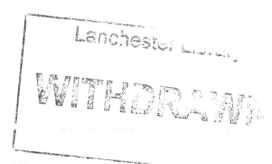

Bioorganic Chemistry:
Carbohydrates

Topics in Bioorganic and Biological Chemistry
A Series of Books in Support of Teaching and Research

Series Editors:
Sidney M. Hecht, University of Virginia
Richard L. Schowen, University of Kansas

Bioorganic Chemistry: Nucleic Acids
    S. Hecht, editor

Bioorganic Chemistry: Peptides and Proteins
    S. Hecht, editor

Bioorganic Chemistry: Carbohydrates
    S. Hecht, editor

Enzyme Catalysis
    R. Schowen

Steady State Enzyme Kinetics
    P. Cook

Bioorganic Chemistry: Carbohydrates

Edited by

Sidney M. Hecht
University of Virginia

New York   Oxford
Oxford University Press
1999

Oxford University Press

Oxford New York
Athens Auckland Bangkok Bogotá Buenos Aires Calcutta
Cape Town Dar es Salaam Delhi Florence Hong Kong Istanbul
Karachi Kuala Lumpur Madrid Melbourne Mexico City Mumbai
Nairobi Paris São Paulo Singapore Taipei Tokyo Toronto Warsaw

and associated companies in
Berlin Ibadan

Copyright © 1999 by Oxford University Press, Inc.

Published by Oxford University Press, Inc.
198 Madison Avenue, New York, New York 10016
http://www.oup-usa.org

Library of Congress Cataloging-in-Publication Data

Bioorganic chemistry : carbohydrates / edited by Sidney M. Hecht.
     p.   cm. — (Topics in bioorganic and biological chemistry)
  Includes bibliographical references and index.
  ISBN 0-19-508469-1 (hardcover)
  1. Carbohydrates. I. Hecht, Sidney M. II. Series.
QD321.B615 1998
572'.56—dc21   98-18287
                                     CIP

Printing (last digit): 9 8 7 6 5 4 3 2 1

Printed in the United States of America
on acid-free paper.

Contents

Foreword

Traditionally, carbohydrate chemistry is considered to relate to the structures and chemical properties of polyhydroxy aldehydes and ketones (the monosaccharides) together with their wide variety of derivatives both natural and synthetic. In fact, saccharides are found as building units of virtually all forms of natural products such as proteins, lipids, alkaloids, steroids, antibiotics, hormones, etc. Thus, for example, many polypeptides that are referred to simply as proteins have complex oligosaccharide side chains; that is, they are in reality glycoproteins. The protein chemist focuses on the structure formed by the amino acids whereas the carbohydrate chemist treats the polypeptide chain as a carrier molecule. Until recently, both disciplines could be practiced independently to good advantage. However, such investigations now must merge in order to enable the achievement of an understanding of the biological properties. For this reason, the unifying concept of glycobiology as a discrete discipline of the biological sciences came into being.

The carbohydrates are the most abundant of the natural products and, by way of postphotosynthetic processes, the source of all biological energy. Because of their importance to nutrition and fermentation, the carbohydrates, especially in the forms of glucose, sucrose, and starch, are of enormous commercial value. Polysaccharides, especially cellulose, are of major industrial importance. For these reasons, a large empirical body of knowledge accumulated in the century prior to 1940. This hotchpotch, composed in large part of singular observations, brought on such jocular remarks as, "the only generalization in carbohydrate chemistry is that no generalization exists." The subject appeared to lack intellectual luster in comparison to other areas of natural-product chemistry and withered in most research-intensive schools of chemistry. A renaissance developed with the discovery of medically important aminoglycoside-type antibiotics in the early 1940s. The further advent in that decade of paper and column chromatographic techniques brought confidence to the planning of carbohydrate research. No longer was progress dependent of fortuitous crystallizations. The concurrently rapidly flourishing field of physical organic chemistry together with the introduction of conformational analysis gave carbohydrate chemistry a vigorous new life. Revolutionary insights came with the introduction of nuclear magnetic resonance spectroscopy in the late 1950s. Within a decade, theoretical carbohydrate chemistry gained a depth of appreciation at the highest levels of organic chemistry. Subsequently, the subject became well integrated within the whole field of chemistry and of basic importance to all the biological sciences.

Meanwhile, the cell biologists became impressed by the ubiquity of complex oligosaccharides covalently attached at cell surfaces and to proteins. Interest surged with the discovery that the antigenic properties of glycoproteins, known as blood-group substances, were produced by the oligosaccharide chains bonded to the core proteins. Indeed, with the always increasing power for isolation and characterization, it became solidly established that the immune response to bacterial infections is in most cases directed to carbohydrate-type structures. Not surprisingly, the structures of cell-surface carbohydrates rapidly attracted widespread attention.

The oligosaccharides proved to be present in minute amounts in highly complex mixtures. Chemical synthetic procedures were developed to provide amounts adequate for detailed assessments of their structural and biological properties. The syntheses required the establishment of a variety of novel procedures. Of these, most outstanding was the need to discover reactions to establish glycosidic linkages in the desired anomeric configuration in useful yields. The breakthrough came with satisfactory methodologies for the synthesis of human blood group determinants in the mid-1970s. Their use in the preparation of artificial antigens and immunoabsorbents provided access to monospecifc antibodies for both clinical and immunochemical research. Significant studies of molecular recognition at the atomic level became possible.

In this regard and of major importance was the ability to synthesize modifications of the natural structures (congeners) to serve as probes for the mapping of the structural features presented (the epitopes) for binding by the receptor sites of proteins (antibody, enzyme, or lectin). Meanwhile, great strides were being made in the X-ray crystallography of large molecules. It had become increasingly possible to derive at high resolution the crystal structures not only of native proteins, but also of their complexes with the stereochemically appropriate synthetic oligosaccharides. It evolved in the 1980s that the combining sites of antibodies and lectins were generally small and shallow cavities that surprisingly were strongly hydrophilic. These epitopes were found to consist most often as small portions of two or three of the monosaccharide building units and most often involving only two or three of the several hydroxyl groups present in the oligosaccharide. Thermodynamic studies of the binding reactions of both the natural oligosaccharides and their congeners led to the conclusion that decreases in the energy of perturbation of water molecules associated with the decrease in surface area exposed to aqueous media made the dominant contribution to the stability of the complex. The term *hydraphobic* was suggested for the phenomenon which, like the hydrophobic effect, involves the stabilization of the system by way of energetically favorable changes in the dynamic of water structure.

The roles played by the wide variety of complex olgiosaccharides that occur as building units of the myriads of glycobiologicals are under intensive investigation at both the molecular and atomic levels. For instance, an international symposium on glycobiology held at the time this book went to press dealt with topics related to inflammation, immunology, infectious diseases, virology, neurology, cell development, cell–cell interactions, fertilization, cancer, and metasthesis. All of these subjects appear orchestrated, in part, by specific molecular recognitions involving oligosaccharides. In view of the widespread occurrence of complex oligosaccharides, such involvements may exist throughout living tissue. It is recognized that, because of their polyfunctionalities, only a few monosaccharides can combine to form a multitude of different oligosaccharides. The number of recognition sites is much greater because a given oligosaccharide can present several different portions of its surface area for recognition. That is, a host of different biological activities can be expressed by a single oligosaccharide. The possibilities are enormous and their delineation represents a major goal for the bioorganic and medicinal chemists of the twenty-first century.

From the above, it is evident that carbohydrate research is now interdisciplinary and, as such, a branch of bioorganic chemistry. This book addresses the subject from a chemist's point of view, and is comprised of contributions from current leaders at a variety of the important frontiers presently under investigation. Thereby, the book sets the stage for plays of the twenty-first century. It will surely be a great and exciting theater.

The book begins with an overview of relevant carbohydrate chemistry for those not already familiar with the subject. There then follow dissertations on syntheses that bring the reader into contact with some of the most spectacular recent accomplishments. The message becomes clear that, given the necessary ingenuity, modern organic chemistry can meet virtually any synthetic goal. Of course, nature was there first. That its methodologies can be practiced *in vitro* to great advantage is documented.

Molecular recognition is of central importance to glycobiology. Being strongly hydrophilic, oligosaccharides readily extend from molecular or cellular surfaces into a surrounding aqueous phase. There, they are well-positioned for detection. Complementarity provides the key to the association of their hydrophilic surfaces. In this regard, the conformational properties of the oligosaccharides are of basic importance and these are well-reviewed. The role of the cell-surface carbohydrates as "biological glues" is also surveyed.

The contents of this book make it evident that glycobiology is in full bloom. The multifacet opportunities are increasingly attracting participation of interdisciplinary groups. The abundant opportunities that exist for research of major significance to human welfare are made evident. It is a truly timely contribution that, along with the companion books on nucleic acids and peptides, will surely have an important influence on the future curricula of research-oriented departments of chemistry.

The molecular biology of the carbohydrates became high science in the twentieth century; the stage is now set for the appreciation of the molecular events at the atomic level.

Raymond U. Lemieux
University of Alberta

Preface

This is the third volume of a series of books in bioorganic and biological chemistry. The first volume, dealing with nucleic acids, appeared in the Spring of 1996. A second volume in the area of peptides and proteins appeared late in 1997.

As noted in the preface to the first volume, the increasingly detailed understanding of the molecular basis by which biological systems operate has dramatically increased the range of studies now considered to be within the domain of organic chemistry. The "core" of expertise required to function effectively in this rapidly expanding field has increased correspondingly. This poses an ongoing educational challenge both for scientists presently working in the field and, especially, for students encountering the subject matter for the first time. This series of books is intended to support the teaching of graduate students in bioorganic chemistry.

As was the case for the first two volumes, *Bioorganic Chemistry: Carbohydrates* consists of a set of chapters, the number of which is roughly equal to the number of weeks in a semester. The subject matter of each chapter is believed to reflect a set of issues and activities that are a focus of current research by bioorganic chemists in the field of carbohydrates. As with the first two volumes, the first part of each chapter contains an overview of basic principles, placed in the context of the area on which that chapter is focused. This is followed by a number of specific examples of current studies which are presented in greater detail. This organization is intended to parallel the typical order of presentation in special topics courses for graduate students and advanced undergraduates. As is the case for the first two volumes, a set of overheads for the figures in the book is available as an aid to classroom presentation.

I thank the authors of the chapters in this volume for their contributions. I also acknowledge the contributions of the graduate and postdoctoral students who participated in a special topics course on carbohydrates that I gave at the University of Virginia, based on the material in this book. I thank Ms. Vickie Thomas for her able assistance in typing the chapters in the volume and Ms. Carolyn Esau for her help in verifying many of the literature references.

Charlottesville   S. M. H.

Contributors

Dominique Armspach
University of Birmingham

Alejandro Aruffo
Bristol-Myers Squibb

Jürgen Bajorath
Bristol-Myers Squibb

Mark Bednarski
Stanford Medical Center

Nicholas J. Bockovich
Scripps Research Institute

David R. Bundle
University of Alberta

Stewart Campbell
Duke University

Bert Fraser-Reid
Duke University

Giuseppe Gattuso
University of Birmingham

Martin Hendrix
Scripps Research Institute

Diane Hollenbaugh
Bristol-Myers Squibb

Daniel Kahne
Princeton University

Rainer Königer
University of Birmingham

Robert Madsen
Duke University

Robert Merritt
Duke University

K. C. Nicolaou
Scripps Research Institute

Jack Preiss
Michigan State University

Florante A. Quiocho
Baylor College of Medicine

Carmichael S. Roberts
Duke University

Anthony S. Serianni
University of Notre Dame

Domingos Silva
Intercardia Research Lab

J. Fraser-Stoddart
University of Birmingham

Marcos Sznaidman
Glaxo Wellcome Co.

François Tropper
Stanford Medical Center

Andrea Vasella
Eidgenössische Technische Hochschule,
  Zurich

Nand K. Vyas
Baylor College of Medicine

Suzanne Walker
Princeton University

Chi-Huey Wong
Scripps Research Institute

Bioorganic Chemistry:
Carbohydrates

1

Introduction to Carbohydrates

Marcos Sznaidman

The word *carbohydrate* was coined during the nineteenth century to describe a family of compounds with the general formula $C_n(H_2O)_n$ (*hydrates of carbon*). With the discovery of new compounds that did not conform to that formula, but still showed the same chemical properties, the term carbohydrate has been modified and broadened. It now includes polyhydroxy aldehydes, ketones alcohols and acids and their simple derivatives as well as their polymers having polymeric linkages of the acetal type. Under this new definition, even those molecules that do not fit the general formula $C_n(H_2O)_n$ are considered to be carbohydrates.

Role of Carbohydrates

Occurrence and Role of Carbohydrates in Nature

Carbohydrates comprise the most abundant group of compounds found in natural sources; they are present both in plants and animals. They perform a wide range of functions and exist in more diverse forms than any other group of natural products.

The main sources of carbohydrates are plants, in which these compounds are synthesized from water and carbon dioxide in a process called *photosynthesis*:

$$\text{energy} + CO_2 + H_2O \rightarrow \text{carbohydrates} + O_2$$

D-glucose is the major carbohydrate produced by photosynthesis. It is utilized by the plant to built polymers such as *cellulose* and *starch*. Cellulose, the most abundant organic substance on earth, is a structural component of plants. Starch serves as energy storage material. Animals, who consume carbohydrates mostly from plants, transform these molecules in the presence of oxygen into carbon dioxide, water, and the energy required for mechanical work. This process is the reverse of photosynthesis:

$$\text{carbohydrates} + O_2 \rightarrow CO_2 + H_2O + \text{energy}$$

Animals can also convert D-glucose into a polymeric form, called *glycogen*, which serves as an energy storage form.

Although carbohydrates constitute a very important group of molecules by themselves, they are also major constituents of diverse natural products: natural antibiotics like *streptomycin* and *puromycin* contain aminosugars as their main constituent;

nucleic acids are carbohydrate-containing polymers that control the biosynthesis of proteins and are responsible for the transfer of genetic information; *adenosine 5'-triphosphate* (ATP) and *adenosine 5'-diphosphate* (ADP) are carbohydrate-containing molecules that are responsible for the energy balance of many metabolic reactions: ATP → ADP + Pi + energy; many proteins contain carbohydrate residues, and are called *glycoproteins*. *Proteoglycans* are components of animal connective tissues; *peptidoglycans* are highly complex macromolecules, which are responsible for the rigidity of the bacterial cell wall; *lipopolysaccharides* are the major constituent of the outer cell envelope of gram-negative bacteria.

Industrial Use of Carbohydrates

Carbohydrates are the basis of many important industries, these include:

1. *The food industry. Sucrose*, a disaccharide of D-glucose and D-fructose obtained from sugar cane, is used as a sweetening agent, as a preservative for food, and as a raw material for the fermentation industry. The manufacturing of baked goods uses large amounts of starch. *Gums* are also used in food processing.
2. *The textile industry.* Despite the advent of synthetic materials, *cotton* (one of the forms of the polysaccharide cellulose) is still used extensively as a material for the production of textiles.
3. *Plastics.* Derivatives of cellulose (e.g., acetate-butyrate, propionate, and acetate-phthalate) are used to make sheets, films, and coating materials.
4. *Packaging.* The manufacturing of paper is one of the most important industries. Other derivatives of cellulose (esters and ethers) are used for the production of transparent wrapping and film.
5. *The pharmaceutical and cosmetic industry.* Antibiotics, intravenous solutions, vitamin C, and the use of carbohydrates as chiral starting materials for the stereoselective synthesis of pharmaceutical agents are examples of the use of carbohydrates in the pharmaceutical industry.

Classification of Carbohydrates

Carbohydrates can be classified into two main groups: *simple carbohydrates*, that is, molecules that contain only carbohydrates in their structure, and *complex carbohydrates* (also called carbohydrate-containing molecules), that is, molecules in which carbohydrates appear covalently bound to lipids, proteins, aglycons, et cetera.

Simple Carbohydrates

Based on their molecular size, simple carbohydrates can be divided into three major groups: *monosaccharides* and monosaccharide derivatives, *oligosaccharides*, and *polysaccharides*. The term saccharide means sugarlike. Sugar in Sanskrit means sweet sand, and is still being used to include the lower members of carbohydrates like monosaccharides and oligosaccharides, which are usually sweet and water soluble. Monosaccharides and monosaccharide derivatives are the lower members of carbohydrates, which can not be degraded by hydrolysis to smaller carbohydrate molecules. They constitute the building blocks of oligosaccharides and polysaccharides, which are polymers of monosaccharides and their derivatives, joined by acetal-type linkages. By definition, oligosaccarides contain between 2 and 10 monosaccharide units, and *polysaccharides* contain more than 10 units. Both, oligo and polysaccharides can be hydrolyzed to their corresponding monosaccharides and monosaccharide derivatives.

Complex Carbohydrates

Many natural products fall into this category, among them complex glycosides, carbohydrate antibiotics, nucleic acids, glycoproteins, proteoglycans, peptidoglycans, lipopolysaccharides, and glycolipids.

The following four sections deal with the structure and natural occurrence of monosaccharides and monosaccharide derivatives, oligosaccharides, polysaccharides, and complex carbohydrates.

Monosaccharides and Monosaccharide Derivatives

Monosaccharides are polyhydroxy aldehydes and polyhydroxy ketones that have the general formula $C_n(H_2O)_n$. Due to their polyfunctionality, they usually exist in many isomeric forms, both *acyclic* and *cyclic* (Fig. 1-1). Most of the cyclic isomers can be isolated in pure form. On the other hand, acyclic isomers have been detected only as very minor components in solution, where they coexist in equilibrium with the cyclic forms.

In order to understand the physical and chemical properties of monosaccharides in a pure state or as equilibrium mixtures in solution, the structures of both the acyclic and cyclic forms should be considered first. (The term *structure* is used here to include constitution, configuration, and conformation [Stoddart, 1971]).

Constitution and Configuration: Acyclic Forms

Figure 1-2 shows the constitution of the acyclic forms of the most common monosaccharides. They all have nonbranched carbon chains with either an aldehydo (*aldoses*) or keto (*ketoses*) functional group. Each one of the remaining carbons has one hydroxyl group. Carbons are numbered such that the carbonyl group has the lowest possible

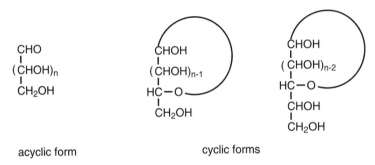

Figure 1–1. Acyclic and cyclic forms of monosaccharides.

Number of carbons	3	4	5	6
General name	**trioses**	**tetroses**	**pentoses**	**hexoses**
Molecular formula	$C_3H_6O_3$	$C_4H_8O_4$	$C_5H_{10}O_5$	$C_6H_{12}O_6$

Aldoses

General constitution

	3	4	5	6
				1 CHO
			1 CHO	*2 CHOH
		1 CHO	*2 CHOH	*3 CHOH
	1 CHO	*2 CHOH	*3 CHOH	*4 CHOH
	*2 CHOH	*3 CHOH	*4 CHOH	*5 CHOH
	3 CH_2OH	4 CH_2OH	5 CH_2OH	6 CH_2OH

General name	**aldotrioses**	**aldotetroses**	**aldopentoses**	**aldohexoses**
Number of chiral carbons	1	2	3	4
Number of possible config. isomers	2	4	8	16

Ketoses[a]

General constitution

	3	4	5	6
				1 CHO
			1 CHO	2 C=O
		1 CHO	2 C=O	*3 CHOH
	1 CHO	2 C=O	*3 CHOH	*4 CHOH
	2 C=O	*3 CHOH	*4 CHOH	*5 CHOH
	3 CH_2OH	4 CH_2OH	5 CH_2OH	6 CH_2OH

General name	**triuloses**	**tetruloses**	**pentuloses**	**hexuloses**
Number of chiral carbons	0	1	2	3
Number of possible config. isomers	0	2	4	8

* Denotes a chiral carbon

[a] only 2-keto compounds are shown since they are the most common ones, but some 3-keto compounds are also known.

Figure I–2. Constitution of acyclic forms of monosaccharides.

number. The number of possible configurational isomers is given by the formula 2^n, where n is the number of chiral carbons. According to their functional group, monosaccharides can be divided into two major groups: aldoses (aldehydo group) and ketoses (keto group). According to the number of carbons, monosaccharides can also be classified as: *trioses* (3 carbons), *tetroses* (4 carbons), *pentoses* (5 carbons), and *hexoses* (6 carbons). Higher carbon chains are known to exist (e.g., *heptoses*), but they are very rare. These two types of classification give origin to many subgroups which unequivocally describe the constitution of a family of configurational isomers. For example: the subgroup of *aldopentoses* includes all configurational isomers of aldoses with a 5-carbon atom chain.

The convention that has been universally adopted to represent the three-dimensional structures of acyclic monosaccharides in two dimensions is the *Fischer projection formula* (Fischer, 1891, 1907; Hudson, 1948). In this convention, the carbon chain is in a vertical arrangement with the carbonyl group at the top (in the case of aldoses), or the nearest to the top (in the case of ketoses) (Fig. 1-3). All the horizontal bonds (e.g., carbon-hydrogen and carbon-oxygen) project toward the viewer, with the vertical bonds pointing away. In this way, the vertical carbon chain forms an arc which is viewed from the convex side. Figure 1-3 shows the Fischer projection formula of one of the configurational isomers of the aldohexoses, D-glucose.

The Fischer projection formulas were created in order to represent three-dimensional structure in two dimensions, and to unequivocally define the configuration of each chiral center. They do not represent any real conformational isomer; if that were the case, they would probably correspond to the least stable conformer of the acyclic monosaccharides, since all the substituents are eclipsed (Fig. 1-3). As discussed below, acyclic monosaccharides adopt a zig-zag conformation, both in solution and in the crystalline state.

The Fischer projection formulas of the aldoses and ketoses of the D-series are represented in Figures 1-4 and 1-5, respectively.

The names of the acyclic monosaccharides shown in Figures 1-4 and 1-5 always start with a capital letter (D or L) followed by a configurational prefix (e.g., *erythro-, threo-, ribo-, gluco-, manno-,* etc.) and the suffix-*se*. The prefixes indicate the number of chiral centers and their *relative configuration*. For example, the prefix *gluco* indicates that there are 4 chiral carbons, and also indicates that in the Fischer projection

perspective representations
of D-glucose

Fischer projection formulas
of D-glucose

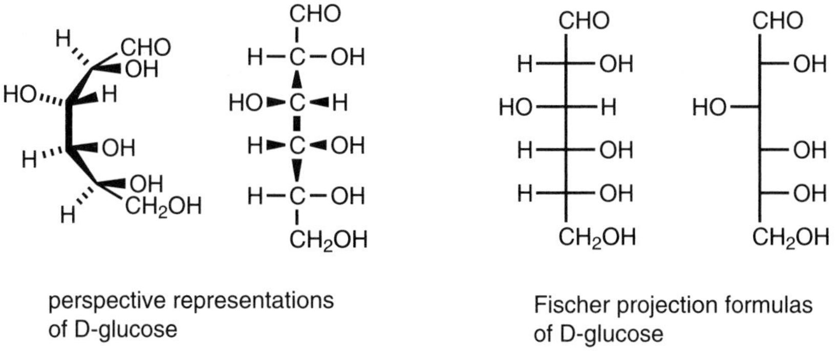

Figure 1–3. Fischer projection formulas.

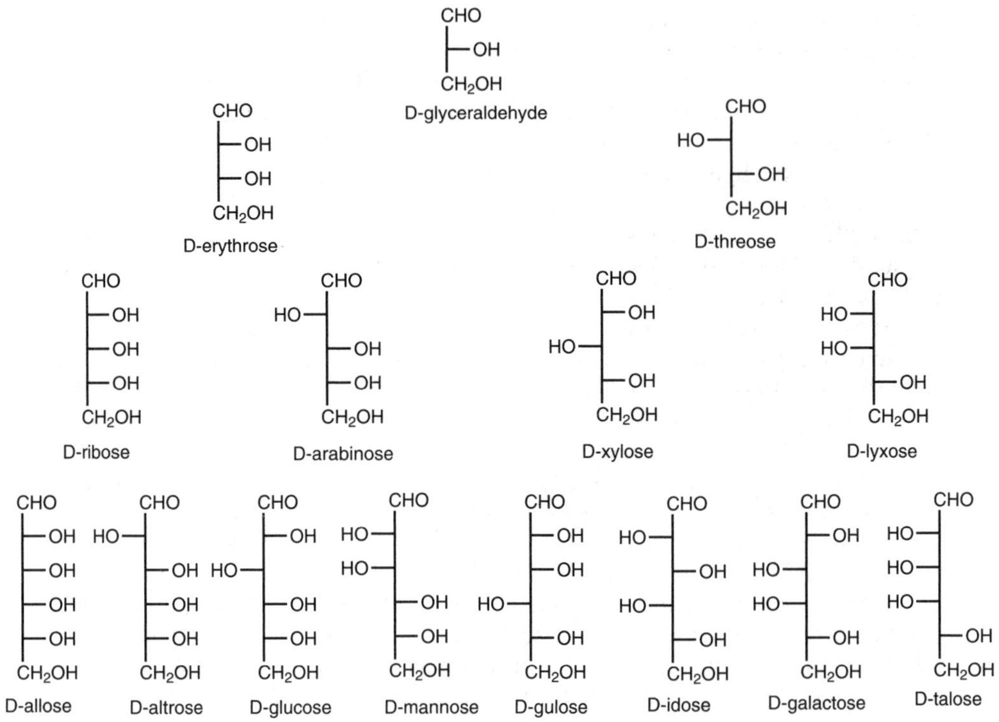

CHO
─OH
CH₂OH
D-glyceraldehyde

CHO
─OH
─OH
CH₂OH
D-erythrose

CHO
HO─
─OH
CH₂OH
D-threose

CHO
─OH
─OH
─OH
CH₂OH
D-ribose

CHO
HO─
─OH
─OH
CH₂OH
D-arabinose

CHO
─OH
HO─
─OH
CH₂OH
D-xylose

CHO
HO─
HO─
─OH
CH₂OH
D-lyxose

CHO
─OH
─OH
─OH
─OH
CH₂OH
D-allose

CHO
HO─
─OH
─OH
─OH
CH₂OH
D-altrose

CHO
─OH
HO─
─OH
─OH
CH₂OH
D-glucose

CHO
HO─
HO─
─OH
─OH
CH₂OH
D-mannose

CHO
─OH
─OH
HO─
─OH
CH₂OH
D-gulose

CHO
HO─
─OH
HO─
─OH
CH₂OH
D-idose

CHO
─OH
HO─
HO─
─OH
CH₂OH
D-galactose

CHO
HO─
HO─
HO─
─OH
CH₂OH
D-talose

Figure 1–4. Fischer projection formulas of D-aldoses.

formula, the hydroxyl groups of carbons 2, 4, and 5 are on the same side and opposite to the hydroxyl group of carbon 3 (Fig. 1-6). The pair of enantiomers shown in Figure 1-6 have the gluco relative configuration. In order to distinguish between these two enantiomers, the Rosanoff convention is adopted (Hudson, 1948; Rosanoff, 1906). In this convention the letters D and L at the beginning of a name indicate the *absolute configuration* of the chiral carbon most remote from the carbonyl group (C-5 in hexoses, C-4 in pentoses, C-3 in tetroses, and C-2 in trioses). Compounds in the D-series have this hydroxyl group to the right in the Fischer projection formula, and those in the L-series have it to the left. According to this convention the structure on the left in Figure 1-6 belongs to the D-series because the hydroxyl group at C-5 is to the right, and hence is called D-glucose. The structure on the right belongs to the L-series because the hydroxyl group at C-5 is to the left, and hence is called L-glucose. Thus, the *absolute configuration* of any monosaccharide is completely determined by combining the proper configurational prefixes (e.g., gluco, manno, etc.) with either D or L as appropriate.

The *Cahn-Ingold-Prelog* convention offers an alternative way of naming monosaccharides. In this convention, the absolute configuration of each chiral carbon is determined by a letter (*R* or *S*) according to a set of rules that assigns a priority number to each substituent (Cahn, 1964; Cahn et al., 1956, 1966). According to this convention the full name of D-glucose is (2*S*, 3*S*, 4*R*, 5*R*)-2,3,4,5,6-pentahydroxyhexanal. Judging from this example, it becomes obvious why this convention has been completely abandoned in carbohydrate chemistry.

Each structure of the D-series has an enantiomer in the L-series. For example, in Figure 1-6 we have already seen the pair of enantiomers of glucose. Figure 1-7 shows the structures of D and L-mannose and their enantiomeric relationship. *Epimers* are di-

CH₂OH

=O

CH₂OH

1,3-dihydroxy-2-propanone
(dihydroxyacetone)

CH₂OH

=O

—OH

CH₂OH

D-*glycero*-tetrulose
(D-erythrulose)

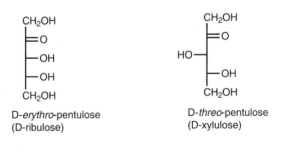

D-*erythro*-pentulose
(D-ribulose)

D-*threo*-pentulose
(D-xylulose)

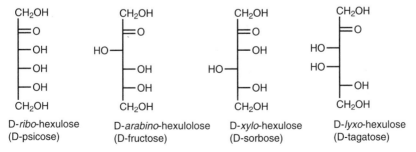

D-*ribo*-hexulose
(D-psicose)

D-*arabino*-hexululose
(D-fructose)

D-*xylo*-hexulose
(D-sorbose)

D-*lyxo*-hexulose
(D-tagatose)

Figure 1–5. Fischer projection formulas of D-ketoses. Trivial names are indicated in parenthesis.

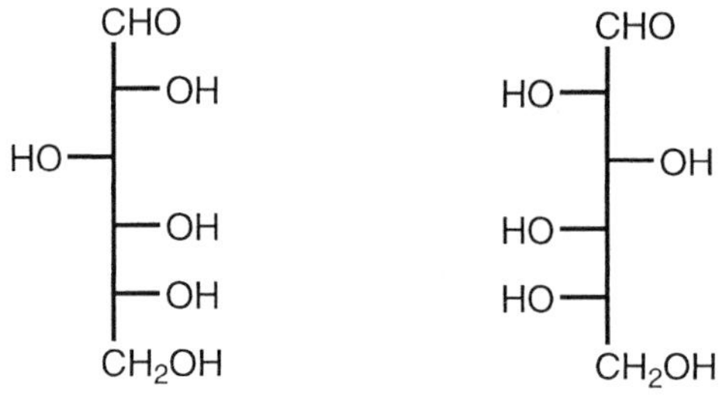

CHO
—OH
HO—
—OH
—OH
CH$_2$OH

CHO
HO—
—OH
HO—
HO—
CH$_2$OH

Figure 1–6. Pair of enantiomers with the *gluco* relative configuration.

astereoisomers that differ in the configuration of only one carbon. For example, D-glucose and D-mannose are epimers because they differ in configuration only at C-2 (cf. Figs. 1-6 and 1-7). Those compounds that are not enantiomers or epimers are simply diastereoisomers.

Constitution and Configuration

Cyclic Forms. Any aldehyde or ketone can react with an alcohol to form a hemiacetal or hemiketal, respectively (Fig. 1-8). If both functional groups are present within the same molecule, then an equilibrium is established between the open chain and the cyclic forms (Fig. 1-9). The position of equilibrium depends on the thermodynamic stability of the acyclic and cyclic forms. Pentoses and hexoses exist predominantly in the cyclic forms.

The Fischer projection formulas can also be used to represent the structures of the cyclic forms of monosaccharides. They are derived from the linear ones by "closing" the ring between the proper oxygen and carbonyl carbon as shown for the 5- and 6-membered rings of D-glucose (Fig. 1-10). Five-membered ring monosaccharides may be considered to be related to tetrahydrofuran (Fig. 1-10), hence called *furanoses*; 6-membered rings are related to tetrahydropyran, and are called *pyranoses*.

When a ring is formed, two configurational isomers can be obtained due to the creation of a new chiral center (denoted with an asterisk in Fig. 1-10), which is called an anomeric center. The two isomers are called *anomers*. In other words, anomers are epimers that differ only in the configuration at the anomeric center. The configuration of the anomeric center is indicated by the prefixes α and β. In the Fischer projection

CHO
HO—
HO—
—OH
—OH
CH$_2$OH

D-mannose

CHO
—OH
—OH
HO—
HO—
CH$_2$OH

L-mannose

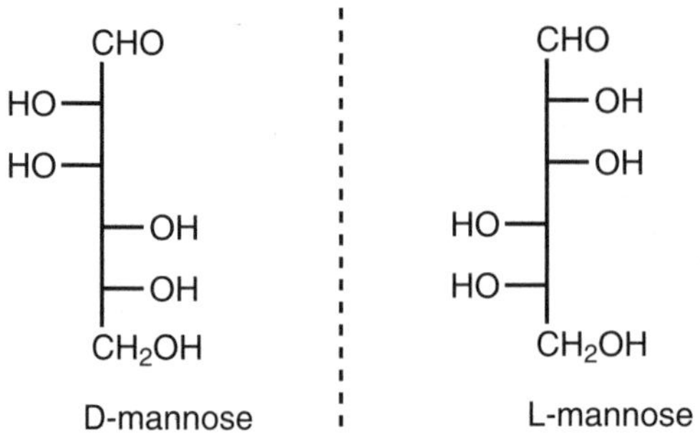

Figure 1–7. Enantiomers of mannose.

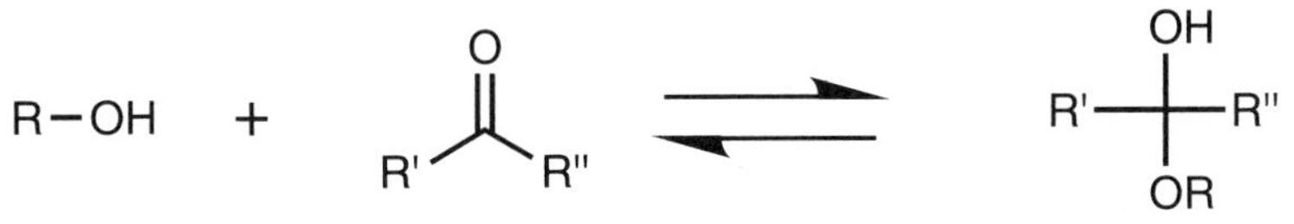

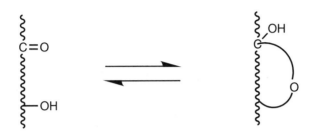

Figure 1–8. Ketal formation from a (carbohydrate) ketone.

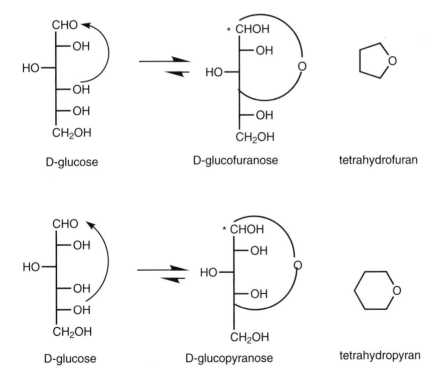

Figure 1–9. Equilibrium between acyclic and cyclic forms of a hydroxy ketone–ketal.

D-glucose → D-glucofuranose tetrahydrofuran

D-glucose → D-glucopyranose tetrahydropyran

* Denotes a new chiral center

Figure 1–10. Cyclic forms of D-glucose.

formulas, the α isomer is the one that has the anomeric hydroxyl group (1-OH group in aldoses, 2-OH group in ketoses) on the same side as the hydroxyl group that determines the D or L configuration (Fig. 1-11). In the β isomer, both hydroxyl groups are on opposite sides.

When drawn side by side, it becomes clear that the enantiomer of α-D-glucopyranose is α-L-glucopyranose, and the enantiomer of β-D-glucopyranose is β-L-glucopyranose (Fig. 1-12). α-D-glucopyranose–β-D-glucopyranose, and α-L-glucopyranose–β-L-glucopyranose are pairs of anomers.

Although the Fischer projection formulas of the ring forms are easily derived from the linear ones, they do not resemble the real geometry of the ring. A more useful way to represent the cyclic forms of monosaccharides, without specifying any conformation, involves the *Haworth perspective formulas* (Drew & Haworth, 1926; Haworth, 1929). The furanoid and pyranoid rings in their Haworth perspective formulas are depicted as planar structures, which are viewed from above at an angle, with the front edge tilted towards the observer (Fig. 1-13). This orientation is emphasized by shading the bonds in the lower edge of the ring. The standard ring orientation has the

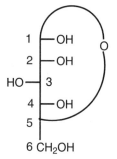

α-D-glucopyranose
OH-1 and O-5 on the same side

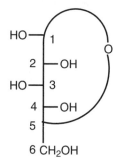

β-D-glucopyranose
OH-1 and O-5 on opposite sides

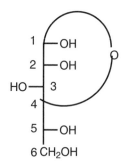

α-D-glucofuranose
OH-1 and OH-5 on the same side

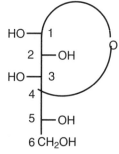

β-D-glucofuranose
OH-1 and OH-5 on opposite sides

Figure 1–11. α and β Anomers of the cyclic forms of D-glucose.

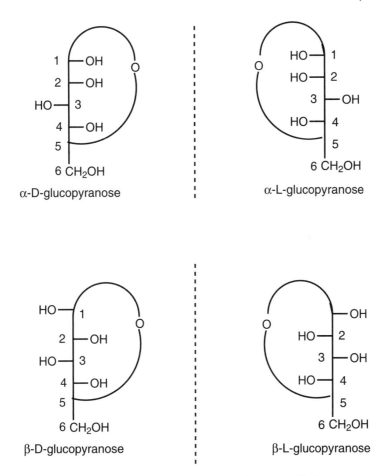

Figure 1–12. Enantiomeric relationships between sugars drawn in cyclic form.

anomeric carbon on the extreme right end of the formula. The numbering of the carbon atoms in the ring increases in a clockwise direction from the anomeric center.

With the Fischer projection formulas, the enantiomeric relationship between α-D-glucopyranose and α-L-glucopyranose, or β-D-glucopyranose and β-L-glucopyranose (Fig. 1-12) is obvious, if the enantiomers are drawn side by side. With the Haworth perspective formulas, the same relationship becomes clear when one structure is on the top of the other (Fig. 1-14).

The enantiomeric relationship of these compounds is clear when they are side by side only if one of the enantiomers is depicted in a "nonstandard" orientation (i.e., the anomeric carbon is on the extreme left of the formula [Fig. 1-15]).

To relate Haworth formulas correctly with Fischer formulas, the following steps should be followed (Fig. 1-16). First, the Fischer projection formula should be written as the cyclic form. Then this representation should be modified so that all of the ring atoms lie on the same vertical plane (this is the modified Fischer formula). Finally, the substituents that were to the right in the modified Fischer projection are written on the

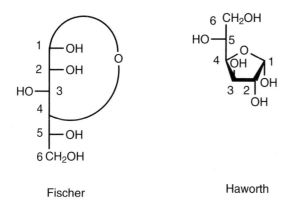

Fischer   Haworth

α-D-glucopyranose

Fischer   Haworth

α-D-glucofuranose

Figure 1–13. Fischer and Haworth perspective formulas of α-D-glucopyranose and α-D-glucofuranose.

bottom plane of the Haworth formula, and all of those that were to the left on the upper plane. This is the equivalent of rotating the plane of the Fischer modified formula by 90°.

A third alternative for representing furanoses and pyranoses is that proposed by Mills (1956). In this system, the ring is projected into the plane of the paper. The orientation of the substituents above or below the ring is represented by solid or broken lines, respectively. In order to be consistent with Haworth perspective formulas, the anomeric carbon should be at the extreme right end of the formula (Fig. 1-17).

While Haworth perspective formulas are suitable for the representation of monocyclic systems and a few simple polycyclic systems, Mills perspective formulas are suitable for both monocyclic and polycyclic systems. In this sense the Mills formulas permit the representation of carbohydrates to be consistent with those of other cyclic

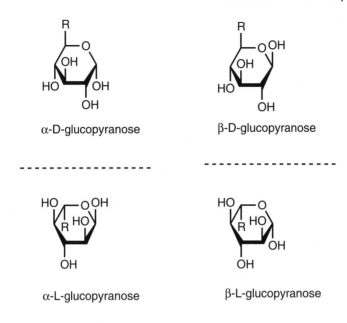

R = CH₂OH

Figure 1–14. Stereochemical relationship between the enantiomers of D and L-glucopyranose, and between epimers of α and β-glucopyranose. The structures are represented as Haworth perspective formulas.

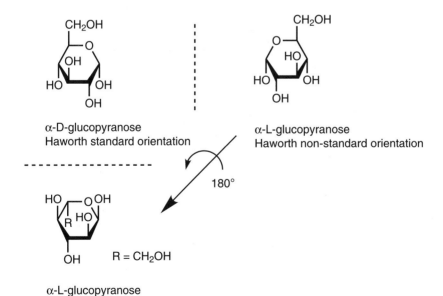

Figure 1–15. Relationship between α-D-glucopyranose and α-L-glucopyranose in Haworth formulas.

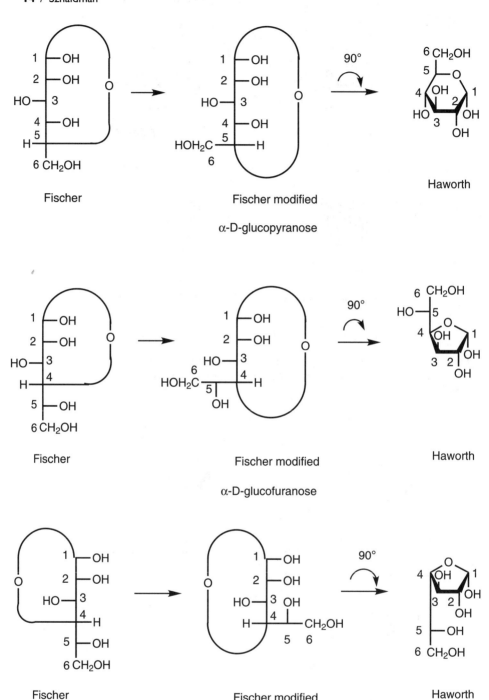

Figure 1-16. Interconversion of Fischer and Haworth formulas.

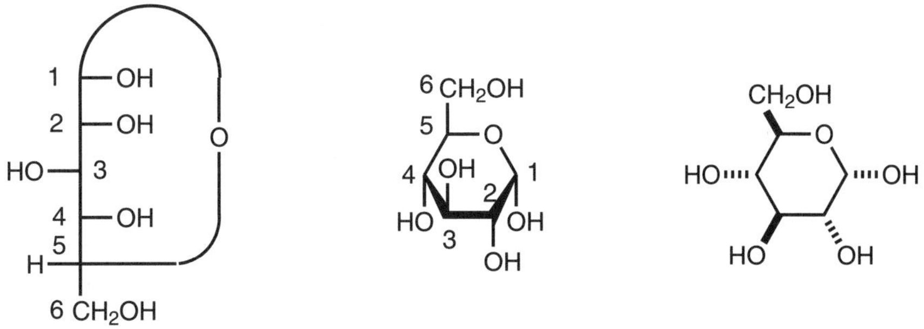

Fischer Haworth Mills

α-D-glucopyranose

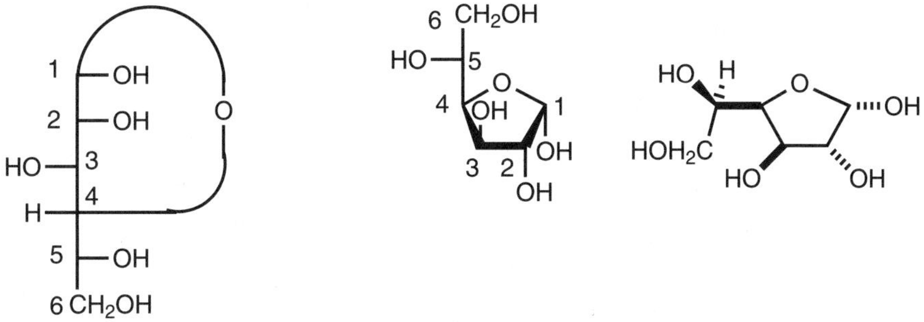

Fischer Haworth Mills

α-D-glucofuranose

Figure 1–17. Mills perspective formulas of *α*-D-glucopyranose and *α*-D-glucofuranose. Fischer projection and Haworth perspective formulas are shown for comparison.

natural products (e.g., steroids, terpenoids, alkaloids, etc.). Nonetheless, Haworth formulas are widely used by carbohydrate chemists to represent monocyclic carbohydrates.

Conformation: Pyranoses

Pyranoid rings can adopt many different conformations similar to those observed in cyclohexane, namely chair, boat, skew and half-chair . They are represented by the conformational formulas, also introduced in carbohydrate chemistry by Haworth (Drew & Haworth, 1926; Haworth, 1929) (Fig. 1-18). These are designated *C* for chair, *B* for boat, *S* for skew, and *H* for half-chair (Schwarz, 1967). The orientation of the conformational formulas is emphasized by shading the bonds nearest to the front edge.

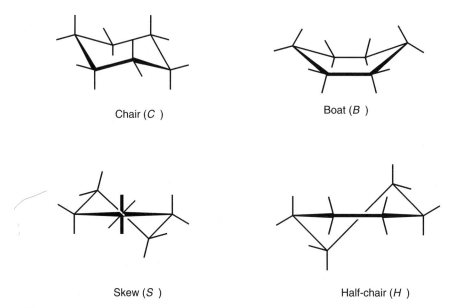

Figure I–18. Conformations of cyclohexane rings.

Pyranoses can adopt two different chair conformations, 6 boat, 6 skew, and 12 half-chair conformations. In order to differentiate and designate each of these, the following convention is adopted. This convention was proposed by Schwarz (Schwarz, 1967; Stoddart, 1971); it is a modification of an earlier proposal (Reeves, 1949, 1951). First, a reference plane is chosen so that it contains four of the ring atoms. When an unequivocal choice is not possible, the reference plane is chosen so that the lowest-numbered carbon in the ring is displaced from the plane. The ring atoms that lie above the reference plane are written as superscripts and precede the letter, while ring atoms that lie below the reference plane are written as subscripts, and follow the letter. The numbering of the carbon atoms increases in a clockwise direction when viewed from above. Figure 1-19 shows two chair ($^1C_4$ and $^4C_1$) and two boat ($^{1,4}B$ and $B_{1,4}$) conformations with their denominations.

In order to describe a conformer fully, reference should be given to whether it belongs to the D or L series. This is illustrated for the conformers of β-D and β-L-xylopyranose (Fig. 1-20). The enantiomeric relationship between $^4C_1$ (D) and $^1C_4$ (L), and between $^1C_4$ (D) and $^4C_1$ (L), is illustrated clearly.

The study of the relative stabilities of different conformers is known as *conformational analysis*. The two chair conformations are so much more stable than the others that conformational analysis of pyranoses is usually restricted to the choice between $^1C_4$ and $^4C_1$ conformers. The proportion of conformers in solution is usually determined experimentally by ^{1}H NMR spectroscopy (Angyal, 1968, 1969; Jochims et al., 1967; Lemieux & Stevens, 1966; Rudrum & Shaw, 1965) and by X-ray crystallography (Capon, 1969; Sweet & Brown, 1968).

^{1}H NMR spectroscopy shows that the most stable conformer of the vast majority of aldohexopyranoses in aqueous solution at room temperature is $^4C_1$ (D). For exam-

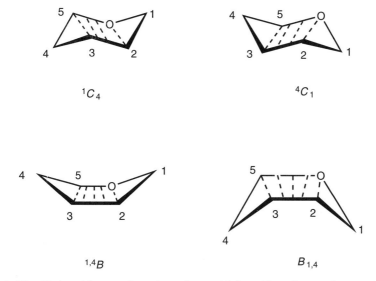

$^1C_4$ $^4C_1$

$^{1,4}B$ $B_{1,4}$

Figure 1-19. Chair and boat conformations of pyranoid rings. The reference planes are indicated.

β-D-xylopyranose

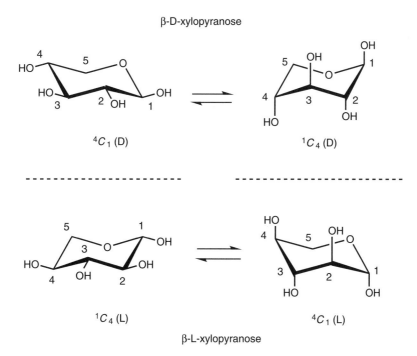

$^4C_1$ (D) $^1C_4$ (D)

$^1C_4$ (L) $^4C_1$ (L)

β-L-xylopyranose

Figure 1-20. Major conformational isomers of β-D and β-L-xylopyranose.

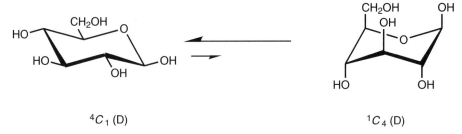

Figure 1–21. Conformational equilibrium of β-D-glucopyranose.

ple, if we consider both conformers of β-D-glucopyranose, the $^4C_1$ conformer with all equatorial substituents is much more stable than conformer $^1C_4$ with all axial substituents (Fig. 1-21).

Another approach for determining the relative stabilities of conformers is the semi-quantitative approach developed by Angyal (1968). In this approach, the relative free energy of each conformer is calculated considering only nonbonding interactions (1,3-diaxial and 1,2-gauche) and the so-called *anomeric effect* (*vide infra*). The predictions made by this method are in good agreement with the experimental data obtained by ^1H NMR spectroscopy.

The preference of an electronegative substituent at the anomeric center for adopting the axial orientation, rather than the expected equatorial position, is known as the anomeric effect (Edward, 1955, Lemieux, 1971; Lemieux & Chu, 1958; Lemieux & Morgan, 1965). For example, 2-chlorotetrahydropyran exists only in the conformation in which the halogen is axial (Fig. 1-22) (Booth & Ouellete, 1966).

The magnitude of the anomeric effect depends, among other factors, on the electronegativity of the anomeric substituent. The anomeric effect decreases through the series: halogen > benzoyloxy > acetoxy > acetylthio > hydroxy > amino > methoxycarbonyl > 4-methylpyridinium. In other words, the more electronegative the substituent, the greater the tendency for its axial disposition. This effect is so powerful that in the case of 2,3,4-tri-O-acetyl-β-D-xylopyranosyl fluoride, the $^1C_4$ conformer is the most stable despite having all axial substituents (Fig. 1-23). On the other hand, electropositive substituents have a tendency to adopt the equatorial position. This is known as the *reverse anomeric effect* (Lemieux & Morgan, 1965). For example, the α-D-*gluco*-pyridinium compound shown in Figure 1-24 (Lemieux & Morgan, 1965) exists in a boat conformation with the positively charged group in an equatorial position. The configuration of the C-2 substituent also effects the magnitude of the anomeric effect: 2-hydroxyl groups oriented axially increase the magnitude of the anomeric ef-

Figure 1–22. Conformational equilibrium of 2-chlorotetrahydropyran.

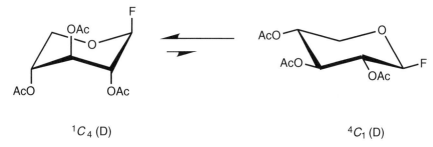

$^1C_4$ (D) $^4C_1$ (D)

Figure 1–23. Conformation equilibrium of 2,3,4-triacetyl-β-D-xylopyranosyl fluoride.

fect with respect to equatorial hydroxyl groups. For example, an equilibrium mixture of D-glucose (2-OH equatorial; Fig. 1-25) will contain 36% of the anomer with the 1-OH axially oriented (α-anomer), while D-mannose (2-OH axial; Fig. 1-26) contains 68% of the anomer with the 1-OH oriented axially. The solvent also influences the magnitude of the anomeric effect; it is larger in solvents of low dielectric constant. For example, methyl 3-deoxy-β-L-*erythro*-pentopyranoside assumes the $^4C_1$ (L) conformation in chloroform, but the $^1C_4$ (L) is the more favorable in water (Fig. 1-27).

At first, the anomeric effect was explained in terms of the interaction between the carbon-oxygen dipole, and the dipole formed by the resulting unshared electron pairs of the ring oxygen atom (Fig. 1-28a) (Edward, 1955). These dipoles form a small angle when the substituent is equatorial, resulting in a large unfavorable dipole moment. The polar nature of the anomeric effect explains the higher tendency of an electronegative substituent to adopt an axial orientation. It can also be seen that, because of dipolar interactions, an axially oriented hydroxyl group at C-2 will destabilize the anomer with the equatorial hydroxyl group even further (Fig. 1-28b). This explains the higher proportion of α-anomer in D-mannose (68%) than in D-glucose (36%).

More recently, the anomeric effect has been explained in terms of stereoelectronic effects (Fig. 1-29) (Deslongchamps, 1983; Romers et al., 1969); the electron density of the substituent (X) located antiperiplanar to one of the nonbonding electron pairs of the ring oxygen will be increased by the HOMO–LUMO interaction between the

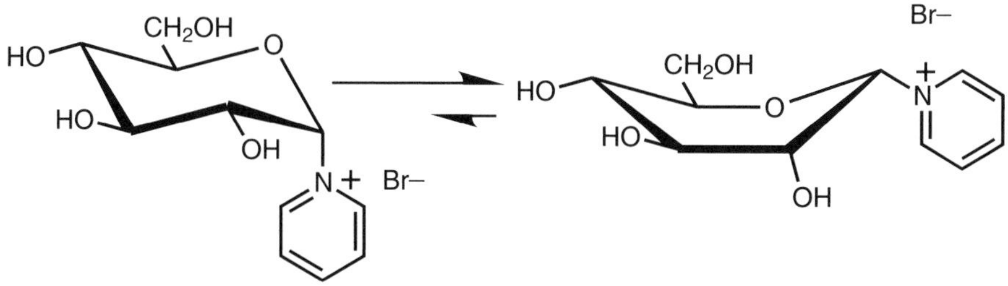

$^4C_1$ (D) $^{1,4}B$ (D)

Figure 1–24. Effect of an electropositive substituent on conformational equilibrium.

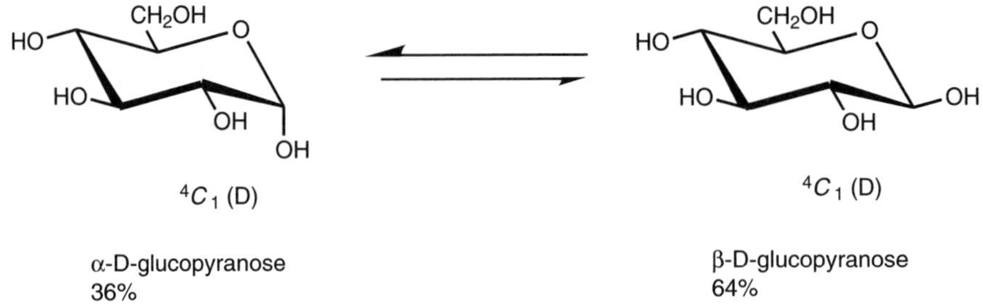

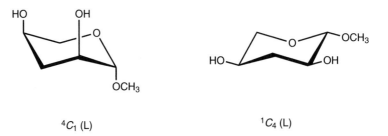

Figure I–25. Anomeric composition of D-glucopyranose.

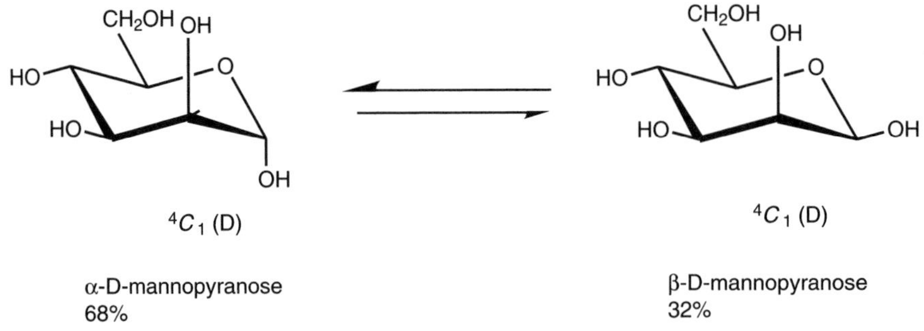

$^4C_1$ (D)

α-D-glucopyranose
36%

$^4C_1$ (D)

β-D-glucopyranose
64%

$^4C_1$ (D)

α-D-mannopyranose
68%

$^4C_1$ (D)

β-D-mannopyranose
32%

Figure I–26. Anomeric composition of D-mannopyranose.

$^4C_1$ (L)

$^1C_4$ (L)

Figure I–27. Conformational isomers of methyl 3-deoxy-β-L-*erythro*-pentopyranoside.

p-orbital of the ring oxygen (HOMO) and the antibonding σ^* orbital of the C—X bond (LUMO). The more electronegative the substituent X, the more favorable the interaction.

Conformation: Furanoses

Conformational analysis of furanoses is much less developed than that of pyranoses. The main reasons for this are that the energy difference between furanose conformers is so low that there are many forms present in equilibrium mixtures, as opposed to the pyranoid rings where one conformer is predominant. Also 5-membered rings are less favorable than 6-membered rings, so they are present in very low concentration. The

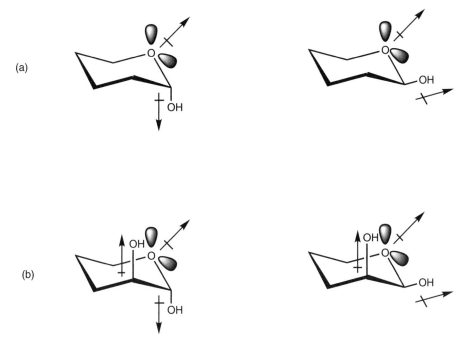

Figure I–28. (a) Polar nature of the anomeric effect. (b) Effect of an axial 2-OH group on the anomeric distribution.

most common conformations that furanoses can adopt are the *envelope* (*E*) or *twist* (*T*) forms. Each one can exist in 10 different arrangements; some of these and their notations are shown in Figure 1-30.

Despite the difficulties in the conformational analysis of furanoses, some generalizations can be made. First, because the interaction between two eclipsing carbon atoms is greater than that between a carbon and an oxygen, the oxygen in the furanoid ring tends to occupy a position along the plane of the ring; hence, usually either C-2 or C-3, or both, will be out of plane. Second, since 1,2-*cis* interactions are very strong, substituted groups will avoid this relationship whenever possible.

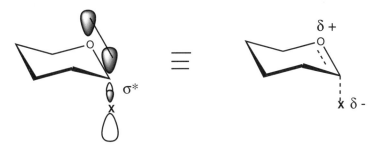

Figure I–29. Stereoelectronic nature of the anomeric effect.

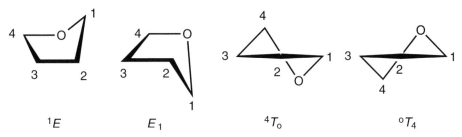

^1E E_1 $^4T_0$ $^0T_4$

Figure I–30. Conformations of furanoses.

Acyclic Conformations

Due to the extremely low proportions of acyclic forms of monosaccharides present in equilibrium mixtures (less than 1%), the study of their conformations is problematic. However, much work has been done with monosaccharides in which the carbonyl group has been either masked with a protecting group, or reduced to an alcohol. In the latter case, a polyol is obtained. The conformations of these polyols have been studied extensively in solution by ^{1}H NMR spectroscopy (El Khadem et al., 1968; Horton & Wander, 1969, 1970) and by X-ray crystallography in the solid state (Jeffrey & Fasiska, 1972; Jeffrey & Kim, 1970). It has been observed that a planar *zig-zag* conformation is the preferred one for these linear acyclic compounds, as in the case of D-mannitol (Fig. 1-31). However, 1,3-interactions between bulky groups can cause deviations from the linear zig-zag conformation. For example, D-glucitol exists predominantly in the *bent* conformation shown in Figure 1-31 because in the zig-zag conformer, 1,3-interactions between 2-OH and 4-OH groups are strongly destabilizing.

Figure 1-31 also illustrates clearly a point mentioned earlier, that is, that the Fischer projection formulas of the acyclic forms of monosaccharides do not represent any real conformation. If that were the case, these representations would suggest the existence of highly hindered conformers.

Monosaccharides in Solution: Equilibrium and Mutarotation

When a monosaccharide is dissolved in a solvent, isomerization takes place and an equilibrium is eventually established. Many constitutional, configurational, and conformational isomers may be present (Angyal, 1984, 1991) as shown in Figure 1-32. The proportion of each isomer is determined by its relative free energy. For example, an equilibrium mixture of D-glucose (Fig. 1-33) contains 36% α-D-glucopyranose and 64% β-D-glucopyranose. The furanoses and the acyclic form together represent less than 1% of the equilibrium mixture. The reason for the higher stability of pyranoses is that in their preferred conformations ($^4C_1$ (D), Fig. 1-25), most of the substituents are equatorially disposed. On the other hand, in the furanoses there are usually strong destabilizing 1,2-*cis* interactions such as those between the 3-OH group and C-4 side chain in α-D-glucofuranose and β-D-glucofuranose (Fig. 1-33). These interactions will be present in any furanose conformer. Since the only two species present to any significant extent at equilibrium are the α and β-D-glucopyranoses in their $^4C_1$ conformations, the representation of the equilibrium mixture of D-glucose can be simplified to the two isomers shown in Figure 1-25. The proportion of the α anomer of

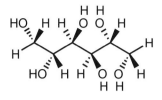

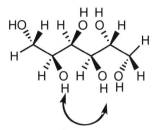

D-mannitol in the preferred
zig-zag conformation

D-glucitol in the less stable
zig-zag conformation

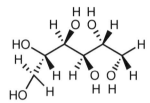

D-glucitol in the preferred
bent conformation

Figure 1–31. Conformations of acyclic forms. Arrows indicate destabilizing interactions.

D-glucose is higher than would be predicted if only nonbonded interactions were considered. This extra stability is due to the anomeric effect, as discussed previously. As mentioned, this effect is even stronger when the 2-OH group is axially disposed. For example, an equilibrium mixture of D-mannose contains 68% α-pyranose and 32% β-pyranose (Fig. 1-26). Again, the equilibrium mixture of D-mannose has less than 1% of furanoses and the acyclic form. Other monosaccharides exhibit more complex mixtures. For example, D-altrose exists in solution as a mixture of pyranoses and fura-

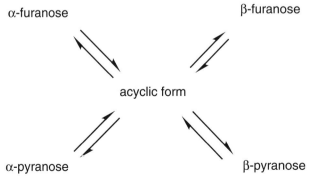

Figure 1–32. Isomers present in an equilibrium mixture of monosaccharides. Each form shown can exist in many different conformations.

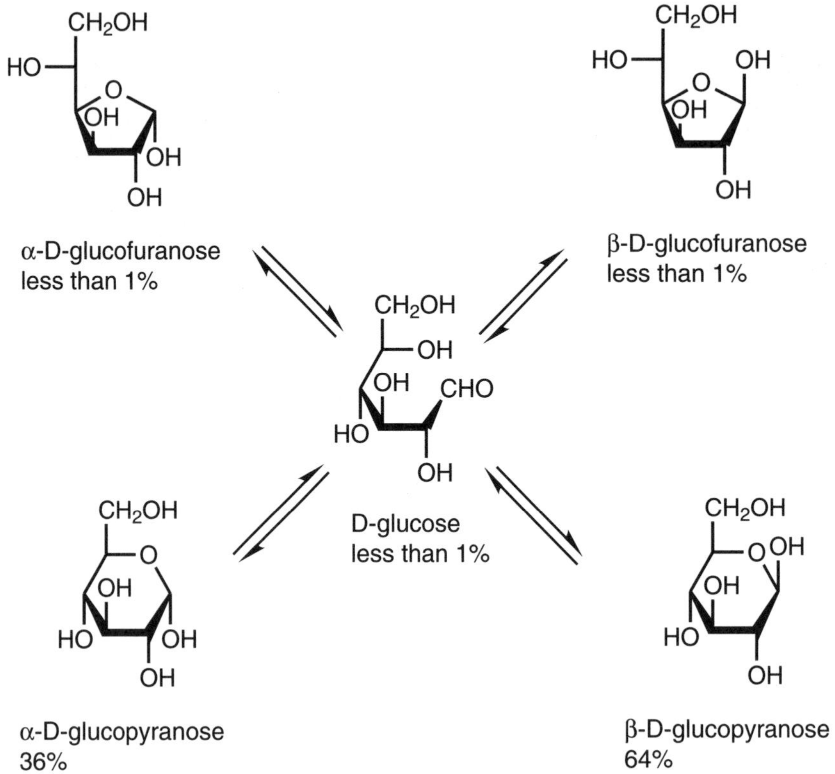

Figure 1–33. Constitutional and configurational isomers of D-glucose present in aqueous solution. Conformational isomers are not shown.

noses as shown in Figure 1-34. The α-D-pyranose form exists as a mixture of conformers. Both have strong 1,3-diaxial destabilizing interactions. Some of these interactions disappear in the β-pyranose form and in the α and β furanoses. Among the furanoses, the β-anomer is less stable due to the strong destabilizing 1,2-*cis* interaction between 1-OH and 2-OH groups.

All the species present in any of these equilibrium mixtures have a particular optical rotation. When a single isomer is dissolved, the optical rotation of the solution will change, reflecting the proportion of isomers present at each particular time. At equilibrium, when the proportion of isomers ceases to change, the optical rotation reaches a constant value. This change of optical rotation with time is known as *mutarotation*. For example, when α-D-glucopyranose is dissolved in water, the optical rotation changes from $+113°$ (pure α-D-glucopyranose) to a final value of $+52°$ at equilibrium. This final rotation reflects the proportion of isomers present according to the following formula: $[\alpha]_{\text{mixture}} = [\alpha]_1 x_1 + [\alpha]_2 x_2 + \ldots\ldots\ldots + [\alpha]_n x_n$, where $[\alpha]_n$ represents the optical rotation of the isomer n, and x_n represents the proportion of that isomer in solution. In the case of D-glucose, we have already seen that there are essentially only two species present at equilibrium: α-D-pyranose ($[\alpha] = +113°$, $x = 0.36$) and

β-D-pyranose ($[\alpha] = +19°$, $x = 0.64$), so $[\alpha]_{\text{mixture}} = 113° \times 0.36 + 19° \times 0.64 = 52°$, which matches the experimentally derived value.

Chemical and Physical Properties of Monosaccharides

Many chemical and physical properties of monosaccharides can be related to their structure. For example, the chemical shift (δ) of a particular proton in the ^{1}H NMR spec-

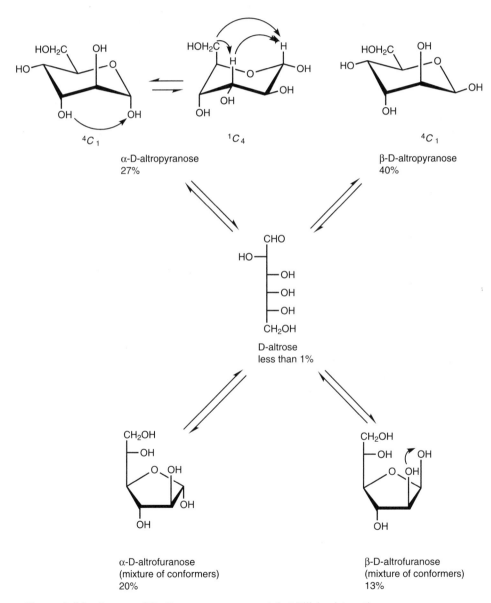

Figure 1–34. Isomers of D-altrose; arrows represent destabilizing interactions.

trum depends on its environment: the presence of electron-withdrawing groups will result in a lower field resonance. For this reason the anomeric proton, with two vicinal oxygens, usually resonates at lower field than other protons in monosaccharides. Coupling constants (J) depend on the number of vicinal protons and their stereochemical relationship. There are equations that correlate the dihedral angle of vicinal protons with their coupling constant (Karplus, 1963). According to these equations, vicinal protons which are *antiperiplanar* (dihedral angle of 180°) exhibit a large coupling constant (7–10 Hz) and vicinal protons which are *syn-clinal* (dihedral angle of 60°) exhibit a small value (1–4 Hz). The ^{1}H NMR spectrum of D-glucose in D_2O has two separate doublets at low fields: one at δ 5.32 with $J = 3.5$ Hz that integrates for 0.36 proton; the second at δ 4.74 with $J = 7.5$ Hz that integrates for 0.64 proton. The first signal corresponds to the anomeric proton of the α anomer. Usually, equatorial protons resonate at a lower field than constitutionally similar axial ones. The low coupling constant indicates a small dihedral angle between H-1 and H-2 ($\sim$60°) (Fig. 1-35). The second signal corresponds to the anomeric proton of the β anomer. The large coupling constant indicates a large dihedral angle ($\sim$180°) between H-1 and H-2 and hence the *trans*-diaxial disposition. The relative ratio of these signals (0.36/0.64) indicates the proportion of these two anomers in solution: 36% of the α-anomer and 64% of the β-anomer (Fig. 1-35).

As opposed to ^{1}H NMR spectral characteristics, it is difficult to correlate the optical rotation with the structure of a monosaccharide. Nevertheless, some entirely empirical methods have been developed (Eliel et al., 1965).

The chemical properties of a particular monosaccharide depend largely on its structure. The reactivity of different hydroxyl groups in pyranoid rings is influenced by a number of factors, including orientation of the hydroxyl groups and steric environment. For example, equatorial hydroxyl groups are more readily esterified than axial ones (Fig. 1-36a). Axial hydroxyl groups at other than the anomeric position are oxidized more rapidly than equatorial ones by O_2 over a platinum catalyst (Fig. 1-36b). Primary alcohols are more reactive toward esterification or etherification than secondary ones. As a result, the primary 6-OH group of aldohexopyranoses and furanoses can be protected selectively in the presence of secondary OH groups (Fig. 1-36c).

Although the acyclic form of sugars is present in small amounts in equilibrium mixtures, it accounts for many of the observed reactions of monosaccharides in aqueous solution. For example, when D-glucose is treated with three equivalents of phenylhy-

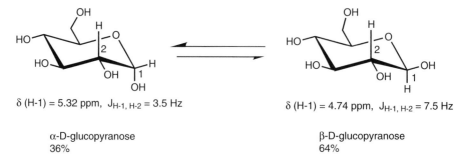

δ (H-1) = 5.32 ppm, $J_{H\text{-}1, H\text{-}2}$ = 3.5 Hz δ (H-1) = 4.74 ppm, $J_{H\text{-}1, H\text{-}2}$ = 7.5 Hz

α-D-glucopyranose β-D-glucopyranose
36%   64%

Figure 1–35. ^{1}H NMR spectral values for the anomeric protons in α and β-D-glucopyranose.

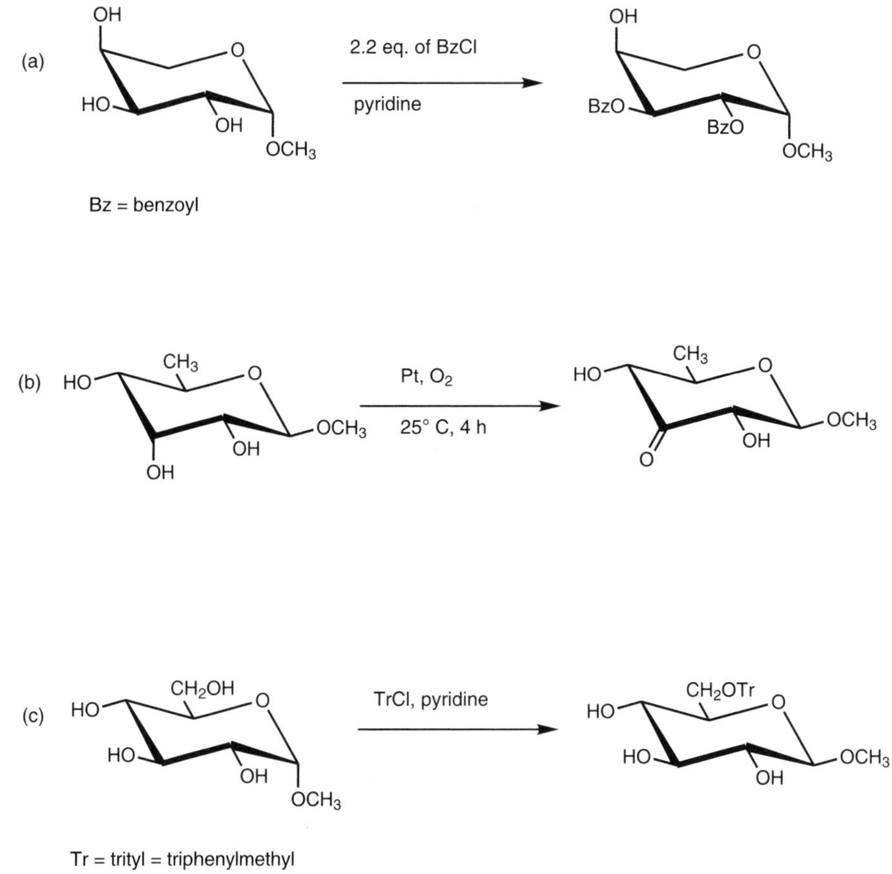
(a)
OH
O
HO
OH
OCH3
2.2 eq. of BzCl
pyridine
OH
O
BzO
BzO
OCH3
Bz = benzoyl
(b)
HO
CH3
O
OCH3
OH
OH
Pt, O2
25° C, 4 h
HO
CH3
O
OCH3
O
OH
(c)
HO
CH2OH
O
HO
OH
OCH3
TrCl, pyridine
HO
CH2OTr
O
HO
OH
OCH3
Tr = trityl = triphenylmethyl

Figure 1–36. **(a)** Selective benzoylation of equatorial hydroxyl groups. **(b)** Selective oxidation of an axial hydroxyl group. **(c)** Selective alkylation of a primary hydroxyl group.

drazine, an osazone is formed. It should be noted that the chirality of C-2 is destroyed, so that D-mannose affords the same osazone (Fig. 1-37).

Finally, it should be mentioned that there seems to be a correlation between the natural abundance of monosaccharides and their relative stabilities. D-glucose has the most stable conformational ring and is the most abundant of the monosaccharides. D-galactose and D-mannose, having each one axial group in their rings, are the next most abundant.

Natural and Synthetic Monosaccharides and Monosaccharide Derivatives

Most monosaccharides occur in nature as part of larger molecules such as simple or complex oligosaccharides or polysaccharides (Schaffer, 1970).

D-glucose occurs free in blood, fruits, honey, plant juices, urine, and cerebrospinal fluid, and is the main component of many oligosaccharides and polysaccharides (in-

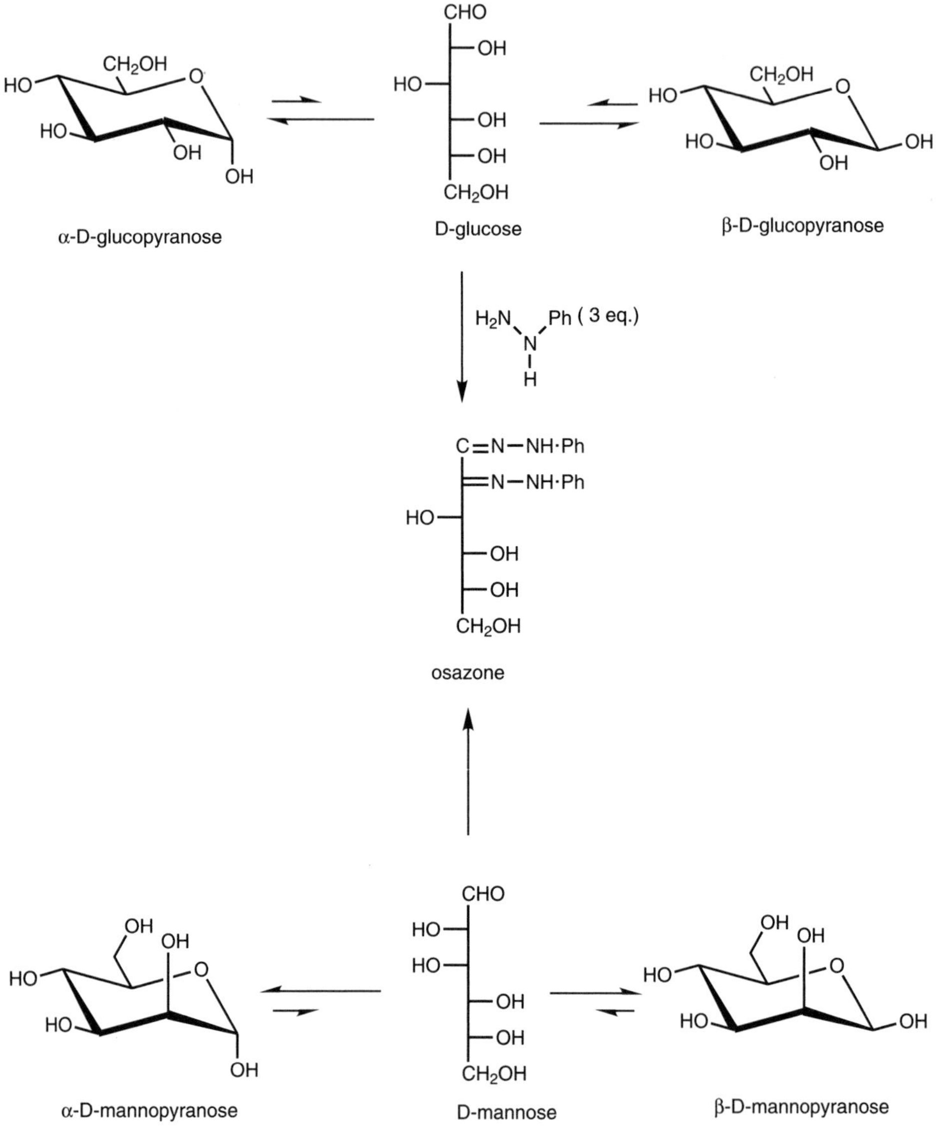

Figure I-37. Formation of an osazone from D-glucose and D-mannose.

cluding *cellulose*, *glycogen*, and *starch*). D-mannose is present in numerous polysaccharides, including some *mucilages* and *hemicelluloses*. Traces of the free sugar are known to exist in apple and peach fruits. *D-fructose* is a constituent of the disaccharide *sucrose*. Certain plants contain polysaccharides of D-fructose known as *inulins*. *D-ribose* is a constituent of *nucleic acids*. *D-galactose* is a constituent of several oligosaccharides: *lactose*, *melobiose*, and *raffinose*. Polysaccharides that yield D-galactose on hydrolysis include *gums* and mucilages. Traces of the free sugar have been detected in fruits.

All monosaccharides that do not fit the formula $C_n(H_2O)_n$ due to the presence of different functional groups (e.g., amino, halogen, acids, thio, etc.) can be considered monosaccharide derivatives. The most important groups of monosaccharides derivatives are described below by mentioning a few characteristic examples, their role and occurrence in nature (Horton & Pigman, 1970) and, in some cases, the most common methods of synthesis (Ferrier & Collins, 1972).

Aldonic acids are monosaccharides in which the aldehydo group has been replaced by a carboxyl group. They are obtained synthetically by oxidation of the corresponding monosaccharide with aqueous bromine. Usually they exist in the lactone form (Fig. 1-38).

They are used in the identification and characterization of monosaccharides. D-gluconic acid occurs as a metabolic intermediate in animals, plants, and microorganisms. L-gulonic acid is an intermediate in the biosynthesis of L-ascorbic acid (vitamin C) (Fig. 1-38).

Uronic acids are monosaccharides in which the primary alcohol is replaced by a carboxylic acid functionality. They are obtained synthetically by oxidation of the corresponding protected monosaccharide with oxygen, using a platinum catalyst or potassium permanganate. Usually they exist in the lactone form (Fig. 1-39).

D-glucuronic acid serves as a detoxifying agent in animals by aiding the excretion of phenols, steroids, and aromatic carboxylic acids in the form of alkyl or aryl D-glucosiduronic acids. Polysaccharides containing D-glucuronic acid include heparin, hyaluronic acid, *Pneumococcus* polysaccharides, and a number of plant polysaccha-

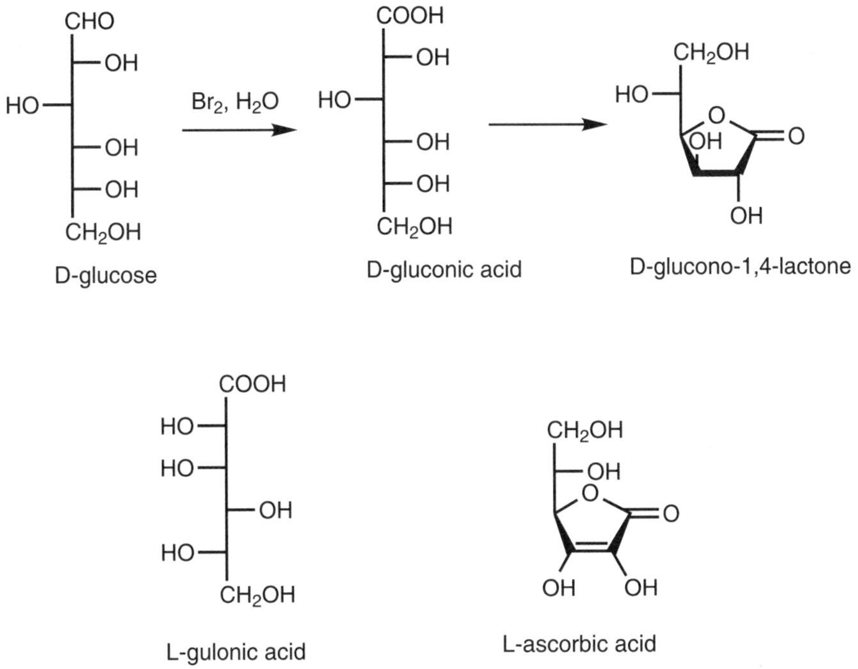

Figure 1–38. Aldonic acids.

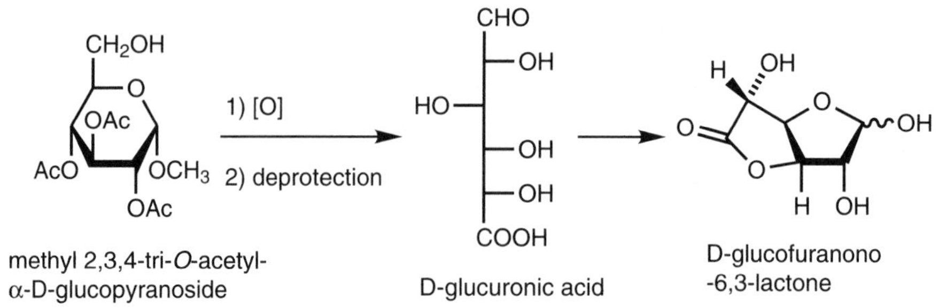

methyl 2,3,4-tri-O-acetyl-
α-D-glucopyranoside

D-glucuronic acid

D-glucofuranono
-6,3-lactone

Figure I–39. Uronic acids.

rides, hemicelluloses, and gums. D-galacturonic acid occurs in fruit pectin. D-mannuronic acid and L-guluronic acid are present in various seaweed polysaccharides.

The reactivity of the carboxyl group makes the identification of these residues in polysaccharides difficult, due to the occurrence of decarboxylation during hydrolysis.

Aldaric acids are monosaccharides in which the aldehydo group and the primary alcohol are replaced by carboxylic groups. These compounds are obtained synthetically by oxidation of the corresponding monosaccharide with nitric acid. They usually exist as dilactones (Fig. 1-40). Because of symmetry properties, D-glucose and L-gulose afford the same aldaric acid. The name D-glucaric acid is preferred to L-gularic acid due to the principle of alphabetic precedence.

Aldaric acids are not known to have any great biological significance.

Alditols are monosaccharide derivatives in which the aldehydo or keto group has been replaced by a hydroxyl group. They are obtained synthetically by reduction of the corresponding monosaccharide with reagents such as sodium borohydride or sodium amalgam. Due to their symmetry properties, reduction of D-glucose, D-fructose, or L-gulose all afford the same alditol (D-glucitol, Fig. 1-41). Reduction of ketoses usually affords the two possible stereoisomers; for example, reduction of D-fructose affords D-glucitol and D-mannitol.

Alditols occur extensively in nature. Glycerol is an essential component of lipids. D-glucitol is found in many fruits, such as apples and plums. D-ribitol occurs in some microbial polysaccharides. D-mannitol is found as a component of polysaccharides in brown seaweeds.

Cyclitols are polyhydroxycyclohexanes and hexenes, and their derivatives. The most studied members of this group are the cyclohexanehexols, commonly known as _inositols_. The structures of all stereoisomers are shown in Figure 1-42.

Myo-inositol is the most widely distributed inositol and occurs free or combined in the tissues of nearly all living species. In animals and microorganisms, _myo_-inositol is in the form of phospholipid. In plants it also occurs as phytate (a hexaphosphate) or as methyl ethers. The next most abundant inositols are derivatives of D-_chiro_-inositol, L-_chiro_-inositol, and _scyllo_-inositol.

Amino sugars are monosaccharide derivatives in which one or more hydroxyl groups, except for the anomeric one, have been replaced by an amino function. When the hydroxyl group that has been replaced is the anomeric one, they are called _glycosylamines_ or _N-glycosides_.

Amino sugars occur widely in nature as simple compounds, as components of larger molecules, and in polysaccharides. One of the most abundant monosaccharides is 2-

amino-2-deoxy-D-glucose. It occurs mainly as its N-acetylated derivative (Fig. 1-43) in polysaccharides, glycoproteins, and proteoglycans. It also occurs as a major constituent in the hard shells of crustaceous and other arthropods. Also occurring in glycoproteins but to a lesser extent are 2-acetamido-2-deoxy-D-galactose, -D-mannose, and -D-talose (Fig. 1-43). One important class of compounds containing amino sugars are the carbohydrate antibiotics (*vide infra*). Amino sugars in which the amino group is at a different position than C-2, and those containing more than one amino group have also been found in nature.

Many methods have been developed for the synthesis of amino sugars. The most common ones involve the displacement of a good leaving group such as the sulfonyloxy group, or the opening of epoxide rings by a nitrogenous nucleophile, typically azide ion, which is then reduced to an amine. Another method involves the reduction of ulose oximes (Fig. 1-44).

The simplest glycosylamines are those in which the aglycon is an amino, alkylamino, or arylamino group. They are generally prepared by direct condensation of the parent monosaccharide with the appropriate amine (Fig. 1-45).

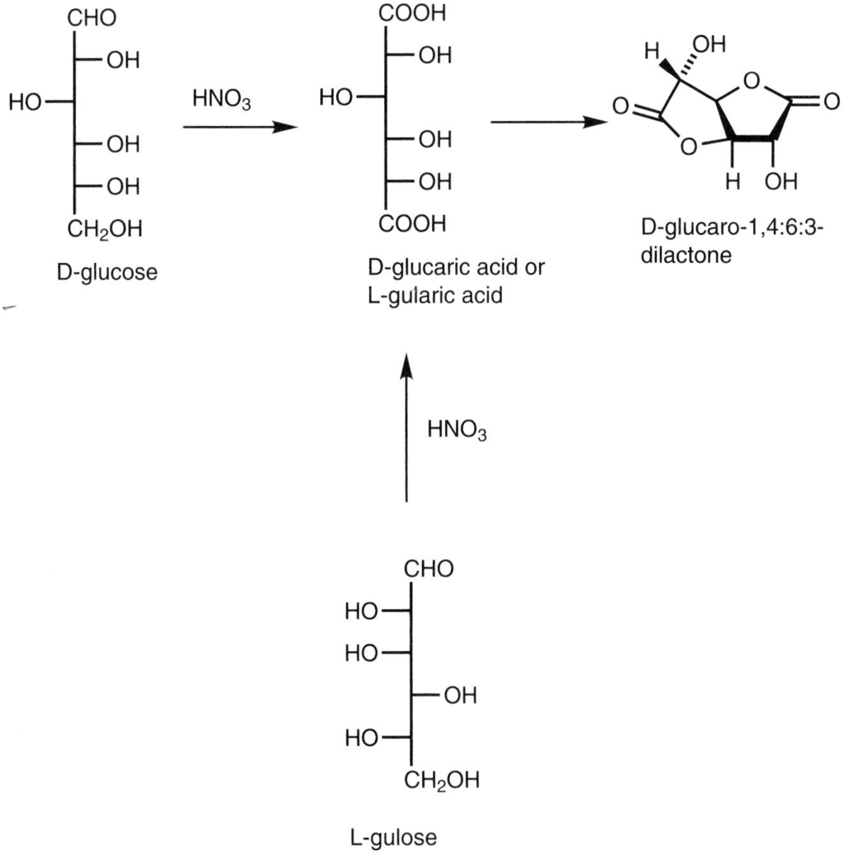

Figure 1–40. Aldaric acids.

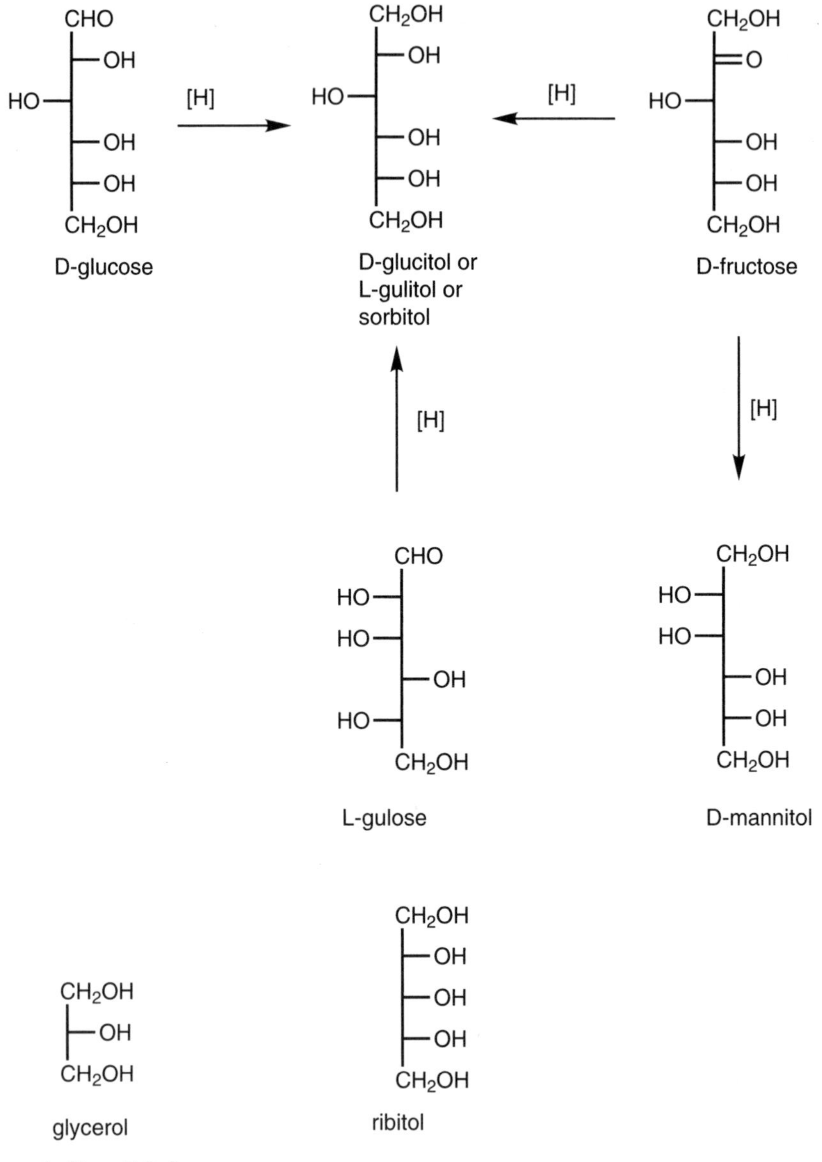

Figure 1–41. Alditols.

The glycosamines of greatest biochemical interest are the *nucleosides*, which are the constituents of nucleic acids (DNA and RNA, *vide infra*). Nucleosides are glycosylamines in which the aglycon is a purine or pyrimidine base. The sugar component is virtually always D-ribofuranose or 2-deoxy-D-ribofuranose (2-deoxy-D-*erythro*-pentofuranose) having a β anomeric configuration (Fig. 1-46). *Nucleotides* are phosphate esters of the nucleosides and represent the monomeric units of nucleic acids. Other nucleotides play important roles as coenzymes, for example adenosine 5′-

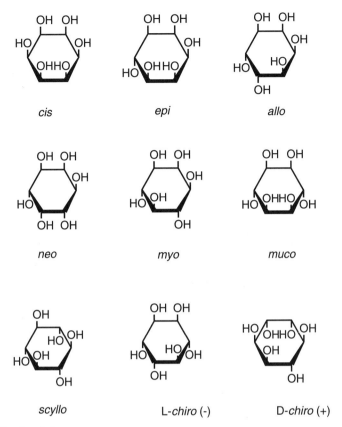

Figure I–42. Inositols.

triphosphate (ATP) and nicotinamide adenine dinucleotide (NAD). ATP acts as a "storage form" for the energy liberated in chemical reactions. The same energy is liberated by hydrolysis of one of the phosphate groups. That energy becomes available to the cell for chemical, mechanical, or electrical work. NAD is involved in numerous enzyme-catalyzed redox processes.

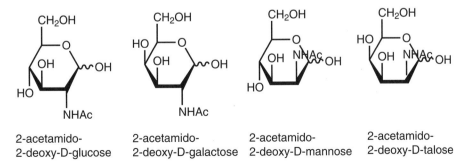

2-acetamido-2-deoxy-D-glucose   2-acetamido-2-deoxy-D-galactose   2-acetamido-2-deoxy-D-mannose   2-acetamido-2-deoxy-D-talose

Figure I–43. Amino sugars.

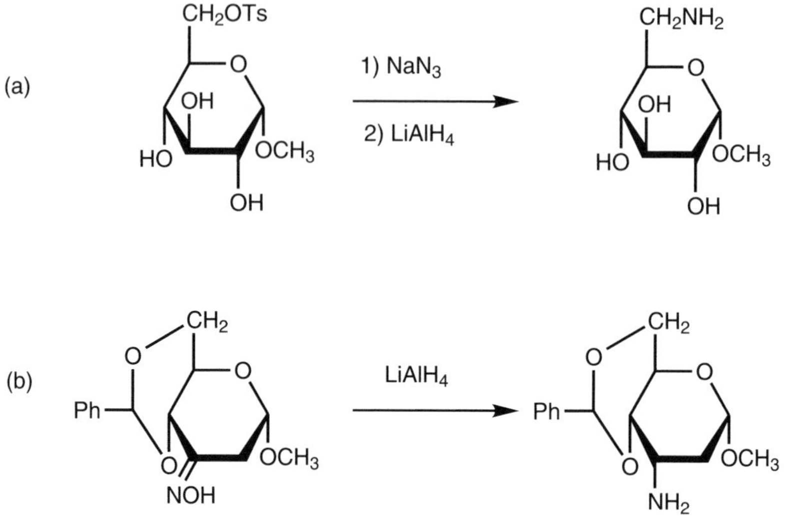

Figure I–44. Synthesis of amino sugars by **(a)** nucleophilic displacement of tosyl group with sodium azide and **(b)** reduction of ulose oxime.

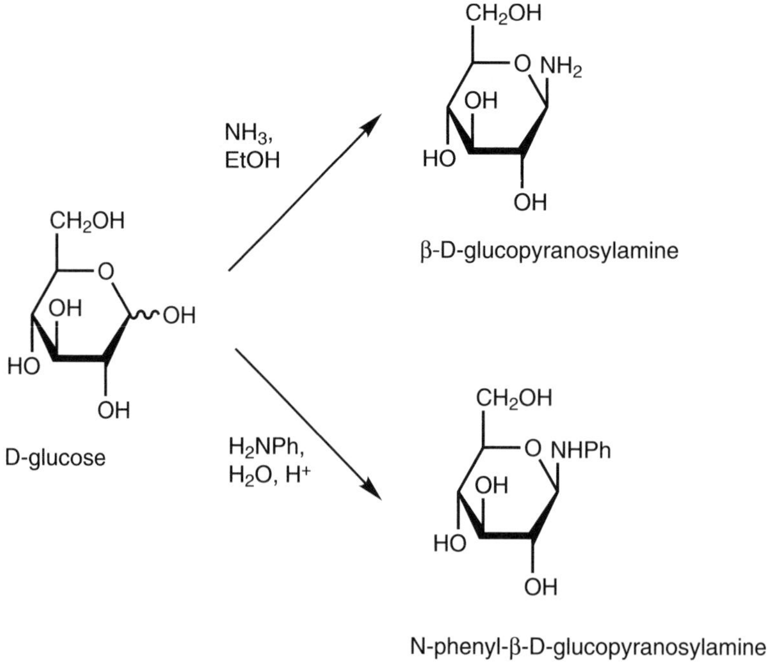

Figure I–45. Synthesis of simple glycosylamines.

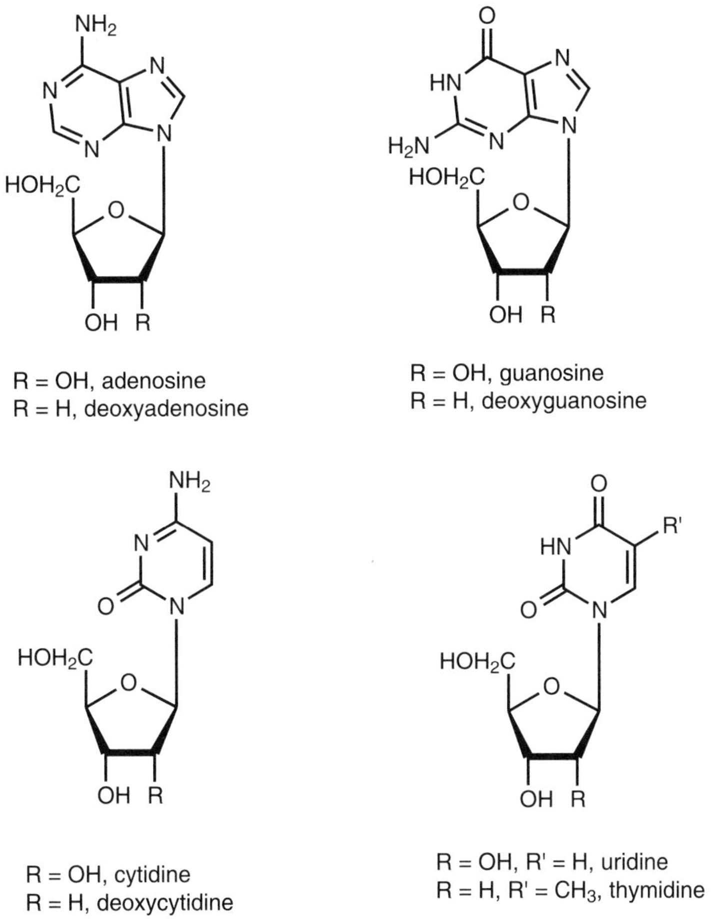

R = OH, adenosine
R = H, deoxyadenosine

R = OH, guanosine
R = H, deoxyguanosine

R = OH, cytidine
R = H, deoxycytidine

R = OH, R' = H, uridine
R = H, R' = CH$_3$, thymidine

Figure 1–46. Nucleosides.

Deoxy sugars are monosaccharide derivatives in which one or more hydroxyl groups, except for the anomeric one, have been replaced by hydrogen. The most common deoxy sugars found in nature are 6-deoxyhexoses (Fig. 1-47). L-Rhamnose (6-deoxy-L-mannose) and L-fucose (6-deoxy-L-galactose) are constituents of many polysaccharides, glycoproteins, and plant glycosides. Occurring in DNA (*vide supra*) is 2-deoxy-D-*erythro*-pentofuranose (2-deoxy-D-ribose).

Among the most common dideoxy sugars are 3,6-dideoxyhexoses, which are constituents of the polysaccharides of the *Salmonella* species, and 2,6-dideoxy-hexoses, which occur as components of some antibiotics (*vide infra*).

As exemplified in Figure 1-48, the most common methods of synthesis of deoxy sugars involve (1) reductive opening of epoxide rings, (2) reductive cleavage of C—X bonds (X = Cl, Br, I, or Ts), and (3) hydrogenation of unsaturated bonds.

Thio sugars are monosaccharide derivatives in which one or more hydroxyl groups have been replaced by a thio group. The most common ones in nature are those in which the anomeric hydroxyl group has been replaced; these are called *1-thioglyco-*

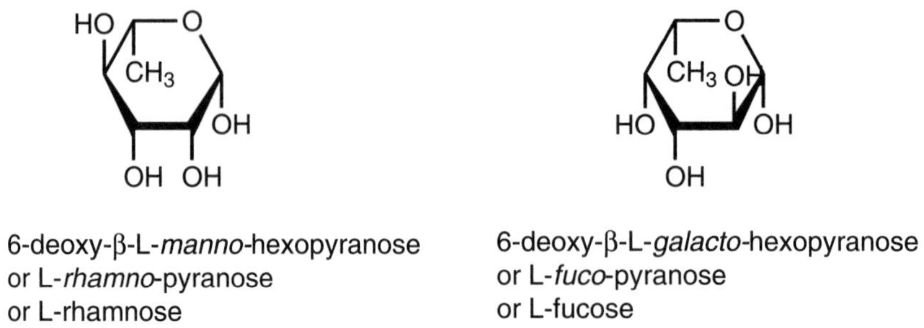

HO
O
CH3
OH
OH OH
6-deoxy-β-L-*manno*-hexopyranose
or L-*rhamno*-pyranose
or L-rhamnose
O
CH3 OH
HO
OH
OH
6-deoxy-β-L-*galacto*-hexopyranose
or L-*fuco*-pyranose
or L-fucose

Figure 1–47. Deoxy sugars.

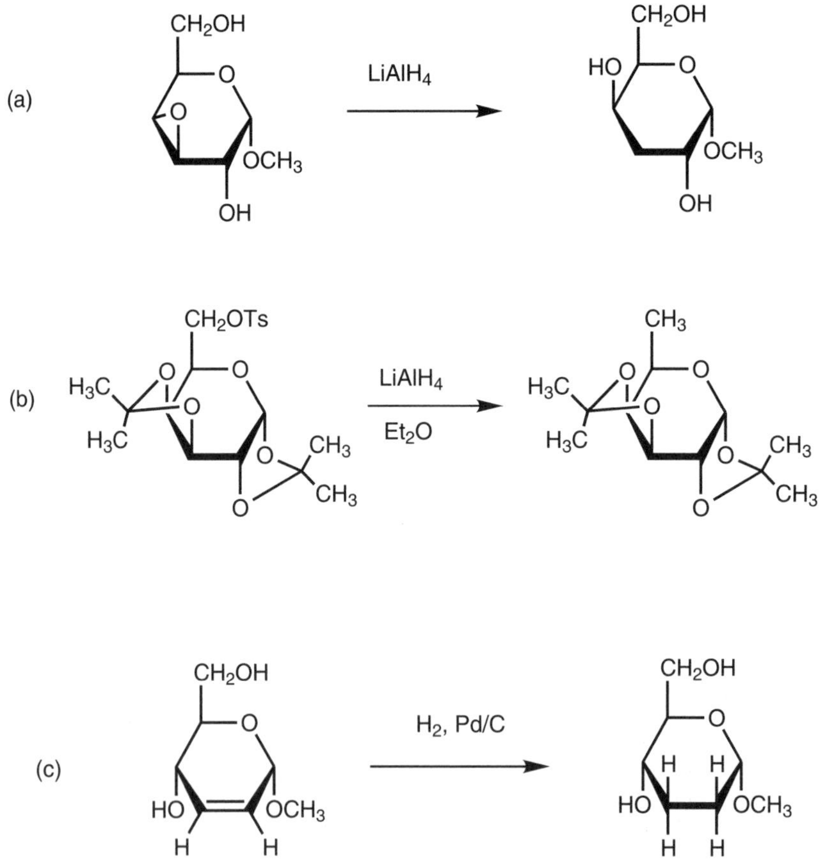
(a)
CH2OH
O
O
OCH3
OH
LiAlH4
CH2OH
HO
O
OCH3
OH
(b)
CH2OTs
H3C
H3C
O
O
O
O
CH3
CH3
LiAlH4
Et2O
CH3
H3C
H3C
O
O
O
O
CH3
CH3
(c)
CH2OH
O
HO
OCH3
H
H
H2, Pd/C
CH2OH
O
H H
HO
OCH3
H H

Figure 1–48. Synthesis of deoxy sugars by **(a)** reductive opening of epoxide rings, **(b)** reductive cleavage of C—X bonds, and **(c)** by hydrogenation of unsaturated bonds.

sides or *S-glycosides*. They occur in the seeds of many plants; e.g., singrin (a derivative of 1-thio-D-glucose, Fig. 1-49a) is isolated from the seeds of black mustard. This and many related 1-thioglycosides are known to be components of the mustard oil glycosides of several plant families, and 1-thioglycosides can be prepared synthetically from peracetylated pyranoses (Fig. 1-49b). They are useful intermediates for the synthesis of glycosyl halides.

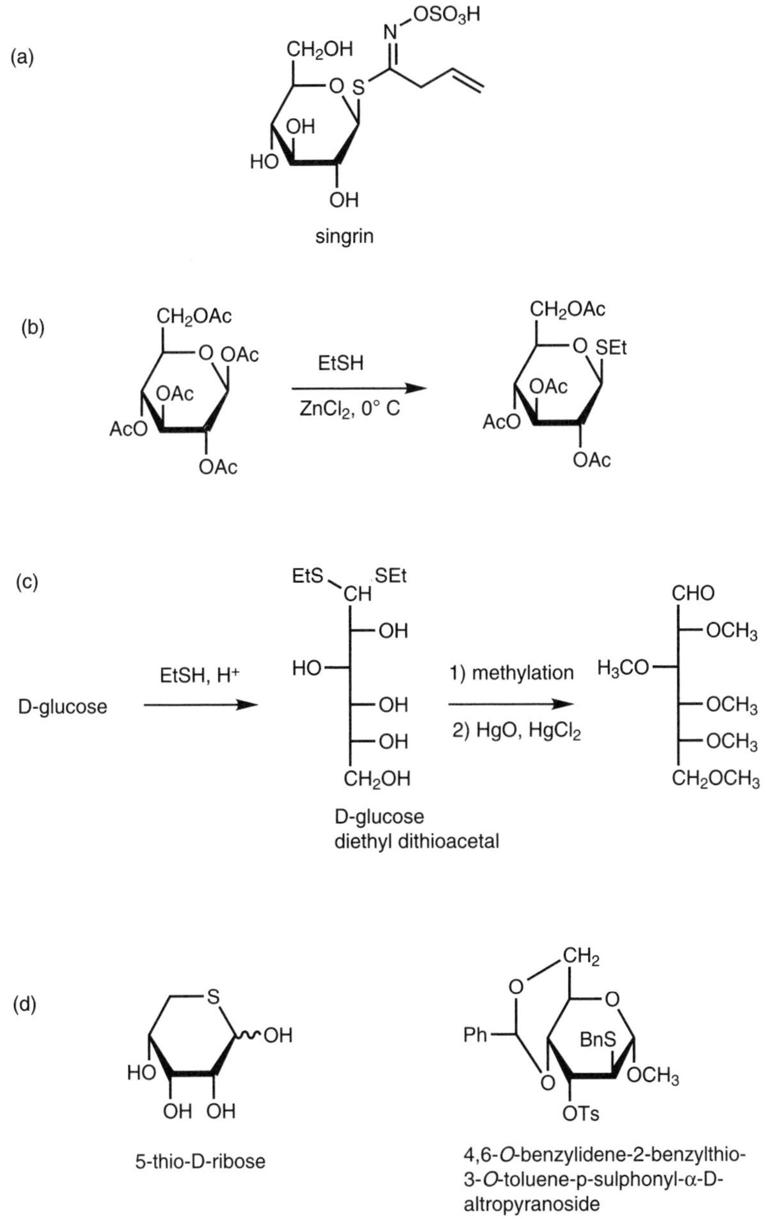

(a)

singrin

(b)

CH$_2$OAc — O OAc, OAc, AcO, OAc   EtSH, ZnCl$_2$, 0° C → CH$_2$OAc — O SEt, OAc, AcO, OAc

(c)

D-glucose   EtSH, H$^+$ →   EtS—CH—SEt, —OH, HO—, —OH, —OH, CH$_2$OH   D-glucose diethyl dithioacetal   1) methylation 2) HgO, HgCl$_2$ →   CHO, —OCH$_3$, H$_3$CO—, —OCH$_3$, —OCH$_3$, CH$_2$OCH$_3$

(d)

5-thio-D-ribose

4,6-*O*-benzylidene-2-benzylthio-3-*O*-toluene-p-sulphonyl-α-D-altropyranoside

Figure I–49. (a–d) Examples of thio sugars and their methods of preparation.

Other thio sugars that have been used extensively as synthetic intermediates are the *thioacetals* (Fig. 1-49c). These are prepared by treating the appropriate monosaccharide with an alkyl or aryl thiol in the presence of aqueous hydrochloric acid. They are very useful intermediates for preparing acyclic derivatives of carbohydrates.

Thio sugars in which some hydroxyl group other than the anomeric one has been replaced by sulfur, have also been prepared synthetically. Examples of these compounds are shown in Figure 1-49d.

Halo sugars are monosaccharides in which one hydroxyl group has been replaced by a halogen atom. Very few halosugars have been obtained from natural sources; they have been obtained by synthesis. The most important are the *1-halosugars*, also known as *glycosyl halides*. These are reactive intermediates that have been used widely for the synthesis of glycosides and oligosaccharides, as shown in Figure 1-50. They are also useful intermediates in the synthesis of *glycals*.

Unsaturated sugars are monosaccharides with a double bond in the carbon skeleton. Examples of natural products containing unsaturated sugars include the antibiotic nucleosides cytosinine and decoyinine, and 2,6-diamino-2,3,4,6-tetradeoxy-D-*glycero*-hex-4-enopyranose, a constituent of the natural antibiotic sisomycin (Fig. 1-51a).

Unsaturated sugars have been found to be valuable intermediates in synthesis. Particularly interesting are the unsaturated sugars having a double bond between C-1 and C-2, also known as *glycals*. These compounds can be prepared by treating the corresponding acetobromo sugars with Zn in AcOH. Thus, acetobromoglucose gives tri-*O*-acetyl-D-glucal (Fig. 1-51b). Glycals are vinyl ethers that readily undergo a variety of addition reactions to the double bond. They have been used as intermediates in the synthesis of 2-deoxy derivatives, glycosides, and 1,2-dihydroxylated compounds.

Branched chain sugars are monosaccharides in which either a hydrogen or a hydroxyl group of the ring have been replaced by an alkyl substituent. Examples of naturally occurring branched chain sugars include D-apiose (3-*C*-hydroxymethyl-D-*glycero*-tetrose), found in parsley, and streptose (5-deoxy-3-*C*-formyl-L-lyxose), a component of the antibiotic streptomycin (Fig. 1-52a).

Branched chain sugars can be obtained synthetically by a variety of methods, including treatment of carbonyl derivatives with diazomethane, Grignard, or Wittig reagents (Fig. 1-52b).

O-Glycosides, simply referred to as *glycosides*, are monosaccharide derivatives in which the anomeric hydroxyl group has been replaced by an alkoxy group. The alkyl portion of the alkoxy group is referred as the *aglycon group* and the corresponding alcohol is called the *aglycon*. The carbohydrate residue is referred to as the *glycosyl*

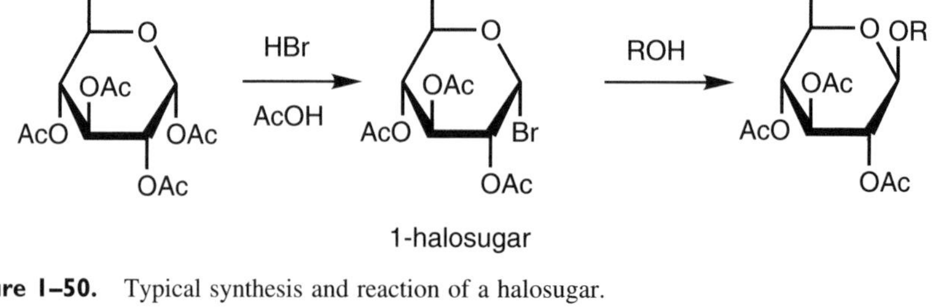

Figure 1–50. Typical synthesis and reaction of a halosugar.

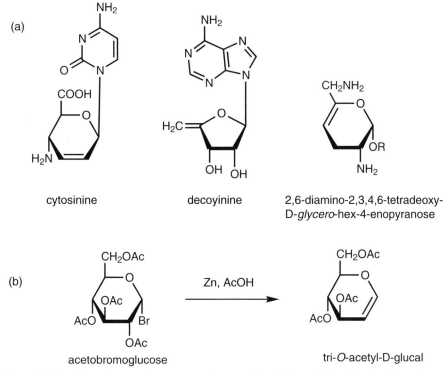

(a)

cytosinine   decoyinine   2,6-diamino-2,3,4,6-tetradeoxy-
                                                            D-*glycero*-hex-4-enopyranose

(b)

Zn, AcOH

acetobromoglucose   tri-*O*-acetyl-D-glucal

Figure 1–51. Naturally occurring unsaturated sugars (**a**) and the synthesis of a glycol from the corresponding acetobromo sugar (**b**).

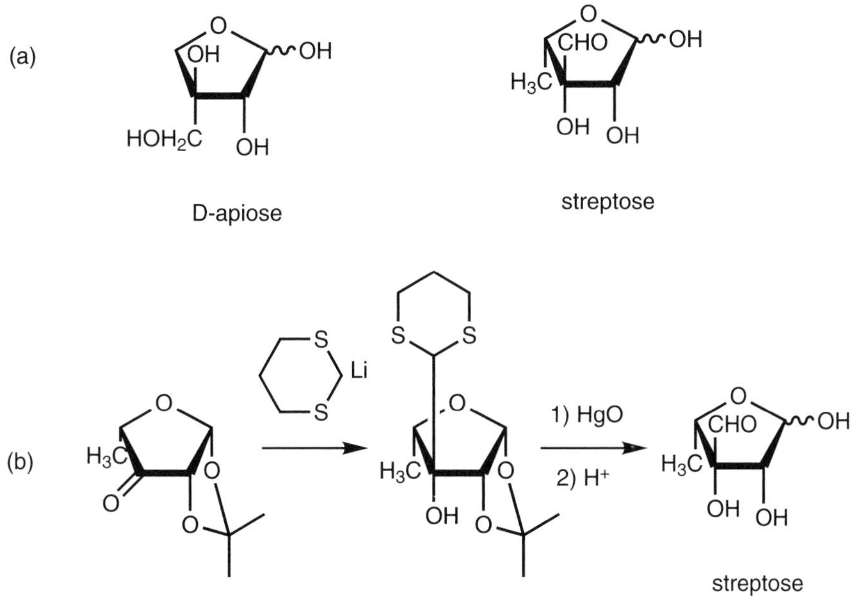

(a)

D-apiose   streptose

(b)

1) HgO
2) H⁺

streptose

Figure 1–52. Naturally occurring branched chain sugars (**a**), and a method for the synthesis of streptose (**b**).

group. For example, methyl α-D-glucopyranoside (Fig. 1-53a) is a glycoside in which the methyl group is the *aglycon group* and D-glucose is the *glycosyl group*. The bond between the sugar and the aglycon is an *acetal linkage*, and is referred to as the *glycosidic* bond.

Glycosides are widely distributed in nature. In many cases, such as complex glycosides, the chemistry of natural glycosides resides to a considerable extent in the aglycon residue (*vide infra*). Simple glycosides like methyl β-D-glucopyranoside and methyl arbutin (Fig. 1-53b) have been isolated from the leaves of certain plants. Salicin is a glycoside found in the bark of the willow. More complex glycosides will be treated in the section on complex glycosides below.

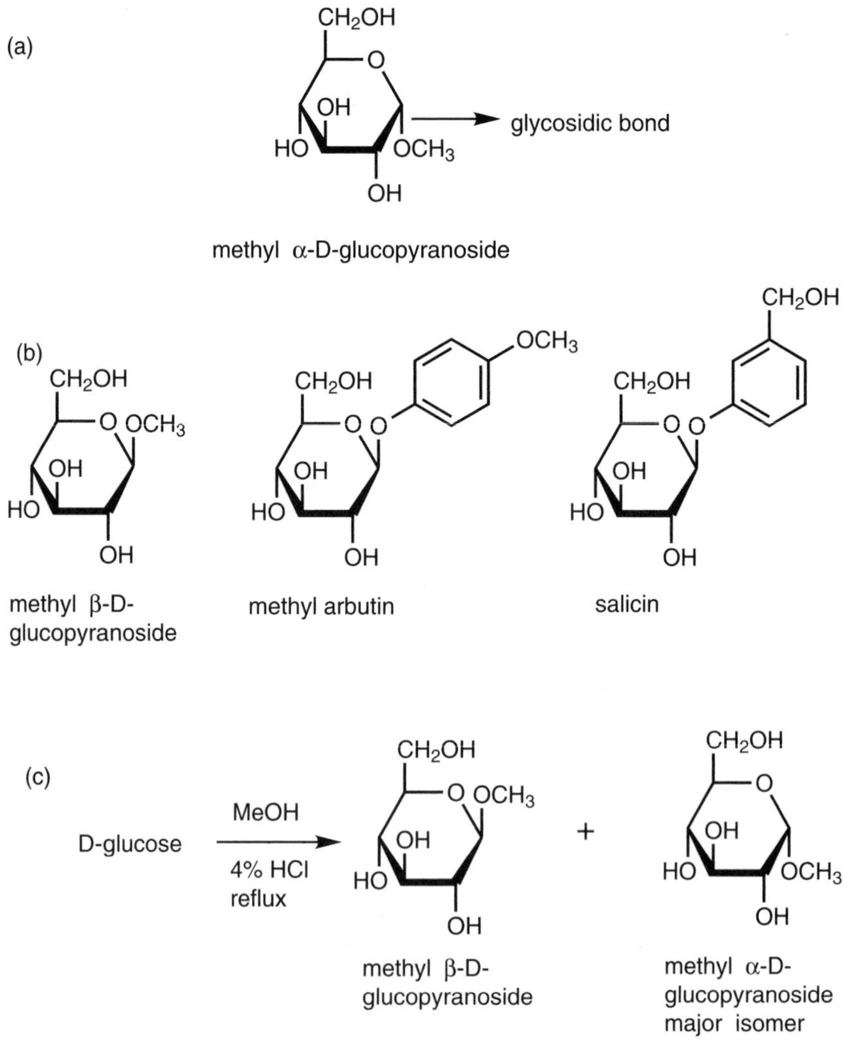

(a)

methyl α-D-glucopyranoside

(b)

methyl β-D-glucopyranoside   methyl arbutin   salicin

(c)

Figure 1–53. Glycosides. **(a)** Methyl α-D-glucopyranoside showing the glycosidic bond. **(b)** Naturally occurring glycosides. **(c)** Synthesis of glycosides.

Simple glycosides, like those containing methyl or ethyl groups, can be prepared by treating the appropriate monosaccharide with an alcohol in the presence of an acid. Usually mixtures of the α and β anomers of furanosides and pyranosides are obtained. However, the conditions can be controlled in most of the cases in order to obtain one major isomer. For example, treatment of D-glucose with hot 4% methanolic hydrogen chloride gives a mixture that is mainly comprised of pyranosides, of which the α-isomer predominates (Fig. 1-53c). Another method involves the treatment of fully acetylated glycosyl halides with the corresponding alcohol in the presence of silver or mercuric salts. This method, known as the Koenigs-Knorr synthesis of glycosides, is also suitable for the synthesis of glycosides having complex aglycons. Glycosides can also be synthesized from glycals, orthoesters, thioacetals, and other suitable precursors.

Acetals, esters, and ethers play an important role as protecting groups in carbohydrate chemistry (Haines, 1976, 1981; Sugihara, 1953). Through selective protection and deprotection of certain hydroxyl groups, complex syntheses have been achieved. The use of these groups has already been illustrated in some of the foregoing examples.

Oligosaccharides

Oligosaccharides (*oligos* means *few* in Greek) are polymers composed of 2 to 10 glycosidically linked monosaccharides (Kennedy & White, 1983). Molecules with more than 10 units are called polysaccharides. This division is somewhat arbitrary. Because of their glycosidic linkages, oligosaccharides are hydrolyzed readily by aqueous acid to their constituent monosaccharides. Oligosaccharides represent a unique type of glycosides, in which the "aglycon" is also a carbohydrate residue instead of a simple alcohol.

Because of their sweet taste, monosaccharides and lower oligosaccharides have been called sugars. Sweetness decreases with increasing degree of polymerization. In general, oligosaccharides having more than four constituent monosaccharides are tasteless.

According to the number of monosaccharide units, oligosaccharides can be classified for example as di, tri, and tetrasaccharides. Each of these groups can be further subdivided into homo-oligosaccharides, consisting of only one type of monosaccharide, or hetero-oligosaccharides, consisting of more than one type of monosaccharide. A second subdivision into *reducing* or *nonreducing* oligosaccharides is based on the presence or absence of a free hemiacetal group. They are called reducing sugars because the hemiacetal function is readily oxidized to a carboxylic acid with concomitant reduction of the oxidizing reagent.

Oligosaccharides comprise a large and important class of polymeric carbohydrates, which are found either free or in a combined form in virtually all living entities. Some oligosaccharides were known before their chemical constitution had been fully elucidated and were given trivial names, which are still being used: examples include sucrose, lactose, and maltose. These names often reflect their source. In the following examples the trivial name, together with the systematic and abbreviated names are presented. For example, the disaccharide with the trivial name lactose has the systematic name 4-O-β-D-galactopyranosyl-D-glucopyranose, which means that a residue of D-galactopyranose is linked glycosidically to the 4-OH of a residue of D-glucopyranose, and the glycosidic bond is β. The abbreviated name is β-D-Gal$p$-(1 → 4)-D-Glc$p$, where Gal and Glc are the abbreviations for galactose and glucose, respectively, and *p* is the abbreviation for pyranose.

Disaccharides

Sucrose (nonreducing hetero-oligosaccharide, Fig. 1-54) is one of the most common carbohydrates, widely distributed throughout the plant world. It is commonly known as table sugar, or just sugar, and it is obtained commercially from sugar beet or cane. It is also the main soluble carbohydrate reserve and energy source and an important component of the human diet.

Lactose (reducing hetero-oligosaccharide, Fig. 1-54), also known as *milk sugar*, occurs mainly in mammalian milk in approximately 5% concentration. Lactose is prepared industrially from whey, a by-product in the manufacture of cheese.

Other disaccharides occur as structural elements of glycosides, polysaccharides, or other oligosaccharides. The most important ones are: *maltose, isomaltose*, and *cellobiose*. Maltose (reducing homo-oligosaccharide, Fig. 1-55) is a structural component of the polysaccharides *starch* and *glycogen*. It is normally prepared by enzymic hydrolysis of starch. Isomaltose (reducing homo-oligosaccharide, Fig. 1-55) is a structural component of the polysaccharides *amylopectin* and *glycogen*; it represents the branch point of these compounds. It is also the major repeating unit of many bacterial polysaccharides of the dextran type. Isomaltose is obtained by partial acid hydrolysis of bacterial dextran. *Cellobiose* (reducing homo-oligosaccharide, Fig. 1-55) is the repeating unit of the polysaccharide *cellulose* and is usually obtained by its partial hydrolysis.

Trisaccharides

Raffinose (nonreducing hetero-oligosaccharide, Fig. 1-56) is the most abundant trisaccharide found in nature. It is widely distributed in plants, but at much lower concen-

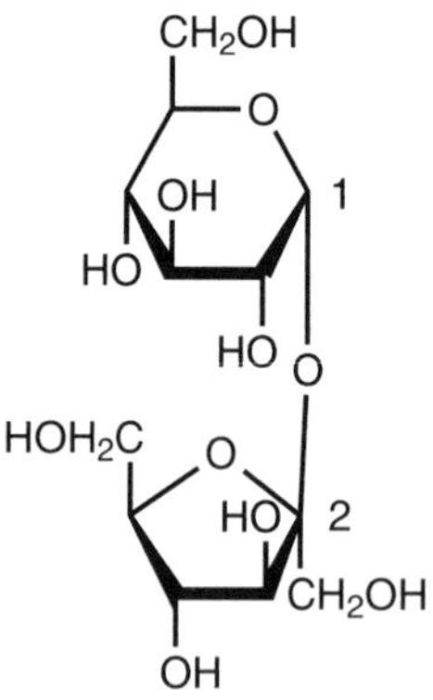

sucrose
β-D-fructofuranosyl α-D-glucopyranoside
β-D-Fru*f*-α-D-Glc*p*

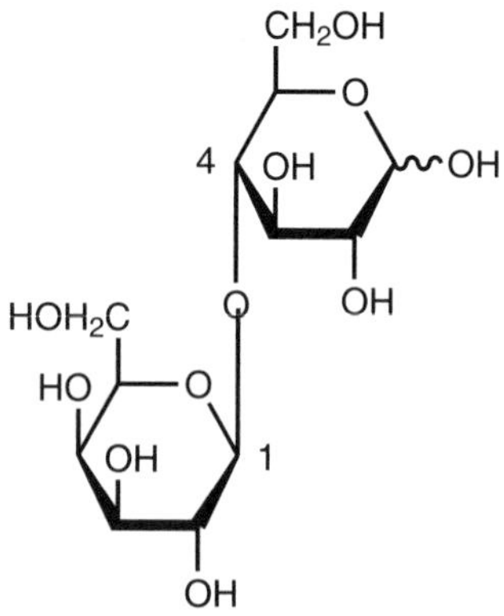

lactose
4-*O*-β-D-galactopyranosyl-D-glucopyranose
β-D-Gal*p*-(1→4)-D-Glc*p*

Figure 1–54. Disaccharides.

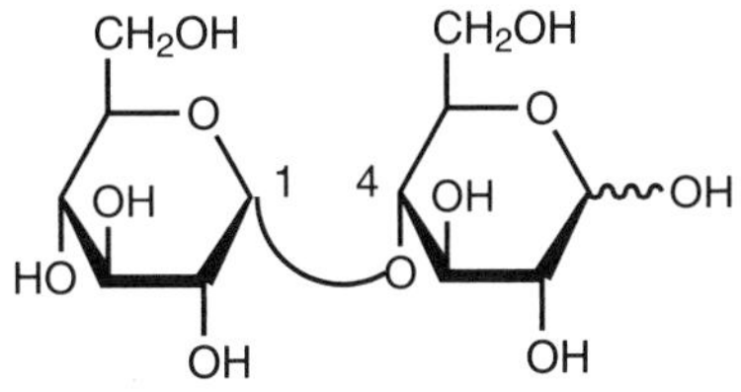

maltose
4-*O*-α-D-glucopyranosyl-D-glucopyranose
α-D-Glc*p*-(1→4)-D-Glc*p*

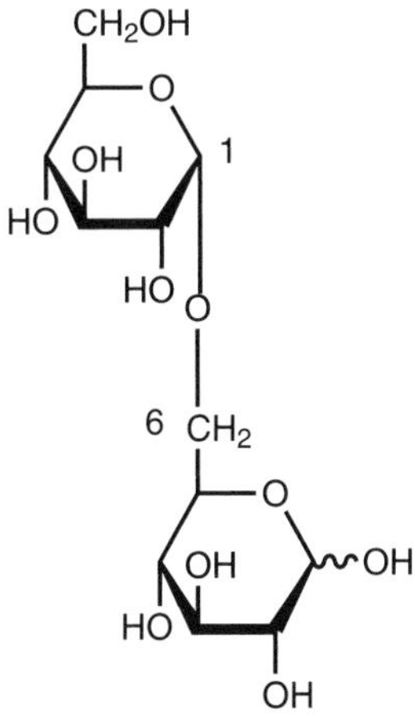

isomaltose
6-*O*-α-D-glucopyranosyl-D-glucopyranose
α-D-Glc*p*-(1→6)-D-Glc*p*

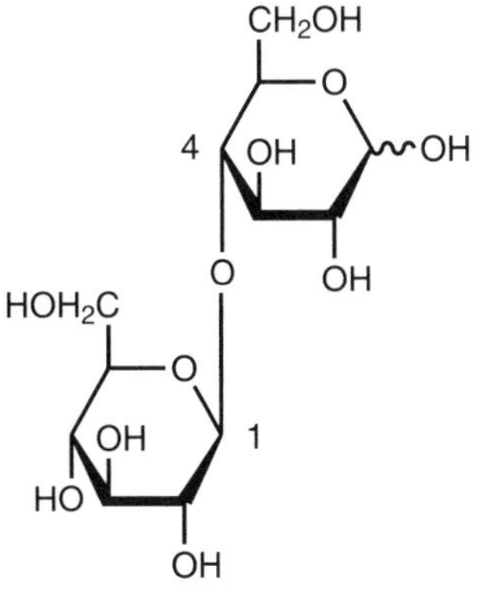

cellobiose
4-*O*-β-D-glucopyranosyl-D-glucopyranose
β-D-Glc*p*-(1→4)-D-Glc*p*

Figure 1–55. Disaccharides.

tration than sucrose. *Panose* (reducing homo-oligosaccharide, Fig. 1-56) is a structural element of the polysaccharides *amylopectin* and *glycogen*.

Higher Oligosaccharides

Oligosaccharides containing between 2 and 10 monosaccharide units, have been isolated from polysaccharides, but few of the larger ones are thought to have significance as discrete entities. One exception is the *Schardinger dextrins* (Clarke et al., 1988;

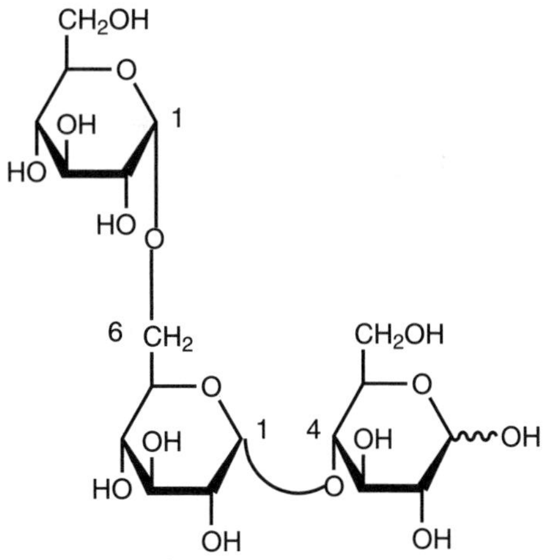

panose

O-α-D-glucopyranosyl-(1→6)-O-α-D-glucopyranosyl-(1→4)-D-glucopyranose

α-D-Glcp-(1→6)-α-D-Glcp-(1→4)-D-Glcp

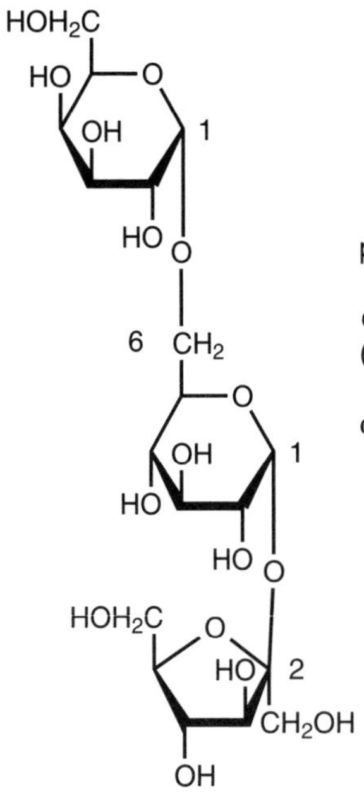

raffinose

O-α-D-galactopyranosyl-(1→6)-O-α-D-glucopyranosyl-(1→2)-β-D-fructofuranoside

α-D-Galp-(1→6)-α-D-Glcp-(1→2)-β-D-Fruf

Figure 1–56. Trisaccharides.

Schardinger, 1911; see also Chapter 13) also known as *cycloamyloses* or *cyclodextrins* (Fig. 1-57). These are cyclic oligosaccharides of glucose which are produced from starch by the action of a hydrolytic enzyme from *Bacillus macerans*. They are all crystalline solids that might contain between 6 and 10 α-D-glucopyranose units per molecule. These compounds bind small molecules within their cyclic structure to form inclusion compounds similar to the polysaccharide *amylose* (*vide infra*). They show the characteristics of hydrolytic enzymes, since they can hydrolyze a number of compounds with which they form inclusion complexes, for example, aromatic esters. These compounds are used in the production of pharmaceutical agents, pesticides, foodstuffs (as protective agents), emulsion stabilizers, and bubbling agents.

Polysaccharides

Polysaccharides (also called *glycans*) are natural polymers composed of more than 10 glycosidically linked monosaccharides (Aspinall, 1982). This number is somewhat arbitrary. Carbohydrates containing 5 to 15 monosaccharide residues rarely occur in nature. A few natural polysaccharides contain 25 to 75 monosaccharide residues, while the vast majority of them contain 80 to 100 residues. Some others, like cellulose, exceed this number, having an average of 3000 residues of D-glucose. Different molecules of a particular polysaccharide will differ in the number of residues, so rather than an exact molecular weight one refers to a distribution of molecular weights or an average molecular weight.

Polysaccharides occur in almost all living organisms. They serve as structural materials (e.g., cellulose) and food storage compounds (e.g., glycogen). They also confer immunological specificity (e.g., polysaccharides of bacterial capsules).

Polysaccharides have names that usually reflect their origin, for example, cellulose is the principal component of cell walls in plants. Other names reflect some property of the polymer, for example, starch is a name derived from the old English word *stercan*, meaning to stiffen. Systematic nomenclature uses the configurational prefix of the parent sugar with the suffix *an* to signify a polymer. The type of glycosidic bond is

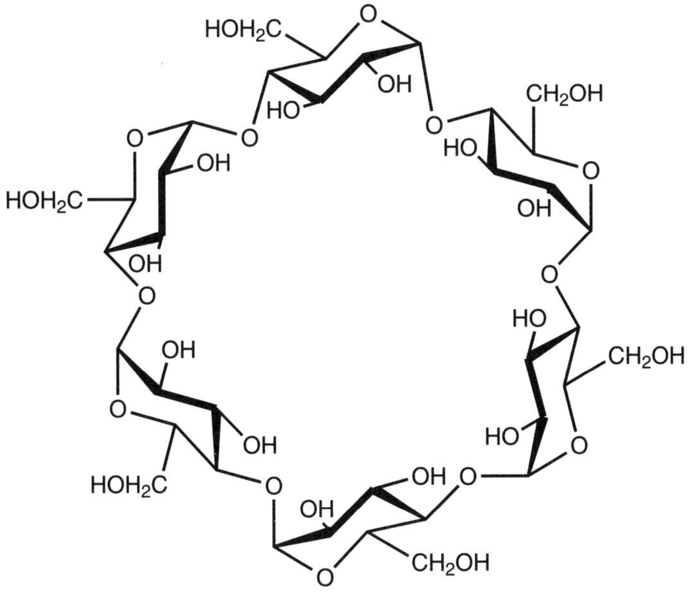

Figure I–57. A cyclodextrin.

also specified, for example, $(1 \rightarrow 2)$-α-D-mannan indicates a polymer of D-mannose, α being the configuration of the anomeric center. Since polysaccharides of the same type differ slightly from one source to another, so it is necessary to specify their origin; an example is maize starch or corn starch. Polysaccharides can be hydrolyzed to their constituent oligo or monosaccharides. As the degree of polymerization increases, their solubility in water decreases and the viscocity increases.

According to their structural features polysaccharides can be classified in two major classes: (1) *homopolysaccharides*, which are constituted by one type of monosaccharide and (2) *heteropolysaccharides*, which are made up of more than one type of monosaccharide. They can be subdivided further into linear or branched. According to their origin they can be classified as plant, animal, bacterial, fungal, or algal polysaccharides. Some representative examples of the plant and animal polysaccharides are presented.

Plant Polysaccharides

Cellulose and starch are the most abundant polysaccharides in the plant kingdom. *Cellulose* (Fig. 1-58) is a linear homopolysaccharide of D-glucose, joined by β-D-$(1 \rightarrow 4)$-linkages, hence its systematic name is $(1 \rightarrow 4)$-β-D-glucan. The average number of D-glucose residues is several thousand; the maximum number is 10,000. These long molecules held in their linear conformation aggregate to form fibrils bound together by hydrogen bonds. This aggregation explains their insolubility in water, their chemical inertness, and their physical strength. Cellulose is much more resistant to acid hydrolysis than other polysaccharides, but in the presence of 70% sulfuric acid it can be converted to D-glucose in good yield. Partial degradation by acetolysis with acetic acid, acetic anhydride, and sulfuric acid gives the disaccharide cellobiose octaacetate. Cellulose occurs throughout the vegetable kingdom as the main constituent of the cell walls of the higher members. It provides their structural strength. Cotton wool is almost pure cellulose (98%). Cellulose is also a constituent of wood (50%), but it is found in close association with other polysaccharides, such as the hemicelluloses and lignin. Other sources of cellulose are flax (80%) and jut (65%).

Starch is the principal energy food reserve in the plant kingdom. It is stored in the roots, tubers, fruits, and seeds in the form of insoluble granules. They form the major source of carbohydrates in the human diet and are, therefore, of great economic importance.

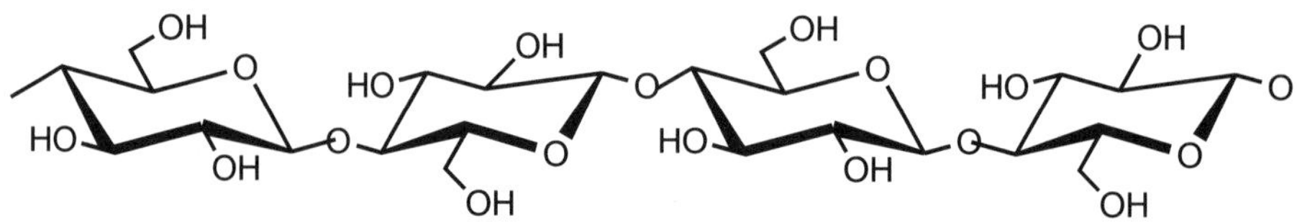

Figure 1–58. Cellulose.

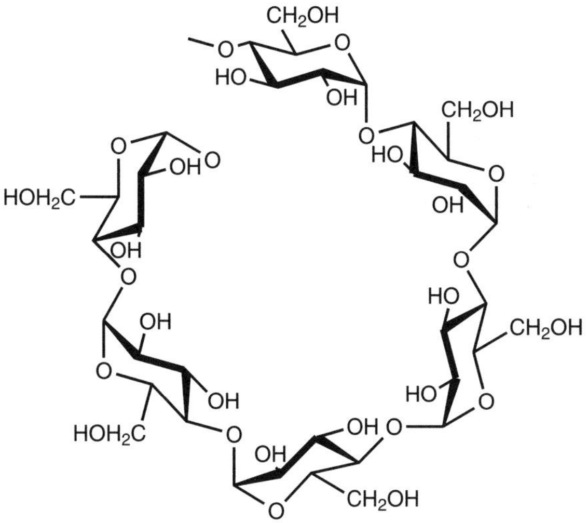

Figure 1–59. Amylose.

Starch is a mixture of two structurally different α-D-glucans: *amylose* and *amylopectin*. The proportion of each one varies depending on the source: waxy maize starch is almost 100% amylopectin, while amylomaize starch is almost 80% amylose. The majority of the starches contain between 15% and 25% of amylose.

Amylose (Fig. 1-59) is a linear homopolysaccharide of D-glucose residues, joined by α-D-(1 → 4) linkages, hence its systematic name is (1 → 4)-α-D-glucan. The average number of D-glucose residues ranges between 1000 and 4000. Amylose adopts a helical structure in solution as a consequence of the steric relationship of adjoining units imposed by the α-bonds. Each turn has approximately six units (Fig. 1-59). Small molecules can occupy the space in the center, as we have seen in the case of *cyclodextrins*. For example, the interaction of amylose with iodine gives an inclusion compound that confers a characteristic blue color. On warming, the color is discharged due to the uncoiling which takes place at high temperatures. When starch solutions are left to stand for a certain period of time, partial precipitation occurs due to the separation of amylose. This phenomenon, called *retrogradation*, occurs when helical molecules align themselves by hydrogen bonding, forming aggregates that precipitate when they exceed colloidal dimensions. Partial hydrolysis of amylose affords the disaccharide maltose.

Amylopectin (Fig. 1-60) is a branched homopolysaccharide of D-glucose. Essentially, it is a (1 → 4)-α-D-glucan with (1 → 6)-α-D linked branch points. Branching occurs randomly. Amylopectin is much larger than amylose; its molecular weight sometimes exceeds one million. It cannot coil effectively like amylose due to the presence of the branching points, hence it does not form inclusion complexes. Amylopectin cannot form aggregates in solution like amylose, so no retrogradation is observed. Partial hydrolysis of amylopectin affords a mixture of the disaccharides maltose and isomaltose and the trisaccharide panose.

Other homopolysaccharides found in the plant kingdom include *laminarin*, a linear β-D-(1 → 3)-linked polymer of D-glucose which occurs in brown seaweeds; *lichenin*, a linear D-glucan consisting of β-D-(1 → 3) and β-D-(1 → 4)-linked units, found in

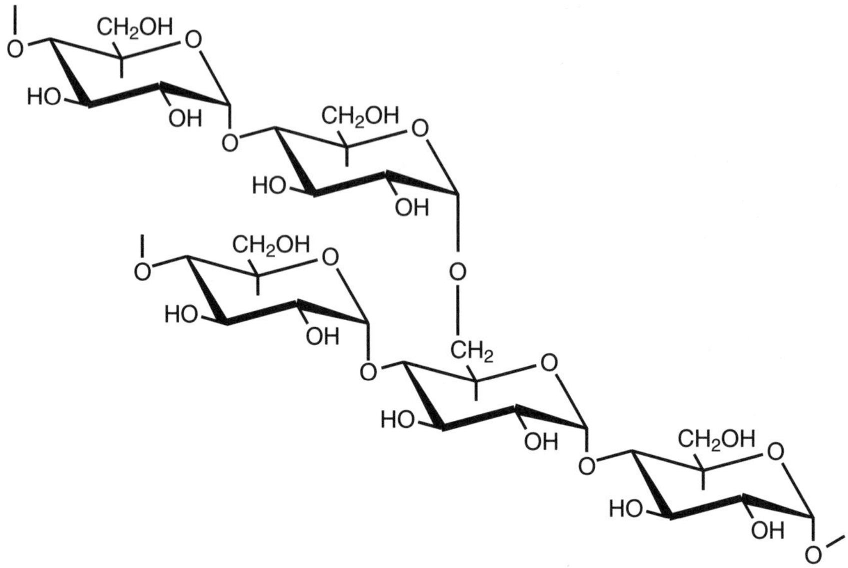

Figure I–60. Amylopectin.

lichens; and *inulin*, a $(2 \to 1)$-β-D-fructan obtained from dahlia and Jerusalem arti-choke tubers.

Some of the most representative heteropolysaccharides in the plant kingdom are *gums, mucilages, polyuronides,* and *hemicelluloses. Gums* constitute a large group of highly branched heteropolysaccharides. They contain hexuronic acid residues in salt form, and several monosaccharide units such as D-galactose, D-xylose, and L-arabinose. These gums are exuded by plants as viscous liquids in order to seal any injury and provide protection from microorganisms. Common examples are gum arabic, gum tragacanth, and mesquite gum.

Mucilages are a group of polysaccharides found in the seeds and barks of plants. They belong to several structural types and may be divided into neutral and acidic polysaccharides. Among the neutral mucilages D-galacto-D-mannans, D-gluco-D-mannans, L-arabino-D-xylans, and D-xylo-L-arabinans are the most common. A characteristic component of acidic polysaccharides is D-galacturonic acid. Typical examples are the acidic polysaccharides from slippery elm bark and cress seeds. It is probable that they function as reservoirs for the retention of water and thus protect the seed against dehydration.

Polyuronides

The basic building blocks of these polysaccharides are uronic acids. *Pectin* is a poly-galacturonic acid methyl ester in combination with arabinan and galactan. They are important cell-wall constituents. These pectins possess considerable gelling power and are widely used for the gelation of fruit juices to form jellies. Another polyuronide is *alginic acid*, which is a main constituent of brown seaweeds. It is a mixed polymer consisting of D-mannuronic and L-guluronic acids, both of which are linked through 1,4-bonds.

Hemicelluloses are a group of polysaccharides found in association with cellulose in plant cell walls. The most common members of this group are the *xylans*, which are formed by β (1 → 4)-linked xylopyranoses and D-glucuronic residues. Other members of this group are D-galacto-D-mannans and D-gluco-D-mannans.

Animal Polysaccharides. *Glycogen* is a (1 → 4)-α-D-glucan with (1 → 6)-α-D-branches. It is the principal reserve polysaccharide in the animal kingdom. Its structure is similar to that of *amylopectin*, except that the degree of branching is greater. It is a material of very high molecular weight with a degree of polymerization of approximately 10^5.

Chitin is a linear homopolysaccharide constituted by 2-acetamido-2-deoxy-D-glucose units with β (1 → 4) linkages . Essentially, it has the same structure as cellulose except that the 2-OH group has been replaced by a 2-acetamido group. It is a structural polysaccharide of the shell of arthropods such as lobsters, crabs, and other curstaceans.

Complex Carbohydrates (Carbohydrate-Containing Molecules)

In the previous sections we have considered molecules that contain only carbohydrates in their structure (mono, oligo, and polysaccharides), and are called *simple carbohydrates*. Other molecules contain carbohydrates as a minor or major constituent. These molecules are called *complex carbohydrates* (Sharon, 1975). In this category we find the *complex glycosides, carbohydrate antibiotics, nucleic acids, glycoproteins, proteoglycans, peptidoglycans, lipopolysaccharides,* and *glycolipids*.

Complex Glycosides

As opposed to simple glycosides, in which the aglycon group was a simple alkyl or phenyl substituent, complex glycosides have more elaborated aglycons such as anthracenes, phenanthrenes, carotenes, and flavones. Usually the chemistry of these molecules resides to a considerable extent in the aglycon residue. Some examples of complex glycosides are shown in Figure 1-61. *Fraxin* is isolated from the bark of the ash tree. *Frangulin* is found in the bark of *Rhamnus frangula*. *Hesperidin* is found in the peels of citrus fruits.

Carbohydrate Antibiotics

Antibiotics are compounds elaborated by microorganisms, that inhibit the growth of other microorganisms; the definition is sometimes extended to include the growth inhibition of animal and plant tumors. The mechanism of action often involves interference with DNA, RNA, or protein synthesis. Although carbohydrate antibiotics are grouped together because of their similar biological activity, structurally they could be classified under many different categories. For example, *Puromycin* can be considered a *glycosamine*, and under this category a *nucleoside. Daunomycin* and *erythromycin* can be considered *complex glycosides*. For this reason, antibiotics are further classified according to their structural features, for example, nucleoside antibiotics, antibi-

fraxin

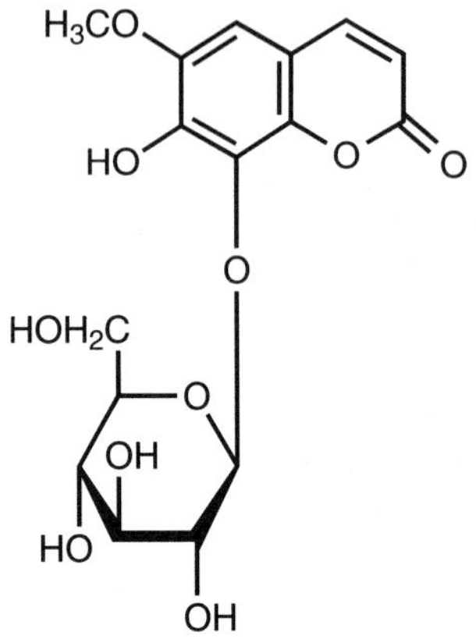

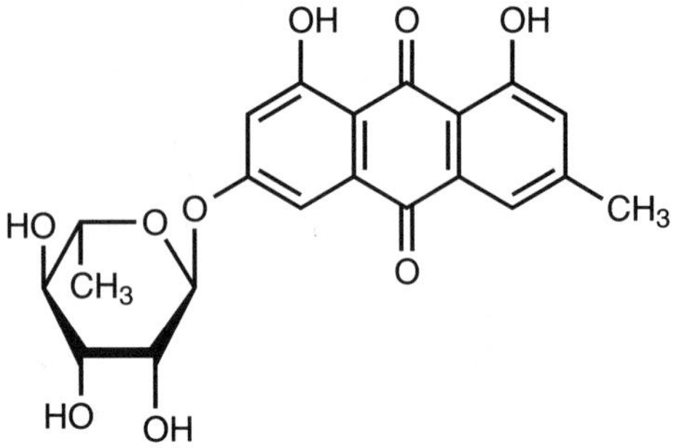

frangulin

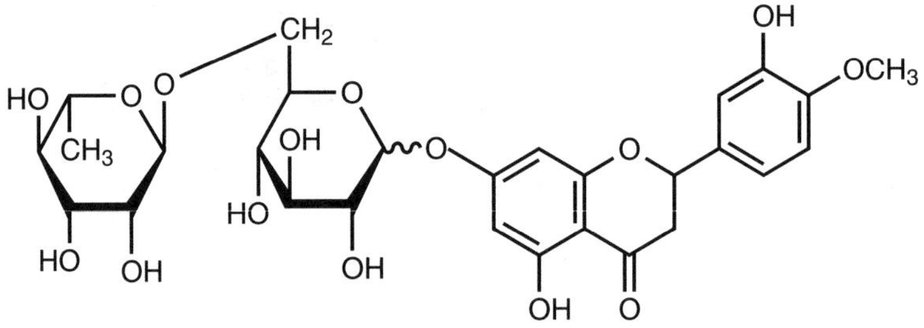

hesperidin

Figure 1–61. Examples of complex glycosides.

otics containing aromatic groups, macrolide antibiotics, and aminoglycoside antibiotics.

Nucleoside antibiotics generally contain a normal base attached to a sugar other than D-ribofuranose, or an unnatural base coupled to β-D-ribofuranose. Puromycin (Fig. 1-62) was the first member of the nucleoside antibiotic group to be discovered. Antibiotics containing aromatic groups (El Khadem, 1982; Umezawa, 1978) contain one or two sugars attached to an aromatic system, generally polycyclic in character. The antibiotic daunomycin (Fig. 1-62) falls in this category. In macrolide antibiotics (Berdy, 1980) the sugar moiety forms a small part of the molecule, the main component being a large lactone ring. However, if the sugar is removed, almost all the biological activity is lost. Erythromycin (Fig. 1-62) is one of the most widely used macrolide antibiotics.

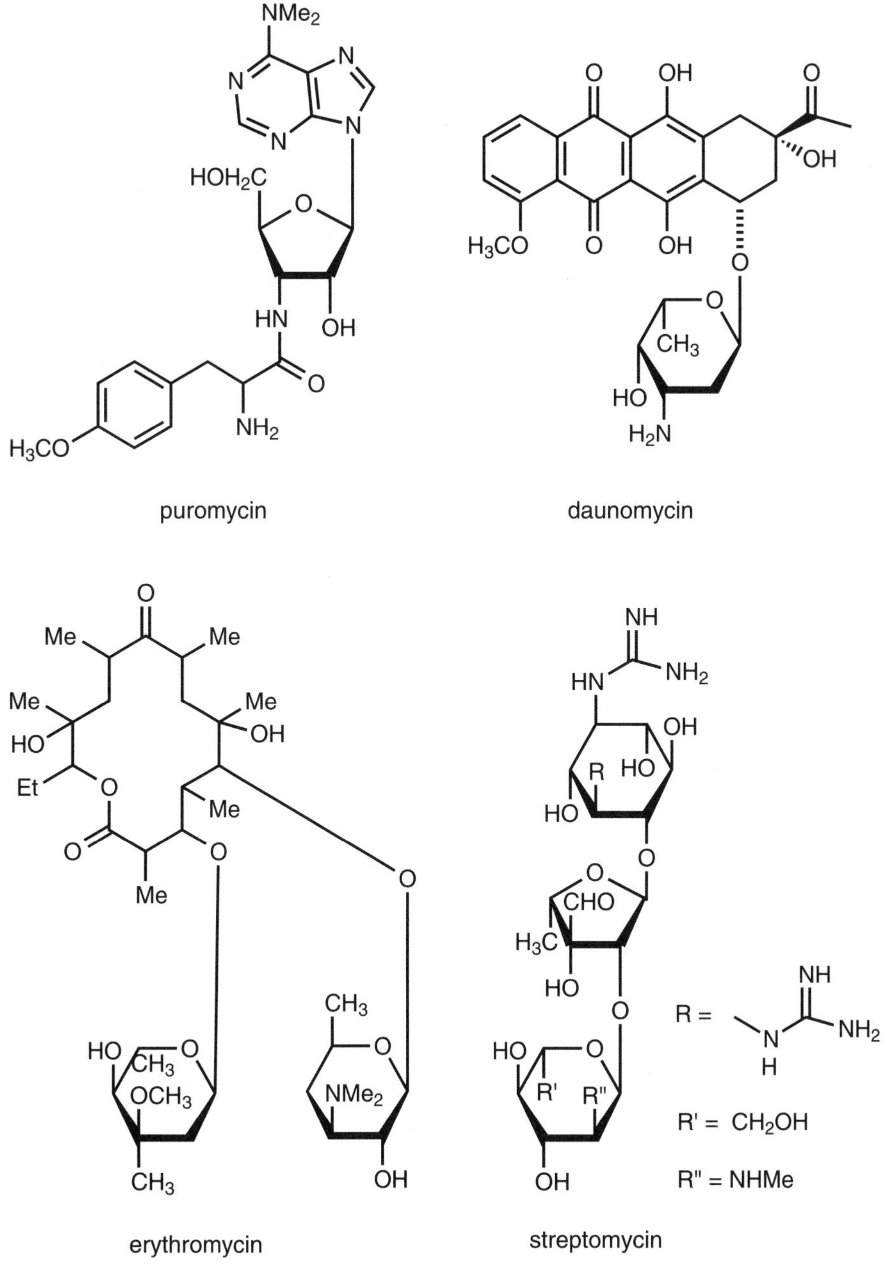

Figure 1–62. Carbohydrate antibiotics.

Aminoglycoside antibiotics (Rinehart & Suami, 1980), contain an amino-cyclitol to which one or more amino-sugars and other sugars are linked. The first natural antibiotic containing a carbohydrate that had an impact on chemotherapy was *streptomycin* (Fig. 1-62).

Nucleic Acids

The structures of *ribonucleic acid* (RNA) and *deoxyribonucleic acid* (DNA) are shown in Figure 1-63. RNA is a polymer of *ribonucleotides* linked by phosphodiester bridges between positions 3' and 5'. DNA is a polymer of *deoxyribonucleotides* linked by phosphodiester bridges between positions 3' and 5'. Both RNA and DNA occupy a central position in molecular biology. DNA is the molecule responsible for storing genetic information. RNA, among other functions, is the molecule responsible for the transfer of genetic information into protein structure.

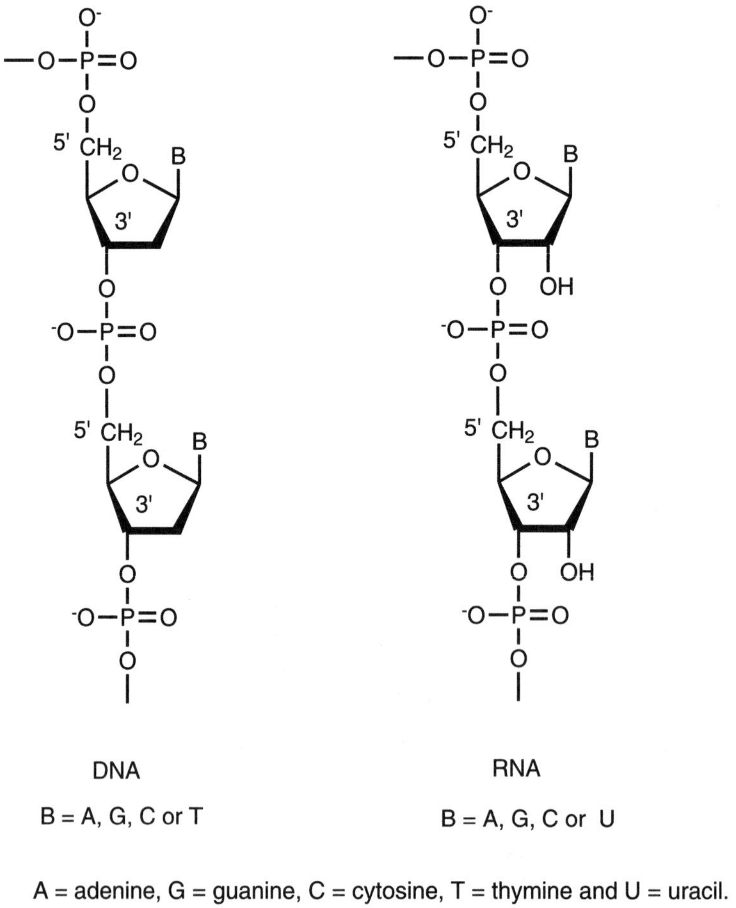

DNA

B = A, G, C or T

RNA

B = A, G, C or U

A = adenine, G = guanine, C = cytosine, T = thymine and U = uracil.

Figure 1–63. Nucleic acids.

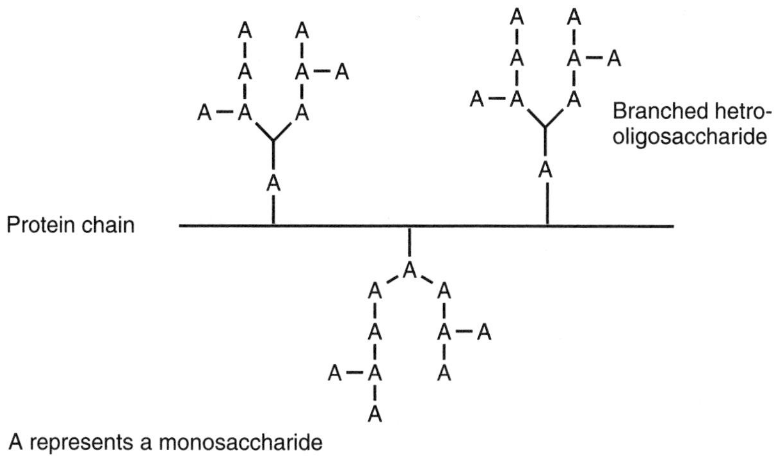

A represents a monosaccharide

Figure I–64. Glycoproteins.

Glycoproteins

There are proteins to which carbohydrates are linked by N- or O-glycosidic bonds (Montreuil, 1982; Sharon & Lis, 1982). They consist of a main protein backbone to which the carbohydrate side chains are attached (Fig. 1-64). These carbohydrate side chains usually consist of branched hetero-oligosaccharides with no more than 15–20 monosaccharide units. The most common monosaccharides present are D-glucose, D-galactose, D-mannose, L-fucose, 2-acetamido-2-deoxy-D-glucose, 2-acetamido-2-deoxy-D-galactose, and N-acetylneuraminic acid (5-amino-3,5-dideoxy-D-*glycero*-D-*galacto*-2-nonulopyranonic acid). The carbohydrate side chains differ widely in carbohydrate content, size and shape. Also, the proportion of total carbohydrate in glycoproteins varies from 0.5% in some collagens, to 85% in blood group substances.

Many proteins are indeed glycoproteins. Their molecular sizes range from 15,000 to over one million. Among the functions of the carbohydrate side chains are stabilization of protein conformation, protection against proteolysis, and classification of human blood types. They are also involved in the adhesion, communication, recognition, antigenic specificity, and regulation of cell growth.

Glycoproteins, which are widely distributed in nature among higher animals, plants, and microorganisms, perform many vital biological functions. Many *hormones* are glycoproteins, for example, human chorionic gonadotrophin and thyroid-stimulating hormone. *Transferrin* is a glycoprotein present in human plasma which is responsible for transporting iron from the tissues to hemoglobin. *Immunoglobulins* are a group of serum glycoproteins that have antibody activity. Blood group substances are a group of structurally related glycoproteins which occur on the surface of red blood cells and are responsible for the determination of particular blood types. Many enzymes are glycoproteins, including porcin ribonuclease B, and horseradish peroxidase. Other glycoproteins like ovalbumin and casein serve as food reserve. The *collagens*, which comprise a quarter to a third of the total protein in human body, are all glycoproteins. *Glycophorin* is the major glycoprotein of the human erythrocyte membrane.

Proteoglycans

Like glycoproteins, *proteoglycans* (Garg & Lyon, 1991; Kuettner & Kimura, 1985; Ruoslathi, 1989) contain a main protein backbone to which carbohydrate side chains are attached glycosidically (Fig. 1-65). In this case, however, each carbohydrate residue is a linear, much larger heteropolysaccharide (about 100 monosaccharide units), made up largely of disaccharide repeating units. In the disaccharide, one sugar unit is always a hexosamine (glucosamine or galactosamine), commonly in its N-acetylated form; the other is a hexuronic acid (glucuronic acid or L-iduronic acid). Sulfate groups may also be present, linked to either the hydroxyl groups or to the amino groups of the hexosamines. Due to the presence of carboxyl and sulfate groups, proteoglycans are highly charged polyanions. Proteoglycans are also called *mucopolysaccharides* and the carbohydrate chains *glycosaminoglycans*.

Proteoglycans occur in many animal tissues and fluids. They are an important component of connective tissues such as skin, bone, cartilage, and ligaments. In the cornea, proteoglycans organize and orientate collagen fibers into the correct array to allow light at visible wavelengths to enter the eye.

Peptidoglycans

Also known as *mureins*, *peptidoglyccans* are highly branched complex macromolecules occurring in bacterial cell walls. They consist of linear polysaccharide strands crosslinked by oligopeptide units. The principal constituents of peptidoglycans are 2-

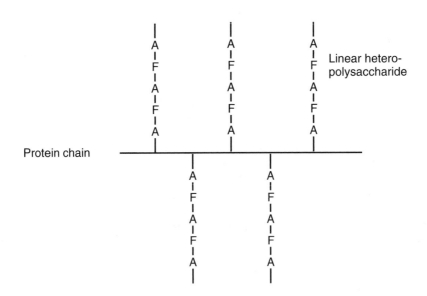

A represents a hexosamine
F represents a hexuronic acid

Figure 1–65. Proteoglycans.

acetamido-2-deoxy-D-glucose and 2-acetamido-2-deoxy-3-*O*-(1-carboxymethyl)-D-glucose, and the amino acids D- and L-alanine, D-glutamic acid, and L-lysine. Their molecular weight may be as high as 5×10^{10}. Peptidoglycans are responsible for the rigidity of the cell wall.

Lipopolysaccharides

These are very complex molecules that are the major constituent of the outer cell envelope of *gram-negative* bacteria. The fatty acid chains are attached to the carbohydrate residues through amide bonds, formed between the carboxylic groups of the fatty acid and the amino group of an amino-sugar. They are immunologically active and they often serve as receptors for bacteriophages.

Glycolipids

Lipids often exist combined with proteins or carbohydrate material, and they are a part of the outer surface of cells. *Glycolipids* (Kochetkov & Smirnova, 1986) are essentially lipids that contain carbohydrates as a minor component. Depending on the nature of the lipid moiety, glycolipids can be subdivided into *glycosphingolipids* and *glycoglycerolipids*. Glycosphingolipids contain a long chain amino alcohol (sphingosine) N-acetylated by a fatty acid, and a glycosyl group bound by a glycosidic linkage to the primary hydroxyl group of the sphingosine. Glycoglycerolipids contain glycerol, acylated fatty acids, and a carbohydrate group glycosidically bound to one of the primary hydroxyl groups of the glycerol. The exact function of glycolipids is still the subject of much speculation. They might transport carbohydrates accross the membrane. They are known to be involved in the biosynthesis of glycoproteins and proteoglycans, and in the inhibition of biological activity of toxins and antiviral agents.

Summary

This introductory chapter gives a general overview of the basic concepts and main areas covered in the field of carbohydrates, starting with a brief discussion of the role of carbohydrates in nature and their industrial use. Carbohydrates are either the main constituent or a part of the following type of molecules: *monosaccharides, monosaccharide derivatives, oligosaccharides, polysaccharides, complex glycosides, carbohydrate antibiotics, nucleic acids, glycoproteins, proteoglycans, lipopolysaccharides,* and *glycolipids.* For each group, the general structure, physical and chemical properties, natural occurrence, and some examples have been given. In this process, basic concepts such as *nomenclature, conformational equilibria, projection formulas, anomeric effect, mutarotation,* et cetera have been introduced. Some of these topics and concepts are discussed in greater detail in the chapters that follow.

Glycosylidene Carbenes

Andrea Vasella

Carbenes have played an important role in preparative organic chemistry for about 30 years (see Aitken, 1991 and earlier volumes; Bethell, 1969; Heydt & Regitz, 1990; Kirmse, 1971; Moss & Jones, 1973, 1975, 1978, 1981, 1985). The study of glycosylidene carbenes, however, is of recent origin; the first rational synthesis of precursors to glycosylidene carbenes was reported in 1989 (Briner & Vasella, 1989), while carbohydrate-derived dialkyldiazirines, possessing an azi function either in the aglycon, or in the glycosidic part, but not at the anomeric center, had been prepared before as photoaffinity reagents (Kuhn et al., 1989, 1990, 1992; Kurz et al., 1985; Lehmann & Thieme, 1986, Lehmann et al., 1988).

Glycosylidene carbenes, derived from glucopyranosylidene carbenes such as **1** and **2** (Scheme 2-1), are cyclic alkoxyalkyl carbenes ("oxacarbenes"), and as such are ambiphilic to nucleophilic singlets (Moss, 1989; Sheridan et al., 1988). They are best represented by a pair of carbene/ylid canonical structures. Alkoxy(alkyl)carbenes have been generated in a number of ways (Heydt & Regitz, 1990), such as by photolysis of cyclobutanones or cyclopentanones (Collins et al., 1974; Pirrung et al., 1989; Yates and Crawford, 1966; Yates & Loutfy, 1975), by cycloelimination (Hoffmann, 1971), from carbonyl-ylides (El-Saidi et al., 1992), by thermolysis or photolysis of N-sulfonyl lactone hydrazones (Agosta & Foster, 1971; Ayral-Kaloustian & Agosta, 1982; Crawford & Raap, 1965a,b), by photolysis of acylsilanes, and by thermolysis or photolysis of alkoxy(alkyl)diazirines (Liu & Stevens, 1987; Moss et al., 1991; Moss & Zdrojewski, 1990).

Glycosylidene carbenes are expected to react by concerted or sequential formation of two bonds at the anomeric center; they should thereby be preparatively useful.

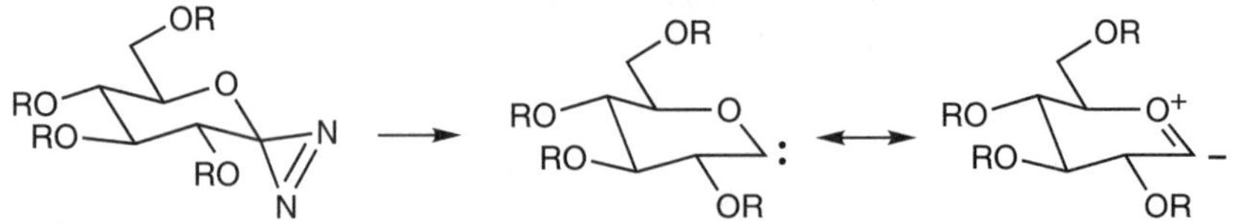

1 R = Bn
2 R = Piv (CO^{t}Bu)

Scheme 2-1.

Moreover, the carbohydrate setting with its *quasi* continuous variation of constitution and configuration of isomers offers a unique opportunity to study structure-reactivity relations in detail, both from the point of view of the carbenes and of the species reacting with them.

Precursors and Generation of Glycosylidene Carbenes

Glycosylidene carbenes are generated in preparatively useful yields by thermolysis or photolysis either of glycosylidene diazirines (i.e., 1-azisaccharides) (Briner & Vasella, 1989; Vasella, 1991, 1993), or of the Na salts of N-sulfonyl glyconolactone hydrazones (Mangholz & Vasella, 1991, 1995; Somsak et al., 1992). Only minor amounts of carbenes are generated from geminal diazides *via* azido nitrenes (Praly et al., 1994). Glycosylidene diazirines form carbenes (and dinitrogen) under mild conditions; significantly harsher conditions are required for the thermolysis or photolysis of N-sulfonyl lactone hydrazone salts; the sulfinate which is simultaneously generated interferes with the reaction of the carbenes and leads to by-products. Although sulfonyllactone hydrazones of type **4** (Scheme 2-2) are conveniently prepared by oxidation of N-glycosyl sulfonylhydrazines (**3**), and are more stable than the diazirines, particularly in the furanose series, studies have concentrated on pyranosylidene diazirines. It has been shown, however, that salts of N-sulfonyl glycopyrano- and -furanolactone hydrazones react similarly to diazirines (but in lower yields) in that they glycosylate phenols and cyclopropanate electron deficient alkenes (Mangholz & Vasella, 1991, 1995).

Typically, glycosylidene diazirines have been prepared from protected aldose oximes *via* hydroximolactones in four generally high yielding steps, involving a modification of the procedure of Schmitz & Ohme (1961), as depicted in Scheme 2-3 (Briner & Vasella, 1989; Vasella et al., 1993b). The intermediate diaziridines are obtained (by predominantly axial attack of ammonia [Mangholz, 1997]) as pairs of *trans*-configured diastereomers, equilibrating in contact with silica gel; under acidic or acylating conditions they readily rearrange to derivatives of lactone hydrazones (Briner, 1995; Mangholz, 1997). It is advantageous to adapt the O-sulfonyl group and the method of oxidizing the hydroxy-oximes to individual cases (Vasella et al., 1993b).

Fully O-benzylated pyranosylidene diazirines are characterized by IR bands at 1560–1700 cm^{-1}, UV maxima around 346 nm, and chemical shift values of about 57 ppm for C(1), and of about 4–4.5 and approximately 2.9 ppm for axial and equatorial

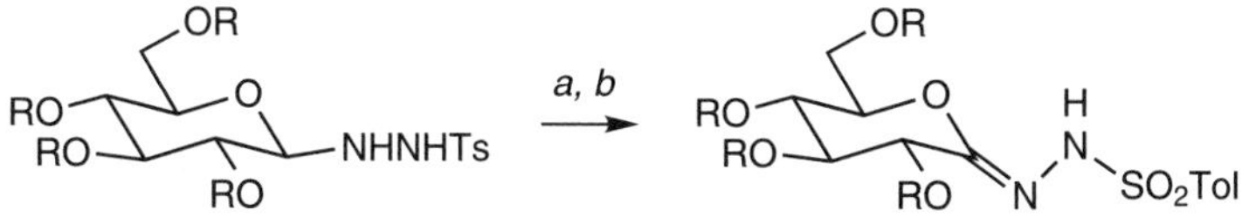

3a R = Bn
3b R = Piv
3c R = Ac

4a R = Bn
4b R = Piv
4c R = Ac

Scheme 2-2. (a) NBS, DBU, DMF, r.t., **4a** (81%) or **4b** (79%). (b) Pyridinium dichromate, CH$_2$Cl$_2$, r.t., **4c** (90%).

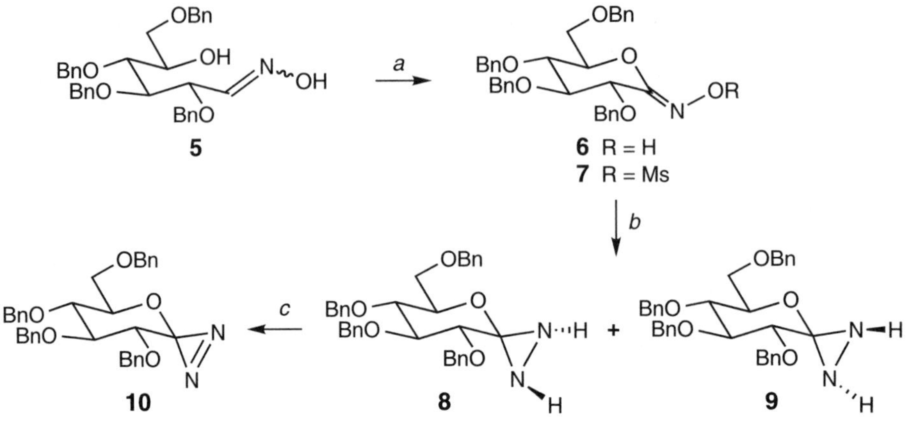

Scheme 2-3. (a)NaIO$_4$, NaOAc, EtOH, 70°–75°C, **6** (86%); MsCl, NEt$_3$, CH$_2$Cl$_2$, r.t., **7** (98%). (b) NH$_3$, MeOH, r.t., 6 bar, 86%. (c) I$_2$, NEt$_3$, MeOH, −25°C, 92%.

HC(2). The X-ray analysis of **12** shows an N,N bond length of 1.258 Å (Linden et al., 1992). Neither of the N-atoms attempts to adopt a truly axial or equatorial position; the mid-point of the N=N bond almost bisects these positions. The N,C(1) bonds differ little in length, the one to the "pseudoaxial" N being slightly longer (1.478 vs. 1.452 Å). This lengthening is not necessarily significant, although it is in agreement with the assumption that this bond should break first.

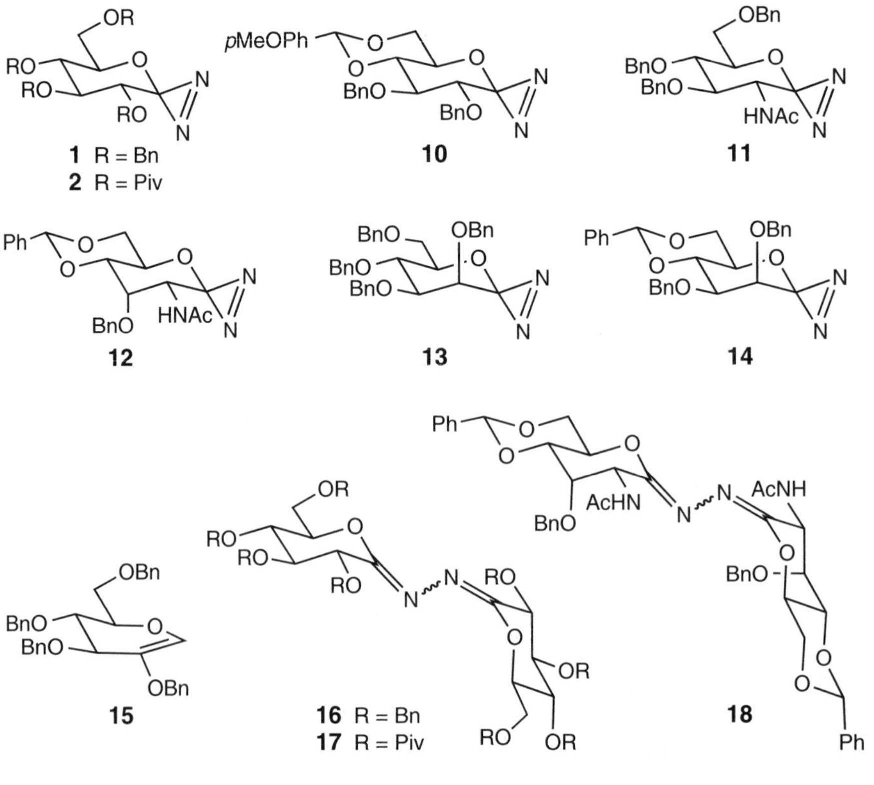

Table 2–1 Kinetic Parameters for the Thermolysis of 1-Aziglycoses in MeOH

Diazirine	τ at 298 K (min)	E_a (kcal/mol)	log A	$\Delta H^{\neq}$ (kcal/mol)	$\Delta S^{\neq}$ (cal/mol·K)
1	33	23.0	13.4	22.4	1.7
2	202	25.0	14.1	24.0	4.8
10	110	22.2	12.4	21.7	−3.2
11	112	22.6	12.6	22.0	−3.1
12	4159	28.1	15.1	27.2	8.1
13	7	23.2	14.2	22.6	5.5
14	23	20.0	11.4	19.5	−8.2

Kinetic data for the thermolysis of glycosylidene diazirines in methanol (Table 2-1) are consistent with a rate determining heterolysis of a C,N bond (Liu, 1982; Liu et al., 1986), quickly followed by loss of nitrogen; evidence for the isomerization to diazo compounds has not been found (Jennings & Liu, 1976; Vasella et al., 1993b). The half-life of a range of glycosylidene carbenes at 298 K varies between 7 and 4159 minutes, corresponding to $\Delta H\ddagger$ and $\Delta S\ddagger$ values of 22.6–27.2 kcal/mol and 5.5–8.1 cal/mol·K, where the extreme values are those for the *manno-* and for the 2-acetamido-2-deoxy-*allo*-diazirines **13** and **12**, respectively. It has been claimed that conformational changes accompanying the thermolysis of diazirines, and the σ-acceptor properties of substituents on the pyranose ring, are responsible for the differences in stability. Thermolysis of glycosylidene diazirines in methanol leads almost quantitatively to a mixture of anomeric methyl glycosides. Thermolysis in aprotic solvents leads mostly to a mixture of diastereomeric lactone azines (cf. Crawford & Raap, 1965b) which equilibrate to the most stable (Z,Z)-isomers **16–18** (Glaser et al., 1993). These lactone azines result from the reaction between the diazirines and the intermediate carbenes (Liu & Ramakrishnan, 1977). They are observed whenever glycosylidene diazirines are thermolyzed or photolyzed in the presence of poorly reactive partners. Derivatives of 2-hydroxyglycals, indicating formal insertion into the adjacent C—H bond, are formed to a minor extent only (3% of **15** by thermolysis of **1** in solution; 18% from solid **1**).

Reaction of Glycosylidene Carbenes with Hydroxy Compounds: Hydrogen Bonds and Regioselective Glycosidation

Glycosidation of Monofunctional Phenols and Alcohols

Glycosylidene diazirines have been prepared with the goal of realizing a promoter-free method for regioselective glycosylation (Scheme 2-4), in which hydroxy compounds should be deprotonated by a glycosylidene carbene (Du et al., 1990; Kirmse, 1963; Kirmse et al., 1981; Kirmse & Kund, 1990; Moss et al., 1988). This would generate an ion pair, simultaneously activating the glycosyl donor and the glycosyl acceptor. Combination of the ions should lead to glycosides (Briner & Vasella, 1989; Vasella, 1993). Regioselectivity may result from a correlation between the (kinetic) acidity of hydroxy groups and the rate of reaction with the carbene, and from differences in the kinetic acidity of individual hydroxyl groups. Proton transfer to the carbene is expected

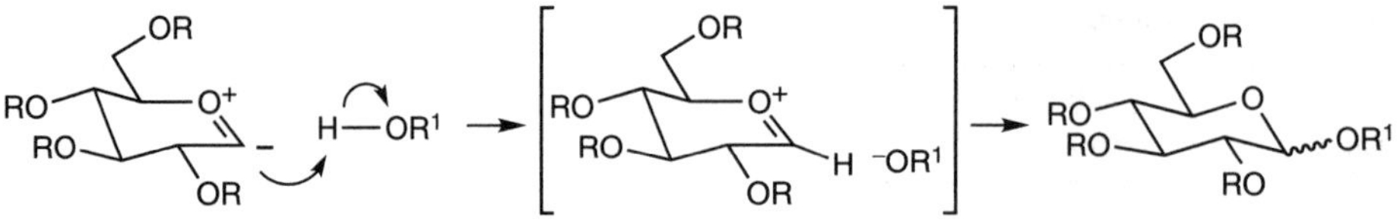

Scheme 2-4.

to take place in the σ-plane of the carbene, while the nascent oxyanion has to attack the glycosyl cation in the π-plane; hence proton transfer and formation of the glycoside cannot be a synchronous process.

The reactivity of glycosylidene diazirines, mostly of **1**, towards a variety of phenols, mono-, di-, and triols has been studied in some detail. Relatively strongly acidic hydroxy compounds, such as phenols (Briner & Vasella, 1990) and fluorinated alcohols (Briner & Vasella, 1992) give mostly 1,2-*trans*-configured glycosides in yields of 70%–80%, particularly using ethers as solvents (Table 2-2). In contradistinction to glycosidations of the Königs-Knorr type (using "classical glycosyl donors"), that is, procedures based on the activation of a potential leaving group at the anomeric center followed by nucleophilic substitution, it is the (kinetic) acidity rather than the nucleophilicity of the alcohol that is the key factor in determining reactivity and yields. This

Table 2–2 Glycosidation of Phenols and Alcohols with Diazirines **1**, **13**, and **22** in CH_2Cl_2 Solution

			Yield (%) of α/β-D-pyranosides		
ROH	pK_{HA}	Diazirine	r.t.	$-70°C$ (hν)	Ratio α-D/β-D
Phenol	9.9	1	70[a]	—	20:80
19	10.2	1	69[b]	—	25:75
20	7.2	1	75	—	40:60
21	12.55	1	81	—	16:84
Phenol	9.9	13	38	—	>95:5
Phenol	9.9	22	58[c]	—	20:80
20	7.2	22	52	—	35:65
CF_3CH_2OH	12.4	1	73	—	30:70
$(CF_3)_2CHOH$	9.3	1	77	—	22:78
$(CF_3)_2MeCOH$	9.6	1	74	—	16:84
$CF_3(CF_2)_5CH_2CH_2OH$		1	85	—	43:57
$CF_3(CF_2)_5CH_2CH_2OH$		1	68[d]	—	24:76
$CHF_2(CF_2)_9CH_2OH$		1	79[d]	—	22:78
MeOH	15.2	1	60	—	50:50
EtOH	15.85	1	55	—	50:50
Me_2CHOH	16.5	1	39	71	50:50
Me_3COH	16.55	1	34	55	50:50

[a] 13% of **23** (1:1).
[b] 16% of **24** (1:1).
[c] 14% of **25** (1:1).
[d] In 1,4-dioxane.

allows the preparation of glycosides of alcohols that are poorly reactive towards classical glycosyl donors, such as highly fluorinated alcohols (Briner & Vasella, 1992). Also, steric hindrance appears to be of relatively little importance.

Formation of 1,2-*trans* glycosides in noncoordinating solvents (e.g., CH_2Cl_2) has been rationalized by postulating an unusual neighboring group interaction of the C(2) O-benzyl group (Briner & Vasella, 1992); the nonstereoselective glycosidation of a 2-deoxy-diazirine is in keeping with this hypothesis (Takahashi & Vasella, 1992). Glycosidation by **1** of weakly acidic alcohols in CH_2Cl_2 at room temperature proceeds in markedly lower yields. Higher yields are obtained when the reaction is performed at low temperatures (the carbene is then generated by photolysis [Frey, 1966; Liu & Stevens, 1987]), but there is no stereoselectivity. There is a strong solvent effect: ethers, particularly THF and dimethoxyethane, increase the diastereoselectivity of the glycosylation of strongly acidic alcohols; thus, hexafluoro-*tert*-butanol in DME yielded the α/β-glucosides in a ratio of 2:98. It is remarkable that photolysis of **1** in the presence of hexafluoro-*tert*-butanol in propionitrile at low temperatures yielded 55% of the iminoether **26** which is almost certainly derived from addition of the alcoholate anion to the intermediate nitrilium ion. Alcohols react with such nitrilium ions by substitution at the anomeric center to give glycosides (Ratcliffe & Fraser-Reid, 1990; Schmidt et al., 1990). Glycosylation of 2-propanol in THF at −78°C also proceeded with good diastereoselectivity, and led to the corresponding β-D-glucoside (d.e. = 84%), in addition to 9% of the anomeric glucosides **27** ($\beta:\alpha$ = 78:22; Briner & Vasella, 1992).

That these glycosylations proceed *via* a glycosyl cation, and thus involve deprotonation of hydroxyl groups, is evidenced by the formation of the C-glycosides **23–25** as byproducts of the glycosidation of electron-rich phenols, by the formation of the imidate **26**, and by the formation of the ethers **27**, derived from ring-opening attack of 2-propanol on the intermediate O-glycosyl tetrahydrofuranylium cations in the glycosidation of 2-propanol in THF. Not surprisingly then, an orthoester (**28**) is the main product of the glycosylation of diisopropylidene-glucofuranose by the pivaloylated diazirine **2**.

19 R = OMe
20 R = NO$_2$

21

22

23 R^1 = R^3 = H, R^2 = OBn
24 R^1 = OMe, R^2 = OBn, R^3 = H
25 R^1 = R^2 = H, R^3 = OBn

26

27

28

The lack of stereoselectivity in the glycosidation of weakly acidic alcohols in non-coordinating solvents has been rationalized by postulating that the intermediate glycosyl cation is solvated more efficiently by weakly acidic and thus relatively nucleophilic alcohols than by the C(2)-benzyloxy group. The anion derived from the protonating unit is located in the σ-plane and cannot directly attack the oxycarbenium centre. However, rapid proton-transfer from a hydrogen bonded neighbor (Crooks, 1975) generates an oxyanion, correctly positioned for attack in the π-plane from above or from below the cationic center, to which it may already be coordinated. In keeping with the postulate that the nucleophilicity, and not the acidity of hydroxyl compounds, is primarily responsible for the lack of stereoselectivity, thiophenols such as 4-methoxy- or 4-nitrothiophenol react with **1** to give approximately 1:1 mixtures of anomeric S-glycosides in good yields (Rajamannar, 1990). The newly generated oxycarbenium ion may also be solvated by a nucleophilic solvent, such as THF, for which axial attack should be stereoelectronically favored. At low temperatures, attack of the oxyanion on the axial C(1)-tetrahydrofuranylium ion leads preferentially to the equatorial glycoside; at higher temperatures, equilibration of the tetrahydrofuranylium ions entails loss of stereoselectivity.

Regioselective Glycosidation, Hydrogen Bonds, and Stereoelectronic Control

Glycosidation of methyl orsellinate (**29**) with equimolar amounts of **1** (Scheme 2-5) proceeds regioselectively, and leads in good yields to the 4-O-glycosides **30** (Briner & Vasella, 1990). The best ratio of anomers (α:β = 15:85, 78%) is realized in toluene. The rate constant for protonation by the chelating hydroxyl (OH) group is reduced by the stability constant of the H-bond (Eigen, 1963); hence OH groups which function as H-bond donors—such as the *ortho* hydroxyl group of **29**—are deactivated towards glycosylidene carbenes, while H-bond accepting OH groups, being partially protonated, are more highly acidic and should be activated. Conversely, H-bond donating OH groups, being partially deprotonated, should be more highly nucleophilic and react preferentially with classical glycosyl donors. Indeed, glycosidation by **1** of catechol, where one hydroxyl group acts as a H-bond acceptor, also yields monoglycosides (68%, α:β = 4:1; Rajamannar, 1990).

Intramolecular H-bonds give rise to IR absorptions which are characteristic for the size and geometry of H-containing rings (see Fig. 2-1; cf. Hönig & Weidmann, 1979; Spedding, 1961; Tichy, 1965), while a Karplus relation may be used to calculate the relevant HCOH dihedral angles (Fraser et al., 1969). In most cases, the FT-IR spectra

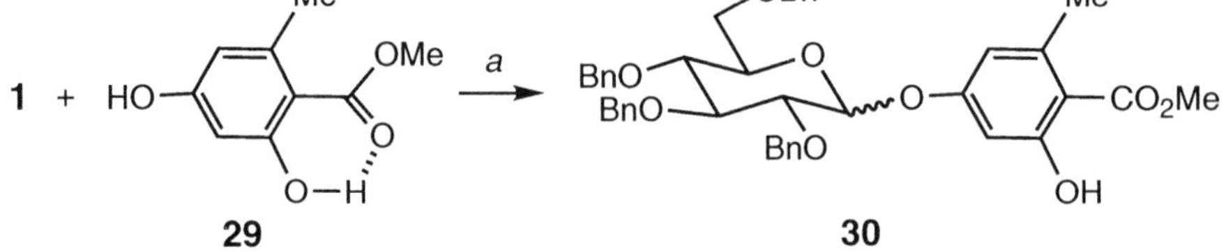

Scheme 2-5. (a) 1 equiv. of **29**, toluene, r.t., **30** (78%, α/β 15:85) *or* 1 equiv. of **29**, CH_2Cl_2, r.t., **30** (79%, α/β 34:66).

Type	OH⋯O$_{Ring}$ or OH (free)	OH⋯O in a *trans-* 5-membered ring	ax. OH⋯O in a *cis-* 5-membered ring
IR (CCl$_4$): IR (CH$_2$Cl$_2$):	3635–3625 3600–3595	3615–3605 3590–3585	3600–3590 cm^{-1} 3575–3565 cm^{-1}

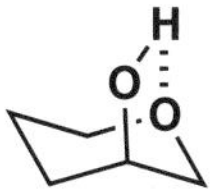

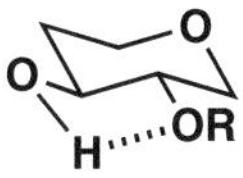

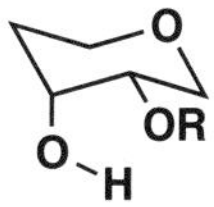

Type	eq. OH⋯O in a *cis-* 5-membered ring	bifurcated OH	OH⋯O in a 6-membered ring
IR (CCl$_4$): IR (CH$_2$Cl$_2$):	3590–3580 3570–3555	3580–3565 3540	3555–3530 cm^{-1} 3525–3500 cm^{-1}

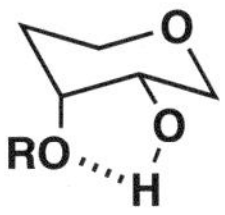

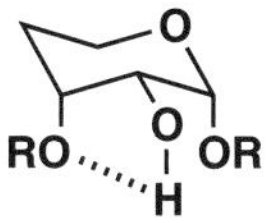

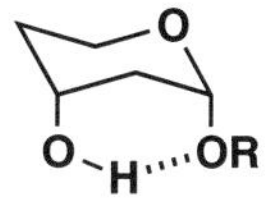

Figure 2–1. Typical IR bands of intramolecularly bonded OH groups in CCl$_4$ and CH$_2$Cl$_2$ solutions.

of solutions of diols and triols in CH$_2$Cl$_2$ or CCl$_4$ at low concentrations show separate and characteristic bands for the intramolecular H-bonds characterizing the individual tautomers; in favorable cases, their proportion can be estimated from the HCOH coupling constants. Alcohols preferentially form intermolecular H-bonds in the solid state (Jeffrey & Saenger, 1991); hence, intramolecular H-bonds in the solid state are quite strong and, as a rule, also found in solution.

The effects of intramolecular H-bonds on regioselective glycosidation are best illustrated by comparing the glycosidation of the orthoester triol **31** (Uhlmann & Vasella, 1992), 1,6-anhydroglucose **34** (Uhlmann & Vasella, 1992), and the methyl α-D-altropyranoside **39** (Bozó & Vasella, 1992) (Scheme 2-6). In the solid state, **31** possesses an intramolecular H-bond between the axial OH groups, characterized by an O,O distance of 2.77 Å; the equatorial OH group is involved in an intermolecular H-bond. In solution, the orthoester **31** possesses two intramolecular H-bonds: a strong one between the two axial OH groups, and a weaker one, where the equatorial OH group functions as an H-bond donor towards the vicinal alkoxy groups and forms a bifurcated H-bond (Fig. 2-2 and Table 2-3; cf. Jochims & Kobayashi, 1976). The two H-bond-donating OH groups are deactivated towards the carbene derived from **1**, while the H-bond accepting, axial OH group is activated. Indeed, glycosidation yielded a 1:1 mixture (90%) of the two diastereomeric, β-D-configured monoglycosides **32** and **33**, derived by interaction of the carbene with one of the enantiotopic, axial OH groups of each of the enantiomeric, rapidly equilibrating H-bonded tautomers. Glycosidation of anhydroglucose **34** by **1** proceeded in lower yields, and gave a 58:42 mixture of a pair

Scheme 2-6. Glc(OBn)$_4$ = 2,3,4,6-tetra-O-benzyl-D-glucopyranosyl. (**a**) 0.8 equiv. of **1**, 1,4-dioxane, r.t., 90% (**32/33** 1:1). (**b**) 0.8 equiv. of **1**, 1,4-dioxane, r.t., **35** (35%, α/β 46:54) and **36** (25%, α/β 40:60). (**c**) 1.1 equiv. of **1**, 1,4-dioxane, r.t., **40** (50%, α/β 18:82), **41** (7%, α/β 18:82), and **42** (ca. 10%, α/β 94:6) *or* 1.1 equiv. of **1**, ClCH$_2$CH$_2$Cl, r.t., **40** (66%, α/β 35:65), **41** (5%, α/β 51:49), and **42** (ca. 10%, α/β 94:6). (**d**) 0.9 equiv. of **37**, 1 equiv. of Et$_4$NBr, 4Å molecular sieves, CH$_2$Cl$_2$, r.t., **40** (9%, α/β 21:79) and **41** (69%, only α). **e**) 1.1 equiv. of **38**, 1. equiv. of BF$_3$·OEt$_2$, ClCH$_2$CH$_2$Cl, −30°, **40** (6%, α/β 46:54) and **41** (61%, α/β 52:48).

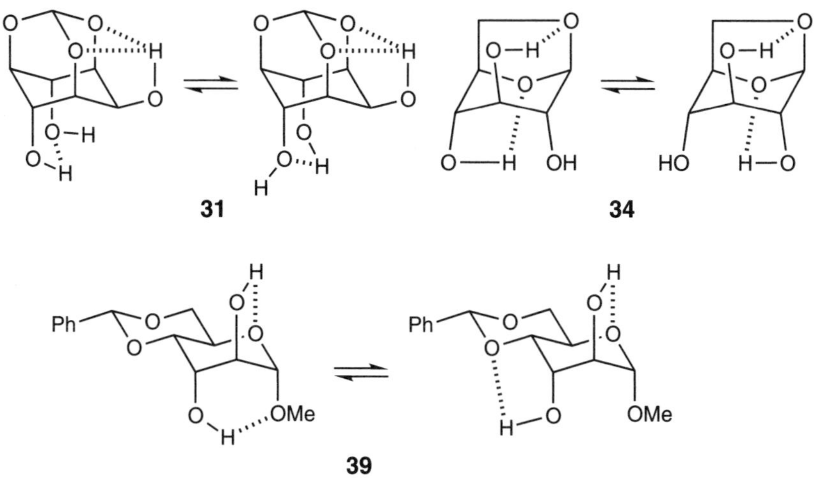

Figure 2–2. Intramolecular hydrogen bonds of **31**, **34**, and **39** in chlorinated solvents.

of anomeric 1,2- and 1,4-linked disaccharides **35** and **36**, with only a small excess of the β-D-anomers. The crystal structure of **34** (Park et al., 1971) shows no intramolecular H-bonds—the distance between the pseudoaxial OC(2 and 4) being too long (3.299 Å). In solution, there is an H-bond between HOC(3) and OC(6), and one formed between HOC(2) or HOC(4) which competes for OC(5) as the H-bond acceptor (Fig. 2-2). Not surprisingly, the yield of **32** and **33** does not depend upon the concentration of the orthoester **31**, while the yield of **35** and **36** decreases from 60% to 0%, when the concentration of levoglucosan **34** is lowered from 0.05 to 0.005 M, showing the importance of intra- or intermolecular H-bonds, respectively, and the use of glycosylidene diazirines in differentiating between them. The altroside **39** exists as a mixture of two H-bonded tautomers: one where HOC(3) forms an H-bond to the anomeric methoxy group, and one where it is H-bonded to OC(4) (Fig. 2-2). In either tautomer, HOC(2) can at best form a weak H-bond, and should preferentially glycosylate a glycosylidene carbene. Conversely, HOC(3) should be more highly nucleophilic than HOC(2). This is confirmed by the glycosidation of **39** by **1**, in contrast to glycosidation by the glucopyranosyl bromide **37** according to the procedure of Lemieux (Lemieux et al., 1975), or by the trichloroacetimidate **38** according to Schmidt (1986). High, and complementary degrees of regioselectivity to afford **40** and **41** are observed; the glycosidation according to Lemieux also proceeds with excellent anomeric selectivity (Scheme 2-6). A slight excess of the β-D-glycosides is observed for the glycosidation of **39** by **1** in chlorinated solvents; it is much improved (but accompanied by a lowered regioselectivity) in THF at −80°C. Glycosidation of the monotosylates derived from **39** does not lead to epoxides, in agreement with the formation of intimate ion pairs.

The simple mechanism of glycosidation by glycosylidene diazirines presented above needed to be specified, considering the nature of the by-products **42** of the glycosidation of **39**, and the results of control experiments, namely the glycosidation of the diol **43** and of the β-D-altroside **48** (to provide **49** and **50**), as detailed below (Scheme 2-7).

Table 2–3 Intramolecular and Free OH-Bands in the IR Spectra (0.005–0.05 M) and Vicinal Coupling Constants J(H,OH) in the ^{1}H-NMR Spectra (0.02–0.05 M in CDCL$_3$) of the Glycosyl Acceptors (Diols and Triols)

Diol or Triol	ν_{OH} in CCl$_4$ (cm^{-1})	ν_{OH} in CH$_2$Cl$_2$ (cm^{-1})	J(2,OH) (Hz)	J(3,OH) (Hz)	J(4,OH) (Hz)
43		3604, 3574	11.9	—	4.3
39	3630, 3601, 3556^a	3598, 3520	6.7	5.7	—
48	3600^a		1.5^b	1.8^b	—
53		3585 (sh), 3568	10.1	<2	—
54	3609, 3582^c		≈8.0	<1	—
55		3590	2.5	2.4	—
56	3615^c		2.2	2.0	—
57		3568	9.1^b	8.0^b	—
58	3615, 3602		1.6	8.9	—
59		3585, 3558, 3500 (sh)	—	9.6	1.7
60		3587, 3575 (sh)	2.5	3.6	—
61		3574	1.7	6.2	—
62		3577, 3490	—	4.1	1.3
63	3610, 3581, 3547	3566, 3540 (sh)	11.7	6.5	
64	3605	3586	6.0	<1.5	
65^d	3599, 3577, 3521	3601, 3568, 3517	8.0	6.4	7.6
67		3595, 3544, 3470 (sh), 3383	—	1.4	10.8
70^e		3400 (broad)	—	<2	—
72^e			—	<2	11.1
68^e		3560, 3480, 3450 (sh)	—	f	6.4
71^e		3590, 3480, 3410	—	1.3	—
69	3618, 3584, 3550, 3515	3600, 3580, 3543, 3500	≈7.5	9.2	8.5

^{a}From Spedding (1961).
^{b}Assignment may be reversed.
^{c}From Hönig & Weidmann, 1979.
^{d}H-NMR in CD$_2$Cl$_2$.
^{e}IR in CHCl$_3$.
^{f}Not determined.

Glycosidation by diazirines of H-bond donating, strongly nucleophilic OH groups, such as HOC(3) of **39**, may proceed not by initial protonation, but by nucleophilic attack on the carbene. This is suggested by the isolation of about 10% of the demethylated (mostly α-D-configured!) 1,3-linked disaccharides **42**. A nucleophilic attack on the carbene by HOC(3) of **39** induces an intramolecular protonation of the methoxy group, followed, on the one hand, by (equatorial!) protonation at C(1′) and, on the other hand, by loss of methanol (methyl tetra-O-benzylglucopyranosides have been identified as by-products), and formation of an unstable 1,2-anhydro derivative. The postulate of an intermediate oxirane is in keeping with the isolation of the epoxides **46** and **47**, among the products of glycosidation of the diol **43** by **1** (followed by acetylation) (Zapata et al., 1996). This diol had been glycosylated to investigate the effects not only of the H-bond between the axial OH groups, but also of the intramolecular

H-bond between the equatorial OH group and the adjacent alkoxy groups. The glyco-sides **44** derived from attack of the axial OH group are indeed the major products (38%, after acetylation), along with 13% of the disaccharides **45**, derived from glycosidation of the equatorial OH group. The formation of the epoxides **51** and **52** is explained by a nucleophilic attack of one of the alkoxy groups on the glycosyl cation, generated upon proton transfer from the equatorial OH group, followed by elimination and in-tramolecular substitution, as illustrated in Scheme 2-8.

Alkylation of one of the alkoxy groups is, however, possible only if the glycosyl cation is properly oriented. Formation of the epoxides therefore suggests that protona-tion is by H-bonded, rather than by free OH groups and, consequently, that the orien-tation of H-bonds determines the angle of attack of the carbene. This angle most prob-

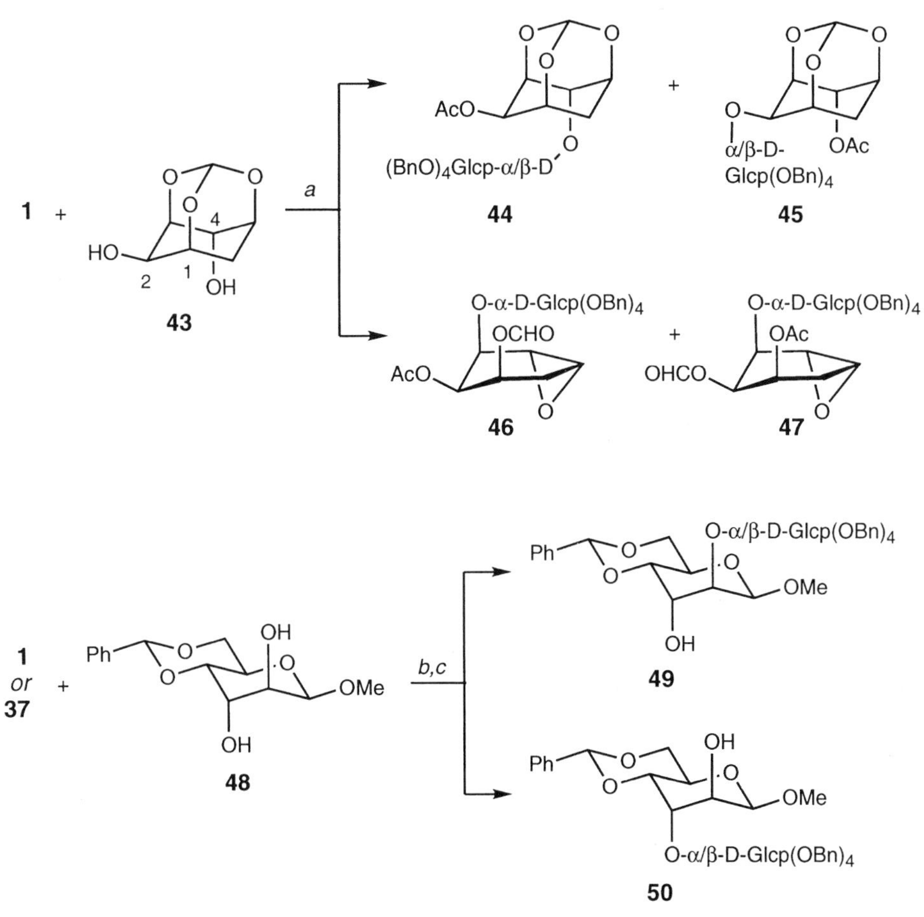

Scheme 2-7. (a) 1.1 equiv. of **1**, 1,4-dioxane, r.t.; Ac$_2$O, pyridine, r.t., **44** (38%, α/β 17:83), **45** (13%, α/β 40:60), **46** (4%), and **47** (3.5%). (b) 1.1 equiv. of **1**, ClCH$_2$CH$_2$Cl, r.t., **49** (28%, α/β 33:67) and **50** (28%, α/β 28:72). (c) 1.2 equiv. of **37**, 1 equiv. of Et$_4$NBr, 4-Å molecular sieves, CH$_2$Cl$_2$, r.t., **49** (71%, α/β 67:33) and **50** (8%, α/β 52:48).

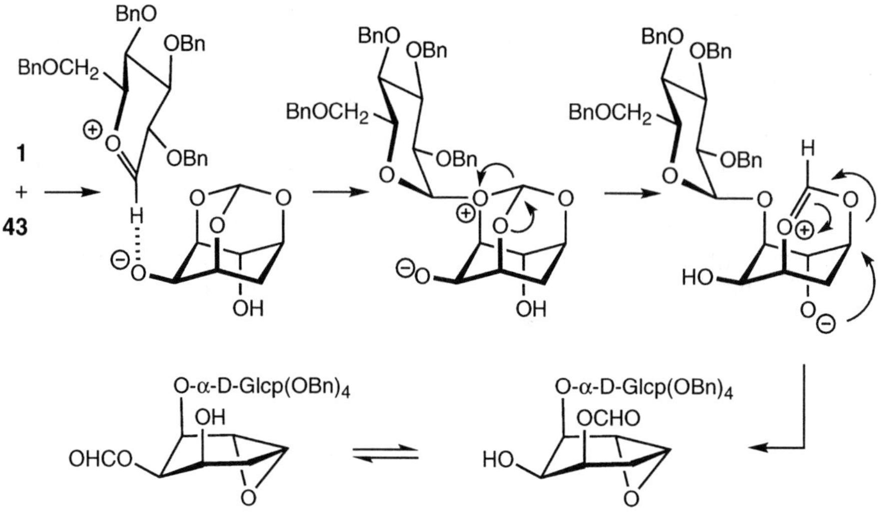

Scheme 2-8.

ably corresponds qualitatively to the position of a third ligand in a bifurcated, or the fourth ligand in a trifurcated H-bond, respectively (Jeffrey & Saenger, 1991).

While glycosidation of the β-D-altroside **48** by **1** proceeds nonregioselectively, as expected on the basis of the similar strength of H-bonds formed by HOC(2 and 3) (both in a *cis*-annulated, five-membered ring), Lemieux glycosidation proceeds with a regioselectivity of 9:1, favoring the 1,2-linked disaccharides. This is surprising as one expects HOC(2), vicinal to the anomeric center, to be less nucleophilic than HOC(3). This regioselectivity thus points to a σ-conjugation between HOC(3) (lone pair or $\sigma_{O,H}$ orbital) through the C(2),C(3) bond with the anomeric methoxy group ($\sigma^*_{C,O}$ orbital). This conjugation lowers the nucleophilicity of OC(3), and should improve the H-bond accepting properties of the anomeric methoxy group.

The 1,2-*trans*-Configured diequatorial diols form weak H-bonds. Glycosidation of the anomeric 4,6-O-benzylidenated *gluco-* and *galacto*-diols **53–58**, and of the 2,6-di-O-benzylated α-D-mannopyranoside **59** by **1** has been studied to evaluate the effect of such weak bonds upon regioselectivity (Muddasani et al., 1994b). No regioselectivity is observed in the glycosidation of any of these diols in dichloroethane (Table 2-4), but apparently for different reasons. The α-D-galactopyranoside **57** is a single tautomer possessing two H-bonds of similar strength (Fig. 2-3, Fig. 2-4A, Table 2-3); protonation should be indiscriminate, and the lack of regioselectivity is no surprise. A similar situation is found for the β-D-glucopyranosides **55** and **56** (Fig. 2-4B). The IR spectra of both glycosides are characterized by a single OH band. The α-D-glucosides **53** and **54** (Fig. 2-4C), the α-D-mannopyranoside **59** (Fig. 2-4E), and the β-D-galactopyranoside **58** (Fig. 2-4C), however, possess H-bonds of differing stability, forming a H-bond arrangement as in Figure 2-3.

The α-D-gluco-, and the β-D-galactopyranosides **53**, **54**, and **58** are single H-bonded tautomers (two OH bands in the IR spectra, separated by 18 and 13 cm^{-1}, respectively), while the mannoside **59** is a mixture of two tautomers, with **59a** dominating

Table 2–4 Glycosidation of *trans*-Diols **53–59** with Diazirine **I** at Room Temperature

Diol (equiv.)	Solvent	Total Yield (%)	Regioselectivity RO—C(2)/RO—C(3)[a]	Stereoselectivity (α-D/β-D)	
				RO—C(2)[b]	RO—C(3)
53 (1.3)	ClCH$_2$CH$_2$Cl	75	46:54	47:53	39:61
53 (0.75)	ClCH$_2$CH$_2$Cl	89	46:54	46:54	38:62
53 (1.3)	1,4-Dioxane	72	44:56	39:61	36:64
54 (1.5)	ClCH$_2$CH$_2$Cl	77	51:49	50:50	37:63
54 (0.9)	ClCH$_2$CH$_2$Cl	73	51:49	46:54	35:65
54 (0.9)	1,4-Dioxane	60	46:54	35:65	26:74
55 (1.3)	ClCH$_2$CH$_2$Cl	80	49:51	45:55	41:59
55 (0.75)	ClCH$_2$CH$_2$Cl	94	52:48	47:53	35:65
55 (1.3)	1,4-Dioxane	69	46:54	32:68	38:62
56 (1.5)	ClCH$_2$CH$_2$Cl	63	52:48	42:58	39:61
56 (0.9)	ClCH$_2$CH$_2$Cl	59	54:46	42:58	35:65
56 (0.9)	1,4-Dioxane	63	59:41	31:69	42:58
57 (1.3)	ClCH$_2$CH$_2$Cl	69	51:49	44:56	52:48
57 (0.75)	ClCH$_2$CH$_2$Cl	80	50:50	44:56	52:48
57 (1.3)	1,4-Dioxane	76	50:50	39:61	59:41
58 (1.5)	ClCH$_2$CH$_2$Cl	61	54:46	40:60	42:58
58 (0.9)	ClCH$_2$CH$_2$Cl	56	56:44	38:62	42:58
58 (0.9)	1,4-Dioxane	51	54:46	32:68	48:52
59 (1.3)	Toluene[c]	75	46:54	48:52	64:36
59 (1.3)	1,4-Dioxane	71	51:49	47:53	59:41
59 (0.75)	1,4-Dioxane	80	51:49	49:51	61:39

[a]RO—C(4)/RO—C(3) for **59**.
[b]RO—C(4) for **59**.
[c]At 70°C.

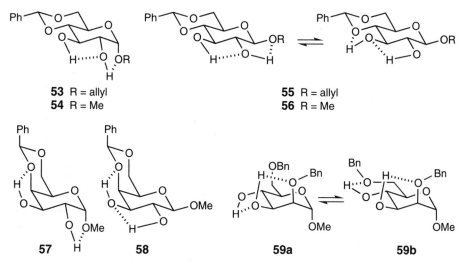

53 R = allyl
54 R = Me

55 R = allyl
56 R = Me

57   **58**   **59a**   **59b**

Figure 2–3. Intramolecular hydrogen bonds of *trans*-diols **53–59** in chlorinated solvents.

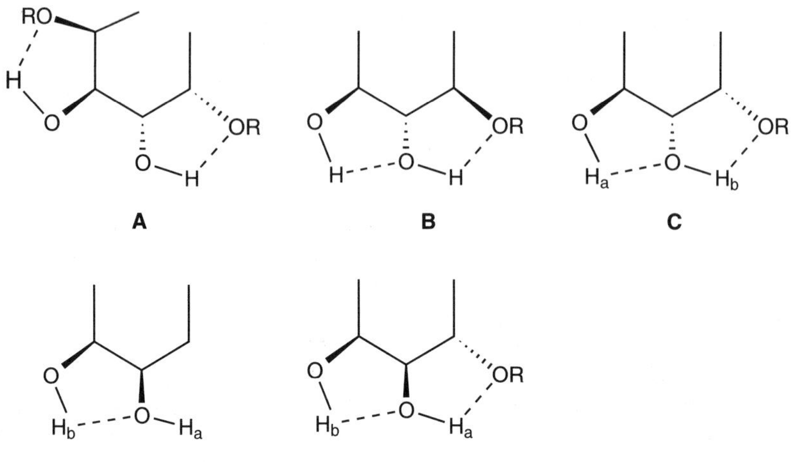

Figure 2–4. (A–E) Types of intramolecular hydrogen bonds in vicinal diols.

(IR at 3585, 3558 cm^{-1}). One has to expect regioselective protonation by H_a in a *trans, cis* configured monoalkylated triol as in (Figure 2-4C), corresponding to HOC(2) in **58**, HOC(3) in **53** and **54**, or HOC(4) in **59a**, respectively. Hence, the lack of regioselectivity of the glycosidation must be the consequence of a nonregioselective combination of the ensuing ions. Regioselective protonation by H_a, that is, by the OH group in the weaker H-bond, positions the glycosyl cation between the vicinal, *trans*-diequatorial hydroxy and oxy groups, which presumably undergo a very rapid proton transfer. This arrangement is quite different from the one in (Fig. 2-4D), where protonation by H_a positions the glycosyl cation close to one oxygen function only. As a rule then, protonation of a glucosylidene carbene by a H-bonded OH group forming a five-membered ring *trans* fused to a pyranosyl ring in a monoalkylated *trans, cis* triol (H_a in the arrangement Fig. 2-4C) is not followed by regioselective combination of ions. It remains to be explored if this rule is also valid for diastereomeric (e.g., *manno*-configured) glycosylidene carbenes.

Several 1,2-*cis* axial, equatorial diols have been glycosylated by **1**. The α-D-mannoside **60** and the α-D-galactoside **62** possess a *cis, trans* configured, monoalkylated triol unit of type **E** (Fig. 2-4). They exist as mixtures of tautomers, as depicted in Figure 2-5. Protonation is expected to be regioselective, by H_a, to position the glycosyl cation close to the oxyanion derived from the equatorial OH group, and to lead to regioselective glycosidation. Preferential glycosidation of the equatorial OH group is indeed observed (Table 2-5); selectivity is poor in (CH$_2$)$_2$Cl$_2$, but respectable in toluene (d.e. = ca. 50% for **60**, and 70% for **62**). Glycosidation of the β-D-mannopyranoside **61** by **1** proceeds without regioselectivity. At a first glance, this result is surprising, as the mannoside **61** is a ca. 1:1 mixture of tautomers (**61a** and **61b**), and the weakest H-bond seems to be between the *trans*-oriented HOC(3) and OC(4) in **61b**; regioselective protonation should lead mostly to 1,3-linked disaccharides. As for **48**, however, σ-conjugation between HOC(3) and C(1)OMe should strengthen the HOC(3) to OC(4) H-bond, and lower the nucleophilicity of HOC(3). This is not only evidenced by the

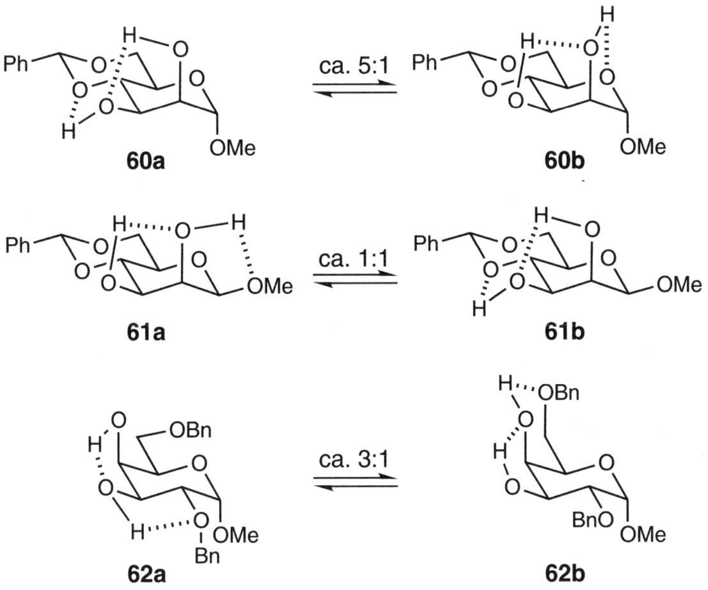

Figure 2–5. Intramolecular hydrogen bonds of *cis*-diols **60–62** in chlorinated solvents.

IR spectrum of **61** which shows a single OH band at 3574 cm^{-1} (CH$_2$Cl$_2$; the α-anomer **60** shows bands at 3587 and 3575 cm^{-1}, Table 2-3), but also by the results of glycosidation by Lemieux's method. Unlike the α-anomers **60** and **62**, where glycosidation with **37** proceeds in good yields and regioselectively at the equatorial hydroxyl group (*via* the tautomers **60a** and **62a**), the β-D-mannoside **61** is unreactive (Table 2-6). Repetition of the Lemieux glycosidation of **60** and **61** by **37**, however, led to a dif-

Table 2–5 Glycosidation of *cis*-Diols **60–62** with Diazirine **I** at Room Temperature

Diol (equiv.)	Solvent	Total Yield (%)	Regioselectivity RO—C(2)/RO—C(3)^a	Stereoselectivity (α-D/β-D)	
				RO—C(2)^b	RO—C(3)
60 (1.3	ClCH$_2$CH$_2$Cl	78	40:60	43:57	41:59
60 (0.75)	ClCH$_2$CH$_2$Cl	96	39:61	48:52	38:62
60 (1.3)	1,4-Dioxane	83	29:71	29:71	39:61
60 (1.3)	Toluene	86	27:73	54:46	39:61
60 (1.3)	Toluene^c	89	24:76	38:62	40:60
61 (0.75)	ClCH$_2$CH$_2$Cl	85	48:52	53:47	56:44
62 (1.3)	1,4-Dioxane	67	26:74	39:61	51:49
62 (0.75)	1,4-Dioxane	74	25:75	39:61	47:53
62 (1.3)	Toluene	73	19:81	53:47	54:46
62 (1.3)	Toluene^c	68	14:86	40:60	54:46

^{a}RO—C(4)/RO—C(3) for **62**.
^{b}RO—C(4) for **62**.
^{c}At 70°C.

Table 2–6 Glycosidation of cis-Diols **60–62** with Bromide **37** at Room Temperature and Trichloroacetimidate **38** at **−30°C**

Donor	Diol	Total Yield (%)	Regioselectivity RO—C(2)/RO—C(3)[a]	Stereoselectivity (α-D/β-D)	
				RO—C(2)[b]	RO—C(3)
37	**60**	80	30:70	93:7	90:10
37	**61**	Trace	—	—	—
37	**62**	75	10:90	90:10	60:40
37[c]	**60**	33	13:87	Only α	Only α
37[c]	**61**	32	27:73	Only α	Only α
38	**60**	93	12:88	40:60	7:93
38	**61**	91	46:54	11:88	8:92
38	**62**	86	20:80	22:78	40:60

[a]RO—C(4)/RO—C(3) for **62**.
[b]RO—C(4) for **62**.
[c]Repetition by Czechtizky, 1996.

ferent result (Czechtizky, 1996); both anomers reacted within 3 days in low yield to mixtures of the α-D-configured 2-O- and 3-O-linked disaccharides in the ratio of 13:87 and 27:73, respectively. The lowering of the nucleophilicity of HOC(3) is also evident from the glycosidation with the trichloroacetimidate **38**, which favors the equatorial, more nucleophilic OH groups of the α-anomers **60** and **62**, but proceeds nonregioselectively for the β-mannoside **61**.

Strong evidence for stereoelectronic control, and the lack of synchronicity and perhaps concertedness of the protonation of a glycosylidene carbene and the combination of the ions thus generated, and for the protonation by H-bonded OH groups, has been derived from the glycosidation of a number of species. These include the anomeric allosides **63** and **64** (Muddasani et al., 1994a), the enantiomeric ribopyranosides **65** and **66** (Uhlmann & Vasella, 1994), and the anomeric glycosides **67** and **68** of N-phthaloylallosamine (Briner et al., 1994). The regioselective glycosidation of the altrotriol **69** has been rationalized by the same mechanism (Bozó & Vasella, 1994).

63   **64**   **65**   **66**

67   **68**   **69**

The α-D-alloside **63** is a mixture of the two tautomers **63a** and **63b** (Fig. 2-6). The bifurcated H-bond is evidenced by the large HCOH coupling constant (Table 2-3; cf. Jochims & Kobayashi, 1976). It is only present in solution; the crystal structure merely shows an intramolecular H-bond between HOC(3) and the anomeric methoxy group. The bond between HOC(3) and OC(4) appears to be of approximately the same strength as the bifurcated H-bond of HOC(2), as evidenced by the single IR OH band. Protonation should occur by either HOC(3) or HOC(2). Protonation by HOC(3) places the glycosyl cation next to OC(3) only, and should lead to 1,3-linked disaccharides; protonation by HOC(2) may lead to glycosidation by either one of the cis-oriented groups (OH and oxy anion). The results of the glycosidation of **63** by **1** depend strongly upon conditions (*see* Table 2-7). HOC(3) is indeed preferentially glycosylated, except in THF at low temperature. The anomeric selectivity is different for the regioisomeric glycosidations: the β-anomer is favored among the 1,2-linked products, most strongly in THF at low temperature, while the a-anomer prevails in the 1,3-linked major products.

These results have been rationalized by pointing out that the OH groups of the diol are located in different planes. Protonation of the carbene in the σ-plane by the H-bonded HOC(2) orients the glycosyl cation in such a way that HOC(3), or the oxy anion derived from it by rapid proton shift, is properly oriented to attack in the π-plane of the cation. The α-anomer is preferentially formed either in consequence of a specific orientation of the carbene or glycosyl cation and the principle of least motion (movement according to a → **A**), or because the ion pair is sufficiently long lived to

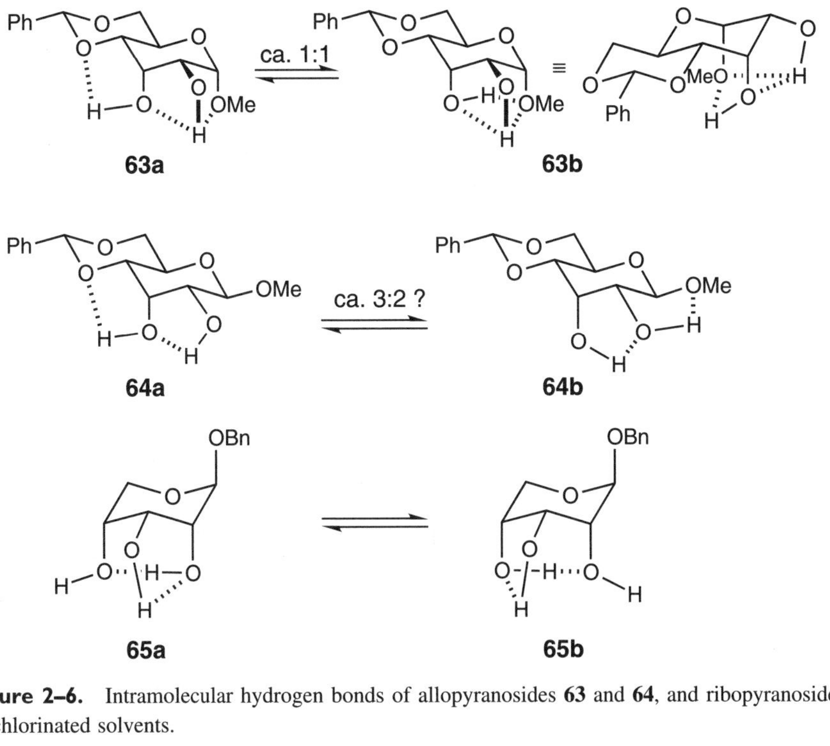

Figure 2–6. Intramolecular hydrogen bonds of allopyranosides **63** and **64**, and ribopyranoside **65** in chlorinated solvents.

Table 2–7 Glycosidation of *allo*-Diols **63** and **64** with Diazirine **I** and Trichloroacetimidate **38**

Glycosyl Donor	Diol (equiv.)	Solvent, Temperature	Total Yield (%)	Regioselectivity RO—C(2)/ RO—C(3)	Stereoselectivity (α-D/β-D)	
					RO—C(2)	RO—C(3)
1	63 (1.3)	CH₂Cl₂, r.t.	90	23:77	41:59	81:19
1	63 (1.3)	CH₂Cl₂, hν, −85°C	82	21:79	44:56	89:11
1	63 (1.3)	ClCH₂CH₂Cl, r.t.	83	32:68	49:51	74:26
1	63 (1.3)	CCl₄, r.t.	30	8:92	49:51	94:6
1	63 (1.3)	Toluene, 70°C	79	20:80	32:68	72:28
1	63 (1.3)	Dioxane, r.t.	81	28:72	39:61	89:11
1	63 (1.3)	THF, r.t.	75	40:60	25:75	66:34
1	63 (1.3)	THF, hν, −85°C	79	72:28	10:90	67:33
38	63 (0.9)	CH₂Cl₂, −30°C	94	28:72	24:76	23:77
1	64 (0.75)	ClCH₂CH₂Cl, r.t.	69	54:46	40:60	45:55
1	64 (0.75)	Dioxane, r.t.	80	51:49	32:68	38:62
38	64 (0.9)	CH₂Cl₂, −30°C	90	46:54	11:89	31:69

allow rotation around the C(1),C(4) axis of the glycosyl cation, leading to stereoelectronically preferred axial attack by the oxyanion (Scheme 2-9). Movement of the cation according to b → **B**, entailing (partial) dissociation of the ion pair, leads to the minor 1,2-linked β-D-anomer. Under conditions of kinetic control, THF solvates the glycosyl cation from the axial side, as discussed above. Substitution by ⁻OC(2) with THF

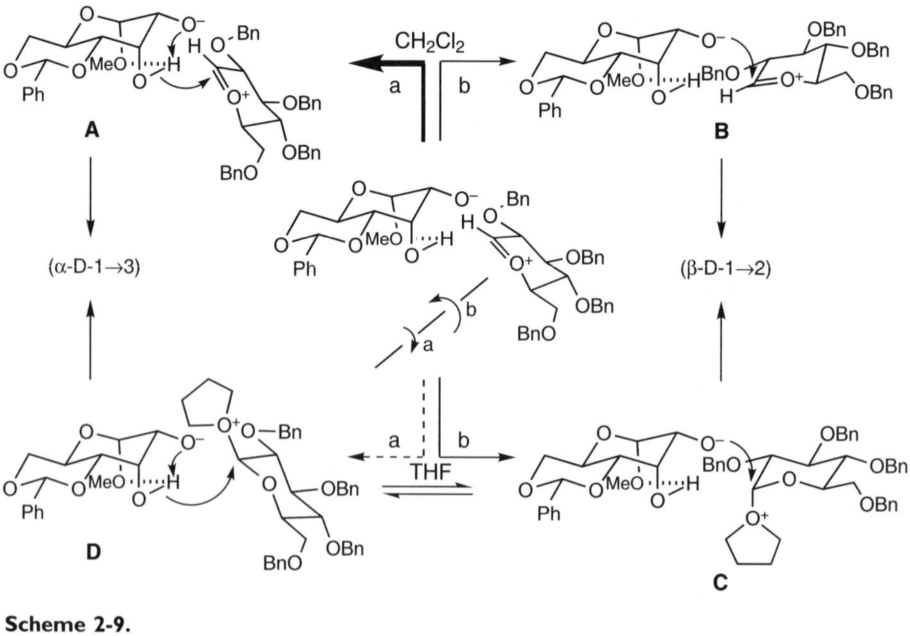

Scheme 2-9.

as the leaving group, and with inversion of the configuration at the anomeric center, appears to be geometrically favorable and prevails at low temperature. At higher temperatures, the tetrahydrofuranylium ions equilibrate, and a competing inverting substitution of the β-anomer by $^-$OC(3) (Scheme 2-9C) leads to increasing amounts of the α-D-1,3-linked disaccharide.

This interpretation has been confirmed by glycosidation of the equatorial glycoside **64**. One expects protonation mostly by the equatorial HOC(2), involved in what should be the weakest hydrogen bond for geometric and electronic reasons, were it not for the effect of the σ-conjugation between OC(3) and the methoxy group. The IR spectrum of **64** (CH$_2$Cl$_2$, Table 2-3) shows only one OH band, indicating that all hydrogen bonds are of similar strength. A barely significant preference for the formation of 1,2-linked disaccharides is indeed observed; the β-anomers are preferred slightly.

A case complementary to the alloside **63**, where protonation of the carbene by an equatorial OH group is followed by interception of the glycosyl cation by an axial OH group, is realized in the glycosidation of the *ribo*-triol **65** and its enantiomer **66** (Uhlmann & Vasella, 1994). The crystal structure of the methyl riboside corresponding to **65** (Hordvik, 1974; James et al., 1978) shows one strong intramolecular H-bond between HOC(2) and HOC(4), and two intermolecular H-bonds. IR and NMR studies of the benzyl riboside **65** (Fig. 2-6) show the presence of an equilibrium between two species, with the same H-bond as in the solid state, but with HOC(3) involved as a H-bond donor to either OC(2) or OC(4), as further evidenced by the partial deuteriation shift. The IR band at 3601 cm^{-1} (Table 2-3) is compatible either with a very weak H-bond between HOC(4) and OC(3) in **65a** or HOC(2) and OC(3) in **65b**, or with a free OH group; the latter interpretation being compatible with the ^{1}H NMR spectra. Independent of this, protonation should be by either HOC(4) or HOC(2), positioning the glycosyl cation close to OC(3), in the π-plane of the cation. Direct attack by OC(3) is then feasible, while attack by HOC(2 or 4) requires partial dissociation of the ion pair. The major regioisomers of glycosidation of **65** by **1** are indeed the anomeric α-D-configured 1,3-linked disaccharides (Table 2-8). Among the minor regioisomers, there is a slight preference for the β-D-1,2-linked, and for the α-D-1,4-linked anomers. Dilution experiments suggest that the 1,3-linked disaccharides derive from the monomeric triol, and the minor products from oligomers. The temperature dependence indicates that partial dissociation is required for attack of the oxycarbenium ion by the protonating HOC(2 or 4). In THF at low temperature, one obtains mostly the 1,2- and the 1,4-linked disaccharides, with a strong preference for the equatorial anomers, showing the importance of solvation of the ion pairs. The results of glycosidation of the benzyl β-L-ribopyranoside **66** differ only slightly, but significantly from those of the enantiomer **65**. The ratio of the 1,2- and 1,4-linked disaccharides is inverted, reflecting the different diastereomeric interactions of the carbene with HOC(2) and HOC(4) of the enantiomeric ribosides. The ratio of anomers of the 1,3-linked products is quite similar for glycosidations in CH$_2$Cl$_2$ at $-78°$C, but consistently lower for glycosidation of the L-enantiomer **66** at 23°C. Considering that the 1,3-linked disaccharides are derived from the oxycarbenium ion formed by protonation by either HOC(2) and HOC(4) (and thus originally in a different orientation), it is significant that the α-D-anomers are clearly the dominating products of glycosidation of both enantiomers in CH$_2$Cl$_2$ and in dioxane. This shows that the C(2) benzyloxy group of the oxycarbenium cation competes at best to a minor extent with HOC(3) of the triol in coordinating with the cationic center of the oxycarbenium ion, and also that the oxycarbenium

Table 2–8 Glycosidation of *ribo*-Triols **65** and **66** with Diazirine **1**

Triol (Conc. (M))	Solvent, Temperature	Total Yield (%)	Regioselectivity RO—C(2)/ RO—C(3)/ RO—C(4)	Stereoselectivity (α-D/β-D) RO—C(2)	RO—C(3)	RO—C(4)
65 (0.05)	Dioxane, r.t.	69	30:40:30	33:67	75:25	60:40
66 (0.05)	Dioxane, r.t.	65	29:41:30	49:51	68:32	43:57
65 (0.05)	CH$_2$Cl$_2$, r.t.	85	19:58:23	47:53	74:26	59:41
66 (0.05)	CH$_2$Cl$_2$, r.t.	86	29:47:24	55:45	64:36	49:51
65 (0.025)	CH$_2$Cl$_2$, r.t.	86	17:61:22	45:55	77:23	61:39
65 (0.005)	CH$_2$Cl$_2$, r.t.	91	14:65:21	31:69	80:20	70:30
66 (0.005)	CH$_2$Cl$_2$, r.t.	90	31:51:18	54:46	65:35	45:55
65 (0.05)	THF, r.t.	80	34:29:37	29:71	69:31	32:68
66 (0.05)	THF, r.t.	83	39:25:36	29:71	54:46	36:64
65 (0.05)	CH$_2$Cl$_2$, -78°C	66	12:62:26	42:58	74:26	69:31
66 (0.05)	CH$_2$Cl$_2$, -78°C	64	21:58:21	62:38	79:21	47:53
65 (0.005)	CH$_2$Cl$_2$, -78°C	81	5:81:14	40:60	77:23	64:36
66 (0.005)	CH$_2$Cl$_2$, -78°C	87	16:74:10	66:34	79:21	58:42
65 (0.05)	THF, -78°C	83	42:19:39	8:92	28:72	5:95
66 (0.05)	THF, -78°C	80	35:21:44	13:87	41:59	12:88

ion is capable of reorienting itself, presumably by rotation around the C(1),C(4) axis. The influence of THF—competition with HOC(3) in the interaction with the oxycarbenium ion—is apparent in both the regio- and stereoselectivity of the glycosidation.

The influence of intramolecular H-bonds, and of the protonation of the glycosylidene carbene by H-bonded OH groups is again clearly evidenced by the glycosidation of the N-phthaloyl-D-allosamine derivatives **67**, **68**, and **70–72** (Briner et al., 1994; Maloisel & Vasella, 1992). Glycosidation of **67** led to a 9:1 mixture of the anomeric 1,3-linked disaccharides. The α-D-configuration of the main product suggests that protonation is by HOC(4). In keeping with this interpretation, the benzylidenated α-D-configured monoalcohol **70**, where HOC(4) is blocked, does not react with **1**, while the β-D-anomer **71** reacts, but yields mostly the β-D-configured disaccharide, and the β-D-diol **68** is nonregioselectively glycosylated.

This behavior is rationalized by a particularly strong intramolecular H-bond between HOC(3) and one of the carbonyl groups of the N-phthalimido substituent of the diol **67**, as evidenced by the chemical shift of the OH groups (HOC[3] at 6.09 ppm!), the coupling constants and the IR bands (Table 2-3), and confirmed by MM3 calculations.

The large value of $J(HOC(4)H)$ (11.1 Hz) shows that there is a weaker H-bond between HOC(4) and OC(3), directing the HOC(4) bond below the plane of the pyranose ring. The result of the glycosidation of the α-D-anomer **67** can only be understood if protonation is by the H-bonded HOC(4), directing the glycosyl cation in front of OC(3). The preferred formation of the α-D-anomer, whenever there was a clear discrepancy between the regioselectivity of the protonation and the ion combination, is more easily reconciled with the idea that the ion pair is sufficiently long lived that the glycosyl cation can rotate around the C(1)–C(4) axis, and that its interception is stereoelectronically controlled, than with the postulate of a specific orientation of the attacking carbene. The β-D-anomer **68** is a mixture of two tautomers, **68a** and **68b**, characterized by different orientations of the phthalimido group; the one reacting with the carbene (**68a**) possesses a relatively weak HOC(3) to OC(4) bond. The orientation of the phthalimido group is also different in the anomeric benzylidene acetals **70** and **71**, with the α-D-anomer **70** possessing a strong H-bond with one of the phthalimido carbonyl groups, while the β-D-anomer **71** forms a weaker H-bond with OC(4). The H-bonding pattern of the diol **67** persists (qualitatively) in the triol **72**, as evidenced by the regioselectivity of the glycosidation, which leads to a mixture of three disaccharides derived from glycosidation of the secondary OH groups only. The main products (Table 2-9) are the 1,3-linked anomers (with an α:β ratio not much different from the one of the 6-O-benzylated analogue **67**) and the β-configured 1,4-linked disaccharide; no products of glycosidation of the primary OH group have been found.

Glycosidation of the altrose-derived triol **69** (Bozo & Vasella, 1994) has been rationalized on the basis of the two H-bonded tautomers **69a** and **69b** (Fig. 2-7) which both possess a weak H-bond involving HOC(2). HOC(3) is H-bonded either to the methoxy group or to OC(4), and HOC(4) either forms a strong H-bond to OC(6), or a weaker one to OC(3). One expects protonation mainly by HOC(2), positioning the glycosyl cation next to OC(2) only, and leading to 1,2-linked disaccharides as major products, followed in importance by protonation either by HOC(4) of the major, or by HOC(3) of the minor isomer. In either case, the glycosyl cation is then positioned between the hydroxy and oxy groups at C(4) and C(3), in such a way that the regioselectivity of protonation differs from that of ion combination; the ratio of the two tautomers has led to the prediction that the second most important regioisomers should be the 1,3-linked disaccharides, and also that both the 1,3- and the 1,4-linked disaccharides should be predominantly α-D-configured. This is indeed found when the glycosidation is performed by **1** in CH_2Cl_2 at $-80°C$ (Table 2-10). This result also shows

Table 2–9 Glycosidation of N-Phthaloyl-allopyranosides **67, 68, 70–72** with Diazirine **I** at Room Temperature in 1,4-Dioxane

Acceptor	Recovered Acceptor (%)	Yield of Regioisomers (%) (α-D/β-D)	
		RO—C(3)	RO—C(4)
70	95	0	—
67	16	80 (90:10)	Trace (only β-D)
72	32	57 (72:28)	9 (only β-D)
71	32	66 (21:79)	—
68	30	29 (45:55)	33 (39:61)

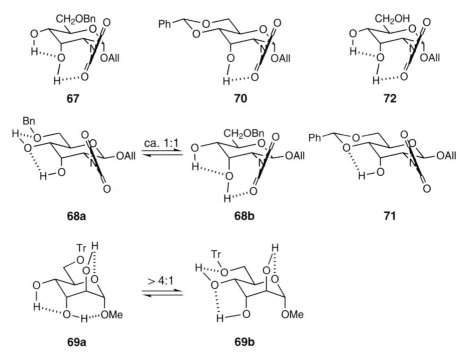

Figure 2–7. Intramolecular hydrogen bonds of 2-phthalimido-allopyranosides **67**, **68**, and **70–72** and of altrotriol **69** in chlorinated solvents.

that the glycosyl cation can be intercepted, not unexpectedly, by a OH group, and not only by an oxyanion. The proportions of the regioisomers remain qualitatively the same, except for glycosidation in THF at low temperatures.

Unfortunately, the disjunction of protonation and ion combination has its negative effects: interaction of **1** with N-Boc protected Asn(OBn) does not lead to an N-

Table 2–10 Glycosidation of Altrotriol **69** with Diazirine **1** at Room Temperature

Solvent	Total Yield (%)	Regioselectivity RO—C(2)/ RO—C(3)/ RO—C(4)	Stereoselectivity (α-D/β-D)		
			RO—C(2)	RO—C(3)	RO—C(4)
ClCH₂CH₂Cl	57	55:31:14	43:57	45:55	61:39
CH₂Cl₂	53	57:30:13	49:51	46:54	64:36
CH₂Cl₂ᵃ	38	55:27:17	35:65	74:26	69:31
Toluene	48	48:36:16	41:59	59:41	83:17
Dioxane	49	50:37:13	62:38	36:64	54:46
THF	54	47:39:13	49:51	28:72	50:50
THFᵃ	50	43:52:5	9:91	13:87	84:16
DME	39	36:46:16	52:48	59:41	79:21

ᵃAt −80°C (hν).

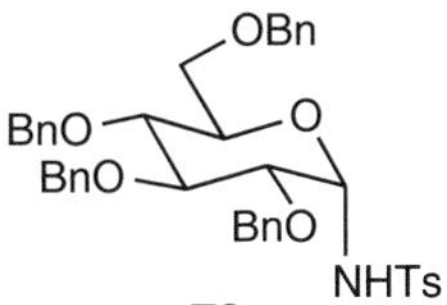

73 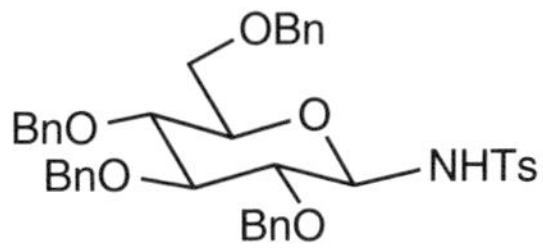   **74**

glycoside, but *via* an unstable intermediate (iminoether?) to a glycosyl ester, while insertion into the N—H bond of toluenesulfonamide yields the N-glycosyl sulfonamides **73** and **74** (50%, α:β = 1:4), along with 16% of the benzyloxyglucal **15** (Rajamannar, 1990).

Glycosidation by Diazirines Derived from 2-Acetamido-2-deoxyaldoses

Glycosidation of hexafluoro-2-propanol and 2-propanol by the N-acetylallosamine derived diazirine **12** (Vasella & Witzig, 1995) shows surprising and significant aspects: the major products are α-D-anomers, the stereoselectivity is higher at higher temperatures, and higher for the less strongly acidic alcohol (Scheme 2-10). The neighboring group activity of 2-acetamido groups of classical glycosyl donors is well known, strong, and leads to β-D-glycosides; it is evidently irrelevant for the glycosidation by the diazirines **11** and **12**.

The results have been rationalized on the basis of two postulates: (1) that the diazirine exists as a mixture of two H-bonded tautomers, where the position of the equilibrium between them is temperature dependent, lower temperatures favoring the in-

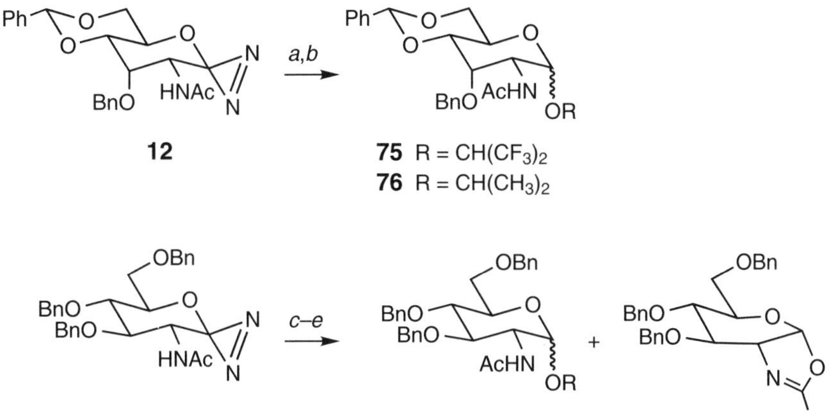

12

75 R = CH(CF$_3$)$_2$
76 R = CH(CH$_3$)$_2$

11

77 R = CH(CF$_3$)$_2$
78 R = CH(CH$_3$)$_2$
79 R = C$_6$H$_4$pNO$_2$

80

Scheme 2-10. (a) (CF$_3$)$_2$CHOH, CH$_2$Cl$_2$, 40°C, **75** (85%, α/β 76:24) *or* (CF$_3$)$_2$CHOH, CH$_2$Cl$_2$, hν, −84°C, **75** (88%, α/β 52:48). (b) (CH$_3$)$_2$CHOH, CH$_2$Cl$_2$, 40°C, **76** (12%, only α) *or* (CH$_3$)$_2$CHOH, CH$_2$Cl$_2$, hν, −84°C, **76** (72%, α/β 78:22). (c) (CF$_3$)$_2$CHOH, CH$_2$Cl$_2$, 0°C → r.t., **77** (72%, α/β 87:13) *or* (CF$_3$)$_2$CHOH, CH$_2$Cl$_2$, hν, −84°C, **77** (59%, α/β 70:30). (d) (CH$_3$)$_2$CHOH, CH$_2$Cl$_2$, hν, −84°C, **78** (70%, only α) and **80** (13%). (e) pNO$_2$C$_6$H$_4$OH, CH$_2$Cl$_2$, 0°C → r.t., **79** (96%, α/β 58:42).

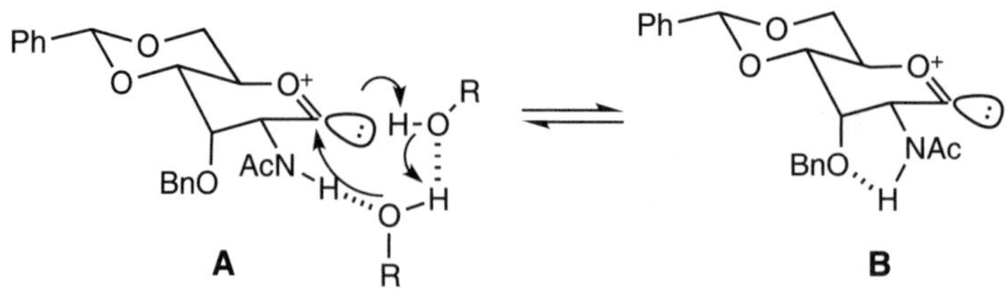

Figure 2–8. (A,B) Hydrogen bonds of the carbene derived from *allo*diazirine **12**.

tramolecular H-bond; and (2) that an intermolecular H-bond between the acetamido group as a H-bond donor and the alcohol as a H-bond acceptor places the glycosyl acceptor below the plane of the pyranose ring, and thereby increases the probability of attack from the α-side (in the π-plane of the glycosyl cation) of this H-bonded alcohol unit (Fig. 2-8). The higher stereoselectivity in glycosidation of 2-propanol is due to its stronger basicity, making it a better H-bond acceptor, while the temperature dependence reflects the position of the equilibrium of tautomers. In keeping with these postulates, the N-acetyl-D-glucosamine derived diazirine **11**, which should at best form a weak intramolecular H-bond, glycosylates 2-propanol and hexafluoro-2-propanol with a higher stereoselectivity than does **12**; also, the diastereoselectivity is proportional to the basic properties of the hydroxy compounds, and lowest for 4-nitrophenol (cf. Kolthoff & Chantooni, 1969). At higher temperatures, 2-propanol yields a number of products, mainly the oxazoline **80**, which is expected from intermolecular protonation of the carbene and neighboring group participation of the acetamido group.

These results demonstrate how 1,2-*cis*-configured glycosides can be prepared from 2-acetamido-2-deoxyglycosyl donors under kinetic control, and show that a stereoselective, *quasi* intramolecular glycosidation is possible by making use of intermolecular hydrogen bonding.

Reaction of Glycosylidene Carbenes with Other Compounds Possessing an HX Group

Three nonconcerted mechanisms have been considered for the formal insertion of a carbene into an X,H bond: deprotonation of XH, as discussed above, nucleophilic addition of XH, leading to an ylid and followed by proton migration, and electrophilic addition of XH, also leading to an ylide, and followed by hydride migration (Bethell, 1969). Nucleophilic attack by an XH group may be expected for phosphines; electrophilic attack is plausible for stannanes. This raised the question of whether glycosylidene carbene insertions might lead to glycosyl phosphines and glycosyl stannanes, and whether the formal insertions would be stereoselective.

Reaction of **1** with diphenylphosphine at room temperature leads to 76% of a 1:1 mixture of anomeric glycosyl phosphines **81** (Vasella et al., 1991a). The α-D-anomer is the product of stereoelectronic rather than steric control, as evident from Scheme 2-11; it is the major product only at low temperatures. The glycosyl phosphines were isolated in the form of the corresponding phosphine oxides **82**. The phosphine oxides, and some analogous pyranose and furanose derivatives, have since been prepared con-

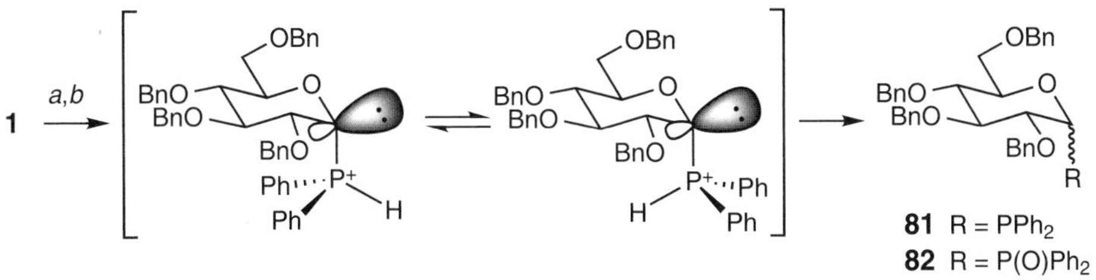

Scheme 2-11. (a) Ph$_2$PH, CH$_2$Cl$_2$, r.t.; H$_2$O$_2$, AcOH, CH$_2$Cl$_2$, r.t., **82** (76%, α/β 50:50). (b) Ph$_2$PH, CH$_2$Cl$_2$, hν, −78°C; H$_2$O$_2$, AcOH, CH$_2$Cl$_2$, r.t., **82** (54%, α/β 75:25).

veniently from glycosyl acetates and methyl diphenyl phosphinite, and reduced to glycosyl phosphines (Lopusinski et al., 1993).

The diazirine **1** reacts with excess Bu$_3$SnH in CH$_2$Cl$_2$ at room temperature to give a mixture of the anomeric glucopyranosyl stannanes **83** with the stannylglucal **85** as a side product (see Scheme 2-12) (Uhlmann et al., 1994b). A similar yield of Sn-glycosides (70%) is obtained in THF, but the ratio of anomers is lowered from α:β = 84:16 to α:β = 79:21. The triphenylstannyl glycosides **84** have been isolated in slightly better yields; no stannylglucal could be isolated. The *gluco*-configured diazirine **1** yields mostly the axial, 1,2-*cis*-configured Sn-glycosides. That the insertion preferentially produces 1,2-*cis* rather than axial Sn-glycosides has been shown by treating the *manno*-diazirine **14** with

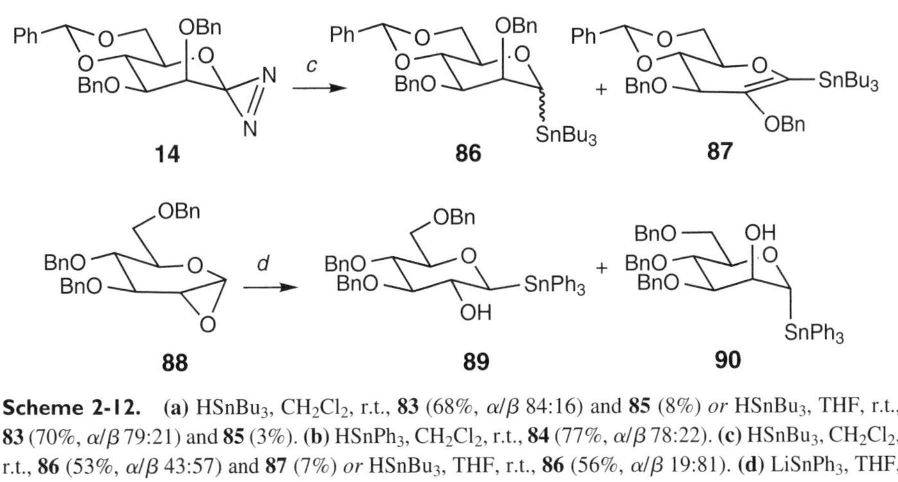

Scheme 2-12. (a) HSnBu$_3$, CH$_2$Cl$_2$, r.t., **83** (68%, α/β 84:16) and **85** (8%) *or* HSnBu$_3$, THF, r.t., **83** (70%, α/β 79:21) and **85** (3%). (b) HSnPh$_3$, CH$_2$Cl$_2$, r.t., **84** (77%, α/β 78:22). (c) HSnBu$_3$, CH$_2$Cl$_2$, r.t., **86** (53%, α/β 43:57) and **87** (7%) *or* HSnBu$_3$, THF, r.t., **86** (56%, α/β 19:81). (d) LiSnPh$_3$, THF, r.t., **89** (65%) and **90** (2%).

Scheme 2-13. (a) 1.2 equiv. of Ph$_3$B, THF, r.t., then H$_2$O$_2$, NaOH, **91** (40%) and **92** (18%). (b) 1.2 equiv. of Bu$_3$B, THF, r.t., then H$_2$O$_2$, NaOH, **93** (45%) and **94** (17%). (c) 1.2 equiv. of Et$_3$B, THF, r.t., then H$_2$O$_2$, NaOH, **95** (60%) and **96** (11%).

Bu$_3$SnH (in CH$_2$Cl$_2$) to yield a mixture of the anomeric *manno*-stannanes **86** in an α:β ratio = 43:57. Yields are hardly affected by using THF, but the preference for the 1,2-*cis*-configured Sn-glycoside is increased significantly to α:β = 19:81.

These results suggest that hydride migration following the nucleophilic attack of the carbene avoids a destabilizing interaction with the C(2)OBn group and that there is a simultaneous favorable interaction between the tetra-ligated Sn center and the C(2) benzyloxy group. Coordination of Sn with THF increases the bulk of the C(1) substituent leading to adoption of the preferred equatorial orientation by the R$_3$Sn group.

The potential insertion of glycosylidene carbenes into the Sn—Sn bond of hexabutyldistannane has also been examined (Uhlmann et al., 1994b). The reaction between **1** and (Bu$_3$Sn)$_2$ gives 17% of the stannylglucal **85** and 60% of a mixture of azines **16**, in addition to benzyloxyglucal **15** and traces of Bu$_3$SnCl. Neither the 1,1-distannane nor Sn-glycosides could be isolated. This shows that the glycosylidene carbene interacts less readily with (Bu$_3$Sn)$_2$ than with **1** (and less readily than with Bu$_3$SnH). Generation of the stannylglucal **85** is not followed by migration of the second Bu$_3$Sn moiety to C(1), but by homolytic fission of the Sn—Sn bond, and radical abstraction from the ensuing glycosyl radical.

Sn-glycosides are of interest as precursors of glycosyl lithium compounds. The α-D-configured dilithium derivatives of 2-hydroxypyranoses, saccharide derivatives with inverted polarity at the anomeric center, have been generated from 2-hydroxy-α-D-glycopyranosyl chlorides (Wittmann & Kessler, 1993); the corresponding β-D-anomers are available from β-D-configured 2-hydroxy Sn-glycosides, which are prepared from 1,2-anhydro pyranoses (Frey et al., 1994).

Reaction with stannanes suggests that boranes should react in a similar way (Graßberger, 1971), by nucleophilic attack of the carbene on the boron center, followed by migration of a boron substituent; this would generate a new (C-glycosyl)borane, so that the process could conceivably be repeated. Exploratory experiments have shown that **1** reacts with boranes to yield, after treatment with NaOH/H$_2$O$_2$, a mixture of chain elongated ketoses and C(1)-arylated or alkylated glycals (Scheme 2-13) (Rajamannar, 1990).

Reaction of Glycosylidene Carbenes with Alkenes, Ketones, and Aldehydes

Glycosylidene diazirines should react with electrophilic or strongly nucleophilic alkenes to yield spiro-cyclopropanes (Brook & Dillon, 1969; Crawford & Raap, 1965b;

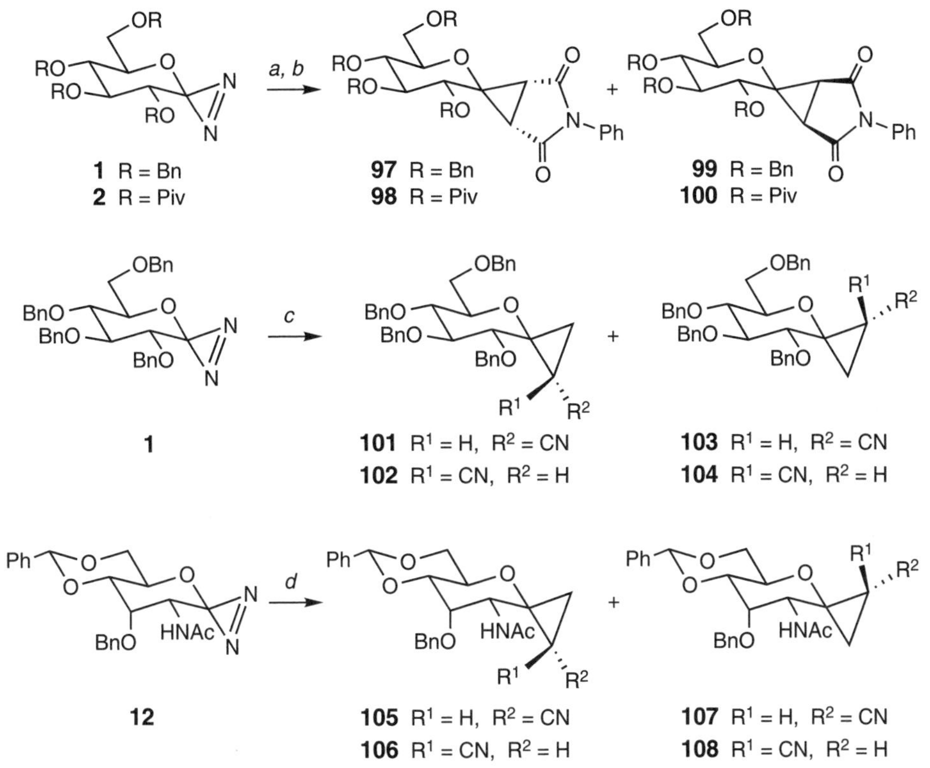

Scheme 2-14. (a) **1**, 10 equiv. of N-phenylmaleimide, 1,4-dioxane, r.t., **97/99** (80%, 90:10). (b) **2**, 10 equiv. of N-phenylmaleimide, 1,4-dioxane, 40°C, **98/100** (60%, 92:8). (c) acrylonitrile, r.t., **101/102/103/104** (70%, 52:31:6:11). (d) acrylonitrile, 50°C, **105/106/107/108** (76%, 46:29:20:5).

Hoffmann et al., 1974; Moss, 1980; Moss et al., 1987). The reaction has been explored to establish a method for making two C—C bonds at the anomeric center, and to explore the possibility of modulating the nucleophilic and electrophilic properties of glycosylidene carbenes by proper choice of protecting groups: stronger σ-acceptors should increase the electrophilicity and lower the nucleophilicity.

N-Phenylmaleimide is cyclopropanated by the O-alkylated **1** to give 80% of a 9:1 mixture of the diastereoisomers **97** and **99** (Vasella & Waldraff, 1991); the analogous reaction with the O-acylated **2** proceeds in lower yields (60%), but with about the same diastereoselectivity (Scheme 2-14) (Waldraff, 1995). A similar diastereoselectivity is observed in the cyclopropanation, by **1** or by **12**, of excess acrylonitrile (cf. Praly et al., 1990, 1994), showing that protonation of the carbene by the NH group of **12** does not interfere with cycloaddition (Vasella et al., 1991b). The four diastereomeric cyclopropanes **101–104** and **105–108** have been isolated in yields of 70% and 76%, respectively, and in similar ratios. In the main products, the cyano group points in the direction of the pyranose ring oxygen and is attached to the methylene group below the average plane of pyranose ring. This has been taken as indicating a two-step cyclopropanation, where Michael addition of the nucleophilic carbene, leading to a dipole, is followed by stereoelectronically preferred axial attack of the carbanionic center on C(1) of the glycosyl cation. The reaction of the Na salt derived from **4a** with N-phenylmaleimide under photolytic conditions (quartz vessel) also yielded the cyclopropanes **97** and **99** (49%, 9:1); reaction with five equivalents of acrylonitrile in THF proceeds in 69% yield to a mixture (53:35:19:3) of the cyclopropanes **101–104** (Mangholz & Vasella, 1991).

Dimethyl fumarate and **1** yield the *trans*-substituted cyclopropanes **109** and **110,** isolated in 43% and 29%, respectively (Scheme 2-15) (Vasella & Waldraff, 1991). The ratio of the products may reflect the relative importance of nonbonding, destabilizing interactions with the methoxycarbonyl groups, as this cyclopropanation is presumably also a two-step process, since the analogous reaction with dimethyl maleate is not stereospecific, but yields the *cis*-configured cycopropanes **111** and **112** and the *trans*-isomers **109** and **110** in 20, 11, 17, and 12% yields, respectively.

The influence of the substituents on the reactivity of the carbenes has also been evidenced by comparing the reactions of **1** and **2** with excess dihydrofuran, 5-methyl-2,3-dihydrofuran, and dihydropyran (Waldraff, 1995). The reaction of **1** with dihydrofuran yields 33% of the single cyclopropanation product **113,** while the yield of the analogous pivaloylated spirocyclopropane **114** from **2** is 72%; both cyclopropanes have been deprotected to the same tetrol **115** (Scheme 2-16). Methyl substitution of the dihydrofuran reduces the yield of the cyclopropane **116** resulting from reaction with **2** to 24%; the (Z,Z)-lactone azine **17** is the main product (50%), showing the reduced reactivity of the enol ether. Dihydropyran is much less reactive than dihydrofuran: even when used in large excess (30 equiv.), it yields only 9% of the benzylated, and 14% of the pivaloylated cyclopropanes **117** and **118,** respectively. Major by-products from **1** are the (Z,Z)- and the (E,E)-lactone azines **16,** which have been isolated in 32% and 28%, respectively; similarly, **2** leads to 55% of (Z,Z)-**17.** 3,5-Bis-O-(triisopropylsilyl)-ribal did not react with **2,** nor did 3,4-O-benzylidene-5-O-trityl-D-glucal. The cycloaddition to enol ethers appears to be a stereospecific process. Although only the (Z)-isomer of the open chain 1-methoxy-1-octenes reacts with **2,** to yield three of the

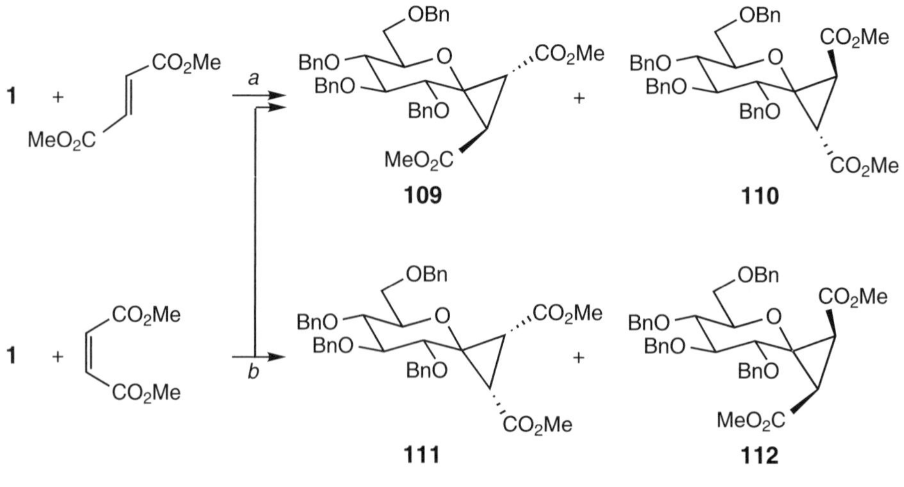

Scheme 2-15. (a) 1,4-dioxane, r.t., **109/110** (72%, 60:40). (b) 1,4-dioxane, r.t., **109/110/111/112** (60%, 29:20:33:18).

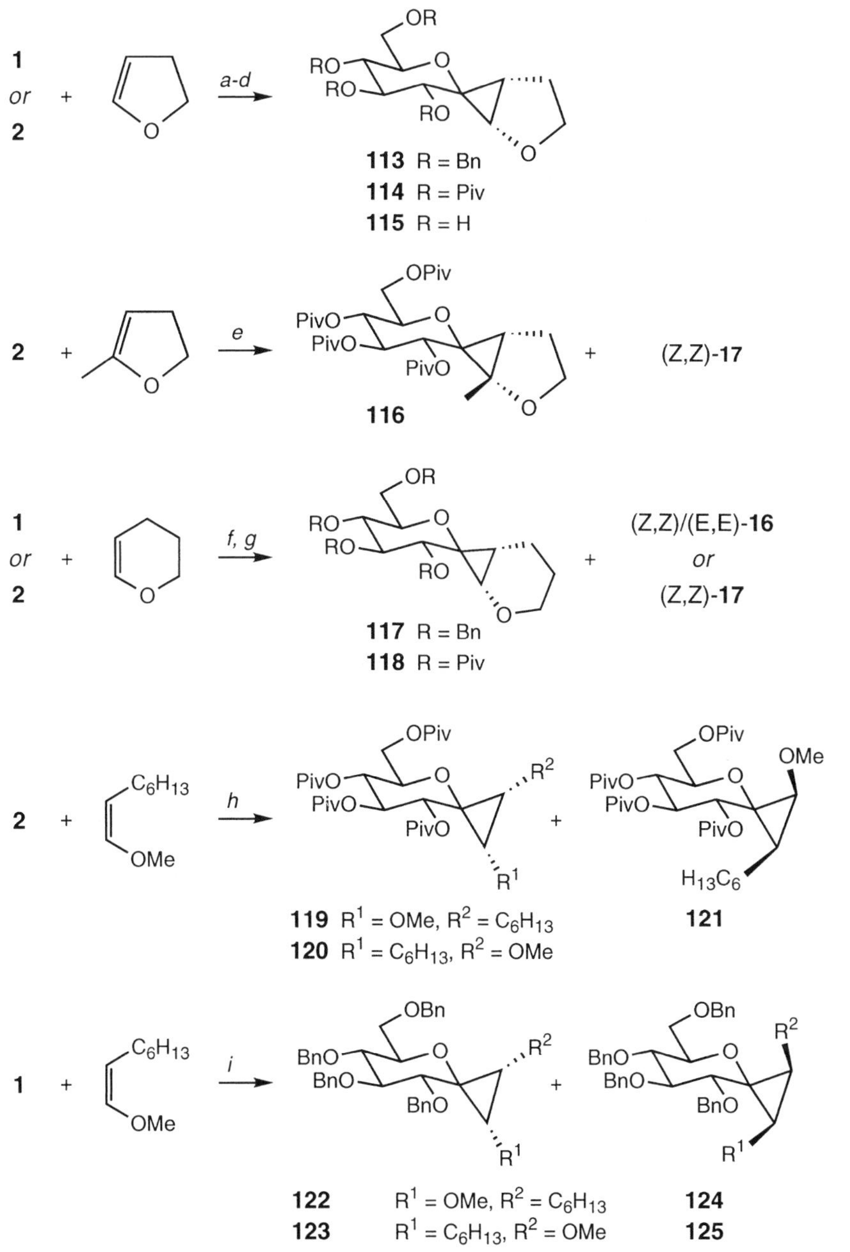

Scheme 2-16. (a) **1**, 1,4-dioxane, r.t., **113** (33%). (b) **2**, 1,4-dioxane, 45°C, **114** (72%). (c) **113**, H₂, Pd(OH)₂/C, MeOH, r.t., **115** (97%). (d) **114**, Bu₄NOH, 1,4-dioxane/H₂O, r.t., **115** (82%). (e) 1,4-dioxane, 45°C, **116** (24%) and (Z,Z)-**17** (50%). (f) **1**, 1,4-dioxane, r.t., **117** (9%), (Z,Z)-**16** (32%), and (E,E)-**16** (28%). (g) **2**, 1,4-dioxane, 45°, **118** (14%) and (Z,Z)-**17** (55%). (h) 1,4-dioxane, 45°C, (Z,Z)-**17** (47%) and **119–121** (17%, 65:23:12). (i) 1,4-dioxane, r.t., (Z,Z)- and (E,E)-**16** (43%, 56:44) and **122–125** (14%, 43:36:14:7).

isomeric *cis*-substituted cyclopropanes **119–121** in addition to 47% of (Z,Z)-**17**, no *trans* cyclopropane has been found. The structure of **119** has been established by X-ray analysis. Similarly, **1** yields 43% of the (E,E)- and (Z,Z)-lactone azines **16**, and the four *cis*-configured cyclopropanes **122–125** in very low yields (Waldraff, 1995).

Reaction of **1** or **2** with C$_{60}$ fullerene has given the first enantiomerically pure fullerene derivatives **126** and **127** (Scheme 2-17) (Vasella et al., 1992). Unfortunately, the protecting groups could not be removed. The benzylidene protected diazirine **14** reacts similarly, to yield **129**. The benzylidene group has been removed by acid hydrolysis, but glycosidation of the resulting diol has not met with success (Uhlmann & Vasella, unpublished results). The 2,3:4,6-di-O-isopropylidenated diazirine **128** also reacts with C$_{60}$ fullerene, to yield 43% of the cyclopropane **130**; here, the isopropylidene groups have been removed with TsOH·H$_2$O in MeOH, to yield a fully deprotected glycosylidene fullerene **131** (Uhlmann et al., 1994a).

The nucleophilic properties of the glycosylidene carbenes have also been probed by treating the diazirines **1** and **12** with benzaldehyde, camphor, acetone, and cyclohexanone (Scheme 2-18) (Vasella et al., 1993a). Reaction of **12** with benzaldehyde gives the two spiro-epoxides **132** and **133** (20% and 7%), the oxazoline **134** (12%), and a

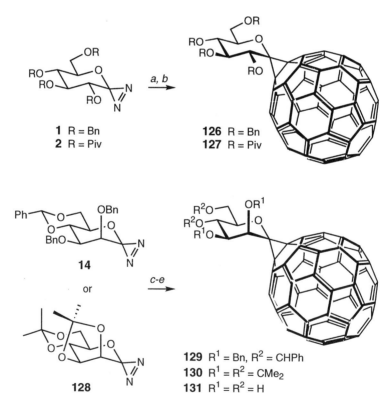

Scheme 2-17. (a) **1**, C$_{60}$, toluene, r.t., **126** (55%) and C$_{60}$ (26%). (b) **2**, C$_{60}$, toluene, 45°, **127** (54%) and C$_{60}$ (28%). (c) **14**, C$_{60}$, toluene, r.t., **129** (44%) and C$_{60}$ (56%). (d) **128**, C$_{60}$, toluene, r.t., **130** (43%). (e) **130**, TsOH·H$_2$O, 9:1 toluene-MeOH, r.t., **131** (73%).

range of by-products from which the acetamidoglycal **135** (3%), the oxazoline **136** (6%), and a mixture of lactone azines **18** (4%) have been isolated (Vasella & Witzig, unpublished results). No reaction with camphor was observed, but thermolysis of **1** in acetone has given 63% of the diastereomeric spiro-epoxides **137** and **138** (70:30), in addition to 5%–10% of a mixture of the diastereomeric lactone azines **16**, the benzyl-

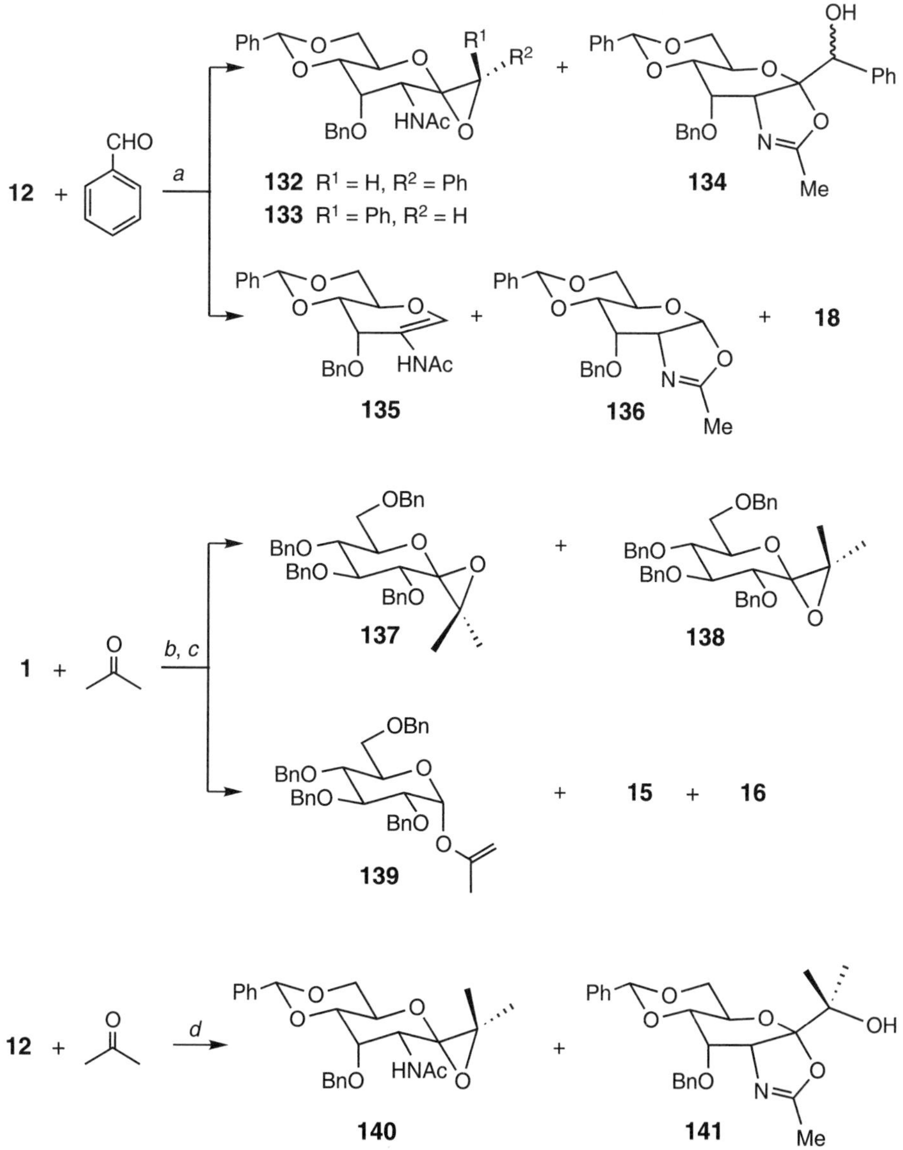

Scheme 2-18. (a) 3 equiv. of PhCHO, MeCN, 60°C, **132/133** (27%, 74:26), **134** (12%), **135** (3%), **136** (6%), **18** (4%). (b) 0.025 M **1** in acetone, r.t., **137/138** (63%, 70:30), **139** (2%), **15** (ca. 5–10%), and **16** (ca. 5–10%). (c) 0.14 M **1** in acetone, r.t., **137/138** (33%, 68:32), **139** (4%), **15** (ca. 5–10%), and **16** (ca. 5–10%). (d) acetone, 60°C, **140** (44%) and **141** (24%).

oxyglycal **15**, and the enol ether **139**. Lower yields of **137/138** result from photolytic generation of the carbenes at lower temperatures. Thermolysis of **1** in cyclohexanone proceeds in a similar way to give 78% of two analogous spiro-epoxides (65:35). Thermolysis of **12** in acetone yields the spiro-epoxide **140** and the dihydro-oxazole **141**, derived either from ring-opening of the diastereomeric spiro-epoxide by the acetamido group, or by nucleophilic attack of the acetamido group on the oxycarbenium center of the zwitterion resulting from the nucleophilic addition of the glycosylidene carbene to acetone. Cyclohexanone reacts in an analogous way, and in higher yields (83 *vs.* 68%).

Summary

Many aspects of the chemistry of glycosylidene carbenes remain to be explored, such as the basicity and solvation of alkoxycarbenes (see e.g., Friedrich et al., 1985; Kirmse et al., 1985, 1991; Kirmse & Kund, 1989; Padwa & Hornbuckle, 1991) and their relevance in the formation of anomeric C,C and C,O bonds; the influence of the nature of the substituents, the protecting groups, and the configuration of the carbenes on the regio- and diastereoselectivity of insertions into O,H bonds; the glycosidation of oligosaccharides, and of amino-, fluoro-, and thiosugars; the insertion into other X,H C,H, and X,Y bonds (boranes, silanes, germanes [Brook & Dillon, 1969], C,H acids, etc.); and addition reactions. Some new preparative methods, which are indirectly related to glycosylidene carbenes (Frey et al., 1994; Li & Vasella, 1993a,b; Lopusinski et al., 1993) have been developed; more are expected to follow, particularly if shorter syntheses of appropriate carbene precursors become available. Glycosylidene diazirines are being applied to the characterization of H-bonds in complex molecules and aggregates. Intramolecular reactions of glycosylidene carbenes are waiting to be explored (cf. Jones et al., 1986; Pirrung et al., 1989), and glycosylidene diazirines commend themselves for the combinatorial synthesis of oligosaccharide libraries.

3

Chemical Synthesis of Oligosaccharides

*Bert Fraser-Reid, Robert Madsen, A. Stewart Campbell,
Carmichael S. Roberts, and J. Robert Merritt*

The explosion of interest in glycoscience during the past decade has helped to draw attention to the difficulties associated with chemical syntheses of oligosaccharides. There have been many spectacular achievements, and these have served to emphasize those areas where improvements in methodologies are still required. Foremost among them are the generation, accessibility, and reactivity of the key reactive intermediate in glycoside coupling, the cyclic oxocarbenium ion. Accordingly, much effort has been properly devoted to the development of new glycosyl donors. In this chapter we confine our attention to four glycosyl donors where the anomeric leaving groups are halide, trichloracetimidate, alkyl or aryl thio, and 4-pentenyloxy (NPG) glycosides. Although many other glycosyl donors have been employed (Danishefsky et al., 1992a,b; Griffith & Danishefsky, 1992; Halcomb & Danishefsky, 1989; Hashimoto et al., 1992; Kahne et al., 1989; Kochetkov et al., 1987, 1991; Lemieux & Huber, 1956; Lemieux et al., 1973a,b; Martin et al., 1993; Raghavan & Kahne, 1993; Sim et al., 1993; Thiem et al., 1978), these four have been chosen because they have been used recently for syntheses of complex oligosaccharides comprising at least five saccharide units.

A recent review on *O*-glycosidation methods by Toshima & Tatsuta (1993) is recommended for additional reading.

General Concepts

In this section some of the concepts and terminologies used frequently throughout this review are briefly defined; references are given to sources where more detailed discussions may be found.

Donors and Acceptors

The synthesis of oligosaccharides involves the coupling of two saccharide units (**1** → **4**) in a process whereby one of them (**5**) functions as a *glycosyl donor* and the other (**3**) as a *glycosyl acceptor*. For the latter, it is necessary that the hydroxyl group of interest be the most reactive or, as is more often the case, the only one that is unprotected. Protecting group strategy is by no means trivial, but an examination of this area is beyond the scope of this chapter (Greene & Wuts, 1991).

With regard to relative reactivities of the glycosyl acceptors, apart from the usual generalizations relating to the reactivity of hydroxyl groups (primary > secondary > tertiary, or equatorial > axial), there are no widely applicable, universal rules. Changes in a remote protecting group can sometimes have a pronounced effect upon a given hydroxyl group. In general, an ester functionality deactivates an adjacent hydroxyl group whereas an ether functionality does not (*vide infra*).

The question of glycosyl donors is even more complex. The key intermediate required is the oxocarbenium ion **2**. Proton-induced loss of the anomeric OH, **1** → **2**, is not a reliable process to enable coupling with the acceptor leading to **4**, because the reverse cleavage, (**4** → **2**) is also proton-induced. Thus it is first necessary to convert the glycose into a derivative (**5**) from which oxocarbenium ion **2** can be generated with specificity.

α and β Configurations

Strictly speaking, the use of α or β by itself lacks precision. Compound **6** is an α-D, not simply an α, derivative. The descriptor D or L refers to orientation, (usually R or S, respectively), of the stereogenic center (usually a secondary alcohol) farthest from the reducing end (usually aldehydic). Thus, although the C1 orientation is the same in hexose **6** and heptoses **7** and **8**, they are designated α-D, β-L, and α-D because the orientations at C5, C6, and C6 are R, S, and R, respectively. Similarly in **9**, the defining center is at C4, which is S, and so it is designated α-L.

However, since most naturally occurring sugars are hexoses of D-configuration (e.g., **6**), the common practice for equating α with axial, and β with equatorial is correct most of the time. However if **6** changes its conformation from $^4C_1$ to the $^1C_4$ (Schwarz, 1973) alternative **6'**, the α-D designation would be maintained even though the aglycon (Y), is now equatorial. Furthermore, caution should be exercised with the hexoses rhamnose and fucose which usually occur in nature as the L-enantiomers.

Anomeric Effect

The anomeric effect is a stereoelectronic phenomenon that was first recognized for substituents at the anomeric center of sugars (Edward, 1955; Kirby, 1983; Lemieux, 1964; Thatcher, 1993). Thus, polar groups (e.g., Y = Br, OAc, OMe [equation ii]) were found to prefer axial orientation *when compared to their cyclohexane counterparts*. The importance of the italicized criterion may be judged from the fact that in D-glucose the population of the anomeric hydroxyl groups at equilibrium is 64% equatorial and 36% axial, which would seem to imply that the anomeric effect is not operating. However in cyclohexyl alcohols, the A-value for OH is 0.9 kcal/mol which predicts 11% axial orientation. The observed value of 36% is, therefore, three times higher than expected. Thus for Y = OH, there is indeed an anomeric effect, albeit not as pronounced as when Y = halogen (Kirby, 1983).

Assignment of Anomeric Configuration

The assignment of configuration of the coupled product can be made readily by studying the proton and carbon NMR spectra. The approximate values for the proton couplings ($J_{1,2}$) shown for **10–13** are diagnostic (Bock & Thogersen, 1982; Vligenthart et al., 1983), but the H1 signal is frequently obscured, particularly when there are sevi)

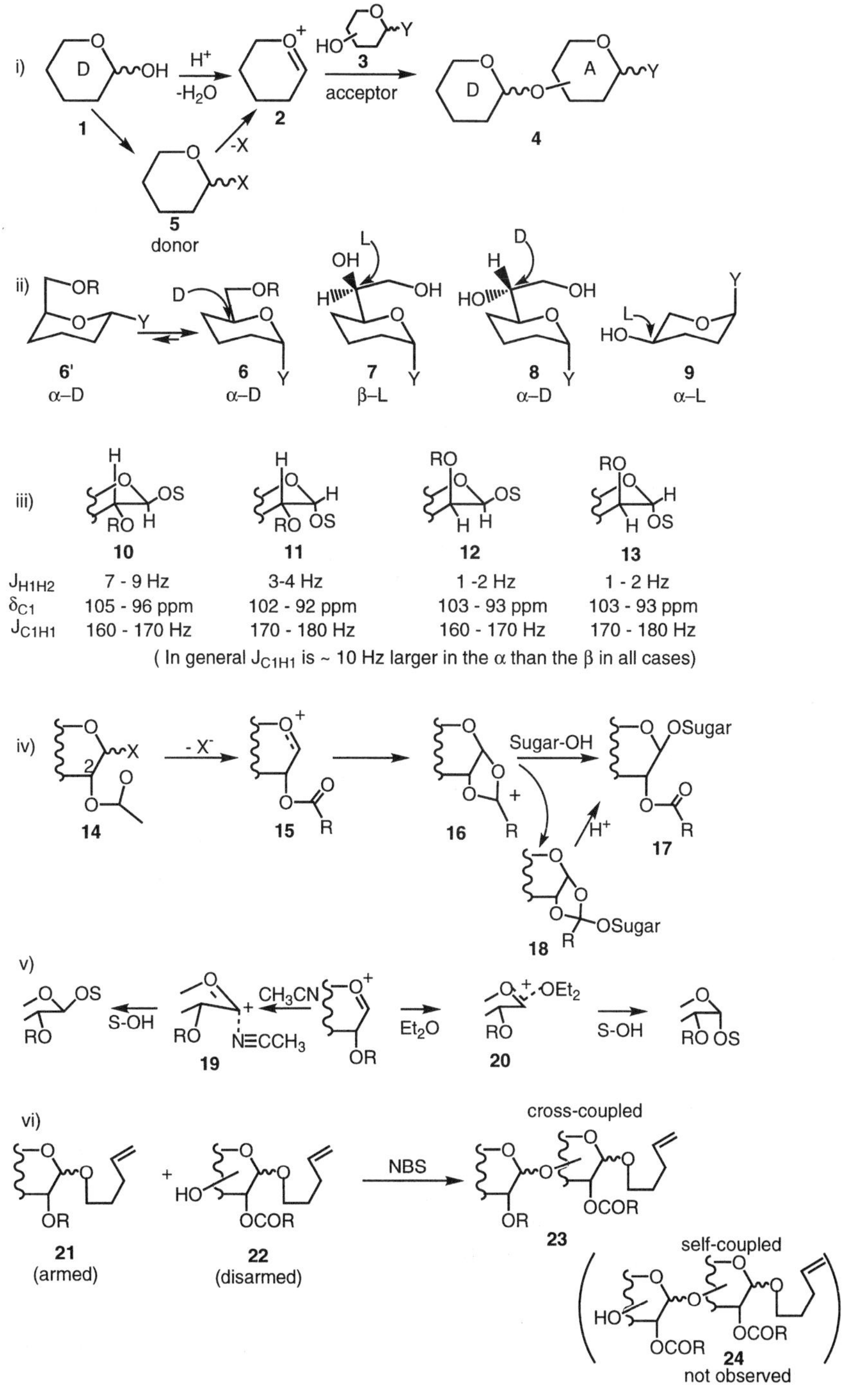

ii)

	6'	**6**	**7**	**8**	**9**
	α–D	α–D	β–L	α–D	α–L

iii)

	10	**11**	**12**	**13**
J_{H1H2}	7 - 9 Hz	3-4 Hz	1 -2 Hz	1 - 2 Hz
δ_{C1}	105 - 96 ppm	102 - 92 ppm	103 - 93 ppm	103 - 93 ppm
J_{C1H1}	160 - 170 Hz	170 - 180 Hz	160 - 170 Hz	170 - 180 Hz

(In general J_{C1H1} is ~ 10 Hz larger in the α than the β in all cases)

iv)

v)

vi)

eral benzyl protecting groups present. The ^{13}C NMR parameters for chemical shifts or the ^1J coupling value for the anomeric CH can be used for assigning α or β orientation (Bock & Pedersen, 1974, 1983; Bock et al., 1984). However, as is usual for ^{13}C NMR, the ranges overlap extensively. Thus, definitive assignments are best made by comparing the parameters for both anomers.

Neighboring Group Participation

In the context of glycoside coupling, the "neighboring group" is always at C2, and the "participation" occurs in that the oxocarbenium ion **15** leads to a dioxolenium ion (**16**) which reacts with the acceptor to give a 1,2-*trans* product (**17**). An orthoester (i.e., **18**) is sometimes also formed; indeed, this can be the major product (Bochkov & Zaikov, 1979). Methods for rearrangement to the glycoside (e.g., **18** → **17**) are well developed (Bochkov & Zaikov, 1979). The product so obtained can be 1,2-*trans*-diequatorial as in **10**, or 1,2-*trans*-diaxial as in **13**.

It is important to note that, as used above, neighboring group participation implies only stereocontrol, and not anchimeric assistance, a distinction which has historical significance. Horace Isbell, an insightful reaction mechanician, rationalized the stereochemistry of the Koenigs-Knorr reaction by postulating participation of the C2-OAc group as outlined in equation iv (Frush & Isbell, 1941). The anchimeric assistance (i.e., rate enhancement) aspect of neighboring group participation was later postulated by Winstein in a manuscript in which the prior contribution by Isbell was duly acknowledged (Winstein & Buckles, 1942).

Solvent Control

In the absence of neighboring group participation, subtle control can be induced by choice of solvent (Fraser-Reid et al., 1988; Igarashi, 1977; Ito & Ogawa, 1987; Wulff & Röhle, 1974). The general rule is that CH_3CN favors equatorial products while Et_2O (usually used along with a halogenated solvent to enhance solubilization) favors axial products. Complexes **19** and **20** may be considered intermediates.

Armed/Disarmed Coupling

This terminology, which was coined by the Fraser-Reid group (Mootoo et al., 1988), is being applied with increasing frequency. The methodology arose from the observation that n-pentenyl glycosides were oxidatively hydrolyzed (NBS/H_2O) much more readily when the C2 protecting group was alkyl (e.g., **21**, armed) than when it was acyl (e.g., **22**, disarmed) (*vide infra*). This observation is an example of the well-known principle that acyl-protecting groups depress anomeric reactivity as compared with ether groups. However, the synthetic advantage was not established until it was shown that these reactivity differences could be exploited for the chemoselective coupling of two *similar* glycosyl donors (Fraser-Reid et al., 1990). Thus, for two *n*-pentenyl glycosides where a free OH group is present on the *disarmed* partner (as in **22**), coupling to the armed donor **21** affords the cross-coupled product **23** in preference to the self-coupled product **24**.

Veeneman and co-workers subsequently extended the armed/disarmed strategy to thioglycosides (Veeneman & van Boom, 1990a,b; Veeneman et al., 1990a) and selenoglycosides (Zuurmond et al., 1993), and more recently gradations of the terminology have been recommended for more specific cases (Roy et al., 1992; Sliedregt et al., 1993). A similar reactivity difference for the C3 protecting group of glycals has been used by Friesen and Danishefsky (1989) to "control the traffic" in electrophilic additions to glycal double bonds.

Glycosyl Bromides and Chlorides

The Koenigs/Knorr reaction (e.g., **29** → **30**) is one of the most widely used methods of glycosidation. However, the first case of glycosidation involving a glycosyl halide was reported in 1879 by Michael when "acetochlorohydrose" (**26**) was allowed to react with "potassium phenate" (Michael, 1879a,b, 1885) (Scheme 3-1). Although Michael was not aware of the correct structure of his starting material nor of the product he obtained, he did recognize the potential of the reaction for glycoside synthesis. Indeed, 22 years later Koenigs and Knorr (1901) resurrected the reaction by use of "acetobrom glucose" (**29**) which they had prepared by treating glucose with acetyl bromide. By treating **29** with methanol and silver carbonate, Koenigs and Knorr obtained the glycoside **30** which, upon deacetylation, gave methyl β-D-glucopyranoside, a compound that had been synthesized previously by Fischer (1893, 1895). As noted above, the exclusive formation of the *trans* 1,2-products **27** and **30** was subsequently attributed to neighboring group participation by Frush and Isbell (1941).

A variety of glycosyl halides were prepared (Stanek et al., 1963) and their reactivities were found to follow the expected trend iodide > bromide > chloride > fluoride. Glycosyl iodides are so reactive that their isolation is problematic; hence, they are rarely used. At the other extreme, the fluorides are very stable and require special activation (*vide infra*).

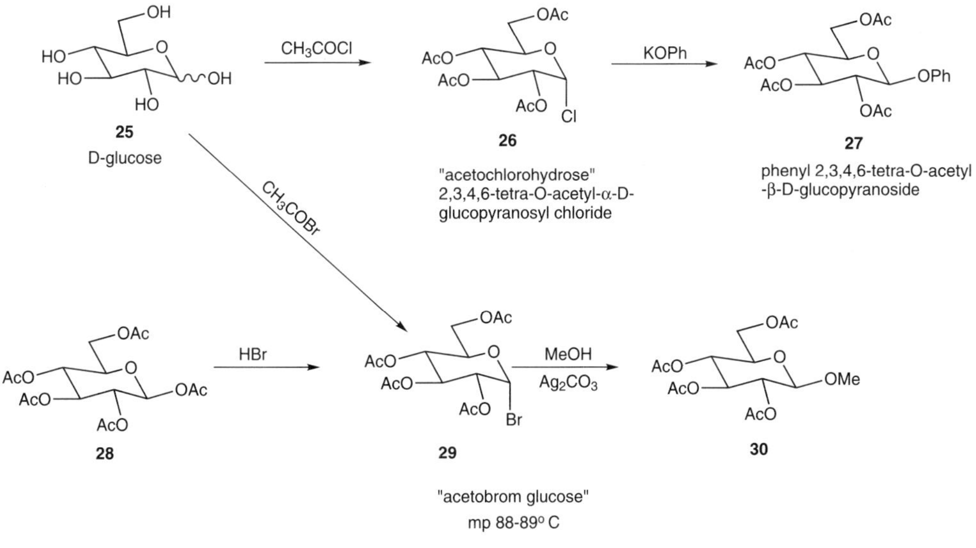

25
D-glucose

CH₃COCl

26
"acetochlorohydrose"
2,3,4,6-tetra-O-acetyl-α-D-
glucopyranosyl chloride

KOPh

27
phenyl 2,3,4,6-tetra-O-acetyl
-β-D-glucopyranoside

CH₃COBr

28

HBr

29
"acetobrom glucose"
mp 88-89° C

MeOH
Ag₂CO₃

30

Scheme 3-1.

The anomeric configuration of "acetobrom glucose" had been originally assigned as β (Stanek et al., 1963); however, the correct designation as **29** was one of the earliest triumphs for Hudson's rules of *iso*rotation (Hudson, 1924)

An anomeric halide can occupy either an axial or equatorial orientation (Scheme 3-2). However, the former is preferred thermodynamically, a circumstance which, in part, led to the discovery of the anomeric effect (Kirby, 1983; Lemieux, 1964; Thatcher, 1993). This is illustrated in Scheme 3-2 with some examples taken from the detailed study of Ishikawa and Fletcher (Ishikawa & Fletcher, 1969). Compound **32** was obtained as a crystalline material from the reaction of the glucosyl *p*-nitrobenzoate **31** with HBr, the β-D structure being assignable from its ¹H NMR spectrum. Upon standing for 1 week, the mother liquors deposited different crystals, assigned as the α-D anomer **33**. According to current rationalizations of the anomeric effect (Kirby, 1983) axial preference should be greater for the more polar bromide than the corresponding chloride; indeed, this is generally true.

Preparation of Glycosyl Halides

Shortly after the above-mentioned work of Koenigs and Knorr (1901), Fischer and Armstrong showed that treatment of glucose pentaacetate (**28**) with hydrogen bromide also led to "acetobrom glucose" (**29**) (Fischer & Armstrong, 1901). This classical method for preparing glycosyl halides by treating the corresponding esters with hydrogen halide (Scheme 3-3, **38** → **39**) is still commonly employed, albeit in modified form (Alais & Veyrieres, 1990; Kartha et al., 1989; Ogawa et al., 1983a,b; Szurmai et al., 1987). Hydrogen bromide is sometimes generated *in situ* by controlled addition of water to phosphorus tribromide (Lemieux, 1963a). The use of Lewis acids (e.g., TiBr₄, and AlCl₃) in organic solvents has a long history and continues to be very popular (Chittenden, 1992; Korytnyk & Mills, 1959; Lemieux, 1963b; Pacsu, 1928; Paulsen et al., 1978).

Direct conversion of the anomeric OH into the halide (Scheme 3-3a) without the intermediacy of an ester (i.e., **37** → **39**), has been accomplished by use of a Lewis acid (Stanek et al., 1963). However, there are now less harsh procedures for this direct con-

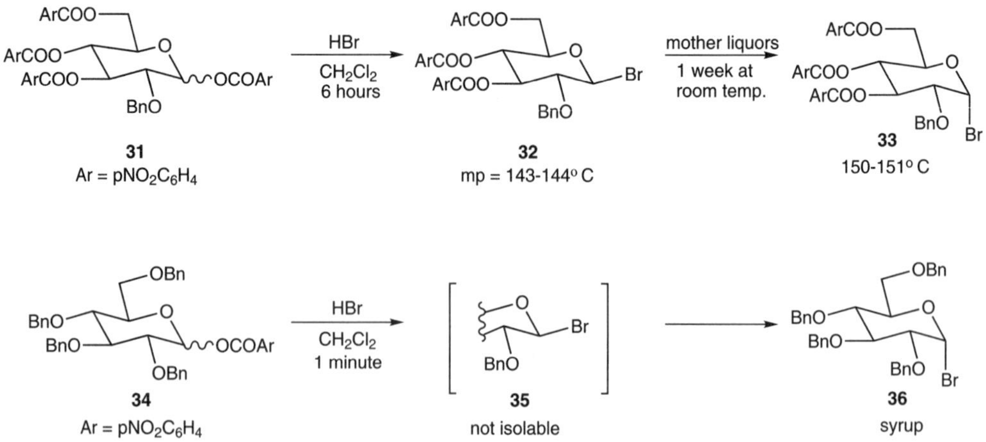

Scheme 3-2.

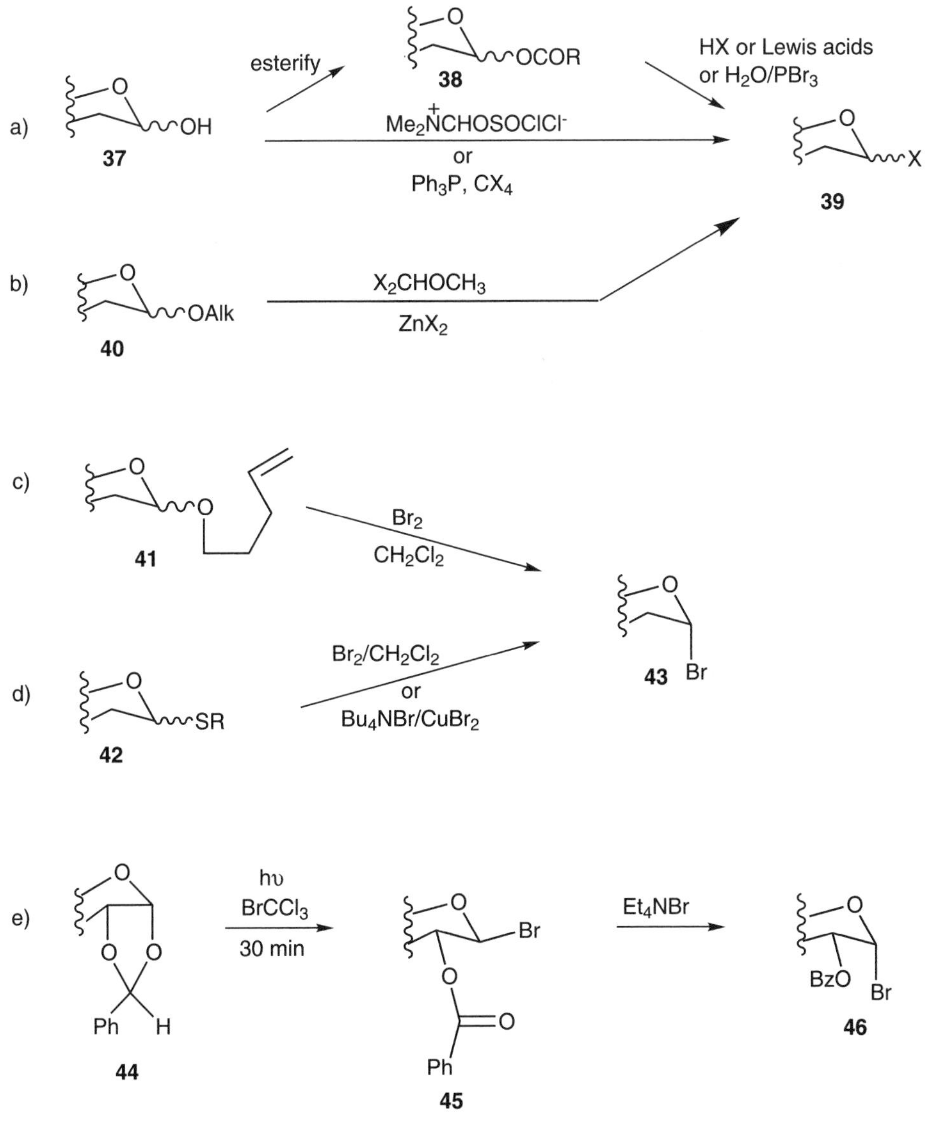

Scheme 3-3.

version. Thus, the use of Vilsmeier-type reagents, generated *in situ* from N,N-dimethylformamide and oxalyl chloride or thionyl chloride, is very popular for direct chlorination (Nukada et al., 1992; Ogawa et al., 1983a,b; Ogawa & Nukada, 1985; Paulsen & Höffgen, 1993; Pinto et al., 1987; Takahashi & Ogawa, 1987). Triphenylphosphine-based halogenation reactions for converting **37** into **39** have also been developed (Hanessian et al., 1972).

Precursors **37** and **38** are comparatively reactive; hence, the use of sturdier substrates is sometimes desirable. Glycosides are ideal substrates; a procedure for converting alkyl glycosides into glycosyl halides by the use of X_2CHOCH_3/ZnX_2 (e.g.,

40 → 39, Scheme 3-3b) **40 → 39** has been developed by Gross et al. (1978). This methodology has been applied with great success to complex oligosaccharides (Iversen & Bundle, 1981, 1982).

Other convenient routes from sturdy precursors utilize *n*-pentenyl derivatives **41** (Fraser-Reid et al., 1992) or alkyl thioglycosides **42** (Fügedi et al., 1987; Weygand & Ziemann, 1962) as starting materials (Scheme 3-3c,d). Titration of either with molecular bromine gives glycosyl bromides **43**. With thioglycosides, Ogawa's group has developed the combination of $CuBr_2/Et_4NBr$ to achieve the same result (*vide infra*) (Sato et al., 1986).

A novel method involving photobromination of 1,2-O-benzylidene derivatives, (Scheme 3-3e) has been introduced by Collins and co-workers (Collins & Ali, 1990; Collins et al., 1988), in a procedure that is reminiscent of the Hanessian-Hular reaction for cleaving O-benzylidene rings to give bromodeoxy sugar benzoates (Hanessian, 1968; Hanessian & Plessas, 1969). A notable advantage of the Collins procedure is the formation of equatorial products such as **45** under nonequilibrating conditions that favor their isolation. In keeping with the examples in Scheme 3-2, conversion to the axial isomer **46** can be achieved readily by addition of Et_4NBr to a solution of **45** (Collins & Ali, 1990; Collins et al., 1988).

Storage of Glycosyl Halides

As noted above the general trend for stability for glycosyl halides is fluoride > chloride > bromide > iodide. In addition, as a rule, stability increases with the number of ester groups present. Thus "acetobrom glucose" (**29**) is a crystalline substance (Lemieux, 1963a), whereas the tetra-O-benzyl analogue **36** is an unstable syrup that must be stored at $-80°C$ (Lemieux et al., 1975). Similarly, in Scheme 3-2 the β-D triester **32** is stable enough to be isolated as a crystalline material, while the β-D tetra-O-benzyl analogue **35** is not isolable (Ishikawa & Fletcher, 1969).

Although glycosyl bromides, and particularly chlorides, are sometimes stable enough to survive chromatographic purification it is usual to prepare and use them directly. When generated from *n*-pentenyl (Fraser-Reid et al., 1992) or thioglycosides (Fügedi et al., 1987; Weygand & Ziemann, 1962) by titration with bromine (Scheme 3-3c,d), bromosugar **43** is conveniently used *in situ* for coupling with an acceptor (*vide infra*).

Coupling

The use of C2-esters to induce *trans* 1,2-coupling by means of neighboring group participation is described above. An elegant method for synthesizing *cis* 1,2 glycosides of type **49** (Scheme 3-4) involves the halide ion-catalyzed protocol developed by Lemieux and co-workers (Lemieux et al., 1975). The procedure depends on the relative stabilities of axial and equatorial halides as revealed by the anomeric effect. Thus, in the presence of highly nucleophilic bromide ion, an axial glycosyl halide (e.g., **47**) reacts quickly to initiate an unfavorable equilibrium involving equatorial isomer **48**. The latter, being more reactive than **47**, undergoes ready displacement to give **49**, the product of overall retention of configuration. Steady state kinetics and LeChatelier's principle combine to optimize this reaction pathway; thus, the halide ion-catalyzed methodology has found widespread success for the preparation of *cis* 1,2-glycosides (Paulsen, 1982; Sadozai et al., 1986a,b) and complements the C2-ester strategy for

halide ion-catalyzed glycosidation

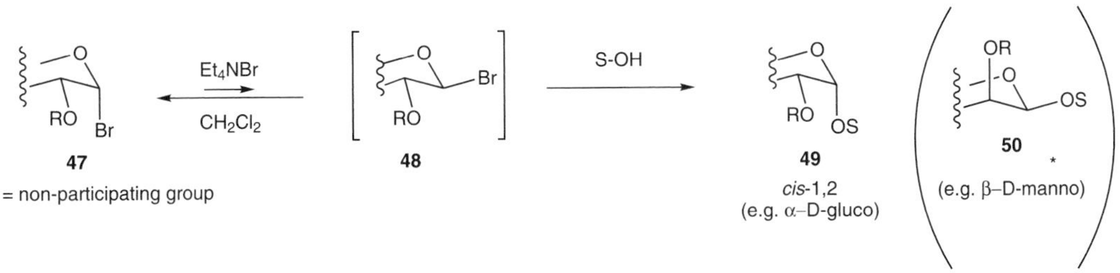

R = non-participating group

Scheme 3-4.

trans 1,2-analogues. An example is shown in Scheme 3-5a (**53** → **54**) (Hronowski et al., 1989).

In view of the mechanistic basis for the halide ion-catalyzed glycosidation, it is not surprising that this methodology cannot be adapted to obtain cis 1,2-glycosides of type **50** (Scheme 3-4) as found in β-D-manno derivatives. Indeed such derivatives pose tremendous synthetic challenges, as discussed below.

In Koenigs-Knorr type glycosidations, the nature of the promoter is of great importance to the outcome of the reaction, as has been described in detail by Paulsen (1982) and Bochkov (Bochkov & Zaikov, 1979). This is illustrated in Scheme 3-5a–d with some examples taken from the recent literature, where glycosyl bromides or chlorides have been used for key coupling reactions. Silver carbonate was used originally by Koenigs & Knorr (1901) to serve both as halophile and for neutralization of HBr; however, the latter reaction produces water which caused hydrolysis of the glycosyl bromide. This problem can be overcome by inclusion of a desiccant, 4Å molecular sieves being the current favorite.

The foregoing and other problems with the original Koenigs-Knorr procedure became apparent as a result of detailed investigations of the reaction by Helferich and Gootz (1929, 1931), and subsequently by others (Helferich et al., 1930; Micheel et al., 1957; Sharp & Stacey, 1951). As a result of these investigations, Hg(CN)$_2$ emerged as the heavy metal salt of choice. It was found that the addition of mercuric bromide had a pronounced effect on the rate and stereoselectivity of the reaction (Flowers, 1972; Newth & Phillips, 1953; Schlubach, 1926; Schlubach & Gilbert, 1930), particularly with glycosyl chlorides (e.g., Scheme 3-5d; **57** + **60b** → **61** Scheme 3-5c). This favorable result may be rationalized by postulating that the added bromide ion induces an S$_N$2 displacement that leads to a more reactive species. The process is, therefore, reminiscent of Lemieux's halide ion-catalyzed protocol (Scheme 3-4) described above.

A return to silver-based promoters was studied from the perspective of making the reaction medium homogenous. Thus, soluble silver salts with nonnucleophilic counter ions were explored (Kronzer & Schuerch, 1973; Marousek et al., 1978). Silver perchlorate, which is soluble in ether, has been found to be particularly useful for preparing a glycosides (Igarashi, 1977; Kovac & Edgar, 1992) as shown in Scheme 3-5b. Silver triflate, introduced by Hanessian and co-workers (Arcamone et al., 1976; Hanessian & Banoub, 1977), now appears to be the soluble salt of choice (e.g., see Scheme 3-5c). Tetramethylurea (Baumann et al., 1983; Lucas et al., 1990; Ogawa et

Scheme 3-5abc.

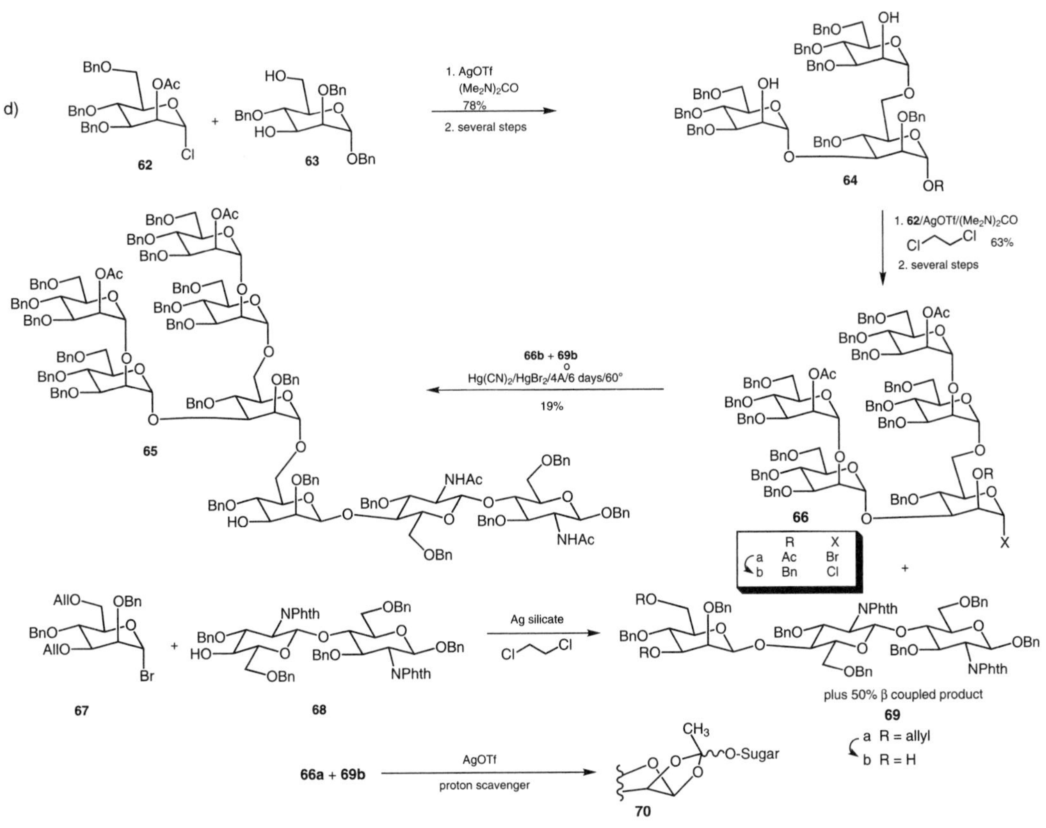

Scheme 3-5d.

al., 1983a,b; Paulsen & Lebuhn, 1984a,b) or a hindered pyridine (Paulsen & Lebuhn, 1983) is usually included as a proton scavenger (Scheme 3-5b–d).

As in other areas of synthetic organic chemistry, the "best method" for a particular transformation usually results from painstaking trial and error. For example, in their approach to high mannose glycoproteins, Paulsen and Peters (1987) used silver triflate for the coupling of diol **56** to glycosyl chloride **57** (Scheme 3-5c). However, for coupling of the same donor **57** to a similar diol (**60b**) to give **61**, the Helferich reagents were found to be preferable.

A similar observation was made by Ogawa's group (Nukada et al., 1992) (Scheme 3-5d). Attempts to couple the glycosyl bromide **66a** and diol **69b** via the agency of silver triflate gave acid labile material that was judged to be orthoester **70**. (By contrast this problem did not arise in Paulsen's coupling of **56** and **57** to give **58** [Scheme 3-5c]). For this reason, the C2 acetate of **66a** was replaced by a benzyl group (**66b**); under the Helferich conditions the octasaccharide **65** was obtained, albeit in the modest yield of 19%. In the words of Ogawa and co-workers, "a reasonable explanation for the subtle differences ... [between] ... **66a** and **66b** is not yet available" (Nukada et al., 1992).

veloped. For example Mukaiyama and co-workers treated 2,3,5-tri-O-benzyl β-D-ribofuranose with 2-fluoro-1-methylpyridinium tosylate at room temperature, which afforded the corresponding ribofuranosyl fluoride in 85% yield (Mukaiyama et al., 1983). More recently the Ishikawa reagent, hexafluoropropene/R_2NH mixtures and Mitsunobu-type methods have been reported from the laboratories of Ishido (Araki et al., 1984) and Kuntz (Kuntz & Sager, 1985), respectively. The use of diethylaminosulfur trifluoride (DAST) now appears to be the fluorinating reagent of choice. This reagent had been introduced by Nicolaou and co-workers for preparing glycosyl fluorides from thioglycosides (Dolle & Nicolaou, 1985; Nicolaou et al., 1984a,b). Its use neat (Rosenbrook et al., 1985) or in solution (Nakahara & Ogawa, 1987; Posner & Haines, 1985) for fluorinating reducing sugars has met with great success.

In view of their extremely high stabilities (*vide supra*), glycosyl fluorides may be stored without problem.

Coupling

As noted above the development of reagents to activate glycosyl fluorides (Scheme 3-7) has been key to their utility as glycosyl donors. Lewis acids such as BF_3 (Voznij et al., 1984), $SnCl_4$ (Nicolaou et al., 1984), $AlMe_3$ (Nicolaou et al., 1984a), TMSOTf (Hashimoto et al., 1984; Noyori & Hayashi, 1987), and TiF_4 (Kreuzer & Thiem, 1986) have been used to good advantage, but the survival of acid-labile functionalities under the reaction conditions is clearly a matter of concern. The use of stannous chloride with silver or trityl perchlorate as promoter was a seminal contribution of Mukaiyama and co-workers (Mukaiyama et al., 1981, 1983). In Ogawa's laboratory it was found that AgOTf could substitute as a promoter (Ogawa & Takahashi, 1985).

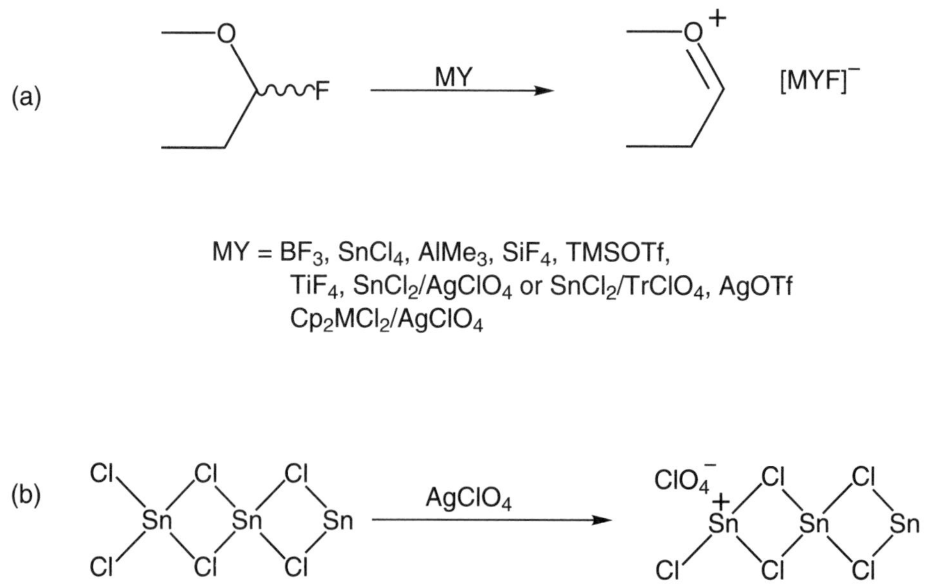

(a)

$$MY \longrightarrow [MYF]^-$$

$$MY = BF_3, SnCl_4, AlMe_3, SiF_4, TMSOTf,$$
$$TiF_4, SnCl_2/AgClO_4 \text{ or } SnCl_2/TrClO_4, AgOTf$$
$$Cp_2MCl_2/AgClO_4$$

(b)

$$\xrightarrow{AgClO_4}$$

Scheme 3-7.

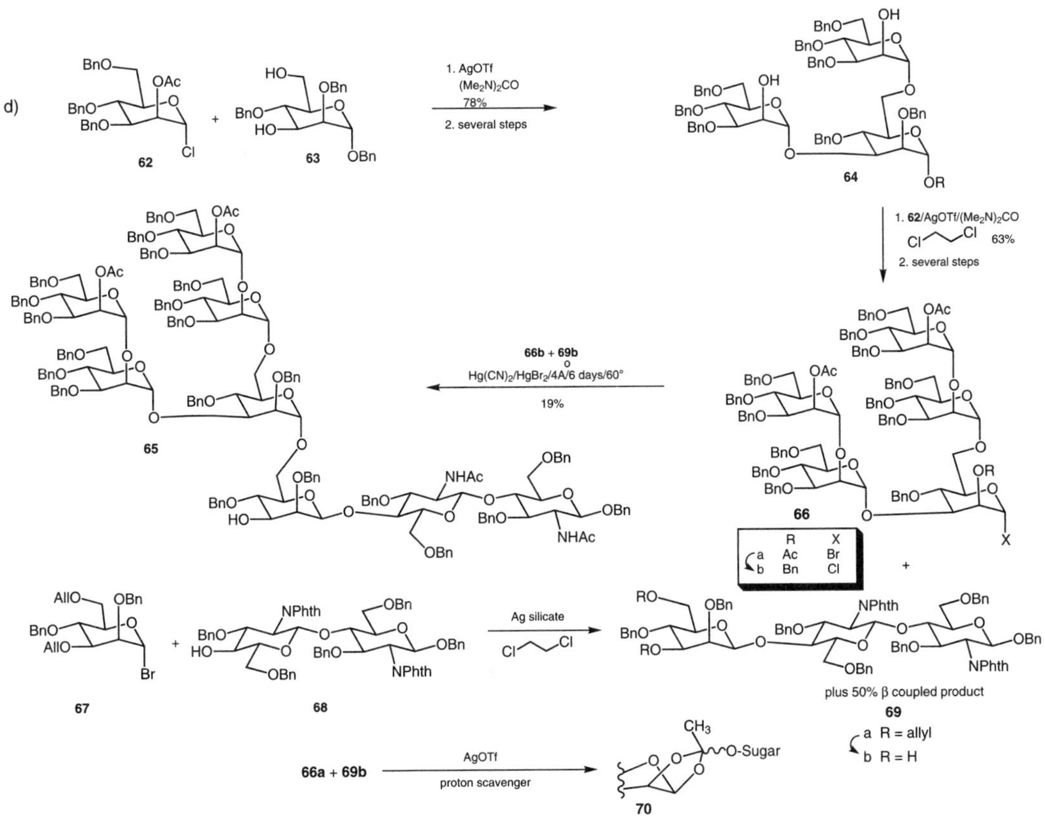

Scheme 3-5d.

al., 1983a,b; Paulsen & Lebuhn, 1984a,b) or a hindered pyridine (Paulsen & Lebuhn, 1983) is usually included as a proton scavenger (Scheme 3-5b–d).

As in other areas of synthetic organic chemistry, the "best method" for a particular transformation usually results from painstaking trial and error. For example, in their approach to high mannose glycoproteins, Paulsen and Peters (1987) used silver triflate for the coupling of diol **56** to glycosyl chloride **57** (Scheme 3-5c). However, for coupling of the same donor **57** to a similar diol (**60b**) to give **61**, the Helferich reagents were found to be preferable.

A similar observation was made by Ogawa's group (Nukada et al., 1992) (Scheme 3-5d). Attempts to couple the glycosyl bromide **66a** and diol **69b** via the agency of silver triflate gave acid labile material that was judged to be orthoester **70**. (By contrast this problem did not arise in Paulsen's coupling of **56** and **57** to give **58** [Scheme 3-5c]). For this reason, the C2 acetate of **66a** was replaced by a benzyl group (**66b**); under the Helferich conditions the octasaccharide **65** was obtained, albeit in the modest yield of 19%. In the words of Ogawa and co-workers, "a reasonable explanation for the subtle differences . . . [between] . . . **66a** and **66b** is not yet available" (Nukada et al., 1992).

A particularly troublesome coupling is that leading to β-D mannopyranosides, and the widespread occurrence of this feature in glycoproteins has prompted considerable study (Alais & David, 1990; Auge et al., 1980; Garegg & Ossowski, 1983; Gunther & Kunz, 1992; Paulsen & Lebuhn, 1984a,b; Taubken et al., 1993; Yamanoi et al., 1993). The innovative approaches reported from the laboratories of Hindsgaul (Barresi & Hindsgaul, 1991) and Stork (Stork & Kim, 1992) are two of the most recent efforts. The use of silver silicate as a heterogeneous catalyst for this coupling has been developed by Paulsen and co-workers (Paulsen, 1982; Paulsen & Lockhoff, 1981). An example is shown in Scheme 3-4c where the β-coupled product **60a** was obtained along with the α-anomer, the ratio being 8:1 (Paulsen & Peters, 1987). Another example comes from Ogawa's laboratory (Nukada et al., 1992), where the glycosyl bromide **67** and disaccharide **68** react in the presence of silver silicate to afford the β (**69a**) and α-coupled products in ~1:1 ratio (Scheme 3-5d). Silver zeolite has been developed in Garegg's laboratory as an alternative insoluble silver promoter that "is more readily prepared than is silver silicate" (Garegg & Ossowski, 1983).

Glycosyl Fluorides

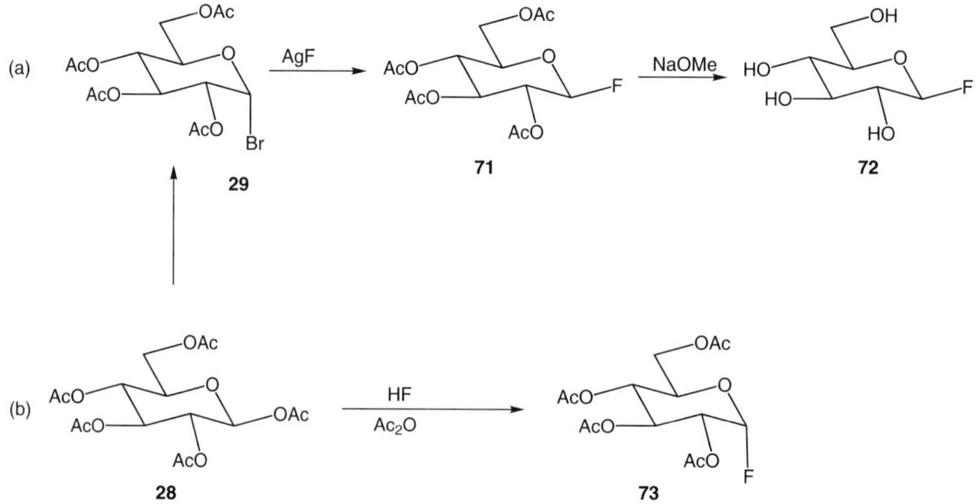

Glycosyl fluorides (Tsuchiya, 1990) were first synthesized as part of the exploratory studies on glycosyl acetates (e.g., **28** → **29**) that also led to the corresponding iodides, bromides, and chlorides (Stanek et al., 1963). The extremely high stability of the product **71** was apparent from the fact that it could be deacetylated in excellent yield with sodium methoxide to give the tetrol **72** (Micheel & Klemer, 1952) under conditions that would have destroyed the other halogenoses. However, the disadvantage of this stability was that special promoters were required to activate such fluorides to serve as glycosyl donors. In the modern era, this challenge was met in 1981 by Mukaiyama and co-workers with the introduction of $SnCl_2/AgClO_4$ as an activating medium for glycosyl fluorides (Mukaiyama et al., 1981, 1983). With the advent of this method of activation, there was a rebirth of interest in glycosyl fluorides as glycosyl donors.

Preparation and Storage

Brauns reported the first preparation of glycosyl fluorides when he treated "acetobrom glucose" with silver fluoride (i.e., **29 → 71**) (Brauns, 1923a,b). The procedure was adopted by Helferich (Helferich & Gootz, 1929), and there have been several modifications since (Brauns, 1923a,b; Hall et al., 1969; Micheel et al., 1955; Micheel & Klemer, 1957; Mukaiyama et al., 1981, 1983; Takahashi & Ogawa, 1987; Wessel et al., 1993). The well studied case of D-glucose illustrates that the α-fluoride (**73**) can be prepared by subjecting the β-glycosyl acetate (**28**) to HF. For the β-anomer (**71**) the α-bromide is a precursor, as described above (Helferich & Gootz, 1929). It may be noted that each reaction occurs with inversion of configuration.

Fluorination by reaction of glycosyl acetates with anhydrous hydrogen fluoride (Micheel et al., 1955; Micheel & Klemer, 1952) has been studied in detail by Bock and Pedersen (Bock & Pedersen, 1976, 1988; Pedersen & Refn, 1978). They have found that extensive rearrangements can occur. The use of commercially available HF/pyridine complex, Olah's reagent (Olah et al., 1979), provides an alternative route that has the advantage of being mild. Modifications from the laboratories of Szarek and Noyori (Hayashi et al., 1984; Jünnemann et al., 1993; Szarek et al., 1984) extend the reagent's usefulness.

However, direct preparation from the glycose (Scheme 3-6) is more expeditious. Although this can be achieved by the use of hydrogen fluoride (Micheel et al., 1955; Micheel & Klemer, 1952; Olah et al., 1979), milder nonprotic methods have been de-

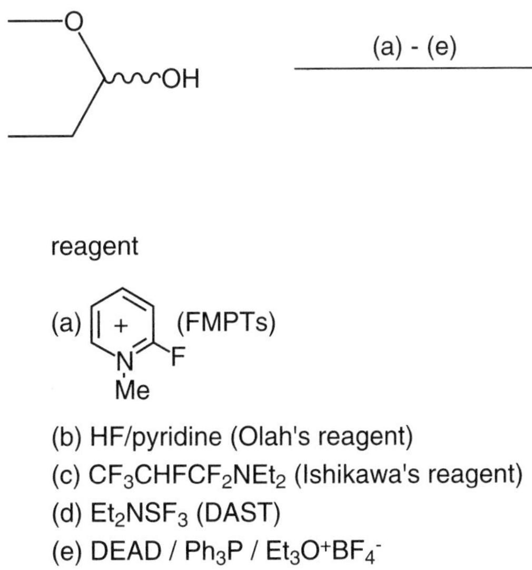

reagent

(a) (FMPTs)

(b) HF/pyridine (Olah's reagent)
(c) $CF_3CHFCF_2NEt_2$ (Ishikawa's reagent)
(d) Et_2NSF_3 (DAST)
(e) DEAD / Ph_3P / $Et_3O^+BF_4^-$

Scheme 3-6.

veloped. For example Mukaiyama and co-workers treated 2,3,5-tri-O-benzyl β-D-ribofuranose with 2-fluoro-1-methylpyridinium tosylate at room temperature, which afforded the corresponding ribofuranosyl fluoride in 85% yield (Mukaiyama et al., 1983). More recently the Ishikawa reagent, hexafluoropropene/R_2NH mixtures and Mitsunobu-type methods have been reported from the laboratories of Ishido (Araki et al., 1984) and Kuntz (Kuntz & Sager, 1985), respectively. The use of diethylaminosulfur trifluoride (DAST) now appears to be the fluorinating reagent of choice. This reagent had been introduced by Nicolaou and co-workers for preparing glycosyl fluorides from thioglycosides (Dolle & Nicolaou, 1985; Nicolaou et al., 1984a,b). Its use neat (Rosenbrook et al., 1985) or in solution (Nakahara & Ogawa, 1987; Posner & Haines, 1985) for fluorinating reducing sugars has met with great success.

In view of their extremely high stabilities (*vide supra*), glycosyl fluorides may be stored without problem.

Coupling

As noted above the development of reagents to activate glycosyl fluorides (Scheme 3-7) has been key to their utility as glycosyl donors. Lewis acids such as BF_3 (Voznij et al., 1984), $SnCl_4$ (Nicolaou et al., 1984), $AlMe_3$ (Nicolaou et al., 1984a), TMSOTf (Hashimoto et al., 1984; Noyori & Hayashi, 1987), and TiF_4 (Kreuzer & Thiem, 1986) have been used to good advantage, but the survival of acid-labile functionalities under the reaction conditions is clearly a matter of concern. The use of stannous chloride with silver or trityl perchlorate as promoter was a seminal contribution of Mukaiyama and co-workers (Mukaiyama et al., 1981, 1983). In Ogawa's laboratory it was found that AgOTf could substitute as a promoter (Ogawa & Takahashi, 1985).

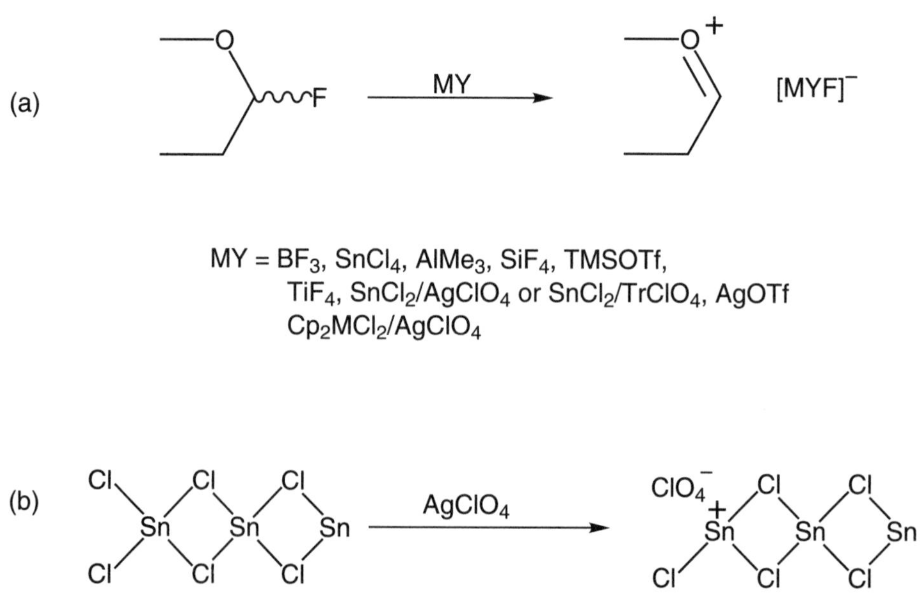

(a)

MY

[MYF]$^-$

MY = BF_3, $SnCl_4$, $AlMe_3$, SiF_4, TMSOTf,
TiF_4, $SnCl_2$/$AgClO_4$ or $SnCl_2$/$TrClO_4$, AgOTf
Cp_2MCl_2/$AgClO_4$

(b)

AgClO$_4$

Scheme 3-7.

(a)

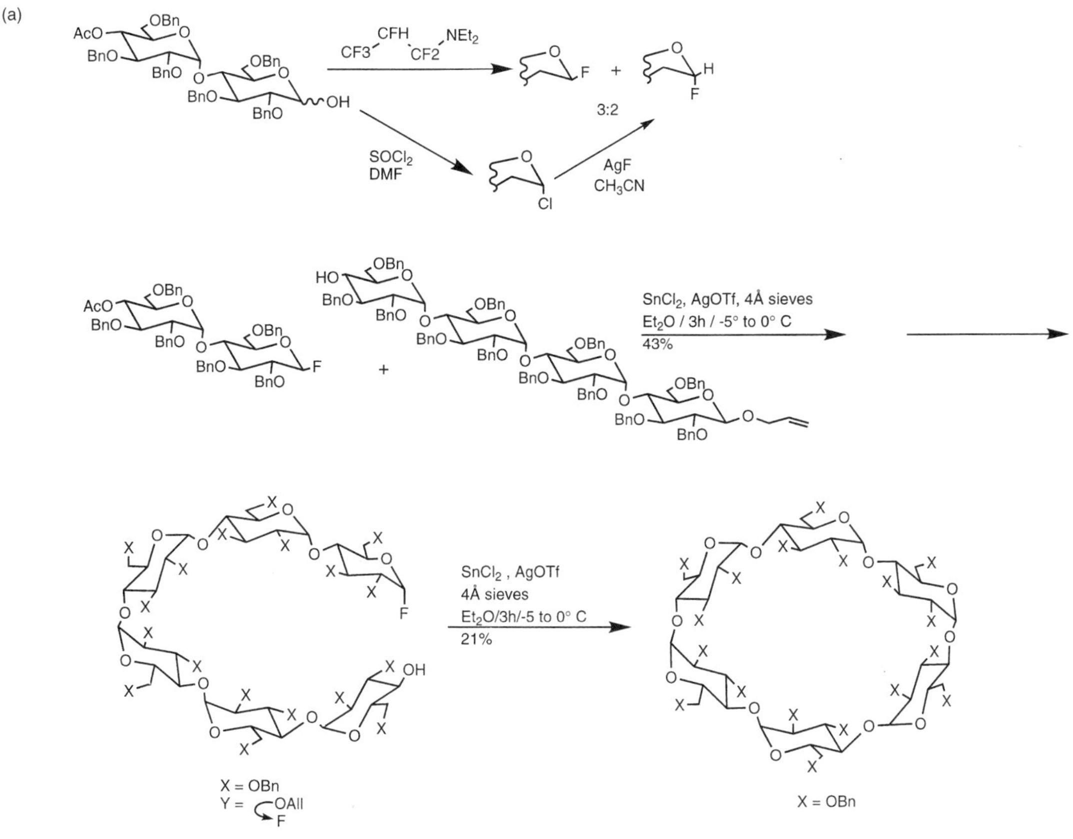

Scheme 3-8a.

Clearly SnCl₂ has specific fluorophilic properties. This has been attributed to the reaction shown in Scheme 3-7b in which polymeric SnCl₂ is ionized by AgClO₄. In this context Suzuki and co-workers have explored the use of metallocenes (Scheme 3-7a) as fluorophilic agents (Cp₂MCl₂/AgClO₄; M = Ti, Zr, Hf) (Matsumoto et al., 1988a,b). Success with O (Matsumoto et al., 1988a,b) and C (Matsumoto et al., 1988a,b, 1989) glycosidation has been demonstrated by Ogawa (Scheme 3-8a) (Takahashi & Ogawa, 1987). They have found great success in the synthesis of complex oligosaccharides (Lucas et al., 1990), and have been particularly favored by Nicolaou (Nicolaou et al., 1988). Schemes 3-8b (Nicolaou et al., 1992) and 3-8c (Nicolaou et al., 1990) show the use of Mukaiyama's and Suzuki's promoters for coupling of glycosyl fluorides in the block synthesis of ceramides.

Glycosyl Trichloroacetimidates

Glycosyl imidates (**75**) were introduced in 1978 as glycosyl donors by Sinaÿ and co-workers (Sinaÿ, 1978) who prepared these substrates from glycosyl halides and N-alkylamides using silver oxide as promoter. Coupling, as illustrated by the preparation

b)

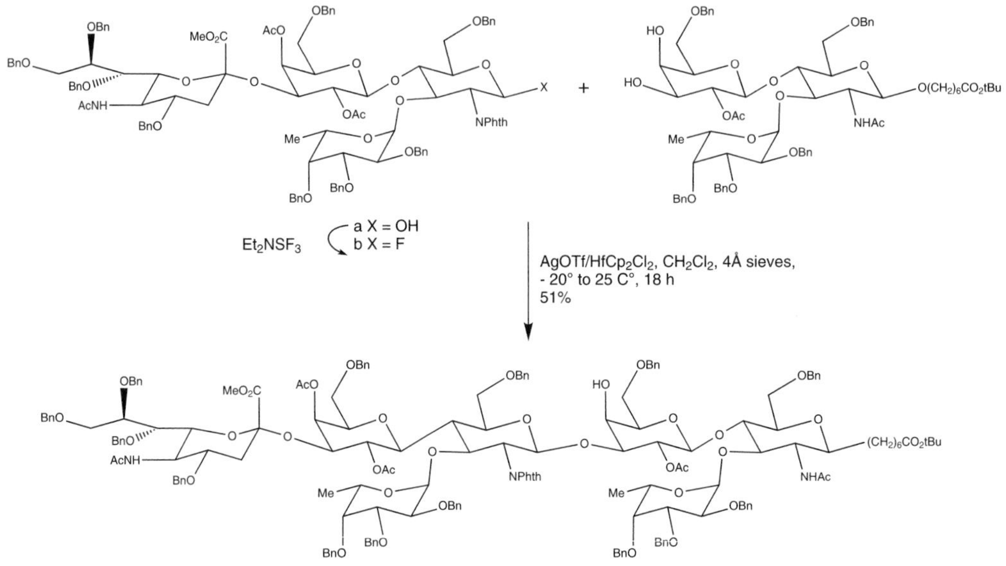

a X = OH
b X = F

Et$_2$NSF$_3$

AgOTf/HfCp$_2$Cl$_2$, CH$_2$Cl$_2$, 4Å sieves,
- 20° to 25 C°, 18 h
51%

c)

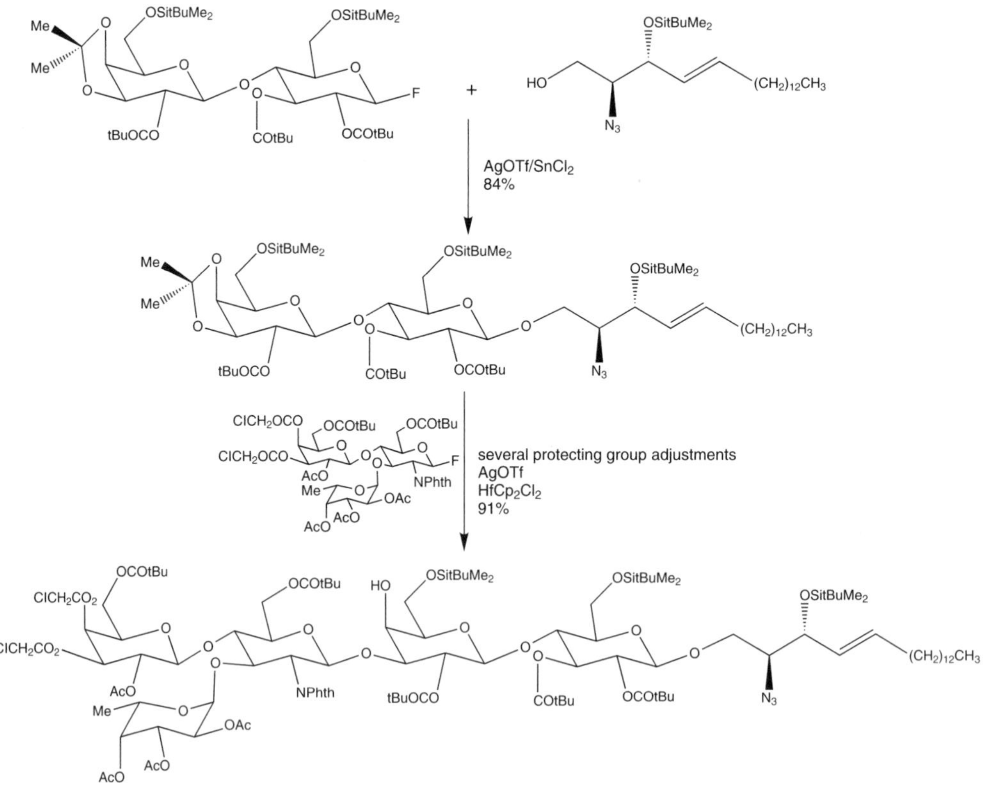

AgOTf/SnCl$_2$
84%

several protecting group adjustments
AgOTf
HfCp$_2$Cl$_2$
91%

Scheme 3-8bc.

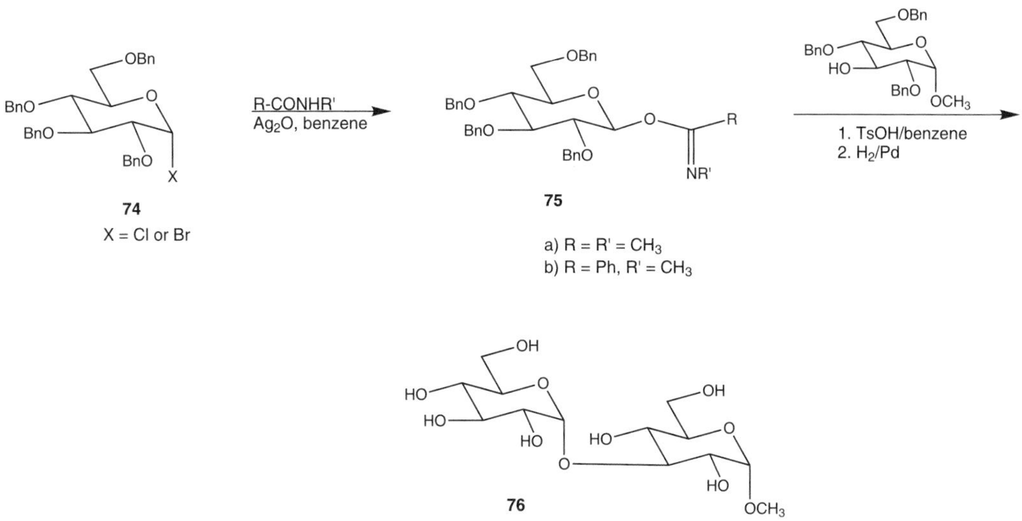

74
X = Cl or Br

75

a) R = R' = CH₃
b) R = Ph, R' = CH₃

76

of a methyl nigeroside (**76**), was effected by the use of *p*-toluenesulfonic acid. Under these conditions if C2-O-acyl groups were present, orthoesters were frequently formed; although these can be processed to give coupling products (Wulff & Röhle, 1974) as illustrated below, their intermediacy was a disadvantage. This problem, combined with the relatively low reactivity of analogues of **75**, and the fact that **75** had to be prepared from glycosyl halide precursors (e.g., **74**) restricted their usefulness. Some years later, Schmidt introduced the trichloromethyl analogue as an alternative to the alkyl counterpart (Scheme 3-9). Schmidt has published several excellent reviews about these powerful glycosyl donors (Schmidt, 1986, 1989).

Preparation and Storage

Trichloroacetimidates are usually prepared from the corresponding hemiacetals (Scheme 3-9a) by base-catalyzed reaction with trichloroacetonitrile in methylene chloride as solvent. The choice of base is critical since it determines the anomeric configuration of the product; this is a manifestation of the equilibria shown in Scheme 3-9. Thus, strong bases such as sodium hydride (Grundler & Schmidt, 1985; Sadozai et al., 1986a), cesium carbonate (Urban et al., 1990) and DBU (Marino-Albernas et al., 1993; Mori et al., 1990; Sadozai et al., 1986b) give the thermodynamically more stable α-trichloroacetimidate, while weaker bases such as potassium carbonate (Grundler & Schmidt, 1984) give the β-trichloro anomers as kinetic products. These conditions do not affect base-labile protecting groups, as is apparent from the examples shown in Scheme 3-10. Schmidt has demonstrated that β-anomers are slowly converted into their α-counterparts by retrograde reaction, oxide anomerization, and recombination (Schmidt, 1986).

As illustrated by the examples in Scheme 3-10, the hemiacetal may be generated from an allyl or benzyl glycoside, or a glycosyl acetate, by chemospecific cleavage. That these procedures can be applied to complex glycosyl donors is evident from Scheme 3-10a and 3-10d.

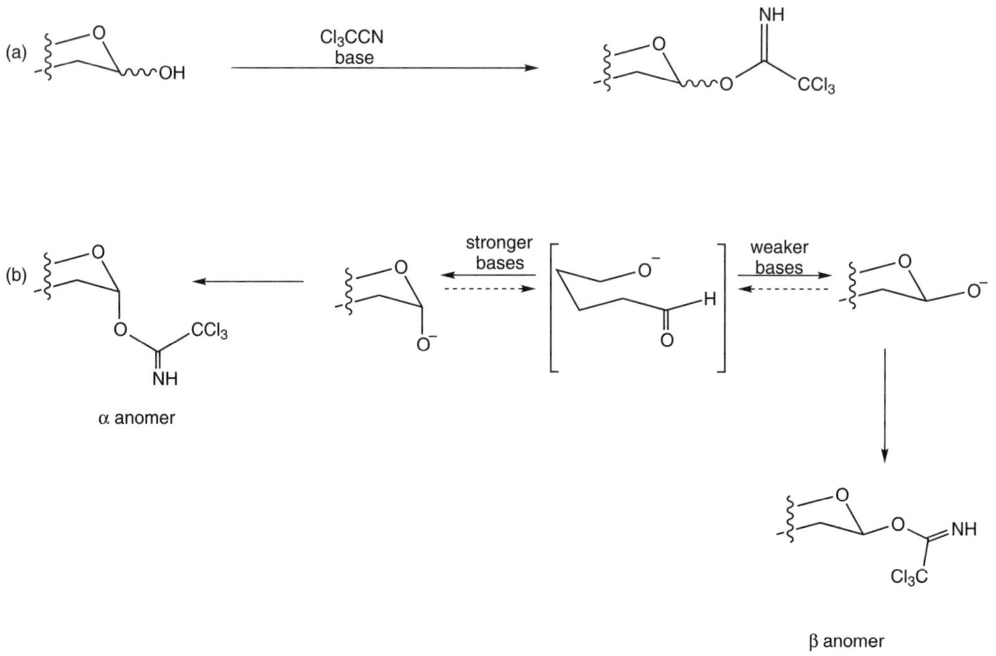

(a)

(b)

α anomer

β anomer

Scheme 3-9.

Glycosyl trichloroacetimidates can be stored below, or at room temperature for hours to months depending on their structures (Grundler & Schmidt, 1984). Some of these species can withstand heating in toluene for several hours (Schmidt et al., 1984) and several are crystalline (Grundler & Schmidt, 1985). However, the usual practice is to prepare the imidates, purify them by silica gel flash chromatography, and use them immediately (Bommer et al., 1991; Marino-Albernas et al., 1993).

Coupling

The coupling of trichloroacetimidates with glycosyl acceptors is usually mediated by Lewis acids, the choice of which influences both the rate and stereochemical outcome of the coupling. Trimethylsilyl trifluoromethanesulfonate (TMSOTf) and boron trifluoride etherate (Schmidt, 1986) are usually the Lewis acids of choice; they allow couplings to occur between −78°C and 0°C over time periods ranging from minutes to hours, depending on the reactivity of the acceptors (Kobayashi et al., 1990; Marino-Albernas et al., 1993; Mori et al., 1990; Paulsen et al., 1992; Sadozai et al., 1986b). Under these conditions a wide range of acid-labile protecting groups are stable, including silyl ethers (Grundler & Schmidt, 1985; Kobayashi et al., 1990; Paulsen et al., 1992), isopropylidene groups (Alais & Veyrieres, 1990; Kinzy & Löw, 1993; Kinzy & Schmidt, 1987), and benzylidene acetals (Greilich et al., 1993; Grundler & Schmidt, 1985; Rio et al., 1993), although caution with *p*-methoxybenzyl ethers is advised according to some reports (Kuyama et al., 1993; Yamazaki et al., 1990a,b). Weaker Lewis acids, such as silver triflate, sometimes work very efficiently (Douglas et al., 1993). With very unreactive acceptors, rearrangement to the corresponding glycosylamide

(Scheme 3-11a) has been observed (Kobayashi et al., 1990; Kinzy & Löw, 1993; Sadozai et al., 1986b; Spijker et al., 1992).

In keeping with the normal trends (*vide supra*) neighboring group participation leads to 1,2-*trans* coupling as exemplified in Scheme 3-11d. In some cases, 2-O-acetylated donors have led to orthoesters, this being reminiscent of the early pioneering work on the alkyl imidates by Sinaÿ's group (Sinaÿ, 1978). However, as illustrated by an example from Ogawa's laboratory (Scheme 3-11e), such orthoesters can be induced to give 1,2-*trans* coupling products by mild Lewis acid treatment (Mori et al., 1990; Zimmermann et al., 1988). The same result can be effected by raising the temperature (Paulsen et al., 1992, 1993).

A valuable facet of the imidate glycosyl donors is that in cases where the C2 substituent is a nonparticipating entity, it is possible to control the stereochemical outcome of the coupling by choice of the Lewis acid catalyst. Thus, when the stronger catalyst trimethylsilyltriflate is used, 1,2-*cis* glycosides are formed preferentially (Kinzy & Löw, 1993; Greilich et al., 1993; Rio et al., 1993; Duchaussoy et al., 1991; Petitou et al., 1991; Lucas et al., 1993). With the weaker catalyst $BF_3 \cdot OEt_2$, coupling usually occurs with inversion of configuration, giving β-glycosides (1,2-*trans*) from α-trichloroacetimidates (Scheme 3-11b) (Bommer & Schmidt, 1989; Schmidt & Michel, 1981; Spijker et al., 1992). With mannopyranosyl trichloroacetimidates, however, coupling usually

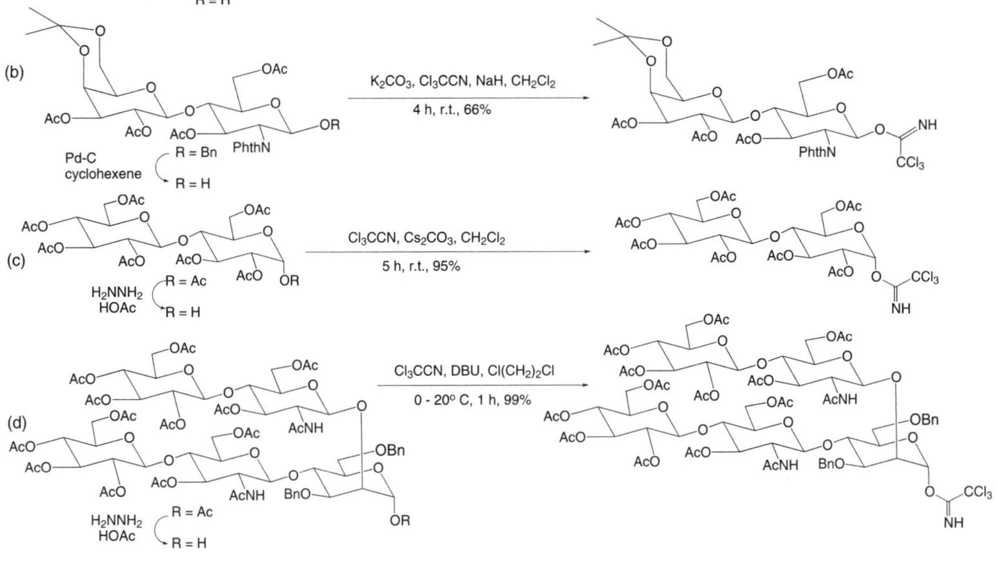

Scheme 3-10.

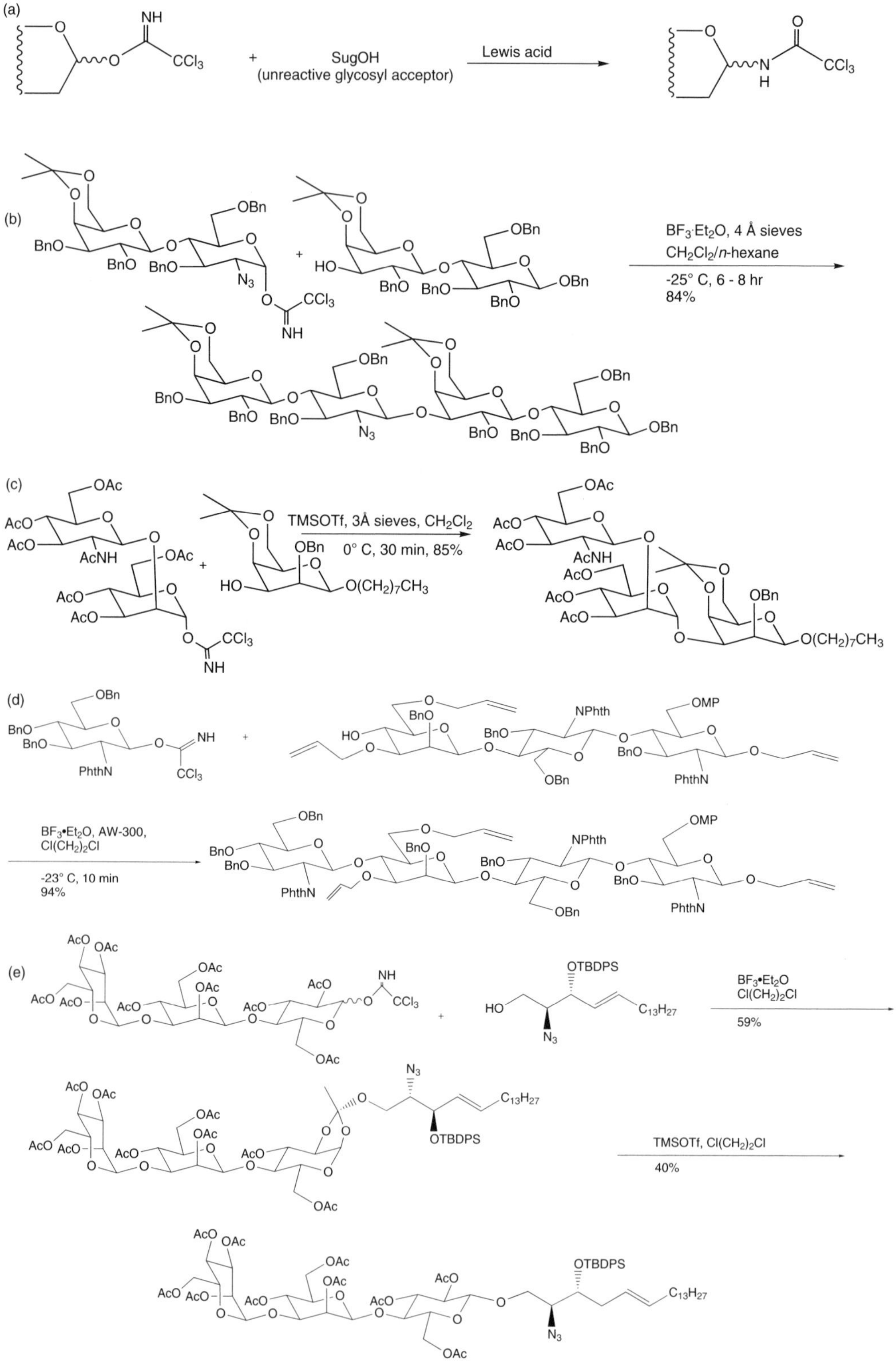

Scheme 3-11.

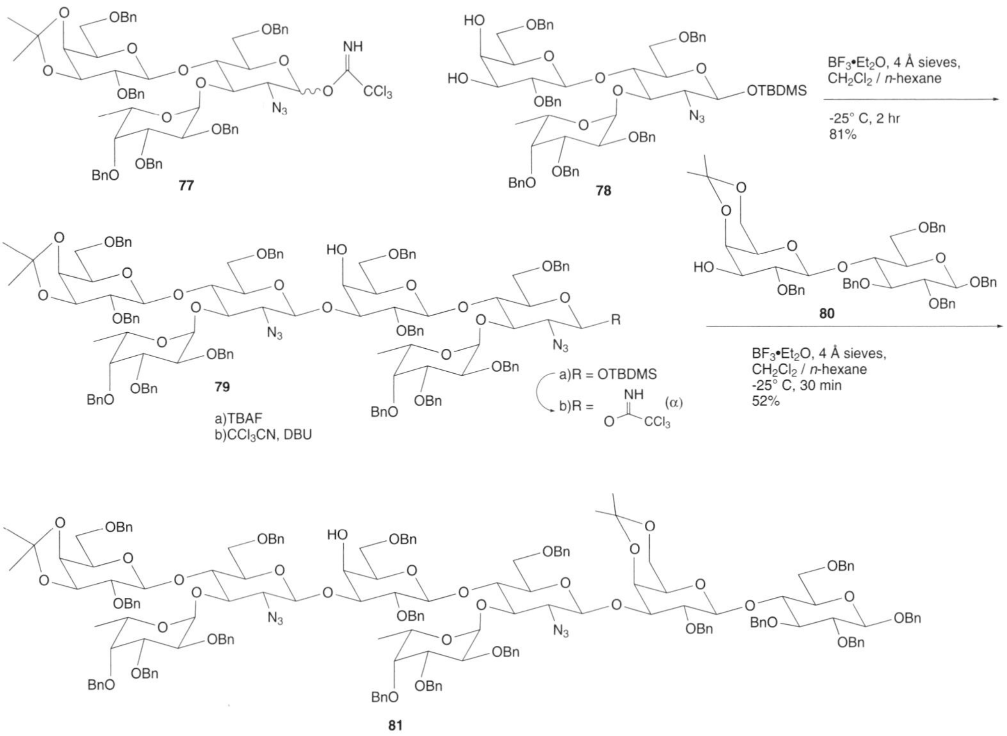

Scheme 3-12.

gives α-D-products regardless of the protecting group at the C2 position (Paulsen et al., 1990, 1993; Ogawa et al., 1986; Sadozai et al., 1986a) (Scheme 3-11c).

Trichloroacetimidates have proved themselves to be impressive glycosyl donors in block syntheses involving large oligosaccharide subunits. This is illustrated by the synthesis of octasaccharide **81** (Scheme 3-12) (Bommer et al., 1991). Although the trisaccharide donor **77** exists as an α/β anomeric mixture, coupling to diol **78** occurs stereoselectively at low temperature to give the β-coupled hexasaccharide **79a** in excellent yield. Two standard operations on the reducing terminus of hexasaccharide **79a** then furnish the α-glycosyl trichloroacetimidate **79b**; reaction with the hindered hydroxyl group of disaccharide **80** occurs rapidly, at low temperature, to give acceptable yields of the β-coupled octasaccharide **81**. The example involving acceptor **78** (Scheme 3-12) shows that high regioselectivity can be found to accompany stereoselective coupling.

2-Amino-2-deoxy gluco- and galactopyranosyl moieties are components of many biologically important oligosaccharides, and hence procedures for their syntheses or incorporation have attracted considerable attention. Protecting the 2-amino-2-deoxy group as its phthalimide (Scheme 3-13) has the advantage of providing for neighboring group participation without the complication that arises with NH-COR groups (Scheme 3-13) (Grundler & Schmidt, 1985). With the latter, participation of the acyl group sometimes results in formation of an oxazoline bridge between C1 and C2. In this context, glycosyl trichloroacetimidates have proved to be very reliable for cou-

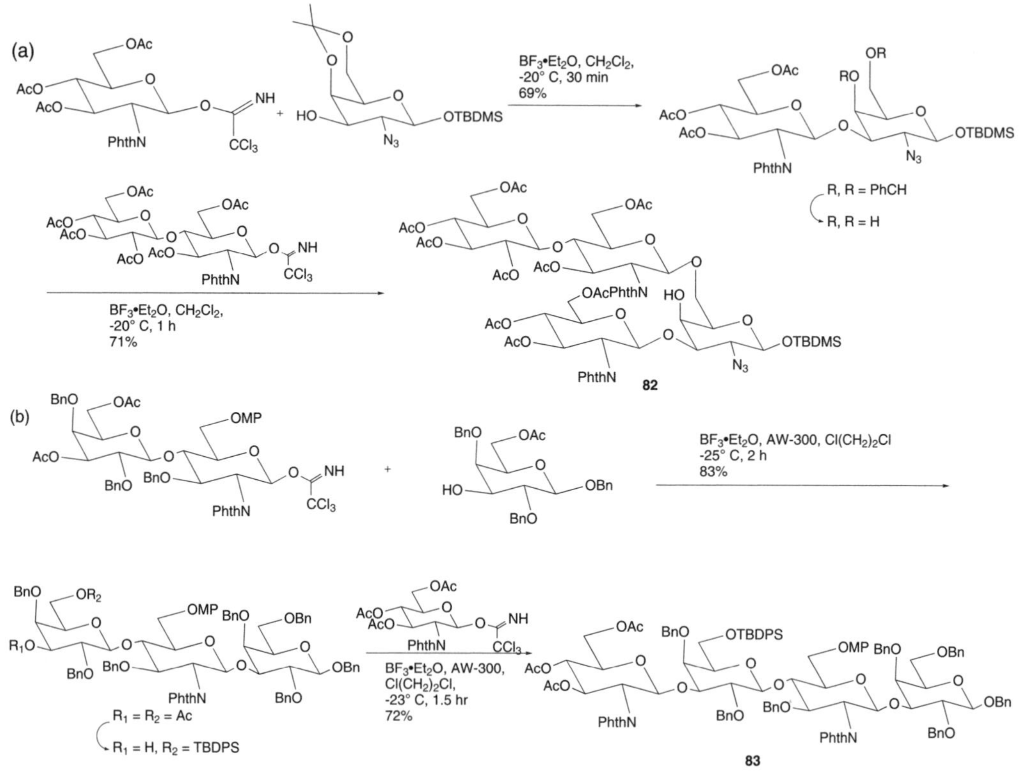

Scheme 3-13.

plings involving 2-deoxy-2-phthalimido glycosyl donors. The examples in Scheme 3-13 (Grundler & Schmidt, 1985; Kobayashi et al., 1990; Paulsen et al., 1992) exemplify the efficiency of these substrates for the formation of the oligosaccharides **82** and **83** (Alais & Veyrieres, 1990; Douglas et al., 1993; Matsuzaki et al., 1992; Sadozai et al., 1986a; Sato et al., 1987; Spijker et al., 1992). Glycosphingolipids, exemplified by compound **84** (Scheme 3-14), are another type of glycoconjugate (Schmidt, 1986, 1989) that have been the focus of much synthetic effort over the past two decades. Early reports on the installation of the ceramide moiety were based on Koenigs-Knorr methodology. Recently, however, two approaches based on glycosyl trichloroacetimidate donors have been explored using protected ceramides or the corresponding azido precursors (pathways I and II, respectively) as glycosyl acceptors (Scheme 3-14) (Schmidt, 1986, 1989). Pathway I suffers from low yields in the coupling reaction, results which have been attributed to steric effects or problems resulting from chelation to the catalyst (Ito et al., 1989; Matsuzaki et al., 1992; Numata et al., 1988; Sato et al., 1987). On the positive side, the glycosphingolipid is easily obtained from the coupled product by a simple deacylation operation. By contrast pathway II gives high yields in the coupling reaction. However, while mostly high yielding, more steps are required for introducing the N-linked fatty acid residue (Hasegawa et al., 1993; Ishida et al., 1993; Prabhanjan et al., 1992; Yoshida et al., 1993).

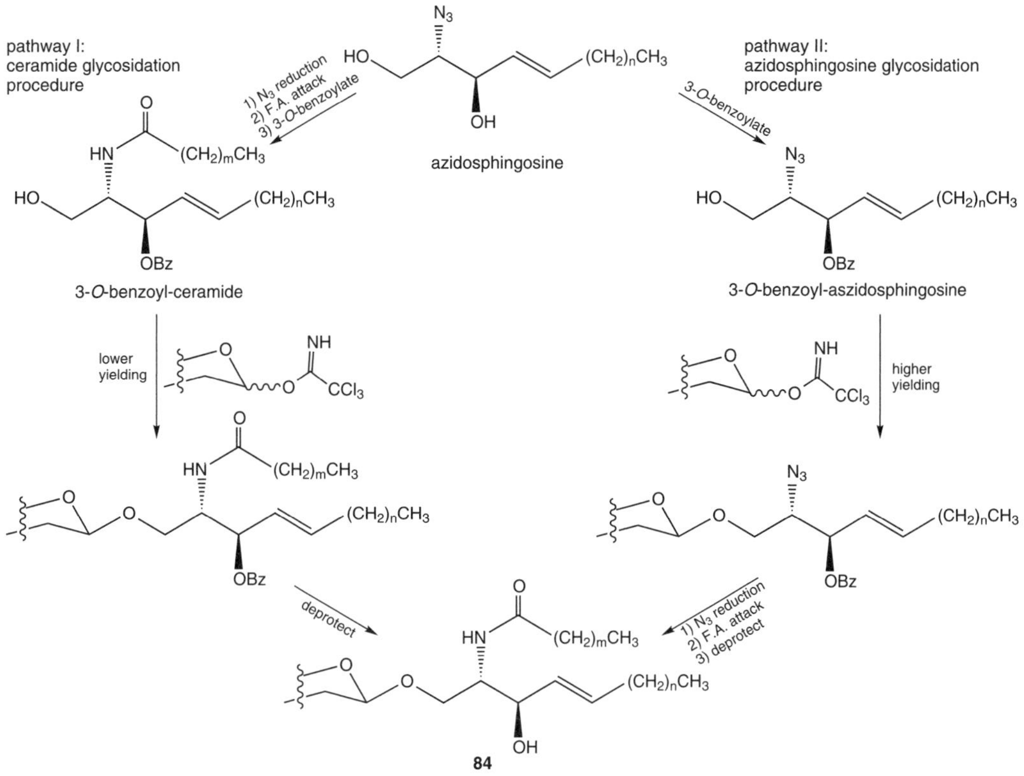

Scheme 3-14.

Thioglycosides

Thioglycosides were the first glycosyl donors to be studied in which the aglycon serves the dual role of anomeric protection and activation. Fischer and Delbrück (1909) reported the first synthesis of a thioglycoside in 1909 when they treated tetra-O-acetyl-α-D-glucopyranosyl bromide with sodium thiophenoxide, that is, **88** → **87** (Scheme 3-15). It should be noted that reaction of an aldose with a thiol under conditions normally used with alcohols for Fischer glycosidation, results in a dithioacetal (i.e., **85** → **86**). This reaction constitutes the most reliable procedure for preparing acyclic aldose derivatives (Wolfrom & Thomson, 1963). The intramolecular cyclization of dithioacetals to give thioglycosides **86** → **87** is, somewhat surprisingly, not an efficient process (Brigl et al., 1939; Pacsu & Wilson, 1939).

Given their current prominence as glycosyl donors (Garegg, 1992), it is remarkable that thioglycosides were first utilized in this role only 20 years ago when Ferrier and co-workers showed that these substances could be activated with mercuric sulfate (Ferrier et al., 1973). Improved methods of synthesis and activation have since made thioglycosides one of the most reliable and versatile glycosyl donors.

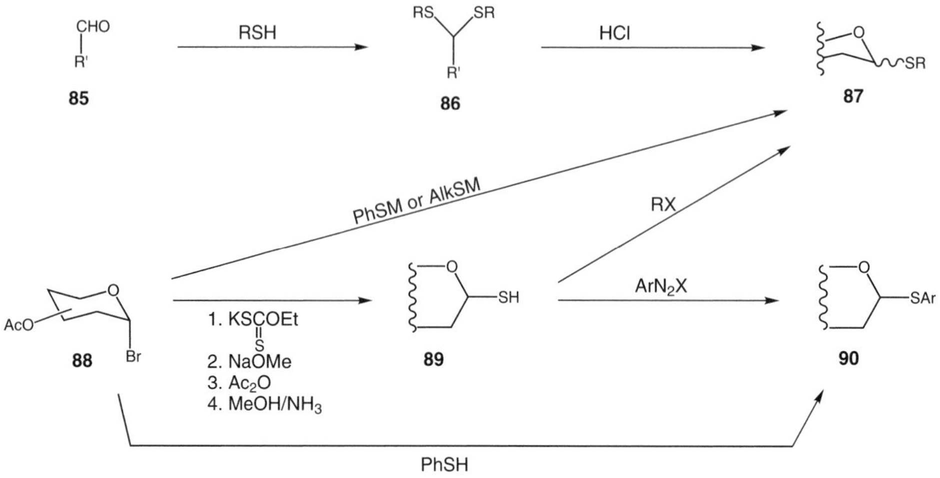

Scheme 3-15.

Preparation

The standard methods now in use for preparation of alkyl thioglycosides involve mercaptolysis of glycosyl bromides, chlorides, or acetates. However the *S*-alkylation of glycosyl mercaptans (**89 → 87**) has been explored by the group of Cerny and Pacak (Cerny et al., 1959; Cerny & Pacak, 1961) (Scheme 3-15). The aromatic analogues (e.g., **90**) have also been prepared by the use of diazonium salts (Cerny et al., 1961). However, the preparation of these precursors involves many steps, as illustrated by the conversion **88 → 89** (Horton, 1963) (Scheme 3-15).

Horton reinvestigated the reaction of glycosyl halides with thiolate anions (Horton, 1963) (Scheme 3-16a); and more recent methodology now allows the replacement to occur without concomitant de-esterification (Scheme 3-16b). In these procedures, the preformed thiolate salts (Apparu et al., 1981; Ding & Liu, 1991) are used with hexamethylphosphoric triamide as solvent. That the reaction is an S_N2-like process is suggested by the result in Scheme 3-16b in which neighboring group participation of the C2-ester is not manifested. However, it should be noted that a variation of this methodology, that is, using Et_3N/RSH instead of RS^-M^+, can lead to a thioorthoester product (*vide infra*).

Under Lewis acid-catalyzed conditions, glycosyl acetates lead to oxocarbenium ions as the expected intermediates and neighboring group participation by a suitable C2 substituent yields the 1,2-*trans* product predominantly (Scheme 3-16c), although not exclusively (Scheme 3-16d). In keeping with an S_N2-like mechanism, substantial synthetic improvements can be achieved by changing the anomeric leaving group or modulating the nucleophilicity of the thiolate anion. With regard to the latter criterion, a major advance occurred with the introduction of tin-mediated thiolates by Ogawa in 1977 (Scheme 3-17e–g) (Matsuzaki et al., 1993; Ogawa et al., 1981; Ogawa & Matsui, 1977; Sadozai et al., 1986b; Smid et al., 1991; Veeneman et al., 1990a; Yamazaki et al., 1990a,b). Without the C2 control element anomeric mixtures of products are obtained (Scheme 3-16f) (Matsuzaki et al., 1993), again in keeping with ample precedent.

Glycosyl bromides can also be used as substrates for the tin-mediated thioalkylation with or without the agency of Lewis acids (Scheme 3-16g) (Ogawa & Matsui, 1977). However, since glycosyl bromides are usually prepared from glycosyl acetates, the latter have emerged as the preferred precursors. Thus their use not only shortens the synthetic route, but has the added advantage that glycosyl esters are extremely stable species, readily amenable to storage.

An alternative class of activating agents which have proved remarkably successful is the trimethylsilyl thioethers first applied by Pozsgay and Jennings (1987), based on the pioneering work of Evans on silylated nucleophiles (Evans et al., 1977). The examples shown (Scheme 3-16) illustrate the efficiency of the procedure (Hasegawa et al., 1991, 1992; Jain, et al., 1993; Jain & Matta, 1992; Kameyama et al., 1990; Khan & Matta, 1993). The ease of preparation of the reagents and the ready availability of glycosyl acetates (Matsuzaki et al., 1993; Pozsgay & Jennings, 1987) probably make

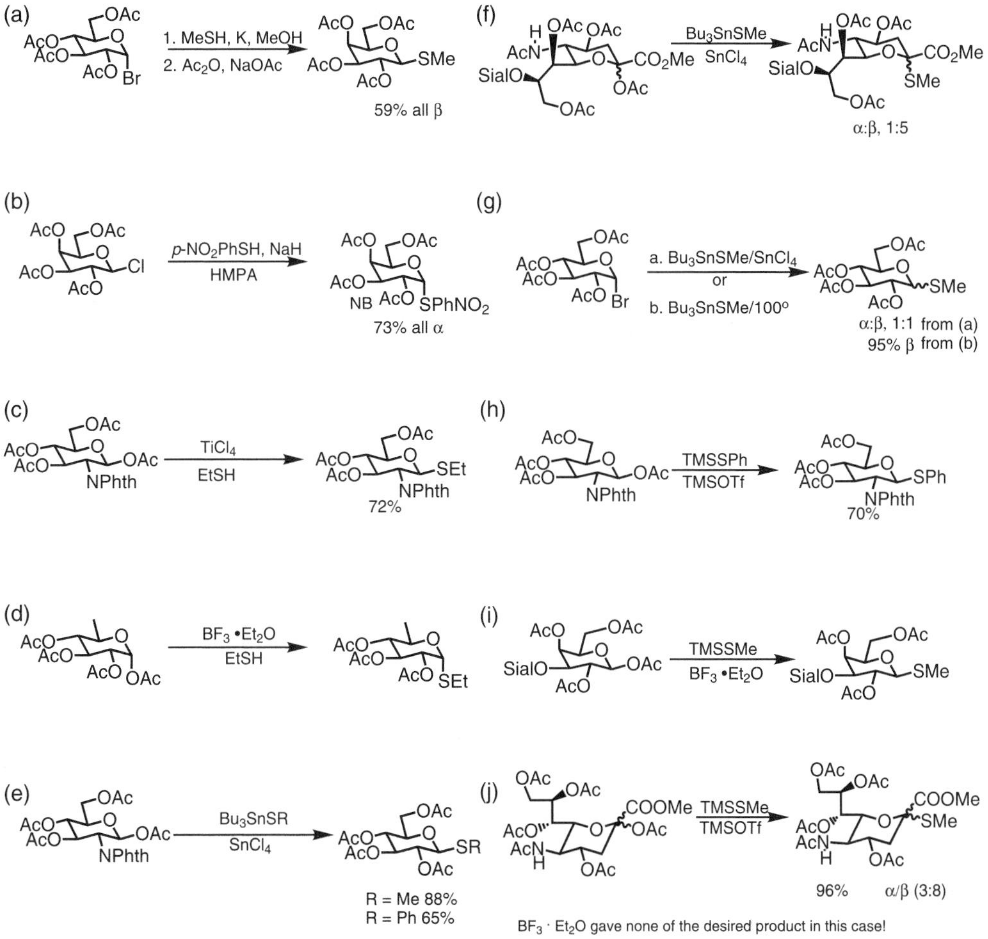

Scheme 3-16. Some procedures for preparing thioglycosides.

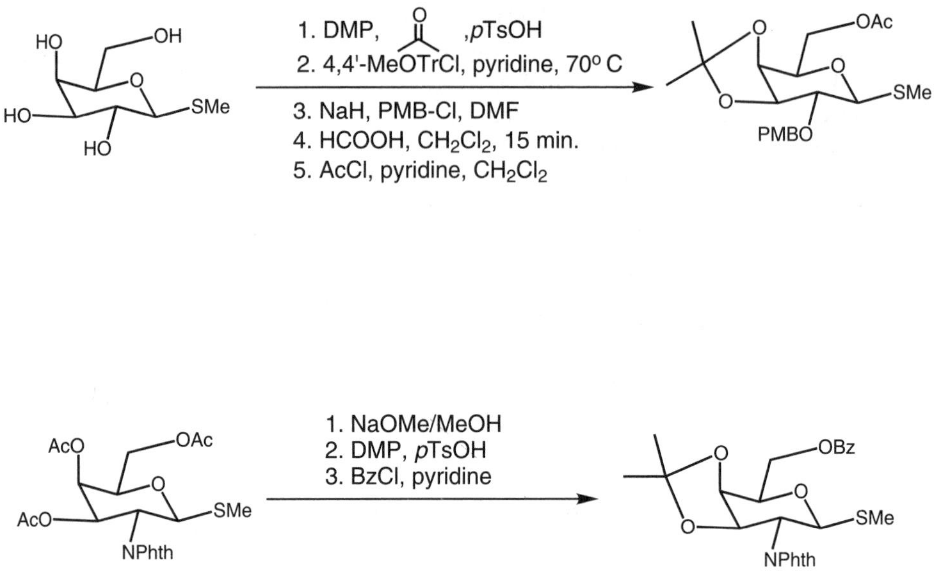

1. DMP, $\overset{O}{\underset{}{\|}}$, pTsOH
2. 4,4'-MeOTrCl, pyridine, 70° C
3. NaH, PMB-Cl, DMF
4. HCOOH, CH$_2$Cl$_2$, 15 min.
5. AcCl, pyridine, CH$_2$Cl$_2$

1. NaOMe/MeOH
2. DMP, pTsOH
3. BzCl, pyridine

(from glycosyl acetate)

Scheme 3-17.

this combination of reactants the current method of choice. However, there are certainly cases where thioglycosides used for block syntheses are prepared readily from glycosyl halides (Leontein et al., 1985). In addition, Schmidt has demonstrated the viability glycosyl trichloroacetimidates as precursors for thioglycosides (Schmidt & Stumpp, 1983).

Storage

Storage of thioglycosides presents few problems since they are stable to a wide variety of reagents and conditions. The examples in Scheme 3-17 (Dmitriev et al., 1975; Garegg & Helland, 1993) indicate the stability to the usual range of protection/ deprotection protocols and, particularly, that alkylation of the sulfide does not occur under the conditions used for O-benzylations. The same holds true for O-methylations (Liptak et al., 1983). Reductive cleavage of benzylidene rings can also be carried out without affecting thioalkylglycoside functionality (Liptak et al., 1983).

Coupling

Given the aforementioned stability of thioglycosides, the key to their successful use has been the development of highly specific methods for S-activation. Heavy metal salts, as used by Ferrier and co-workers in their pioneering experiment (Ferrier et al., 1973) are of limited utility since the metal can also react with glycosyl halides. Thus, in the reaction shown in Scheme 3-18a, aglycon exchange occurs with either silver or mercuric promotors (Kihlberg et al., 1991). Use of n-pentenyl glycosides as donors (Scheme 3-18b) is also not possible, as shown by van Boom and co-workers (Smid et al., 1991; Veeneman et al., 1990a) since the halonium ion reacts preferentially with sulfur. Thus, instead of the intended cross-coupled product, the self-coupled product was obtained. As illustrated in Scheme 3-18c, trichloroacetimidates are excellent donors for use with thioglycoside acceptors (Ziegler et al., 1993).

Given the wealth of literature on the use of glycosyl halides in oligosaccharide synthesis, the ability to convert thioglycosides into these well-explored glycosyl donors is of great strategic value (Hällgren & Widmalm, 1993; Kihlberg et al., 1990; Peters & Bundle, 1989; Srivistava & Hindsgaul, 1991). Ogawa's group was one of the first to carry out this transformation efficiently and to effect *in situ* coupling to a glycosyl acceptor (Sato et al., 1986) (Scheme 3-19a). Thus, treatment of **91** with copper (II) bromide and tetraethylammonium bromide effected the first step affording the glycosyl bromide; direct addition of **92** led to trisaccharide **93**.

The groups of Bundle and Hindsgaul have carried out Br_2/AgOTf-catalyzed couplings to give pentasaccharide **94** (Kihlberg et al., 1991) and disaccharide **97** (Kihlberg et al., 1990) (Scheme 3-20b,c). However, it has not been established that such reactions proceed through glycosyl bromides. Indeed, bromine and silver triflate should react to generate Br^+, which could function as a thiophilic agent leading to an oxocarbenium ion which could be captured by the acceptor (Scheme 3-20a). In this connection the use of N-bromosuccinimide as an electrophile for activating thioglycosides was initiated by Hanessian (Hanessian et al., 1980) and Nicolaou (Nicolaou et al., 1982).

In the reaction leading to **97** the intermediacy of the imidic cation **98** would account for the high stereoselectivity observed in the coupling reaction, by a process which is reminiscent of Lemieux's halide ion-catalyzed reaction to give 1,2-*cis* glycosides (Lemieux et al., 1975) (see Scheme 3-4). It may be noted that such an ion has been characterized as an intermediate by Bock (Bock et al., 1992).

One of the most successful methods of *S*-activation involves the use of methyl triflate as an alkylating agent (Scheme 3-21a) (Lönn, 1985). The merits of this method

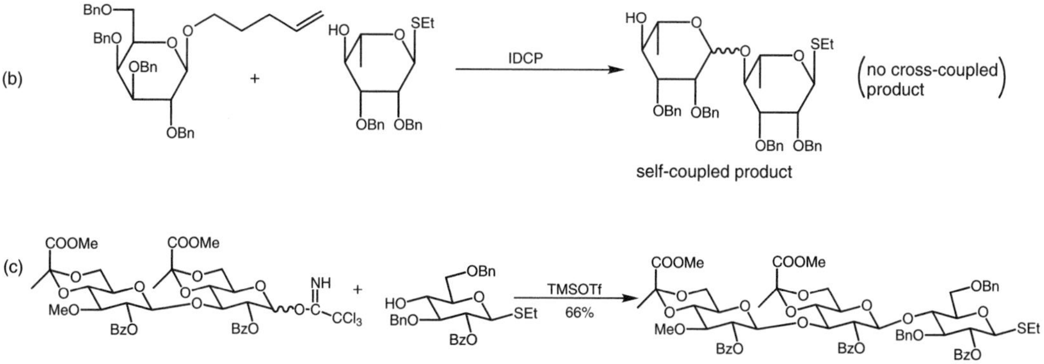

Scheme 3-18.

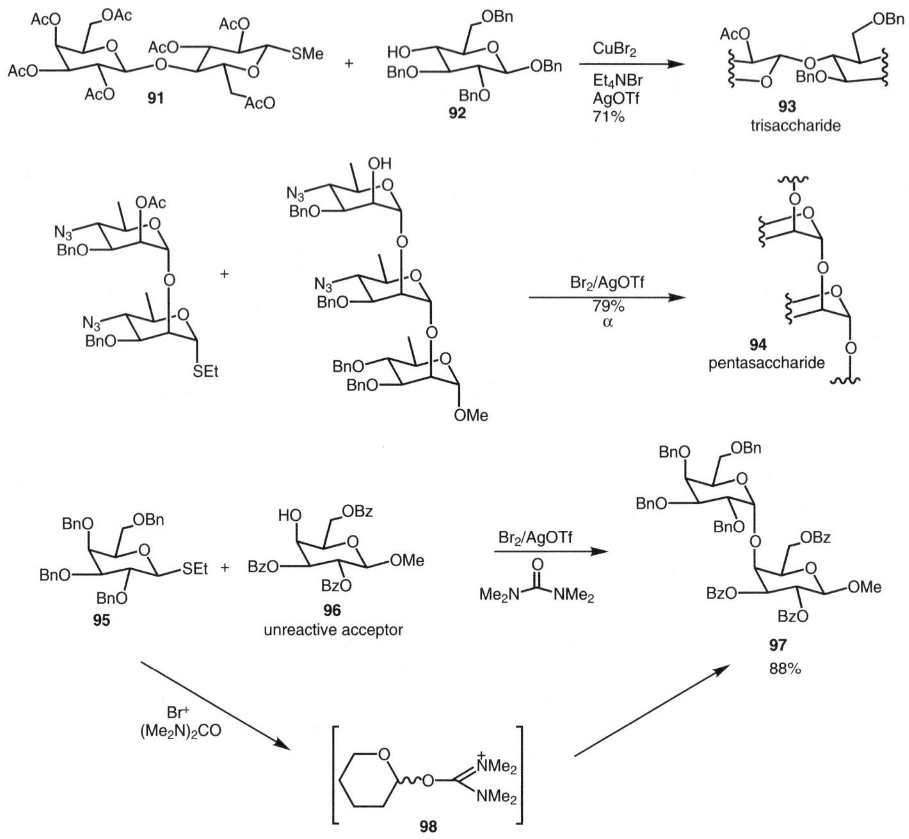

Scheme 3-19.

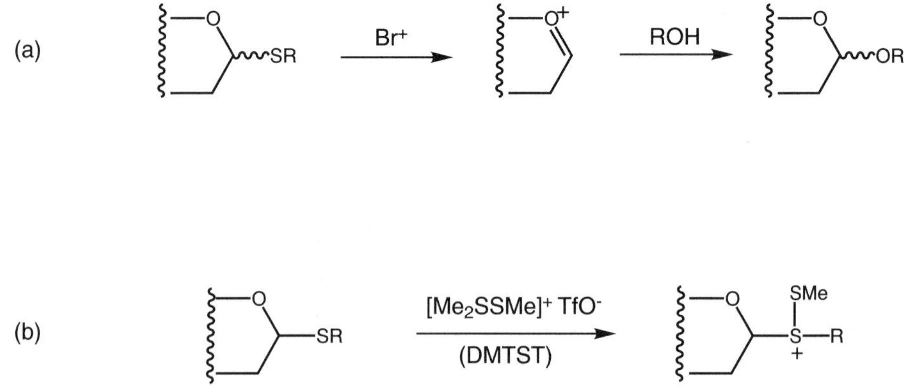

Scheme 3-20.

may be exemplified by the block synthesis of tetrasaccharide **99** (Nilsson et al., 1991) and more impressively decasaccharide **100** (Scheme 3-21b) (Birberg et al., 1989). In spite of its exceedingly high toxicity, which constitutes a severe limitation, this reagent still finds widespread use in some laboratories. The reagent can also cause O-methylation and this may present a problem in that alkylation of hydroxyl groups of unreactive acceptors might occur.

The recent development by Fügedi and Garegg (1986) of the use of dimethyl-(methylthio)sulfonium triflate (DMTST) as a methyl sulfenylating agent has emerged as a superior method (Murase et al., 1988). The trialkylated sulfenium intermediate (Scheme 3-20b) provides an excellent leaving group. Scheme 3-22 presents a representative selection of results in which the stereoselectivity in coupling was controlled either by the use of ether as solvent to induce 1,2-*cis* selectivity (Scheme 3-22a) (Garegg & Helland, 1993), or by neighboring group participation to induce 1,2-*trans* coupling (Scheme 3-23b,c) (Hasegawa et al., 1993; Nilsson & Norberg, 1990). Alternatively, the Lemieux halide ion-catalyzed protocol (Lemieux et al., 1975) (see Scheme 3-4) can be adapted to induce a 1,2-*cis* configuration, as the results in Scheme 3-22d with and without added tetra-*n*-butylammonium bromide illustrate (Andersson et al., 1986; Hasegawa et al., 1986). The examples in Scheme 3-22 indicate that the use of DMTST generally leads to good yields even with unreactive acceptors (Scheme 3-22a), and that a wide range of protecting groups in the reactants can be tolerated.

Perhaps the most significant application of DMTST has been for the glycosidation of N-acetylneuraminic acid (sialic acid), exemplified particularly in the synthesis of gangliosides, an area of intense activity during the past 5 years. Initially it appeared that a pure anomer was necessary to obtain stereoselective coupling (Scheme 3-22e) (Murase et al., 1988), a requirement which would have presented a serious limitation for large scale preparations. However, prodigious effort by the Hasegawa group has

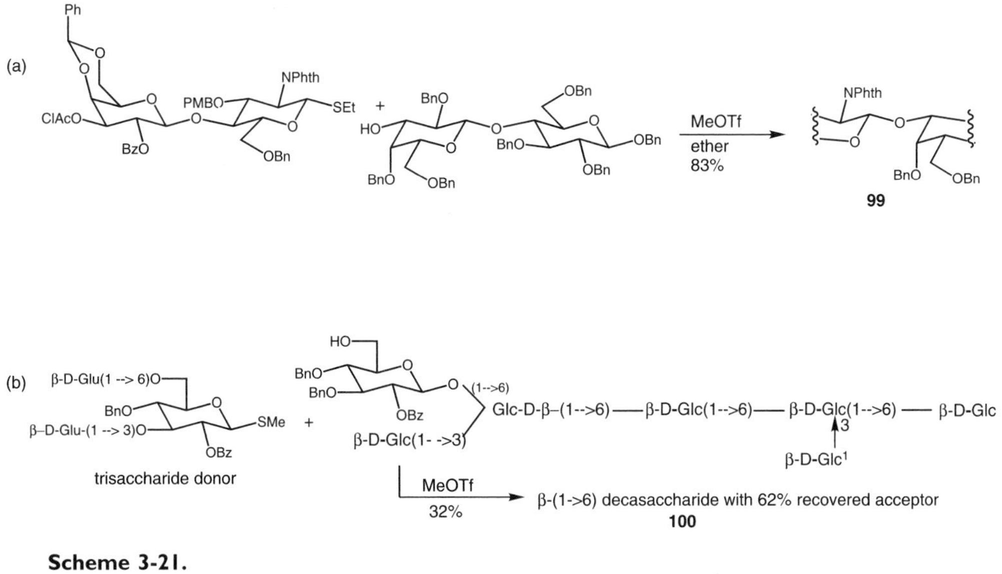

Scheme 3-21.

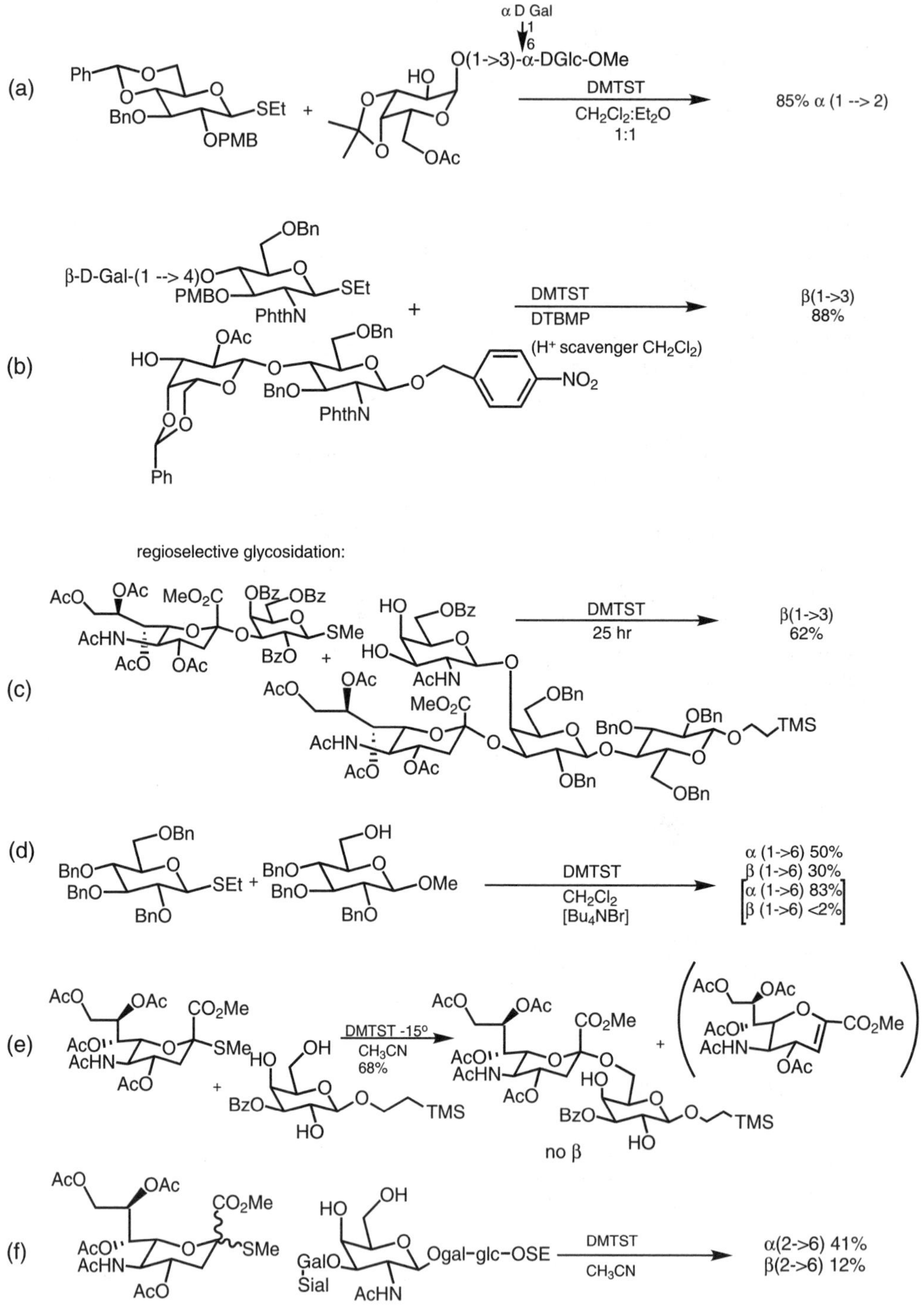

Scheme 3-22.

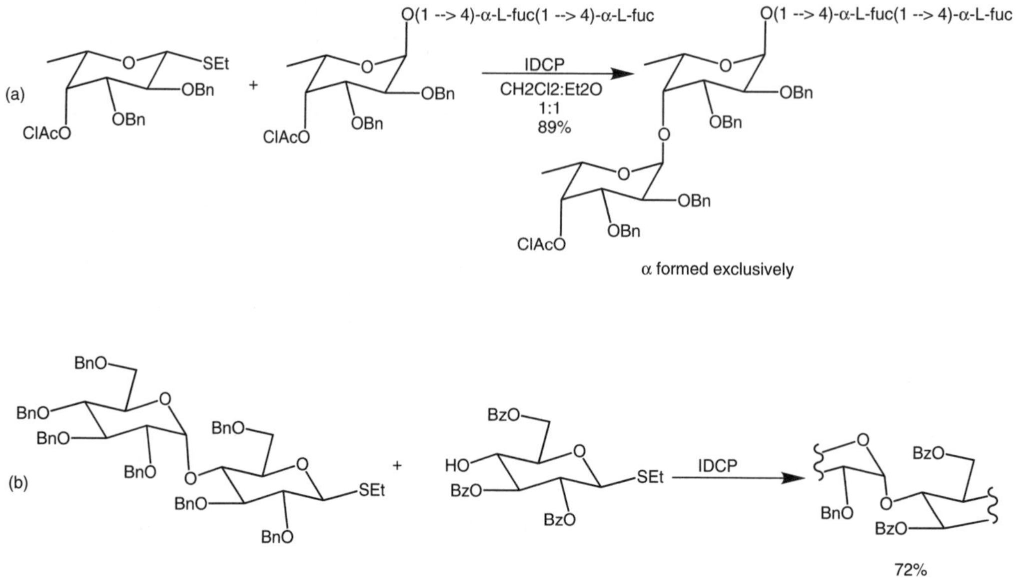

Scheme 3-23.

shown that the anomeric configuration of the thioglycoside donor is not important (Scheme 3-22f) (Murase et al., 1988; Prabhanjan et al., 1991; Zuurmond et al., 1991). Thus the easily accessible α/β mixture of donors, if used under kinetically controlled conditions, gives only the α-sialylated product in many cases. Some glycal formation has been observed as a by-product under these conditions. This development has made it possible for ready routes to specific gangliosides (Hasegawa et al., 1993).

In Scheme 3-20a it was shown that thioglycosides can be activated by halonium ions. Following the use of iodonium dicollidine perchlorate (IDCP) for activating n-pentenyl glycosides by Fraser-Reid and co-workers (Fraser-Reid et al., 1988), van Boom's laboratory applied this reagent to thioglycosides (Veeneman & van Boom, 1990a,b). The results in Scheme 3-23 show that high yields and high stereoselectivities can be achieved (Smid et al., 1991). This may be attributable to the effect of the ethereal solvent. Additional influence may arise from the fact that IDCP is not a very potent iodonium ion source and so greater stereoselectivity may be obtainable when it is employed for coupling.

In 1990 the combination of N-iodosuccinimide and triflic acid was introduced independently by the laboratories of van Boom (Veeneman et al., 1990a,b,c) and Fraser-Reid (Konradsson et al., 1990) as a potent source of iodonium ion. With these reagents, the reactions of thioglycosides could be effected much more rapidly, typically from 5 minutes to 1 hour depending on the temperature. In these rapid reactions, stereoselectivity may suffer somewhat, except when a participating group is present at C2.

n-Pentenyl Glycosides

The use of *n*-pentenyl glycosides (NPGs) as glycosyl donors originated from a serendipitous observation in Fraser-Reid's laboratory, as reviewed elsewhere (Fraser-Reid et

al., 1992). For the commonly used glycosyl donors described in the preceding sections, it is first necessary to convert the glycose into a suitable derivative (e.g., a glycosyl ester or bromide) or to protect the "other" hydroxyl groups before the desired activating group can be installed at the anomeric center. However, the *n*-pentenyl group can be attached directly to the unprotected aldose by a modification of the Fischer glycosidation procedure for preparation of simple alkyl glycosides (*vide infra*). Being a simple alkyl residue itself, the *n*-pentenyl group ably protects the anomeric center during a wide variety of chemical transformations, yet can be chemospecifically triggered to serve as a leaving group at the appointed time. Hydrogen cannot be used for reactions of NPGs, for example, for cleavage of benzyl protecting groups, but this result can be achieved by Birch reduction protocols (Fraser-Reid et al., 1992).

Preparation and Storage

Since NPGs are normal alkyl glycosides, methods for their preparation are borrowed or adapted from standard protocols (Scheme 3-24). Thus, the most direct methods of preparation involve treating an aldose with 4-pentenyl alcohol and camphorsulfonic acid as catalyst, that is, **101** → **102** (Fraser-Reid et al., 1992; Konradsson et al., 1991). At the completion of the reaction, the unreacted 4-pentenyl alcohol can be recovered by distillation and saved for use in future glycosidations. As expected, the reaction gives a mixture of anomers, the composition of which is approximately the same as that of corresponding methyl glycosides prepared by standard Fischer glycosidation methods.

The above analogy to simple alkyl glycosides can also be exploited for alternative methods of preparation from glycosyl esters or halides (**103** or **104**, respectively) as illustrated in Scheme 3-24. Resort to this prior art is particularly useful when either an α or β anomer is specifically required, or when a special pattern of protecting group deployment is necessary. In such cases the published procedure for preparing the methyl

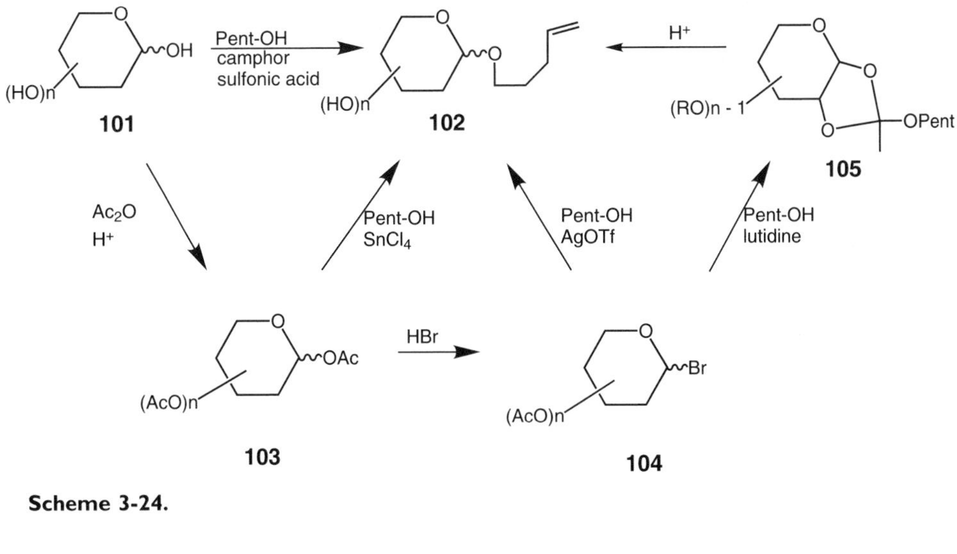

Scheme 3-24.

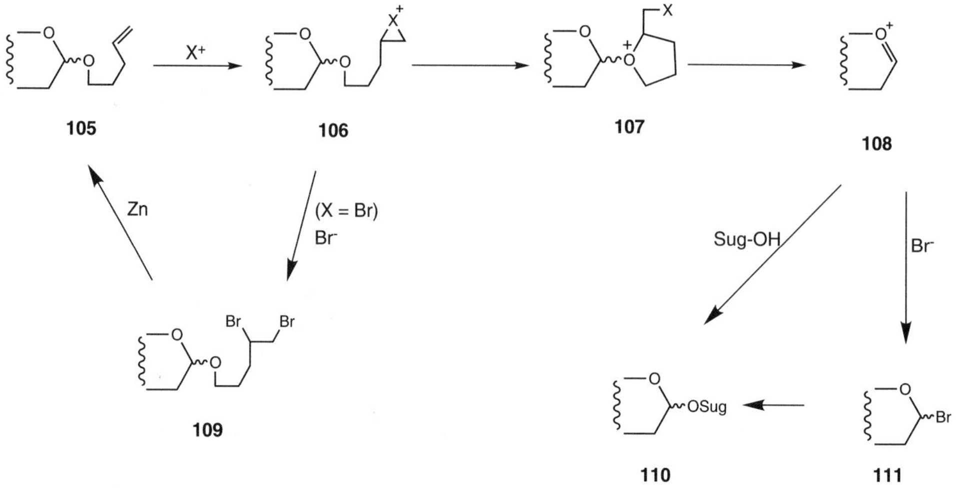

Scheme 3-25.

(or simple alkyl) analogue can usually be adopted with minor adjustments. *n*-Pentenyl orthoesters (**126**) can also be prepared under standard conditions (*vide infra*). NPGs are as stable as other alkyl glycosides and, therefore, can be stored indefinitely at room temperature whether in crystalline or syrupy form.

Coupling

Electrophilic activation of an NPG produces a cascade of ionic intermediates culminating in the formation of the key oxocarbenium ion **108**, as illustrated in Scheme 3-25 (Fraser-Reid et al., 1992). N-Halosuccinimides have been the most widely investigated electrophiles; used by themselves, the reaction of NPGs is comparatively slow, sometimes requiring hours or days to proceed to completion. However, addition of a protic or Lewis acid enhances the rate of reaction. For example, with NIS/Et$_3$SiOTf the reaction of NPGs, even when disarmed (cf. **21** and **22**) is usually complete within the time it takes to sample the mixture by thin layer chromatography (Konradsson et al., 1990).

Of intermediate potency is iodonium dicollidine perchlorate (IDCP), as illustrated in Scheme 3-26a (Ratcliffe et al., 1990, 1991). Thus coupling of the armed and disarmed partners, **112** and **113** respectively, via the agency of IDCP afforded the coupled product **114** after 10 hours in 63% yield. The resulting disaccharide **114** is a disarmed glycosyl donor; however, by use of NIS/TfOR (R=H or Et$_3$Si), coupling to acceptor **115** occurred within minutes to give trisaccharide **116**.

The first stage of the foregoing reaction relies on the fact that IDCP reacts more slowly with the disarmed donor **113** in keeping with the protocol described above. On the other hand, such disarmed donors react rapidly with NIS/TfOR (R=H or Et$_3$Si). However, it must be emphasized that with the latter promoter, armed/disarmed selectivity is still observed, as exemplified in Scheme 3-26b (unpublished). Thus both stages of the reaction in Scheme 3-26a could conceivably have been carried out, if so desired, by the use of NIS/TfOR.

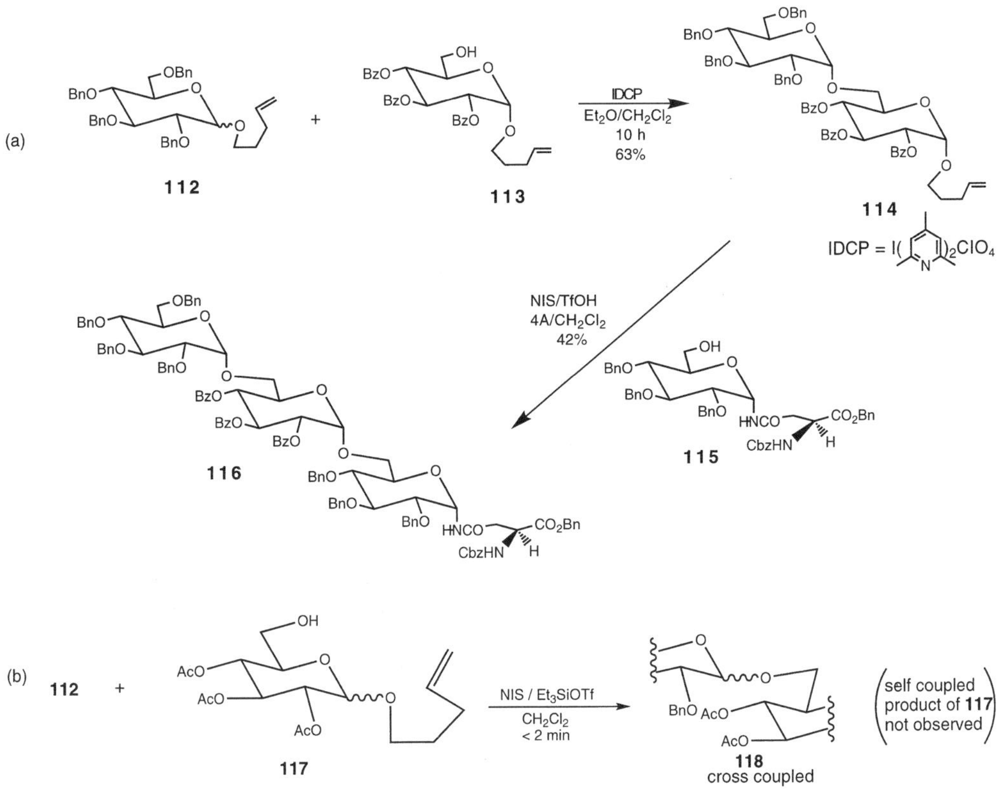

(a)

112 + **113** → IDCP / Et$_2$O/CH$_2$Cl$_2$ / 10 h / 63% → **114**

IDCP = I(collidine)$_2$ClO$_4$

NIS/TfOH 4A/CH$_2$Cl$_2$ 42%

115

116

(b)

112 + **117** → NIS / Et$_3$SiOTf / CH$_2$Cl$_2$ / < 2 min → **118** cross coupled

(self coupled product of **117** not observed)

Scheme 3-26.

The use of this promoter for rapid assembly of oligosaccharides is demonstrated for preparation of the blood group substance β tetrasaccharide (**122b**). Thus NIS/Et$_3$SiOTf was used for *all* couplings whether the donor was armed (**120** and **121**) or disarmed (**119**), the individual reactions being complete in 10–15 min (Udodong et al., 1992). One advantage of this development is that assembly of tetrasaccharide **122a** involves virtually iterative practices: couple ... deprotect ... couple ... deprotect ... couple. ...

A problem using IDCP is that the coupling tends to "stop," leaving quantities of un-reacted glycosyl donor, and addition of further amounts of IDCP does not drive the re-action to completion. This is probably due to an inhibiting effect of the collidine lib-erated during the reaction, a circumstance that might account for the potency of NIS/Et$_3$SiOTf system since there are no ligands present to curtail the reactivity of I$^+$ (Fraser-Reid et al., 1992).

Formation of the oxocarbenium ion **108** in Scheme 3-25 requires that the cyclic bromonium ion **106** undergo intramolecular cleavage leading to the furanylium ion **107**. If a sugar alcohol, Sug-OH, is present, coupling to give **110** is observed. However, if molecular bromine is used as the electrophile, the Br$^-$ that is liberated becomes at-tached to the anomeric center (i.e., **108** → **111**).

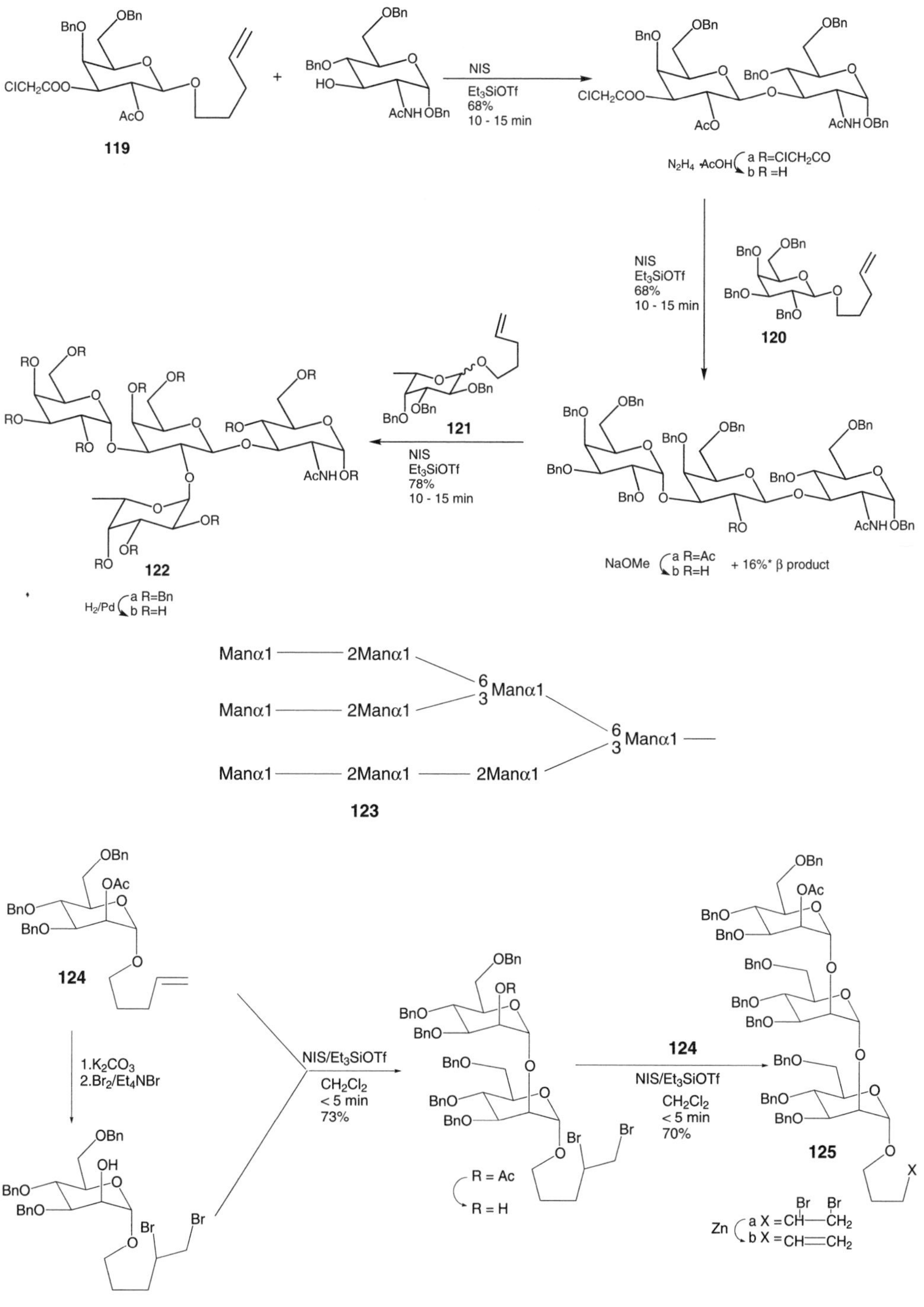

Scheme 3-27.

In addition, if an excess of Br⁻ is added at the start of the reaction, there is bimolecular competition for intermediate **106** which results in the formation of the vicinal dibromide **109**. The double bond can be readily restored from the latter by reductive elimination using zinc (Fraser-Reid et al., 1990). This development has great strategic value since it means that a given NPG can be used as a donor immediately (e.g., **102** → **108**) or subsequently by blocking the double bond as in **109**, exposing a hydroxyl group to produce a glycosyl acceptor, and after coupling regenerating the double bond (**109** → **102**) at the appropriate time.

The advantage of this strategy for oligosaccharide synthesis is that it reduces the number of starting materials that must prepared, as exemplified in Scheme 3-27. Thus Merritt and Fraser-Reid (1992) prepared the nonamannose component **123** of high mannose glycoprotein from two pentenyl glycosides, one of them being **124**. The use of the latter for preparing **125**, which is the trimannan of the lowest antenna, is summarized.

A different strategy for reducing the number of ultimate starting materials involves the use of *n*-pentenyl orthoesters, for example, **126** (Scheme 3-28). 1,2-O-Orthoesters are valuable synthetic intermediates which undergo stereocontrolled proton-induced re-

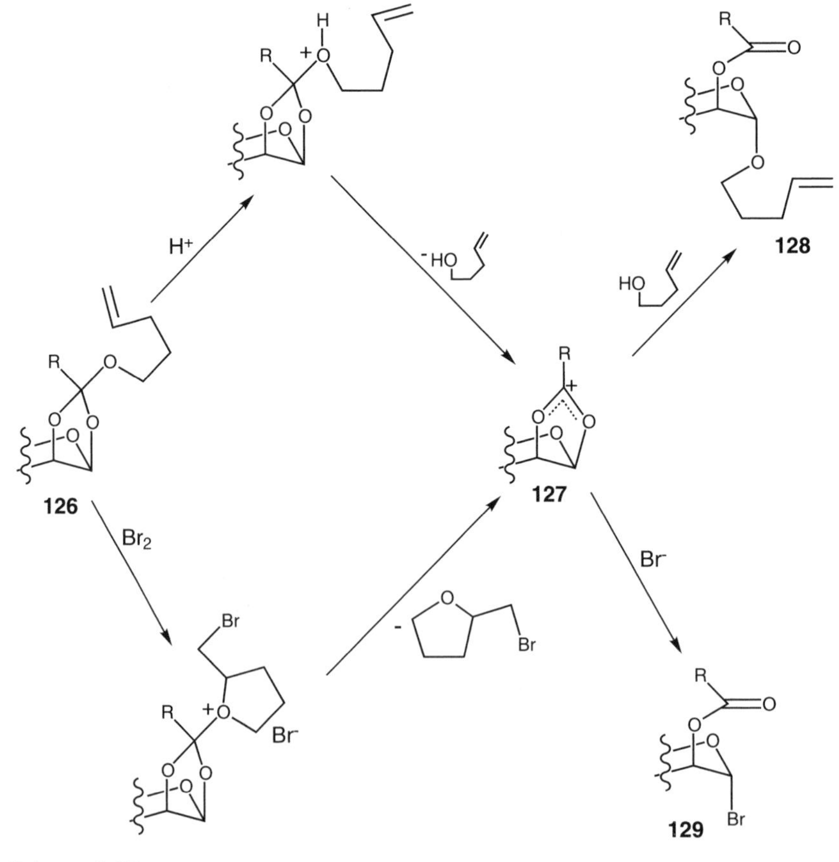

Scheme 3-28.

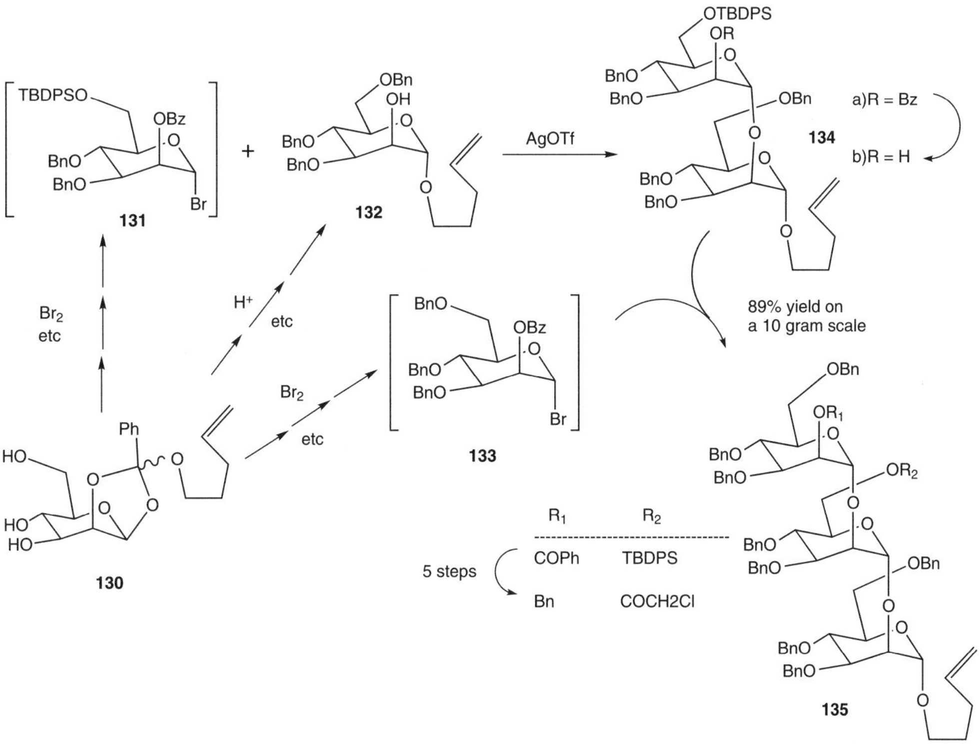

Scheme 3-29.

arrangement to give 1,2 -*trans* glycosides (Bochkov & Zaikov, 1979). Thus with **126**, the extruded 4-pentenyl alcohol is transferred to the anomeric center by nucleophilic attack upon the dioxolenium ion **127** to give the *n*-pentenyl glycoside **128**. If orthoester **126** is titrated with molecular bromine, the same dioxolenium ion **127** is produced; however, the extruded by-product is now the nonnucleophilic 2-bromomethyl tetrahydrofuran. Ion **127** is therefore quenched with the only available nucleophile, Br⁻, leading to glycosyl bromide **129**, the latter procedure being reminiscent of the brominolysis **105** → **111** described in Scheme 3-25.

Thus a given orthoester can be processed to give glycosyl donors **128** and **129** having vastly different reactivities. The use of this strategy for simplifying the number of starting materials needed for oligosaccharide synthesis is illustrated in Scheme 3-29. By standard protection strategies orthoester **130** is converted into donors **131**, **132**, and **133**. Each bromide is generated and used *in situ*. Thus **131** undergoes *in situ* coupling to give **134a** which can be debenzoylated to obtain the acceptor **134b**. The latter then couples with bromide **133** to give trimannan **135** which is a component of the Thy-1 glycoprotein (Udodong et al., 1990).

It was noted that the β-selective stereocontrol which is observed when acetonitrile is used as solvent is consistent with formation of a transient complex **19** (*vide supra*). Indeed, evidence of a relatively long-lived complex such as **137** (Scheme 3-30) was

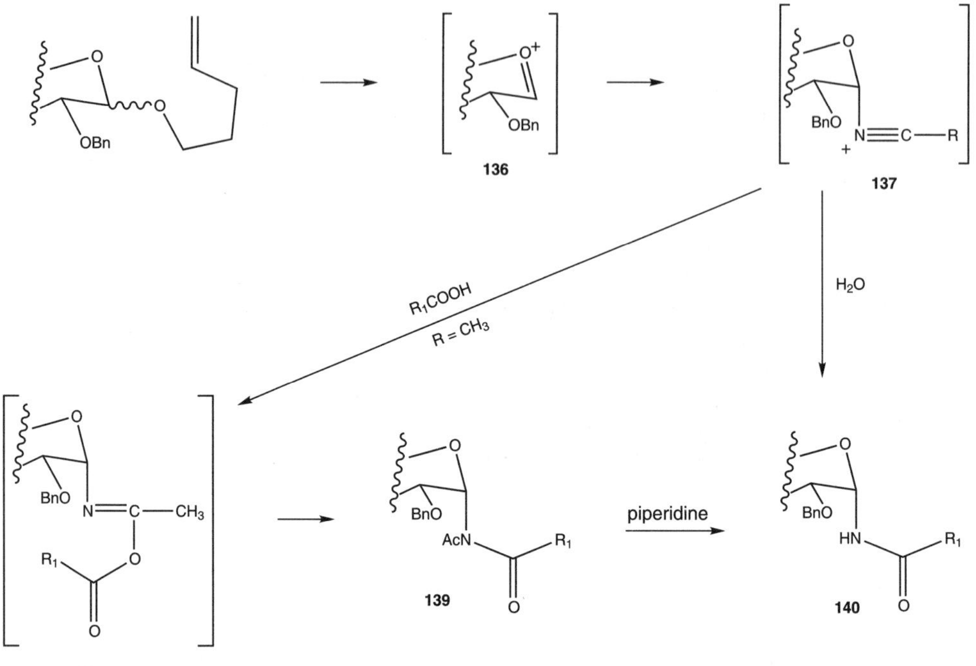

Scheme 3-30.

apparent from the isolation of small amounts of acetamide **140** (Ratcliffe et al., 1990, 1991). Further, such intermediates have been trapped by Sinaÿ (Sinaÿ & Pougny, 1976) and Schmidt (Schmidt & Michel, 1985) by including aromatic carboxylic acids in the solution.

Ratcliffe and Fraser-Reid established the correct stereochemistry of the process (Ratcliffe & Fraser-Reid, 1990) and further showed that amino acids could be trapped to give *N,N*-diacyl derivatives (e.g., **139**) (Rao et al., 1993; Ratcliffe et al., 1990, 1991). The development of a chemospecific procedure for the *N*-deacetylation by the use of piperidine made it possible to convert **139** into glycopeptide units, for example, **140** (R_1 = amino acid residue). This strategy has now been extended to obtain the chito-biosyl-*N*-glycopeptide moiety (Handlon & Fraser-Reid, 1993).

Solid Phase Oligosaccharide Synthesis

The development of the glycosyl donors discussed above has had such a remarkable effect on solution phase oligosaccharide synthesis that glycosidation has now become a relatively standard transformation that can be achieved in very high yield and frequently with good stereocontrol. Thus preparation of oligosaccharides comprising up to five sugars may be considered routine, contrary to 20 years ago when the synthesis of a trisaccharide was still a major task. The primary obstacles to rapid oligosaccharide assembly are the need for strategic protecting group deployment, and the constant labor of chromatographic purification. The latter can be overcome by solid phase methods, as was first shown by Merrifield (1963) for oligopeptides and extended soon thereafter to oligonucleotides.

Application of solid phase methods to oligosaccharide synthesis was first explored in the early 1970s (Excoffier, et al., 1972; Frechet & Schuerch, 1971; Zehavi & Patchornik, 1973). The elegant, 20-year-old synthesis of isomaltose in Scheme 3-31 (Chiu & Anderson, 1976) was a notable, early achievement. However, momentum could not be sustained owing to shortcomings of then available glycosyl donors, and the absence of appropriate analytical techniques. Recent advances in both of these areas have reinvigorated efforts in solid phase oligosaccharide synthesis, and the advent of combinatorial libraries (Fodor et al., 1991; Geysen, et al., 1984) as an adjunct to drug development is providing a powerful driving force.

Typically, the solid support is a polymer, the most popular being (a variation of) the Merrifield resin. A more rigid material like pure glass or silica (Hermkens, et al., 1996, 1997) has been recently introduced. An inert linker is usually used so as to suppress steric interactions between the support and the glycosyl reactant.

Two modalities for oligosaccharide assembly are conceivable (Scheme 3-32) wherein either the glycosyl donor or glycosyl acceptor is immobilized on the solid support. Typically oligosaccharide assembly begins by attaching the first monosaccharide to the support, and in each cycle a fully protected glycosyl donor or acceptor is coupled to the solid bound counterpart under the agency of an appropriate promoter. A protective group is then removed and the polymer-bound sugar is ready for the next coupling. After each step all components in the solution are conveniently removed by filtration and washing of the solid support.

The first modality (Scheme 3-32a) is patterned after the Merrifield (1963) technique in which the "reactive species" is used in solution, so that side products accumulate in solution rather than on the solid support. However Danishefsky, a major exponent in the rejuvenation of solid phase methods, has explored the modality in Scheme 3-32b (Danishefsky et al., 1993). He and his co-workers have achieved remarkable success using bound glycal epoxides, as glycosyl donors, and glycals as acceptors, as illustrated in Scheme 3-33 (Randolph et al., 1995; Roberge et al., 1995).

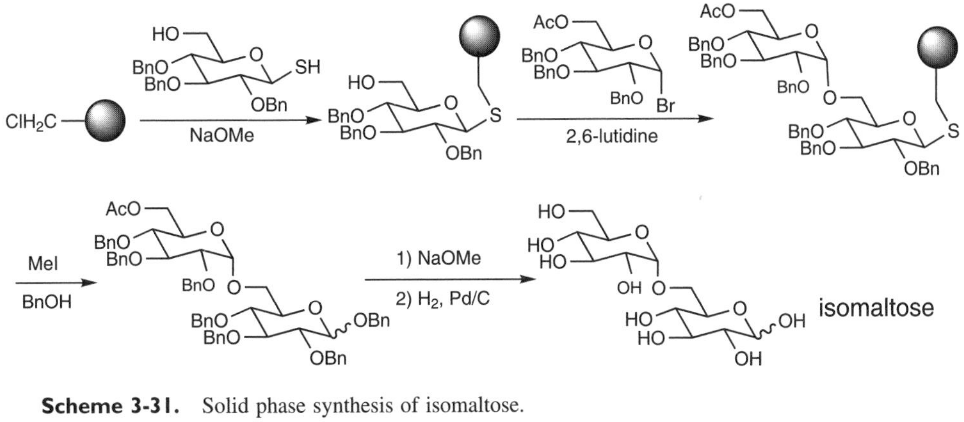

Scheme 3-31. Solid phase synthesis of isomaltose.

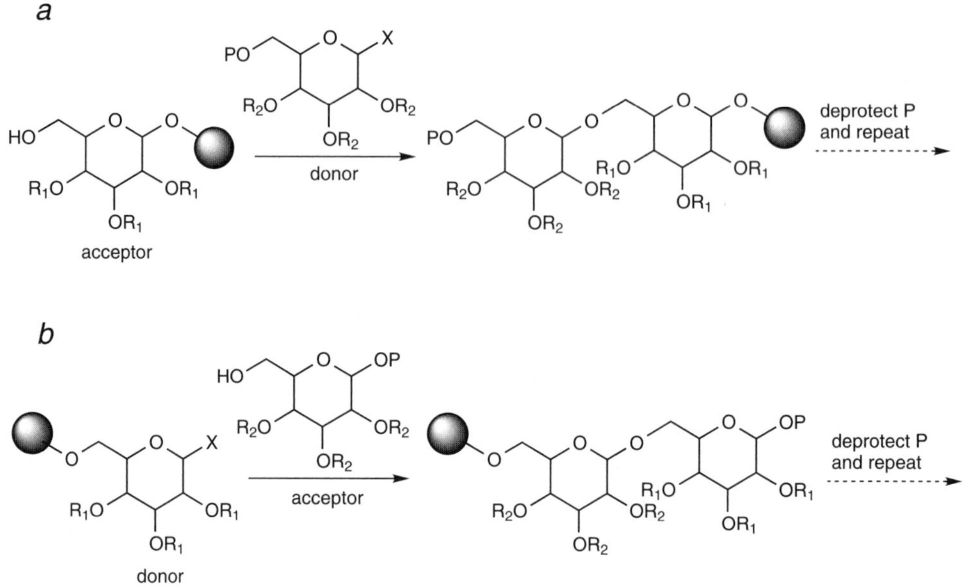

a

b

Scheme 3-32. (a) Polymer-bound glycosyl donor. (b) Polymer-bound glycosyl acceptor.

The modality in Scheme 3-32a has found many proponents, geared frequently to associated developments in glycosyl donor methodology. An iterative procedure that requires only one operation on the solid bound material in readiness for the next coupling event, is ideal. The burden of the synthesis would then be shifted to the solution phase preparation of an appropriate glycosyl donor.

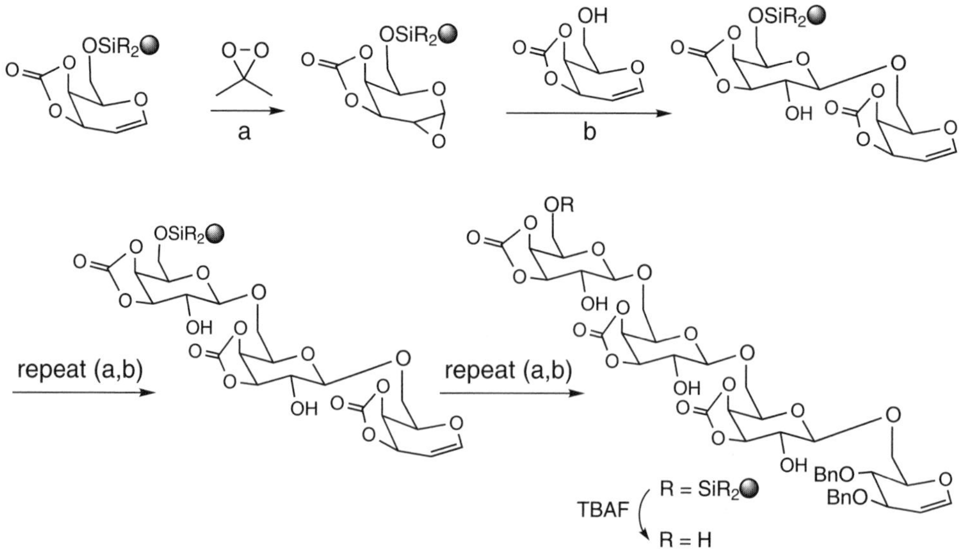

Scheme 3-33.

Scheme 3-34 summarizes a recently developed iterative protocol for rapid assembly of 1-6-gluco-oligosaccharides (Rademann & Schmidt, 1996). Merrifield resin was used as the solid support and propanedithiol as the linker. Each glycosidation was shown to proceed in over 95% yield with three equivalents of the trichloroacetimidate donor. Progress of the glycosidation could be monitored by MALDI-TOF mass spectrometry.

Branched oligosaccharides are more demanding than their linear counterparts in view of the need for selective deprotection at some stage. An example using *n*-pentenyl glycosides is illustrated in Scheme 3-35 (Joshi et al., 1997). An impressive synthesis of a highly branched phytoalexin elicitor achieved in Nicolaou's laboratory (Nicolaou et al., 1997) is summarized in Scheme 3-36.

To be effective for the preparation of larger oligosaccharides, the solid phase strategy places some severe demands on each coupling reaction. Since fractionation of the solid material is not possible, each glycosidation must ideally proceed in very high yield and with complete stereocontrol. Excess glycosyl donor or repetitive processing can be used to drive reactions to completion. This is sometimes necessary because glycosidations in the heterogeneous environment of a solid support are significantly slower than in solution.

An alternative procedure, that overcomes the problem of kinetics, is to use soluble polymers that can be precipitated after each coupling (Douglas et al., 1991;

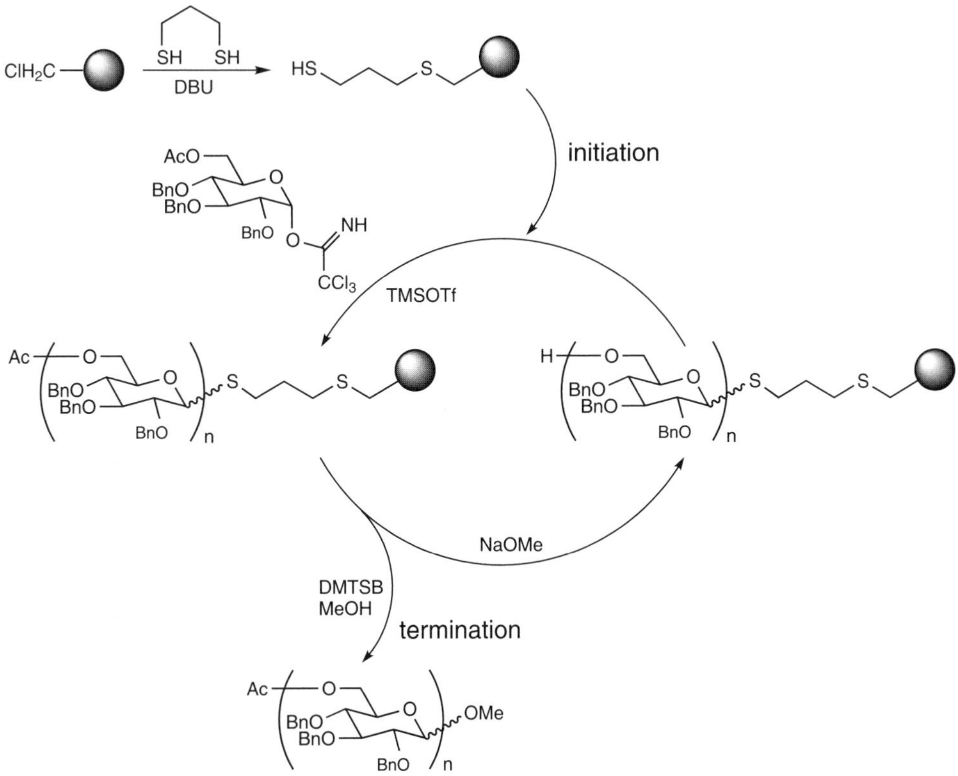

Scheme 3-34. Solid phase synthesis by trichloroacetimidate method.

Guthrie et al., 1973). The glycosidations occur in solution and therefore follow, approximately, standard reactivities. However, a disadvantage is in the precipitation of the oligosaccharide bound polymer. Thus the physical properties of the soluble polymer change as the bound oligosaccharide becomes larger, making precipitation more difficult.

A linear pentamannan has been produced through an iterative couple, deprotect, couple strategy using a trichloroacetimidate donor and the MPEG-DOX-OH polymer by Krepinsky and co-workers (Douglas et al., 1995). As illustrated in Scheme 3-37, steps a, b, c, and d were repeated five times. After the polymer-bound pentasaccharide had been assembled, treatment with Raney nickel at reflux for 16 hours afforded the totally deprotected pentasaccharide.

The development of solid phase oligosaccharide synthesis opens up new opportunities that are not shared by solution phase alternatives. The most important is the possibility of making combinatorial libraries of oligosaccharides which can be screened against biological targets for possible discovery of lead compounds. On the assumption that the epitope for binding is relatively small, a di- or trisaccharide may be all that is required. Fully deprotected oligosaccharides have been prepared in Fraser-Reid's laboratory (Joshi et al., 1997) bound to polystyrene-grafted "crowns," so that they are reusable for screening (Scheme 3-38).

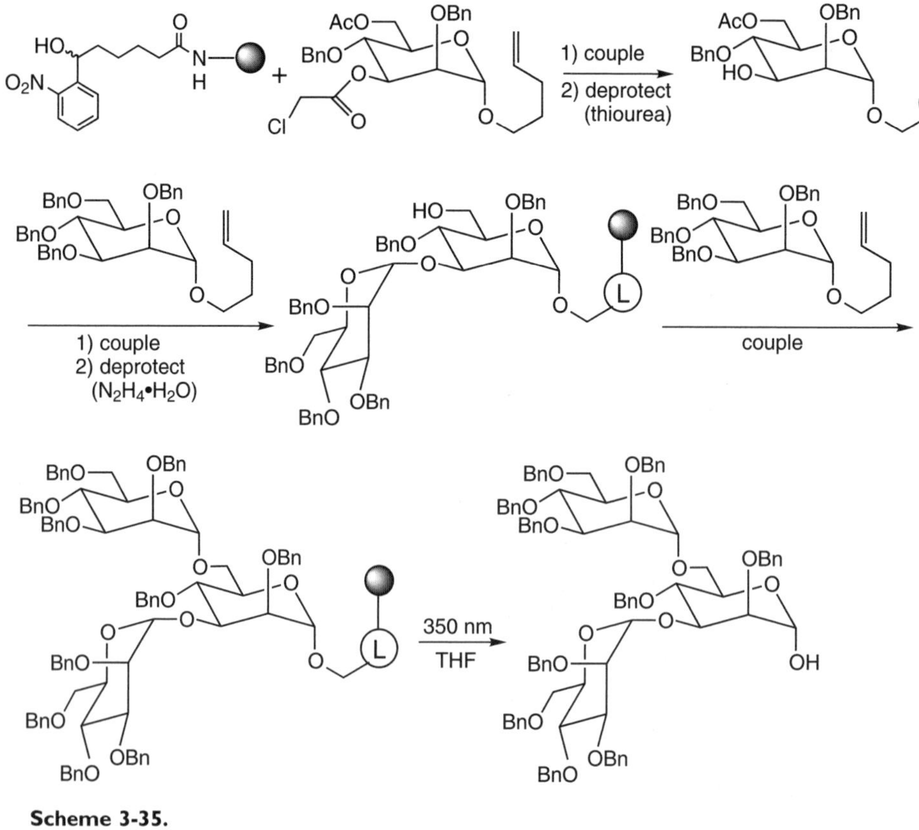

Scheme 3-35.

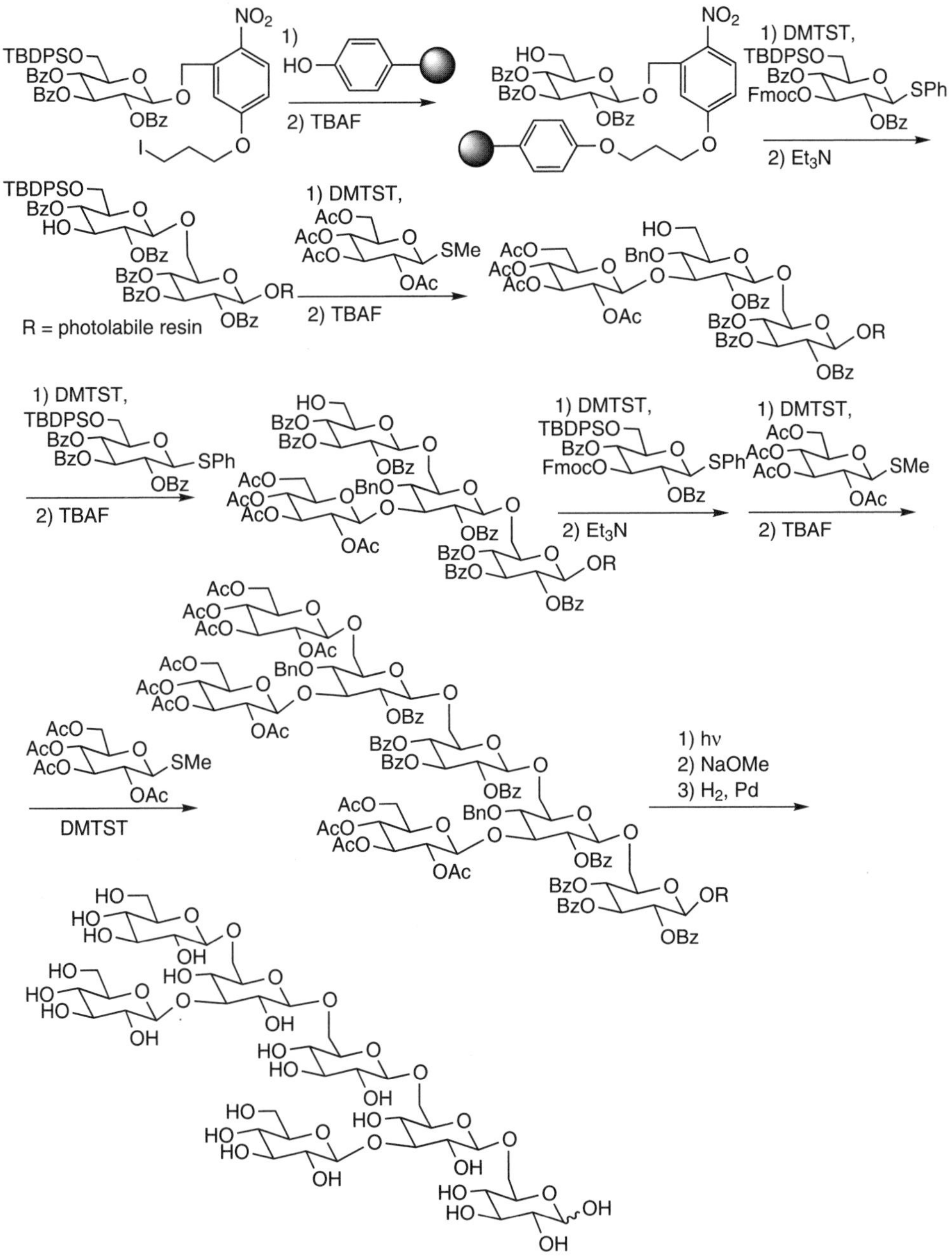

Scheme 3-36.

Kahne and co-workers have developed the solid phase synthesis of a combinatorial library of about 1300 di- and trisaccharides (Liang et al., 1996). The library was synthesized with chemical encoding on a TENTA GEL resin using a "split and mix" strategy. The library was screened by binding to a lectin, and several strong ligands were thereby discovered.

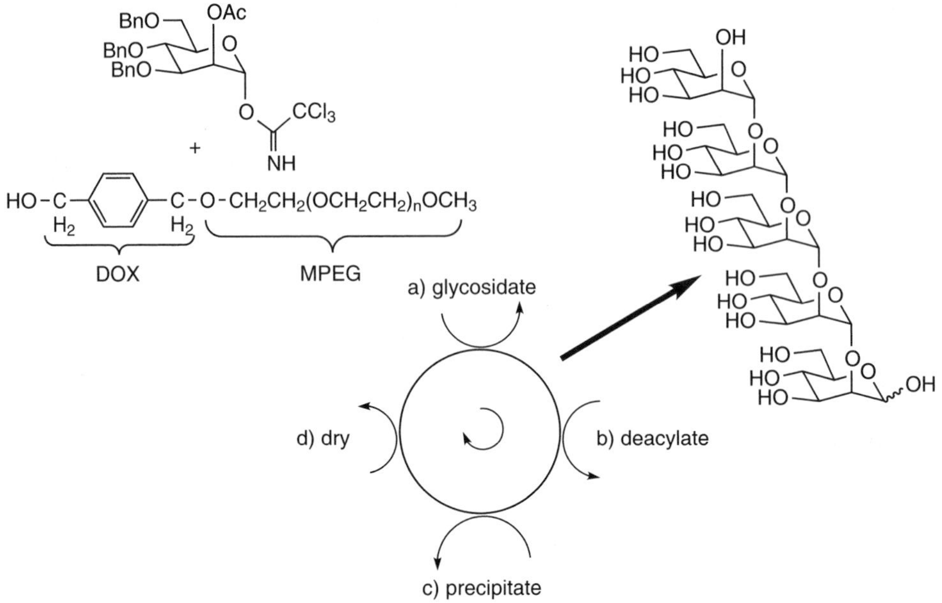

Scheme 3-37.

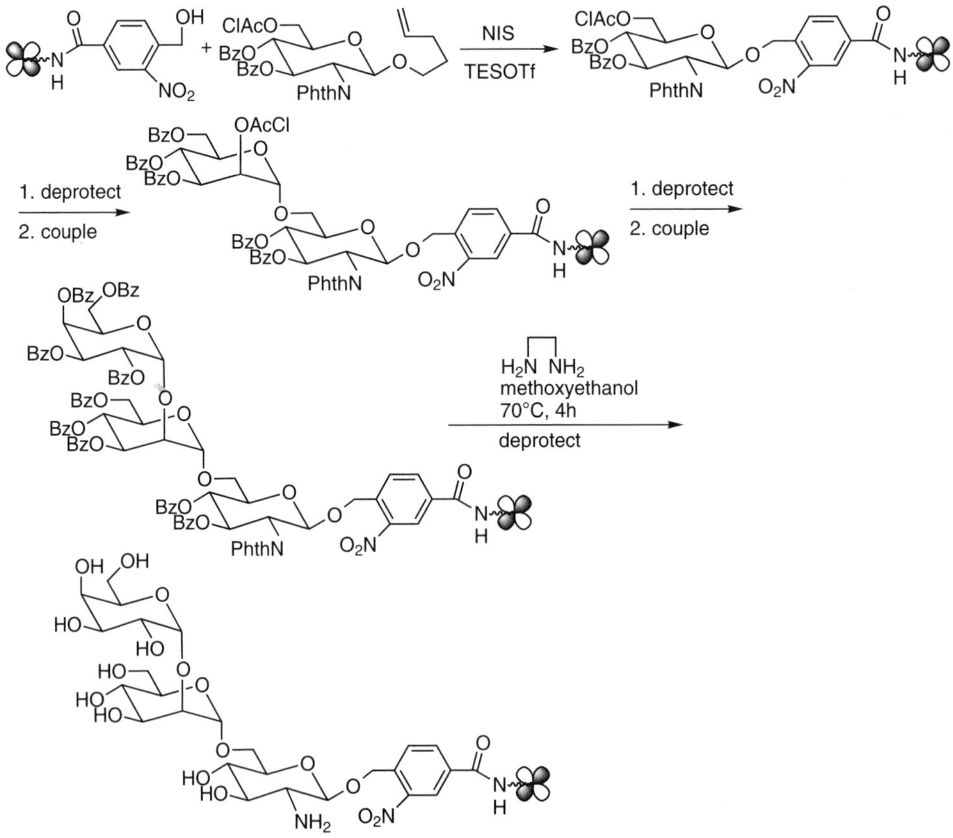

Scheme 3-38.

In spite of the growth of solid phase synthesis of oligosaccharides, there are major problems to be solved, and it will take a major breakthrough to reach the level of efficiency currently enjoyed by oligopeptide and oligonucleotide syntheses. However, the ability to make preparative amounts of oligosaccharides by iterative procedures and also for preparation of combinatorial libraries for biological testing are powerful inducements for improving the present state of the art.

Summary

In 1982, the first comprehensive overview concerning complex oligosaccharide synthesis was published by Paulsen (1982). Anticipating the probable question "Is there a best procedure to follow?", Paulsen wrote "Although we have now learned to synthesize oligosaccharides, it should be emphasized that each oligosaccharide synthesis remains an independent problem—There are no universal reaction conditions for oligosaccharide syntheses."

That statement is much less true today than it was then. Nevertheless, it is impossible to draw comparisons between various donors or accceptors because synthetic chemists do not, indeed cannot, perform the same coupling using a variety of glycosyl partners. As in much of organic synthesis, the skill of the experimentalist in "optimizing" yields, selectivity, et cetera will never be replaced. Such expert "know-how" and prior art, as exemplified in Hasegawa's finely tuned glycosidation of sialic acid (Schemes 3-22e,f) can become adopted as state-of-the-art.

At the conclusion of the 1982 article, Paulsen surveyed "future prospects" by noting that "the area of carbohydrate chemistry will increase in significance in the coming years—[and]—novel synthetic methods should be used to prepare larger oligosaccharide units; thereby, prefabricated synthetic blocks, for example, disaccharide units, presumably will be used in future to a greater extent."

This chapter demonstrates that progress has been as good as, if not greater than, Paulsen had predicted. Indeed the developments in oligosaccharide synthesis are of such quantity and quality that full coverage in a single article may no longer be possible. Since this chapter was written, some excellent overviews have appeared (Boons, 1996a,b; Hanessian,1997; Khan & O'Neil, 1996).

The area of solid phase oligosaccharide synthesis has seen vigorous activity during the last 3 years, and may be viewed as a period of transition. Accordingly, a brief survey of some recent developments has been included within the chapter.

4

Chemical Synthesis of Complex Carbohydrates

K.C. Nicolaou and Nicholas J. Bockovich

The importance of carbohydrates in biological systems is now well established (Sharon & Lis, 1993), with new exciting findings arriving everyday. The study of the chemistry and biology of these molecules is, therefore, of crucial importance to our understanding of living organisms and to biomedical research. From the chemistry point of view, both structural elucidation and synthesis currently play major roles in pushing the field forward. This chapter focuses mainly on chemical synthesis (Barresi & Hindsgaul, 1995; Boons, 1996; Hanessian, 1997; Paulsen, 1982; Schmidt, 1986; Toshima & Tatsuta, 1993) of complex oligosaccharides, while other chapters expand on the equally important and emerging enzymatic approach (Toone et al., 1989; David et al., 1991; Ichikawa et al., 1992a,b) to oligosaccharide synthesis.

As evidenced by the ever increasing complexity of targeted oligosaccharides, the synthetic methods available for glycosidic bond formation are becoming quite powerful and efficient. Despite the great strides seen in this field in recent years, however, even more effective methods are needed. Relatively low yields and stereochemical issues still prevent adoption of solid phase synthetic techniques in the carbohydrate field in the same way that solid phase methods are utilized routinely today in the fields of peptide (Kent, 1988) and oligonucleotide (Gait, 1984) synthesis.

Below, we first summarize the main synthetic methods available for glycosidic bond formation and then proceed to discuss a selected number of complex oligosaccharide total syntheses. We apologize to those researchers whose work was not included here due to space limitations; we hope that the examples cited will assist the reader in appreciating the state of the art in this field.

Chemical Methods for Oligosaccharide Synthesis

Koenigs–Knorr Method

The classical Koenigs–Knorr method of oligosaccharide synthesis dates back to 1901 (Koenigs & Knorr, 1901). This method, along with its modern variations, achieves activation through the use of glycosyl halides (e.g., bromides, chlorides) in the presence of heavy metal salts (generally silver or mercury). Several reviewers have already sum-

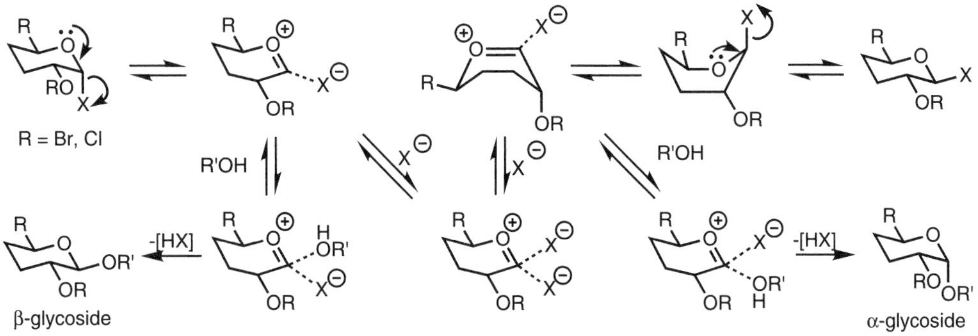

Scheme 4-1. Mechanism of the Koenigs–Knorr glycosidation reaction.

marized applications of this method for complex oligosaccharide synthesis (Paulsen, 1982; Schmidt, 1986; Toshima & Tatsuta, 1993).

The general mechanism of activation in this method involves *in situ* conversion of an α-halide to the more reactive β-halide and subsequent glycosidation, accomplished by inversion of configuration to form an α-glycoside (Lemieux & Haymi, 1965) (Scheme 4-1). The synthesis of β-glycosides typically involves the utilization of 2-*O*-acyl substituents in neighboring group participation (Igarashi, 1977) (Scheme 4-2).

Despite the efficiency of several of the variants of the Koenigs–Knorr method, some severe disadvantages remain. These include (1) the harsh conditions generally required to produce the glycosyl halides; (2) the relative instability of glycosyl halides, and the need to prepare and react them at relatively low temperatures; (3) the high sensitivity of glycosyl halides to aqueous hydrolysis; and (4) the often high costs and toxicity of the heavy metal salts used in the reactions.

New catalysts for this glycosidation reaction are, therefore, constantly sought in order to improve the overall efficiency of the process.

Glycosyl Fluoride Method

The use of glycosyl fluorides in oligosaccharide synthesis was pioneered by the Mukaiyama group. They demonstrated that in the presence of silver perchlorate and tin dichloride, glycosyl fluorides could be activated for glycosidation (Mukaiyama et al., 1981) with alcohols (Scheme 4-3).

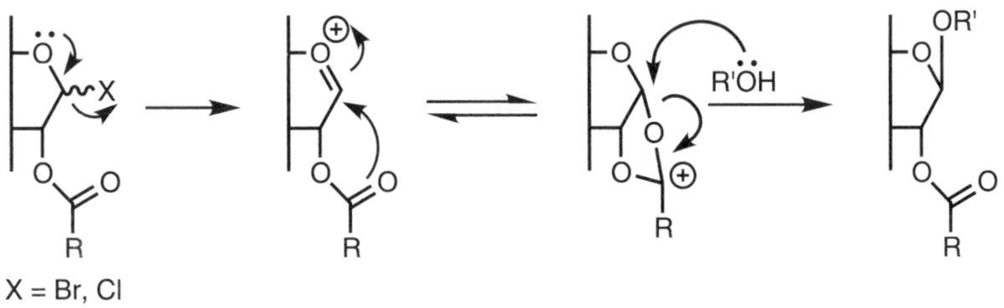

X = Br, Cl

Scheme 4-2. Neighboring group participation in the formation of 1,2-*trans*-glycosides (often β-glycosides).

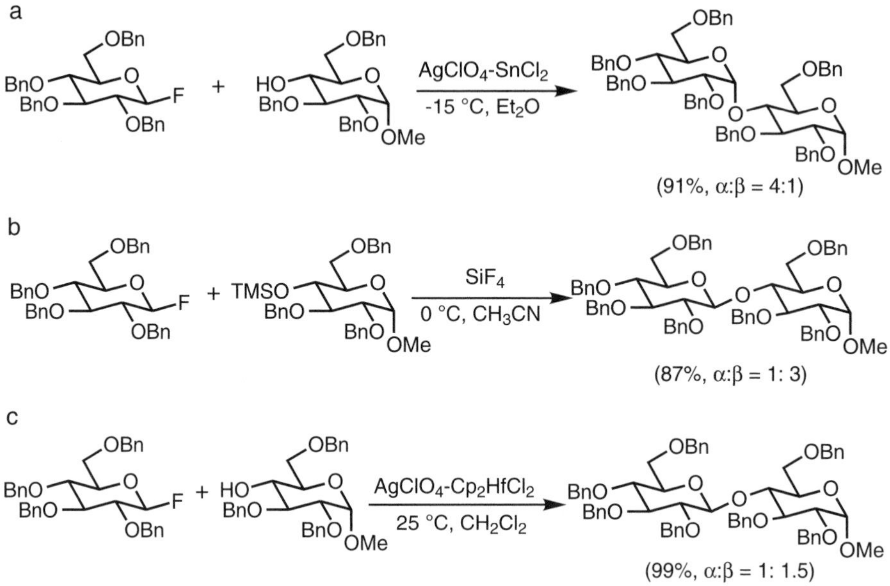

a

$$(91\%, \alpha:\beta = 4:1)$$

b

$$(87\%, \alpha:\beta = 1:3)$$

c

$$(99\%, \alpha:\beta = 1:1.5)$$

Scheme 4-3. The Mukaiyama glycosyl fluoride glycosidation reaction (**a**) (Mukaiyama et al., 1981). Other activating reagents include SiF_4 (**b**) (Hashimoto et al., 1984) and Cp_2HfCl_2-$AgClO_4$ (**c**) (Suzuki et al., 1989). Bn = benzyl, TMS = trimethylsilyl.

The mechanism of activation of the glycosyl fluoride functionality is the same as in the Koenigs–Knorr method (Scheme 4-1), so without a participating group at the 2-position, α-glycosides are formed preferentially. Analogously, one can form β-glycosides through neighboring group participation strategies.

Often, glycosyl fluorides have several advantages over their bromide and chloride counterparts. These advantages include (1) the ease by which glycosyl fluorides can be prepared via a variety of mild and selective conditions (Scheme 4-4); (2) glycosyl

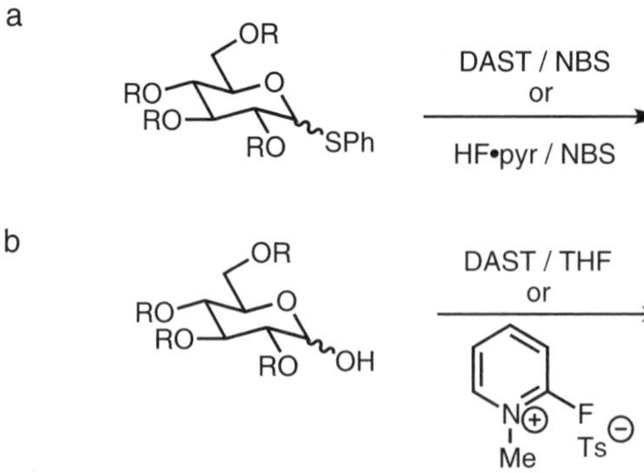

a

b

Scheme 4-4. Methods for glycosyl fluoride formation using thioglycosides (**a**) (Nicolaou et al., 1984) and lactols (**b**) (Mukaiyama et al., 1981; Posner & Haines, 1985; Rosenbrook et al., 1985). DAST = diethylaminosulfur trifluoride, NBS = N-bromosuccinimide.

fluorides are generally more stable than bromides or chlorides, and they can be purified by chromatography on silica gel and stored for extended periods of time without decomposition; and (3) new methods for the coupling of glycosyl fluorides with hydroxy compounds that have been developed recently (Hashimoto et al., 1984; Suzuki et al., 1989).

Glycosyl Trichloroacetimidate Method

Glycosyl trichloroacetimidates (Scheme 4-5) were first developed as glycosyl donors by Schmidt in 1980 (Schmidt & Michel, 1980). Prior to this development, Sinaÿ had utilized glycosyl acetimidates (Scheme 4-5) as donors in what turned out not to be the optimum glycosidation procedure for this activating group (Pougny et al., 1977). Trichloroacetimidates are stable, frequently (but not always) isolable intermediates that can be activated through acid catalysis without the use of heavy metal salts. Boron trifluoride etherate at low temperature is generally the reagent of choice for coupling reactions, although other Lewis or protic acids have been used as well. This method of coupling generally gives products with inversion of configuration at the anomeric center (e.g., α-trichloroacetimidates give β-products and vice versa) (Scheme 4-5). There are exceptions to this rule, however. For example, the presence of 2-O-acyl substituents leads invariably to the product of neighboring group participation (1,2-*trans*) (Amvam-Zollo, 1983). Furthermore, glycosyl donors with an axial substituent at the 2-position (e.g., mannose, rhamnose) generally give α-products owing to the strong anomeric effect in these systems (Fügedi et al., 1982).

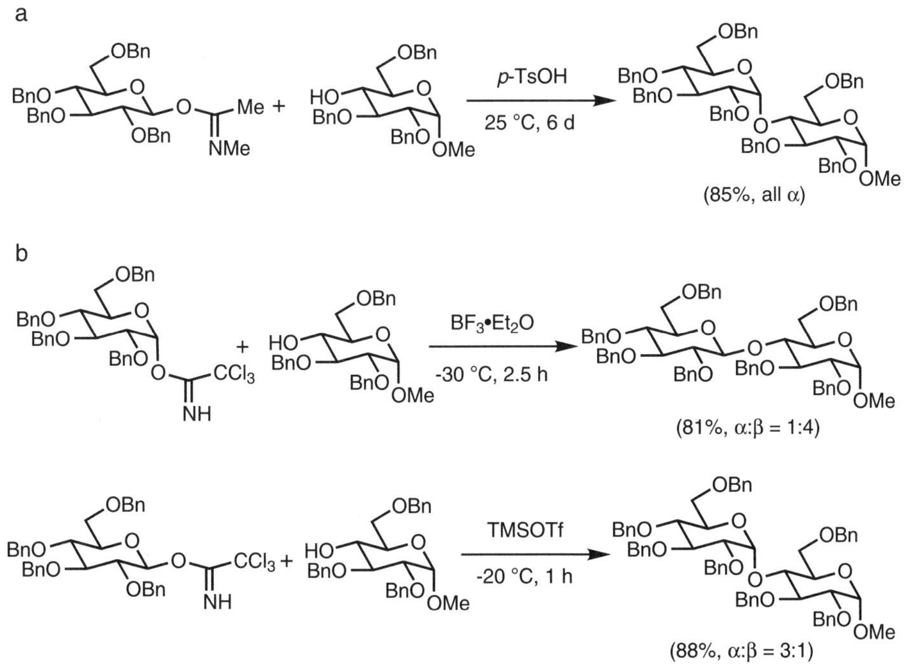

Scheme 4-5. (a) The Sinaÿ imidate coupling method (Pougny et al., 1977). (b) The Schmidt trichloroacetimidate method (Schmidt & Michel, 1985). TMSOTf = trimethylsilyl triflate.

The ability to obtain a glycosidation product whose anomeric configuration depends on the anomeric configuration of the glycosyl donor is an advantage, provided that one is able to synthesize both the α- and β-trichloroacetimidates selectively. This issue has been explored by the Schmidt group (Schmidt & Michel, 1984) who found that under kinetically controlled conditions the β-isomer is produced, whereas thermodynamic control leads preferentially to the α-isomer (Scheme 4-6). If a strong base is used (e.g., NaH), then conversion of the β-isomer to the α-isomer is favored due to reversibility of the reaction, leading to the α-anomer. On the other hand, if a weaker base is used (e.g., K$_2$CO$_3$), then interconversion is much slower and the β-anomer can be isolated as the kinetically favored product (Schmidt & Michel, 1984). The trichloroacetimidate method has several advantages, including (1) control of anomeric configuration of the product by selecting the required donor precursor; and (2) avoidance of heavy metal salts. The method is not without disadvantages, however. Specifically, strong bases, such as NaH, may not always be compatible with complex oligosaccharides.

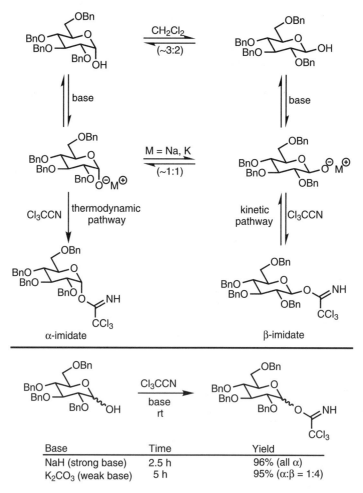

Base	Time	Yield
NaH (strong base)	2.5 h	96% (all α)
K$_2$CO$_3$ (weak base)	5 h	95% (α:β = 1:4)

Scheme 4-6. Selective formation of α- and β-trichloroacetimidates (Schmidt & Michel, 1984).

Furthermore, trichloroacetimidates must be synthesized from free lactols, a condition that requires selective deprotection at the reducing end, thus resulting in a three-step procedure for coupling that can become troublesome in complex situations. However, it is fair to state that, perhaps more than any other method, the trichloroacetimidate glycosidation procedure has enjoyed considerable success in complex glycosidations, as we will see later with specific examples.

Thioglycoside Method

Thioglycosides are highly versatile and useful intermediates in oligosaccharide synthesis. A distinct advantage they have is that their thio group can be used for temporary protection, then employed for almost any coupling procedure currently in use. Thioglycosides react with bromine or chlorine to give glycosyl bromides (Weygand et al., 1958) or chlorides (Wolfrom & Groebke, 1963), respectively, diethylaminosulfur trifluoride (DAST) and N-bromosuccinimide (NBS) to give fluorides (Nicolaou et al., 1985), or water and NBS to produce anomeric lactols (Nicolaou et al., 1983). Furthermore, the thioglycosidic group can be introduced early in the synthesis (Nicolaou et al., 1983; Hanessian & Guindon, 1980) and maintained throughout long sequences, being quite stable toward most protecting groups and other chemical manipulations often encountered in oligosaccharide synthesis.

Extensive usage of thioglycosides as glycosyl donors is a relatively new development (Fügedi et al., 1987) compared to the Koenigs–Knorr reaction. The mechanism of activation in this process is generally similar to that of the Koenigs–Knorr reaction in that a thiophile activates the sulfur first, which is then released to give the corresponding oxonium ion. Glycosidation then follows by trapping this species with a hydroxy compound in the normal way as shown in Scheme 4-7a.

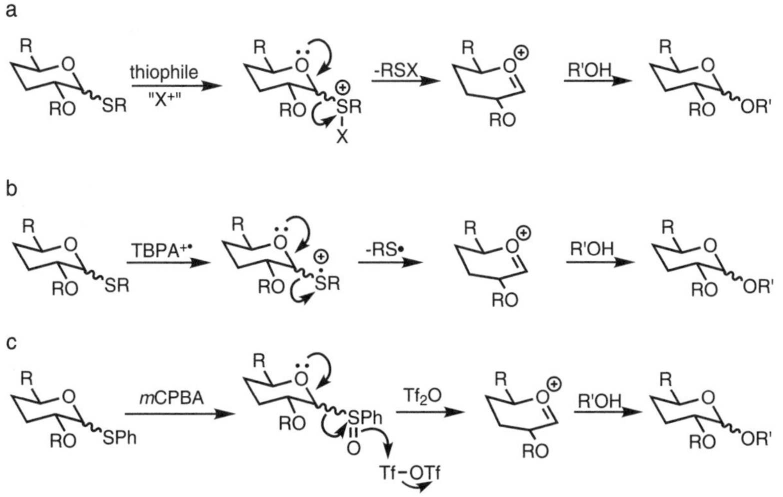

Scheme 4-7. Thioglycoside-based glycosidation reactions. (**a**) thiophile = N-bromosuccinimide (Nicolaou et al., 1983); methyl triflate (Lönn, 1985); dimethyl(methylthio)sulfonium triflate (Fügedi & Garegg, 1986); N-iodosuccinimide-triflic acid (Veeneman et al., 1990). (**b**) TBPA$^{+\cdot}$ = tris-(4-bromophenyl)ammoniumyl hexachloroantimonate (Sinaÿ, 1991). (**c**) mCPBA = m-chloroperbenzoic acid (Kahne et al., 1989).

Thioglycosides were first introduced as glycosyl donors by Ferrier (Ferrier et al., 1973; Ferrier & Haines, 1984) who utilized the ethylthio group. Subsequently, Hanessian (Hanessian et al., 1980) demonstrated the use of the 2-pyridylthio group. Recently, the phenylthio and ethylthio groups have become the favored choices.

The thiophile "X$^+$" can be quite varied. The initial studies employed heavy metal salts such as mercury, copper, or lead salts (Fügedi et al., 1987). These methods, however, suffer from low reactivity and have not been extensively utilized. More recently NBS has been used (Nicolaou et al., 1983), but this method also has disadvantages, such as variable reactivity and somewhat poor selectivity. More effective activating agents introduced recently include methyl triflate (Lönn, 1985), dimethyl(methylthio)sulfonium triflate (DMTST) (Fügedi & Garegg, 1986), N-iodosuccinimide-triflic acid (NIS-TfOH) (Veeneman et al., 1990b), and iodonium di-sym-collidine perchlorate (IDCP) (Veeneman et al., 1990a). Activation by methyl triflate or DMTST involves methylation of the anomeric sulfur, followed by elimination of the sulfur species to give the oxonium ion. The NIS-TfOH activation method involves iodination of the sulfur, followed by replacement with triflic acid to give the highly reactive glycosyl triflate that is then glycosidated.

A novel method of thioglycoside activation has been developed by Sinaÿ (1991). This method involves a single-electron transfer path to achieve activation utilizing tris-(4-bromophenyl)ammoniumyl hexachloroantimonate (TBPA$^{+\cdot}$) as the activating agent. TBPA$^{+\cdot}$ is a radical cation that can accept an electron from sulfur, thereby generating a glycosyl radical cation which cleaves to give a thiyl radical and an oxonium ion as depicted in Scheme 4-7b.

This method has been applied to the synthesis of various disaccharides (Sinaÿ, 1991) and exhibits neighboring group characteristics similar to other glycosidation reactions.

Another coupling method based on thioglycosides has been developed by Kahne (Kahne et al., 1989). This method involves oxidation of the thioglycoside to a sulfoxide and activation with triflic anhydride to yield saccharides as shown in Scheme 4-7c.

Two-Stage Activation Method

The strategy of combining glycosyl fluorides and thioglycosides for the synthesis of complex oligosaccharides was developed by the Nicolaou group in 1984 (Nicolaou et al., 1984). Known as the two-stage activation procedure, this method, depicted in Scheme 4-8, uses the stable phenylthioglycosides as key building blocks. In activation stage I, the phenylthio group of a carbohydrate unit is converted to the fluoride by treatment with DAST and NBS. Activation stage II involves coupling of the glycosyl fluoride with a glycosyl acceptor carrying a phenylthio group to give a disaccharide (Scheme 4-8). This disaccharide can then be either deprotected to give a glycosyl acceptor or treated with DAST and NBS to give a glycosyl fluoride. Further couplings can then be performed to extend the saccharide chain, leading to higher oligosaccharides in a rapid manner. This strategy has the advantages of both the thioglycoside and glycosyl fluoride methods. The phenylthioglycosides are introduced early in the synthesis and are stable to the glycosidation conditions as well as many protecting group manipulations. The glycosyl fluorides are synthesized under mild and selective conditions, and are coupled without affecting other glycosidic linkages present in the mol-

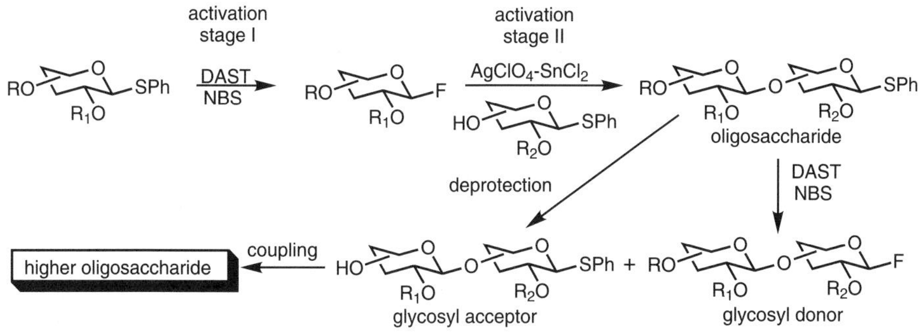

Scheme 4-8. The two-stage activation procedure for glycoside bond formation (Nicolaou et al., 1984).

ecule. This iterative process has been used successfully by our group to synthesize several complex oligosaccharides as will be discussed later.

Glycal Method

Glycals have been used in glycoside synthesis for some time. Originally, they were utilized in the synthesis of 2-deoxyglycosides (Ito & Ogawa, 1987; Janrand et al., 1981; Lemieux & Morgan, 1965; Thiem et al., 1978). These methods exploit the electron-rich olefin of the enol ether present in the glycal by treatment with an electrophilic reagent (E^+) to form a 3-membered "onium" ion species, which is then glycosidated by an alcohol (Scheme 4-9a). The electrophile (E) can then be removed reductively to form the 2-deoxyglycoside. A variety of electrophilic reagents (E^+) have been introduced for glycal-coupling reactions. Thus, Lemieux and Morgan (1965) utilized I(collidine)$_2$ClO$_4$ to generate an iodonium species as the reactive intermediate, whereas Thiem et al. (1978) exploited the same iodonium species by generating it with N-iodosuccinimide (NIS). Sinaÿ (Janrand et al., 1981) has used phenylselenyl chloride (PhSeCl) to generate an episelenonium ion species, which undergoes subsequent gly-

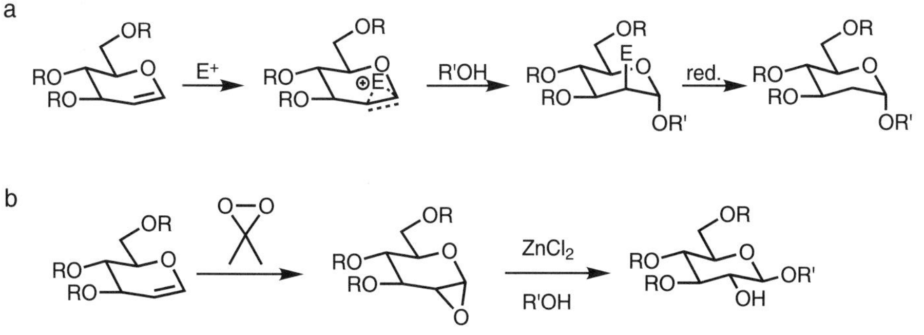

Scheme 4-9. Glycal-based glycosidation reactions. (**a**) $E^+ = $ I(collidine)$_2$ClO$_4$ (Lemieux & Morgan, 1965); N-iodosuccinimide (Thiem et al., 1978); PhSeCl (Janrand et al., 1981); PhSCl (Ito & Ogawa, 1987). (**b**) Danishefsky method (Halcomb & Danishefsky, 1989).

cosidation in the presence of hydroxy compounds. Ito and Ogawa (1987) introduced phenylsulfenyl chloride as a promoting agent for glycosidation through an episulfonium ion. In most cases, α-glycosides are formed in these reactions due to preferential top-face attack by the electrophile (E^+) on the glycal.

Danishefsky has recently developed a different approach for activation of glycals (Halcomb & Danishefsky, 1989). In this instance, the electrophilic reagent (E^+) is dimethyldioxirane that transfers an oxygen atom to the glycal, generating a glycosyl epoxide (Scheme 4-9b). The epoxide can then be activated with a Lewis acid (e.g., $ZnCl_2$) and the derived reactive species trapped with an acceptor to form the glycosidic bond. There are several advantages to Danishefsky's glycal method. They include (1) the relative stability of glycosyl epoxides that often allows their isolation, (2) the use of oxygen as the electrophile, providing entries to both 2-oxo and 2-deoxyglycosides; and (3) the formation of β-glycosides due to preference for α-attack by the electrophile, followed by inversion of configuration at the glycosidation step.

n-Pentenyl Glycoside Method

The use of *n*-pentenyl glycosides in oligosaccharide synthesis was introduced by Fraser-Reid (Fraser-Reid et al., 1988). The development of *n*-pentenyl glycosides resulted from a search for a stable alkyl glycoside that could be used both as a glycosyl donor and a glycosyl acceptor. Activation of these intermediates is achieved by electrophilic addition to the olefin, followed by intramolecular trapping by the anomeric oxygen and eventual expulsion of the side chain to produce an oxonium ion, which is glycosidated by the hydroxy component (Scheme 4-10).

A useful feature of this method is the ability to modulate the reactivity of the glycosyl donor by varying the protecting groups in the molecule. Thus, glycosides with ether-protecting groups at the 2-position are much more reactive donors than those with ester-protecting groups. This observation led to the development of "armed" and "disarmed" *n*-pentenyl glycosides (Mootoo et al., 1988). In this process, an esterified *n*-pentenyl glycoside can be glycosylated by an *O*-alkylated *n*-pentenyl glycoside. The esters in the new product can then be exchanged for ethers, and the *n*-pentenyl glycoside which was originally "disarmed" now becomes "armed."

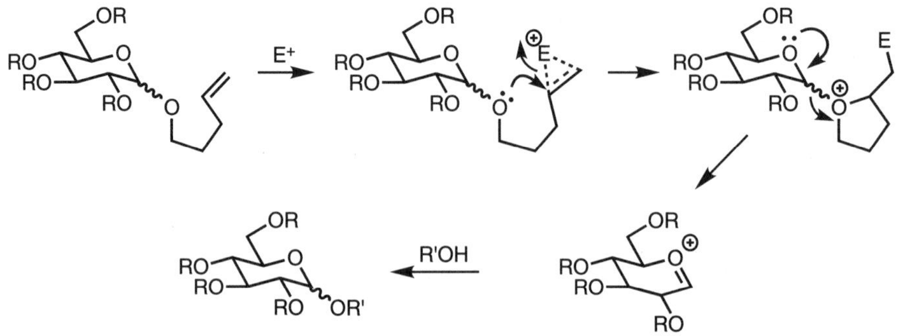

Scheme 4-10. The *n*-pentenyl glycoside-based glycosidation method (Fraser-Reid et al., 1988). E^+ = I(collidine)$_2$ClO$_4$.

a

b

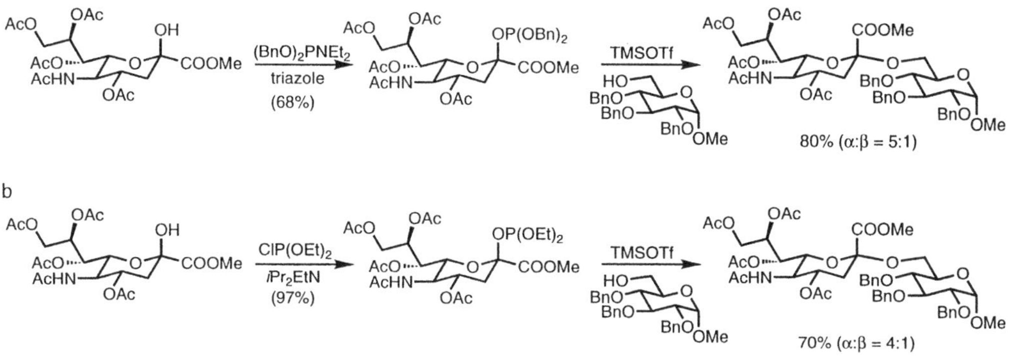

Scheme 4-11. Phosphite-based glycosidation methods. (**a**) C.-H. Wong (Kondo et al., 1992; Sim et al., 1993). (**b**) Schmidt (Martin & Schmidt, 1992). Ac = acetyl, Bn = benzyl.

Glycosyl Phosphite Method

Glycosyl phosphites are generally encountered as intermediates in the synthesis of glycosyl phosphates. The glycosyl-donating abilities of phosphites have only recently been explored (Kondo et al., 1992; Martin & Schmidt, 1992; Sim et al., 1993). These intermediates can be synthesized from free anomeric lactols and a phosphitylating agent, such as diethyl chlorophosphite (Martin & Schmidt, 1992) or dibenzyl N,N-diethylphosphoramidite (Kondo et al., 1992; Sim et al., 1993).

Activation of glycosyl phosphites can be achieved by treatment with catalytic amounts of trimethylsilyl triflate (TMSOTf). In the presence of a glycosyl acceptor, this activation method leads to glycosidation (Scheme 4-11). While this procedure appears to be quite general, the most interesting examples so far reported are those using sialic acid phosphites as sialylating agents (Kondo et al., 1992; Martin & Schmidt, 1992; Sim et al., 1993).

The Enzymatic Approach

Obviously, nature uses enzymes to construct complex carbohydrates. The use of such enzymes (glycosyl transferases) in the laboratory has resulted in a spate of recent successes; these have, in certain cases, proven advantageous over chemical synthesis. Advantages include (1) the high substrate specificity of glycosyl transferases, which allows for high chemoselectivity in the glycosidation reaction, (2) the highly stereoselective glycosidation step that provides either the α or the β-anomer exclusively and, (3) no required complex schemes for protection and deprotection.

The use of glycosyl transferases is not without drawbacks, however. For example, glycosyl transferases are not always readily available. Furthermore, the required glycosyl donors are usually sugar nucleotides which tend to be rather expensive and, therefore, prohibitively impractical as stoichiometric reagents. Also, these enzymatic reactions are sometimes inhibited by nucleoside phosphates, which are by-products of the reaction. Wong and his collaborators have recently developed a system which can obviate some of these problems (Ichikawa et al., 1991). In this new process, sugar nucleotides are prepared *in situ* and then coupled, allowing the use of inexpensive sugars as starting materials. Furthermore, the problem of product inhibition was addressed by enzymatic conversion of the nucleoside diphosphates to nucleoside triphosphates which are necessary for the generation of the sugar nucleotides. Examples of complex oligosaccharide construction by this enzymatic method (Ichikawa et al., 1992b) include members of the highly publicized sialyl Lewis x (Le[x]) family of compounds.

Solid Phase Oligosaccharide Synthesis

Given the successes of solid phase synthesis in the areas of polypeptides (Kent, 1988) and polynucleotides (Gait, 1984), much effort has been expended for the development of similar technology for oligosaccharide synthesis. There are, however, several inherent problems that have to be dealt with in order to arrive at a practical solution in this field. The following points and comparisons amplify the main issues facing solid phase oligosaccharide synthesis; namely (1) several nearly equivalent acceptor hydroxyl groups are present in most glycosyl acceptors, as opposed to the easily differentiated groups in amino acids and peptides, and 2-deoxynucleosides and nucleotides; (2) the stereochemistry of the glycosidic bond (α- and β-anomers) must be controlled, whereas there is no stereochemical issue in the amide and phosphate bonds; and (3) the coupling technology for oligosaccharide construction often suffers from modest yields, whereas coupling technology for nucleotides and peptides is generally high yielding and reliable (often approaching 100% yields!).

There are two conceptually different paths that can be exploited for solid phase synthesis (Scheme 4-12). The first involves attaching the glycosyl acceptor to the polymer support followed by addition of an excess of glycosyl donor and a promoter. Initial attempts to achieve solid phase synthesis of oligosaccharides followed this approach (Frechet & Schuerch, 1972; Veeneman et al., 1987). A more recent application of this strategy was employed by Krepinsky and his group (Scheme 4-13) (Douglas et al., 1991). According to this method, the glycosyl acceptor is attached to a poly(ethylene glycol) monomethyl ether support via a succinate linker, and glycosylated by repeated additions of glycosyl donor. Several disaccharides were prepared by this method (Douglas et al., 1991). The second approach for a solid phase oligosaccharide synthesis involves attachment of the glycosyl donor to the polymer support and then adding an excess of glycosyl acceptor to achieve glycosidation. Such an approach is utilized

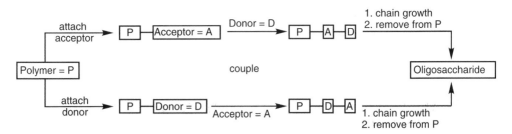

Scheme 4-12. Alternative strategies for solid phase oligosaccharide synthesis.

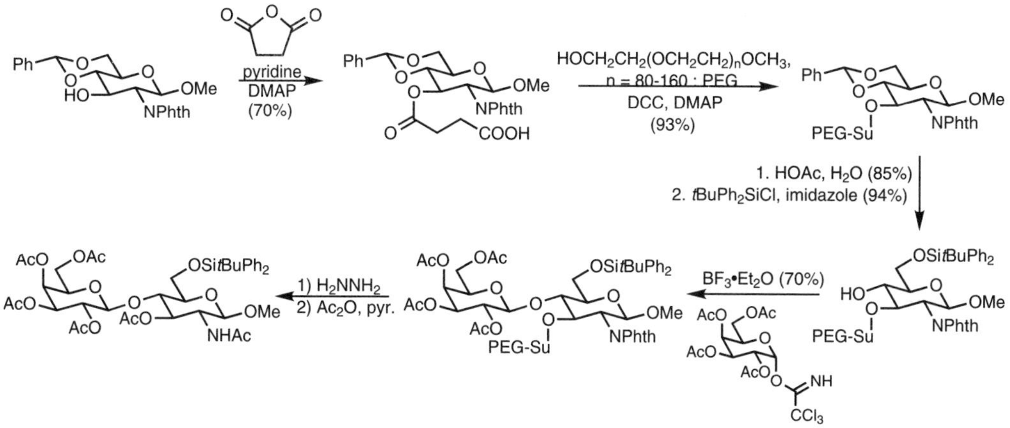

Scheme 4-13. The Krepinsky method of solid phase oligosaccharide synthesis (Douglas et al., 1991). Ac = acetyl, Bn = benzyl, Phth = phthaloyl, Su = succinoyl, DCC = dicyclohexylcarbodiimide, DMAP = 4-dimethylaminopyridine.

in the strategy recently reported by Danishefsky and his group, who developed a solid phase version of their glycal-based methodology for oligosaccharide synthesis (Scheme 4-14) (Danishefsky et al., 1993). In this method, a glycal intermediate is attached to a polymer support followed by epoxidation with 3,3-dimethyldioxirane. The glycosyl epoxide so obtained is then treated with $ZnCl_2$ and a glycal acceptor to achieve activation and coupling. The polymer is then washed to remove excess acceptor and any hydrolyze unreacted epoxide followed by reiteration to extend the chain to the desired length. The process can be terminated by adding a conventional sugar rather than a glycal as acceptor. Oligosaccharides containing as many as four carbohydrate units have been synthesized by this method (Danishefsky et al., 1993).

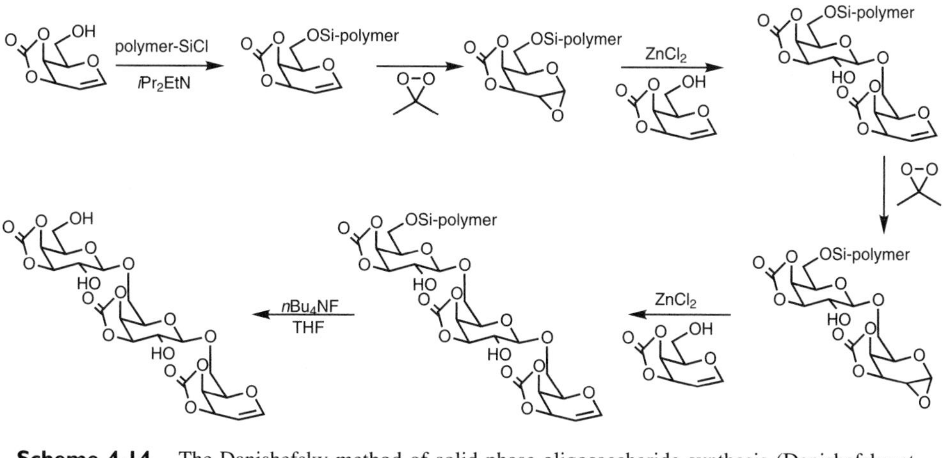

Scheme 4-14. The Danishefsky method of solid phase oligosaccharide synthesis (Danishefsky et al., 1993).

Total Synthesis of Complex Oligosaccharides

In the following sections the total syntheses of a selected number of oligosaccharides are described. These syntheses exemplify the power of several methods for the construction of glycosidic bonds, and highlight a number of different strategies for building such molecular assemblies. Space limitations did not allow, unfortunately, the inclusion of the many noteworthy examples in this highly active field.

We begin with the elegant work of Lemieux and co-workers in the mid-1970s which culminated in the first syntheses of naturally occurring oligosaccharides utilizing the halide ion-catalyzed approach to glycosidic bond formation (Lemieux & Driguez, 1975a,b). This strategy involved the *in situ* anomerization pathway, where an α-glycosyl halide is initially converted to its more reactive β-anomer and then glycosidated.

Lewis a Blood Group Antigenic Determinants

The Lewis a (Le^a) blood group determinant (**1**, Scheme 4-15) is a trisaccharide consisting of *N*-acetylglucosamine at the reducing end carrying a β-linked-D-galactose at the 3-position and an α-linked-L-fucose at the 4-position. The retrosynthesis of this target corresponding to the Lemieux synthesis is shown in Scheme 4-15. Thus, protections and disconnections as indicated lead to key intermediates **3**, **4**, and **5** as starting materials. The utilization of a benzyl and an acetate group at the 2-position of carbohydrate units **3** and **4**, respectively, ensure the desired stereochemical outcome of the two glycosidic bond forming reactions (*vide infra*).

Scheme 4-16 details the synthesis of **1**. Coupling of benzylidene derivative **5** with galactosyl bromide **4** in the presence of Hg(CN)$_2$ led to disaccharide **6** in 91% yield. Removal of the benzylidene group followed by selective acetylation of the primary hydroxyl group with acetylimidazole furnished derivative **2** in 76% overall yield. Attachment of fucose using bromide **3** was achieved in the presence of Et$_4$NBr and Hunig's base to afford trisaccharide **7** (83% yield). Deacetylation (Et$_3$N, MeOH, H$_2$O) followed by reductive elimination (Zn, AcOH) and hydrogenolysis (H$_2$, Pd on carbon) yielded the Le^a determinant **1** in 56% overall yield.

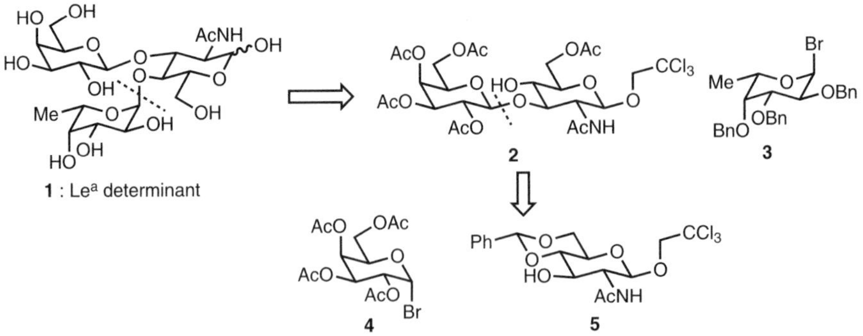

Scheme 4-15. Retrosynthetic analysis of Le^a determinant (**1**) (Lemieux & Driguez, 1975a). Ac = acetyl, Bn = benzyl.

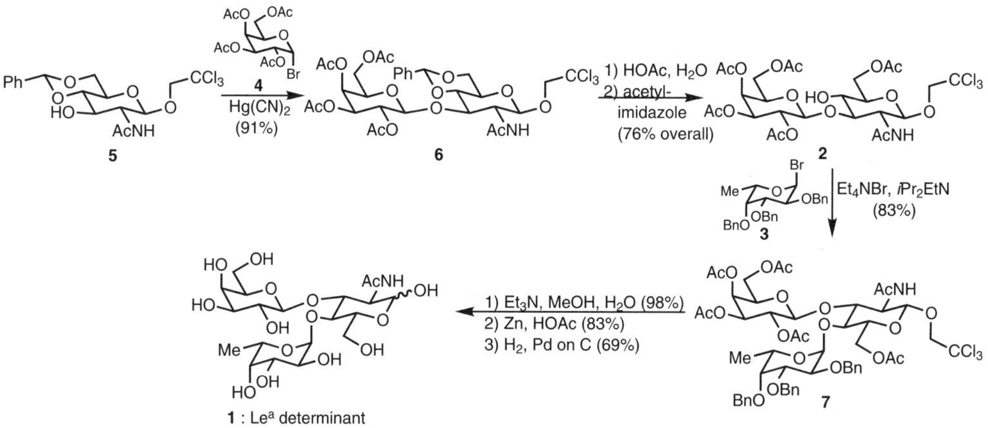

Scheme 4-16. Total synthesis of Le^a determinant (**1**) (Lemieux & Driguez, 1975a).

Blood Group B Antigenic Determinant

The blood group B determinant **8** (Scheme 4-17) is a trisaccharide consisting of galactose at the reducing end with an α-L-fucose at the 2-position and an α-D-galactose at the 3-position. Retrosynthetic analysis of **8** (Scheme 4-17) involves protection and glycosidic bond disconnections as indicated leading to the three carbohydrate units **3**, **10**, and **11**.

The Lemieux synthesis of blood group B determinant **8** (Scheme 4-18) began with isopropylidenation of 2,2,2-trichloroethylglycoside **12** with dimethoxypropane under acidic conditions (79%). Selective acetylation at the primary position using acetylimidazole afforded alcohol **11** in 73% yield. Coupling of the free hydroxy group in **11** with fucosyl bromide **3** in the presence of Et$_4$NBr yielded disaccharide **13** in 66% yield via a glycosidation reaction that involves the *in situ* anomerization of the C—Br bond.

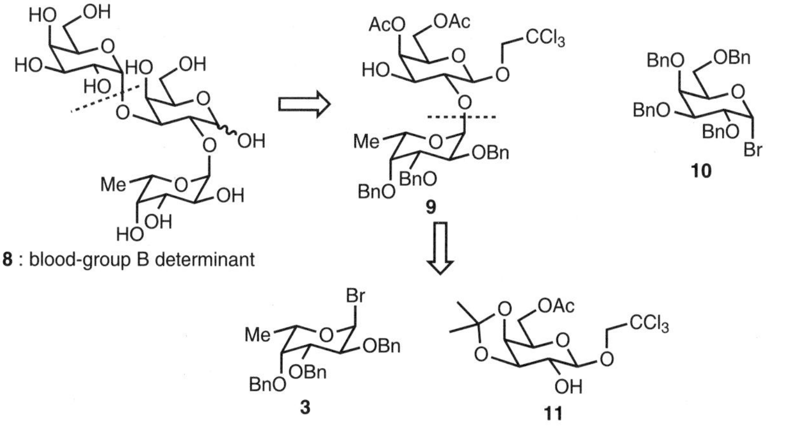

Scheme 4-17. Retrosynthetic analysis of blood group B determinant (**8**) (Lemieux & Driguez, 1975b). Ac = acetyl, Bn = benzyl.

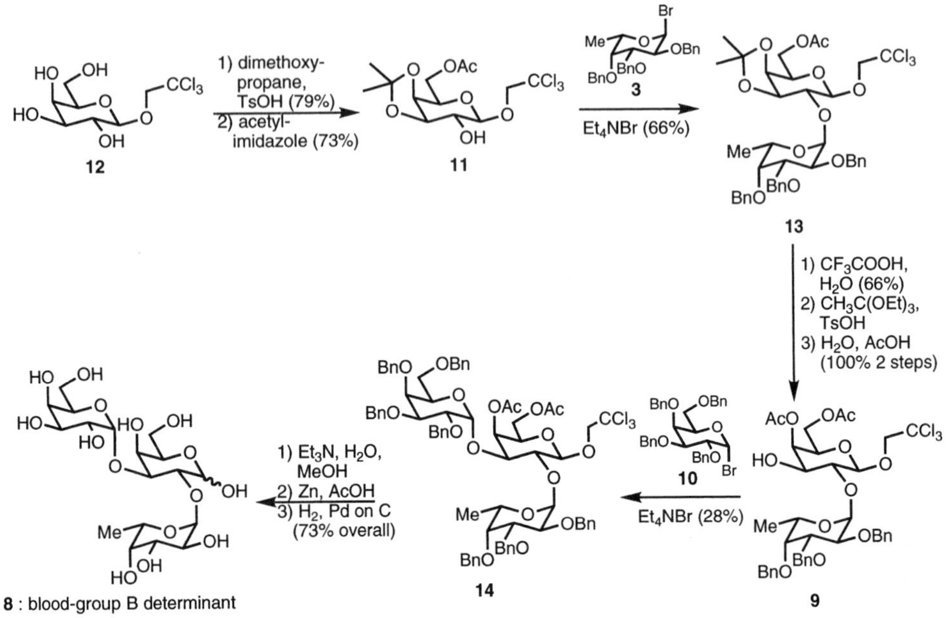

Scheme 4-18. Total synthesis of blood group B determinant (**8**) (Lemieux & Driguez, 1975b).

Removal of the acetonide group (CF$_3$COOH, H$_2$O, 66%) followed by selective axial acetylation [MeC(OEt)$_3$, H$^+$, AcOH, 100%] yielded alcohol **9**. Attachment of galactosyl bromide **10** to alcohol **9** proceeded smoothly in the presence of Et$_4$NBr, furnishing trisaccharide **14** (28%). Finally, deacetylation (Et$_3$N, MeOH, H$_2$O) followed by reductive elimination (Zn, AcOH) and hydrogenolysis (H$_2$, Pd—C) gave the target compound **8** in 73% overall yield.

Pectinioside E Hexasaccharide

Pectinioside E is a steroidal glycoside isolated from starfish. It is a member of the family of steroidal glycosides known as the asterosaponins. These compounds are part of the saponin family of molecules that are responsible for the toxicity of sea cucumbers and starfish. An elegant total synthesis of the oligosaccharide segment of pectinioside E was recently reported by Schmidt and co-workers (Jiang & Schmidt, 1992). This synthesis involves the use of the trichloroacetimidate technology developed earlier by the same group for glycosidic bond formation.

The retrosynthesis of this target molecule (Scheme 4-19) involves initial removal of the 6-deoxyglucose (quinovose) residue to yield α-imidate **16** and pentasaccharide **17** as potential precursors for the construction. Further disconnection of one of the β-glucosidic junctions in **17** affords α-imidate **18** and disaccharide **19**. The disaccharide fragment is then disconnected to two quinovose units, α-imidate **20** and alcohol **21**, as shown. The trisaccharide imidate (**18**) is disconnected to give D-fucose imidate **22**, xylose imidate **23**, and glucose diol **24**.

The synthesis of the hexasaccharide **15** begins with the TMSOTf-induced coupling of xylose imidate **23** and glucose diol **24** to yield disaccharide **25** as the major product (51%) (Scheme 4-20). A minor product (19%) resulting from coupling at the 4-position of **24** is also observed in this reaction. The remaining hydroxyl group in **25** is then similarly coupled with D-fucose imidate **22** to afford trisaccharide **26** in 95% yield. The silyl glycoside is then converted into the trichloroacetimidate **18** by treatment with *n*-Bu₄NF in THF (95%) followed by Cl₃CCN and 1,8-diazabicyclo-

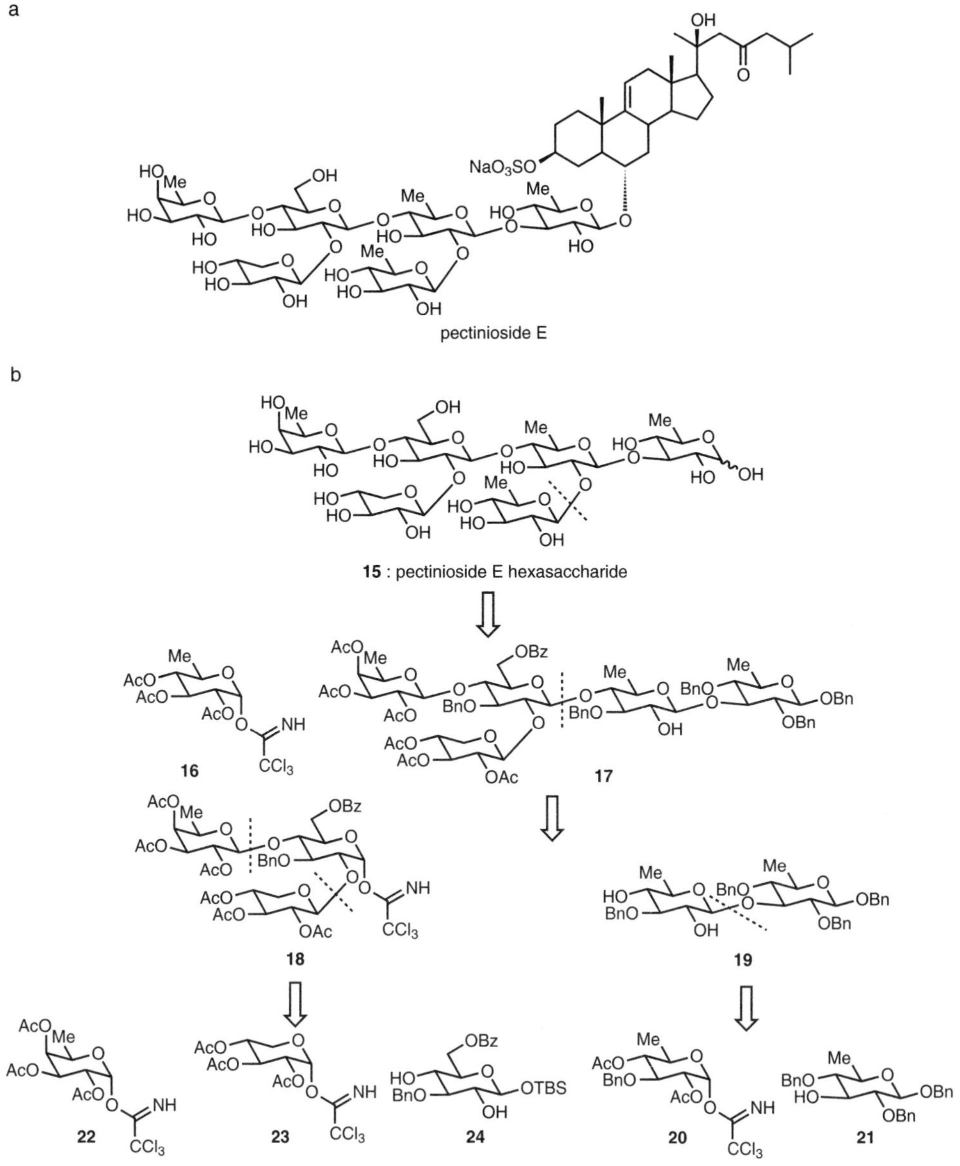

a

pectinioside E

b

15 : pectinioside E hexasaccharide

Scheme 4-19. (a) Structure of pectinioside E and (b) retrosynthetic analysis of pectinioside E hexasaccharide (**15**) (Jiang & Schmidt, 1992). Ac = acetyl, Bz = benzoyl, Bn = benzyl, TBS = *t*-butyldimethylsilyl.

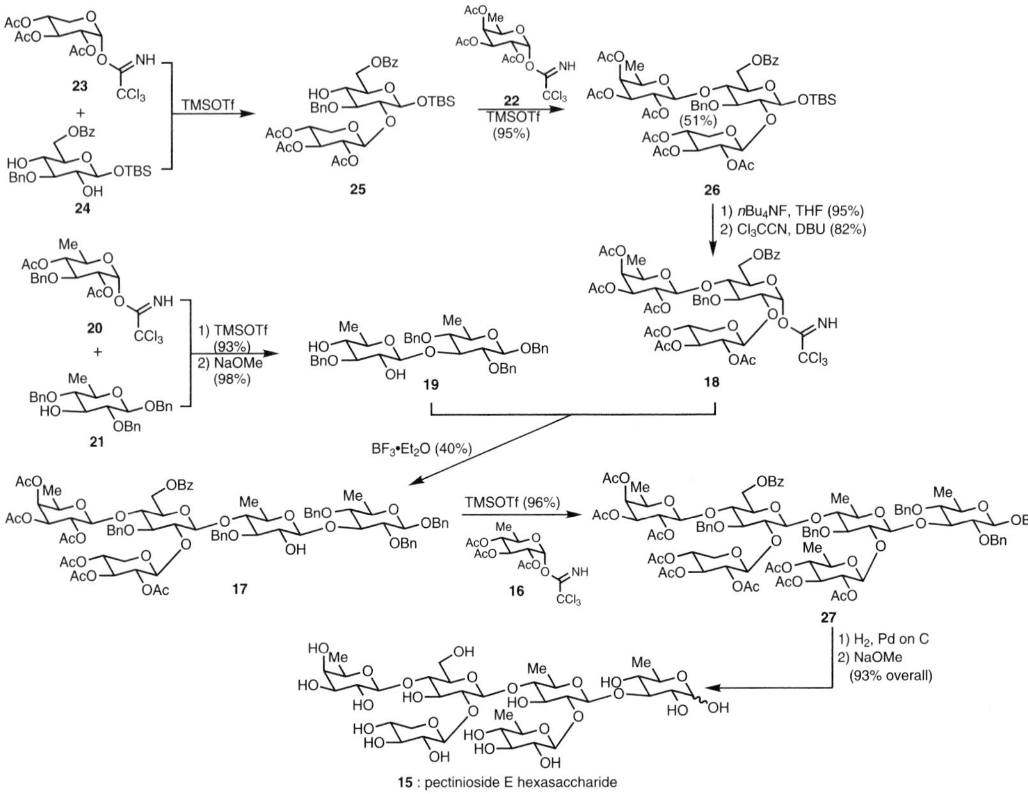

Scheme 4-20. Total synthesis of pectinioside E hexasaccharide (**15**) (Jiang & Schmidt, 1992). DBU = 1,8-diazabicyclo[5.4.0]undec-7-ene.

[5.4.0]undec-7-ene (DBU) (82%). The disaccharide acceptor is synthesized by first coupling quinovose imidate **20** and quinovose alcohol **21** using TMSOTf as catalyst to furnish the corresponding disaccharide (93%), followed by cleavage of the acetates with NaOMe (98%) to furnish diol **19**. Coupling of the two fragments **18** and **19** (BF$_3$·Et$_2$O, −10°C) proceeded exclusively at the 4-position to afford pentasaccharide **17** in 40% yield. The α-coupled isomer was obtained as a minor product (11% yield) in this reaction. The remaining hydroxyl group was then coupled with quinovose imidate **16** using TMSOTf as a catalyst to yield hexasaccharide **27** in 96% yield. Finally, deprotection by hydrogenolysis and treatment with NaOMe-furnished pure pectinioside E hexasaccharide **15** in 93% overall yield.

Bisected Glycoprotein Core Sequence

N-Linked glycoproteins are a large family of bioconjugates whose carbohydrate regions tend to have similar structural characteristics. These carbohydrates generally have a "core" structure which consists of a mannose unit, β-linked to one or more *N*-acetyl-glucosamine units. The 3- and 6-positions of this β-mannose unit are substituted with α-mannose units. These α-mannose residues can be substituted in one or more places by *N*-acetyl lactosamine units. "Bisected" structures, as these compounds are called, also have an *N*-acetylglucosamine unit at the 4-position of the β-mannose core.

An impressive synthesis of a nonasaccharide sequence of a "bisected" structure has been reported by Paulsen and co-workers (Paulsen et al., 1990). Retrosynthesis of this target molecule **28** (Scheme 4-21) starts with the disconnection of the two α-mannosidic linkages to yield a trisaccharide donor (**29**, which is used twice) and a trisaccharide acceptor (**30**). Further disconnection of the donor yields lactosaminyl bromide **31** and mannose derivative **32**. Disconnection of the acceptor yields glucosaminyl bromide **33**, mannosyl bromide **34**, and 1,6-anhydroglucosazide derivative **35**.

The synthesis of the trisaccharide acceptor (**30**, Scheme 4-22) began with the coupling of mannosyl bromide **34** and 1,6-anhydroglucosazide derivative **35**, which pro-

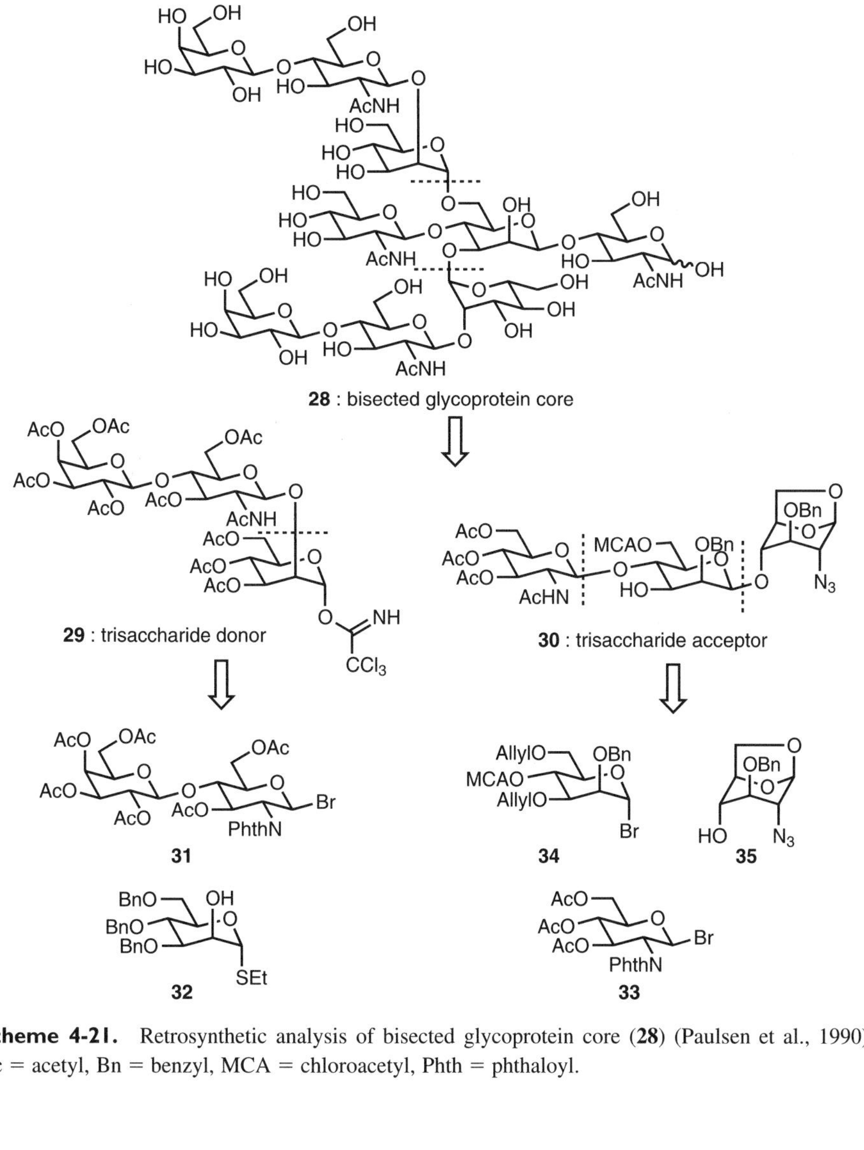

Scheme 4-21. Retrosynthetic analysis of bisected glycoprotein core (**28**) (Paulsen et al., 1990). Ac = acetyl, Bn = benzyl, MCA = chloroacetyl, Phth = phthaloyl.

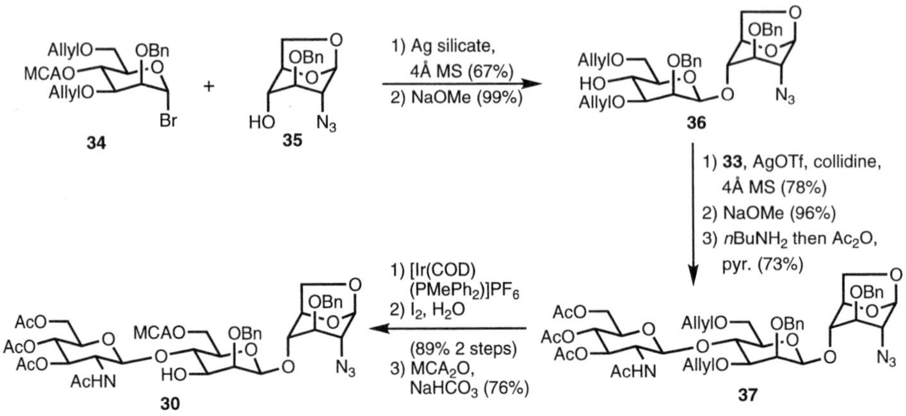

Scheme 4-22. Construction of trisaccharide acceptor (**30**) (Paulsen et al., 1990).

ceeds under the influence of insoluble silver silicate to yield the challenging β-mannosidic linkage in 67% yield. The chloroacetate was then removed with NaOMe (99%) and the resulting alcohol **36** was coupled with bromide **33** (AgOTf, 78%) to give the corresponding trisaccharide. The acetate groups were then removed (NaOMe, 96%) and the phthalimide was converted to an N-acetyl group by treatment with n-butylamine followed by acetylation (Ac$_2$O, pyridine, 73% overall) to yield trisaccharide **37**. The allyl groups were then removed in a two-step procedure, involving isomerization to the enol ether using an iridium catalyst, followed by hydrolysis with iodine and water (89% overall). The diol was then selectively chloroacetylated at the primary position with chloroacetic anhydride to yield the trisaccharide acceptor **30** in 76% yield.

The synthesis of the trisaccharide donor (**29**, Scheme 4-23) involved coupling of lactosaminyl bromide **31** with mannose alcohol **32** (AgOTf, 81%) to yield compound **38**. The phthalimide moiety in **38** was then converted to an N-acetyl group by sequential treatment with NaOMe, hydrazine hydrate, and acetic anhydride in pyridine to yield trisaccharide **39** (89% overall). The benzyl ethers were replaced by acetates (H$_2$, Pd on carbon, followed by Ac$_2$O in pyridine, 82%) and the thioglycoside was then converted to the trichloroacetimidate system **29** by a three-step procedure involving (1) conversion to the corresponding bromide with Br$_2$ in CH$_2$Cl$_2$ (99%); (2) hydrolysis of the bromide to the lactol (90%); and (3) conversion of the lactol to trichloroacetimidate by treatment with Cl$_3$CCN and DBU (87%) to provide the requisite trisaccharide donor **29**.

The completion of the synthesis was accomplished as outlined in Scheme 4-24. Thus, BF$_3$·Et$_2$O-promoted coupling of the trisaccharide donor **29** with trisaccharide acceptor **30**, produced hexasaccharide **40** in 69% yield. The chloroacetate group was removed from **40** using hydrazine dithiocarbonate (54%), and the resulting alcohol was coupled again with the trisaccharide donor **29** (BF$_3$·Et$_2$O) to yield nonasaccharide **41** (58%). Removal of the benzyl ether groups in **41** (H$_2$, Pd on carbon) was accompanied by concomitant azide reduction to give the corresponding amino diol, which was fully acetylated with acetic anhydride (62% overall yield). The 1,6-anhydro linkage was then cleaved with trifluoroacetic acid in acetic anhydride (61%) to yield the cor-

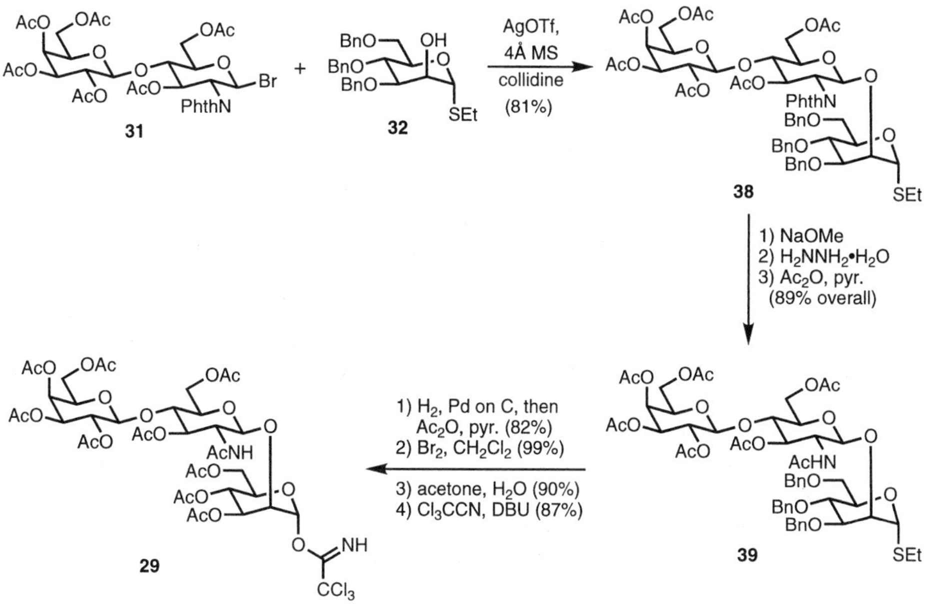

Scheme 4-23. Construction of trisaccharide donor (**29**) (Paulsen et al., 1990).

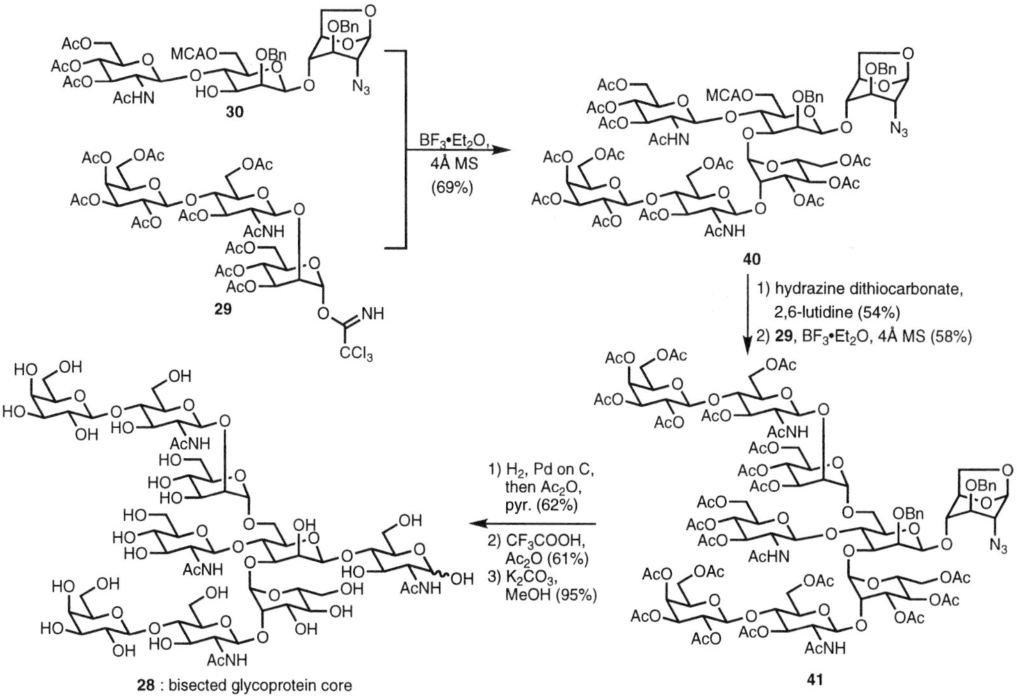

Scheme 4-24. Total synthesis of bisected blycoprotein core (**28**) (Paulsen et al., 1990).

responding peracetylated nonasaccharide. Finally, complete deacetylation of the hydroxyl groups by treatment with K_2CO_3 in methanol yielded the targeted nonasaccharide **28** in 95% yield.

Trimeric Le^x

The Le^x-type glycosphingolipids have been recognized as tumor cell markers due to their presence in higher concentrations on the surface of tumor cells relative to their concentrations in normal cells. Their structure consists of a ceramide segment (sphingosine acylated at nitrogen with a fatty acid) and a carbohydrate component which includes a lactose group and a number of Le^x moieties. The most complex member of the group is trimeric Le^x (**42**) (Scheme 4-25). The first total synthesis of this molecule (Nicolaou et al., 1990a), involving the two-stage activation procedure, is summarized below.

Retrosynthetic analysis (Scheme 4-25) of trimeric Le^x (**42**) involves initial disconnection of the molecule at the indicated positions revealing three units of Le^x trisaccharide (**43**) and lactosyl sphingosine acceptor (**44**) as potential precursors. Further disconnections of the Le^x fluoride lead to carbohydrate units (**45–47**), whereas dismantling of the lactosyl sphingosine acceptor (**44**) reveals lactosyl fluoride **48** and sphingosine precursor **49** as potential starting points for this construction.

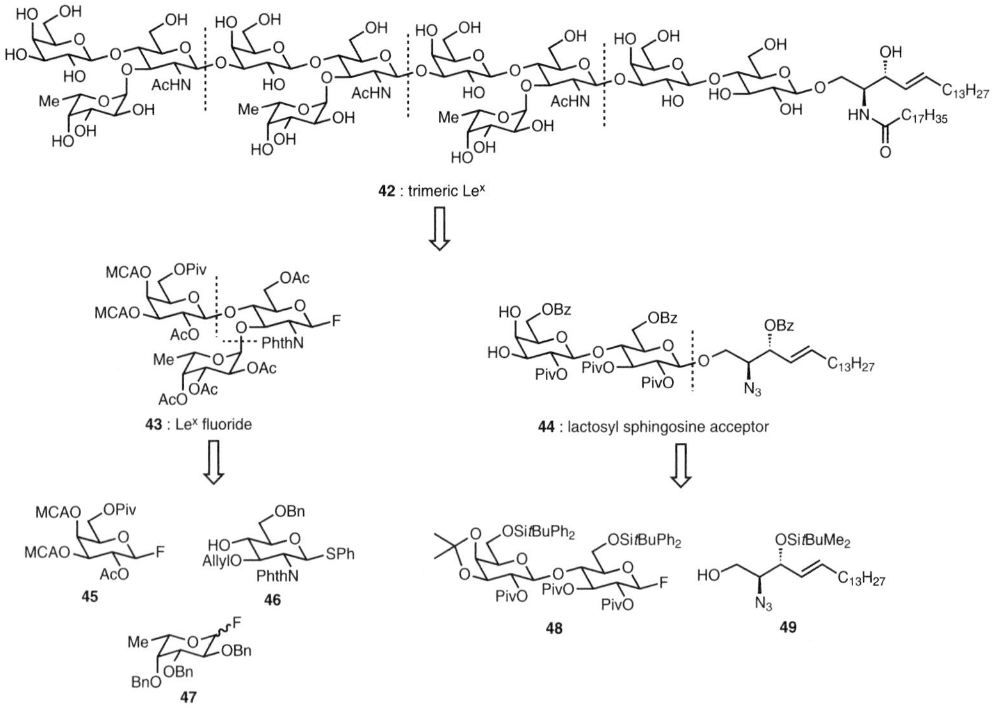

Scheme 4-25. Retrosynthetic analysis of trimeric Le^x (**42**) (Nicolaou et al., 1990a). Ac = acetyl, Bz = benzoyl, Bn = benzyl, MCA = chloroacetyl, Phth = phthaloyl, Piv = pivaloyl.

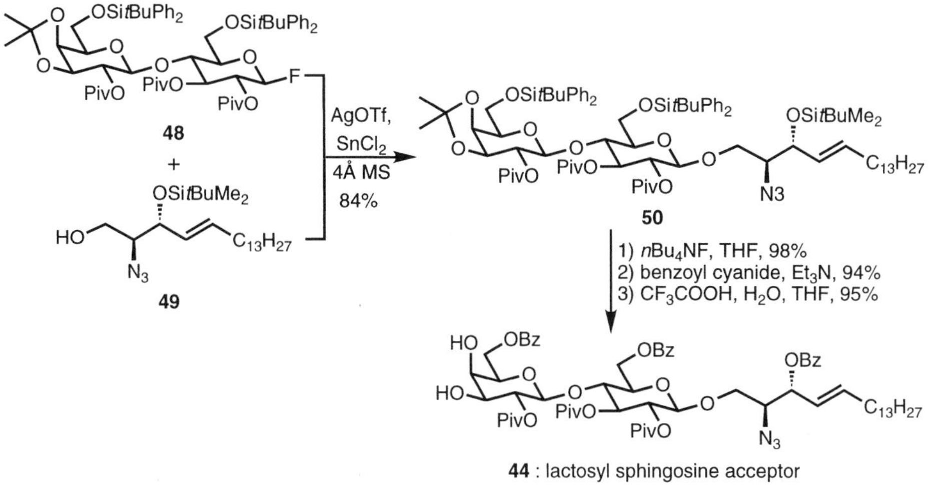

Scheme 4-26. Construction of lactosyl sphingosine acceptor (**44**) (Nicolaou et al., 1990a).

Scheme 4-26 details the synthesis of the lactosyl sphingosine acceptor (**44**). Thus, coupling of lactosyl fluoride **48** and sphingosine precursor **49** (AgOTf, SnCl$_2$, 84%) yields compound **50**. A pivaloate ester is used to direct the glycosidation and to avoid orthoester formation (which prevails when the acetate group is used). Compound **50** was then converted to acceptor **44** via desilylation with n-Bu$_4$NF, (98%), followed by benzoylation (PhCOCN, 94%) and acetonide removal (CF$_3$COOH, 95%).

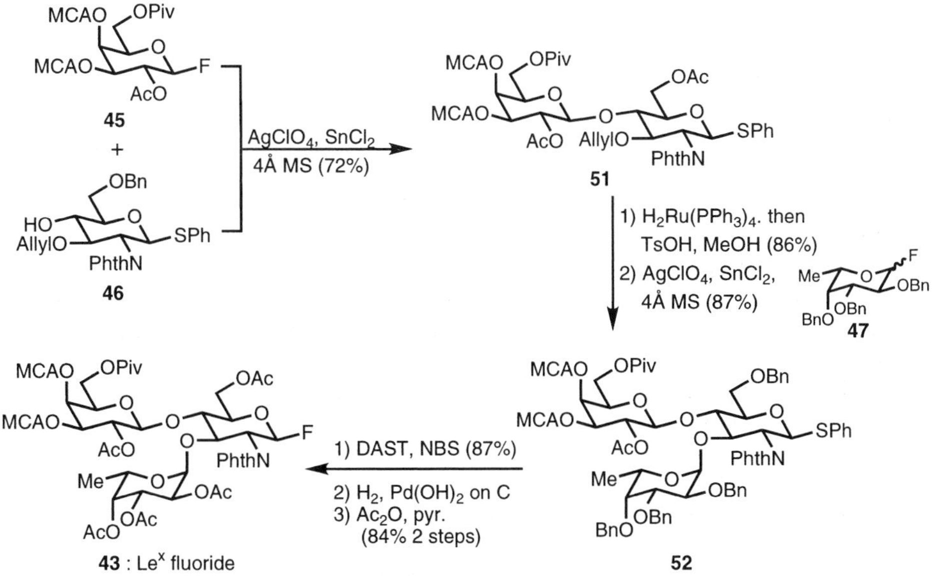

Scheme 4-27. Construction of the Le^x fluoride (**43**) (Nicolaou et al., 1990a).

The synthesis of the Le^x fluoride (**43**, Scheme 4-27) began with the coupling of galactosyl fluoride **45** with glucosamine acceptor **46** (AgClO$_4$, SnCl$_2$, 72%) to yield disaccharide **51**. Removal of the allyl group (H$_2$Ru(PPh$_3$)$_4$, then TsOH, 86%) followed by coupling with fucosyl fluoride **47** (AgClO$_4$, SnCl$_2$, 87%) gave Le^x derivative **52**. A three-step process then converted **52** to the requisite Le^x donor **43**. First, the phenylthio group was converted to a fluoride by treatment with NBS-DAST (87%), second, the benzyl ethers were cleaved by hydrogenolysis [Pd(OH)$_2$-H$_2$], and finally, the liberated hydroxyl groups were acetylated (Ac$_2$O-pyr, 84% overall).

Scheme 4-28 summarizes the completion of the total synthesis of trimeric Le^x (**42**). Thus, coupling of Le^x fluoride **43** with acceptor **44** (AgOTf, Cp$_2$HfCl$_2$, 91%) furnished pentasaccharide **53**. Two iterations of a two-step process involving monochloroacetate removal (thiourea, 2,6-lutidine) and Le^x fluoride coupling, led to undecasaccharide **55** *via* octasaccharide **54**. Removal of chloroacetate groups, followed by acetylation and azide reduction (Sn[SPh]$_2$, PhSH, Et$_3$N, 89%) furnished the corresponding amine which was coupled with stearic acid (bis(2-oxo-3-oxazolidinyl)phosphinic chloride [BOP-Cl], Et$_3$N, 85%) to afford ceramide **56**. Ester cleavage with NaOMe, followed by phthalimide removal (MeNHNH$_2$), peracetylation (Ac$_2$O-pyr, 68% overall yield for three steps) and base-induced deprotection gave the targeted molecule, trimeric Le^x (**42**) in 96% yield.

This synthesis demonstrated the power of the two-stage activation procedure in the synthesis of complex oligosaccharides and rendered these compounds available in pure form for the first time.

Triantennary Blood Group I Antigen

Triantennary blood group I antigen **57** (Scheme 4-29) belongs to a class of glycosyl ceramides found on cell surfaces of rabbit erythrocyte and human placenta cells. Its molecular structure contains three identical arms (*N*-acetyl lactosamine) attached onto the linear hexasaccharide chain consisting of two *N*-acetyl lactosamines and a lactosyl ceramide. The following elegant synthesis has recently been accomplished by Ogawa and his group in Japan (Matsuzaki et al., 1992).

Scheme 4-29 presents a retrosynthetic analysis of this complex oligosaccharide. Disconnection of the bonds joining the three "arms" and the ceramide unit with the central hexasaccharide, leads to lactosamine fluoride **58** (to be used three times), hexasaccharide **59**, and ceramide derivative **60**. Further disconnection of the hexasaccharide fragment as shown, leads to lactosamine fluoride derivative **61** (to be used twice) and lactose acceptor **62**.

The Ogawa synthesis of the triantennary compound (Scheme 4-30) began with the coupling of disaccharides **61** and **62** (AgOTf, Cp$_2$HfCl$_2$) to give tetrasaccharide **63** in 91% yield. Removal of both levulinate groups with hydrazine acetate (92%), followed by selective acylation of the primary alcohol with levulinic acid (2-chloro-1-methylpyridinium iodide, 1,4-diazabicyclo[2.2.2] octane [DABCO], 100%) gave a monohydroxy compound which was glycosylated with fluoride **61** (AgOTf-Cp$_2$HfCl$_2$) to give hexasaccharide **64** in 96% yield. The levulinate groups were removed with hydrazine acetate (94%) and the generated hydroxyl groups were glycosylated with fluoride donor **58** as above to afford dodecasaccharide **65** in 92% yield. The benzyl ethers were removed from **65** by hydrogenolysis and the generated hydroxyl groups were acetylated using acetic anhydride in pyridine. The more reactive anomeric acetate was then removed by treatment with hydrazine acetate and the generated lactol was con-

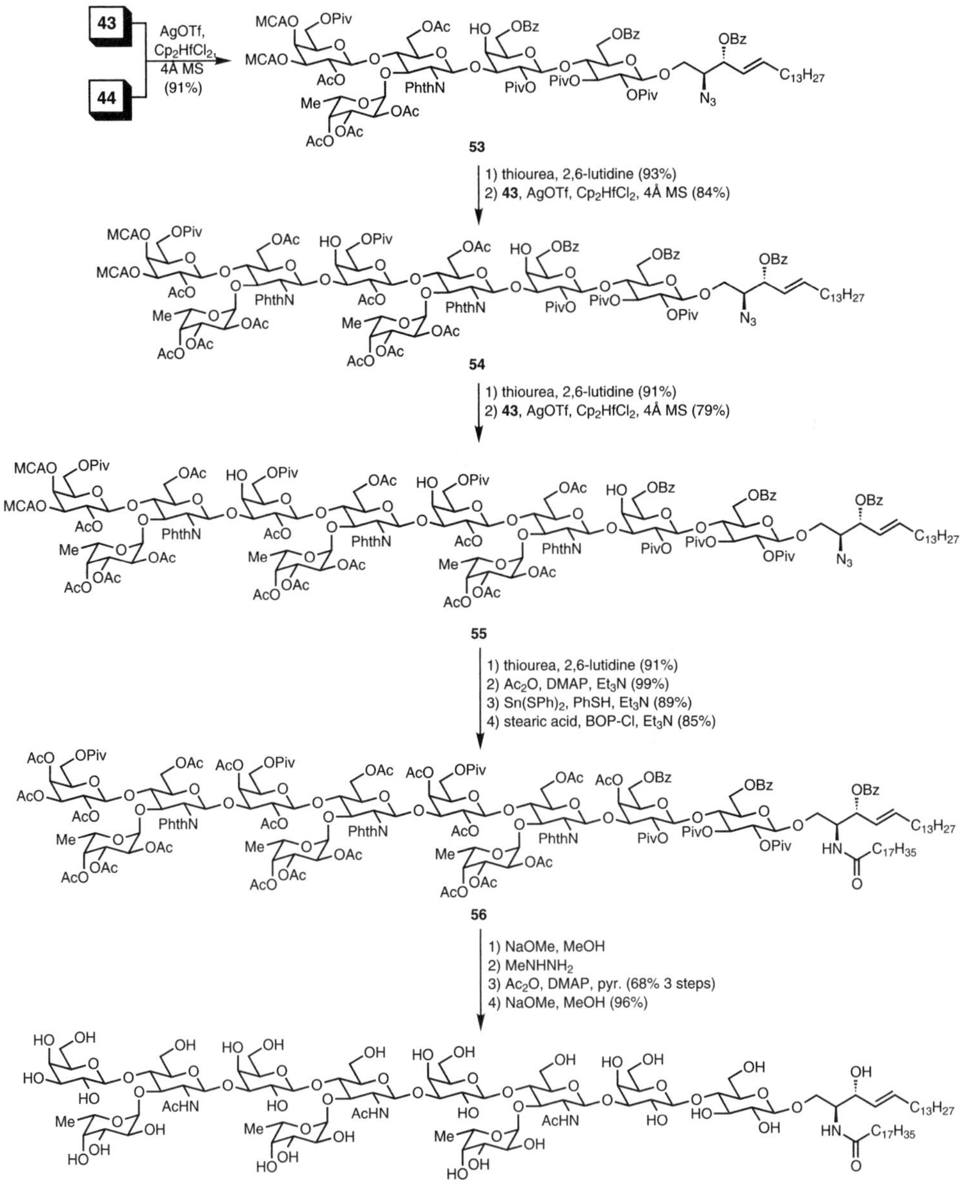

42 : trimeric Le^x

Scheme 4-28. Total synthesis of trimeric Le^x (**42**) (Nicolaou et al., 1990a). DMAP = 4-dimethylaminopyridine, BOP = bis(2-oxo-3-oxazolidinyl)phosphinic chloride.

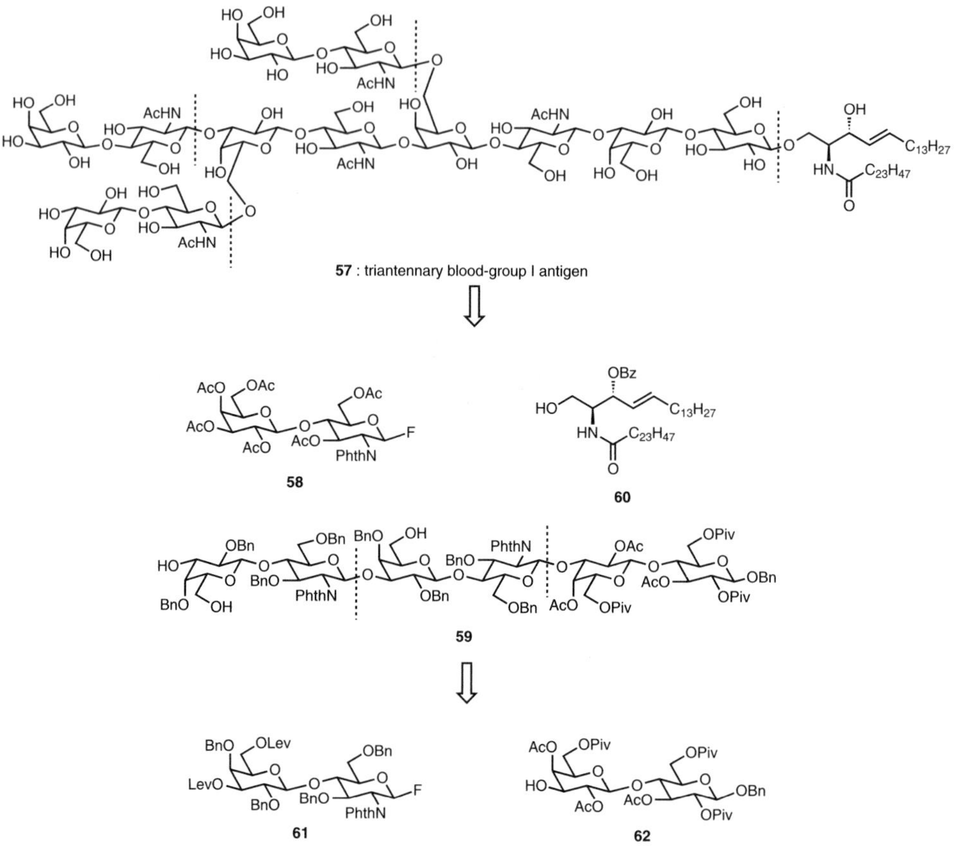

Scheme 4-29. Retrosynthetic analysis of triantennary blood group I antigen (**57**) (Matsuzaki et al., 1992). Ac = acetyl, Bz = benzoyl, Bn = benzyl, Lev = COCH$_2$CH$_2$COCH$_3$, Phth = phthaloyl, Piv = pivaloyl.

verted to trichloroacetimidate **66** with Cl$_3$CCN and DBU (56% overall for four steps). The trichloroacetimidate **66** was then coupled with ceramide derivative **60** under the influence of TMSOTf to afford the corresponding glycosyl ceramide (45%), which was deprotected by a three-step sequence. First, the phthalimide and acetate groups were removed with methylamine; second, the product was peracetylated (Ac$_2$O-pyr); and finally, the O-acetyl groups were cleaved by exposure to NaOMe (54% overall for three steps).

Repeating Unit of Salmonella Thompson

The structure of the *Salmonella thompson* serogroup C$_1$ O-antigen lipopolysaccharide contains the repeating oligosaccharide unit shown in structure **67** (Scheme 4-31). This structure includes two challenging β-mannosidic linkages and was recently synthesized by the Garegg group using two different strategies (Garegg & Hällgren, 1992).

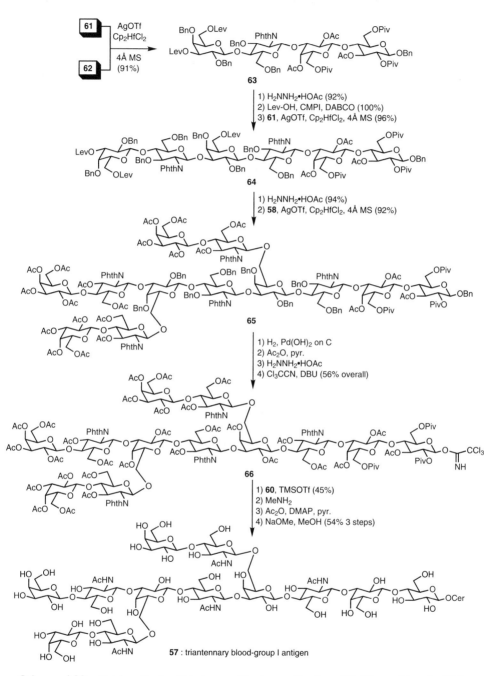

Scheme 4-30. Total synthesis of triantennary blood group I antigen (**57**) (Matsuzaki et al., 1992). CMPI = 2-chloro-1-methylpyridinium iodide, DABCO = 1,4-diazabicyclo[2.2.2]octane.

The retrosynthetic analysis of the Garegg synthesis is shown in Scheme 4-31. Thus, disconnection of the α-mannosidic linkages as shown led to three disaccharide components: β-mannose donor **68**, disaccharide donor **69**, and β-mannose acceptor **70**. Further disconnection of the β-mannose donor **68** led to mannose derivatives **71** and **72** as potential precursors for this fragment, whereas dismantling of the disaccharide **69** led to units **73** and **74**. Retrosynthesis of the β-mannose acceptor **70** included inversion of the hydroxyl-bearing carbon to afford **75**, followed by glycosidic bond rupture, leading to precursors **76** and **77**.

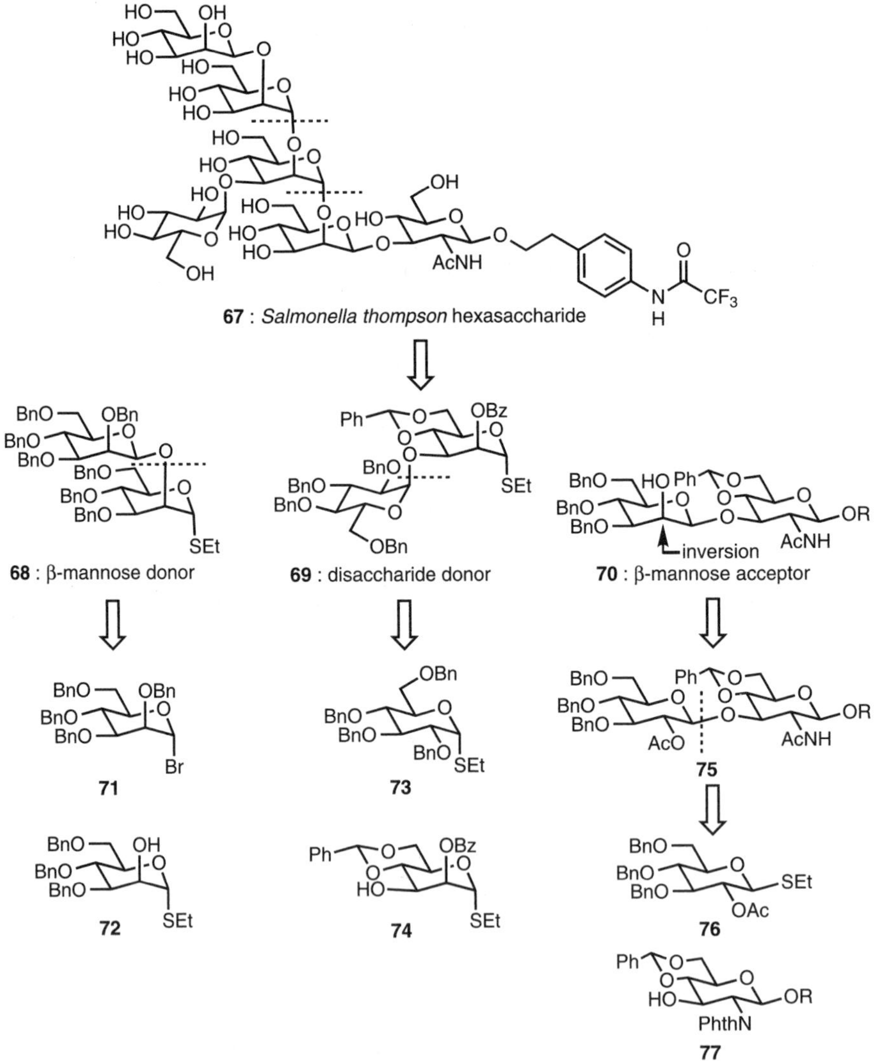

67 : *Salmonella thompson* hexasaccharide

68 : β-mannose donor

69 : disaccharide donor

70 : β-mannose acceptor

71

72

73

74

75

76

77

Scheme 4-31. Retrosynthetic analysis of the *Salmonella thompson* hexasaccharide (**67**) (Garegg & Hällgren, 1992). Ac = acetyl, Bz = benzoyl, Bn = benzyl, Phth = phthaloyl.

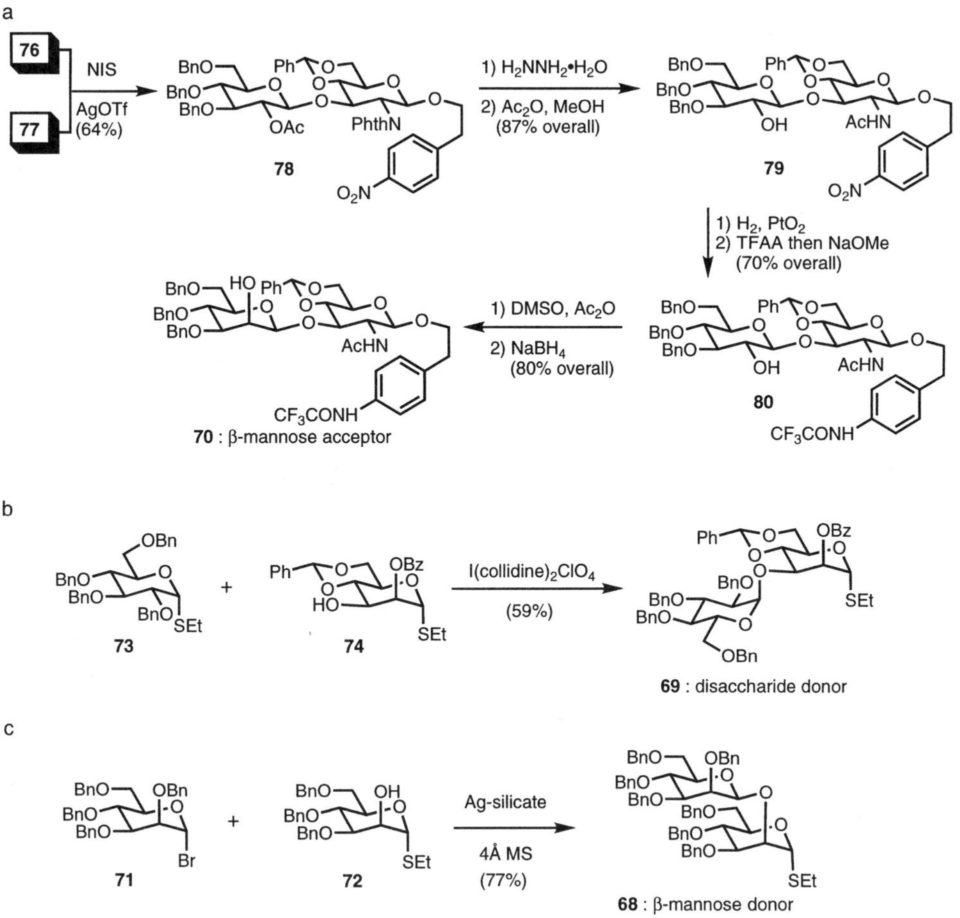

Scheme 4-32. Construction of disaccharide fragments **70 (a)**, **69 (b)**, and **68 (c)** (Garegg & Hällgren, 1992).

Scheme 4-32 outlines the construction of the requisite disaccharides **68–70**. Coupling of **76** and **77** (*N*-iodosuccinimide, AgOTf) gave disaccharide **78** in 64% yield (Scheme 4-32a). Removal of the phthalimido and acetate groups using hydrazine, followed by acetylation, afforded disaccharide **79** in 87% overall yield. The nitro group was then reduced and trifluoroacetylated on nitrogen, affording compound **80** in 70% overall yield. Oxidation of the secondary alcohol followed by reduction gave the β-mannose acceptor **70** in 80% yield. The synthesis of the disaccharide donor **69** (Scheme 4-32b) proceeded via coupling of key intermediates **73** and **74** in 59% yield by treatment with (I[collidine]$_2$ClO$_4$). The β-mannose donor **68** was obtained as the major product by direct β-glycosylation of alcohol **72** with glycosyl bromide **71** in the presence of silver silicate (77% yield, plus 18% of the α-anomer).

The total synthesis of the target molecule (**67**) was completed as indicated in Scheme 4-33. Coupling of fragments **69** and **70** in the presence of DMTST afforded tetrasaccharide **81** in 75% yield; **81** was debenzoylated (NaOMe) and coupled with disaccha-

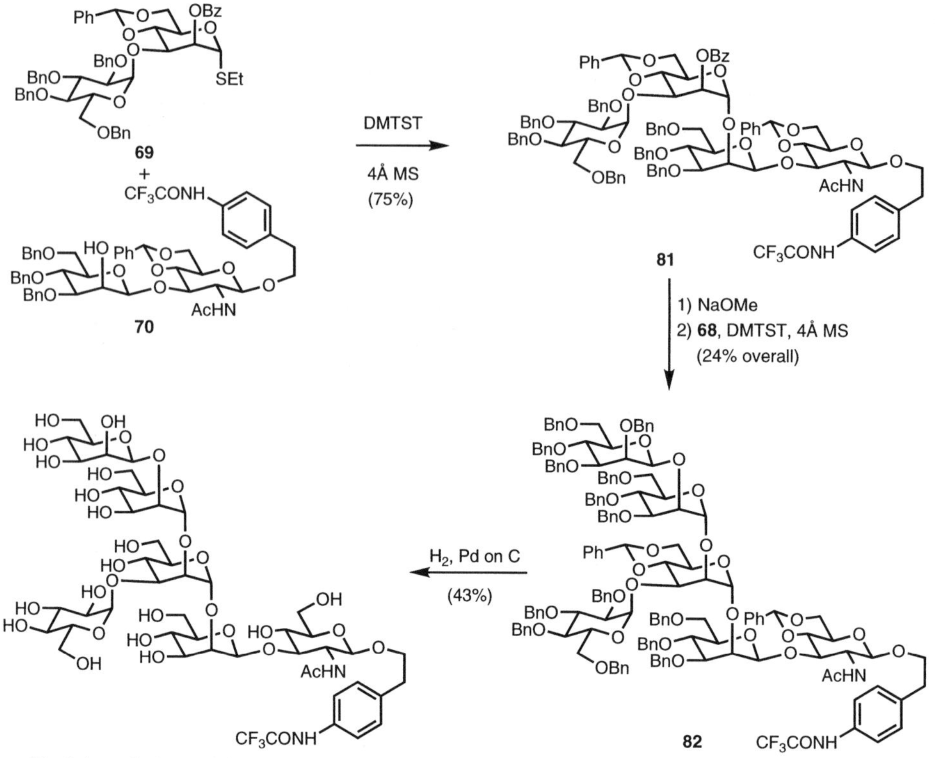

Scheme 4-33. Total synthesis of *Salmonella thompson* hexasaccharide (**67**) (Garegg & Hällgren, 1992). DMTST = dimethyl(methylthio)sulfonium triflate.

ride **68** (DMTST) to furnish hexasaccharide **82** in 24% overall yield. Final deprotection then gave the targeted *Salmonella thompson* repeating unit (**67**) in 43% yield.

Nodulation Factor NodRm-IV (Ac,S)

Nodulation factors are important signaling molecules involved in the rhizobium-legume symbiosis. They are excreted by the bacteria and cause the root of the plant to form nodules that house the bacteria and, thereby, ensure nitrogen fixation for the plant. Despite their structural similarities, these molecules are highly specific. For example, NodRm-IV (Ac,S) (**83**, Scheme 4-34) is active only on alfalfa, whereas the nonsulfated compound NodRm-IV (Ac) is only active on pea and vetch. The total synthesis of NodRm-IV (Ac,S) (**83**) reported by our group (Nicolaou et al., 1992a) is discussed below.

A retrosynthetic analysis of NodRm-IV (Ac,S) (**83**) is depicted in Scheme 4-34. Disconnection of the terminal glycosidic bond and removal of the fatty acid chain leads to unsaturated acid **84**, glycosyl fluoride **85**, and trisaccharide **86** as key intermediates. Further disconnection of the trisaccharide **86** at the indicated glycoside bonds gives glycosyl fluoride **87** (two units) and alcohol **88**.

The synthesis of NodRm-IV (Ac,S) (**83**, Scheme 4-35) began with the coupling of fluoride **87** and alcohol **88** (AgOTf, Cp_2ZrCl_2, 2,6-di-t-butyl-4-methylpyridine) to yield disaccharide **89** in 64% yield, plus 30% recovered **88**. The acetate group was then cleaved (K_2CO_3, MeOH, 90%) and the alcohol was coupled again with fluoride **87** (AgOTf, Cp_2HfCl_2, 2,6-di-t-butyl-4-methylpyridine) to afford trisaccharide **90** in 61% yield, along with 35% recovered alcohol. The phthalimido and acetate groups were then cleaved with hydrazine hydrate, and the amino groups generated were acylated with acetic anhydride in MeOH to give triacetamide **86** in 72% yield. The alcohol **86** was then coupled with fluoride **85** (AgOTf, Cp_2HfCl_2, 2,6-di-$tert$-butyl-4-methylpyridine, 50%) to yield the corresponding tetrasaccharide. Removal of the phthalimido group with hydrazine hydrate (87%) led to the corresponding amine, which was acylated, with carboxylic acid **84** (2-chloro-1-methylpyridinium iodide [CMPI], Et_3N, 73%) to yield tetrasaccharide **91**. The $tert$-butyldimethylsilyl group was selectively cleaved (pyridinium p-toluenesulfonate [PPTS], EtOH, 85%) and the derived hydroxyl group was acetylated (Ac_2O, 4-dimethylaminopyridine [DMAP], Et_3N, 90%). The $tert$-butyldiphenylsilyl ether was then removed (n-Bu$_4$NF, THF, 88%), and the alcohol was sulfated ($SO_3 \cdot NMe_3$, pyridine, 85%) to afford the corresponding sulfate. The latter compound was converted to NodRm-IV (Ac,S) (**83**) by treatment with ceric ammonium nitrate in 30% yield.

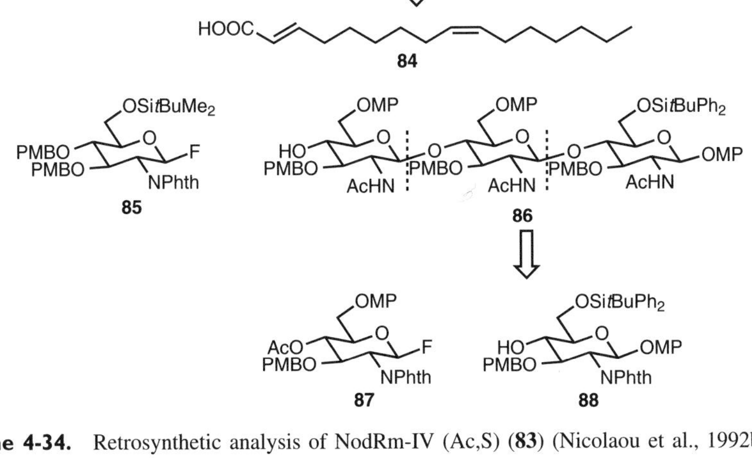

Scheme 4-34. Retrosynthetic analysis of NodRm-IV (Ac,S) (**83**) (Nicolaou et al., 1992b). Ac = acetyl, MP = 4-methoxyphenyl, PMB = 4-methoxybenzyl, Phth = phthaloyl.

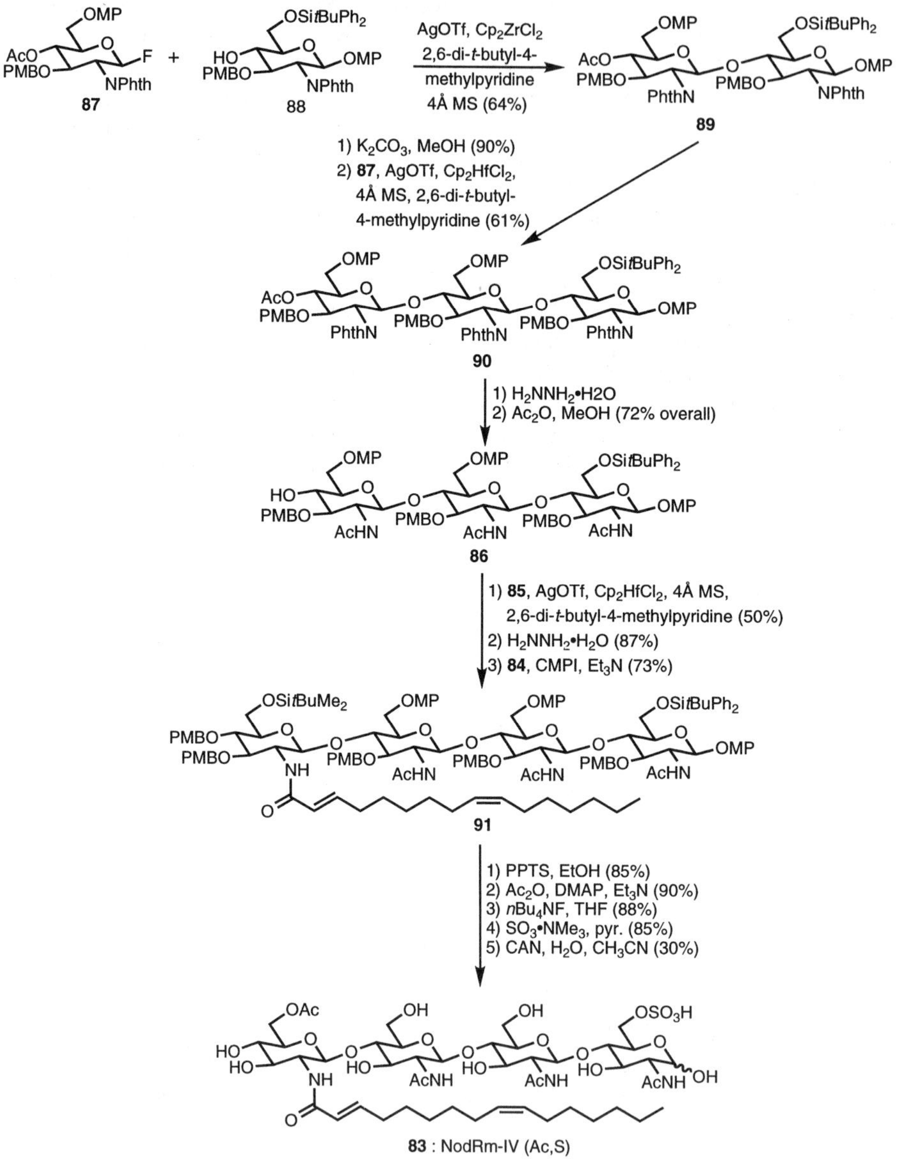

Scheme 4-35. Total synthesis of NodRm-IV (Ac,S) (**83**) (Nicolaoou et al., 1992a). PPTS = pyridinium p-toluenesulfonate, CAN = ceric ammonium nitrate, CMPI = 2-chloro-1-methylpyridinium iodide.

Sialyl Le^x Ganglioside

Sialyl Le^x-containing molecules are quite abundant in nature and have a wide variety of functions. Oligosaccharide **92** (Scheme 4-36) is an example of a ganglioside which serves as a tumor cell marker. Others, such as **93** (Scheme 4-39, *vide infra*) are ligands of E-selectin and are involved in the cell adhesion process of leukocytes. A synthesis of sialyl Le^x ganglioside **92** has recently been reported by Hasegawa's group (Kameyama et al., 1991) and is highlighted below.

Retrosynthesis (Scheme 4-36) of **92** begins with the disconnection of the sialyl-galactose and ceramide segments to give sialyl-galactose derivative **95**, tetrasaccharide

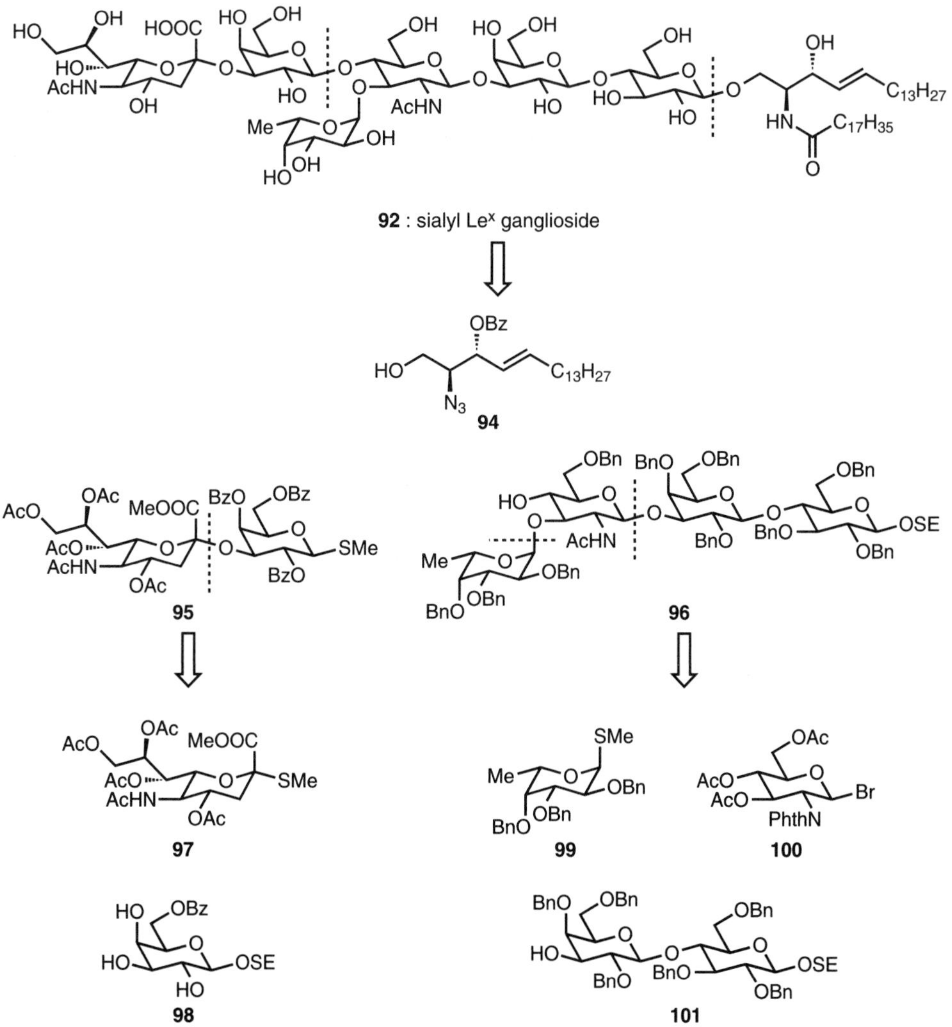

Scheme 4-36. Retrosynthetic analysis of sialyl Le^x ganglioside (**92**) (Kameyama et al., 1991). Ac = acetyl, Bz = benzoyl, Bn = benzyl, Phth = phthaloyl, SE = 2-trimethylsilylethyl.

96, and sphingosine derivative **94**. Further disconnection of sialyl-galactose derivative **95** leads to sialic acid derivative **97** and galactose derivative **98**. Disconnection of the tetrasaccharide **96** leads to fucose thioglycoside **99**, glucosamine bromide **100**, and lactose derivative **101** as potential starting materials.

The total synthesis of tetrasaccharide **96** (Scheme 4-37a) begins with the coupling of alcohol **101** with bromide **100** (Ag$_2$CO$_3$, AgClO$_4$, 4Å molecular sieves) to give trisaccharide **102** in 94% yield. The acetates were then cleaved (NaOMe, MeOH) from **102** to give the triol, which was treated with hydrazine hydrate to remove the phthalimido group. The derived amino group was acetylated selectively to give triol **103** in 88% overall yield. Compound **103** was then treated with benzaldehyde dimethyl acetal and acid to give the corresponding 6-membered ring benzylidene (83%). The remaining secondary alcohol was coupled with fucose thioglycoside **99** (DMTST, 4Å molecular sieves) to afford the expected tetrasaccharide in 86% yield. The benzylidene moiety was then reductively cleaved with NaCNBH$_3$ in the presence of acid, furnishing the desired compound **96** in 75% yield.

The synthesis of the requisite sialic acid fragment **95** (Scheme 4-37b) commenced with the coupling of sialic acid thioglycoside **97** with galactose derivative **98** (DMTST, 4Å molecular sieves, 46% yield) to afford a disaccharide which was fully benzoylated (BzCl, 60%) and then exposed successively to Ac$_2$O-BF$_3$·Et$_2$O and MeSSiMe$_3$-BF$_3$·Et$_2$O, leading to methyl thioglycoside **95** in 75% overall yield.

Continuing with the synthesis (Scheme 4-38), fragments **95** and **96** were coupled under the influence of DMTST to afford hexasaccharide **104** in 41% yield. The benzyl ethers were then removed (H$_2$, Pd on carbon) and replaced with acetate groups

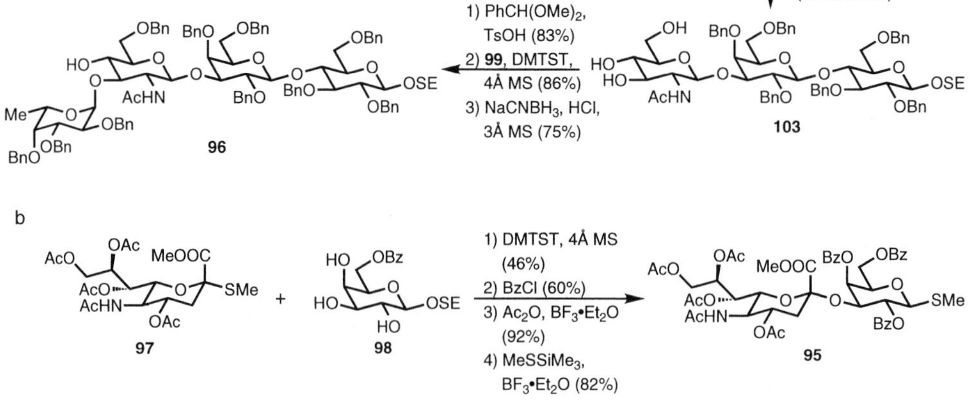

Scheme 4-37. Construction of tetrasaccharide fragment **96** (a) and disaccharide fragment **95** (b) (Kameyama et al., 1991). DMTST = dimethyl(methylthio)sulfonium triflate.

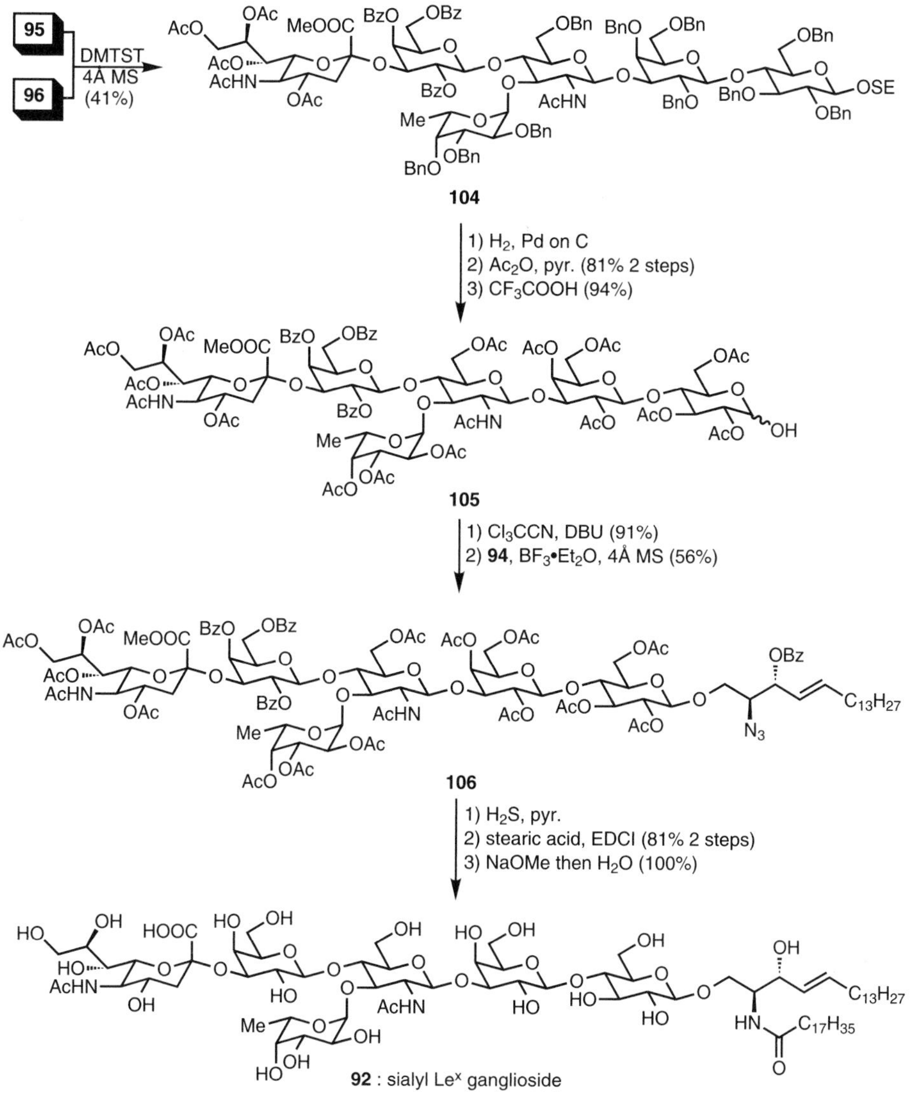

Scheme 4-38. Total synthesis of sialyl Le^x ganglioside (**92**) (Kameyama et al., 1991). EDCl = 1-(3-dimethylaminopropyl)-3-ethylcarbodiimide hydrochloride.

(Ac$_2$O, pyridine, 81% overall). The trimethylsilylethyl glycoside was then cleaved by treatment with aqueous CF$_3$COOH to yield lactol **105** in 94% yield. The lactol was then converted to the corresponding trichloroacetimidate (Cl$_3$CCN, DBU, 91% yield) and coupled with the sphingosine derivative **94** (BF$_3$·Et$_2$O, 4Å molecular sieves) to give hexasaccharide **106** in 56% yield. Reduction of the azido group (H$_2$S, pyridine) followed by coupling with stearic acid (EDCl) gave the expected ceramide (81% overall yield), which was converted to the target compound **92** by removal of the ester groups under basic conditions (100% yield).

Sialyl Dimeric Le^x

Sialyl dimeric Le^x (**93**, Scheme 4-39) is a ligand for the cell adhesion molecule E-selectin and it may play an important role in inflammation. A total synthesis of this complex oligosaccharide has recently been reported from these laboratories (Nicolaou et al., 1992b).

The retrosynthetic analysis of the sialyl dimeric Le^x derivative **93** involved the straightforward disconnections shown in Scheme 4-39 by which intermediates **107–111** are defined as potential precursors.

The syntheses of the two advanced intermediates **107** and **108** were achieved as outlined in Scheme 4-40. Thus, coupling of Le^x derivative **110** with o-nitrobenzyl alcohol in the presence of AgOTf-Cp$_2$HfCl$_2$ gave the corresponding o-nitrobenzyl glycoside in 94% yield. Ester cleavage (NaOMe), followed by coupling with sialic acid derivative **109** (Hg[CN]$_2$, HgBr$_2$) led to compound **112** in 56% yield. Conversion of **112** to glycosyl fluoride **107** was achieved by (1) acetylation of the free hydroxyl groups (Ac$_2$O, DMAP, pyridine, 86%); (2) photo-induced removal of the o-nitrobenzyl group (82%); (3) reductive cleavage of the PhS group (Ph$_3$SnH-AIBN, 42% yield plus 46% recovered starting material); and (4) exposure to DAST (78% yield). The Le^x triol **108**

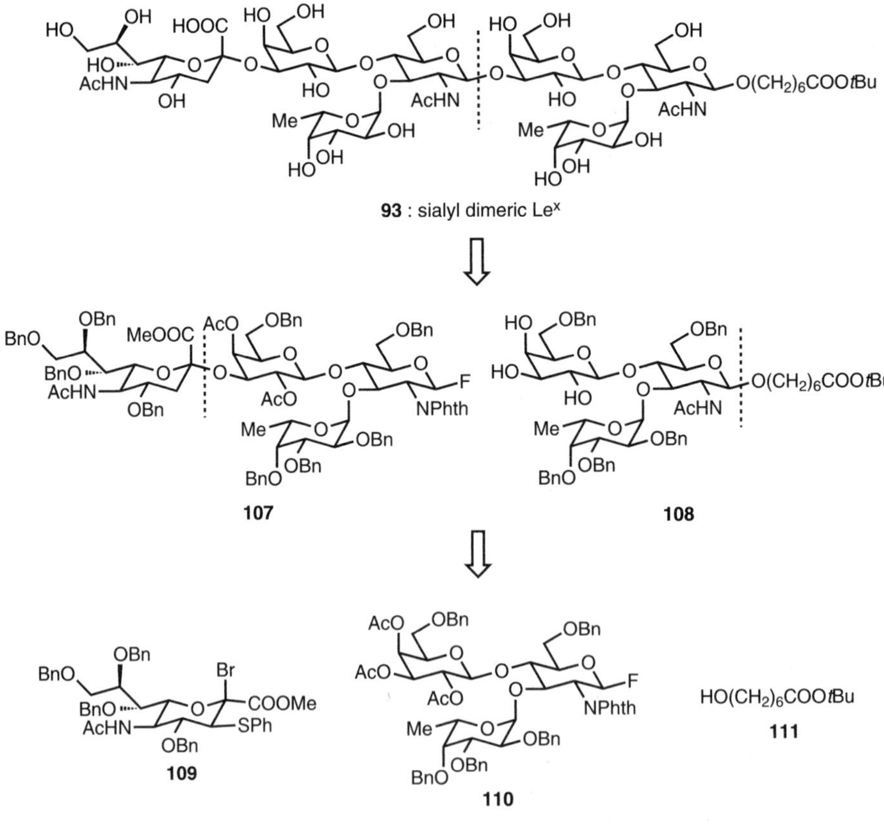

Scheme 4-39. Retrosynthetic analysis of sialyl dimeric Le^x (**93**) (Nicolaou et al., 1992b). Ac = acetyl, Bn = benzyl, Phth = phthaloyl.

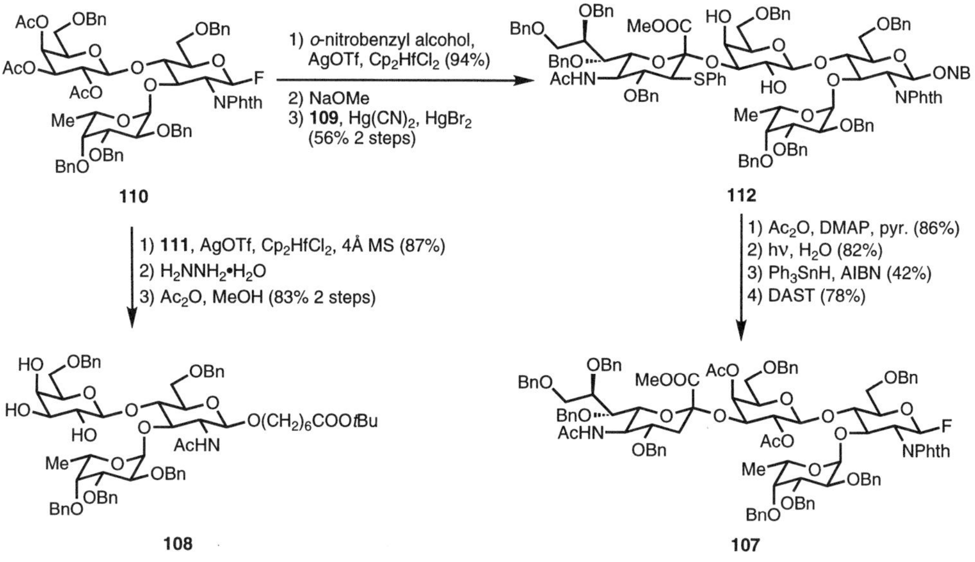

Scheme 4-40. Synthesis of trisaccharide acceptor (**108**) and tetrasaccharide glycosyl fluoride (**107**) (Nicolaou et al., 1992b). AIBN = 2,2'-azobisisobutyronitrile, NB = o-nitrobenzyl.

was synthesized from fluoride **110** by coupling with hydroxy ester **111** (AgOTf-Cp₂HfCl₂, 87%) followed by treatment with hydrazine hydrate to remove the acetate and phthalimido groups, and selective acetylation of the derived primary amine (83% for two steps).

Scheme 4-41 summarizes the completion of the synthesis of sialyl dimeric Le^x (**93**). Thus, coupling of fluoride **107** and triol **108** (AgOTf-Cp₂HfCl₂) proved regioselective, leading to heptasaccharide **113** in 51% yield. Removal of the methyl ester (LiI, pyridine, Δ, 80%), followed by deprotection-selective protection (hydrazine hydrate, then Ac₂O) led to a mixture of carboxylic acid **114** and lactone **115** (ca 1:1) in 75% total yield. Carboxylic acid **114** was converted directly to sialyl dimeric Le^x (**93**) by hydrogenolysis (95%), whereas lactone **115** required first hydrogenolysis (95%) and then standing in aqueous solution to generate the target molecule (**93**) in quantitative yield.

Calicheamicin γg₁^I Oligosaccharide

Calicheamicin γ₁^I (Scheme 4-42a) is a prominent member of the enediyne family of antitumor anticancer antibiotics (Lee et al., 1991). The carbohydrate fragment of this compound is thought to be responsible for its DNA binding properties. This oligosaccharide has several intriguing structural features. Included among these are two 2-deoxyglycosidic bonds, a glycosidic hydroxylamino linkage, a thioester linkage, and a hexasubstituted benzene ring carrying an iodine substituent. The synthesis of the calicheamicin γ₁^I oligosaccharide as its methyl glycoside **116** (Scheme 4-42b) was first reported by our group in 1990 (Nicolaou et al., 1990b; Groneberg et al., 1993). This construction is discussed briefly below.

Retrosynthetic analysis of the target compound **116** (Scheme 4-42b) began with disconnection of the thioester bond to give acid chloride **117** and thiol **118**. Further elab-

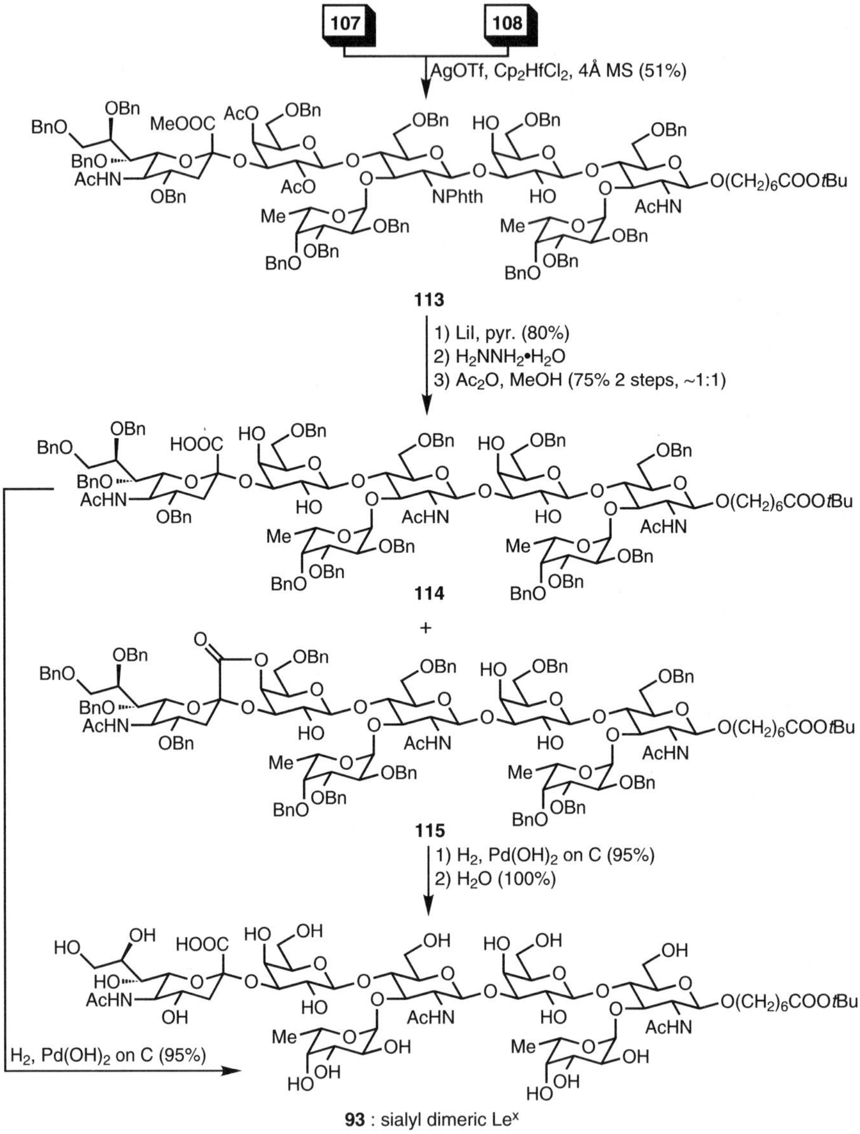

Scheme 4-41. Total synthesis of sialyl dimeric Le^x (**93**) (Nicolaou et al., 1992b).

oration and disconnection of the thiol trisaccharide **118** gave hydroxylamino glycoside **119**, D-fucose derivative **120**, and amino sugar **121** as potential precursors. The oxygen substituent at the 2-position of hydroxylamino glycoside **119** was to be used to introduce the thiol in **118** *via* a 3,3-sigmatropic rearrangement (*vide infra*).

Synthesis of the oligosaccharide **116** (Scheme 4-43) began with coupling of fluoride **121** and alcohol **120** (AgClO$_4$-SnCl$_2$, 4Å molecular sieves) to give disaccharide

122 in 70% yield, along with 16% of the corresponding β-isomer. The carbonate group was then removed (NaH, ethylene glycol, 93%), and the axial alcohol was oxidized selectively by treatment with n-Bu₂SnO, followed by Br₂ and n-Bu₃SnOMe, to give ketone **123** in 70% yield. Ketone **123** was treated with hydroxylamino glycoside **119** in the presence of acid (PPTS) to form the oxime in 83% yield. The free alcohol was silylated (TESOTf, 2,6-lutidine, 100%) and the m-chlorobenzoate group was removed

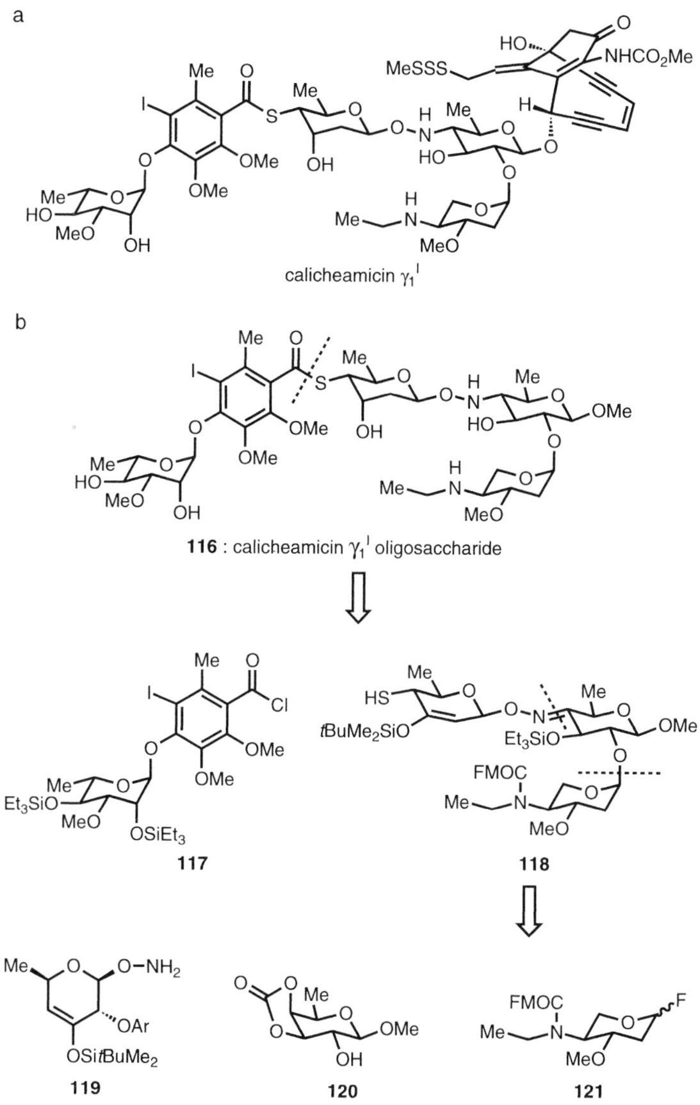

Scheme 4-42. (a) Structure of calicheamicin γ_1^I and (b) retrosynthetic analysis of calicheamicin γ_1^I oligosaccharide **116** (Nicolaou et al., 1990b). Ar = m-chlorobenzoyl, FMOC = 9-fluorenyl-methoxycarbonyl.

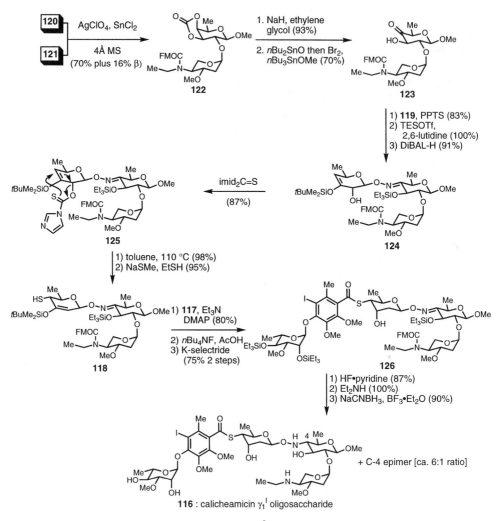

Scheme 4-43. Total synthesis of calicheamicin γ_1^I oligosaccharide (**116**) (Nicolaou et al., 1990b; Groneberg et al., 1993). Imid = imidazole, PPTS = pyridinium *p*-toluenesulfonate, DIBAL = di-*iso*-butylaluminum hydride.

by treatment with DIBAL-H to yield trisaccharide **124** in 91% yield. The free alcohol was acylated with thionocarbonyldiimidazole to give thionoimidazolide **125** (87% yield), which underwent a 3,3-sigmatropic rearrangement as indicated (toluene, 110°C, 98% yield). The resulting thioester was cleaved (NaSMe, EtSH) to give the free thiol **118** in 95% yield. The free thiol was acylated by acid chloride **117** (obtained in quantitative yield by treatment of the corresponding acid with oxalyl chloride) in the presence of Et₃N and DMAP to give the expected aryl tetrasaccharide in 80% yield. The enol silyl ether so obtained was selectively hydrolyzed by treatment with 1.0 equivalent of *n*-Bu₄NF and acetic acid, leading to the corresponding ketone. The ketone was reduced to the axial alcohol **126** in 75% yield using K-selectride. Compound **126** was then converted to the final product via a three-step sequence. First, cleavage of the silyl ethers (HF·pyridine, 87%); second, removal of the 9-fluorenylmethoxycarbonyl (FMOC) group (Et$_2$NH, 100%); and finally, reduction of the oxime bond with NaCNBH$_3$ in the presence of BF$_3$·Et$_2$O to give the calicheamicin γ_1^I oligosaccharide **116** as a 6:1 mixture of C-4 epimers in 90% total yield. The total synthesis of calicheamicin γ_1^I was achieved in 1992 (Nicolaou, et al., 1992c).

Summary

It is hoped that the above survey conveys both the state of the art in the chemical synthesis of complex oligosaccharides, and the excitement in the field. The impressive strides made in the last few years in the area of chemical synthesis of complex oligosaccharides complemented by enzymatic approaches will, no doubt, aid further investigations into the chemistry, biology, and medicinal properties of these compounds. At the same time, however, the realization of the need for further improvements in synthetic methodology in the carbohydrate field should serve as a continuing challenge to the synthetic chemist.

Acknowledgments

It is a pleasure to acknowledge our many talented collaborators, whose names appear in the original articles listed in this chapter for their contributions to this research effort. This work was supported financially by the National Institutes of Health, Merck Sharp & Dohme, Pfizer Research, Glaxo, Rhone-Poulenc Rorer, Schering Plough, and The Scripps Research Institute.

5

DNA-Binding Glycoconjugates

Daniel Kahne, Domingos Silva, and Suzanne Walker

DNA plays a central role in biology. It encodes the information that gives all organisms their particular identities. Access to this information involves a complex series of events requiring specific interactions between the DNA and various proteins. Microorganisms produce an impressive variety of cytotoxic compounds that bind in the minor groove of DNA and interfere with its ability to interact with processing enzymes such as DNA polymerase, RNA polymerase, and topoisomerases I and II (Chiang et al., 1994). These cytotoxic compounds are of great interest to the research community for several reasons. First, some of them have enormous potential as antitumor agents because they kill cells. Second, it is possible to obtain very detailed information on their interactions with DNA because they are relatively small. Knowledge of how these small natural molecules recognize DNA may lead to the ability to design synthetic DNA-binding ligands with potentially useful properties. Finally, studies on small molecule–DNA recognition can provide information about sequence-dependent variations in DNA structure.

DNA Interactive Agents

Minor groove binders have traditionally been classified according to the way in which they interact with DNA. The major classes of minor groove binders include intercalators (Waring, 1981a,b), noncovalent groove binders (Zimmer & Wähnert, 1986), covalent groove binders (Hurley et al., 1984), and DNA-cleaving agents (Goldberg, 1986; Hecht, 1986; Lee et al., 1991; Stubbe & Kozarich, 1987). The minor groove binders within each of these classes tend to share certain structural features. For example, intercalators typically have a polycyclic planar aromatic ring system which can insert between adjacent base pairs in double-stranded DNA. DNA-cleaving agents, on the other hand, often contain a structural unit capable of generating a radical under the appropriate conditions. These radicals initiate DNA cleavage by abstracting hydrogen atoms from the ribose–phosphate backbone. Covalent groove binders contain functional groups that react with nucleophiles in the minor groove of DNA. Thus, it is frequently possible to anticipate the mode of action of new DNA-binding antitumor agents by examining their structures and identifying the defining motifs. However, many of the details of how various molecules interact with DNA are determined not by these defining motifs, but by other groups on the molecules. To understand DNA binding, one must understand how these other groups function.

A large number of DNA binders contain one or more sugars that are known to play a role in DNA binding. These sugar-containing DNA binders can be classified under the general rubric *DNA-binding glycoconjugates*. There are important DNA-binding glycoconjugates in almost every category of antitumor agent. For example, many of the anthracyline antibiotics, which are intercalators, have one or more sugars attached to the anthraquinone core (Lown, 1988; Fig. 5-1). Several well-known DNA-cleaving agents, including the ene-diynes calicheamicin and esperamicin, also contain func-

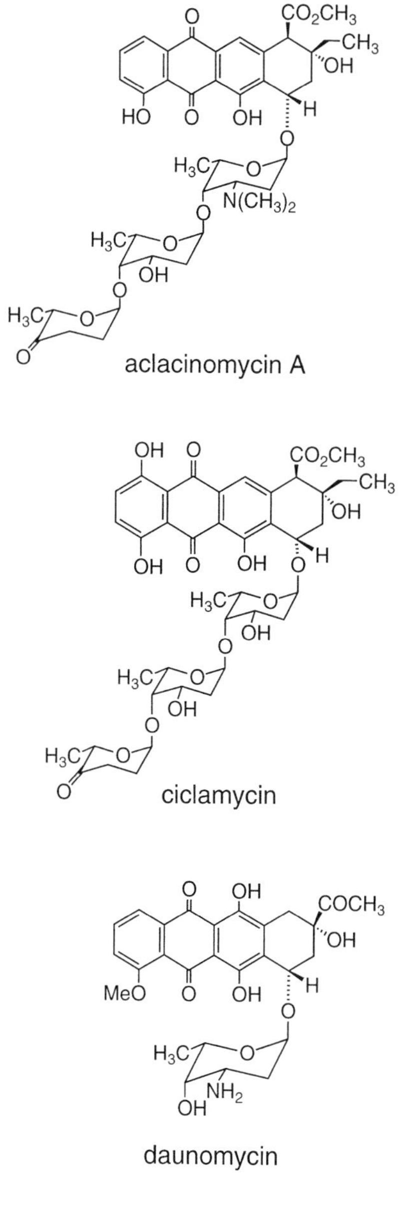

aclacinomycin A

ciclamycin

daunomycin

Figure 5–1. Glycoconjugates of the anthracycline group.

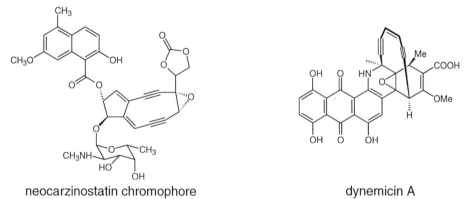

calicheamicin

esperamicin

neocarzinostatin chromophore

dynemicin A

Figure 5–2. Representative members of the ene-diyne antibiotics.

tionally important oligosaccharides (Lee et al., 1991) (Fig. 5-2). In addition, the aureolic acid antibiotics, an important class of noncovalent groove binders, contain saccharide chains which are essential for effective DNA binding (Skarbek & Speedie, 1981) (Fig. 5-3). In spite of their functional importance, however, the sugars in most DNA-binding glycoconjugates have typically received less attention than the aglycone. Past neglect of DNA-binding sugars can be attributed in part to technical difficulties that limited their study. Recent advances in synthetic methods, particularly in methods to synthesize carbohydrates, have made it easier to make derivatives of various glycoconjugates in order to probe the role of the different groups in DNA binding. Advances in techniques for studying DNA binding have also facilitated efforts to delineate the role of the sugars in DNA-binding glycoconjugates.

In this chapter, we examine the function of the sugars in two very different DNA-binding glycoconjugates. Our aim is to use specific molecules to discuss some of the

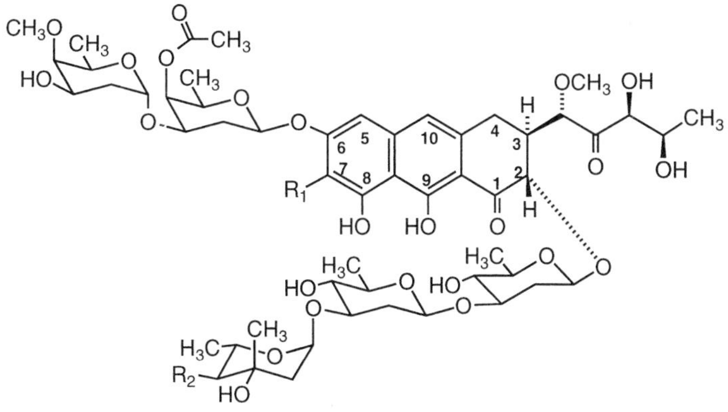

chromomycin A_3 (R_1=CH_3, R_2=acetyl)

olivomycin A (R_1= H, R_2=isobutyryl)

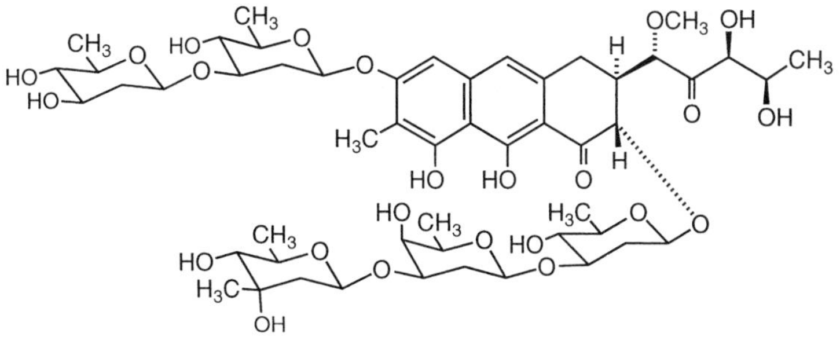

mithramycin

Figure 5–3. Glycoconjugates of the aureolic acid group.

different roles that oligosaccharides may play in DNA binding. One of the molecules we focus on, chromomycin A_3 (CRA_3) (Fig. 5-3), belongs to the aureolic acid class of antibiotics. CRA_3 has been known for over 30 years, but until recently almost nothing was known about how it interacts with DNA or the function(s) of the sugars. The other molecule, the ene-diyne anitiotic calicheamicin (Fig. 5-1), contains an aryl-tetrasaccharide chain that binds site selectively in the minor groove of DNA. More than any other DNA-binding glycoconjugate, calicheamicin has been responsible for drawing the attention of the scientific community to the central role that sugars can play in DNA binding. One question to keep in mind as we examine these molecules is why nature has used sugars in the design of so many DNA binders. Is there something about their structures and properties that makes them good DNA-binding elements? Do the sugars found in DNA-binding glycoconjugates differ from sugars attached to other biologically important molecules? If so, do the structural differences provide insight into their functions? Answers to these kinds of questions are important to make sense of carbohydrate–DNA interactions. However, they may also be relevant to how we think about the role of carbohydrates in other molecular recognition events.

Chromomycin A_3

Many DNA-binding glycoconjugates have been identified by screening the fermentation broths of microorganisms for antibiotic and antitumor activity. Chromomycin A_3 (Fig. 5-3), one of the first glycoconjugates used in cancer chemotheraphy, was isolated from fermentation broths of *Streptomyces griseus* in 1960 (Shibata et al., 1960). Several related compounds, including mithramycin A and olivomycin A, were isolated from the fermentation broths of other micoorganisms shortly thereafter (Skarbek & Speedie, 1981) (Fig. 5-3). Chromomycin A_3, mithramycin A, olivomycin A, and related compounds were collectively named the *aureolic acid* antibiotics because of their vivid golden color. All the aureolic acid antibiotics contain a tricyclic aromatic chromophore that is glycosylated at two positions. The various members of the class differ with respect to the substituents on the chromophore and the structures of the attached oligosac charides.

   CRA_3 attracted a great deal of interest at the time of its discovery because of its notable activity against deadly cancers such as testicular cancer, sarcomas, leukemias, and advanced stomach cancer (Skarbek & Speedie, 1981). Unfortunately, CRA_3 also proved to be highly toxic, presumably because it does not exhibit much selectivity between tumor cells and normal cells. The toxicity of CRA_3 has limited its therapeutic use primarily to the treatment of advanced stomach cancer. Nevertheless, interest in CRA_3 has remained high because it is an unusual DNA binder. Moreover, it may ultimately be possible to modify CRA_3 to reduce its toxicity without greatly diminishing its efficacy.

   Investigations into the mode of action of CRA_3 began in the 1960s. Ward and co-workers showed that CRA_3 binds tightly to double-stranded DNA, but only in the presence of Mg^{2+} (Itzhaki et al., 1990; Kamiyama, 1968; Ward et al., 1965). Wakisaka and co-workers showed that CRA_3 inhibits the action of both DNA and RNA polymerases *in vitro* (Wakisaka et al., 1963; Ward et al., 1965). These results suggested that CRA_3 kills cells by binding to DNA and preventing replication or transcription. The ability of CRA_3 to bind to DNA and inhibit polymerase activity stimulated interest in understanding the molecular basis for the interaction between the drug and DNA.

Initial studies on the binding properties of CRA_3 focused on defining its binding site preferences. Spectroscopic experiments carried out by Kersten and co-workers in 1966 showed that CRA_3 has a preference for GC-rich DNA (Kersten et al., 1966). Many years later, both enzymatic and chemical footprinting experiments defined the binding selectivity more precisely (Fox & Howarth, 1985; Stankus et al., 1992; van Dyke & Dervan, 1983). The footprints revealed that CRA_3 binds to DNA sites that are three base pairs long and contain at least two contiguous G·C base pairs. The affinity of the drug for a given binding site was found to be dependent on the nature of the flanking sequence as well, suggesting a complex relationship between base sequence and binding affinity.

Early studies on CRA_3 also included efforts to identify which parts of the molecule are critical for DNA binding. For example, in 1967 Kaziro and Kamiyama showed that the aglycone of CRA_3, chromomycinone (CRN) (Fig. 5-4), does not inhibit RNA polymerase (Kaziro & Kamiyama, 1967). The authors concluded that the aglycone itself does not bind to DNA. This conclusion was supported by studies showing that CRN, unlike CRA_3, does not precipitate with calf thymus DNA in the presence of Mg^{2+} (Kamiyama, 1968). Studies on other degradation products of CRA_3 lacking various sugars (Behr et al., 1969; Hayasaka & Inoue, 1969; Koschel et al., 1966) showed that inhibition of RNA polymerase activity decreased as the sugars were removed sequentially. Thus, a comparison of CRA_3 with its degradation products in various assays directly implicated the chromomycin sugars in DNA binding.

The early results on the involvement of the chromomycin sugars in DNA binding were of great importance. There had been a widely held belief that the sugars in biologically active natural products rarely interact directly with the therapeutic targets of the natural products. Instead, the sugars were commonly regarded as appendages that contribute to biological activity through some beneficial effect on pharmacokinetics (such as absorption, bioavailability, and distribution). Although this may be true in many cases, the example of chromomycin showed that the sugars in DNA-binding glycoconjugates can play a fundamental role in DNA binding. Unfortunately, there was no physical basis for understanding how the chromomycin sugars contributed to DNA binding because there was no structural information on the complex formed between CRA_3 and DNA. The discovery of chromomycin preceded by many years the development of the NMR techniques that permit the routine elucidation of high-resolution structures of drug–DNA complexes.

Finally, in 1989, almost 30 years after the initial discovery of CRA_3, Gao and Patel published a landmark paper on the NMR structure of CRA_3 bound to the self-complementary duplex d($T_1T_2G_3G_4C_5C_6A_7A_8$) (Gao & Patel, 1989; Gao et al., 1992; see also Banville et al., 1990a,b). The authors showed that CRA_3 binds in the minor groove of

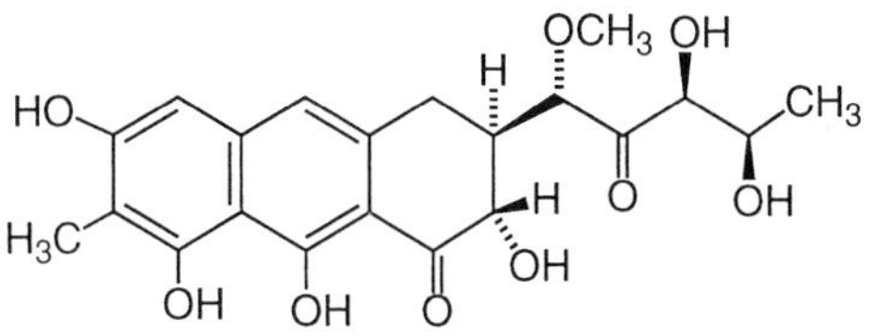

Figure 5–4. The aglycone of CRA_3, chromomycinone (CRN).

DNA as a dyad-symmetric dimer. There is a Mg^{2+} ion at the center of the dimer, co-ordinated by the C1 ketone oxygen and the ionized C9 phenolate of each chromophore (Fig. 5-5a). An X-ray structure of the CRA_3–DNA complex obtained at about the same time showed that the Mg^{2+} ion occupies an octahedral site, with the O9 phenolate atoms in the apical positions (Hendrickson, 1990) (Fig. 5-5b). The two other ligands in the Mg^{2+} complex are water molecules. The orientation of the rigid CRA_3 ligands around the metal center is such that the complex has a right-handed twist which is complementary to the twist of the DNA (Fig. 5-5a). However, the width of the complex is larger than the average width of the minor groove in B form DNA. Hence, the DNA undergoes a binding-induced structural transition in order to accommodate the 2:1 $CRA_3:Mg^{2+}$ complex. The DNA unwinds and the minor groove becomes much wider and shallower, with many features characteristic of A form DNA. The drug dimer fits into the groove with the hydrophilic edges of the chromophores (where the phenols are located) contacting the floor of the minor groove. The trisaccharide on each CRA_3 molecule is folded against the tricyclic chromophore on the other CRA_3 molecule, contacting the walls of the minor groove. There are also some contacts from the A ring of the disaccharide to the ribose phosphate backbone.

Based on the structure of the CRA_3–DNA complex, Gao and Patel proposed that the selectivity of CRA_3 for GC-rich regions of DNA is due to hydrogen bonds between the G_3 and G_4 guanine bases in the duplex $d(T_1T_2G_3G_4C_5C_6A_7A_8)$ and the CRA_3 dimer (Gao et al., 1992). They suggested that the C8 phenol on each CRA_3 molecule in the dimer forms a set of hydrogen bonds to each G_4 base (Fig. 5-6). They also proposed a hydrogen bonding interaction between the ring ether oxygen of each E-ring sugar and the G_3 amino proton. These interactions may explain the sequence selectivity in part. However, given that the DNA must undergo large conformational changes to accommodate the CRA_3–Mg^{2+} complex, it is likely that sequence-dependent DNA flexibility also plays a role in the binding selectivity. It is surely not coincidental that GC-rich regions of DNA, and in particular GpG sites, undergo B-form to A-form

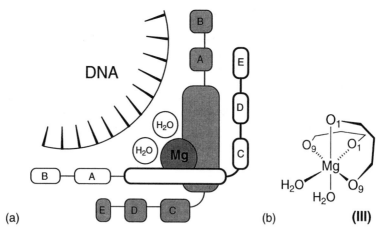

(a)   (b)   (III)

Figure 5–5. (a) Schematic representation of the 2:1 $CRA_3:Mg^{2+}$ complex in the minor groove of DNA. (b) Schematic representation of the coordination geometry of Mg^{2+} in the 2:1 CRA_3-metal complex bound to the minor groove of DNA. The ligand CRA_3 is represented by its chelating O1 and O9 oxygens.

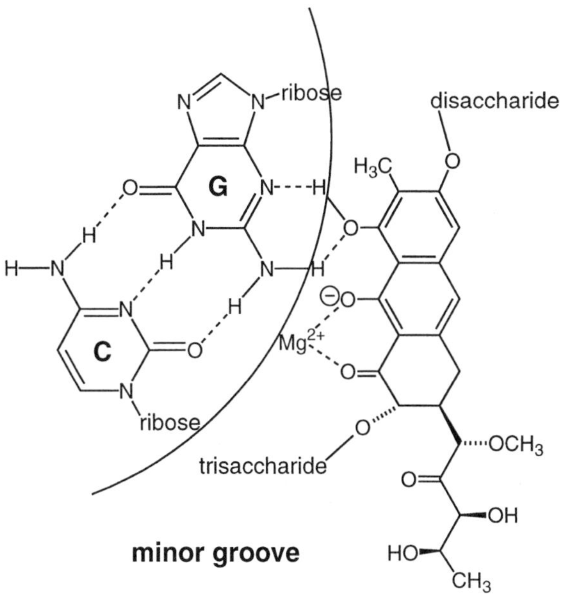

Figure 5–6. Details of the 2:1 complex showing putative hydrogen bonds established to the recognition site in the minor groove of DNA. Only the shaded CRA$_3$ molecule (Fig. 5-5a) is shown for clarity.

conformational transitions more readily than other sites (Heinemann et al., 1990). A model in which sequence-dependent DNA flexibility plays a large role in binding selectivity would also explain the influence of the flanking sequences on binding affinity.

Gao and Patel's nuclear magnetic resonance (NMR) study provided critical information about the molecular basis for site selectivity as well as the role of the metal ion. However, an examination of the structure of the CRA$_3$–DNA complex did not suggest an obvious role for the sugars in DNA binding. Although a few intermolecular contacts were observed between the sugars and the DNA, this information did not immediately lead to a clear understanding of how the sugars contributed to DNA binding.

In considering the relationship between structure and function in a complex biological molecule, it is sometimes helpful to try to reduce the molecule to some essential element. For example, if one regards CRA$_3$ as a ligand for a metal ion, it becomes apparent that the β-keto phenolate coordination site resembles an acetoacetone (*acac*) ligand (Fig. 5-7). While little may be known about CRA$_3$ itself, a great deal is known about other *acac* ligands and the complexes that they form with various metal ions. Acetoacetone ligands have been studied extensively by inorganic chemists and are known to form both 2:1 and 1:1 octahedral complexes with Mg^{2+} in solution, that is, Mg(*acac*)$_2$(solvent)$_2$ or (Mg[*acac*][solvent]$_4$)$^+$. The major factor determining the preferred stoichiometry of the complex formed between an *acac* ligand and Mg^{2+} is the

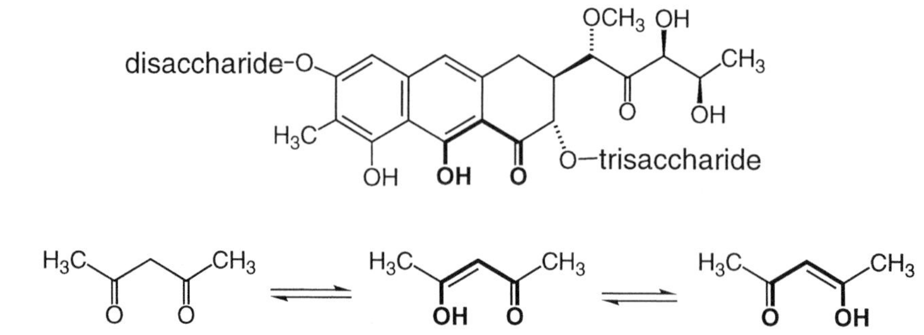

Figure 5–7. Structures of CRA$_3$ and acetoacetone (*acac*).

size of the substituents on the *acac* ligand. The more sterically hindered the *acac* ligand, the less stable the 2:1 octahedral complex (Graddon, 1969). CRA$_3$ resembles a hindered *acac* ligand and one might, therefore, expect it to form a 1:1 complex rather than a 2:1 complex. However, the NMR structure showed that CRA$_3$ clearly binds to DNA as a 2:1 complex. The unusual behavior of CRA$_3$ as a ligand for Mg^{2+} led us to wonder whether the DNA somehow stabilizes the more hindered 2:1 complex, or if there are structural elements in the CRA$_3$ molecule itself that override the effects of steric hindrance and stabilize the 2:1 complex.

The Gao and Patel NMR structure showed that the trisaccharide of one CRA$_3$ molecule and the chromophore of the other are in close proximity in the 2:1 complex that binds to DNA. This proximity suggested that the CRA$_3$ sugars might be involved in stabilizing the 2:1 complex. Accordingly, we decided to investigate the interaction of CRA$_3$ and its aglycone, CRN, with Mg^{2+} in methanol to determine whether they form 2:1 or 1:1 complexes (Silva et al., 1993; Silva & Kahne, 1993). (Although it is generally preferable to study biologically active molecules in an aqueous environment similar to the conditions in which they normally function, there are cases where this is not possible. CRA$_3$ and its complexes aggregate even at low concentrations in water, making it impossible to establish the preferred stoichiometry. In such a case, altering the conditions can permit one to obtain useful information.) The results of several spectroscopic studies of the behavior of CRA$_3$ and its aglycone in the presence of increasing concentrations of Mg^{2+} showed that CRA$_3$ forms a stable 2:1 complex with Mg^{2+} while its aglycone forms a 1:1 complex. Thus, the CRA$_3$ sugars somehow stabilize the 2:1 complex in solution, even though the 1:1 complex is less hindered.

The aforementioned spectroscopic investigations established the stoichiometry of the CRA$_3$–Mg^{2+} complex in methanol but not the structure. The formula Mg(AB)$_2$C$_2$, where AB is a chiral bidentate ligand like CRA$_3$ and C is an achiral monodentate ligand like water, can give rise to eight different diastereoisomeric complexes (Fig. 5-8). Analysis of CPK models of the eight possible 2:1 complexes formed by CRA$_3$ indicates that most of them can be excluded on steric grounds (Fig. 5-9). For example, isomers VII and VIII can be excluded because the two CRA$_3$ chelating rings cannot come close enough to form a complex if they are located in the same plane. Isomers I and V can be excluded because unfavorable intermolecular interactions between the trisaccharide of one molecule and the chromophore of the other should prevent their formation. Finally, in isomers II and VI the CDE trisaccharides create so much steric congestion around the metal center that it is not possible to coordinate two water and two

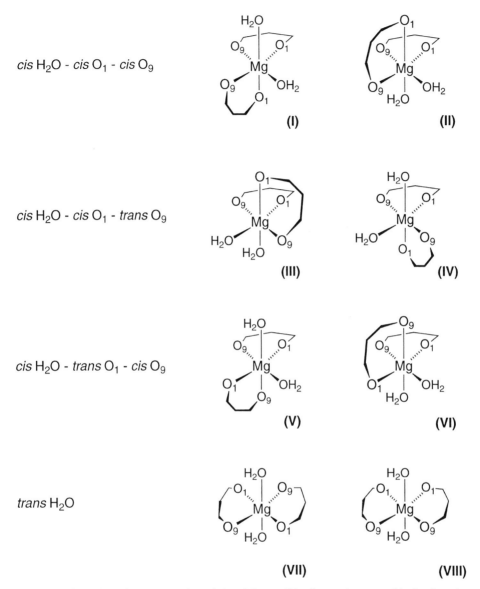

cis H_2O - cis O_1 - cis O_9

(I) (II)

cis H_2O - cis O_1 - trans O_9

(III) (IV)

cis H_2O - trans O_1 - cis O_9

(V) (VI)

trans H_2O

(VII) (VIII)

Figure 5–8. Schematic representation of the eight possible diastereoisomers with the formula $Mg(CRA_3)_2(H_2O)_2$. The bidentate ligand CRA_3 is represented by its chelating O1 and O9 oxygens.

CRA_3 molecules simultaneously. Only isomers III and IV do not present visible steric problems. An NMR study of the 2:1 $CRA_3:Mg^{2+}$ complex in methanol showed that the predominant complex (> 95%) is identical to isomer III (Silva & Kahne, 1993). In this isomer, the CDE trisaccharide of one CRA_3 molecules folds against the chromophore of the other CRA_3 molecule *exactly* as in the complex that binds to DNA in water (Gao et al., 1992). Favorable intermolecular interactions between the CDE trisac-

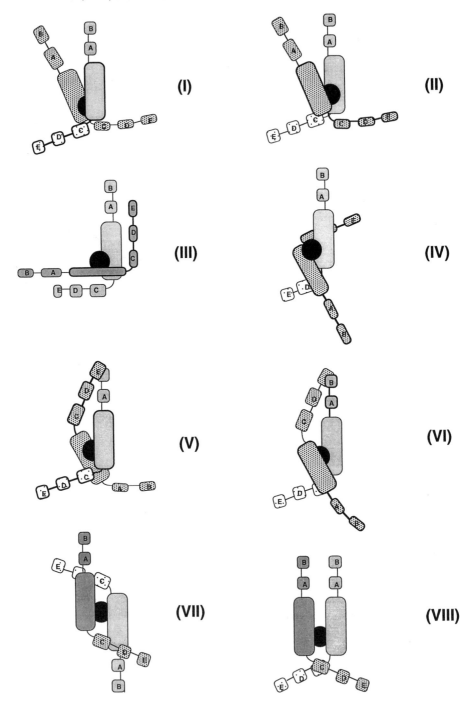

Figure 5–9. Block representation of the eight possible diastereoisomers of the complex $Mg(CRA_3)_2(H_2O)_2$. In the drawings the water molecules have been omitted for clarity. The numbering is the same as in Figure 5-8.

charide and the chromophore apparently stabilize this 2:1 complex relative to the less congested 1:1 complex. Hence, along with the Mg^{2+} ion, the sugars in CRA_3 play a critical role in the assembly of the diastereoisomeric complex that binds to DNA.

Further studies on a series of acidic degradation products of CRA_3 showed that the CDE trisaccharide is essential for stabilizing the 2:1 complex with Mg^{2+}. The disaccharide and the hydrophilic side chain do not appear to be involved in dimer formation. We have, therefore, concluded that the CRA_3 chromophore and the CDE trisaccharide contain all the information required for assembly of a diastereoisomeric metal complex with a coordination geometry identical to that in isomer III and an overall shape with a right-handed twist. This conclusion can be tested by making the appropriate CRA_3 derivative and studying the complexes it forms with Mg^{2+}. It remains to be seen whether the ability to form the appropriate 2:1 complex with Mg^{2+} is sufficient for DNA binding.

The preceding studies have shown that one of the key roles of the chromomycin sugars in DNA binding is to promote assembly of the dimeric complex that binds to DNA. These studies have laid the groundwork for designing new molecules to investigate further structural requirements for DNA binding (Silva et al., 1994). Using the accumulated structural information, it should ultimately be possible to design new DNA binding metal complexes that are simpler than CRA_3 or have other desirable properties. At the same time, the studies on CRA_3 and the role of the sugars in complex formation have provided valuable information about metal complexes in general and how one might go about designing diastereoisomeric metal complexes for purposes *other* than DNA binding.

Calicheamicin

Calicheamicin (Fig. 5-2) is a member of a rapidly growing family of natural products called the ene-diyne antibiotics (Doyle & Borders, 1995; Lee et al., 1991). The ene-diyne antibiotics are a structurally diverse group of DNA cleaving agents that share an ene-diyne motif and have potent antitumor activity. Long before the isolation of any ene-diyne-containing natural products, the behavior and reactivity of simple ene-diynes had already been studied by Sondheimer, Masamune, and Bergman (Bergman, 1973; Darby et al., 1971; Jones & Bergman, 1972; Mayer & Sondheimer, 1966). Bergman and co-workers showed in the early 1970s that acylic ene-diynes cyclize spontaneously to form 1,4-benzenoid diradicals (Fig. 5-10a). Upon the discovery of the first ene-diyne-containing antitumor agents, it was suggested that these agents might undergo a so-called "Bergman cyclization" to produce a diradical intermediate that damages DNA. Subsequent work has confirmed this idea (Goldberg, 1991; Sugiura & Matsumoto, 1993; Sugiura et al., 1989; Zein et al., 1988).

The mechanism for cleavage of DNA by calicheamicin is shown in Figure 5-10b (Lee et al., 1991; Zein et al, 1988). The ene-diyne unit in calicheamicin is contained within a bicyclic ring system with a bridgehead double bond. Because of this bridgehead double bond, the bicyclic ring system is extremely strained. Cyclizaton of the ene-diyne is virtually impossible because it would increase the strain (Snyder, 1990). Therefore, the first step of the cleavage mechanism involves "triggering" calicheamicin to generate a molecule that is capable of cyclizing. Triggering is accomplished by attack of a sulfhydryl or other suitable nucleophile on the methyl trisulfide to produce

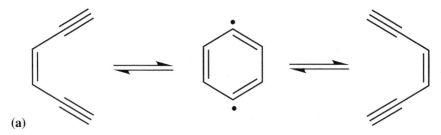

(a)

Figure 5–10a. The spontaneous Bergman cyclization of a simple ene-diyne.

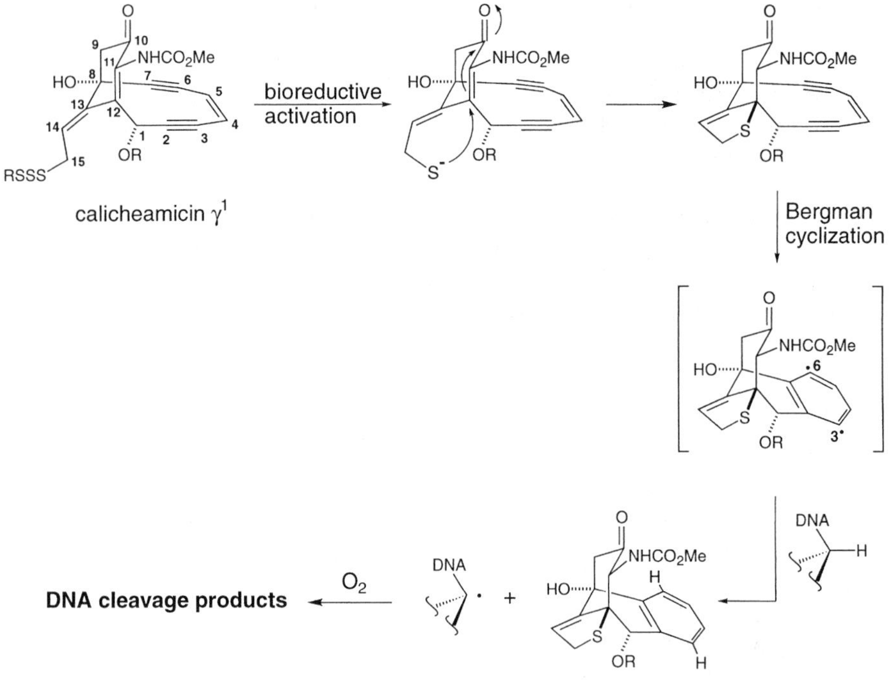

calicheamicin γ^1

bioreductive activation

Bergman cyclization

DNA cleavage products ← O_2 — DNA + HO

DNA

(b)   calicheamicin ε

Figure 5–10b. Proposed mechanism by which calicheamicin γ^1 effects DNA cleavage.

a free thiolate. The thiolate undergoes a 1,4-conjugate addition into the bridgehead double bond to form a tricyclic dihydrothiophene intermediate. The ene-diyne then cyclizes rapidly to produce an aromatic *para* diradical. This diradical abstracts a hydrogen atom from each strand of the ribose phosphate backbone of the DNA. The carbon radicals on the DNA react with oxygen to produce peroxy radicals, which undergo a further series of reactions resulting in DNA strand scission.

All of the ene-diyne antibiotics include a stabilized ene-diyne structural motif as well as a triggering mechanism that activates them for cyclization (Fig. 5-2). However, there are notable differences in the structures and the cleavage behaviors of different ene-diynes. Calicheamicin causes more double-stranded lesions than any other ene-diyne; furthermore, it is the most site-selective ene-diyne cleaving agent identified to date. The highly selective cleavage activity of calicheamicin has been the focus of many studies aimed at understanding how this unusual molecule interacts with DNA.

A great deal can be learned about the interaction of DNA-cleaving agents with DNA from the sizes of the cleavage fragments that are produced. For example, by analyzing the products from the cleavage of a 3' or 5' end-labeled DNA restriction fragment on a polyacrylamide gel in parallel with a marker lane containing radiolabeled fragments of known sizes, one can identify the preferred cleavage sites. Based on this kind of cleavage fragment analysis, TCCT was identified as the preferred cleavage site for calicheamicin (Fig. 5-11). Several other sites, including CTCT, ACCT, TCCA, and TCTC, were identified as secondary cleavage sites. All of these sites contain at least three pyrimidines on one strand, with a corresponding number of purines on the other strand. The cleavage analysis further showed that calicheamicin cleaves at position $+2$ on the pyrimidine strand and position $-2'$ on the purine strand, producing a 3 base-pair stagger in the 3' direction between cleavage sites (Lee et al., 1991; Zein et al., 1988). A stagger in the 3' direction between cleavage sites on opposite DNA strands is characteristic of molecules that cleave from the minor groove (Zein et al., 1988). Subsequent experiments have confirmed the suggestion that calicheamicin, like most small molecule groove binders, interacts with the minor groove of DNA.

The initial cleavage studies also provided information concerning which hydrogen atoms are abstracted by calicheamicin. The 1', 4', and 5'R/5'S hydrogen atoms on the ribose sugars of the DNA are accessible from the minor groove. The cleavage chemistry that occurs following a hydrogen atom abstraction event depends on which of these hydrogen atoms is abstracted (Dedon & Goldberg, 1992). It is often possible to deduce which hydrogen atom is abstracted by treating the DNA cleavage fragments with various reagents and monitoring any changes in mobility of the treated cleavage products compared to the initial cleavage product. From these types of experiments, Zein et al. (1988) concluded that calicheamicin abstracts a 5' hydrogen atom from the $+2$ position of the pyrimidine strand (Fig. 5-12). The data indicated that a different hydrogen atom was abstracted from the purine strand, and Zein suggested that the 4' hydrogen atom was the most likely target.

An elegant series of isotope transfer experiments have identified specifically which hydrogen atoms are abstracted from the DNA as well as where they end up in calicheamicin (DeVoss et al. 1990b; Hangeland et al., 1992). The isotope transfer experiments involved synthesizing DNA duplexes containing deuteriums at the proposed hydrogen

Figure 5–11. A TCCT recognition site in an octamer duplex. **N** can be any nucleotide. The calicheamicin cleavage sites, at nucleotides $+2$ and $-2'$, are indicated by **arrows**.

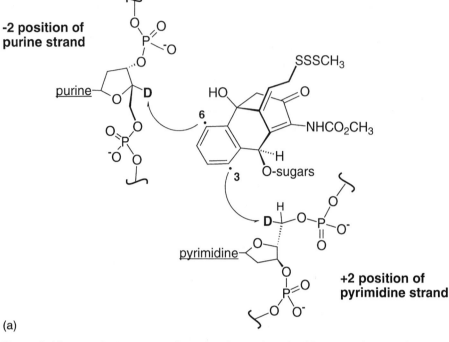

**-2 position of
purine strand**

**+2 position of
pyrimidine strand**

(a)

Figure 5–12a. The hydrogen atom abstraction sites as determined from deuterium transfer experiments. The C3 radical on calicheamicin abstracts a 5′ hydrogen from the pyrimidine strand and the C6 radical abstracts a 4′ hydrogen atom from the purine strand.

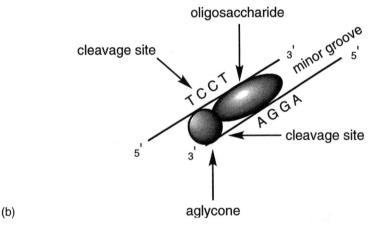

(b)

Figure 5–12b. The orientation of calicheamicin with respect to the recognition site in the minor groove. The oligosaccharide-aryl tail directly contacts the four base-pair recognition sequence.

atom abstraction sites. The deuterium-labeled duplexes were cleaved with calicheamicin, and the fate of the deuterium labels was determined. The results showed that the C3 radical on calicheamicin abstracts a 5' deuterium from the +2 pyrimidine, while the C6 radical abstracts a 4' deuterium from the −2' purine (Fig. 5-12a). These findings indicated that calicheamicin binds to DNA with the oligosaccharide-aryl tail lying in the 3' direction along the pyrimidine strand relative to the aglycone (Fig. 5-12b). Therefore, the oligosaccharide-aryl tail contacts the recognition sequence, which suggested that it may play a significant role in DNA recognition.

It should be noted, however, that molecules that cleave DNA can achieve specificity either through kinetic effects or through thermodynamic binding. In the case of calicheamicin, Townsend pointed out that there could be differences in the rates of Bergman cyclization at different locations in the minor groove. If the Bergman cyclization were the rate limiting step in hydrogen atom abstraction, then the cleavage selectivity might reflect kinetic factors rather than thermodynamic binding (Chatterjee et al., 1994; De Voss et al., 1990a). Therefore, before committing significant energy to studying the interactions of the carbohydrate-aryl tail with the minor groove of DNA, it was important to determine whether the cleavage selectivity was due to kinetics or thermodynamics.

Initially, a degradation product of calicheamicin provided some information on the likely origins of the cleavage selectivity (Walker et al., 1990, 1992). This degradation product, calicheamicin T (Fig. 5-13), which contained the aglycone, the A ring, and the E ring of calicheamicin, was shown to cleave DNA from the minor groove and cause double-stranded lesions in the DNA just like the parent molecule; however, there was virtually no selectivity to the cleavage. This meant there were no differences in the rates of cleavage at different sites in the minor groove. Since it is difficult to imagine how an intact carbohydrate tail could influence the rate of cyclization of the aglycone without binding selectively, it was suggested that the cleavage selectivity observed in the parent molecule reflects selective binding in the minor groove. Subsequent experiments confirmed that the calicheamicin oligosaccharide-aryl tail does bind selectively to DNA in the absence of the ene-diyne (Aiyar et al., 1992; Mah et al., 1994; Nicolaou et al., 1992). More recently, Townsend and co-workers have provided quantitative data on DNA cleavage at different sites that show that there are no differences in cleavage rates; observed differences in the *extent* of cleavage at different sites reflect differences in the binding affinity of calicheamicin for those sites (Chatterjee et al., 1994, 1995). Thus, calicheamicin is the first DNA binding glycoconjugate in which the carbohydrate portion is known to play a primary role in determining the binding selectivity. An important goal now is to understand how the carbohydrate tail of calicheamicin recognizes DNA.

Schreiber and co-workers proposed the first model for how binding selectivity might be achieved by calicheamicin (Hawley et al., 1989). They suggested that the oligosaccharide-aryl tail binds over the TCCT recognition sequence with the iodine on the aromatic ring interacting favorably with the N2 guanine amino groups at positions +2 and +3. This intriguing proposal has been tested by several other groups (Bailly & Waring, 1995; Chatterjee et al. 1995; Li et al., 1994). For example, Joyce, Nicolaou, and co-workers have compared the binding affinities of the native calicheamicin oligosaccharide-aryl tail and several derivatives in which the iodine is substituted by other atoms or groups, including Cl, Br, Me, and H (Li et al., 1994). They found that the native oligosaccharide binds better than any of the derivatives and that the hydro-

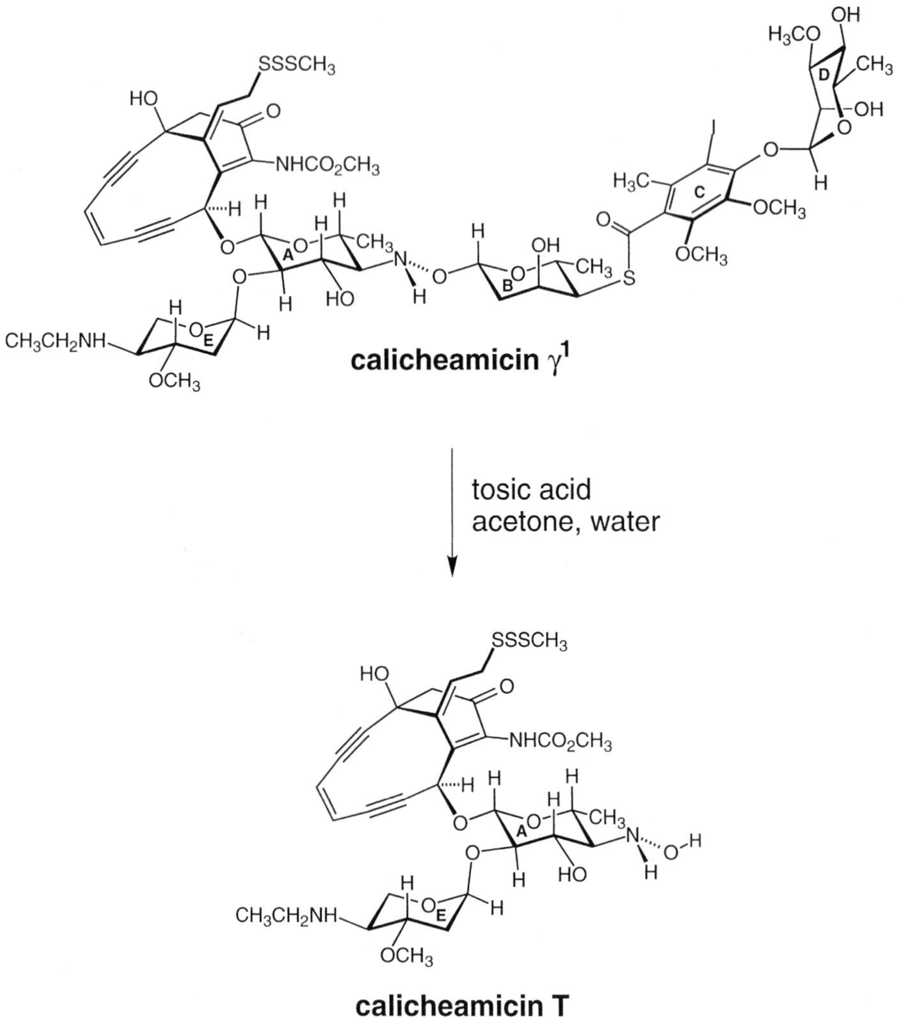

Figure 5–13. Preparation of calicheamicin T (Cal T) by acidic hydrolysis of calicheamicin γ^1.

gen analogue binds the worst. To try and evaluate whether the effect is related to an interaction with one or both N2 amino groups, the guanines were replaced with inosines. There was no effect on binding when the guanine at the $+2'$-position was replaced with inosine, but there was a significant effect when the guanine at the $+3'$ position was replaced. Townsend and Tullius have also used inosine replacements to examine the role of the guanine amino groups in recognition; they, too, have concluded that the amino group at the $+3'$-position plays an important role in binding (Chatterjee et al., 1995). Their studies indicate that the interaction of the aryl-iodide with this amino group involves approximately 1 kcal/mol of binding energy. Hence, Schreiber's proposal that the binding selectivity of calicheamicin involves interactions between the aryl-iodide and the N2 amino groups in the minor groove has proven to be partially correct.

Unfortunately, the origins of the binding selectivity are not that simple. When Schreiber made his proposal, it was thought that calicheamicin recognizes TCCT sequences and a few other GC-containing pyrimidine–purine sites. Subsequent studies, however, showed that calicheamicin also recognizes TTTT sites (Mah et al. 1994; Sugiura & Matsumoto, 1993; Walker et al. 1992). In fact, calicheamicin appears to recognize a large number of sequences that contain three or more pyrimidines. Townsend and Tullius have quantified the differences in affinity of calicheamicin for TCCT and TTTT sites, and the $\Delta\Delta G$ is only about 0.5 kcal/mol (Chatterjee et al. 1995). Furthermore, they have shown that calicheamicin binds to TTTT sites, which have no guanine amino groups, with virtually the same affinity as to CTCT sites, which contain a guanine amino group at position $+3'$. Hence, calicheamicin seems to have a strong preference for pyrimidine–purine tracts in general (with the exception of poly dG-poly dC tracts). A favorable interaction between the aryl iodide of calicheamicin and an N2 amino group in the minor groove is neither sufficient nor necessary for selective binding.

To put the unusual site selectivity of calicheamicin in context, a slight digression is in order. Prior to about 1990, all site-selective minor groove binders were classified according to whether they recognized AT-rich regions of DNA or GC-rich regions of DNA (Yanagi et al., 1991). The polypyrrole antibiotics represented the prototype AT-selective binders; the aureolic acid antibiotics represented the prototype GC-selective binders. The division between AT and GC selective binders seemed inviolable because there are some fundamental differences in the minor groove at AT-rich regions and GC-rich regions. The principal difference is that guanine contains an amino group at the 2-position which protrudes into the minor groove. This guanine amino group makes the minor groove at GC steps very different both sterically and electronically from the minor grove at AT steps (Fig. 5-14). The discovery of a minor groove binder which recognizes pyrimidine–purine tracts rather than AT or GC tracts is interesting because it suggests that pyrimidine–purine tracts share some heretofore unrecognized conformational features that small molecules can recognize. Therefore, studying how calicheamicin interacts with DNA may provide new insight into DNA structure.

To understand the molecular basis for recognition better, several NMR studies of calicheamicin bound to DNA have been carried out (Ikemoto et al., 1995; Kumar et al., 1997; Paloma et al., 1994; Walker et al., 1993). The first NMR study identified the position and location of calicheamicin in the minor groove of an octamer duplex containing an ACCT recognition sequence. This study showed that the oligosaccharide-aryl tail fits closely into the minor groove at the recognition sites. The close fit is achieved because the diameters of the individual rings in the oligosaccharide-aryl tail roughly match the width of the minor groove. Moreover, all of the internal rings in calicheamicin (the A ring, B ring, and C ring) are 1–4 linked to produce an extended chain (Hawley et al., 1989). This chain of subunits has a curvature that follows the curvature of the minor groove. The curvature is caused primarily by the presence of an unusual hydroxylamine bond between the A and B sugars. N—O bonds are rare in biological molecules and they have very different conformational preferences from C—C bonds or C—O bonds (Walker et al., 1991, 1994a). Whereas C—C and C—O bonds prefer a staggered conformation in which the bulky substituents are *anti* to one another, N—O bonds prefer a conformation in which the substituents are oriented at a 120° angle (Fig. 5-15). This is one of only two conformations that avoids the destabilizing four-electron interactions that occur whenever lone pairs on adjacent atoms

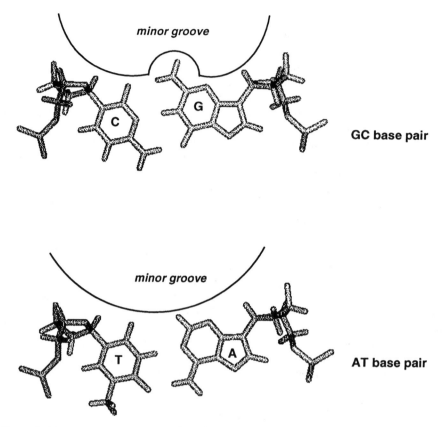

Figure 5–14. A schematic representation of the steric differences in the minor groove at GC versus AT base pairs.

eclipse one another or are *anti* to one another. The curvature that results from holding the two halves of calicheamicin at a 120° angle permits the molecule to maintain contact with the minor groove over several base pairs.

The first NMR study of calicheamicin bound to DNA also defined the contacts made between calicheamicin and the DNA. The A ring was shown to contact the pyrimidine at position +2 and the purine at position −2′; the B ring was shown to contact the pyrimidines at positions +2 and +3; the methyl group on the C ring contacts the pyrimidine at the +3-position, while the iodo substituent contacts the N2 amino group of the corresponding guanine. The aromatic face of the C ring itself is close to the 4′ hydrogen of the purine at +1′, while the D ring contacts the purine at +2′. The E ring protrudes out of the minor groove, with the positively charged ethylamino group making an electrostatic contact to the phosphate of the +2 pyrimidine (Fig. 5-16). Subsequent NMR studies on calicheamicin bound to two different TCCT-containing duplexes have confirmed this general picture for how calicheamicin binds to DNA (Ikemoto et al., 1995; Kumar et al., 1997; Paloma et al., 1994). In addition, one of these studies provided information on how the aglycone fits into the minor groove, showing that it is aligned so that the C3 carbon atom is close to the 5′ methylene at the +2-position of the pyrimidine strand, while the C6 carbon atom is close to the 4′ hydrogen atom at the −2′ position of the purine strand (Ikemoto et al., 1995; Kumar et al., 1997). Thus, binding of calicheamicin in the minor groove precisely orients the aglycone so that it is positioned to abstract the appropriate hydrogen atoms from the DNA backbone. This precise alignment, which is determined primarily by the oligosaccharide-aryl tail, undoubtedly accounts for the ability of calicheamicin to cause primarily double stranded lesions.

Armed with structural and biochemical information pertinent to the way in which calicheamicin contacts DNA, several groups have been trying to design analogues of the oligosaccharide to test the roles of various structural features in determining affinity and specificity. Nicolaou and co-workers have achieved notable success in this area, designing dimers of the oligosaccharide that bind much more tightly than the oligosaccharide itself (Bifulco et al., 1996; Nicolaou et al., 1995, 1996).

Unfortunately, although the NMR studies provided critical information that can be used as a basis for the design of calicheamicin analogues, they have not provided much insight into the origins of the binding specificity. There are a few hydrogen bonds from calicheamicin to the DNA backbone and to certain bases; there is also a close contact between the iodine and the N2 amino group at the +3′-position. These interactions as well as van der Waals contacts between calicheamicin and the DNA are presumably energetically favorable. However, they do not explain why calicheamicin binds preferentially to pyrimidine–purine sequences. Many of the interactions, such as the hydrogen bonds between calicheamicin and the DNA backbone, could presumably form to any DNA sequences. Other interactions, such as the iodine–amino group contact,

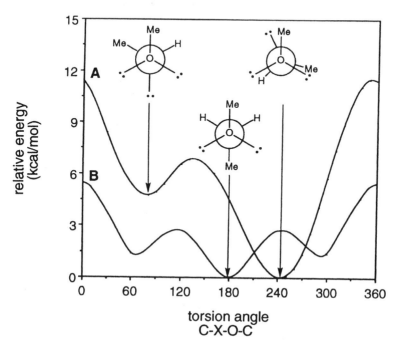

Figure 5–15. Rotational profiles for MeNHOMe (**A**) and MeCH₂OMe (**B**).

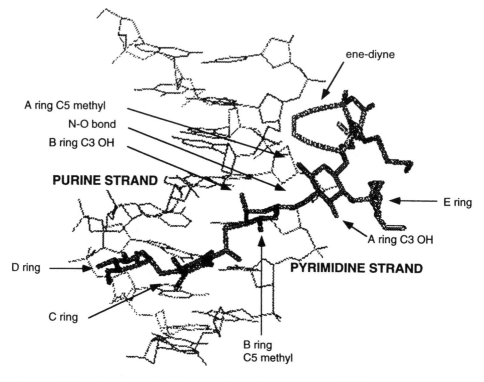

A ring C5 methyl
N-O bond
B ring C3 OH

ene-diyne

PURINE STRAND

E ring

A ring C3 OH

D ring

PYRIMIDINE STRAND

C ring

B ring
C5 methyl

Figure 5–16a. Front view of calicheamicin bound in minor groove of DNA.

are present only in some pyrimidine–purine sequences. Thus, the direct contacts that are observed between calicheamicin and the DNA do not adequately explain the selectivity. What, then, can account for the preference of calicheamicin pyrimidine–purine sites?

One possible explanation was proposed based on the initial NMR studies on calicheamicin bound to DNA (Walker et al., 1993, 1994b). It was observed that calicheamicin causes some distortions in the DNA upon binding. For example, although the global conformation of the DNA in the complex is B form, and the Watson–Crick base pairing is maintained, the conformation of the ribose sugar at the +3 pyrimidine of the recognition sequence distorts significantly upon binding calicheamicin (Kumar et al., 1997; Walker et al. 1993). In addition, there are some notable changes in the twist angles between base pairs upon binding (Kumar et al., 1997). In contrast, the conformation of calicheamicin itself does not change significantly upon DNA binding (Ikemoto et al., 1995; Paloma et al., 1994; Walker et al., 1993). These findings raise the possibility that calicheamicin binds to pyrimidine–purine tracts because they can adapt more easily to the shape of calicheamicin. Sequence-dependent DNA flexibility is an accepted mechanism by which proteins can achieve binding selectivity, and small molecules should also be able to exploit differences in the ability of various DNA sequences to adapt upon binding. For sequence-dependent DNA flexibility to play a role in binding, the small molecule must be relatively rigid and have a shape that is not

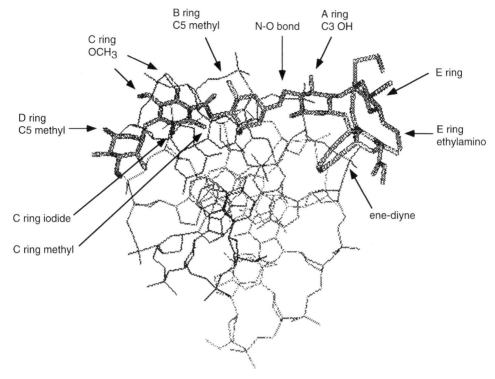

C ring
OCH₃

B ring
C5 methyl

N-O bond

A ring
C3 OH

E ring

D ring
C5 methyl

E ring
ethylamino

C ring iodide

ene-diyne

C ring methyl

Figure 5–16b. Side view of calicheamicin bound in the minor groove of DNA. The functional groups that point out of the groove are indicated. The ene-diyne and the C ring iodo and methyl substituents, which are also noted, point in towards the floor of the groove.

perfectly complementary to the DNA. The DNA is then forced to adapt to maximize the complementarity of fit to the small molecule. Calicheamicin is a relatively rigid molecule and it is clear from the NMR studies that the DNA does undergo some conformational changes in order to bind calicheamicin.

Unfortunately, it is not a simple matter to evaluate the role of sequence-dependent DNA flexibility in binding. There is a great deal of research focused on understanding the energetic cost of distorting DNA at different sequences, but at this point there have been no studies addressing whether pyrimidine–purine tracts undergo certain types of conformational adjustments more readily than other sequences. Nevertheless, there are some notable differences between pyrimidine–purine tracts and all other sequences that at least provide a basis for speculation. Because one strand contains only purines and the other strand contains only pyrimidines, one would expect the energy available from stacking interactions to be different along the two DNA strands. Stacking interactions are believed to make a large contribution to the overall stability of DNA. Although it has not been possible to define the conformation of the DNA upon binding calicheamicin precisely using NMR, there is evidence that the twist angles in the recognition sequence are smaller than for idealized B form DNA. Perhaps smaller twist angles can be accommodated more readily by sequences containing a run of purines because the stacking interactions between purines at adjacent steps are more favorable than in a mixed sequence. Alternatively, Dedon and co-workers, noting that most calicheamicin cleavage sites have a pyrimidine flanking the purine strand on the 3' side, have suggested that there may be a transient kink at the pyr–pur step (Yu et al., 1995). This structural perturbation may, in turn, make it possible for the bulky aglycone to insert into the minor groove. Although the DNA conformation cannot be established with sufficient precision using NMR to determine if there is a local kink, there is evidence for conformational changes in some of the ribose sugars in this region of the DNA.

Further work on calicheamicin is clearly necessary to understand its selectivity better. One question that has not been addressed yet is whether, in fact, calicheamicin binds in a similar mode to all its recognition sites. Although cleavage evidence suggests that it does, there is no direct structural proof, and it has been suggested that different conformations of calicheamicin could recognize different sequences of DNA (Langley et al., 1994). A second question is whether all the calicheamicin recognition sites will show similar DNA distortions. It is likely that the selectivity of calicheamicin reflects a combination of sequence-dependent flexibility in conjunction with some base-specific contacts. The relative importance of the two factors probably varies from site to site, and it may be possible to alter the selectivity of calicheamicin by manipulating the base-specific contacts. A full understanding of how calicheamicin achieves selectivity will require a combination of more structural work with additional biochemical studies on oligosaccharide analogues designed to test various hypotheses about the recognition event.

Why Sugars?

In the introduction to this chapter we wondered why nature has used sugars as building blocks in the design of so many DNA binders. A comparison of CRA_3 and calicheamicin shows that the sugars in DNA-binding glycoconjugates can play very different roles. In CRA_3 the sugars promote the assembly of a diastereoisomeric metal complex with a right-handed twist. This metal complex induces a large conformational change in the DNA that turns the minor groove into a complementary binding pocket. In calicheamicin, the oligosaccharide-aryl tail is a linear molecule with a shape tailored for a tight fit in the minor groove of DNA. At first glance there would seem to be few structural similarities between the CRA_3–Mg^{2+} complex and calicheamicin. However, a more general consideration of the role of the sugars reveals some key similarities between these two DNA binding glycoconjugates. The most obvious similarity is that the sugars in both molecules are hexoses, or six-membered ring sugars. Unlike other rings, six-membered rings are not conformationally flexible; they typically adopt single conformations in solution. Thus, six-membered rings make relatively rigid building blocks. The NMR structures of the DNA complexes of both CRA_3 and calicheamicin show that the individual sugar rings bind to DNA in the same conformation they adopt in solution. Hence, it can be said that the individual sugars in these two DNA-binding glycoconjugates are *preorganized* for binding. In calicheamicin, the hexoses are connected using linkages that produce a shape complementary to the minor groove. In CRA_3, the hexoses actually play a key role in the assembly of two CRA_3 molecules to form a higher order structure that binds to DNA. Thus, the CRA_3 sugars, which are themselves preorganized, help organize the CRA_3–Mg^{2+} complex that binds to DNA.

The sugars in CRA_3, calicheamicin, and most other DNA binding glycoconjugates have another striking similarity: they are unusually hydrophobic. They are deoxygenated at one or more positions around the ring. Moreover, the oxygen substituents are often found as ethers or esters rather than free hydroxyls. The hydrophobicity of the sugars promotes DNA binding out of water (Ding & Ellestad, 1991).

CRA_3, calicheamicin, and most other DNA binding glycoconjugates are complicated molecules. Nevertheless, studies such as the ones described above are beginning to shed light on the many roles that sugars can play and reveal the relationships between structure and function. It should be possible to abstract the essential structural features from molecules such as CRA_3 and calicheamicin and use them in the design of new carbohydrate-based DNA binders. In the case of CRA_3, studies on the role of the sugars are also leading to new strategies for the design of self-assembling diastereoisomeric metal complexes. Although there is still a great deal to learn about how to exploit carbohydrates in the design of DNA binders or other useful molecules, we have clearly come a long way from the view that the sugars in most bacterial glycoconjugates are necessary only because they promote membrane transport or otherwise improve the pharmacokinetics of a biologically active aglycone.

6

Enzymatic Synthesis of Carbohydrates

Martin Hendrix and Chi-Huey Wong

Complex carbohydrates and their conjugates in biological systems are either structural or informational molecules. Many cell-surface carbohydrates of glycoconjugates are involved in various types of biochemical recognition processes (Carver, 1993; Hakomori, 1985; Kobata, 1993; Lasky, 1992; Lemieux, 1989; Paulson, 1992; Weiss et al., 1992), including growth, development, immune responses, infection, cell adhesion, metastasis, and numerous signal transductions. These compounds often exist in minute quantities and are difficult to isolate, characterize, and synthesize in amounts large enough for biological study and therapeutic evaluation. Although only 7–8 monosaccharides are commonly found as building blocks in mammalian systems (Beyer et al., 1981), the multifunctionality of these monomers allows the assembly of an immense variety of complex structures. Different tetrasaccharide structures numbering several million, for example, can be assembled from this small number of building blocks, when considering the branching, stereochemistry of glycosylation, and modifications. Oligosaccharides are therefore effective in coding for the vast amounts of information required in various biological recognition processes. The enzymes (e.g., glycosyltransferases) involved in the biosynthesis and degradation of oligosaccharides have been valuable reagents for the synthesis of saccharides (Ichikawa et al., 1992b; Nilsson, 1988b). These glycoenzymes are also interesting targets for inhibition as they may cause metabolic disorders and diseases (Look et al., 1993a; Palcic et al., 1989; Winchester & Fleet, 1992). Figure 6-1 lists a number of naturally occurring carbohydrate structures and glycoconjugates which possess important biological activities. Given the complexity and multifunctional nature of these structures, the synthesis of these and related compounds or designed mimetics in a practical manner remains a challenging task. The following describes various types of enzymatic reactions which combined with simple chemical transformations may provide a practical route to numerous sugar-like compounds.

Enzymatic Synthesis of Monosaccharides and Related Structures Based on Aldolases

Aldolases catalyze the stereospecific aldol condensation of an aldehyde acceptor with a ketone donor. Two types of aldolases are known: type I aldolases are found primarily in animals and higher plants. Their catalysis proceeds through a covalent interme-

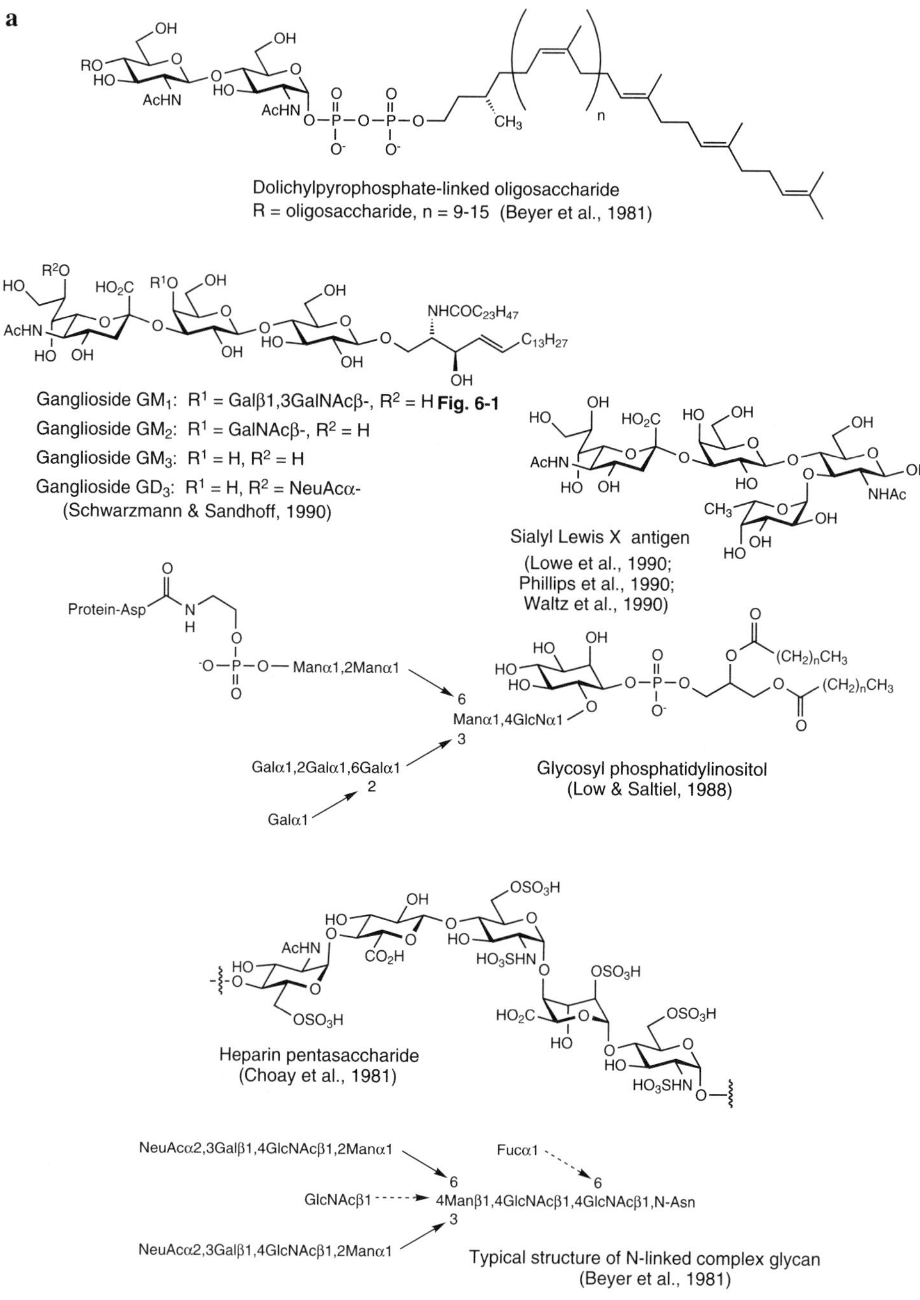
a
Dolichylpyrophosphate-linked oligosaccharide
R = oligosaccharide, n = 9-15 (Beyer et al., 1981)
Ganglioside GM1: R1 = Galβ1,3GalNAcβ-, R2 = H
Fig. 6-1
Ganglioside GM2: R1 = GalNAcβ-, R2 = H
Ganglioside GM3: R1 = H, R2 = H
Ganglioside GD3: R1 = H, R2 = NeuAcα-
(Schwarzmann & Sandhoff, 1990)
Sialyl Lewis X antigen
(Lowe et al., 1990;
Phillips et al., 1990;
Waltz et al., 1990)
Protein-Asp
Manα1,2Manα1
Manα1,4GlcNα1
Galα1,2Galα1,6Galα1
Galα1
Glycosyl phosphatidylinositol
(Low & Saltiel, 1988)
Heparin pentasaccharide
(Choay et al., 1981)
NeuAcα2,3Galβ1,4GlcNAcβ1,2Manα1
Fucα1
GlcNAcβ1
4Manβ1,4GlcNAcβ1,4GlcNAcβ1,N-Asn
NeuAcα2,3Galβ1,4GlcNAcβ1,2Manα1
Typical structure of N-linked complex glycan
(Beyer et al., 1981)

Figure 6–1. Biologically active carbohydrates and glycoconjugates.

b

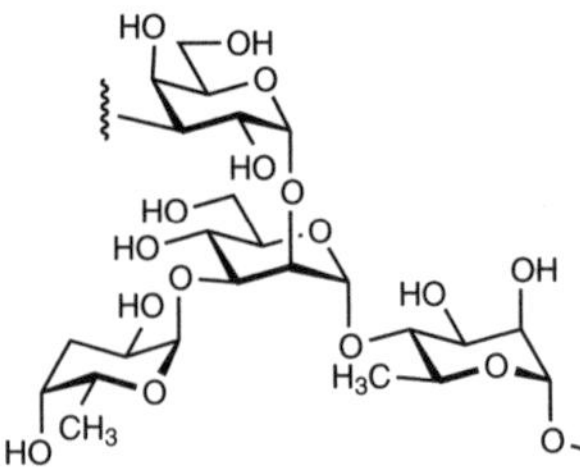

Salmonella O-antigen (Cygler et al., 1991)

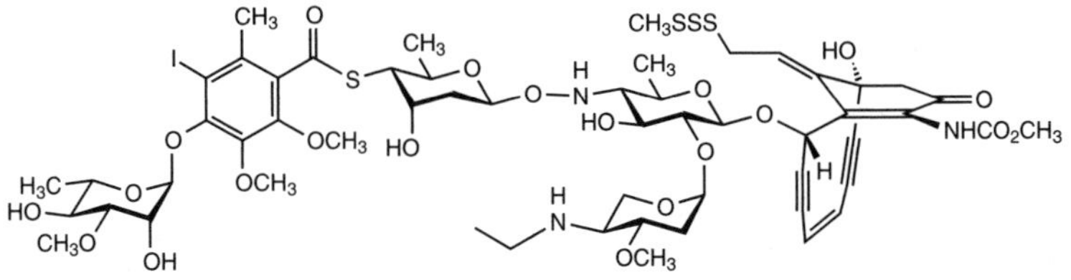

calicheamicin $\gamma_1{}^I$ (Nicolaou & Dai, 1991)

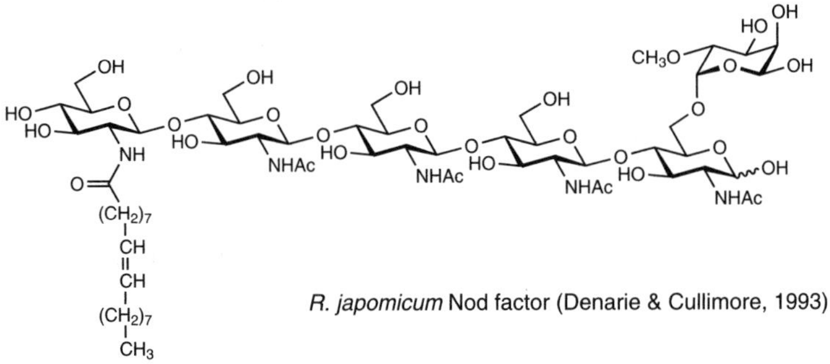

R. japomicum Nod factor (Denarie & Cullimore, 1993)

c

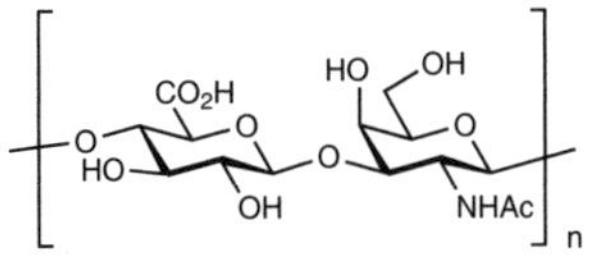

hyaluronic acid
(van Brunt, 1986)

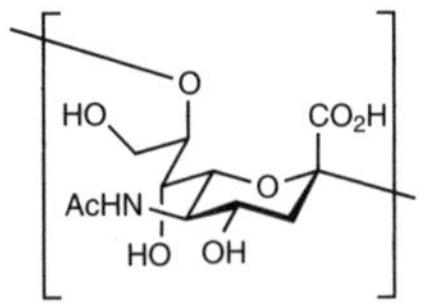

poly-$\alpha2,8$-linked sialic acid
(Baumann, 1993)

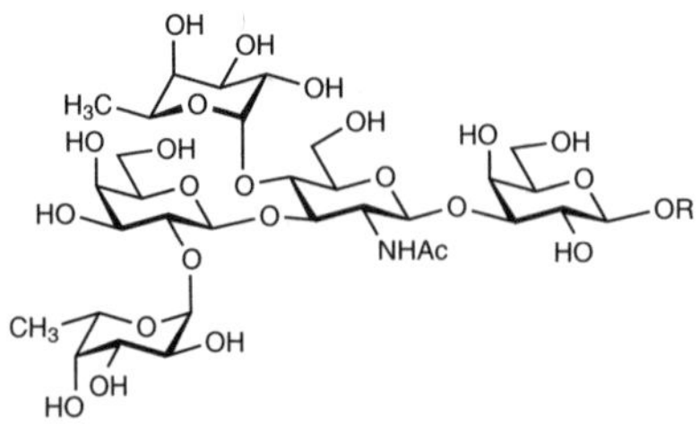

Blood Group O Antigen
and receptor for ulcer-causing *H. pylori*

Figure 6–1. (continued)

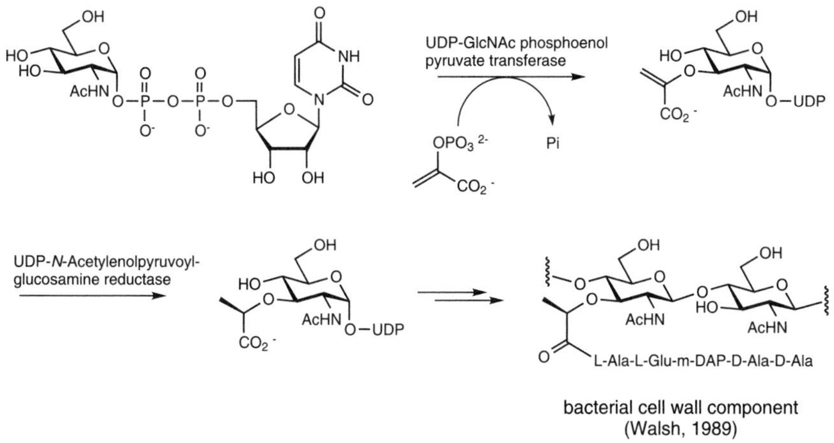

Figure 6–1. (continued)

diate with a Schiff base formed between the donor and the enzyme. They do not require a metal cofactor. In contrast type II aldolases, found predominately in microorganisms, require a Zn^{2+} cofactor as a Lewis acid in the active site. No covalent enzyme intermediate is formed. Generally, aldolases are very specific for their donors but accept a variety of unnatural acceptor substrates. Often the stereoselectivity of the reaction is controlled by the enzyme and is highly predictable. Figure 6-2 indicates the aldolases that have been reported. The arrows indicate the C—C bond associated with aldol reactions. These enzymes have been employed in the synthesis of natural as well as nonnatural monosaccharides. We shall now discuss the synthetically most useful members of this class in more detail.

Fructose 1,6-Diphosphate (FDP) Aldolase (E.C. 4.1.2.13)

FDP aldolase catalyzes the reversible aldol addition reaction of dihydroxyacetone phosphate (DHAP) and D-glyceraldehyde 3-phosphate (G3P) to form D-fructose 1,6-diphosphate (FDP) and has a $K_{eq} \approx 104$ M^{-1} in favor of FDP formation (Fig. 6-3). Both type I and II enzymes have been isolated from a variety of eukaryotic and prokaryotic sources. Most of the mechanistic studies have been carried out on FDP aldolases from rabbit muscle (Kuo & Rose, 1985; Rose & Warms, 1985) or yeast (Belasco & Knowles, 1983; Kadonaga & Knowles, 1983). The X-ray structure (2.7 Å resolution) (Sygusch et al., 1987) of the enzyme from rabbit muscle (RAMA) indicates that Lys-229 is responsible for Schiff base formation with DHAP. The distance between the Schiff base nitrogen and the phosphate on the G3P (glyceraldehyde-3-phosphate) in the active site of RAMA has been calculated to be 8.3 Å (Lees & Whitesides, 1993). Another lysine residue in the active site (Lys107), approximately 8.9 Å from the site of Schiff base formation, is thought to stabilize the negatively charged phosphate of G3P (Fig. 6-3). The enzyme RAMA has been cloned and overexpressed, and site-directed mutagenesis studies have identified Asp-33 as another critical residue in catalysis (Morris & Tolan, 1993).

Generally, the type I FDP aldolases exist as tetramers (molecular weight [MW] ca. 160 kDa), while the type II enzymes are dimers (MW ca. 80 kDa). The sequences of

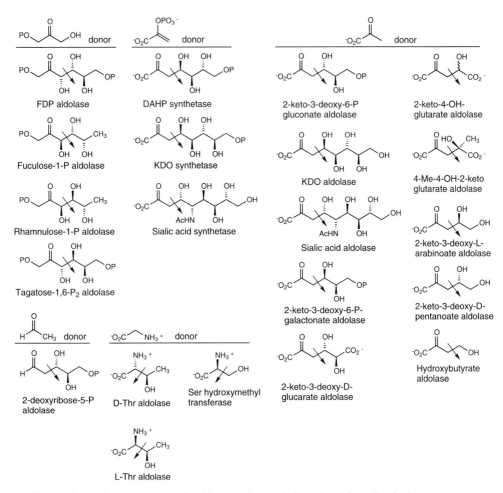

Figure 6–2. Natural substrates for aldolases. **Arrows** indicate bond formed or broken.

the type I enzymes have a high degree of homology (>50%) with the active site sequence being conserved throughout evolution (Sygusch et al., 1987). However, significant differences in the C-terminal regions have been identified which may be important in mediating substrate specificity. Some of the type I aldolases are commercially available and have useful specific activity (ca. 60 units/mg). These enzymes are not particularly air sensitive, although there is an active site thiol group which may be oxidized. The half-life of the free enzyme, which is approximately 2 days in aqueous solution at pH 7.0 (von der Osten et al., 1989b), can be increased by immobilization or by enclosure in a dialysis membrane (Bednarski et al., 1989). The type II aldolase from yeast is commercially available. Other microbial aldolases have been cloned recently (Alefounder et al. 1989; Schwelberger et al., 1989; von der Osten et al., 1989a,b). Recent X-ray (Dreyer & Schulz, 1993) and inhibition studies of a type II fuculose-1-phosphate aldolase and other DHAP-dependent type II aldolases suggest that the reaction proceeds through a Zn^{2+}-coordinated enediolate intermediate (Fessner et al., 1996). The enzyme from *Escherichia* (*E.*) *coli* has no thiol group in the active site, and has a half-life of approximately 60 days in 0.3 mM Zn^{2+} at pH 7.0 (von der Osten et al., 1989b). Despite the small degree of homology in primary sequence between the enzymes from *E. coli* (type II) and rabbit muscle (type I), studies have shown that they possess almost the same substrate specificity.

To date FDP aldolase, especially RAMA, has been the most widely used aldolase in organic synthesis. This enzyme accepts a wide range of aldehyde acceptor substrates with DHAP as the donor to generate 3S, 4R vicinal diols, stereospecifically, with D-threo stereochemistry (Bednarski et al., 1989). Suitable acceptors include unhindered aliphatic aldehydes, and monosaccharides and their derivatives. Phosphorylated aldehydes react more rapidly than their unphosphorylated analogues (Bednarski et al., 1986), but aromatic, sterically hindered aliphatic and α,β-unsaturated aldehydes are generally not substrates. The specificity for the donor substrate is much more strin-

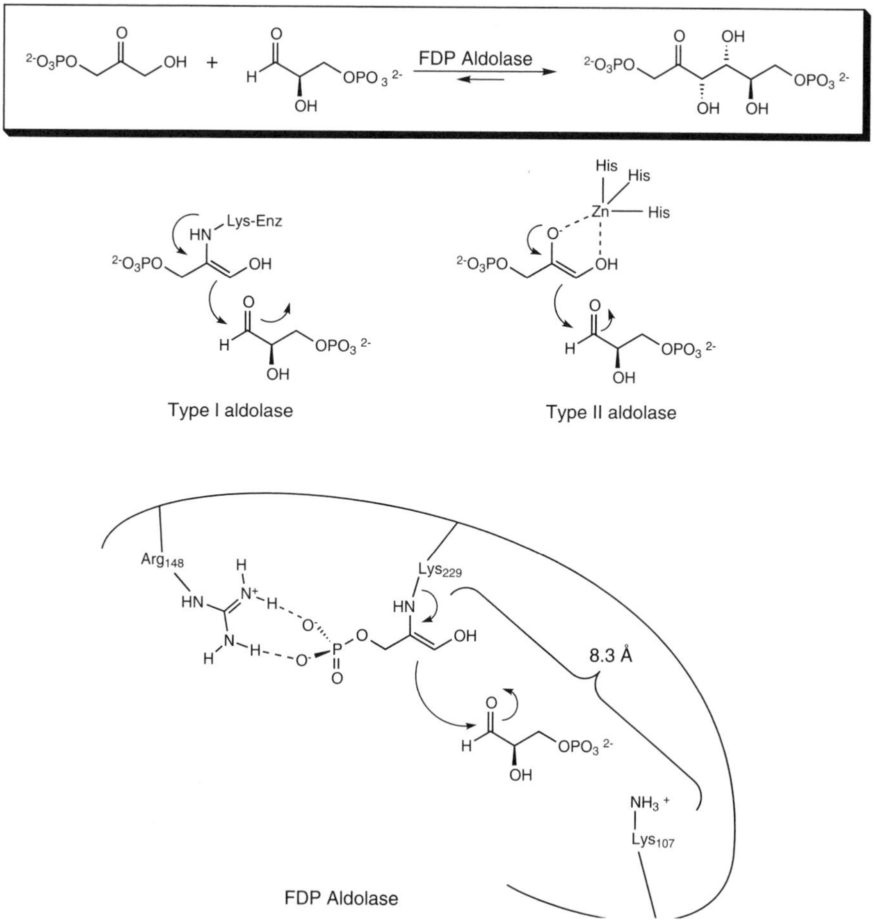

Figure 6–3. Mechanism of aldolase-catalyzed aldol reactions.

gent, allowing only isosteric replacements on the phosphate side of the molecule (Bednarski et al., 1989; Bischofberger et al., 1988; Fessner & Sinerius, 1994).

The diastereoselectivity exhibited by FDP aldolase was found to be dependent on the reaction conditions. In a kinetically controlled experiment, the D-enantiomer of G3P was accepted by the enzyme with a 20:1 preference over the L-enantiomer (Bednarski et al., 1989), whereas no selectivity was observed for the unphosphorylated aldehyde (Lees & Whitesides, 1993). Racemic mixtures of unnatural aldehyde acceptors may be partially resolved in a kinetically controlled reaction if they bear a suitably positioned anionic group. This kinetic selectivity has been rationalized by fixing the conformation of the aldehyde through binding of both the carbonyl group and the anionic substituent. The latter interaction would involve the lysine residue that normally interacts with the phosphate on G3P (Lees & Whitesides, 1993). When the substrate is constrained by this two-point binding, nucleophilic *si* face attack onto the *R* aldehyde is faster since the *S* aldehyde has a bulky group on the same face where the nucleophile approaches. A study with a series of carboxylated substrates supports this model (Lees & Whitesides, 1993). When racemic α-hydroxyaldehyde carboxylates ($n = 1$–3, 5) were treated with DHAP and RAMA, the substrates which had carboxylates that could best overlap with the second lysine residue gave the highest enantiomeric excess.

In the cases where the product can cyclize to form a six-membered ring hemiketal, racemic aldehydes are stereoconvergently transformed into one adduct if the enzymatic reaction is under thermodynamic control. Since the reaction is reversible, the product with the fewest 1,3-diaxial interactions will predominate after equilibration. For example, with racemic β-hydroxybutyraldehyde (Bednarski et al., 1989; Durrwachter & Wong, 1988) or 2-hydroxymethyl-4-pentenal (Durrwachter & Wong, 1988) as substrates, only a single diastereomer was obtained with the methyl or allyl group in the equatorial position (Fig. 6-4).

Examples Using FDP Aldolase

The numerous examples using FDP aldolase in organic synthesis that have appeared recently include the production of ^{13}C-labeled sugars (Serianni et al., 1982; Wong & Whitesides, 1983; Wong et al., 1991b), nitrogen-containing sugars (Look et al., 1993a), deoxysugars (Durrwachter et al., 1986; Durrwachter & Wong, 1988; Wong et al., 1983b), fluorosugars (Durrwachter et al., 1986; Wong et al., 1988), and 7, 8, and 9-carbon sugars (Bednarski et al., 1986; Durrwachter & Wong, 1988; Wong & Whitesides, 1983; Wong et al., 1983b). In general, two groups have been used as "masked" aldehydes in the synthesis of the necessary substrates: a terminal olefin that can be cleaved by ozonolysis (Bischofberger et al., 1988), or an acetal that is deprotected through acid hydrolysis (von der Osten et al., 1989a). The α-chiral substituent can be prepared by nucleophilic ring opening of readily-available (*R*)- and (*S*)-glycidaldehyde acetal, or the corresponding thiirane and aziridine (Pederson et al., 1990). Both enantiomers of glycidaldehyde acetal can be prepared by lipase-catalyzed resolution of 3-chloro-2-hydroxypropanal diethyl acetal (Pederson et al., 1990; von der Osten et al., 1989b). Following resolution, the epoxide is simply generated by treating the chloro-alcohol with base. Examples of enzymatic syntheses with some of these nonnatural aldehydes using FDP aldolase are discussed below.

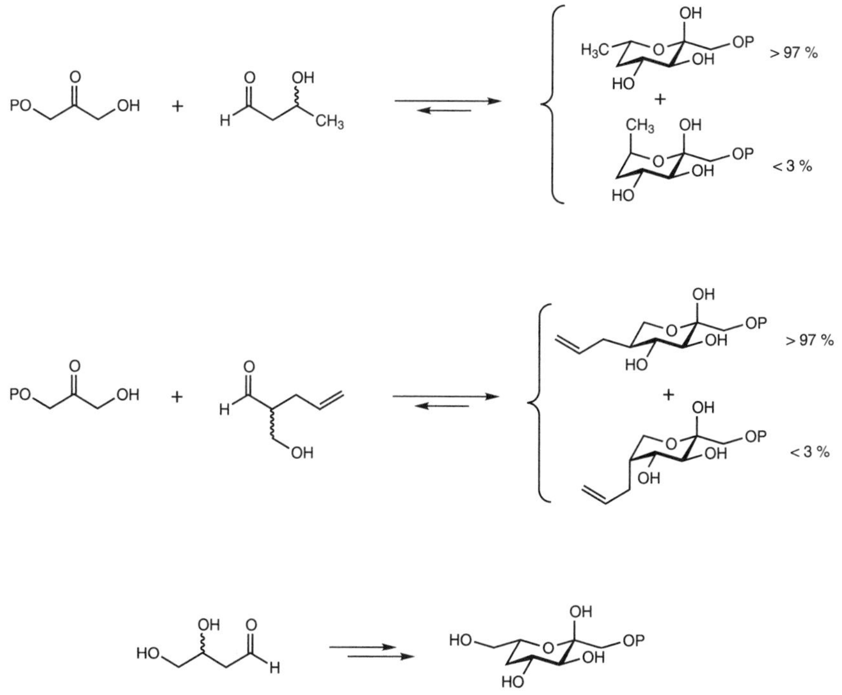

Figure 6–4. Thermodynamically-controlled resolution of racemic aldehydes using FDP aldolase from rabbit muscle (RAMA).

The FDP aldolase is useful for the synthesis of iminocyclitols, which have become increasingly important targets since they have been found to be inhibitors in glyco-processing (Look et al., 1993a). Two potent glycosidase inhibitors, deoxynojirimycin and deoxymannojirimycin, were prepared readily in three steps with RAMA being used in the key C—C bond forming step (Pederson et al., 1988; Ziegler et al., 1988). Using racemic 3-azido-2-hydroxypropanal and DHAP, diastereomeric 6-azido ketoses were formed. Following the acid phosphatase-catalyzed removal of phosphate and subsequent reductive amination (Fig. 6-5), the products were isolated in a 4:1 ratio favoring

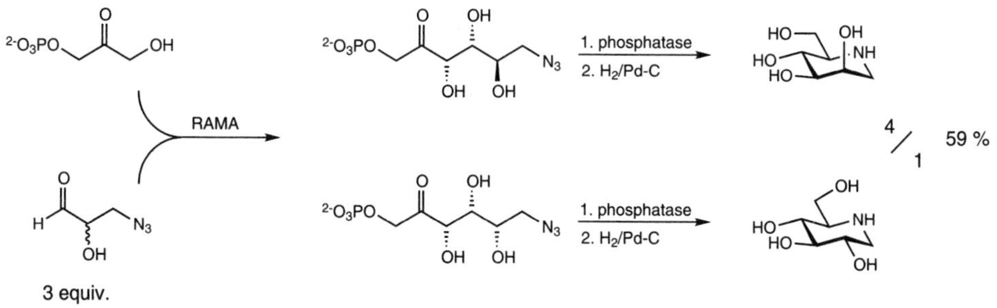

Figure 6–5. Synthesis of azasugars from racemic aldehydes using RAMA.

the manno derivative, which indicates that the D-aldehyde is a better substrate for the enzyme. A similar result was obtained with FDP aldolase from *E. coli* (von der Osten et al., 1989b). Deoxynojirimycin and deoxymannojirimycin can also be prepared exclusively by utilizing the respective optically pure azidoaldehydes (von der Osten et al., 1989b). Both (*R*)- and (*S*)-3-azido-2-hydroxypropanal were obtained in >98% ee by resolution of the racemic acetal precursor catalyzed by LP-80 (Fig. 6-6) (Pederson et al., 1990; von der Osten et al., 1989b). Using a similar strategy several other 3-substituted-2-hydroxypropanal acetals were prepared which were precursors to a variety of 6-substituted ketoses (Pederson et al., 1990).

RAMA has also been used in the synthesis of iminocyclitols corresponding to *N*-acetylglucosamine and *N*-acetylmannosamine from (*S*) and (*R*)-3-azido-2-acetamidopropanal, respectively, using the analogous RAMA-catalyzed aldol reaction/reductive amination procedure (Fig. 6-6) (Kajimoto et al., 1991b).

When 2-azidoaldehydes were used as substrates for the RAMA-catalyzed aldol reaction and subsequent reductive amination, a number of polyhydroxylated pyrrolidines were synthesized (Fig. 6-7) (Hung et al., 1991; Kajimoto et al., 1991a; Liu et al., 1991b). Both (2*R*,5*R*)- and (2*S*,5*R*)-bis(hydroxymethyl)-(3*R*,4*R*)-dihydroxypyrrolidine were synthesized from racemic 2-azido-3-hydroxypropanal, and 1,4-dideoxy-1,4-imino-D-arabinitol was prepared from CBz-protected α-aminoacetaldehyde (Pederson et al., 1988) or azidoacetaldehyde (Hung et al., 1991). In the latter case, the kinetic product of the aldol addition was converted to the (2*R*,5*R*) pyrrolidine (Hung et al., 1991), while the thermodynamic product gave the (2*S*,5*R*) stereoisomer (Kajimoto et al., 1991a; Liu et al., 1991a). Similar transformations using 2-azido-3-acetamidopropanal gave the iminocyclitols structurally related to *N*-acetylglucosamine (Takaoka et al., 1993). Using

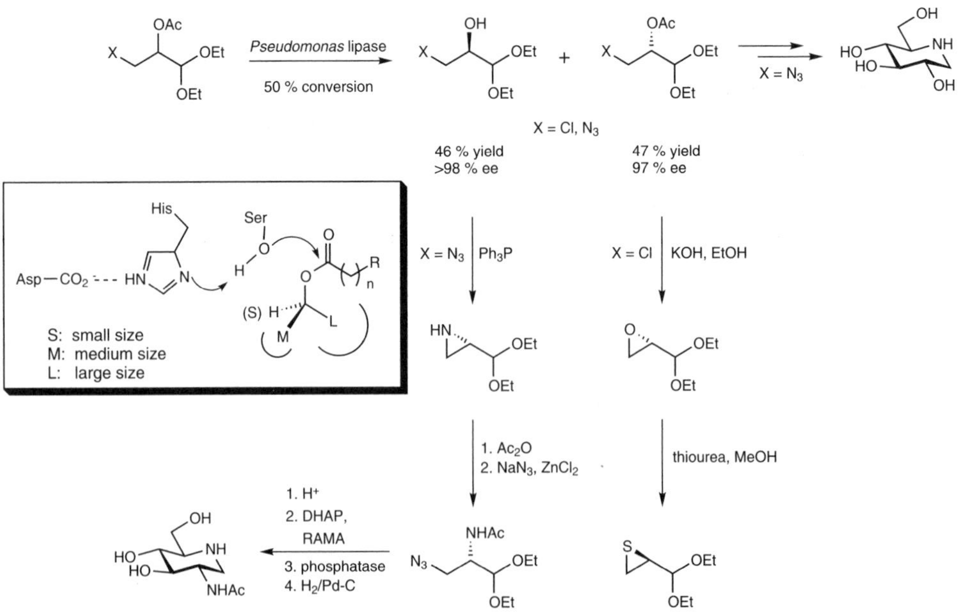

Figure 6–6. Synthesis of azasugars from enantiomerically pure aldehydes using RAMA. Indicated in the box is the mechanism and the stereochemical preference of the lipase reaction.

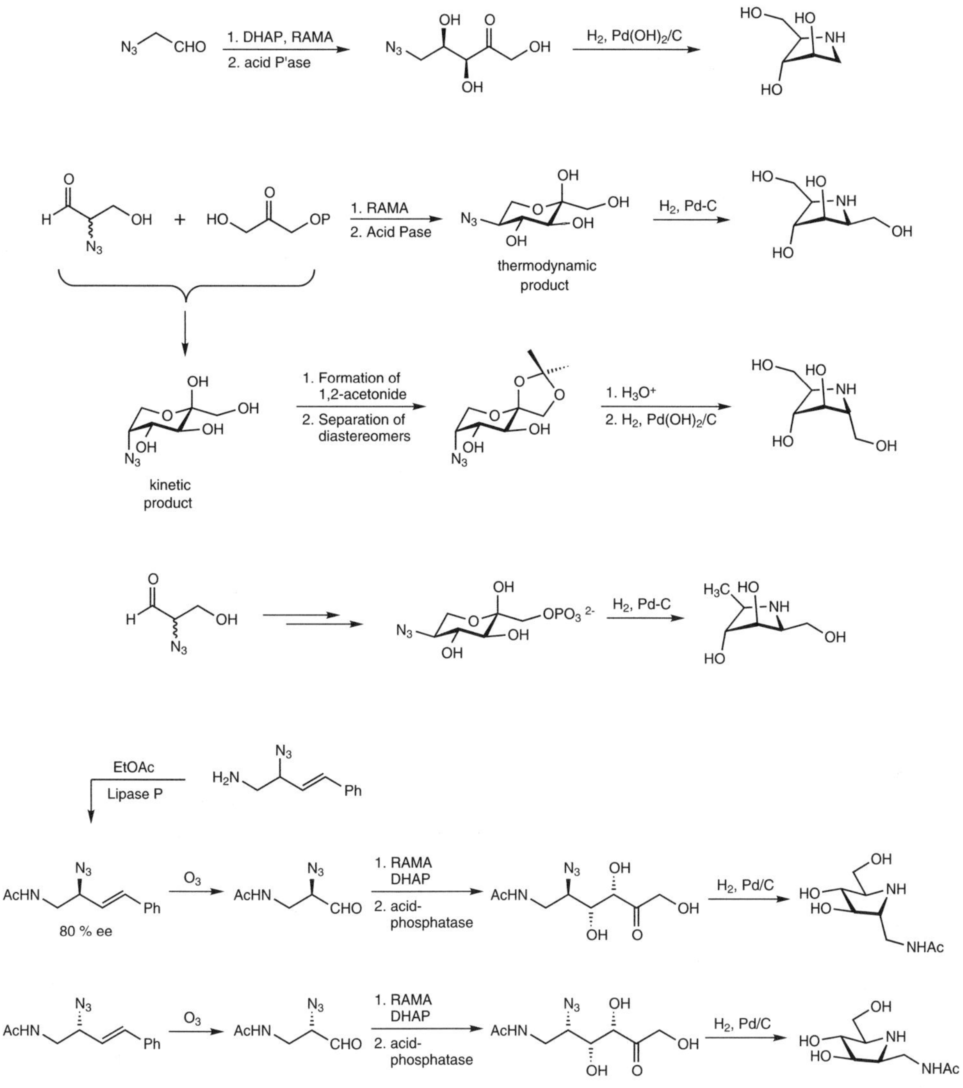

Figure 6–7. Synthesis of polyhydroxylated pyrrolidines using RAMA.

a four-carbon azidoaldehyde, a homo-aza sugar has been prepared (Fig. 6-8) (Henderson et al., 1994a). However, attempts to prepare seven-membered iminocyclitols *via* the same reductive amination process were unsuccessful (Liu et al., 1991b). Isomerization of the azidoketose to aldose followed by reductive amination, however, provides a seven-membered iminocyclitol (Moris-Varas et al., 1996). The Pd-catalyzed reductive amination of azido-ketoses is stereoselective. Reduction presumably proceeds through an imine intermediate which is further reduced to form a *trans*-product. When an axial-OH group is present in the imine intermediate, hydrogen is delivered to the opposite face due to steric reasons.

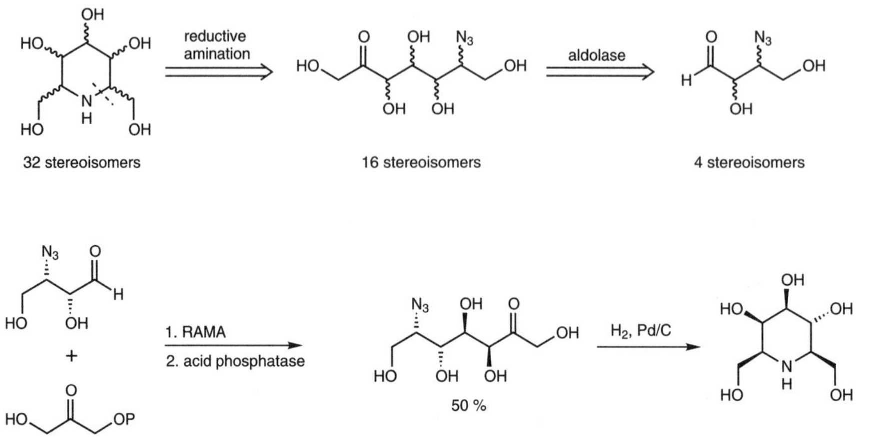

Figure 6–8. Aldolase-mediated synthesis of six-membered homoazasugars.

Prior to removal of the phosphate group, 6-deoxy iminocyclitols and their analogues can also be prepared by direct reductive amination of the aldol products (Kajimoto et al., 1991b). Again, reduction is thought to occur through an imine intermediate (Kajimoto et al., 1991a; Liu et al., 1991a).

In addition to its application to iminocyclitols, RAMA has been used in the synthesis of a variety of oxygen heterocycles. For example, 3-deoxy-D-arabinoheptulosonic acid (DAHP), which is an important intermediate in the shikimate pathway for the biosynthesis of aromatic amino acids in plants, has been synthesized using RAMA (Fig. 6-9) (Turner & Whitesides, 1989). Using N-acetylaspartate β-semialdehyde as a substrate for RAMA, DAHP was synthesized in four steps and 13% overall yield. The enzymatic step produced the desired D-*threo* stereochemistry, and chemical reduction of the ketone gave the desired (6R)-stereoisomer in 60% diastereomeric excess. Other analogues of DAHP are also potentially available by this route since RAMA utilizes a number of aldehyde acceptors.

Products from the RAMA-catalyzed aldol reactions have also been used to synthesize carbocycles (Schmid & Whitesides, 1990). The protected chloro-sugar which is generated from the aldolase reaction between DHAP and chloroacetaldehyde, undergoes nucleophilic attack at the carbonyl carbon with allylmagnesium bromide. Radical cyclization conditions then yield a cyclitol (Fig. 6-10). When the Grignard reagent is

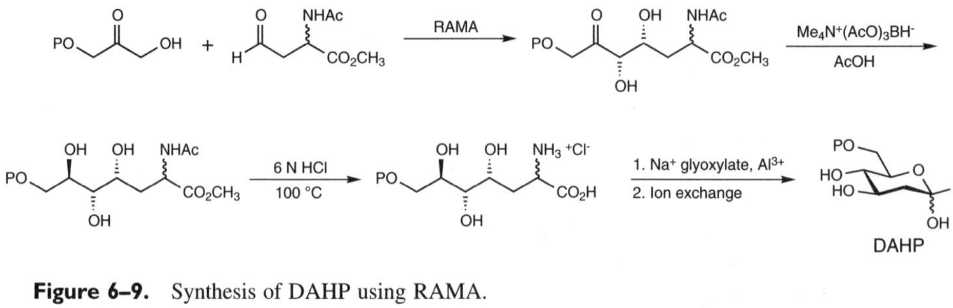

Figure 6–9. Synthesis of DAHP using RAMA.

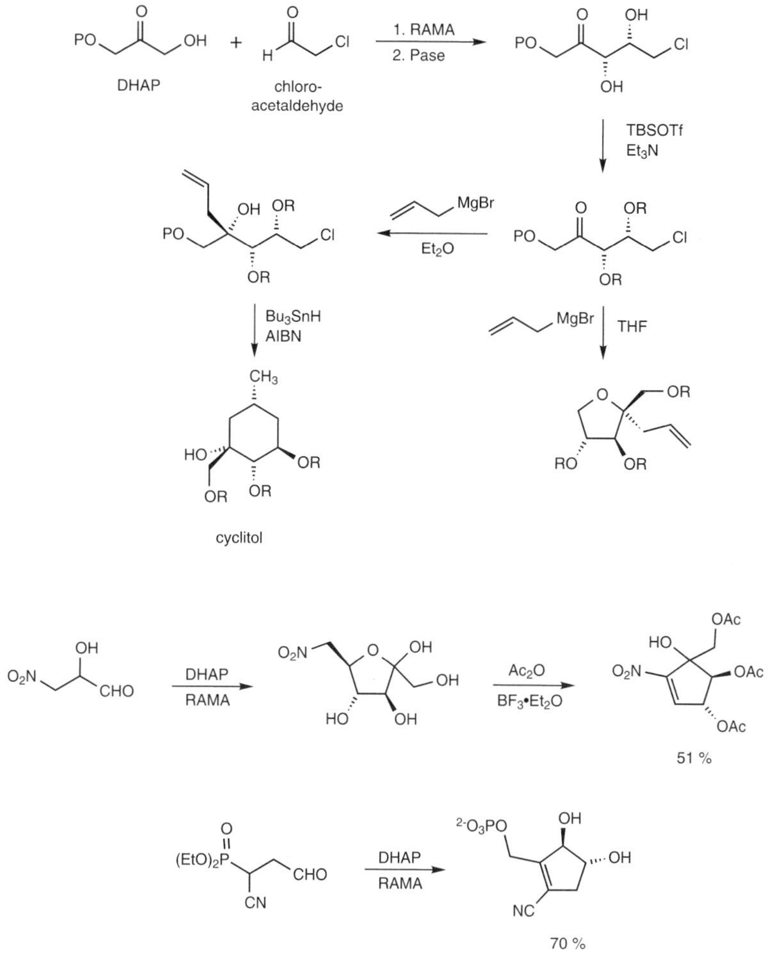

Figure 6–10. Synthesis of carbocycles and a *C*-glycoside using RAMA.

added in tetrahydrofuran, addition to the ketone is followed by displacement of the chloride, giving the *threo*-pentulose-C-allylglycoside as the major product. In principle, this technique can also be applied to the products of other aldolase-catalyzed reactions to provide stereoisomeric derivatives. Using nitroaldehydes or phosphonoaldehydes as acceptors the products undergo intramolecular nitroaldol reactions or Horner-Wadsworth-Emmons reaction spontaneously to give carbocycles (Chou et al., 1995; Gijsen & Wong, 1995a).

The discovery that RAMA also accepts pentose and hexose phosphates as viable substrates provided a new route to novel high-carbon sugars. Several of these compounds have been synthesized, including analogues of sialic acid and 2-keto-3-deoxyoctanoate (KDO) (Bednarski et al., 1986).

Synthesis with RAMA is not limited to carbohydrate derivatives. In fact, during the synthesis of the naturally occurring beetle pheromone (+)-*exo*-brevicomin the aldolase

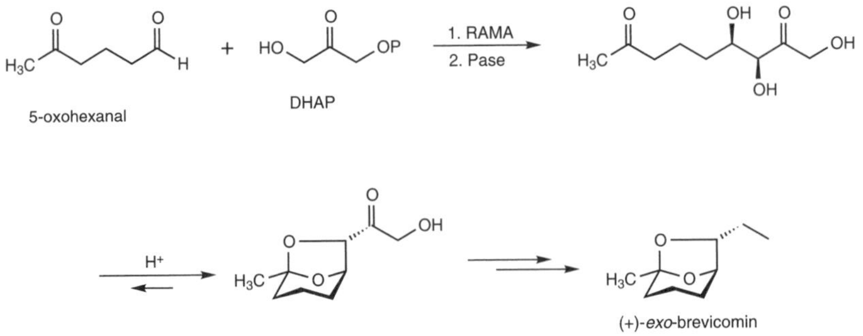

Figure 6–11. Synthesis of (+)-*exo*-brevicomin using RAMA.

reaction was used to establish the two chiral centers in the target molecule (Fig. 6-11) (Schultz et al., 1990). Another novel use of RAMA is exemplified in the synthesis of homo-C-nucleosides (Fig. 6-12) (Liu & Wong, 1992).

As shown in the many examples above, FDP aldolase generates several types of ketose monosaccharides. However, many of the important naturally occurring carbohydrates are aldoses. One way to generate aldoses from FDP-aldolase products is by using glucose isomerase (GI) or xylose isomerase (E. C. 5.3.1.5) (Liu & Wong, 1992). This enzyme catalyzes the isomerization of fructose to glucose and is used in the food industry for the production of high fructose corn syrup. GI also accepts fructose analogues, which are modified at the 3, 5, and 6 positions (Fig. 6-13) (Durrwachter et al., 1986). Various FDP aldolase products can be equilibrated to a mixture of the ketose and aldose forms; subsequently, the two isomers can be separated using Ca^{2+} or Ba^{2+}-treated cation exchange resins. Aldose analogues including 6-deoxy, 6-fluoro, 6-O-methyl, and 6-azidoglucose have been synthesized using this FDP aldolase–GI methodology. Not all products from FDP aldolase reactions give the desired aldose when

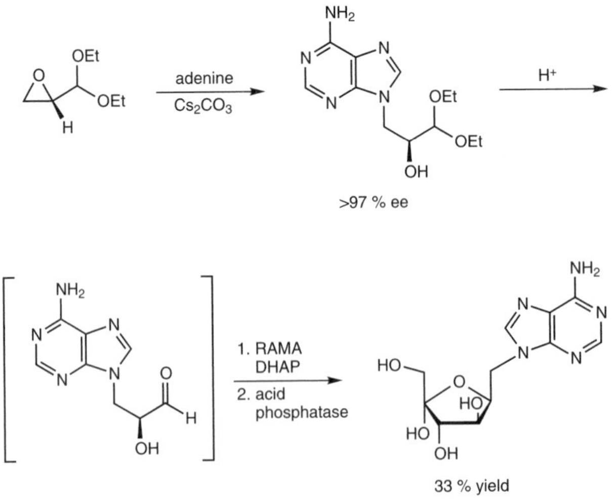

Figure 6–12. Synthesis of homo-C-nucleosides.

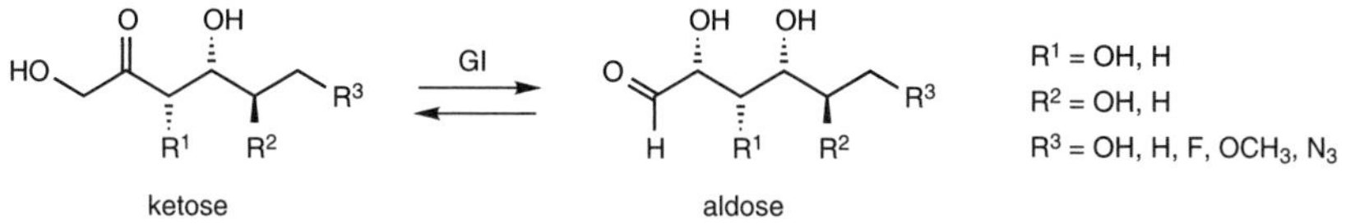

Figure 6–13. Substrate specificity of glucose isomerase (GI).

subjected to GI. In some cases, for example, 5-deoxy-D-fructose, the equilibrium lies completely in favor of the ketose, while in other cases the products from the aldolase reaction are simply not substrates.

An alternative method for aldose generation from the FDP aldolase products is through the so-called "inversion strategy" (Borysenko et al., 1989). Here, monoprotected dialdehydes are used as substrates for FDP aldolase, affording protected aldehyde ketoses. The carbonyl group is then reduced stereoselectively, either chemically with borohydride or enzymatically with iditol dehydrogenase, and the aldehyde is subsequently deprotected to afford the aldose (Fig. 6-14). The NADH-dependent iditol dehydrogenase (IDH) from *Candida utilis* (also known as sorbitol or polyol dehydrogenase (E.C. 1.1.14)) has been used previously to reduce the carbonyl group of ketoses to give the (S)-alcohol (Wong et al., 1983b). The corresponding (R)-alcohol was obtained by reduction of the ketone with $NaBH(OAc)_3$ giving a 2:1 mixture of diastereomers from which the minor (S)-epimer was selectively removed through oxidation catalyzed by IDH.

While most of the FDP aldolase reactions use the commercially available aldolase from rabbit muscle, there is still considerable interest in determining the utility of FDP aldolase from other sources. In fact, the type II aldolases from *E. coli* (von der Osten et al., 1989b) and yeast (Alefounder et al., 1989), which have been subcloned and overexpressed, have the potential to become useful synthetic catalysts since they have enhanced stability compared to RAMA (*vide supra*) and are available from a microbial as opposed to a mammalian source.

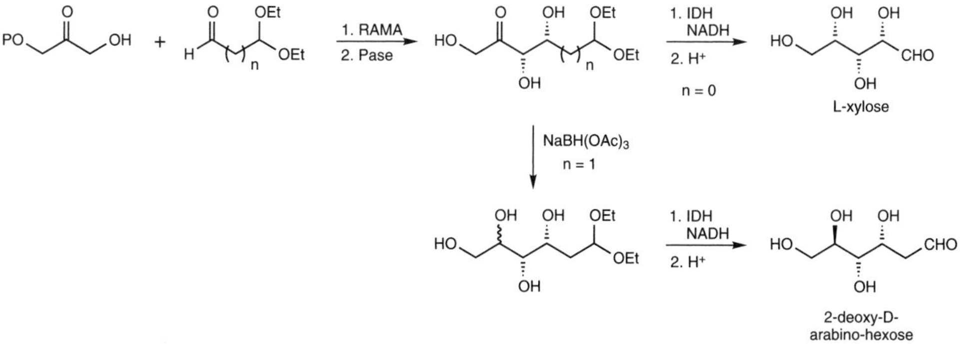

Figure 6–14. Use of the "inversion strategy" to synthesize L-xylose and 2-deoxy-D-*arabino*-hexose.

Fuculose 1-Phosphate (Fuc 1-P) Aldolase (E.C. 4.1.2.17), Rhamnulose 1-Phosphate (Rha 1-P) Aldolase (E.C. 4.1.2.19), and Tagatose-1,6-diphosphate (TDP) Aldolase

Besides FDP aldolase, there are three other known aldolases that use DHAP as the donor in the aldol reaction: Fuc 1-P aldolase, Rha 1-P aldolase, and TDP aldolase. Fuc 1-P aldolase catalyzes the reversible condensation of DHAP and L-lactaldehyde to give L-Fuc-1-P, and with the same substrates Rha-1-P aldolase produces L-Rha-1-P (Fig. 6-15). Both of these enzymes are type II aldolases and are found in several microorganisms. Fuc-1-P aldolase (Fessner et al., 1991; Ozaki et al., 1990) and Rha-1-P aldolase (Fessner et al., 1991) have been cloned, overexpressed, and purified. Tagatose 1,6-diphosphate (TDP) aldolase, a type-I aldolase involved in the galactose metabolism of *cocci*, catalyzes the reversible condensation of G3P with DHAP to give D-TDP.

Both Fuc-1-P and Rha-1-P aldolase accept a variety of aldehydes generating vicinal diols with D-*erythro* and L-*threo* configurations, respectively. While the enzymes yield products with the (3*R*)-configuration stereospecifically, the stereoselectivity at C-4 is somewhat diminished for a few substrates (Ozaki et al., 1990). However, sterically unhindered 2-hydroxyaldehydes normally give very high diastereoselectivities.

These two aldolases also show significant kinetic preference for the L-enantiomer of 2-hydroxyaldehydes (>95:5), so they facilitate the kinetic resolution of a racemic mixture of these compounds (Fessner et al., 1992). Both enzymes have been used in the synthesis of rare ketose 1-phosphates (Fessner et al., 1992) and several aza and deoxyazasugars (Kajimoto et al., 1991a; Lees & Whitesides, 1992; Liu et al., 1991a; Wang et al., 1993b). Figure 6-16 illustrates the use of Fuc-1-P aldolase in the synthesis of five-membered azasugars and Figure 6-17 demonstrates the enantiocomplementary nature of Rha-1-P aldolase and FDP aldolase.

Fuc-1-P and Rha-1-P aldolases have also been utilized in whole-cell systems with DHA and catalytic inorganic arsenate (Drueckhammer et al., 1989). With L-lactaldehyde and DHA-arsenate as the substrates in the Rha-1-P aldolase reaction, the aldol

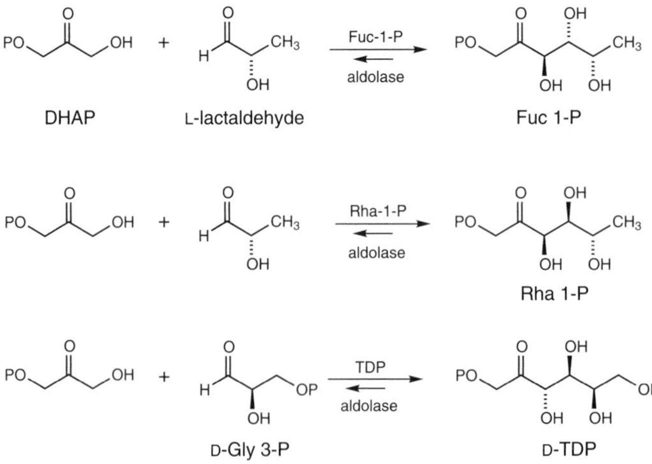

Figure 6–15. Aldol addition reaction catalyzed *in vitro* by Fuc-1-P aldolase, Rha-1-P aldolase, and TDP aldolase.

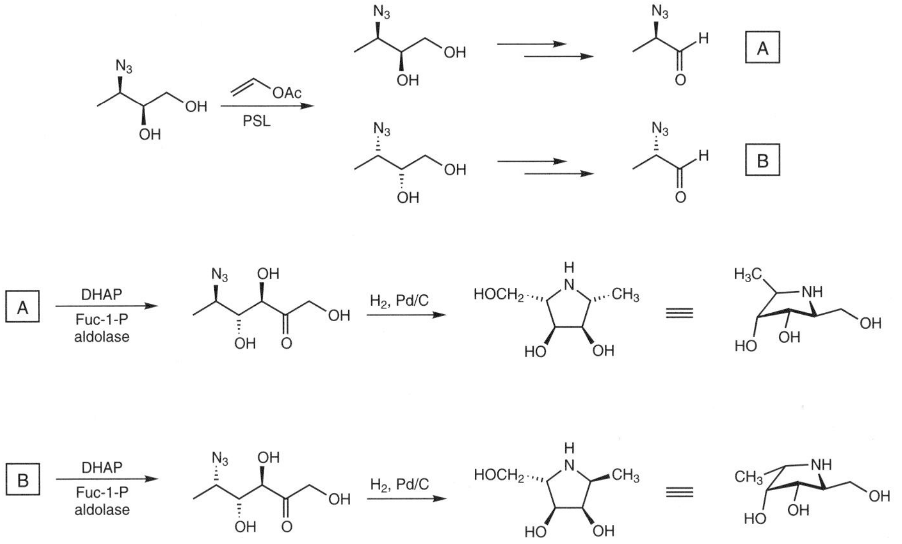

Figure 6–16. Synthesis of azasugars using fuculose-1-phosphate aldolase and *Pseudomonas* lipase (PSL).

product L-rhamnulose was subsequently isomerized to L-rhamnose catalyzed by rhamnose isomerase present in the cell. No such isomerization was observed with L-xylulose when the corresponding aldol product using glycoaldehyde was the substrate. Studies have since shown that both rhamnose and fucose isomerase accept only substrates with a (2R,3R) and (2S,3R) stereochemistry, respectively (Fessner et al., 1992).

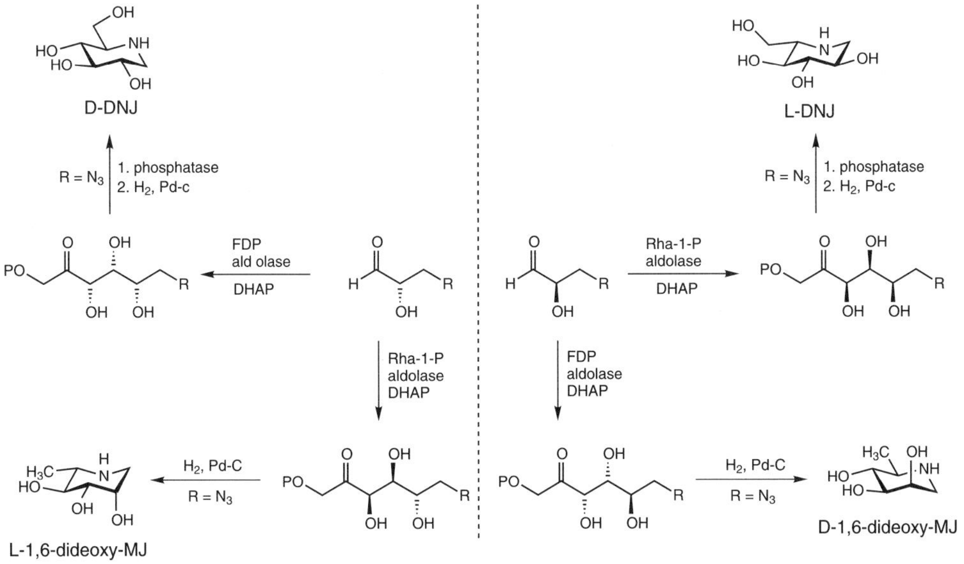

Figure 6–17. Aldolase-mediated synthesis of each enantiomer of deoxymannonojirimycin and 1,6-dideoxymannonojirimycin.

An alternative method for preparing the same ketose 1-phosphates available from the Rha-1-P and Fuc-1-P aldolases is to use the rhamnose and fucose isomerases (Fessner et al., 1992). The isomerase only partially converts the aldose to the ketose, so a second step which phosphorylates the ketose drives the reaction to completion. Several ketose 1-phosphates were prepared using rhamnulose kinase in the second step to give overall yields of 72%–90%.

TDP aldolase has been isolated from several sources (van Rooijen et al., 1991). In particular, the enzyme from *E. coli* has a narrow pH profile with an optimum at pH 7.5, but it still displays acceptable activity within the pH range 6.5-7. Like the other DHAP aldolases, TDP aldolase accepts a variety of acceptor substrates for the aldol reaction, including glycoaldehyde, D- and L-glyceraldehyde, acetaldehyde, and *iso*-butyraldehyde (Fessner & Eyrisch, 1992). However, in all cases investigated so far, a diastereomeric mixture of products was formed. Surprisingly, instead of exhibiting the expected D-tagatose-like *erythro* configuration, >90% of each product possessed the *threo* configuration similar to D-fructose. Only with the natural substrate D-glyceraldehyde, does the major product (D-TDP) have the tagatose configuration. Due to this lack of stereoselectivity, TDP aldolase is not yet synthetically useful; however, with suitable protein engineering, this may change in the future.

Each of the DHAP-dependent aldolases yields a product whose stereochemistry at C3 and C4 is complementary to the other products; that is, all four stereochemical permutations are provided by these aldolases. FDP, Rha-1-P, Tag-1-P, and Fuc-1-P aldolases provide four diastereomers from a variety of nonnatural aldehydes. In principle, any one of the C3/C4 stereoisomeric ketoses is obtainable using these enzymes. Examples have been shown in the synthesis of enantiomeric D- and L-fructose and their derivatives (Fig. 6-18) (Henderson et al., 1994b).

N-Acetylneuraminate (NeuAc) Aldolase (E.C. 4.1.3.3) and NeuAc Synthetase (E.C. 4.1.3.19)

The enzyme NeuAc aldolase, also known as sialic acid aldolase, catalyzes the reversible condensation of pyruvate with D-*N*-acetylmannosamine (ManNAc) to form *N*-acetyl-5-amino-3,5-dideoxy-D-*glycero*-D-*galacto*-2-nonulosonic acid (NeuAc or sialic acid) (Fig. 6-19) (Comb & Roseman, 1960; Uchida et al., 1985). Although the β-anomer predominates in solution, the α-anomer of NeuAc is the substrate for the enzyme, and the initial products of aldol cleavage are α-D-ManNAc and pyruvate (Deijl & Vliegenthart, 1983). The enzyme has a catabolic function *in vivo*, with an equilibrium constant for the retro-aldol reaction of 12.7 M^{-1}. However, for synthetic purposes, an equilibrium favoring the aldol product can be achieved by using excess pyruvate (Uchida et al., 1985). NeuAc aldolase has been isolated from both bacteria and animals; in both cases, it is a Schiff base-forming type I aldolase. The optimum pH for activity is 7.5, but it is still active between pH 6 and 9 and is stable under oxygen (Uchida et al., 1985). The enzymes from *Clostridia* and *E. coli* are now commercially available (Toyobo), and the enzyme from *E. coli* has been cloned and overexpressed (Aisaka et al., 1987). NeuAc aldolase has been used in both free and immobilized form (Auge & Gautheron, 1987; Kim et al., 1988; Liu et al., 1992); in some instances, it has been enclosed in a dialysis membrane (Bednarski et al., 1987).

Synthetic studies have shown that a high conversion (ca. 90%) of ManNAc to NeuAc could be achieved using the isolated enzyme, although several equivalents of pyruvate

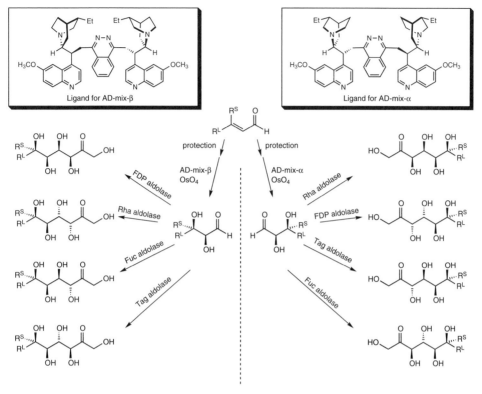

Figure 6–18. Strategy for the enantioselective synthesis of all possible isomers of the 2-ketohexoses.

were required. To eliminate the need for excess pyruvate and to aid in the isolation of products, the NeuAc synthesis can be coupled to a more thermodynamically favorable process. One example using ManNAc as the starting material involved the coupling of the NeuAc aldolase reaction with a sialyltransferase reaction to produce sialylsaccharides (Ichikawa et al., 1991a). Instead of pure ManNAc, a mixture of ManNAc and GlcNAc may be used whereby the GlcNAc is epimerized to ManNAc chemically (Simon et al., 1988) or enzymatically (Kragl et al., 1991) with *N*-acetylglucosamine 2-epimerase.

Extensive substrate specificity studies have been carried out on this enzyme. Only pyruvate is acceptable as the donor; 3-fluoropyruvate, acetylphosphonate, 3-hydroxybutanoate, 2-oxobutyrate, and 3-bromopyruvate are not substrates (Kim et al., 1988). In the case of the acceptor substrate the enzyme is more flexible. Substitutions at C-2, -4, and -6 of ManNAc are allowed with only a slight preference for the absolute stereochemistry at C-4, -5, or -6 being the same as the stereochemistry in ManNAc (Auge et al., 1988, 1990; David et al., 1991). Some pentoses and their analogues are also substrates, but two and three carbon molecules are not accepted.

Recently, there has been significant interest in the synthesis of NeuAc analogues due to their role in cellular biology. NeuAc and derivatives are found at the termini of mammalian glycoconjugates and play an important role in biochemical recognition

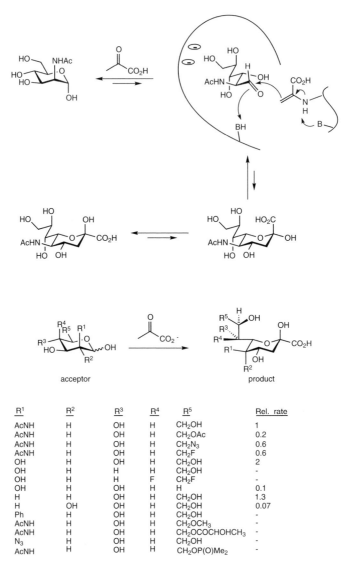

R[1]	R[2]	R[3]	R[4]	R[5]	Rel. rate
AcNH	H	OH	H	CH2OH	1
AcNH	H	OH	H	CH2OAc	0.2
AcNH	H	OH	H	CH2N3	0.6
AcNH	H	OH	H	CH2F	0.6
OH	H	OH	H	CH2OH	2
OH	H	H	H	CH2OH	-
OH	H	H	F	CH2F	-
OH	H	OH	H	H	0.1
H	H	OH	H	CH2OH	1.3
H	OH	OH	H	CH2OH	0.07
Ph	H	OH	H	CH2OH	-
AcNH	H	OH	H	CH2OCH3	-
AcNH	H	OH	H	CH2OCOCHOHCH3	-
N3	H	OH	H	CH2OH	-
AcNH	H	OH	H	CH2OP(O)Me2	-

Figure 6–19. Substrate specificity of NeuAc aldolase.

(Paulson, 1989). Polysialic acids are also found in bacteria and mammalian tissues and may be involved in cell adhesion and cell–cell communication (Lasky, 1992; Paulson, 1992). Because NeuAc aldolase accepts a wide variety of substrates and is readily available, many sialic acid derivatives have been prepared that may be used in biological studies.

Of the many NeuAc derivatives synthesized so far, most give products with S-configuration at C-4. Recent observations indicate, however, that under thermodynamically controlled reaction conditions with certain sugars, the absolute stereochemistry at C-4 can become reversed. For example, in the NeuAc aldolase-catalyzed synthesis of KDO a mixture of (S)-C-4 and (R)-C-4 products were isolated when D-arabinose was the substrate (Auge et al., 1990). Also, NMR studies with several other sugars (e.g., L-mannose, D-gulose, 2-azido-2-deoxymannose) showed that over time, products with a C-4 equatorial group predominated, in some cases disregarding the normal stereochemical preference for the enzyme (Gautheron-Le Narvor et al., 1991; Lin et al., 1992). Apparently, pyruvate attacks the acceptor sugar to give the thermodynamically more stable product, and the facial selectivity is merely a consequence of the preference to form C-4 equatorial product. Several biologically interesting L-sugars were synthesized using this method including L-NeuAc, L-KDO, and L-KDN. The excess pyruvate can be decomposed with pyruvate decarboxylase to simplify the product isolation (Lin et al., 1992).

A facile synthesis of 9-O-acetylNeuAc was achieved by regioselective irreversible acetylation of ManNAc catalyzed by subtilisin, followed by NeuAc aldolase-catalyzed condensation of the resulting 6-O-acetylManNAc with pyruvate (Kim et al., 1988). This two-step enzymatic synthesis provided 9-O-acetyl-NeuAc in approximately 80% yield. Some other 9-O-acylated NeuAc derivatives may also be prepared in this fashion (Liu et al., 1992).

The synthesis of NeuAc *in vivo* is accomplished using NeuAc synthetase (Blacklow & Warren, 1962), which catalyzes the irreversible condensation of phospoenol pyruvate (PEP) and *N*-acetylmannosamine. Although, this enzyme has not yet been isolated and characterized, it may prove synthetically useful in the future since the forward reaction is favored kinetically.

3-Deoxy-D-manno-2-octulosonate Aldolase (E.C. 4.1.2.23) and 3-Deoxy-D-manno-2-octulosonate 8-Phosphate Synthetase (E.C. 4.1.2.16)

3-Deoxy-D-manno-2-octulosonate aldolase, also known as 2-keto-3-deoxyoctanoate (KDO) aldolase, catalyzes the reversible condensation of pyruvate with D-arabinose to form KDO with an equilibrium constant in the cleavage direction of 0.77 M^{-1} (Fig. 6-20). KDO and its activated form 2-chloro-1-methylpyridinium (CMP)-KDO are key intermediates in the synthesis of the outer membrane lipopolysaccharide (LPS) of gram-negative bacteria, so analogues of KDO may inhibit LPS biosynthesis or LPS binding protein (Andersson et al., 1990; Schumann et al., 1990). KDO aldolase has been isolated and purified from *E. coli* (Ghalambor & Heath, 1966a) and *Aerobacter cloacae* (Ghalambor & Heath, 1966b). Preliminary investigations of this enzyme showed high specificity for KDO in the direction of cleavage, whereas the condensation reaction proceeded with some flexibility; several unnatural substrates, including D-ribose, D-xylose, D-lyxose, L-arabinose, D-arabinose 5-phosphate, and *N*-acetylmannosamine were reported to be weak substrates (relative rate <5% cf. D-arabinose) (Ghalambor & Heath, 1966a). More recent studies on the substrate specificity showed that the KDO aldolase from *Aureobacterium barkerei*, strain KDO-37-2, accepted an even wider variety of substrates which included trioses, tetroses, pentoses and hexoses (Sugai et al., 1993). The enzyme is specific for substrates having (R)-configuration at C-3, but the stereochemical requirements at C-2 are less stringent. Under kinetic control, the C-2 S-configuration is favored while the C-2 R configuration is favored thermodynamically. Several aldol addition reactions were conducted on a preparative scale including the synthesis of KDO itself which was obtained in a 67% yield. In each case, at-

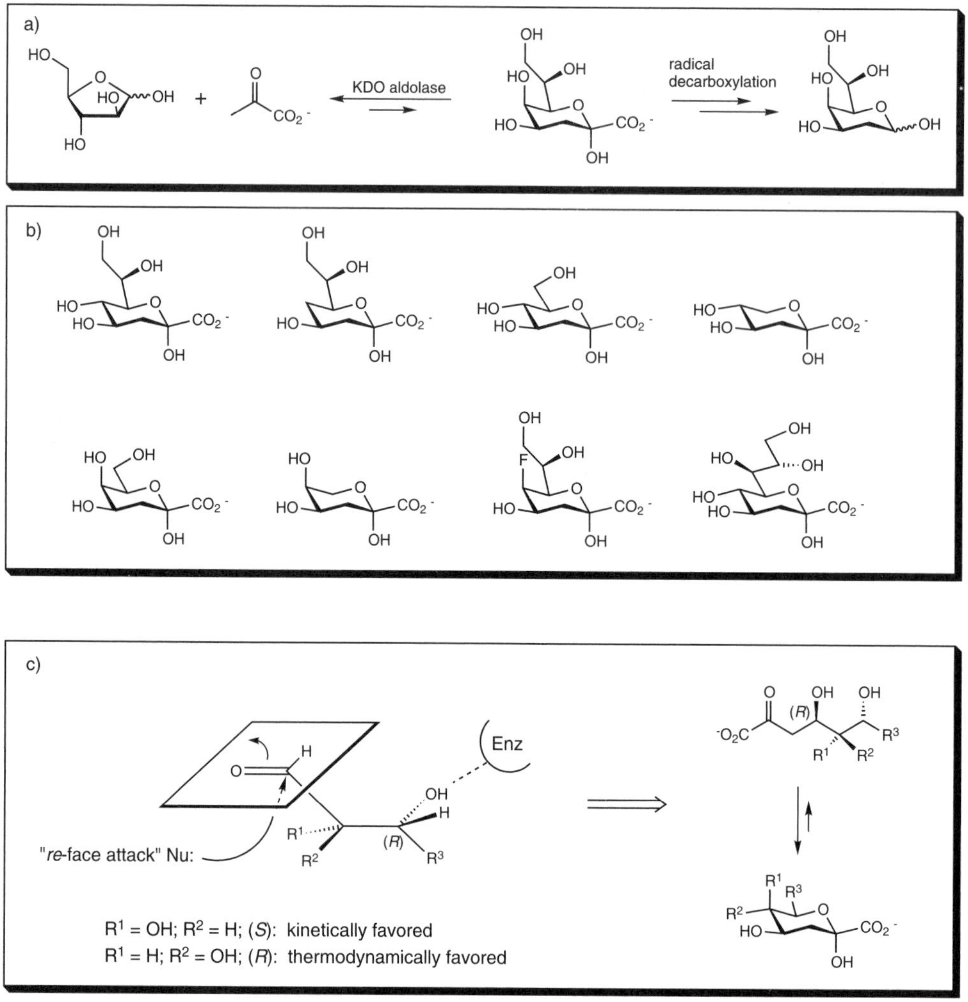

Figure 6–20. (a) Synthesis of higher 2-deoxysugars with KDO aldolase. (b) KDO derivatives synthesized with KDO aldolase. (c) Mechanism of KDO aldolase.

tack of the pyruvate took place on the *re* face of the carbonyl group of the acceptor substrate.

3-Deoxy-D-*manno*-2-octulosonate 8-phosphate synthetase, also known as phospho-2-keto-3-deoxyoctanoate (KDO 8-P) synthetase, catalyzes the irreversible aldol reaction of PEP and D-arabinose 5-phosphate to give KDO 8-P (Ray, 1980, 1982). The enzyme isolated from *E. coli* B has been used in the synthesis of KDO 8-P (Bednarski et al., 1988). To date, the substrate specificity of KDO 8-P synthetase has not been well investigated, but initial studies indicate this enzyme is very specific for its natural substrates.

2-Deoxyribose 5-Phosphate Aldolase (DERA) (E.C. 4.1.2.4)

The enzyme DERA (Racker, 1952) is unique among the aldolases in that the donor of the aldol reaction is an aldehyde. *In vivo*, this enzyme catalyzes the reversible condensation of acetaldehyde and G3P to form D-2-deoxyribose 5-phosphate with an equilibrium constant for deoxyribose phosphate of 2×10^{-4} M (Fig. 6-21). DERA is a type I aldolase. The enzyme from *E. coli* has been overexpressed (Barbas et al., 1990; Chen et al., 1992). At 25°C and pH 7.5, this enzyme is fairly stable and retains 70% of its activity after 10 days.

A number of unnatural substrates are accepted by DERA, and the newly generated chiral center always has the (*S*)-configuration. The enzyme from *L. plantarum* accepts various acceptor substrates including L-G3P, D-erythrose 4-phosphate, glycoaldehyde phosphate, D-ribose 5-phosphate, D,L-glyceraldehyde, D-erythrose, and D-threose, but not D-ribose or glycoaldehyde (Rosen et al., 1965). Only propionaldehyde could weakly replace acetaldehyde as the donor. The *E. coli* enzyme, on the other hand, is able to utilize several other donor substrates including acetaldehyde, propionaldehyde, acetone, and fluoroacetone (Barbas et al., 1990; Chen et al., 1992). A number of aliphatic aldehydes, sugars, and sugar phosphates are acceptor substrates. However, in some cases, the rates of the aldol reactions are very slow (0.4%–1% of the natural substrates). The application of DERA in the synthesis of deoxygenated azasugars is shown in Figure 6-21. Sequential enzymatic aldol reactions with more than two substrates and one or two aldolases have also been achieved (Gijsen & Wong, 1995b).

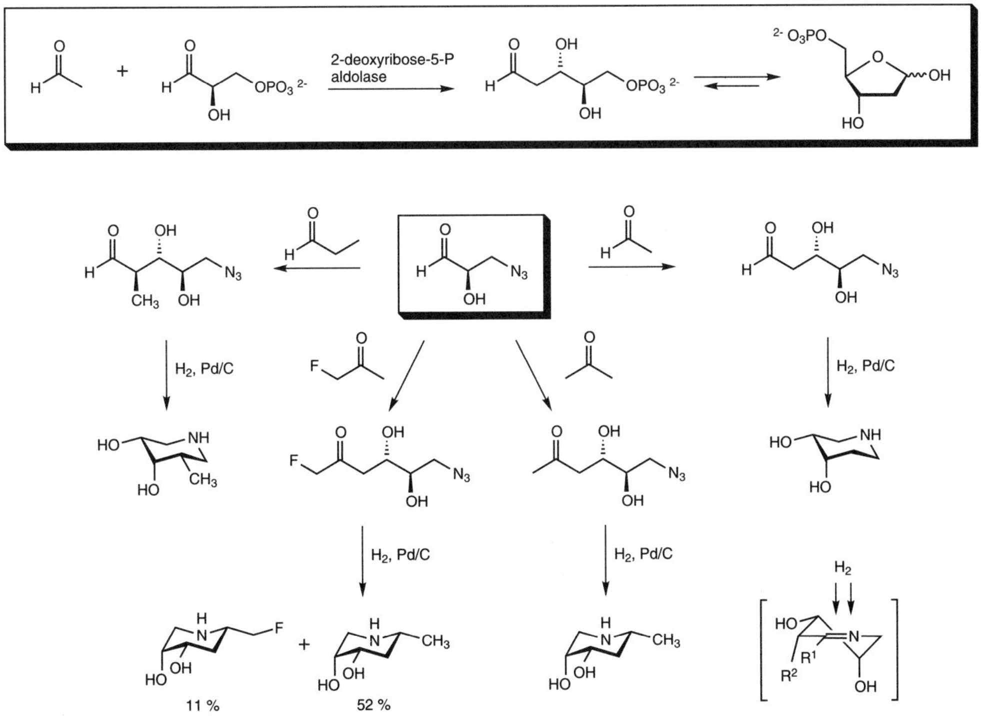

Figure 6–21. Synthesis of azasugars using 2-deoxyribose-5-phosphate aldolase.

Transketolase (TK) (E.C. 2.2.1.1)

One of the enzymes in the pentose phosphate pathway is transketolase (TK) (Racker, 1961) which reversibly transfers the C1–C2 ketol unit from D-xylulose 5-phosphate to D-ribose 5-phosphate to generate D-sedoheptulose 7-phosphate and G3P. The enzyme relies on two cofactors for activity: thiamine pyrophosphate (TPP) and Mg^{2+}. The enzyme isolated from yeast shows a slightly higher diastereoselectivity (ca. 100%) (Demuynk et al., 1990) than the enzyme from spinach (ca. 95%), with the newly formed hydroxymethine chiral center always possessing (S)-configuration. Besides D-xylulose 5-phosphate, D-erythrose 4-phosphate also acts as a ketol donor that produces Fru-6-P and G3P (Fig. 6-22); β-hydroxypyruvic acid (HPA) also acts as a ketol donor that is transferred to an aldose acceptor with an activity 4% compared to that of D-xylulose 5-phosphate (Bolte et al., 1987; Hobbs et al., 1993). One valuable feature of HPA

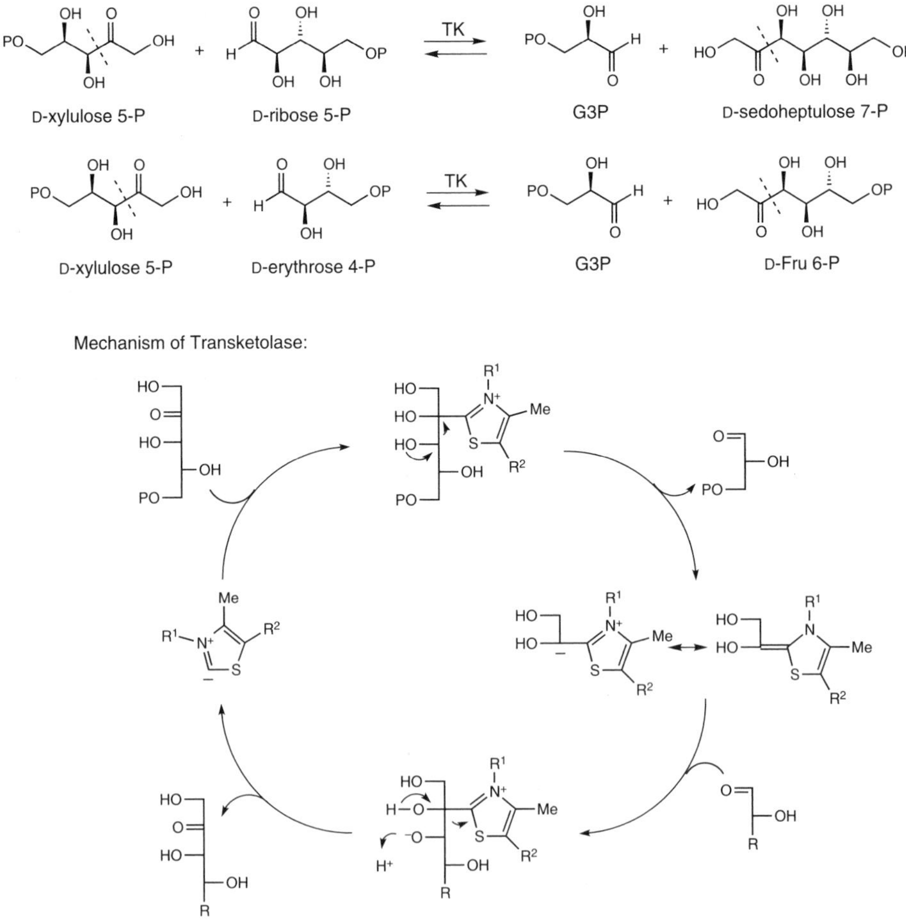

Figure 6–22. Ketol transfer reactions in the oxidative pentose phosphate pathway catalyzed by transketolase (TK).

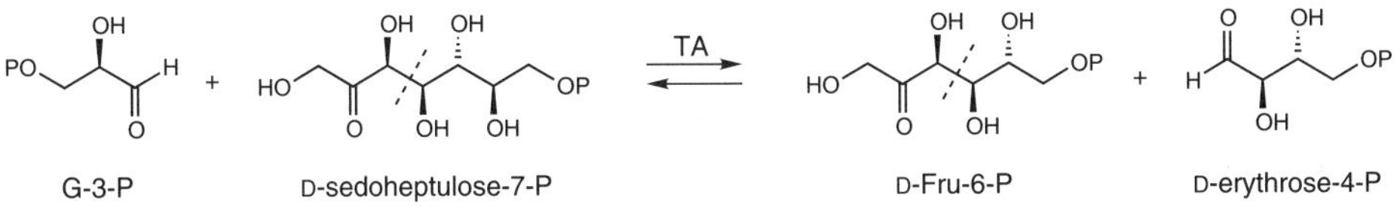

Figure 6–23. Aldol transfer reaction in the oxidative pentose phosphate pathway catalyzed by transaldolase (TA).

is that once the ketol unit is transferred by TK, carbon dioxide is lost, rendering the overall reaction irreversible. Other analogues of HPA, for example, 2,3-dioxopropionic acid, 2-oxo-3-hydroxybutyric acid, 2-oxomalonic acid, and 2-ketogluconic acid, are not substrates, but a wide range of aldehydes are ketol acceptors including aliphatic, α,β-unsaturated, aromatic, and heterocyclic aldehydes (Demuynk et al., 1991; Kobori et al., 1992). The presence of a hydroxy group at C-2 or C-3 has a positive effect on the rate, while steric hindrance near the aldehyde decreases the rate. β-Hydroxyaldehydes epimeric at C-2 can be resolved efficiently by TK since only the D-enantiomers are substrates, giving D-*threo* products (Effenberger et al., 1992; Kobori et al., 1992). Beyond the C-2 configuration, the enzyme appears to have no preferred stereochemistry.

A few synthetic examples that employ TK are illustrated in the synthesis of (+)-*exo*-brevicomin (Myles et al., 1991) and the azasugar 1,4-dideoxy-1,4-imino-D-arabinitol (Ziegler et al., 1988). Both syntheses involve the condensation of HPA with racemic 2-hydroxyaldehydes whereby the ketol unit is diastereoselectively transferred only to the D-enantiomer of the aldehyde.

Transaldolase (TA) (E.C. 2.2.1.2)

Like transketolase, transaldolase (TA) is an enzyme in the pentose phosphate pathway (Racker, 1961). TA, which operates through a Schiff base intermediate, catalyzes the transfer of the C1–C3 aldol unit from D-sedoheptulose 7-phosphate to G3P to produce D-Fru 6-P and D-erythrose 4-phosphate (Fig. 6-23). Although it is commercially available, TA has rarely been used in organic synthesis, and no detailed substrate specificity study has yet been performed. In one application, TA was used in the synthesis D-fructose from starch (Moradian & Benner, 1992). The aldol moiety was transferred from Fru-6-P to D-glyceraldehyde in the final step of this multi-enzyme synthesis of D-fructose.

Enzymatic Formation of Glycosidic Bonds

Nature employs two groups of enzymes in the biosynthesis of oligosaccharides: the enzymes of the Leloir pathway (Leloir, 1971) and those of non-Leloir pathways. The Leloir pathway enzymes are responsible for the synthesis of most *N*- and *O*-linked glycoproteins and other glycoconjugates in mammalian systems. The *N*-linked glycoproteins are characterized by a β-glycosidic linkage between a GlcNAc residue and the δ-amide nitrogen of an asparagine. The less common *O*-linked glycoproteins contain an α-glycosidic linkage between a GalNAc (or xylose) and the hydroxyl group of a serine or threonine. The addition of oligosaccharide chains to glycoproteins occurs cotranslationally for both O-linked and N-linked types, and occurs in the endoplasmic reticulum and the Golgi apparatus (Beyer et al., 1981; Kornfeld & Kornfeld, 1985). The biosynthesis of the N-linked type involves an initial synthesis of a dolichol pyrophosphoryl oligosaccharide intermediate in the endoplasmic reticulum by the action of GlcNAc-transferases and mannosyltransferases. This structure is further glucosylated, and then the entire oligosaccharide moiety is transferred to an Asn residue of the growing peptide chain by the enzyme oligosaccharyltransferase (Kornfeld & Kornfeld, 1985). The Asn is typically part of the amino acid sequence Asn-X-Ser(Thr), where X ≠ Pro or Asp (Imperiali et al., 1992a,b; Lennarz, 1987). Before transport into the Golgi apparatus, the glucose residues and some mannose residues are removed by the action of glucosidase I and II and a mannosidase, in a process called trimming, to reveal a core pentasaccharide (peptide-Asn-[GlcNAc]$_2$-[Man]$_3$). The resulting core structure is further processed by mannosidases and glycosyltransferases present in the Golgi apparatus to produce either the high mannose type, the complex type, or the hybrid type oligosaccharides. Monosaccharides are then added sequentially to this core structure to provide the fully elaborated oligosaccharide chain.

In contrast to the dolichol pyrophosphate-mediated synthesis of N-linked oligosaccharides, for the synthesis of O-linked oligosaccharides in the Golgi apparatus, monosaccharide residues are added sequentially to the growing oligosaccharide chain (Kornfeld & Kornfeld, 1985).

The glycosyltransferases of the Leloir pathway in mammalian systems utilize as glycosyl donors monosaccharides that are activated as glycosyl esters of nucleoside mono- or diphosphates (Kornfeld & Kornfeld, 1985). Non-Leloir transferases typically utilize glycosyl phosphates as activated donors. The Leloir glycosyltransferases utilize primarily eight nucleoside mono- or diphosphate sugars as monosaccharide donors for the synthesis of most oligosaccharides: UDP-Glc, UDP-GlcNAc, UDP-Gal, UDP-GalNAc, GDP-Man, GDP-Fuc, UDP-GlcUA, and CMP-NeuAc. Many other monosaccharides, such as the anionic or sulfated sugars of heparin and chondroitin sulfate, are also found in mammalian systems, but they usually are a result of modification of a particular sugar after it is incorporated into an oligosaccharide structure. A very diverse array of monosaccharides (e.g., xylose, arabinose, KDO, deoxysugars) and oligosaccharides is also present in microorganisms, plants, and invertebrates (Grisebach, 1978; Hash, 1975; Oths et al., 1990; Russell & Liu, 1991; Vara & Hutchinson, 1988). The enzymes responsible for their biosynthesis , however, have not been extensively exploited for synthesis, though they follow the same principles as do those in mammalian systems.

The glycosyltransferases from the Leloir and non-Leloir pathways have been used for the synthesis of oligosaccharides and glycoconjugates (Hindsgaul, 1991; Ichikawa et al., 1992b). Glycosidases have also been exploited for synthesis (Nilsson, 1988b). The function of glycosidases in vivo is to cleave glycosidic bonds; however, under appropriate conditions they can be useful synthetic catalysts. Each group of enzymes has certain advantages and disadvantages for synthesis. Glycosyltransferases are highly specific in the formation of glycosides; however, the availability of many of the necessary transferases is a limiting factor. Fortunately, the recent advances in genetic engineering and recombinant techniques rapidly alleviate this drawback. Glycosidases have the advantage of wider availability and lower cost; however, they are not as specific or high yielding in synthetic reactions. Several other enzymatic methods have also been used to synthesize N-glycosides, such as nucleosides.

Glycosyltransferases of the Leloir Pathway

Leloir pathway glycosyltransferases utilize glycosyl esters of nucleoside phosphates as activated monosaccharide donors. Most of these sugar nucleoside phosphates are biosynthesized *in vivo* from the corresponding monosaccharides. The initial step is a kinase-mediated phosphorylation to produce a glycosyl phosphate. This glycosyl phosphate then reacts with a nucleoside triphosphate (NTP), catalyzed by a nucleoside diphosphosugar pyrophosphorylase, to afford an activated nucleoside diphosphosugar [Eq. 6-1]. Other sugar nucleoside phosphates, such as GDP-Fuc and UDP-GlcUA, are biosynthesized by further enzymatic modification of these existing key sugar nucleotide phosphates. Another exception is CMP-NeuAc, which is formed by the direct reaction of NeuAc with CTP [Eq. 6-2]. Some of the enzymes involved in the biosynthesis of sugar nucleotides also accept unnatural sugars as substrates. In general, however, the rates are quite slow, thus limiting the usefulness of this approach.

$$\text{Sugar-1-P} + \text{NTP} \rightarrow \text{NDP-Sugar} + \text{PPi} \tag{6--1}$$

$$\text{NeuAc} + \text{CTP} \rightarrow \text{CMP-NeuAc} + \text{PPi} \tag{6--2}$$

The appropriate nucleoside triphosphates are utilized as substrates for the biosynthesis of sugar nucleoside phosphates. Therefore, biosynthesis-based enzymatic preparation of these donors for use in glycosylations requires a practical scale synthesis of NTPs.

Most preparative scale enzymatic syntheses of NTPs use commercially available nucleoside monophosphates (NMPs) as starting materials. Alternatively, all of the NMPs can be obtained from yeast RNA digests at low cost (Leuchs et al., 1979), or can be easily prepared chemically (Heidlas et al., 1992). In general, these methods involve the sequential use of two kinases to transform NMPs to NTPs, *via* the corresponding NDPs. Any of three kinases may be used to synthesize NTPs from the corresponding NDPs, each of which uses a different phosphoryl donor: pyruvate kinase (EC 2.7.1.40) uses PEP (Hirschbein et al., 1982; Simon et al., 1989) as a phosphoryl donor, acetate kinase (EC 2.7.2.1) uses acetyl phosphate, and nucleoside diphosphate kinase (EC 2.7.4.6) uses ATP. Pyruvate kinase is generally the enzyme of choice because it is less expensive than nucleoside diphosphate kinase (Heidlas et al., 1992), and because PEP is more stable and provides a more thermodynamically favorable driving force for phosphorylation than does acetyl phosphate (Fig. 6-24).

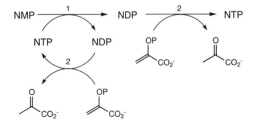

1. Adenylate kinase (EC 2.7.4.3, N = A, C, U)

   Guanylate kinase (EC 2.7.4.8, N = G)

   Nucleoside monophosphate kinase
   (EC 2.7.4.4, N = U)

2. Pyruvate kinase (EC 2.7.1.40)

Figure 6–24. Synthesis of nucleoside triphosphates (NTPs).

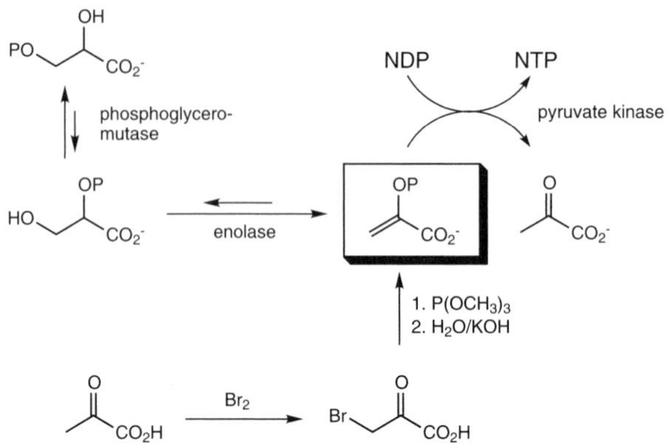

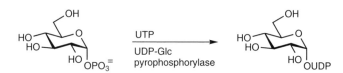

Figure 6–25. Synthesis of phospoenol pyruvate (PEP).

The preparation of NDPs from NMPs is more complicated, and requires different enzymes for each NMP. Adenylate kinase (EC 2.7.4.3) phosphorylates AMP and CMP, and also slowly phosphorylates UMP (Heidlas et al., 1992). Guanylate kinase (EC 2.7.4.8) catalyzes the phosphorylation of GMP. Nucleoside monophosphate kinase (EC 2.7.4.4) uses ATP to phosphorylate AMP, CMP, GMP, and UMP; however, the enzyme is relatively expensive and unstable. Both CMP and UMP kinases exist but are not commercially available. For those kinases requiring ATP as a phosphorylating agent, ATP is usually used in a catalytic amount and recycled from ADP using pyruvate kinase-PEP or acetate kinase-acetyl phosphate (Heidlas et al., 1992). Phosphoenolpyruvate may be prepared chemically from pyruvate or generated enzymatically from D-3-phosphoglyceric acid (Simon et al., 1989) (Fig. 6-25).

Comparisons of chemical and enzymatic methods for the synthesis of NTPs conclude that enzymatic methods provide the most convenient route to CTP and GTP (Heidlas et al., 1992). Chemical deamination of CTP is the best method for preparing UTP (Simon et al., 1990). ATP is relatively inexpensive from commercial sources, although it has been synthesized enzymatically from AMP on a 50 mmol scale. Mixtures of NTPs can be prepared from RNA by sequential nuclease P1, polynucleotide phosphorylase, and pyruvate kinase-catalyzed reactions (Wong et al., 1983a). This mixture can be selectively converted to a sugar nucleotide using a particular sugar nucleoside diphosphate pyrophosphorylase. Figures 26–31 illustrate the synthesis of different sugar nucleotides (Heidlas et al., 1992; Ichikawa et al., 1992b).

Figure 6–26. Synthesis of UDP-Glc.

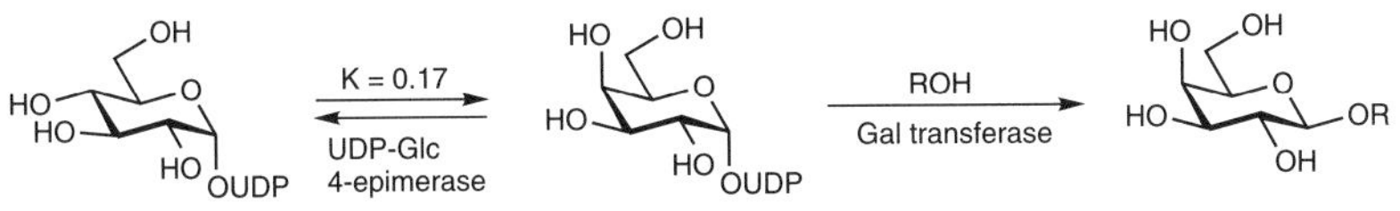
K = 0.17
UDP-Glc
4-epimerase
ROH
Gal transferase
OUDP
OR

Figure 6–27. Synthesis of UDP-Gal and its use in glycosylations.

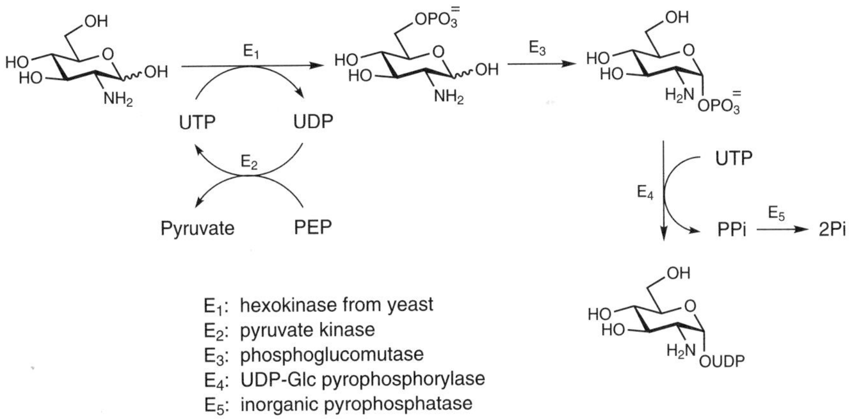
E1
E2
E3
E4
E5
UTP
UDP
Pyruvate
PEP
PPi
2Pi
E1: hexokinase from yeast
E2: pyruvate kinase
E3: phosphoglucomutase
E4: UDP-Glc pyrophosphorylase
E5: inorganic pyrophosphatase

Figure 6–28. Synthesis of UDP-glucosamine that can be converted to UDP-GlcNAc via chemical acetylation.

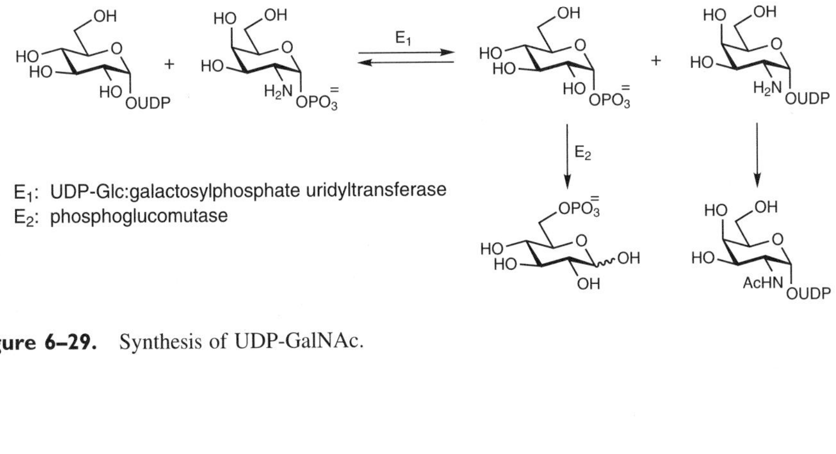
E1
E2
E1: UDP-Glc:galactosylphosphate uridyltransferase
E2: phosphoglucomutase

Figure 6–29. Synthesis of UDP-GalNAc.

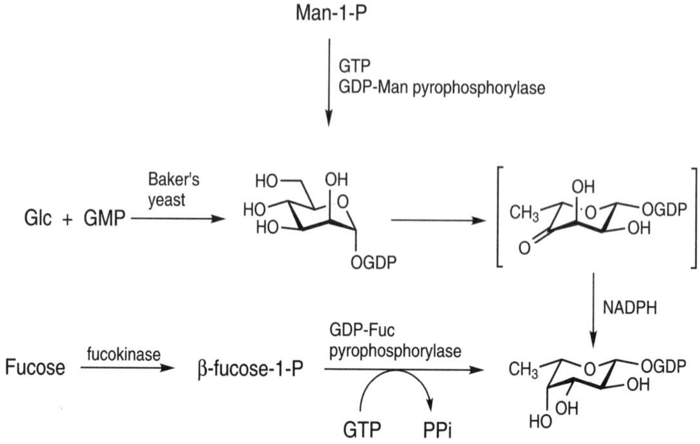

Figure 6–30. Synthesis of GDP-Fuc.

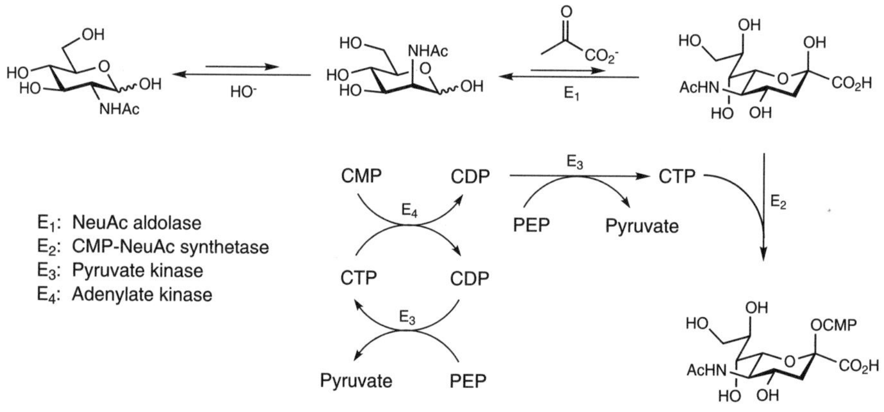

E$_1$: NeuAc aldolase
E$_2$: CMP-NeuAc synthetase
E$_3$: Pyruvate kinase
E$_4$: Adenylate kinase

Figure 6–31. Synthesis of CMP-NeuAc.

Substrate Specificity and Synthetic Applications of Glycosyltransferases

In general, many glycosyltransferases exist for each sugar nucleotide glycosyl donor, each transfering the particular donor to different acceptors. These enzymes are generally considered to be specific for a given glycosyl donor and acceptor, as well as for the stereochemistry and the linkage position of the newly formed glycoside bond. This specificity has led to the "one enzyme-one linkage" concept (Beyer et al., 1981). The specificity of the glycosyltransferases ensures fidelity in oligosaccharide sequences *in vivo* without the use of a template scheme. Though systematic investigations of the *in vitro* substrate specificity of most glycosyltransferases have not been carried out, some deviations from this picture of absolute specificity have already emerged, both in the glycosyl donors and acceptors. Additionally, studies toward the design of inhibitors of glycoprotein biosynthesis (Schwarz & Datema, 1982) have shown that the specificities of glycosyltransferases are not absolute.

Galactosyltransferase

Because of its availability, β1,4-galactosyltransferase (UDP-Gal:N-acetylglucosamine β1,4-galactosyltransferase, EC 2.4.1.22) (Beyer et al., 1981; Ichikawa et al., 1992b) is one of the most extensively studied mammalian glycosyltransferases with regard to synthesis and substrate specificity. This enzyme catalyzes the transfer of galactose from UDP-Gal to the 4-position of β-linked GlcNAc residues to produce the Galβ1,4GlcNAc substructure. In the presence of lactalbumin, however, glucose is the preferred acceptor, resulting in the formation of lactose, Galβ1,4Glc. The enzyme has been employed in the *in vitro* synthesis of N-acetyllactosamine and glycosides thereof, as well as other galactosides.

Galactosyltransferase utilizes as acceptor substrates N-acetylglucosamine and glucose and β-glycosides thereof, 2-deoxyglucose, D-xylose, 5-thioglucose, N-acetylmuramic acid, and *myo*-inositol (Berliner et al., 1984; Nunez & Barker, 1980; Schanbacher & Ebner, 1970). Modifications at the 3- or 6-position of the acceptor GlcNAc are also tolerated. For example, Fucα1,6GlcNAc and NeuAcα2,6GlcNAc are substrates (Auge et al., 1984; Palcic et al., 1987). Acceptor substrates that are derivatized at the 3-position include 3-O-methyl-GlcNAc, 3-deoxy-GlcNAc, 3-O-allyl-GlcNAcβOBu, and 3-oxo-GlcNAc (Wong et al., 1991a). All glycosides of GlcNAc that are reported to be substrates for the galactosyltransferase have β-glycosidic linkages. Both α and β-glycosides of glucose are acceptable; however, the presence of lactalbumin is required for galactosyl transfer onto α-glycosides. Neither D-mannose, D-allose, D-galactose, D-ribose, nor D-xylose are substrates. Monosaccharides that have a negative charge, such as glucuronic acid and α-glucose-1-phosphate, are also not accepted as substrates. Figure 6-32 illustrates several disaccharides that have been synthesized with galactosyltransferase (Nishida et al., 1992, 1993a,b; Srivastava et al., 1993; Thiem & Wiemann,

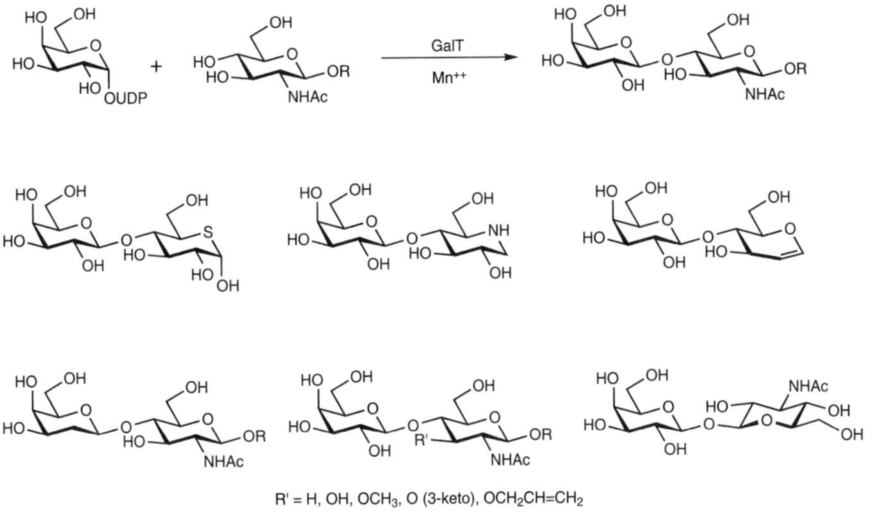

R' = H, OH, OCH$_3$, O (3-keto), OCH$_2$CH=CH$_2$

Figure 6–32. Some disaccharides synthesized using β1,4-galactosyltransferase.

1990; Wong et al., 1991a). A particularly interesting example is the β,β-1,1-linked disaccharide, in which the anomeric hydroxyl of 3-acetamido-3-deoxyglucose serves as the acceptor moiety. The acetamido function apparently controls the position of glycosylation.

β1,4-Galactosyltransferase has also been employed in solid phase oligosaccharide synthesis, and has been used to effect the galactosylation of *gluco* or *cellobio* subunits of polymer-supported oligosaccharides and polysaccharides (Zehavi & Herchman, 1984; Zehavi et al., 1983). The resulting oligosaccharides can then be removed from the support by either a photochemical cleavage or a chymotrypsin-mediated hydrolysis. The types of polymer supports employed include polyacrylamide and a water soluble poly(vinyl alcohol). N-Acetylglucosaminyl amino acids and peptides have also been used as substrates for galactosyltransferase to afford galactosylated glycopeptides (Auge et al., 1989; Thiem & Wiemann, 1990; Unverzagt et al., 1990). The carbohydrate chain can then be extended further with other transferases, for example, sialyltransferase. Similarly, the synthesis of a ceramide glycoside that was subsequently enzymatically sialylated provided a GM$_3$ analogue (Guilbert et al., 1992).

With regard to the donor substrate, the β-galactosyltransferase also transfers glucose, 4- and 6-deoxygalactose, arabinose, glucosamine, galactosamine, N-acetylgalactosamine, 2-deoxygalactose, and 2-deoxyglucose from their respective UDP-derivatives, providing an enzymatic route to oligosaccharides that terminate in 1,4-linked residues other than galactose (Table 6-1) (Berliner & Robinson, 1982; Lowary & Hindsgaul, 1993; Palcic & Hindsgaul, 1991; Yuasa et al., 1992). An example worthy of note is the transfer of 5-thiogalactose to an acceptor. Although the rate of the enzyme-catalyzed transfer of many of these unnatural donor substrates is quite slow, this method is useful for milligram-scale synthesis. The α1,3-galactosyltransferase (and α1,3-GlcNAc transferase) involved in the synthesis of the B blood–group antigen (and A blood–group antigen) has also been studied (Berliner et al., 1984).

Sialyltransferase

Several sialyltransferases, classified as either α2,6- and α2,3-sialyltransferases, have been used for oligosaccharide synthesis (Auge & Gautheron, 1988; Sabesan & Paulson, 1986; Thiem & Treder, 1986). These sialyltransferases generally transfer N-acetylneuraminic acid to either the 3- or 6-position of terminal Gal or GalNAc residues. The 2,8-sialyltransferase is involved in the synthesis of 2,8-linked polysialic acids (Finne, 1985; McCoy et al., 1985). Some sialyltransferases have been shown to accept CMP-NeuAc analogues that are derivatized at the 9-position of the sialic acid–side chain (Conradt et al., 1984; Gross & Brossmer, 1988; Petrie et al., 1989), such as those in which the hydroxyl group at C-9 is replaced with an amino, fluoro, azido, acetamido, or benzamido group, and at the 5-position with different acyl groups (Gross & Brossmer, 1995). Analogues of the acceptors Galβ1,4GlcNAc and Galβ1,3GalNAc, in which the acetamido function is replaced by an azide, phthalimide, carbamate, or pivaloyl functionality are also substrates for the enzymes (Ito et al., 1993). A recent synthesis of a GM$_3$ analogue started with a disaccharyl ceramide derivative in which the fatty acid amide group was replaced with the azide group (Liu & Danishefsky, 1993). To incorporate sialic acid analogues into sialosides is, however, problematic because sialyltransferases are very specific for their natural substrates.

Table 6.1 Relative Rates of β1,4-Galactosyltransferase Catalyzed Transfer of Donor Substrates

Donor Substrate		Relative Rate	Reference
UDP-Gal		100	Berliner & Robinson, 1982
UDP-Glc		0.3	Palcic & Hindsgaul, 1991
UDP-4-deoxy-Glc		5.5	Berliner & Robinson, 1982
UDP-Ara		4.0 R = H 1.3 R = CH$_3$ 0.2 R = CH$_2$F	Berliner & Robinson, 1982
UDP-GalNAc		4.0	Palcic & Hindsgaul, 1991
UDP-GlcNAc		0.00	Palcic & Hindsgaul, 1991
UDP-GlcN		0.09	Palcic & Hindsgaul, 1991
UDP-5-thio-Gal		5.0	Yuasa et al., 1992
UDP-2-deoxy-Gal		90	Srivastava & Hindsgaul, 1993

Fucosyltransferase

Fucosyltransferases are involved in the biosynthesis of many oligosaccharide structures such as blood-group determinants and cell-surface and tumor-associated antigens. Fucosylation is one of the last modifications of oligosaccharides *in vivo*. Several fucosyltransferases have been isolated and used for *in vitro* synthesis (Gokhale et al., 1990; Ichikawa et al., 1992a; Palcic et al., 1989). For example, α1,3-fucosyltransferase has been used to effect the L-fucosylation of the 3-position of the GlcNAc moiety of *N*-acetyllactosamine and of sialyl α2,3-*N*-acetyllactosamine to provide the Lewis x (Le^x) and sialyl Le^x structural motifs, respectively (Ichikawa et al., 1992a; Palcic et al., 1989). Several other acceptor substrates with modifications in the GlcNAc residue can also be used as substrates for fucosylation (Fig. 6-33) (Hubbard, 1981). Galβ1,4Glc, Galβ1,4Glucal, and Galβ1,4(5-thioGlc) are all substrates. A similar enzyme, α1,3/4-fucosyltransferase, has also been used for synthesis. This enzyme effects the fucosylation either on the GlcNAc 3-position of Galβ1,4GlcNAc or on the GlcNAc 4-position of Galβ1,3GlcNAc to afford Le^x or Lewis a (Le^a), respectively (Palcic et al., 1989). The corresponding sialylated substrates have also been employed as acceptors.

The Le^a α1,4fucosyltransferase has been shown to transfer unnatural fucose derivatives from their GDP esters. 3-Deoxyfucose and L-arabinose are transferred to Galβ1,4GlcNAcO(CH$_2$)$_8$CO$_2$CH$_3$ at a rate 2.3% and 5.9%, respectively, that of L-fucose (Gokhale et al., 1990). Furthermore, this enzyme will transfer a fucose residue which is substituted on C-6 by a very large sterically demanding substituent (Srivastava et al., 1992). This approach has been used to alter the antigenic properties of cell surface glycoproteins.

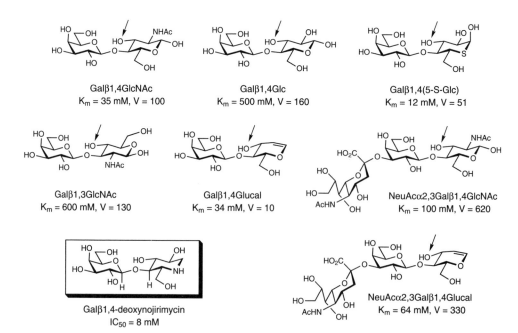

Galβ1,4GlcNAc
K_m = 35 mM, V = 100

Galβ1,4Glc
K_m = 500 mM, V = 160

Galβ1,4(5-S-Glc)
K_m = 12 mM, V = 51

Galβ1,3GlcNAc
K_m = 600 mM, V = 130

Galβ1,4Glucal
K_m = 34 mM, V = 10

NeuAcα2,3Galβ1,4GlcNAc
K_m = 100 mM, V = 620

Galβ1,4-deoxynojirimycin
IC$_{50}$ = 8 mM

NeuAcα2,3Galβ1,4Glucal
K_m = 64 mM, V = 330

Figure 6–33. Substrates and inhibitors of fucosyltransferase.

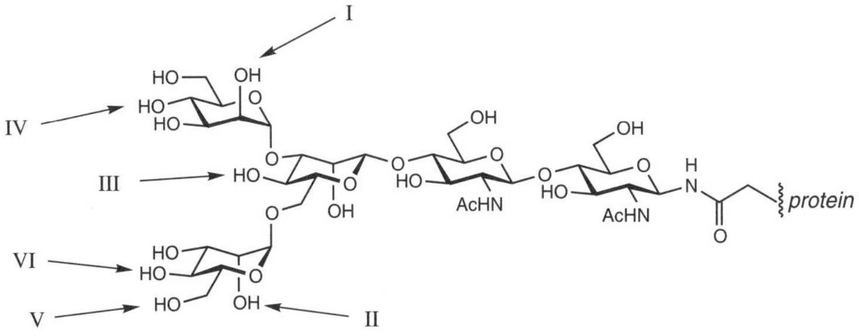

Figure 6–34. Specificity of GlcNAc transferases I–VI.

N-Acetylglucosaminyltransferase

In vivo, the N-acetylglucosaminyl transferases control the branching pattern of N-linked glycoproteins (Schachter, 1986). Each of the enzymes transfers a β-GlcNAc residue from the donor UDP-GlcNAc to a mannose or other acceptor. The GlcNAc transferases I–VI, which catalyze the addition of the GlcNAc residues to the core pentasaccharide of asparagine glycoproteins as outlined in Figure 6-34 have been identified and characterized (Brockhausen et al., 1988, 1989; Schachter, 1986). These, as well as other GlcNAc transferases, have been exploited for purposes of oligosaccharide synthesis (Kaur et al., 1990; Lindh & Hindsgaul, 1991; Linker et al., 1993; Srivastava et al., 1990).

GlcNAc transferases have also been utilized to transfer non-natural residues onto oligosaccharides. In addition to transferring GlcNAc, N-acetylglucosaminyl transferase I from human milk catalyzes the transfer of 3-, 4-, or 6-deoxy-GlcNAc from its respective UDP derivative to Manα1,3(Manα1,6)ManβO(CH$_2$)$_8$CO$_2$CH$_3$ (Srivastava et al., 1990). The 4- and 6-deoxy-GlcNAc analogues can also be transferred by GlcNAc transferase II; however, UDP-3-deoxy-GlcNAc is not a substrate for this enzyme. In addition to the synthetic applications of GlcNAc tranferases, a GlcNAc transferase has been used to attach the terminal GlcNAc of GlcNAcβ1,4GlcNAcα dolichol pyrophosphate, a substance employed in the study of oligosaccharyl transferase (Imperiali & Zimmerman, 1990). A mouse kidney GlcNAc transferase was used in the synthesis of a sialyl Le^x containing hexasaccharide. This enzyme catalyzes the transfer of GlcNAc to 6-OH of Galβ1,3GlcNAc (Oehrlein et al., 1993).

Mannosyltransferase

Various mannosyltransferases have been shown to transfer mannose and 4-deoxymannose from their respective GDP adducts to acceptors (McDowell et al., 1987). The α1,2-mannosyltransferase from yeast has been cloned and overexpressed in *E. coli* ($\sim$ 1unit/L) and was employed to transfer mannose to the 2-position of various derivatized α-mannosides and α-mannosyl peptides to produce the Manα1,2Man structural unit (Wang et al., 1993a). A recent report indicates that mannosyltransferases from pig liver accept GlcNAcβ1,4GlcNAc phytanyl pyrophosphate, an analogue of the natural substrate in which the phytanyl moiety replaces dolichol (Flitsch et al., 1991a,b, 1992).

Sucrose Synthetase

The fructose derivatives 1-azido-1-deoxy-, 1-deoxy-1-fluoro-, 6-deoxy-, 6-deoxy-6-fluoro-, and 4-deoxy-4-fluorofructose have been used as glycosyl acceptors in the sucrose synthetase-catalyzed synthesis of sucrose analogues (Fig. 6-35) (Card & Hitz, 1984; Card et al., 1986). 6-Deoxy- and 6-deoxy-6-fluorofructose were generated *in situ* from the corresponding glucose derivatives under catalysis by glucose isomerase. Because of the reversible nature of the reaction, the sucrose synthetase from rice was used for the preparation and regeneration of UDP-Glc. The enzyme also accepts TDP, ADP and GDP (Elling et al., 1993).

Oligosaccharyltransferase

As mentioned earlier, oligosaccharyltransferase catalyzes the transfer of an oligosaccharide consisting of two GlcNAc, nine mannose, and three glucose units from a dolichol pyrophosphate intermediate to an Asn residue of a nascent peptide or protein (Kornfeld & Kornfeld, 1985). The enzyme also transfers the minimal structure GlcNAcβ1,4GlcNAc from the corresponding dolichyl pyrophosphate donor or from a derivative in which the lipid component is truncated or simplified (Imperiali & Shannon, 1991). The minimal peptide structure that will serve as an acceptor is the tripeptide Asn-X-Ser/Thr. Oligosaccharyltransferase has been utilized for the *in vitro* synthesis of several peptides containing glycosylated Asn residues. The glycopeptides Bz-Asn-(oligosaccharyl)-Leu-Thr-NH$_2$ (Clark et al., 1990), Bz-Asn(GlcNAc$_2$)-Leu-Thr-NH$_2$ (Lee & Coward, 1992), and Ac-Asn(GlcNAc$_2$)-Leu-Thr-OCH$_3$/NHCH$_3$ were synthesized, as were glycosylated cyclic peptides such as oxid(Cys-Als-Asn(GlcNAc$_2$)-Cys)-Thr-Ser-Ala (Imperiali et al., 1992a,b).

In Situ Cofactor Regeneration

Though analytical and small scale synthesis using glycosyltransferases is extremely powerful, the high cost of sugar nucleotides and the product inhibition caused by the released nucleoside mono- or diphosphates present major obstacles to large scale synthesis. A simple solution to both of these problems is to use a scheme in which the sugar nucleotide is regenerated *in situ* from the released nucleoside diphosphate. The first example of the use of such a strategy was the galactosyltransferase-catalyzed synthesis of *N*-acetyllactosamine (Wong et al., 1982) (Fig. 6-36). A catalytic amount of

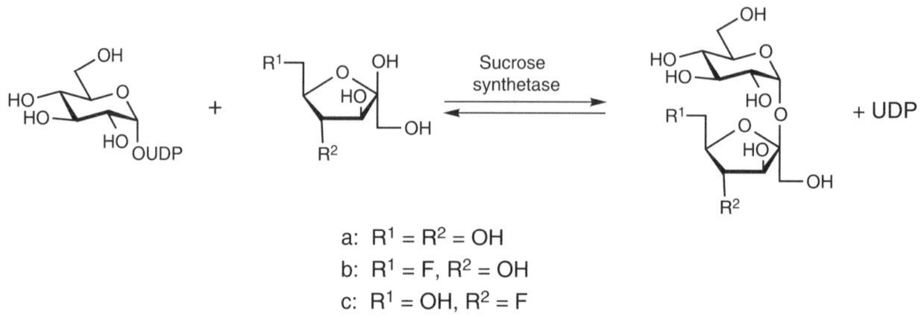

a: R^1 = R^2 = OH
b: R^1 = F, R^2 = OH
c: R^1 = OH, R^2 = F

Figure 6–35. Synthesis of sucrose analogues using sucrose synthetase.

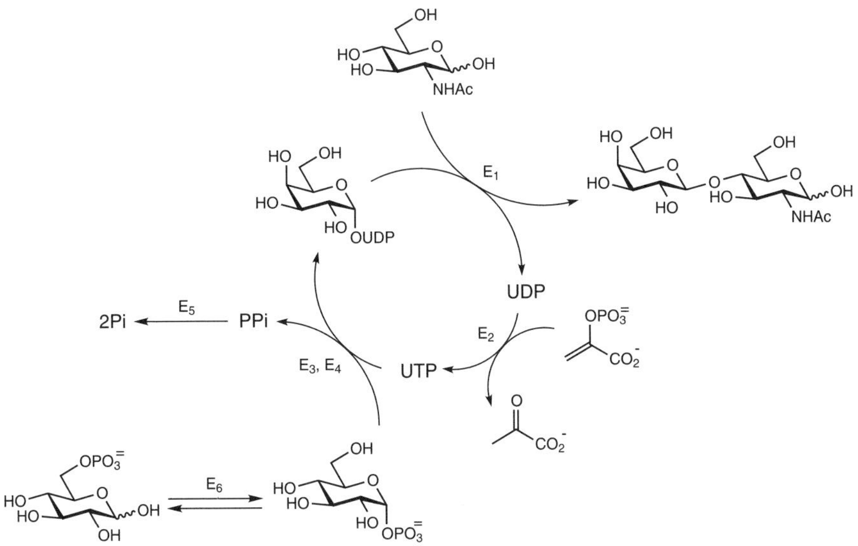

E$_1$: β1,4-galactosyltransferase; E$_2$: pyruvate kinase; E$_3$: UDP-Glc pyrophosphorylase
E$_4$: UDP-Glc epimerase; E$_5$: pyrophosphorylase; E$_6$: phosphoglucomutase

Figure 6–36. Enzymatic transfer of galactose with galactosyltransferase.

UDP-Gal was used initially to glycosylate GlcNAc; UDP-Gal was regenerated from the product UDP and galactose using an enzyme-catalyzed reaction sequence which required stoichiometric amounts of a phosphorylating agent. A second regeneration system for UDP-Gal, which is based on the use of galactose-1-phosphate uridyltransferase, has also been developed (Wong et al., 1992), and has been used in the preparation of analogues such as 2'-deoxy-LacNAc and 2'-amino-2'-deoxy-LacNAc (Fig. 6-37). A third regeneration method for UDP-Gal is based on sucrose synthetase which catalyzes the formation of UDP-Glc from sucrose and UDP (Elling et al., 1993).

In situ cofactor regeneration offers a few advantages. First, a catalytic amount of nucleoside diphosphate and a stoichiometric amount of monosaccharide can be used as starting materials rather than a stoichiometric quantity of sugar nucleotide, thus reducing costs tremendously. Second, product inhibition by the released NDP is minimized due to its low concentration in solution. Third, isolation of the product is greatly facilitated.

A regeneration system for CMP-NeuAc has also been developed, and is illustrated in Figure 6-38 (Ichikawa et al., 1991a,b). The UDP-Gal and CMP-NeuAc regeneration schemes have been combined in a one-pot reaction (Fig. 6-39) and applied to the synthesis of sialyl Le^x (Ichikawa et al., 1992a). The development of these regeneration systems, as well as the more recent development of regeneration schemes for UDP-GlcNAc (Look et al., 1993b), GDP-Man (Wang et al., 1993a), GDP-Fuc (Ichikawa et al., 1992a), and UDP-GlcUA (Gygax et al., 1991) should facilitate the more widespread use of glycosyltransferases for oligosaccharide synthesis. One example is the recent synthesis of sialyl Le^x (Fig. 6-40) (Ichikawa et al., 1992a).

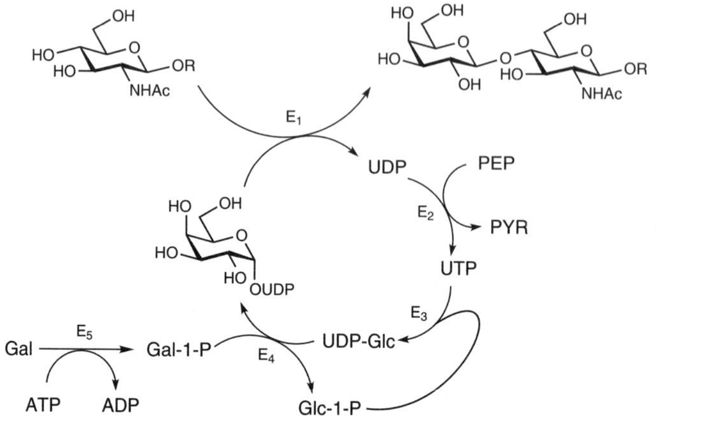

E_1: β1,4-galactosyltransferase; E_2: pyruvate kinase; E_3: UDP-Glc pyrophosphorylase
E_4: Galactose-1-phosphate uridyltransferase; E_5: galactokinase

Figure 6–37. Galactosyltransferase-catalyzed glycosylation with *in situ* regeneration of UDP-Gal.

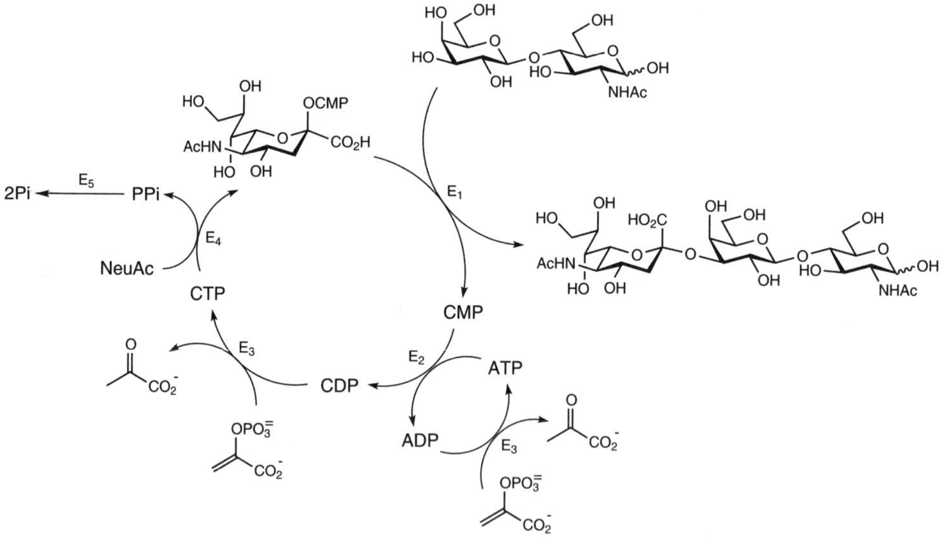

E_1: α2,3-sialyltransferase; E_2: nucleoside monophosphate kinase or adenylate kinase;
E_3: pyruvate kinase; E_4: CMP-NeuAc synthetase; E_5: pyrophosphatase

Figure 6–38. Enzymatic sialylation with *in situ* regeneration of CMP-NeuAc.

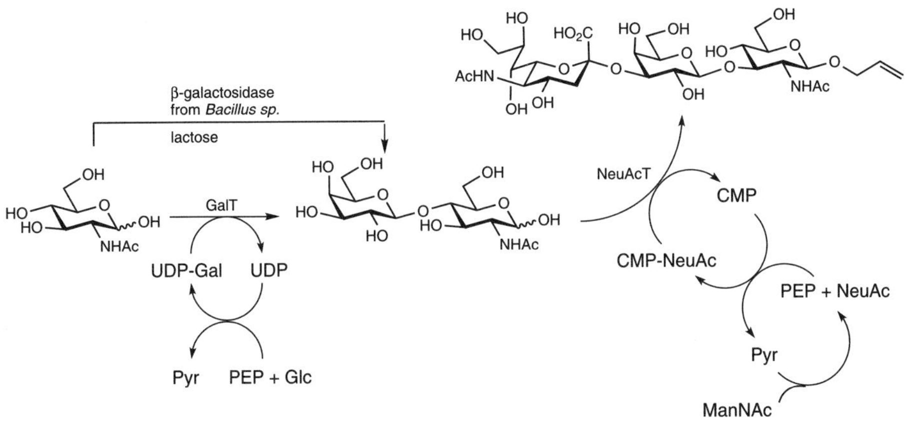

Figure 6–39. Synthesis of a trisaccharide using a galactosidase/sialytransferase enzyme system.

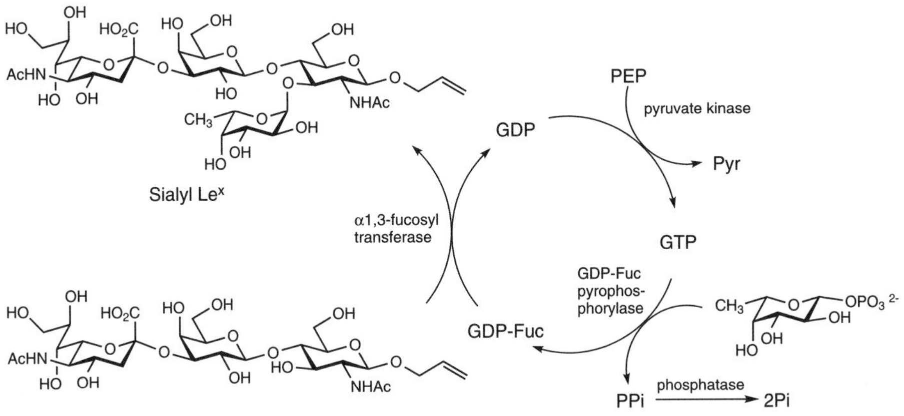

Figure 6–40. Enzymatic fucosylation with *in situ* regeneration of GDP-fucose.

Cloning and Expression of Glycosyltransferases and Sugar Nucleotide Synthetases

While many glycosyltransferases catalyze similar reactions and in many cases use the same donor substrate, there appears to be little sequence homology among the different transferases. There is, however, a significant cross species homology between the same enzymes. For instance, one finds an 86% identity in comparing the $\beta1,4$-galactosyltransferase from humans to the protein sequence from rats. The different glycosyltransferases do exhibit some similarity in that all the cDNA sequences determined to date encode regions consistent with a short N-terminal tail, a hydrophobic transmembrane sequence, a short-stem sequence, and a large C-terminal catalytic domain (Paulson & Colley, 1989). In addition to the membrane-bound form of the glycosyltransferases, soluble forms have also been identified in various body fluids such as the blood, milk, and colostrum. Indeed, these fluids have been the sources of protein for the purification of some of these enzymes. A comparison of the cDNA sequences of these soluble enzymes with the N-terminal protein sequence of the glycosyltransferases that have been sequenced suggests that the stem region has been cleaved to release the large catalytic domain from the membrane. Presumably, this theme of signal-sequence cleavage is consistent for all the glycosyltransferases.

The amount of a glycosyltransferase that can be isolated from a natural source is often limited by the low concentrations of these enzymes present in most tissues and body fluids. The purification of glycosyltransferases is further complicated by the relative instability of this group of enzymes (Beyer et al., 1981). For this reason, the cloning of the glycosyltransferase genes into convenient expression systems has been of great interest.

To date, very few glycosyltransferases have been cloned, expressed, and produced in quantities sufficient for enzymatic synthesis (Lowe, 1991). However, given the advantages of enzymatic synthesis of oligosaccharides over traditional schemes, research into the overexpression of glycosyltransferases will continue to flourish. The most practical expression systems are those based on baculovirus (Ichikawa et al., 1992a; Williams et al., 1995) and yeast (Krezdorn et al., 1993).

Non-Leloir Glycosyltransferases: Transfer of Glycosyl Donors from Glycosyl Phosphates and Glycosides

Oligosaccharides can also be prepared using non-Leloir glycosyltransferases. Phosphorolysis of polysaccharides is catalyzed by a group of these enzymes called glucan phosphorylases. The reaction is reversible and can be used in the synthesis of oligo or polysaccharides. Two particularly important examples are the syntheses of sucrose and trehalose, catalyzed by sucrose phosphorylase and trehalose phosphorylase (Haynie & Whitesides, 1990), respectively (Fig. 6-41). Examples of other enzymes of this class are those involved in synthesis of dextrans and levans (Dedonder, 1966).

Non-Leloir transferases have also been used to synthesize a variety of polysaccharides. The synthesis of modified polysaccharides may provide materials with more de-

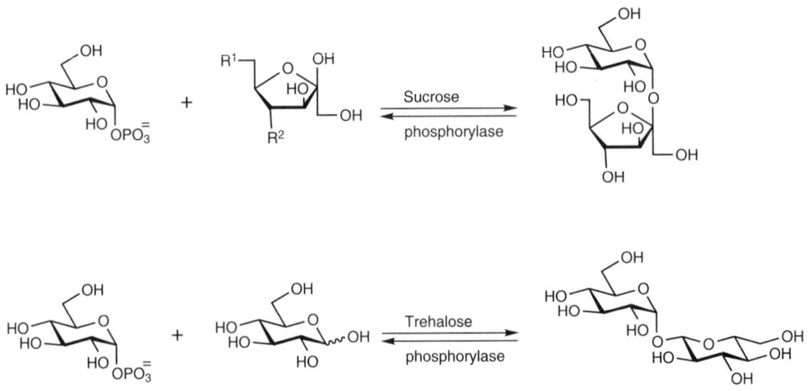

Figure 6–41. Phosphorylase-catalyzed synthesis of sucrose and trehalose.

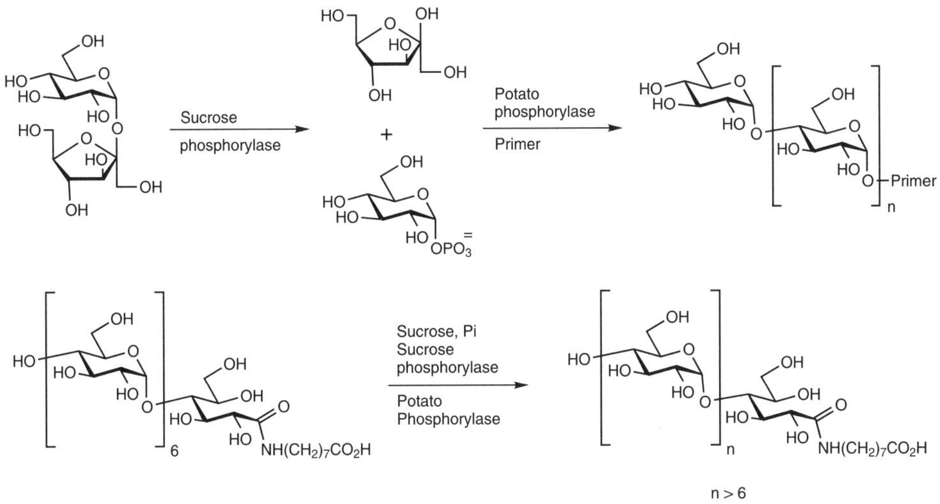

Figure 6–42. Phosphorylase-catalyzed polysaccharide synthesis.

sirable physical and biological properties than their natural counterparts. Approaches to controlling the characteristics of polymers include the control of genes encoding the enzymes responsible for their production, regulation of the activity of these enzymes, or the influence of their *in vivo* synthesis (Carlson, 1987). Potato phosphorylase (E.C. 2.4.1.1) has been used *in vitro* to prepare oligomeric maltose, and a family of linear, star- and comb-shaped polymers (Ziegast & Pfannemuller, 1987).

Improvement of this system has been accomplished by the use of a coupled enzyme system in which glucose-1-phosphate is generated *in situ* from sucrose and inorganic phosphate catalyzed by sucrose phosphorylase (Waldmann et al., 1986). The inorganic phosphate liberated by potato phosphorylase is used by sucrose phosphorylase to drive the formation of polymer, thereby increasing the yield. This coupled-enzyme system also allows for regulation of the molecular weight of the polysaccharide product by control of the concentration of the primer. Unnatural primers bearing functional groups can also be used to prepare tailor-made polysaccharides for further manipulation, for example, attachment to proteins or other compounds (Fig. 6-42).

Cyclodextrin 1,4-glucosyltransferase (EC 2.4.1.19) from *Bacillus macerans* catalyzes the cyclization of oligomaltose to form α, β, and δ-cyclodextrin, and the transfer of sugars from cyclodextrin to an acceptor to form oligosaccharides (Bender, 1980; French, 1957; Saenger, 1980; Wallenfels et al., 1978). It was established recently that the enzyme was able to transform α-glucosyl fluoride into a mixture of α- and β-cyclodextrins and malto-oligomers in almost equal amounts (Treder et al., 1986). When immobilized on a silica gel support that was functionalized with glutaraldehyde, the enzyme was very stable, and no loss of activity was observed after 4 weeks of storage at 4°C. With an appropriate choice of acceptor, this type of enzymatic catalysis may provide a new route to cyclodextrin analogues and novel oligosaccharides.

Unnatural sugar acceptors that are structurally similar to glucose are also substrates for cyclodextrin-glucosyltransferase. Using deoxynojirimycin as acceptor and cyclodextrin as donor, an oligoglucosyl deoxynojirimycin was produced that was sub-

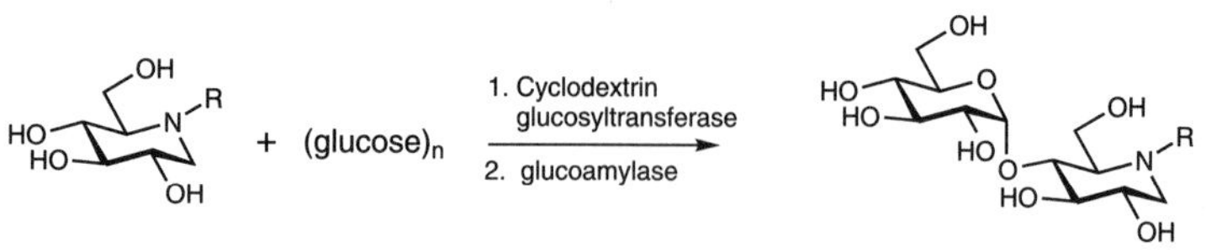

Figure 6–43. Synthesis of glycosyl *N*-alkyldeoxynojirimycin derivatives.

sequently hydrolyzed by glucoamylase to give 4-*O*-α-D-glucopyranosyl deoxynojiri-mycin in ~60% yield (Fig. 6-43) (Ezure, 1985; Ezure et al., 1989). A number of *N*-substituted derivatives of deoxynojirimycin were also good substrates for the trans-ferase, and the products can be degraded with glucoamylase to afford glucosylazasugars. One of these glucosylazasugars, 4-O-α-D-glucopyranosyl-*N*-methyldeoxynojirimycin, was reported to be a potent inhibitor of glucosidase.

A key step in the biosynthesis of lipid A is the glycosylation of the 6-position of a 2,3-diacylglucosamine-1-phosphate by the donor UDP-2,3-diacylglucosamine to pro-duce a lipid A precursor. This transformation is catalyzed by the enzyme lipid A syn-thetase. Lipid A synthetase has been cloned and overexpressed (Crowell et al., 1986), and used for the *in vitro* synthesis of the lipid A precursor and analogues thereof. Some examples include *C*-glycoside (Vypel et al., 1991) and phosphate analogues (Scholz et al., 1992) (Fig. 6-44).

Glycosidases and Transglycosidases

Glycosidase-catalyzed glycoside synthesis may be used under either equilibrium or ki-netically controlled conditions (Nilsson, 1988b). Although operationally straightfor-ward, the equilibrium approach provides poor yields, generally not exceeding 15%. Also, the low yield of the desired product and the formation of side products gener-ally make purification difficult.

Kinetically controlled synthesis relies on the trapping of a reactive intermediate gen-erated from an activated glycosyl donor with exogenous nucleophiles to form a new glycosidic bond. Suitable glycosyl donors for this transglycosylation reaction include di- or oligosaccharides, aryl glycosides, and glycosyl fluorides. This reaction must be

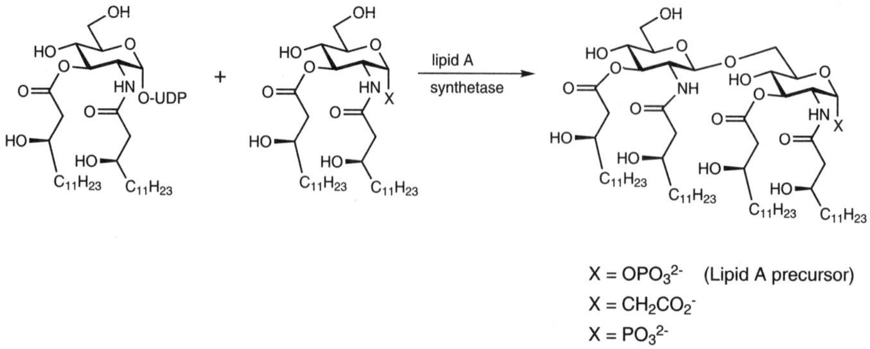

$X = OPO_3^{2-}$ (Lipid A precursor)
$X = CH_2CO_2^-$
$X = PO_3^{2-}$

Figure 6–44. Synthesis of lipid A analogues using lipid A synthetase.

monitored carefully and stopped when the glycosyl donor is consumed in order to minimize glycoside hydrolysis.

In general, the primary hydroxyl group of the acceptor reacts preferentially over secondary hydroxyl groups, resulting in a 1,6-glycosidic linkage. Some control of selectivity has been demonstrated by the selection of an appropriate donor/acceptor combination (Crout et al., 1992; Lehmann & Schroter, 1979; Nilsson, 1987, 1988a). For example, the α-galactosidase catalyzed reactions of α-Gal-OPh-p-NO$_2$ with α-Gal-OMe (or allyl α-galactoside) and β-Gal-OMe form predominantly α-1,3 and α-1,6 linkages, respectively. The configuration of the anomeric center of the acceptor controls, to some extent, the position of glycosylation. Gal-OPh-p-NO$_2$ acting both as donor and acceptor, forms preferentially the α-1,3 linkage whereas the o-nitrophenyl glycoside reacts in a similar fashion to form predominantly the α-1,2 linkage. With β-galactosidase, the β1,3-linked disaccharides were formed predominantly when benzyl or allyl β-galactoside was used as acceptor (Nilsson, 1987, 1988a, b). The use of glycals as acceptors has also been employed as a means of controlling selectivity (Look & Wong, 1992), giving the 1,3-linked glycoside as the major product.

One can also use glycosidases from different species to control regioselectivity. For example, the β-galactosidase from testes catalyzes the formation of Galβ1,3GlcNAc or Galβ1,3GlcNAcβSEt (Hedbys et al., 1989) from lactose and either GlcNAc or GlcNAcβSEt, respectively. The minor products produced in this preparation were then hydrolyzed by the *E. coli* β-galactosidase that preferentially hydrolyzes β1,6-linked galactosyl residues. The overall yield of the β1,3-linked disaccharides was around 10%–20%. Synthesis of polysaccharides based on kinetically controlled glycosidase reactions have been accomplished, as exemplified by the cellulase-catalyzed polymerization of β-cellobiosyl fluoride to form cellulose, with a degree of polymerization >22 (Kobayashi et al., 1991). The β-galactosidase from *Bacillus circulans* was used in the large scale synthesis of N-acetyllactosamine (Herrmann et al., 1993b) from lactose and GlcNAc. When this reaction was coupled with a sialyltransferase reaction *in situ*, a sialyl lactosamine was prepared (Herrmann et al., 1993a) (Fig. 6-39), thus preventing the secondary hydrolysis of the galactosidase product.

Glycosyl transfer to nonsugar acceptors has also been demonstrated. These reactions are especially interesting with chiral-, racemic-, or meso-alcoholic acceptors, as one might expect some degree of diastereoselectivity due to the asymmetric microenvironment of an enzyme active site. Such selectivity has indeed been observed, with diastereoselectivities ranging from moderate to exceptional (Ichikawa et al., 1992b).

Transglycosidases are related to glycosidases in that they cleave glycoside bonds; however, they differ in that they usually transfer the glycosyl moiety to another acceptor with a minimal amount of hydrolysis. Transglycosidases have been found to be useful catalysts for glycosylation. For example, a β-fructofuranosidase from *Antherobacter sp.* K-1 has been used to transfer fructose from sucrose to the 6-position of the glucose residues of stevioside and rubusoside (Ishikawa et al., 1990). Also, a sucrase from *Bacillus subtilis* catalyzes the reversible transfer of fructose from sucrose to the 6-hydroxyl group of a fructose unit at the nonreducing end of a levan chain (Dedonder, 1966; Kunst et al., 1974). Several unnatural sucrose derivatives have been prepared by taking advantage of this process (Rathbone et al., 1986).

A transsialidase from *Trypanosoma cruzi* has been shown to transfer sialic acid reversibly to and from the 3-position of terminal β-Gal residues (Schenkman et al., 1991). Chains terminating in α-linked galactose are not substrates. A number of oligosaccharides containing the NeuAcα2,3Galβ substructure have been synthesized using this transsialidase (Vanderkerckhove et al., 1992). Additionally, this enzyme has been shown to resialylate the terminal galactose units of the cell surface glycoproteins and glycolipids of sialidase-treated erythrocytes (Tomlinson et al., 1991) as well as a synthetic galactoside (Ito & Paulson, 1993). Thus, the *T. cruzi* transsialidase potentially provides a useful alternative to α 2,3-sialyltransferase.

Glycopeptide Synthesis and Glycoprotein Remodeling

A number of the proteins of interest as human pharmaceuticals (tissue plasminogen activator, juvenile human growth hormone, CD4) are glycoproteins. There is substantia interest in developing methods that will permit modification of oligosaccharide structures on these glycoproteins by removing and adding sugar units ("remodeling") and in making new types of protein-oligosaccharide conjugates (Conradt et al., 1987; Little et al., 1984; Livingston et al., 1990). The motivation for these efforts is the hope that modification of the sugar components of naturally occurring or unnatural glycoproteins might increase serum lifetime, increase solubility, decrease antigenicity, and promote uptake by target cells and tissues.

Enzymes are plausible catalysts for manipulating the oligosaccharide content and structure of glycoproteins. The delicacy and polyfunctional character of proteins, and

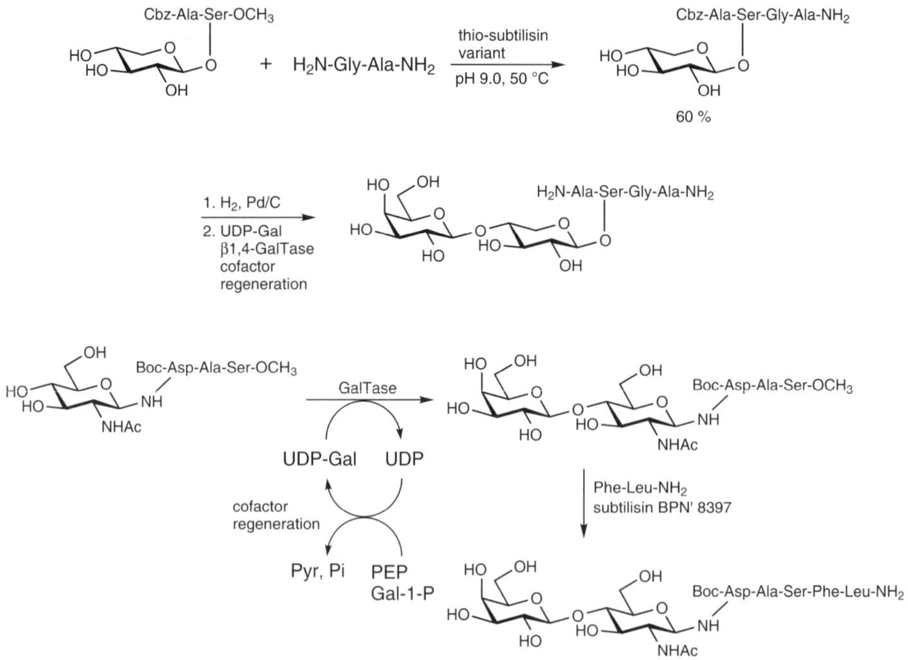

Figure 6–45. Synthesis of glycopeptides using a modified subtilisin and galactosyltransferase. The thio-subtilisin variant contains the following changes: Met50Phe, Asn76Asp, Gly169Ala, Asn218Ser, Ser221Cys.

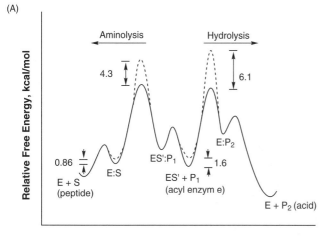

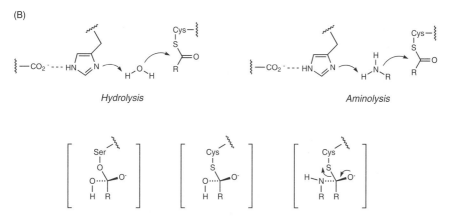

Figure 6–46. (a) Enhancement of the ratio of hydrolysis/aminolysis by thiosubtillisin. (b) Mechanism of subtilisin and thiosubtilisin.

the requirement for high selectivity in their modification, predict that classical synthetic methods will be of limited use. A major problem in the widespread use of enzymes in glycoprotein remodeling and generation is that many of the glycosyltransferases that are plausible candidates for this area are not available. Furthermore, glycosyltransferases likely act on unfolded or partially folded proteins *in vivo* and might not be active at the surface of a completely folded protein.

A useful method for glycopeptide synthesis involves the incorporation of glycosylamino acids into oligopeptides chemically (Bielfeldt et al., 1992; Kunz, 1993) or enzymatically (Wong et al., 1993), followed by introduction of additional sugars by glycosyltransferases. Enzymatic formation of peptide and glycoside bonds in certain cases is quite effective as both procedures can be carried out in aqueous solution, thus minimizing protection/deprotection steps in peptide synthesis. Glycosyl amino acids can be used as the P_2, P_3, P_2', and P_3' residues in subtilisin-catalyzed glycopeptide segment condensation. Using a thermostable variant developed by site-directed mutagenesis, with the active-site Ser converted to Cys, the enzymatic coupling of glycopeptide segments can be carried out effectively at 60°C in aqueous solution (Fig. 6-45) (Wong et al., 1993). The enzyme prefers aminolysis to hydrolysis by a factor of approximately 10,000, and kinetic studies indicate that the selectivity comes from the acyl-enzyme intermediate reacting more selectively with the amine nucleophile than that from the wild-type enzyme (Fig. 6-46). Alternatively, the solid phase chemical synthesis of glycopeptides followed by enzymatic glycosylation, works well on aminopropyl-silica support (Schuster et al., 1994). A key element is to attach a proper acceptor-spacer group with a selectively cleavable bond. This strategy allows a rapid iterative formation of peptide and glycosidic bonds in organic and aqueous solvents, respectively, and enables the release of glycopeptides from the support by enzymatic means (Fig. 6-47).

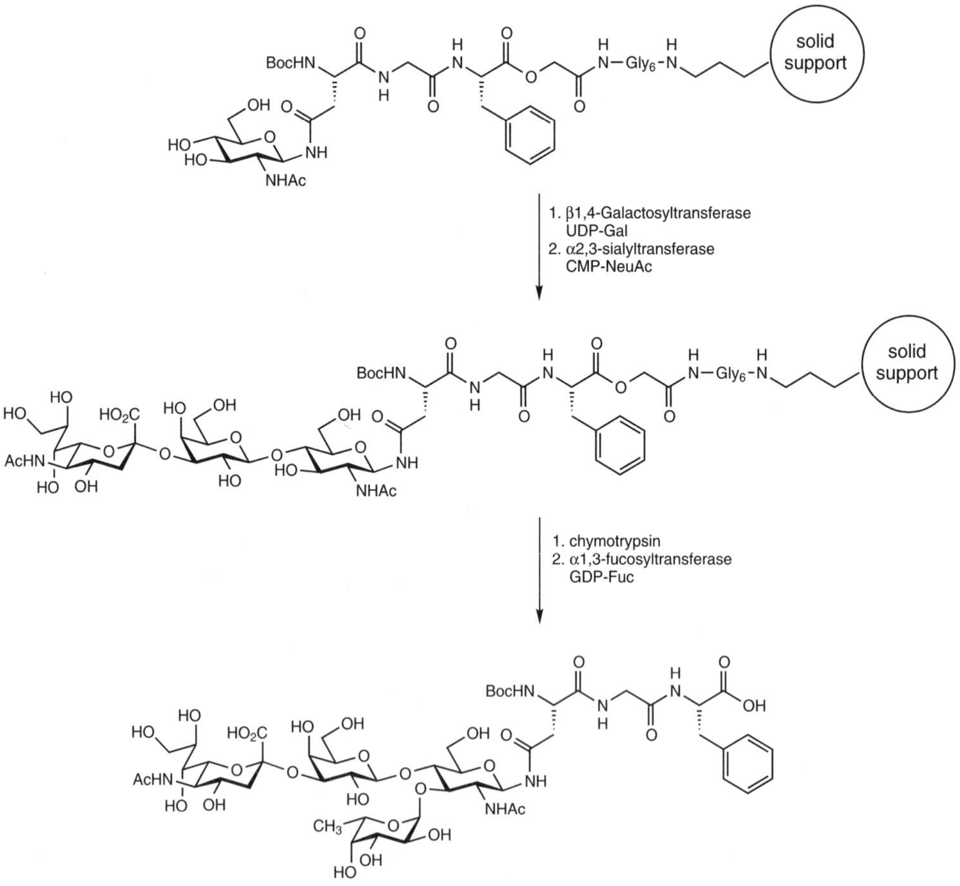

Figure 6–47. Solid-supported enzymatic synthesis of a glycopeptide containing sialyl Lewis x.

Summary

The pace of development of carbohydrate-derived pharmaceutical agents has, in general, been slower than that of more readily accessible classes of molecules. The difficulties in the synthesis and analysis of carbohydrates have undoubtedly contributed to this slow pace, but at least three areas of biology and medicinal chemistry have redirected attention to carbohydrates. First, interfering with the assembly of bacterial cell walls (Rietschel & Brade, 1992; Walsh, 1989) remains one of the most successful strategies for the development of antimicrobials. As bacterial resistance to penams and cephams becomes more widespread, there is increasing interest in interfering with the biosythesis of the characteristic carbohydrate components of the cell wall, especially KDO, heptulose, lipid A, and related materials. Second, cell surface oligosaccharides play a pivotal role in cell communication, cell adhesion, infection, differentiation, and development, and may be relevant to abnormal states of differentiation, such as those characterizing some malignant tumors. Third, the broad interest in diagnostics has finally begun to generate interest in carbohydrates as markers of human health. In addition, there are a number of other possible applications of carbohydrates that warrant attention, for example as dietary constituents, as antivirals agents, or as components of liposomes. By rendering carbohydrates more accessible, enzymatic methods of synthesis will contribute to further research in all of these areas.

7

Carbohydrate Structure, Conformation, and Reactivity: NMR Studies with Stable Isotopes

*Anthony S. Serianni*

Carbohydrate molecules have complex personalities. Being polyfunctional in nature, these molecules participate in a multitude of chemical and biochemical reactions, making them ideal scaffolds for a wide range of *in vitro* and *in vivo* applications. Biological systems have selected simple monosaccharides such as D-glucose (**1**) as a primary fuel source, oligosaccharides as specific ligands for cell-surface recognition, and insoluble polysaccharides to maintain structural rigidity and global shape. Carbohydrates play key roles as raw materials in the chemical and pharmaceutical industries. In the latter regard, fermentation of carbohydrates derived from plants (e.g., corn) is used to produce ethanol as an alternative to fossil fuels, while there is increasing use of carbohydrate precursors in the synthesis of complex organic compounds having useful therapeutic properties.

The many functional roles manifested by carbohydrates are intimately related to their structures. The term "structure" is a hierarchical concept. In its elementary form, structure describes which atoms are bound to which, defined here as primary structure. Since carbohydrates contain one or more chiral centers, absolute configuration must be addressed; this is defined as secondary structure in this chapter. At the highest structural level is three-dimensional (3D) shape or conformation (tertiary structure), dictated by bond rotations within the molecule which give rise to numerous potential 3D structures. These 3D structures are not all equally probable in solution (i.e., they do

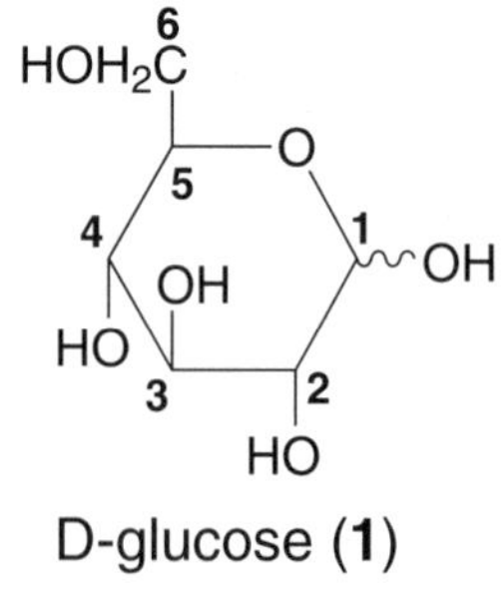

D-glucose (**1**)

not have the same energies). As a consequence, usually only a small subset of the total ensemble of 3D forms is preferred in solution. A number of experimental techniques are available to determine, with different degrees of confidence, the primary, secondary, and tertiary structures of carbohydrates.

In this chapter, we examine how stable isotopes, primarily ^{13}C and ^{2}H, can be used in conjunction with NMR spectroscopy to evaluate various aspects of carbohydrate structure, conformation, and reactivity in solution. The aim is not to provide a comprehensive treatment of the literature, but rather *to provide a general discussion of how isotopic enrichment of carbohydrate structures can provide useful information complementary to that obtained by other methods.* The use of isotopically labeled compounds cannot provide answers to all of the problems enumerated above, but can provide supplemental information on which to draw firmer structural and conformational conclusions. Given the complexity of carbohydrate structure elicidation, these conclusions should be based on maximal data derived from varied experimental and computational methods.

Defining the Problems

As monosaccharides are linked together to form oligosaccharides and polysaccharides, the ability to fully define structure at all three levels decays rapidly. The origin of the problem may be appreciated by first considering the ways by which monosaccharide building blocks are used to construct more complex carbohydrates. Consider the construction of a disaccharide using a single monosaccharide, D-glucose (**1**), an example of an aldohexose. Examination of **1** reveals the presence of a hemiacetal carbon at C1 resulting from an intramolecular reaction between the aldehyde functional group and a specific hydroxyl group in the acyclic (aldehydic) form (Scheme 7-1). Hemiacetal formation occurs *via* attack on either planar face of the carbonyl carbon, producing two different configurations at C1. Cyclic forms of a monosaccharide that differ only in configuration at the hemiacetal (or hemiketal) carbon are defined as anomers, one having the α-configuration (α-anomer) and the other the β-configuration (β-anomer). In general, if ring closure of the acyclic form of a monosaccharide is possible as described (i.e., if the carbonyl carbon and an hydroxyl group are properly disposed for ring closure), then ring formation will occur and cyclic forms will be preferred over acyclic forms. The latter include the carbonyl form and its corresponding hydrate (gem-diol) in aqueous solution. Cyclic forms are limited mainly to five- (furanose) and six- (pyranose) membered rings, with the latter generally more thermodynamically favored than

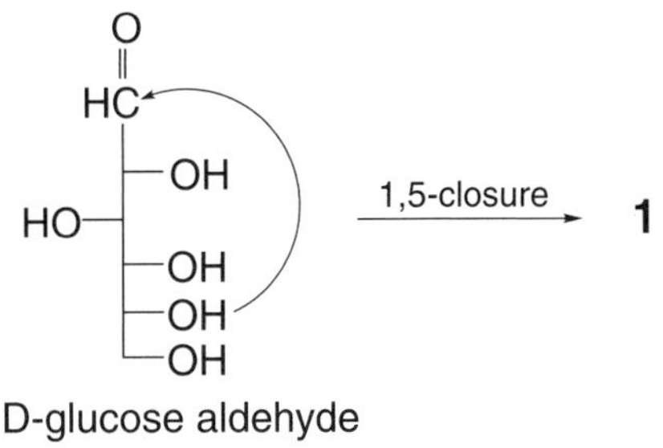

Scheme 7–1. Ring-closure of the acyclic aldehyde form of D-glucose, forming pyranose anomers.

the former. Thus, D-glucose (**1**) in aqueous solution occurs predominantly in the pyranose form, with the β-pyranose (65%) more abundant (Angyal, 1969); very little furanose and acyclic forms are detected. It should be appreciated that monosaccharides like D-glucose (**1**) spontaneously ring-open and ring-close in solution, establishing a dynamic equilibrium between several forms depending on structure. This reaction is responsible for the observed rotation of plane-polarized light upon dissolution (known as mutarotation) of crystalline anomers, while the reaction itself is referred to as anomerization (Scheme 7-2) (Isbell & Pigman, 1969; Pigman & Isbell, 1968).

The relative stabilities of anomers is, in part, determined by the anomeric effect, first described by Lemieux (1963). This stereoelectronic effect was originally proposed to explain the enhanced preference of substituents to orient in the axial position at the anomeric carbon of aldopyranosyl rings as these substituents become more electronegative. Thus, for example, halogen substituents at the anomeric carbon (i.e., glycosyl halides) highly favor the axial position. The precise origin of the effect remains unresolved, although two mechanisms have been proposed, one based on electrostatic (dipolar) interactions and the other on molecular orbital interactions. More detailed discussions of the anomeric effect and its origin have been published (Juaristi & Cuevas, 1995; Kirby, 1983; Perrin & Armstrong, 1993; Perrin et al., 1994; Salzner & Schleyer, 1993; Tvaroska & Bleha, 1989) and are also described in Chapters 1 and 3 of this volume. The effect is not only important in pyranosyl rings, but also apparently plays a key role in dictating ring geometry in furanosyl rings. For example, in the latter rings, the C1—O1 bond can assume a quasi-axial or quasi-equatorial orientation *regardless of the configuration at C1* due to ring flexibility (see below). Nonplanar forms having the C1—O1 bond quasi-axial will thus be stabilized by the anomeric effect, although

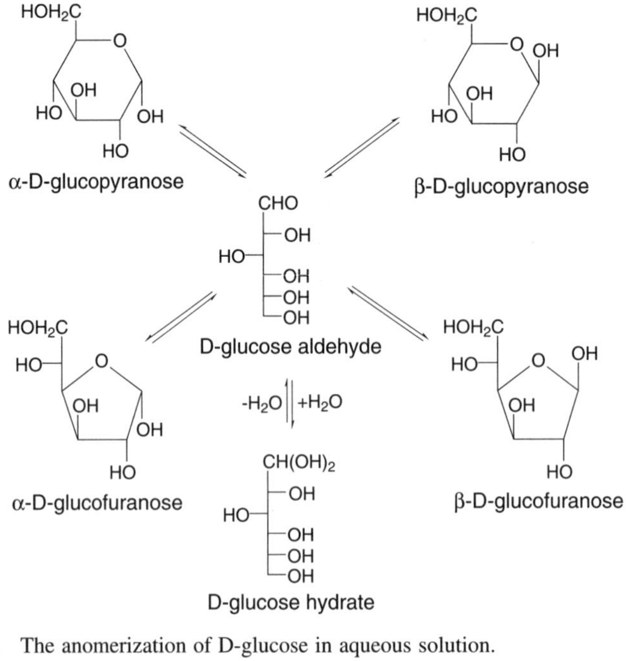

Scheme 7–2. The anomerization of D-glucose in aqueous solution.

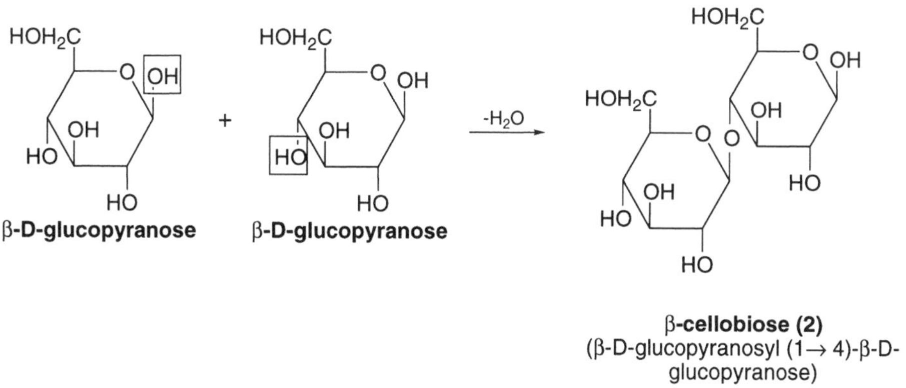

β-cellobiose (2)
(β-D-glucopyranosyl (1→ 4)-β-D-
glucopyranose)

Scheme 7–3.

these forms may not necessarily predominate, since additional conformational factors within these rings compete with stereoelectronic effects at C1 (see below).

Having established that D-glucose (**1**) can exist in four relatively stable cyclic forms, we now examine how these forms combine to produce a glucose–glucose disaccharide. This combining reaction involves the formation of a new bond between each monosaccharide known as a glycosidic linkage. This linkage is formed *via* a dehydration reaction, that is, two hydroxyl groups, one from each sugar, are brought together with the loss of H_2O to produce the new bond. In the construction of a glycosidic bond, however, *one of the participating hydroxyl groups must be attached to an anomeric carbon*. Thus, for example, O1 of β-D-glucopyranose may be linked to C4 of β-D-glucopyranose to give the disaccharide, β-cellobiose (**2**) (Scheme 7-3). A completely different product, β-maltose (**3**), having its own chemical and physical properties, results when O1 of α-D-glucopyranose is linked to C4 of β-D-glucopyranose (Scheme 7-4). It should be apparent that a number of different glucose–glucose disaccharides can be produced in this fashion. Thus, structural methods must be sufficiently precise to distinguish between these many forms. The problem becomes more acute when different monosaccharides or more than one glycosidic linkage are involved, as is often found in biologically important oligo- and polysaccharides.

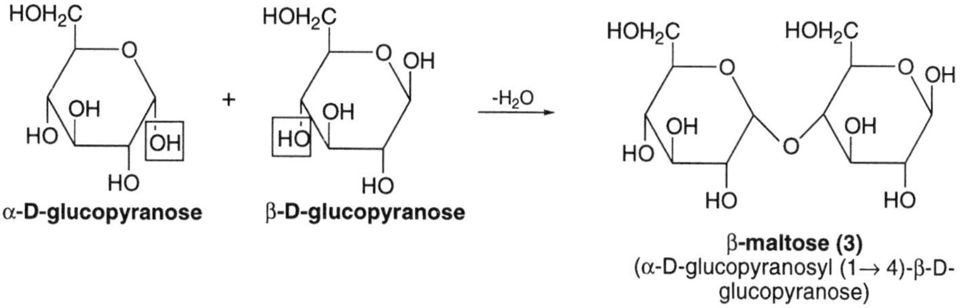

β-maltose (3)
(α-D-glucopyranosyl (1→ 4)-β-D-
glucopyranose)

Scheme 7–4.

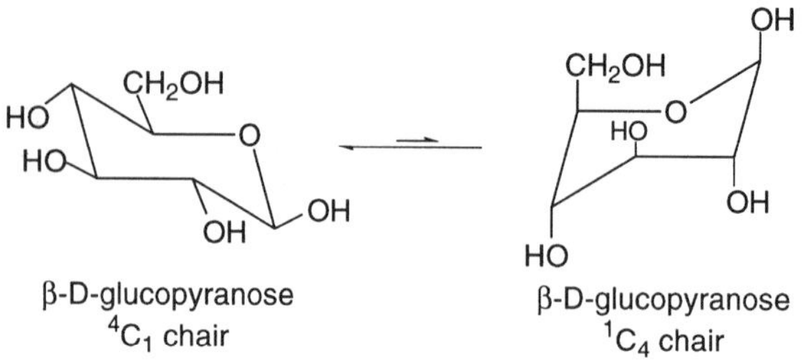

β-D-glucopyranose
$^4C_1$ chair

β-D-glucopyranose
$^1C_4$ chair

Scheme 7–5.

Once the forms of glucose in a particular glucose–glucose disaccharide are identified and the linkage established, the problem of structure takes on new complexity. Let's examine a particular compound, namely, α-D-glucopyranosyl-(1 → 4)-β-D-glucopyranose (**3**) (β-maltose). (A recent update on carbohydrate nomenclature can be found on the world wide web at the following address: http://www.chem.gmw.ac.uk/iupac2/2carb/carb.html.) This molecule contains several structural elements that must be considered in order to establish overall conformation in solution: (1) the conformations of the constituent monosaccharides; (2) the conformations of the exocyclic hydroxymethyl (CH$_2$OH) groups; (3) the conformation of the glycosidic linkage; and (4) the conformations of the exocyclic C—O bonds.

Ring conformation of pyranoses is generally easy to establish, especially those having the *gluco, galacto,* and *manno* configurations. In general, chair forms ($^4C_1$ and $^1C_4$) are preferred (Scheme 7-5) (Angyal, 1969), although other ring forms (i.e., skew and boat forms) are possible. For simple monosaccharides, a general rule is that equatorial substituents confer stability to a given chair form, whereas axial substituents are destabilizing due to the presence of 1,3-diaxial interactions. The magnitude of the latter effect is particularly strong for exocyclic hydroxymethyl groups. Thus, the preferred chair form of both anomers of D-glucopyranose is $^4C_1$ in which no (β-anomer) or only one (O1, α-anomer) substituent is axial (Scheme 7-5). As the number of axial and equatorial substituents within a given chair form become comparable, the difference in the stabilities of both chair forms decreases. In this situation, both chair forms will be present in measurable quantities in solution. This situation is encountered for the aldohexose, α-D-idohexopyranose (**4**), and its derivatives which in solution can exist in more than one conformation (Rao et al., 1995; Snyder & Serianni, 1986). In disaccharide **3**, we assume that both glucose monomers assume the $^4C_1$ conformation. If, on the other hand, furanose forms are involved, the assignment of ring conformation is more complicated. This complication arises from the ability of these rings to assume

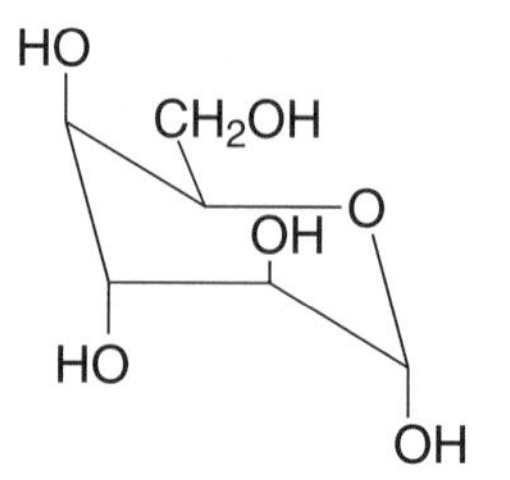

α-D-idopyranose (**4**)

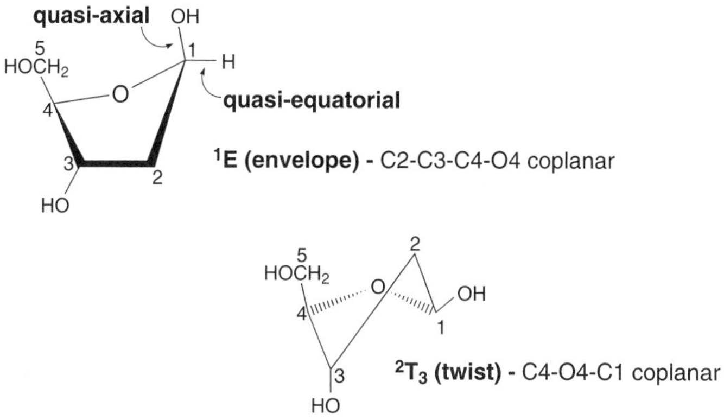

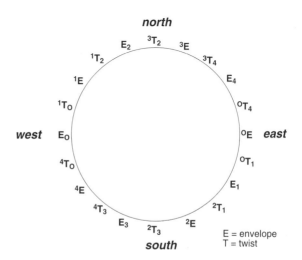

Scheme 7–6. Two nonplanar forms of 2-deoxy-β-D-*erythro*-pentofuranose.

various nonplanar geometries in solution characterized by two generic types, the envelope (E) and twist (T) forms. The former contains four contiguous coplanar atoms with the remaining atom out-of-plane, whereas the latter contains three contiguous coplanar atoms and two out-of-plane atoms (one above and one below the plane) (Scheme 7-6). Twenty idealized nonplanar forms are possible (10 E and 10 T); in fact, many more are possible if consideration is given to the extent to which the out-of-plane atom(s) deviate from planarity (puckering amplitude). These nonplanar forms interconvert readily *via* pseudorotation (Altona & Sundaralingam, 1972) (Scheme 7-7) or inversion, the latter involving the planar form as an intermediate. Studies have suggested that the energies of nonplanar forms are very similar, and thus it is expected that two or more conformers may be present in solution for a given furanose ring. The nature of the conformational equilibrium and dynamics is likely to be very much configuration dependent, so no general rule can be readily applied at the present time to predict preferred geometries. However, several factors influence furanose conformation, namely, the anomeric effect (see above), the preferred quasi-equatorial orienta-

Scheme 7–7. Pseudorotational itinerary of an aldofuranose ring.

tion of side chains, staggered orientation of ring substituents, the gauche effect, and a preference for tetrahedral carbon angles in the ring (Serianni & Barker, 1984). The stereoelectronic gauche effect (Kirby, 1983; Wolfe, 1972; Zefirov et al., 1978) describes the preferred geometry about X-C-C-X molecular fragments; as the substituents X become more electronegative, the gauche conformation becomes more preferred.

The conformation of exocyclic hydroxymethyl groups involves assessing the torsional behavior of the C5—C6 bond in aldohexopyranoses (or the C4—C5 bond in aldopentofuranoses). One can address this problem in several ways (Bock & Duus, 1994). In the first, only three idealized (i.e., perfectly staggered) rotamers (gg, gt, tg) are considered as potential forms in solution (Scheme 7-8). All other rotamers are ignored; in fact, they are implicitly treated as transient intermediates in the rapid interconversion of the three staggered forms and thus have lifetimes very short compared to the experimental measurement time. As a consequence of the latter, they do not make a contribution to measured experimental parameters. In an intermediate approach, the same three rotamers are considered, but they may not be perfectly staggered. Finally, in the most complex and less constrained treatment, rotation about the C—C bond is treated as a continuum, that is, all rotamers are considered in the analysis (Poppe, 1993).

Conformation about the glycosidic linkage is defined by two or three torsion angles depending on the nature of the linkage. In disaccharide **3**, only two torsion angles are involved, namely C1—O1 (phi, ϕ) and O1—C4' (psi, ψ) (Scheme 7-9). The ϕ torsion angle is affected by the exoanomeric effect (see below), whereas both ϕ and ψ will be influenced by nonbonded interactions and potential interresidue H-bonding. For linkages involving the oxygen of an exocyclic CH_2OH group, a third torsion (ω) is required (e.g., C5—C6) to define linkage geometry.

The remaining conformational feature, exocyclic C—O torsions, is the most subtle and thus the most difficult to assess (in practice, this feature is virtually ignored at the present time for nonanomeric C—OH groups). It is important to recognize, however, that the C—O torsion at the anomeric carbon of monosaccharides is subject to stereoelectronic control, unlike C—O torsions at nonanomeric carbons. The so-called exoanomeric effect, first described by Lemieux (Lemieux, 1971; Lemieux et al., 1979), has important implications in the conformational analysis of oligosaccharides. In simple aldopyranoses, the preferred C1—O1 torsion orients OH-1 gauche with respect to the ring oxygen and H1, whereas in the least stable conformation OH-1 is oriented anti to the ring oxygen. Similar effects are observed in simple glycosides. In oligosaccharides, *O*-glycoside linkage conformation is controlled in part by this effect.

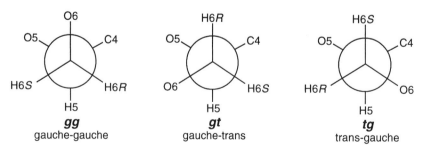

Scheme 7–8. Three perfectly staggered rotamers about the exocyclic C5—C6 bond of aldohexopyranoses.

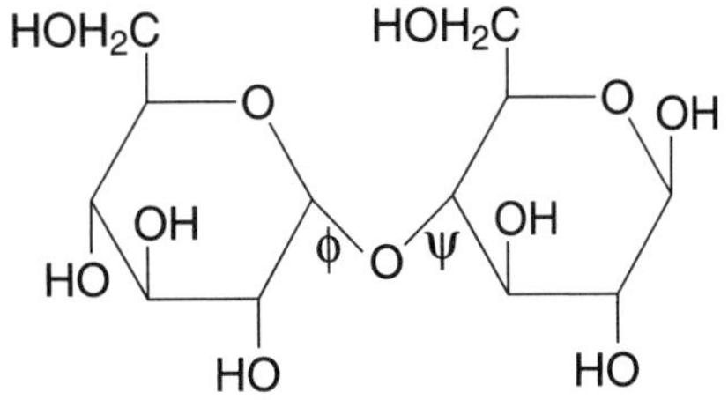

Scheme 7-9. Definition of the phi (ϕ) and psi (ψ) glycosidic torsion angles in a disaccharide (β-maltose [3]).

The problem of tertiary structure determination would be simplified if each of the above-noted conformational features displayed one highly preferred state in a given oligosaccharide. This, unfortunately, is not the case. In some or all of these conformational domains, more than one conformational state appears to be sampled in solution, and thus experimental NMR parameters will be averaged to reflect these populations. It is possible, in fact probable, that conformational domains within an oligosaccharide are interdependent, as has been observed in nucleosides (Saenger, 1984), with a particular conformational state in one domain either stabilizing or destabilizing a particular state in another. Thus, for example, intramolecular hydrogen bonding involving the hydroxymethyl oxygen of a given oligosaccharide residue and a hydroxyl group on another residue could dictate a preferred geometry about the C5—C6 bond and geometry about one or both C—O bonds of the glycosidic linkage.

The conformational behavior of a particular oligosaccharide in solution is determined not only by intramolecular interactions but also by potential interactions that exist between the oligosaccharide and its environment. Thus, the nature of the solvent, or more generally the environment in which the compound finds itself, can profoundly affect its geometry. Two such environments are particularly noteworthy, namely the fully solvated state (e.g., the molecule free in solution) and the molecule "solvated" by a macromolecular receptor (i.e., the molecule bound to the active site of a protein). As hydrophilic (polar) molecules, carbohydrates naturally interact with other polar substances, most notably water, in an organized fashion (Liu & Brady, 1996; Schmidt et al., 1996). With water, the common interaction involves the formation of hydrogen bonds. The nature of the hydrogen-bonding network around a carbohydrate is probably related to its structure. The type of aqueous solvent cage surrounding an aldohexopyranosyl ring is expected, for example, to be sensitive to the configuration of hydroxyl groups on the ring. It might be expected that some configurations "fit" better into a hydrogen-bonding network with water, whereas others are less well accommodated. Thus, the strength and number of hydrogen bonds that solvent water makes with a carbohydrate is likely to vary with structure.

The strength and nature of water interactions with sugars has important implications for understanding how these substances interact with macromolecular receptors (Scheme 7-10). As the carbohydrate reacts with the binding site, water molecules bound to the sugar are stripped away. This loss of water results in a positive enthalpy change (breaking H-bonds) and a positive entropy change (more freedom for the water molecules). Of course, the released water molecules are not actually "free" but become bulk water, thereby forming new H-bonds with other water molecules instead of the sugar. Thus, the overall enthalpy change may not be significant, assuming that the strengths and number of the hydrogen bonds broken are similar to those newly formed. This en-

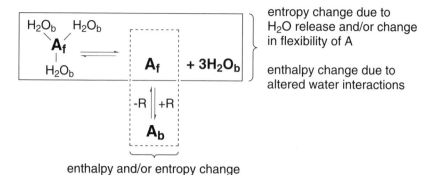

Scheme 7–10. Enthalpy and entropy changes that accompany the binding of substrate A to receptor R.

thalpic balance is likely to be influenced by the structure of the carbohydrate. As the unsolvated sugar binds to the receptor, it experiences new interactions with the receptor, some being hydrophobic in nature, others electrostatic (H-bonding). These interactions are accompanied by a negative enthalpy change. Furthermore, in some cases the conformational flexibility of the carbohydrate may be altered, resulting in either a positive (more flexible when bound) or negative (less flexible when bound) entropy change. Similar thermodynamic considerations must also be made for the receptor. For example, binding of the carbohydrate substrate may result in the rearrangement or expulsion of receptor-bound water, processes which can significantly affect the overall enthalpy and entropy changes for the binding reaction. Additional considerations must be made if the receptor or carbohydrate are ionic. These factors, and others not discussed here, come into play to determine the overall binding affinity (or free energy of binding, ΔG_b) for carbohydrate-receptor interactions, as the sign and magnitude of ΔG_b will be determined by competition between ΔH_b and ΔS_b. Much of what determines the specificity and strength of such recognition is not fully understood. It should, nevertheless, be appreciated that the number, kinds, and strengths of interactions that a given carbohydrate experiences differ when solvated by water and by the receptor, and thus it is not to be expected that the conformational behavior of the molecule free in solution will be identical to that when bound to a receptor. It should also be appreciated that, in either state (free or bound), the carbohydrate may be sampling more than one geometry (i.e., be conformationally flexible or mobile), and this sampling may or may not result in global topological changes.

It should be apparent from the above discussion that the determination of carbohydrate structure and conformation has many varied elements, and some of these cannot be readily evaluated by current experimental methods. Fortunately, some general principles have emerged from structural studies conducted over the past 2–3 decades that provide, in the least, an intellectual framework to address these problems in greater depth.

NMR Spectroscopy of Carbohydrates

To appreciate the effects of isotopic substitution on NMR spectra of carbohydrates in solution, we first review some basic spectroscopy. The NMR experiment supplies several experimental parameters that may be used to assess the solution properties of molecules: chemical shift (δ), spin–spin coupling constants (J), spin–lattice relaxation times (T_1), spin–spin relaxation times (T_2), and nuclear Overhauser enhancement (NOE). It is beyond the scope of this review to treat each of these in detail, and the reader is referred to several reviews for more thorough accounts (Ernst et al., 1987; Homans, 1990; Kessler et al., 1988; Serianni, 1992; Wüthrich, 1986). Here we discuss the salient features of chemical shift, spin–spin-coupling constants and nuclear spin relaxation, illustrated with simple carbohydrate models.

Chemical Shift

The chemical shift (δ) provides information on the NMR signal position (resonance frequency) of a specific NMR-sensitive nucleus in a molecule. Signal (or resonance) position is determined by several factors, most notably the bonding environment of the nucleus. Were this not the case, all protons in a molecule would have the same chemical shift and NMR spectroscopy would be a relatively useless structural tool. The range of chemical shifts (usually reported in parts per million, or ppm, of the applied rf field) accessible to different nuclei is not the same. For ^1H, the range extends roughly over a 10 ppm region, whereas for ^{13}C the range is about 200 ppm. Thus, the likelihood of encountering signal overlap is greater for ^1H than for ^{13}C, and the interpretation of ^1H NMR spectra is commonly plagued by this problem. For methyl β-D-galactopyranoside (**5**), the ^1H chemical shifts (in ppm relative to the standard reference, sodium 2,2-dimethyl-2-silapentane-5-sulfonate [DSS]) are as follows: H1, 4.413; H2, 3.598; H3, 3.741; H4, 4.019; H5, 3.792; H6, 3.893; H6′, 3.848; CH3, 3.670 (Fig. 7-1) (Podlasek et al., 1995). As observed for most simple aldopyranosyl rings, the H1 signal is lo-

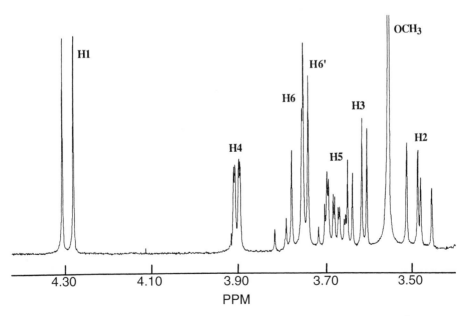

Figure 7–1. The 300 MHz ^1H NMR spectrum of methyl β-D-galactopyranoside (**5**) in ^2H_2O, showing signal assignments. Reprinted from Kline et al., 1990, with permission.

cated significantly downfield (i.e., shifted to higher δ) of the remaining protons, being connected to a carbon having two oxygen substituents (a hemiacetal or acetal carbon) rather than one (an alcohol carbon). These so-called anomeric protons are useful diagnostic probes of carbohydrate chemical structure, being usually well resolved in ^{1}H spectra. As discussed above and illustrated by **5**, the remaining ^{1}H chemical shifts are very similar, covering the range 3.598–4.019 ppm, for a total dispersion of only 0.421 ppm. On the ppm scale, this dispersion is independent of rf field strength (e.g., H1 is located at 4.413 ppm at an observe frequency of 100 *and* 500 MHz), but the *separation in Hz* is rf field strength dependent (e.g., ^{1}H NMR signals separated by 1 ppm on 300 and 500 MHz spectrometers are separated by 300 Hz and 500 Hz, respectively). Consequently, signals which overlap on a lower field spectrometer will often be better resolved and more readily assigned on higher field spectrometers (Figs. 7-2 and 7-3). The higher field spectrometer also provides the added benefit of enhanced sensitivity, allowing analysis of sample solutions of lower concentration.

Carbon chemical shifts (in ppm relative to DSS) for **5** are as follows: C1, 105.3; C2, 72.2; C3, 74.3; C4, 70.2; C5, 76.6; C6, 62.5; CH$_3$, 58.7 (Fig. 7-4). Again we observe a well resolved, downfield signal for C1. Secondary alcohol carbons (C2—C5) usually resonate between 70 and 80 ppm, whereas primary alcohol carbons (C6) resonate at around 62 ppm. Note, however, the greater dispersion of the nonanomeric carbons (17.9 ppm range) compared to the nonanomeric protons (0.421 ppm).

^{1}H and ^{13}C chemical shifts of carbohydrates are affected significantly by ring substitution, the effect being the greatest at the site of substitution. Thus, for example, the ^{1}H chemical shift of H4 of methyl β-D-glucopyranoside (**6**) is 3.36 ppm, whereas the corresponding value in the β-Glc moiety of the disaccharide, methyl β-lactoside (**7**),

methyl β-D-galactopyranoside (**5**)

methyl β-D-glucopyranoside (**6**)

methyl β-lactoside (**7**)
(β-D-galactopyranosyl-(1→ 4)-methyl-
β-D-glucopyranoside)

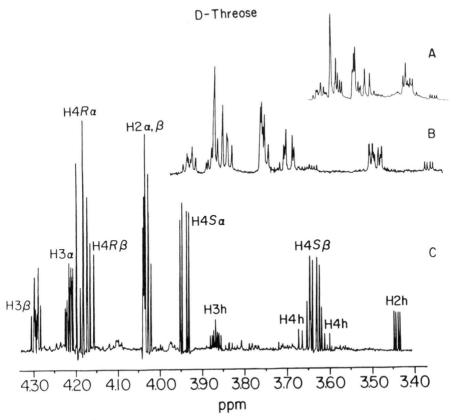

Figure 7–2. ^{1}H NMR spectra of D-threose in ^{2}H$_2$O showing the effect of increasing spectrometer field strength on spectral dispersion. (**A**) 180 MHz. (**B**) 300 MHz. (**C**) 600 MHz. Reprinted from Serianni & Barker, 1979, with permission.

N-acetyl-β-D-glucosamine (**8**)
(2-deoxy-2-acetamido-β-D-glucopyranose)

N-acetyl-β-lactosamine (**9**)
(β-D-galactopyranosyl-(1→ 4)-2-deoxy-2-acetamido-β-D-glucopyranose)

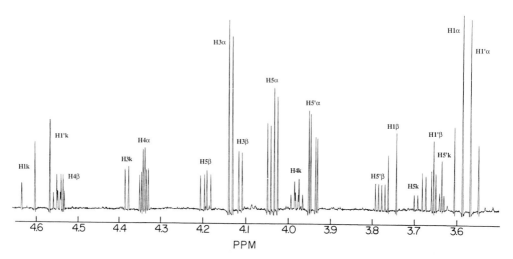

Figure 7–3. The 620 MHz ¹H NMR spectrum of D-*erythro*-2-pentulose (D-ribulose **23**), showing signal assignments for the α and β-ketofuranoses and the acyclic carbonyl form. Resolution-enhancement was employed to improve the measurement of ¹H—¹H spin-couplings. Reprinted from Vuorinen & Serianni, 1990b, with permission.

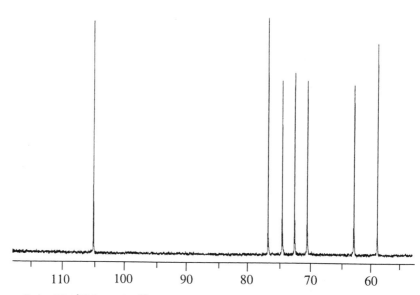

Figure 7–4. The ¹H-decoupled ¹³C NMR spectrum of methyl β-D-galactopyranoside (**5**) in ²H₂O obtained at 75 MHz. Signal assigments (in ppm) are as follows: C1, 105.3; C2, 72.2; C3, 74.3; C4, 70.2; C5, 76.6; C6, 62.5; CH₃, 58.7.

is 3.63 ppm (Hayes et al., 1982b). The C4 chemical shift of β-GlcNAc (**8**) is 71.2 ppm, whereas the corresponding shift in the β-GlcNAc of the disaccharide, β-Gal(1 $\rightarrow$ 4)-β-GlcNAc (**9**), is 79.7 ppm (Rosevear et al., 1982). These substitution effects can be used as a qualitative tool to determine sites of substitution in oligosaccharides, although structural assignments based solely on chemical shift arguments should be avoided since such shifts are not completely reliable.

Determinations of the types of residues in oligosaccharides and the manner in which they are linked together can sometimes be assisted by assessing ^{1}H chemical shifts. This method makes use of structural "reporter" groups, as first suggested by Vliegenthart and co-workers (Vliegenthart et al., 1983), that permit the identification of specific residues. For example, in N-linked oligosaccharides, several reporter groups have been identified: (1) anomeric protons, (2) mannose H2 and H3 protons, (3) sialic acid H3 protons, (4) fucose H5 and CH$_3$ protons, (5) galactose H3 and H4 protons, and (6) aminosugar N-acetyl CH$_3$ protons.

Other than the global relationships between ^{1}H and ^{13}C chemical shift and mono-saccharide structure discussed briefly above, there does not exist at present a full understanding of how these parameters are influenced by structure. Some rules have been proposed that predict with reasonable accuracy the ^{13}C chemical shifts of simple aldohexopyranosyl rings (King-Morris & Serianni, 1987). It appears that the chemical shift of ring carbons bearing equatorial OH groups are much more sensitive to configuration at other ring carbons than are those of ring carbons bearing axial OH groups. This observation seems consistent with the fact that axial ring *protons* (i.e., those bound to carbons having equatorial OH groups) are more sensitive to ring configuration than equatorial protons (Podlasek et al., 1995). Empirical rules based on these observations can be used to provide a rough prediction of ^{1}H and ^{13}C chemical shifts in aldopyranosyl rings having different configurations. Nevertheless, much work remains to be done in this area. In the absence of precise rules to predict ^{1}H and ^{13}C chemical shifts, the unequivocal assignment of the signals, especially for ^{13}C (in the absence of isotopic enrichment), can only be made *via* experimental methods, as discussed below.

Mention should be made of the matter of stereochemical assignments of prochiral protons (diastereotopic protons) in carbohydrates and their derivatives. For example, in **5**, signals due to H6 and H6$'$ can be identified *via* ^{1}H—^{1}H spin-coupling patterns (see below), but the assignments of which signal is the pro-R and which the pro-S proton (Scheme 7-8) cannot be made without further information. In simple carbohydrate systems (C,H,O-containing), these stereochemical assignments can be made in several ways: (1) *via* chemical shifts, (2) *via* selective substitution with deuterium, (3) *via* complementary vicinal ^{1}H—^{1}H and ^{13}C—^{1}H spin-couplings, and (4) *via* $^2J_{CH}$ values. For example, consider the structure of methyl α-D-threofuranoside (**10**). Compound **10** con-

methyl α-D-threofuranoside (**10**)

tains a prochiral carbon at C4, and its ^{1}H NMR spectrum (Serianni & Barker, 1979) yields two distinct signals for H4 and H4', the latter defined as the more shielded C4 proton. Empirical rules (Anteunis & Danneels, 1975) predict that furanosyl ring protons *cis* with respect to an electronegative substituent will be more shielded than protons *trans* to the same substituent. Based on this "syn-upfield" rule, the pro-*S* proton (that which is *cis* to O3) is assigned as the more shielded C4 proton. This assignment may be tested by observing $^3J_{CH}$ values for complementary ^{13}C—^{1}H coupling pathways in **10** involving the ring oxygen (Serianni & Barker, 1984) (see more detailed discussion of ^{13}C—^{1}H spin-couplings below). Thus, $^3J_{C1,H4} = 4.0$ Hz, $^3J_{C1,H4'} = 1.6$ Hz, and $^3J_{C4,H1} = 6.0$ Hz. The relatively large coupling between C4 and H1 establishes that, on average, the dihedral angle between C4 and H1 is large ($\sim 160°$). This being the case, structural constraints impose dihedral angle limits on the C1—O4—C4—H4 and C1—O4—C4—H4' torsion angles, with the former on average being larger ($\sim 160°$) than the latter ($\sim 75°$). Since the more shielded proton (H4') exhibits the smaller coupling (due to the smaller dihedral angle), it may be assigned as the pro-*S* proton. The magnitudes of two-bond ^{13}C—^{1}H spin-couplings ($^2J_{CH}$) involving C3 and H4/H4' of **10** were also used to assign the pro-*R* and pro-*S* protons at C4 (Serianni & Barker, 1984).

Stereospecific deuteration of **10** at C4 can be achieved by treatment with Raney nickel in ^2H_2O (Wu et al., 1983). This method replaces the H4*S* proton, thus causing its signal in the ^{1}H NMR spectrum to disappear, thereby giving direct evidence for its assignment. $^2J_{CH}$ has also been used to assign the ^{1}H signals of the diastereotopic protons at C2' of 2'-deoxyribonucleosides (Bandyopadhyay et al., 1993). Thus, for example, $^2J_{CH} = -5.7$ Hz and 0.4 Hz for the C2' protons *trans* (H2'*S*) and *cis* (H2'*R*) to O3' in [1'-^{13}C]2'-deoxyadenosine (**11**). Using model compounds to calibrate these couplings, it is possible to assign H2'*S* as the more shielded C2' proton. In a similar vein, $^2J_{CH}$ values can be used to assign the hydroxymethyl protons of β-D-ribofuranosyl rings in RNA oligomers (Hines et al., 1993, 1994; Marino et al., 1996), and in methyl aldofuranosides, although little has been reported to date on their application in other carbohydrate-containing molecules.

Spin–Spin Coupling Constants

Inspection of the ^{1}H NMR spectrum of **5** reveals multiple signals (multiplets) for most of the ^{1}H nuclei in the molecule. The magnetic dipole of a given ^{1}H nucleus, H_a, is, of course, affected primarily by the strong uniform magnetic field imposed by the spectrometer magnet (denoted as B_o). However, H_a also senses local fields induced by the spin orientations of other nearby ^{1}H dipoles. The actual magnetic field experienced by

$$^2J_{C1',H2'S} = -5.7 \text{ Hz}$$
$$^2J_{C1',H2'R} = 0.4 \text{ Hz}$$

[1'-^{13}C]2'-deoxyadenosine (**11**)

H_a will be determined by both B_o and the latter local fields. Thus, for example, consider a simple case where H_a experiences only one interaction with another *nonequivalent* ^1H nucleus, H_b. Since the ^1H nucleus has a spin-quantum number of 1/2, the spin-states available to H_b are $+1/2$ or $-1/2$ (also referred to as α and β states, respectively). Both of these spin states are felt by H_a, and thus its signal appears as a doublet, with both signals having equal intensity. If H_a were affected by three nearby nonequivalent protons, the H_a signal would be composed of eight equally intense lines (not all of these may be observed experimentally as we shall see) because H_a senses eight different combinations of spin-states from the three neighboring ^1H nuclei ($\beta\beta\beta$, $\beta\beta\alpha$, $\beta\alpha\beta$, $\beta\alpha\alpha$, $\alpha\beta\beta$, $\alpha\beta\alpha$, $\alpha\alpha\beta$, and $\alpha\alpha\alpha$). This effect is known as spin–spin-coupling, and the magnitude of the splitting observed is the (indirect or scalar) spin–spin-coupling constant, denoted as J. A general equation used to predict the multiplicity of a given proton (or ^{13}C) signal is $2nI + 1$, where n = number of equivalent coupled nuclei and I = spin quantum number of these nuclei. For three coupled nonequivalent protons, the multiplicity would be $[2(1)(0.5) + 1] \times [2(1)(0.5) + 1] \times [2(1)(0.5) + 1] = 8$ (all signals of equal intensity). The signal multiplicity of a proton J-coupled to two equivalent methylene protons would be $2(2)(0.5) + 1 = 3$ (a triplet with the center line having twice the area as each of the outer lines).

Spin–spin-coupling is essentially a through-bond phenomenon. (Through-space contributions to J can also be important. For a discussion of this phenomenon and other factors affecting spin–spin coupling, see Ando and Webb [1983]). As the number of chemical bonds increases between H_a and a nearby ^1H nucleus H_b, the effects of the spin-states of H_b on H_a are truncated and eventually fall to zero. For $^1H—^1H$ spin-coupling in most carbohydrates, the number of bonds over which couplings are routinely observed is 2 and 3, although, in some cases, $^1H—^1H$ spin-coupling through four bonds can be detected and used for structure determinations (Snyder et al., 1989). These couplings are denoted as $^nJ_{HH}$, where the variable superscript "n" denotes the number of intervening bonds separating the coupled nuclei and the subscripts the type of coupling interaction (HH, an example of a *homo*nuclear coupling). The trivial names "geminal" and "vicinal" couplings are associated with ^2J and ^3J values, respectively, regardless of the nature of the nuclei involved.

Of the two types of J_{HH} routinely observed in carbohydrates, $^3J_{HH}$ are by far the more valuable for structure elicidation. As originally shown by Lemieux (Lemieux et al., 1957) and Karplus (1959), the magnitude of $^3J_{HH}$ depends on the value of the torsion angle ϕ between the coupled protons; the original Karplus equation took the form $^3J_{HH} = A + B \cos \phi + C \cos 2\phi$. This general relationship (Karplus curve) is illustrated in Figure 7-5. Coupling is maximal when the two coupled protons are eclipsed ($\phi = 0°$) or antiperiplanar ($\phi = 180°$), and minimal when they are approximately orthogonal ($\phi = 90°$). As a consequence of this dependence, $^3J_{HH}$ has been used widely to examine the conformational properties of many different types of molecules in solution (Altona, 1996), including carbohydrates. In the latter case, three $^3J_{HH}$ types are encountered in pyranosyl rings, $^3J_{He,He}$, $^3J_{He,Ha}$ and $^3J_{Ha,Ha}$, where H_e and H_a denote equatorial and axial protons, respectively. $^3J_{He,He}$ and $^3J_{He,Ha}$ correspond to the same dihedral angle ϕ ($\sim 60°$, gauche) and thus have similar magnitudes (~ 3.5 Hz or less), and these two coupling types are not always readily distinguished (Podlasek et al., 1995). Both, however, can be distinguished from $^3J_{Ha,Ha}$, which corresponds to a $\phi = \sim 180°$ and typically has a value of 8–10 Hz (Podlasek et al., 1995).

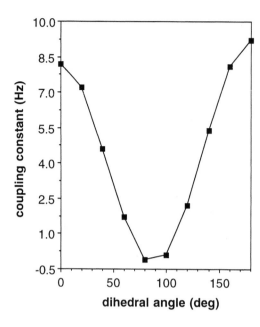

Figure 7–5. A "Karplus curve" relating the dihedral angle for a H_aC—CH_b fragment to the vicinal ^1H—^1H spin-coupling constant, $^3J_{Ha,Hb}$. Coefficients A, B, and C used in the original Karplus equation (see text) were 4.22, −0.5 and 4.5, as described by Karplus for a C—C bond length of 1.543 Å (sp^3-hybridized carbons).

Despite the success of $^3J_{HH}$ values as conformational probes, several limitations must be appreciated when applying them in structural studies. The exact form of the Karplus equation depends highly on structure of the H—C—C—H fragment; the values of A, B, and C in the original Karplus equation reflect this dependency. In particular, the presence of electronegative substituents on or near the H—C—C—H fragment, and their orientation with respect to the coupling pathway, significantly affect the amplitude, and to a lesser extent, the shape of Karplus curves (Altona, 1996). Since the original report by Karplus (1959), who himself recognized this limitation, several more sophisticated Karplus equations have been proposed that take substitution effects into account. Most notable is that proposed by Altona and co-workers (Haasnoot et al., 1980). *Thus, $^3J_{HH}$ values cannot be interpreted without access to a Karplus equation appropriate for the coupling pathway under consideration.* Furthermore, the interpretation of $^3J_{HH}$ values in a molecule (assuming a valid Karplus equation is available) depends on its dynamical properties which are often unknown or assumed in many analyses. The simplest interpretation assumes one and only one conformation in solution. $^3J_{HH}$ Values are then assessed to identify this single conformation. However, more often the molecule is not conformationally fixed, that is, it is sampling two or more conformations in solution with exchange between these states rapid compared to the frequency of observation. Under these circumstances, the observed $^3J_{HH}$ values will be averaged to reflect this heterogeneity, and this averaging can be complex depending on the dynamics of the system and the NMR parameter involved (Jardetzky, 1980). In this case, interpretation of $^3J_{HH}$ is not straightforward, especially when little knowledge is available to assess the structures and relative energies of those conformers most likely to occur. We will return to this problem later.

The above discussion now permits an analysis of the ^1H NMR spectrum of methyl β-D-galactopyranoside (**5**) (Fig. 7-1). The H1 signal appears as a doublet since this proton is coupled only to H2 ($^3J_{H1,H2}$). The magnitude of this coupling is 7.9 Hz, indicative of the antiperiplanar relationship between H1 and H2. The H2 signal is a doublet of doublets, being coupled both to H1 and H3. Since $\phi_{H1,H2}$ and $\phi_{H2,H3}$ are ~ 180°, $^3J_{H1,H2}$ and $^3J_{H2,H3}$, which are responsible for the multiplicity of the H2 signal, are of comparable magnitude (7.9 and 9.9 Hz, respectively). (The fact that they are not exactly the same magnitude reflects small differences in ϕ and/or different dependencies of $^3J_{HH}$ on ϕ for these two different coupling pathways.) The H4 signal, again a doublet of doublets due to coupling to both H3 and H5, shows $^3J_{H3,H4} = 3.4$ Hz and $^3J_{H4,H5} = 1.1$ Hz, consistent with $\phi = \sim 60°$. The most complex multiplet is observed for H5, which is coupled to three nonequivalent protons (H4, H6, and H6') and thus appears as an octet with $^3J_{H4,H5} = 1.1$ Hz, $^3J_{H5,H6} = 7.9$ Hz and $^3J_{H5, H6'} = 4.4$ Hz. The only geminal J_{HH} in **5** occurs between H6 and H6', giving a value of -11.7 Hz (the negative sign is explained below). It should be apparent that the assignment of ^1H signals can be made with certainty by first identifying the well resolved H1 signal, which has a characteristic chemical shift, and then establishing redundant and internally consistent couplings within the remaining multiplets. Thus, $^3J_{H1,H2}$ observed from the H1 signal must also be present in the H2 signal. Likewise, $^3J_{H2,H3}$ observed in the H2 signal must be observed in the H3 signal, and so forth. If the ^1H—^1H coupling network for a given monosaccharide can thus be deciphered and internal consistency established, then signal assignments can be made with confidence.

A critical assumption made in the above analysis of the ^1H NMR spectrum of **5** is that peak positions measured directly from the spectral data can be used to determine J_{HH} values. This assumption is not always valid, especially for carbohydrates. The effect we are considering is referred to as non-first-order behavior (or strong coupling). In first-order spectra, the chemical shift difference in Hz between any pair of coupled nuclei is at least 10 times greater than the coupling constant between the two nuclei in these pairs. As this factor is reduced, significant distortion in the spectral signals results and coupling constants (and chemical shifts) can no longer be measured reliably by direct analysis of the spectrum. These non-first-order effects not only generate distortions in the signals of the coupled nuclei, but also at remote signals in the spectrum (Fig. 7-6). The latter effects, sometimes called virtual coupling, produce splittings that are *not* caused by spin–spin-coupling and, if not recognized, can lead to erroneous interpretations. Since a characteristic feature of carbohydrate ^1H NMR spectra is a clustering of signals over a small spectral range as discussed above, non-first-order effects are very common. Access to higher field spectrometers sometimes can reduce or eliminate these effects, but it is best to resort to spectral simulation if any uncertainty exists. These spin simulation programs are often supplied as part of the operating software accompanying commercial spectrometers (e.g., the VNMR software on modern Varian NMR spectrometers includes an iterative 8-spin simulation program based on the LAOCOON program originally described by Bothner-by and Castellano [1964]).

The above example illustrates the assignment of ^1H signals using one-dimensional (1D) data. Such an analysis is feasible for small molecules having a limited number of protons giving signals that are reasonably dispersed in the spectrum. For more complex structures such as oligosaccharides, 1D analyses have been replaced by two-dimensional (2D) approaches, the most common being the homonuclear (^1H—^1H) correlation experiment (COSY, COrrelation SpectroscopY) (Bax et al., 1981) and its descendant, DQF–COSY (Double Quantum Filtered COSY) (Piantini et al., 1982). In

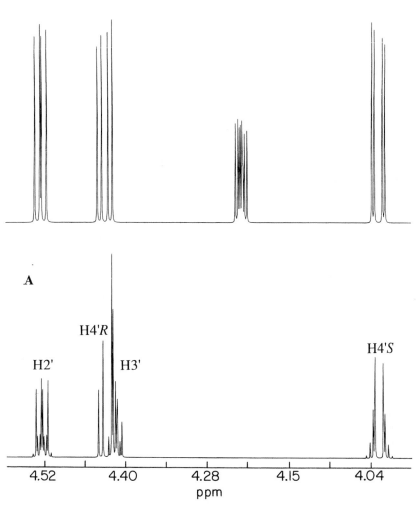

Figure 7–6. An illustration of strong coupling (non-first-order behavior) in the ^{1}H NMR spectrum of the nucleoside, erythrouridine. In **A** (bottom), the chemical shifts of H3′ and H4′R are very similar, leading to non-first-order behavior in the signals arising from these nuclei and those of H2′ and H4S. The simulated spectrum in **A** is virtually identical to the actual spectrum obtained at 620 MHz. In **B** (top), the chemical shift of H3′ has been artificially displaced upfield, thereby removing the strong coupling and providing simple first-order patterns for all four multiplets. In **A** and **B**, the H1′ signal appears downfield (5.84 ppm) and is not shown. Reprinted from Kline & Serianni, 1992, with permission.

this experiment, the 1D spectrum appears along the diagonal (signals are usually represented as contours) (Fig. 7-7), and off-diagonal elements (crosspeaks) denote which protons are coupled to which. This single experiment provides spin-coupling connectivities for all protons within the molecule and thus permits a rapid assignment of signals. DQF-COSY differs from simple COSY in that the former suppresses the detection of singlets (these often intense signals obscure nearby signals and do not supply useful coupling information) and provides better resolution of crosspeaks near the diagonal, and is thus usually preferred over COSY (Fig. 7-7). Normally COSY and DQF-COSY spectra are used for assignment purposes only, that is, no attempt is made to use the fine structure of the crosspeaks to measure ^{1}H—^{1}H spin-couplings quantitatively. However, such measurements are possible (Emsley et al., 1993; Majumdar & Hosur, 1992; Schmitz et al., 1990) provided that sufficient digital resolution is employed to detect the signal components of the crosspeaks and that line widths are relatively narrow.

In comparison to the ^{1}H signal assignments, signal assignments in a ^{13}C NMR spectrum are less straightforward. To understand why this is so, we need to compare the abundances of the ^{1}H and ^{13}C isotopes found in nature. The most abundant isotope of hydrogen is ^{1}H (>99 atom %), and fortunately this isotope is NMR-sensitive (I =

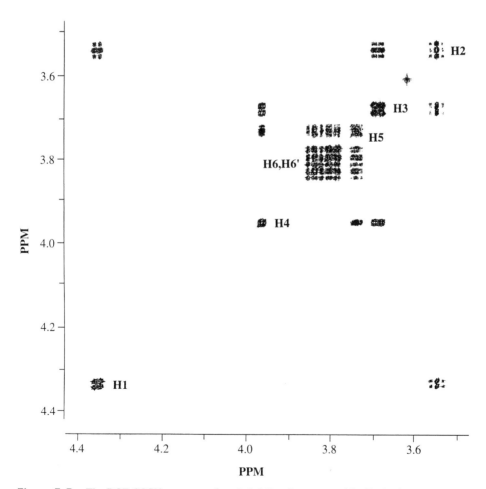

Figure 7–7. The DQF-COSY spectrum of methyl β-D-galactopyranoside (**5**). Assignments of the on-diagonal signals are indicated. The methyl proton signals of the aglycone group ($\sim$ 3.6 ppm) are almost fully suppressed.

1/2). In contrast, ^{12}C is the most abundant isotope of carbon ($\sim$ 98.9 atom %) but it is not NMR-sensitive (I = 0). The ^{13}C isotope can be detected by NMR (I = 1/2), but it is present in natural compounds at low abundance ($\sim$ 1.1 atom %). Consequently, on the basis of abundance alone, the detection of ^{13}C will be more difficult than ^1H. As it turns out, the ^1H nucleus is intrinsically more sensitive towards detection than ^{13}C as well, making the differential between them even greater. (This intrinsic sensitivity is embodied in a parameter known as the gyromagnetic ratio, γ, which we will encounter again later.) All of this means that, for a given ^{13}C nucleus in a molecule, *spin-coupling interactions with nearby ^1H nuclei are highly probable, since the latter are present in high abundance, whereas spin-coupling interactions with nearby ^{13}C nuclei are very rare* (e.g., given the natural isotopic abundance of 1.1%, there is only a 1 in 8264 chance of encountering two contiguous ^{13}C nuclei in a molecule). Thus, in ^{13}C NMR spectra of natural compounds, it will be difficult (but not impossible) to measure homonuclear ^{13}C—^{13}C spin-couplings, but straightforward, in principle, to measure ^{13}C—^1H spin-couplings.

The low abundance of ^{13}C is, however, a mixed blessing. While it presents problems for structure determination with natural compounds, it also allows for selective or uniform isotopic enrichment which can be extremely valuable. It is interesting to note that ^2H nuclei are often substituted into molecules to *reduce* the information content (i.e., remove signals or ^1H—^1H spin-couplings) of ^1H NMR spectra, whereas ^{13}C isotopes are incorporated most often to *increase* information content (e.g., permit access to ^{13}C—^{13}C or ^{13}C—^1H spin-couplings, or to facilitate nuclear relaxation measurements).

In carbohydrates, the carbons are almost always bound to at least one proton or are in the vicinity (within 2 or 3 bonds) of others. Consider C1 of methyl β-D-galactopyranoside (**5**). This carbon is attached directly to H1, separated from H2 by two bonds, and separated from H3 and H5 by three bonds. Since both ^{13}C and ^1H are NMR-sensitive nuclei, each with I = 1/2, spin–spin-coupling occurs between them, and the magnitude of this coupling depends on the bonding relationship between the ^{13}C and ^1H. Thus, for directly bonded protons, the coupling is large, typically between 130 and 170 Hz in carbohydrates depending on structure; this coupling is defined as $^1J_{CH}$. Protons two bonds removed from the carbon produce two-bond (geminal) ^{13}C—^1H couplings ($^2J_{CH}$), whereas ^{13}C and ^1H separated by three bonds generate three-bond (vicinal) ^{13}C—^1H spin-couplings ($^3J_{CH}$) (Scheme 7-11). These ^{13}C—^1H spin-couplings are valuable structural probes (Marshall, 1983), as we shall discuss later. However, with regard to the *assignment* of signals in 1D ^{13}C spectra, these couplings are of little value *in the direct sense*, that is, they cannot be used as ^1H—^1H spin-couplings

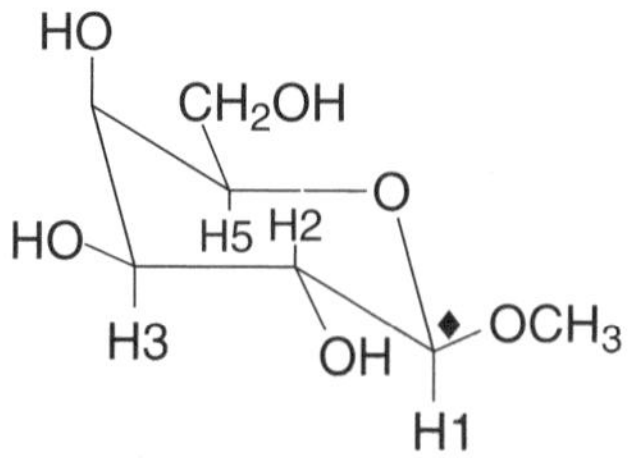

$$^1J_{C1,H1} = 160.7 \text{ Hz}$$
$$^2J_{C1,H2} = -6.1 \text{ Hz}$$
$$^3J_{C1,H3} = 1.3 \text{ Hz}$$
$$^3J_{C1,H5} = 2.5 \text{ Hz}$$
$$^3J_{C1,Me} = 4.5 \text{ Hz}$$

Scheme 7–11. ^{13}C—^1H spin-coupling constants in methyl β-D-[1-^{13}C]galactopyranoside (**5**).

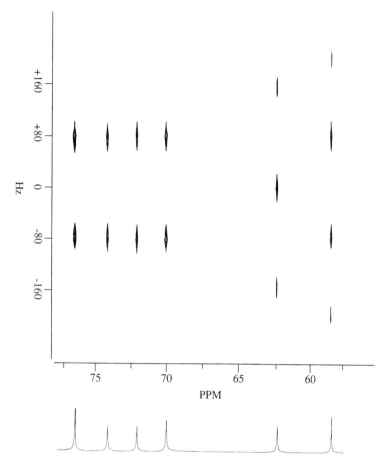

Figure 7–8. The partial heteronuclear *J*-spectrum of methyl β-D-galactopyranoside (**5**), showing only the nonanomeric carbons (C2—C6). ^{13}C Signals between 70 and 77 ppm appear as doublets, indicating carbons bound to a single proton. The C6 signal (62 ppm) appears as a triplet (two attached protons), whereas the methyl signal (59 ppm) appears as a quartet (three attached protons).

were used above to assign the signals in these spectra. In contrast to ^{1}H—^{1}H spin-couplings, which are *homo*nuclear, ^{13}C—^{1}H spin-couplings are *hetero*nuclear and thus cannot provide straightforward connectivities between the carbon nuclei. While use of homonuclear J_{CC} can be made to establish C—C connectivities in a molecule (Bax et al., 1980; Mareci & Freeman, 1982), the method (INADEQUATE; Incredible Natural Abundance Double QUAntum Transfer Experiment) is neither sensitive nor simple to perform. The number of protons attached to a carbon, however, can be determined readily by 1D (e.g., APT [Patt & Shoolery, 1982]) and two dimensional (2D) (e.g., heteronuclear *J*-spectroscopy [Müller et al., 1975]; Fig. 7-8) methods, and this information provides some assignment information.

Since J_{CH} cannot be used directly for complete assignment purposes, ^{13}C NMR spectra are typically obtained under conditions of broadband ^{1}H-decoupling, accomplished at present with the use of composite pulses (e.g., WALTZ [Shaka & Keeler, 1987; Shaka et al., 1983]). In the presence of this decoupling, the carbons sense only a single average energy level for each coupled proton (rather than distinct α and β states), and thus each carbon produces a single resonance. Collapse of the multiplicity caused by ^{1}H-decoupling significantly improves sensitivity (along with other factors, as we shall see), but of course all coupling information is lost. In practice, for most molecules including carbohydrates, a full assignment of the ^{1}H-decoupled ^{13}C spectrum cannot be made with certainty in the absence of additional NMR data of the multipulse or -dimensional variety, or through selective ^{13}C-labeling. The most useful 2D NMR method for the purpose of ^{13}C signal assignment is ^{13}C—^{1}H shift-correlation spectroscopy which was first introduced as a ^{13}C—detected method (HETCOR for heteronuclear correlation) (Bodenhausen & Freeman, 1977) and latter refined as a ^{1}H-detected method (HMQC for Heteronuclear Multiple-Quantum Coherence) (Bax & Subramanian, 1986). The latter modification results in significantly enhanced sensitivity, thereby allowing analyses of smaller sample quantities, and is frequently the method of choice. Compared to HETCOR, however, HMQC is more challenging to implement in the laboratory. In both HETCOR and HMQC, crosspeaks are detected between carbons and their *directly bonded protons*. This correlation is mediated by $^1J_{CH}$ values for each ^{13}C—^{1}H bond. Therefore, if the *proton* spectrum of a carbohydrate sample can be interpreted and the signals firmly assigned (usually *via* 2D ^{1}H—^{1}H COSY or related methods), then the signals in the ^{13}C spectrum may be assigned *via* HETCOR or HMQC, as illustrated for methyl β-D-galactopyranoside (**5**) (Fig. 7-9). A related method, heteronuclear multiple-bond correlation (HMBC), provides longer

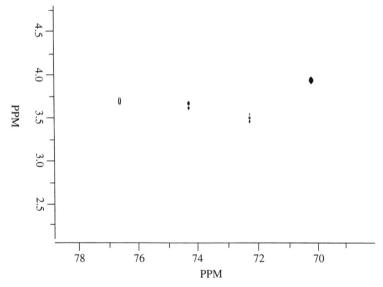

Figure 7–9. The partial HETCOR spectrum of methyl β-D-galactopyranoside (**5**), showing only the methine (CH) carbons (C2, C3, C4, and C5). Contours correlate ^{1}H chemical shifts of H2, H3, H4, and H5 assigned from the 1D proton spectrum (Fig. 7-1) with those of carbons directly bound to each, thereby permitting unequivocal signal assignments of the latter.

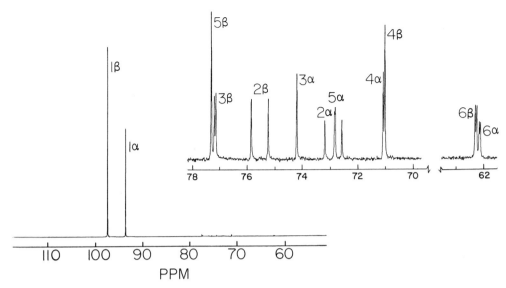

Figure 7–10. The ^{1}H-decoupled ^{13}C NMR spectrum (75 MHz) of D-[1-^{13}C]glucose in ^{2}H$_2$O, showing signal assignments for the α and β-pyranoses, and splitting of specific signals due to ^{13}C—^{13}C spin-coupling. Reprinted from King-Morris & Serianni, 1987, with permission.

range ^{13}C—^{1}H correlation information which is mediated by $^2J_{CH}$ and $^3J_{CH}$ values, rather than $^1J_{CH}$ (Bax & Summers, 1986; Summers et al., 1986). HMBC spectra thus give signal assignment information complementary to that obtained from HETCOR and HMQC, and have also been used to establish linkage positions in oligosaccharides (via $^3J_{COCH}$ values across the O-glycoside bonds).

Assignment information can also be obtained by enriching the molecule with ^{13}C at one or more carbons. For example, consider the effect of enhancing the ^{13}C content at C1 of D-glucose (**1**) from natural abundance (1.1%) to 99% on the ^{13}C spectrum, leaving all other carbons at natural abundance (Fig. 7-10). Firstly, the detection of C1 of both α- and β-pyranoses is significantly improved, resulting in a much greater signal intensity for the labeled carbon compared to the natural abundance carbons. Thus its assignment can be made by inspection. In addition, the *unlabeled* carbons now sense the ^{13}C isotope at C1 with essentially 100% probability, and thus ^{13}C—^{13}C couplings can be easily detected between these carbons and the labeled C1. As observed for ^{13}C—^{1}H spin-coupling, ^{13}C—^{13}C spin-couplings decrease in magnitude as the number of intervening bonds between the coupled carbons increases. Coupling across one bond ($^1J_{CC}$) is large (in **5**, $^1J_{C1,C2} = 46.7$ Hz) (Scheme 7-12) and the observation of this coupling allows a firm assignment of the C2 signal. Couplings across two bonds ($^2J_{C1,C3}$, $^2J_{C1,C5}$) may or may not be observed, depending on molecular structure (Scheme 7-12), and some empirical rules have been proposed that allow their use for some assignment purposes (Church et al., 1996; King-Morris & Serianni, 1987; Serianni et al., 1996; Walker et al., 1976; Wu et al., 1992), although at present this application is rare. The magnitude of vicinal ^{13}C—^{13}C spin-couplings (e.g., $^3J_{C1,C6}$) is sensitive to torsion angle (i.e., $^3J_{CC}$ appear to have Karplus-like dependencies as do $^3J_{HH}$ and $^3J_{CH}$) (Wu et al., 1992). For example, in **5** the torsion angle between C1 and C6 is

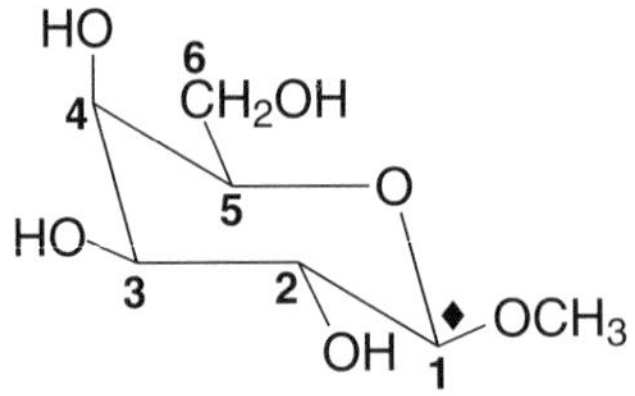

The scheme shows a galactopyranose structure with labels HO, 6, CH₂OH (CH$_2$OH), 4, 5, O, HO, 3, 2, OH, 1, OCH₃ (OCH$_3$), alongside:

$$^1J_{C1,C2} = 46.7 \text{ Hz}$$
$$^2J_{C1,C3} = 4.6 \text{ Hz}$$
$$^2J_{C1,Me} = 1.9 \text{ Hz}$$
$$^2J_{C1,C5} = 0 \text{ Hz}$$
$$^3J_{C1,C6} = 4.3 \text{ Hz}$$

Scheme 7–12. ^{13}C—^{13}C spin-coupling constants in methyl β-D-[1-^{13}C]galactopyranoside (**5**).

$\sim 180°$, and a large value for $^3J_{C1,C6}$ is observed (4.3 Hz) (Scheme 7-12). In **5**, however, the observation of $^3J_{C1,C6}$ has no value *from an assignment standpoint*, since C6 can be readily distinguished from C2, C3, C4, and C5 due to its characteristic chemical shift or different number of attached protons (Fig. 7-8).

From an assignment standpoint, ^{13}C signal assignments are commonly mediated *via* $^1J_{CH}$ (through HETCOR and HMQC) and $^1J_{CC}$. $^1J_{CC}$ can be used to assign carbon signals in natural abundance compounds *via* INADEQUATE (Bax et al., 1980; Mareci and Freeman, 1982), although the method is comparatively insensitive and not employed routinely. In addition, $^2J_{CH}$, $^3J_{CH}$, $^2J_{CC}$, and $^3J_{CC}$ have value not only for assignment purposes, but also as probes of 3D structure. While these couplings can be measured in unlabeled compounds, precise measurements are best made with ^{13}C-labeled samples. As discussed above, $^2J_{CC}$ and $^3J_{CC}$ are obtained from 1D ^{13}C spectra of selectively ^{13}C-labeled compounds (Walker et al., 1976) or *via* 1D INADEQUATE spectra (King-Morris & Serianni, 1987; Wu et al., 1992). The advantage of the latter approach is that those natural abundance carbon signals not spin-coupled to the labeled carbon are suppressed (i.e., only those signals due to the coupled carbons are detected), which can be helpful when analyzing more complex molecules (Fig. 7-11). ^{13}C—^{13}C COSY methods have also been used to measure J_{CC} values in uniformly ^{13}C-labeled carbohydrates (Bossennec et al., 1990).

The measurement of J_{CH} values is accomplished *via* several approaches. The simplest involves use of 1D ^{1}H-coupled ^{13}C NMR spectra (Fig. 7-12) (Garrett & Serianni, 1990). The splittings observed at each carbon are caused by the ^{13}C—^{1}H couplings experienced by that carbon. The trick is to decipher which splitting belongs to which coupling, since there is no internal consistency to identify specific couplings as was available in ^{1}H NMR spectra. Sometimes this problem can be addressed with the use of selective ^{1}H-decoupled ^{13}C spectra (Snyder & Serianni, 1986). In this experiment, specific ^{1}H signals are irradiated selectively while observing the ^{13}C spectrum. By noting which splitting is lost upon irradiation, specific splittings can be assigned to specific ^{13}C—^{1}H couplings. It should be appreciated, however, that non-first-order effects commonly occur in ^{1}H-coupled ^{13}C spectra of carbohydrates due to the non-first-order behavior in their ^{1}H NMR spectra (Cyr et al., 1974). When such effects are encountered, it is difficult to extract ^{13}C—^{1}H couplings directly from the ^{13}C spectrum.

Several 1D and 2D NMR methods have been devised to measure ^{13}C—^{1}H spin-couplings in natural abundance molecules, and some of these have been applied to carbohydrate systems (Mulloy et al., 1988; Nuzillard & Freeman, 1994; Varela et al., 1996; Zhu et al., 1994). Some of these methods, however, are restricted to comparatively small oligosaccharides (up to tetrasaccharides).

One-dimensional ^{1}H NMR spectra of moderately small ^{13}C-labeled carbohydrates can be used directly to measure ^{13}C—^{1}H couplings (Angelotti et al., 1987; Duker & Serianni, 1993; Serianni & Barker, 1984; Vuorinen & Serianni, 1990a,b). In natural compounds, the abundance of ^{13}C is so low that the detection of ^{13}C—^{1}H splittings (satellites) in ^{1}H spectra is difficult. Upon enhancing the ^{13}C content at a given carbon to 99 atom %, those protons near the labeled carbon sense the additional spin-1/2 nucleus and may exhibit a coupling to the labeled carbon. Thus, additional splittings (relative to those observed in the spectrum of the unlabeled compound) appear in those multiplets arising from protons that are coupled to the labeled carbon (Fig. 7-13). Again, non-first-order effects can complicate the measurement of these couplings; in these cases, spectral simulation is required to make reliable determinations.

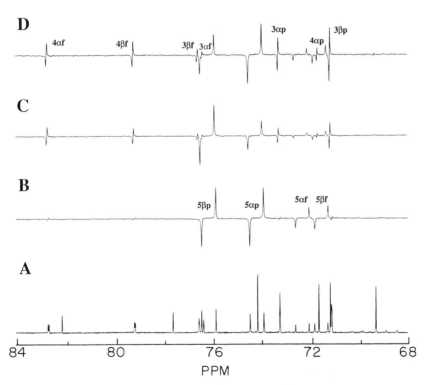

Figure 7–11. (Read bottom up) (**A**) The partial ^{1}H-decoupled ^{13}C NMR spectrum (75 MHz) of L-[6-^{13}C]idose. (**B-D**) INADEQUATE ^{13}C NMR spectra of L-[6-^{13}C]idose obtained with a 6 (**B**), 50 (**C**), and 100 ms (**D**) mixing time. Shorter mixing times enhance the detection of carbons directly bonded to the enriched site, while longer mixing times suppress the one-bond correlations and enhance the detection of coupled carbons further removed from the enriched site. Close inspection of data in **D** reveals small couplings to C3 of the α-furanose ($\sim$ 76.4 ppm) and to C4 of the α-pyranose ($\sim$ 71.7 ppm) that were not detected in the conventional 1D spectrum in **A**. Reprinted from Wu et al., 1992, with permission.

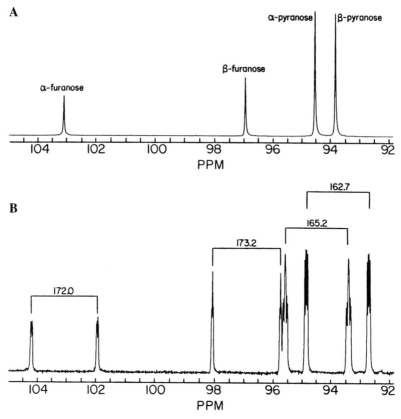

Figure 7–12. (**A**) The anomeric carbon region of the 75 MHz ^{1}H-decoupled ^{13}C NMR spectrum of D-[1-^{13}C]idose, showing signal assignments. (**B**) The same region as in **A** of the ^{1}H-coupled (with NOE) ^{13}C NMR spectrum, processed with resolution enhancement. The one-bond ^{13}C—^{1}H coupling in each form is as indicated; longer range ($^{2}J_{CH}$ and $^{3}J_{CH}$) couplings can also be observed for each form. Reprinted from Snyder & Serianni, 1986, with permission.

For larger molecules, 2D and 3D NMR methods can be applied to ^{13}C-labeled compounds to extract accurate J_{CH} values (Biamonti et al., 1994). The most useful 2D method is homonuclear correlation spectroscopy which includes COSY (Bax & Freeman, 1981; Piantini et al., 1982), TOCSY (total correlation spectroscopy) (Bax & Davis, 1985; Braunschweiler & Ernst, 1983; Davis & Bax, 1985), and NOESY (Nuclear Overhauser Effect SpectroscopY) (Bodenhausen & Ernst, 1982; Macura & Ernst, 1980). We confine our discussion to 2D TOCSY here, with the realization that the same concepts apply to related homonuclear methods.

In TOCSY spectra, signals (contours) along the diagonal correspond to the 1D ^{1}H spectrum. Crosspeaks denote correlations between specific proton pairs. These correlations arise due to scalar coupling interactions (i.e., spin–spin-coupling between proton pairs) *and* longer-range interactions mediated solely through spin-propagation induced by the spin-lock in the TOCSY pulse sequence. Thus, for example, the TOCSY spectrum of methyl β-D-allopyranoside (**12**) shows two crosspeaks correlating H1 with

B

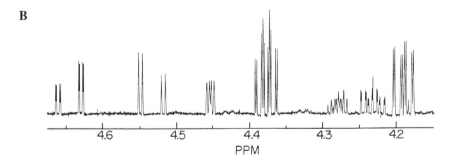

A

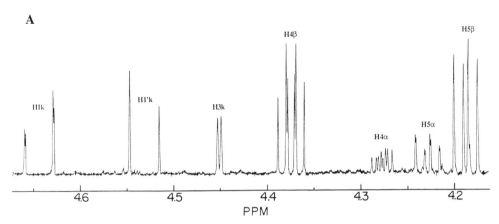

Figure 7–13. (Read bottom up) (**A**) Partial ^{1}H NMR spectrum (620 MHz, resolution-enhanced) of D-*threo*-2-pentulose, showing signal assignments. (**B**) Partial ^{1}H NMR spectrum (620 MHz, resolution-enhanced) of D-[2-^{13}C]*threo*-2-pentulose. Comparison with the spectrum in **A** reveals the additional signal splittings (J_{CH}) caused by ^{13}C-enrichment. Reprinted from Vuorinen & Serianni, 1990b, with permission.

H2 and H3. The H1—H2 crosspeak arises from $^3J_{H1,H2}$ (8.3 Hz). In contrast, the H1—H3 crosspeak is *not* mediated by $^4J_{H1,H3}$ (this coupling is zero in **12**) but rather is induced by the spin-lock. The TOCSY experiment thus permits a tracing of intact spin systems through spin propagation. *A major determinant of TOCSY crosspeak intensity is the magnitudes of $^3J_{HH}$ between the protons in the intact spin system.* The existence of a small $^3J_{HH}$ within the pathway will severely reduce the efficiency of magnetization transfer between protons and will truncate crosspeak intensities. In al-

methyl β-D-allopyranoside (**12**)

dopyranosyl rings, such small couplings are commonly encountered for $^3J_{He,Ha}$ and $^3J_{He,He}$; specific examples include $^3J_{H4,H5}$ in methyl β-D-galactopyranoside (**5**) (1.1 Hz), and $^3J_{H1,H2}$ in methyl β-D-mannopyranoside (**13**) (0.9 Hz). In addition, *crosspeak intensity falls off as the number of bonds between the correlated protons increases.* Typically, two to four correlations are observed, although more can be detected in some instances. With unlabeled molecules, TOCSY data are obtained along with DQF–COSY data to make signal assignments, the latter providing correlations between scalar-coupled protons, and the former providing longer-range correlations to confirm assignments.

TOCSY spectra obtained on ^{13}C-labeled carbohydrates may be used to extract J_{CH} values (Serianni & Podlasek, 1994). For example, consider the partial TOCSY spectrum of methyl β-D-[1-^{13}C]allopyranoside (**14**) (Fig. 7-14), showing only the crosspeaks involving H1. The on-diagonal H1 signal is split by the large $^1J_{C1,H1}$ (163.4 Hz) and by the smaller $^3J_{HH}$ (8.3 Hz). Thus, each correlation (to H2 and H3) appears as two paired crosspeaks due to the large $^1J_{CH}$ splitting. Inspection of the *relative alignment* of each crosspeak pair reveals a displacement when coupling exists between C1 and the respective proton. Thus, the displacement of the H1—H2 paired crosspeaks is 6.5 Hz, which corresponds to the coupling constant between C1 and H2. Likewise, the H1—H3 paired crosspeaks are displaced by 6.0 Hz, which is equal to $^3J_{C1,H3}$. Thus, provided that TOCSY data are obtained with sufficient digital resolution, this ^1J-resolved displacement method provides a straightforward way to measure J_{CH} values. In practice, this method provides better resolution of small J_{CH} (< 0.5 Hz) than can be achieved by the simpler 1D methods. Similar displacements are observed in COSY (Fig. 7-15), DQF–COSY, and related 2D spectra, but these data are often more limited due to the smaller number of crosspeaks observed in these spectra.

In addition to measuring the *magnitudes* of J_{CH} values from TOCSY spectra of ^{13}C-labeled compounds, information on coupling *sign* is also readily obtained. Sign information is difficult to assess *via* other methods, and is particularly important when interpreting the structural significance of $^2J_{CH}$ values. The simplified Dirac model of coupling signs (Harris, 1986) predicts that, for nuclei having positive γ and for satu-

methyl β-D-mannopyranoside (**13**)

methyl β-D-[1-^{13}C]allopyranoside (**14**)

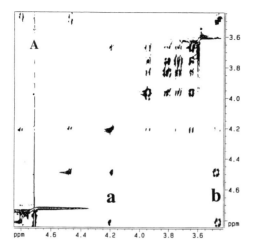

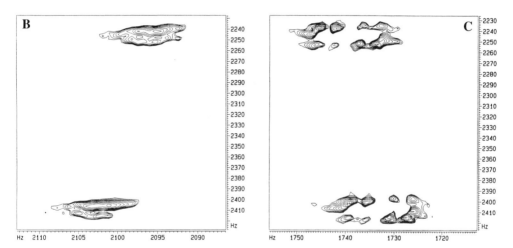

Figure 7–14. (A) The 500 MHz TOCSY spectrum of methyl β-D-[1-^{13}C]allopyranoside (**14**), showing the H1—H3 (**a**) and H1—H2 (**b**) crosspeaks. (B) Expanded H1—H3 crosspeaks, showing the displacement from which $^3J_{C1,H3}$ was measured (6.0 Hz). The relative displacement of these signals (downfield signal displaced to the left) is correlated with a J_{CH} having a positive sign. (C) Expanded H1—H2 crosspeaks, showing a displacement from which $^2J_{C1,H2}$ was determined (-6.5 Hz). The relative displacement of these signals has the opposite sense of that observed in **B**, indicating that $^2J_{C1,H2}$ has a negative sign. Reprinted from Serianni & Podlasek, 1994, with permission.

rated systems, one-bond (e.g., $^1J_{CC}$, $^1J_{CH}$) and three-bond couplings (e.g., $^3J_{CH}$, $^3J_{HH}$, $^3J_{CC}$) will have positive signs, whereas two-bond couplings will be negative (e.g., $^2J_{HH}$, $^2J_{CH}$, $^2J_{CC}$). In carbohydrates, these rules are generally observed for ^1J and ^3J, but not for ^2J. For example, two-bond ^{13}C—^{1}H couplings *often have positive signs* (Bock & Pedersen, 1977; Schwarcz & Perlin, 1972; Schwarcz et al., 1975), and thus their use in structural analyses must take sign into account. Indeed, coupling sign predictions

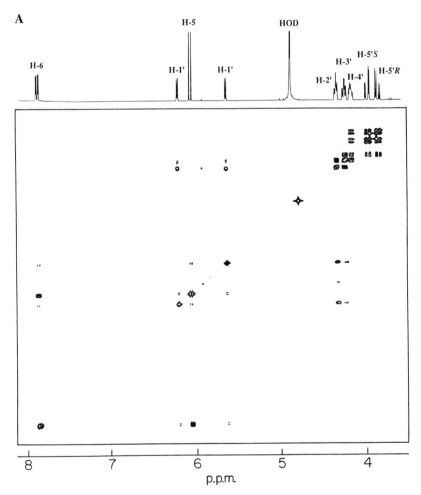

Figure 7–15. (A) The ¹H—¹H COSY spectrum of [1′-¹³C]cytidine in ²H₂O showing signal assignments. Expansion of the paired crosspeaks in **A** for (**B**) H1′—H2′, H1′—H3′, and (**C**) H1′—H6. The shifts in these paired crosspeaks yield $^2J_{C1',H2'}$ (-1.7 Hz), $^3J_{C1',H3'}$ (3.0 Hz) and $^3J_{C1',H6}$ (2.8 Hz). The weak signal between the H1′—H2′ crosspeaks arises from the small population of molecules not enriched with ¹³C at C1′. Reprinted from Wu & Serianni, 1992, with permission.

based on the Dirac model are not borne out experimentally even for the one-bond case (Harris, 1986).

TOCSY data provide J_{CH} coupling sign information in the following manner (Biamonti et al., 1994). An inspection is made of the *relative orientation* of the displaced paired crosspeaks for a $^3J_{CH}$ interaction (e.g., the H1—H3 crosspeaks of methyl β-D-[1-¹³C]allopyranoside [**14**], Fig. 7-14). This orientation is correlated with a positive coupling ($^3J_{CH}$ is assumed to have a positive sign). All other displaced paired crosspeaks having the same *relative* orientation must therefore relate to J_{CH} values having positive signs. In addition, those having the opposite relative orientation, such as that observed between H1 and H2 in **14** (Fig. 7-14), must relate to J_{CH} values having a neg-

B

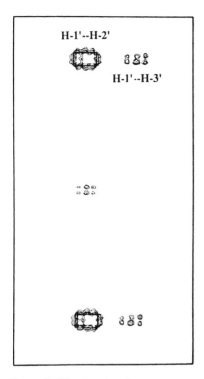

C

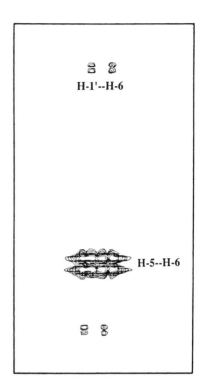

Figure 7–15. (continued)

ative sign; thus $^2J_{C1,H2}$ in **14** $= -6.5$ Hz. By using the relative orientation of displaced paired crosspeaks in TOCSY spectra of ^{13}C-labeled compounds, *both the magnitudes and signs of J_{CH} values may be ascertained.* Related E.COSY-based 3D NMR methods have been developed to determine accurate J_{CH} magnitudes and signs in larger biomolecules such as RNA (Marino et al., 1996).

Although homonuclear correlation 2D spectra are perhaps the most efficient way to determine J_{CH} values in labeled carbohydrates, other 2D methods have been used to accomplish the same end. For example, homonuclear *J*-spectroscopy has been used to extract J_{CH} values (without sign information) by examining the data projected along the chemical shift axis (Fig. 7-16) (Vuorinen & Serianni, 1990a). Normally, these data correspond to the "proton-decoupled" ^{1}H spectrum, with each ^{1}H signal appearing as a singlet. Thus, any multiplicity observed in the projected signals must arise from *heteronuclear couplings*, such as J_{CH} (which may exist in ^{13}C-labeled molecules) or J_{HP} in phosphorylated structures. The accuracy and sensitivity of this approach, however, are considerably less than those of the displacement methods discussed above.

The above discussion has focused on the use of singly ^{13}C-labeled compounds and 2D NMR methods to evaluate J_{CH} values. It should be appreciated, however, that multiply ^{13}C-labeled compounds can be employed to make similar measurements, but of course the 2D spectra arising from these compounds will be significantly more com-

A

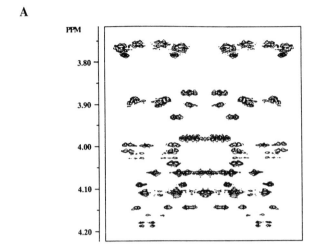

B

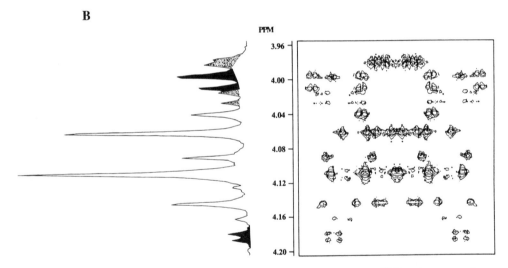

Figure 7–16. (A) Partial homonuclear 2D *J*-spectrum (500 MHz) of D-[1-^{13}C]*erythro*-pentos-2-ulose (**26**) at 30°. (B) An expanded region of **A**, showing additional contour splittings caused by ^{13}C—^{1}H spin-coupling. The projected spectrum reveals four doublets from protons spin-coupled to ^{13}C at C1. Reprinted from Vuorinen & Serianni, 1990a, with permission.

plex. This complexity can be reduced by adding a third dimension to the experiment, an approach embodied in 3D HMQC–TOCSY (Wijmenga et al., 1989) where the ^{13}C nucleus is used to edit the 2D TOCSY spectrum. In other words, the spectral dispersion of ^{13}C is used to collect TOCSY subspectra showing only correlations to a specific labeled carbon. This application is illustrated with use of a simple aqueous sample containing equimolar quantities of [1'-^{13}C]adenosine, [1'-^{13}C]uridine, [1'-^{13}C]cytidine, and [1'-^{13}C]ribothymidine. The 2D TOCSY spectrum of this mixture showing only H1′ correlations is crowded and thus is not amenable to analysis to extract J_{CH} values. However, if the 2D spectrum is edited with respect to the *unique* C1′ chemical shifts of the four nucleosides in a 3D HMQC–TOCSY experiment, separate TOCSY subspectra of each of the four compounds are obtained, allowing extraction of J_{CH} values in the same manner used for the 2D TOCSY spectrum (Fig. 7-17). In this illustration, we are editing only with respect to C1′ chemical shifts, since only these carbons are labeled, but other carbons could be labeled and used as editing nuclei. It is clear that this method can be very valuable in carbohydrate conformational analysis, not only in studies of nucleic acids, but for oligo and polysaccharides, provided that ^{13}C-labeled compounds can be obtained. However, it should be appreciated that the presence of high atom % ^{13}C-enrichment on contiguous carbons may complicate 3D HMQC–TOCSY data due to multiplicity arising from homonuclear $^{13}C{-}^{13}C$ couplings, and thus such studies are normally conducted with molecules labeled at lower atom % ^{13}C ($\sim$ 30 atom % has been reported to be optimal) (Hines et al., 1994) when the aim is to measure J_{CH} values.

The above discussion focuses exclusively on ^{1}H and ^{13}C nuclei, although ^{2}H substitution is commonly employed in carbohydrate studies. In contrast to ^{1}H and ^{13}C, ^{2}H is a quadrupolar nucleus (I = 1). Thus, when a single ^{2}H is coupled to ^{1}H or ^{13}C, signal multiplicity will be 2(1)(1) + 1 = 3 (triplet). In addition, the signals of scalar-coupled ^{1}H or ^{13}C nuclei are shifted upfield in the ^{2}H-substituted molecule due to the isotope effect (Fig. 7-18) (Kline & Serianni, 1988, 1990b; Serianni & Barker, 1979).

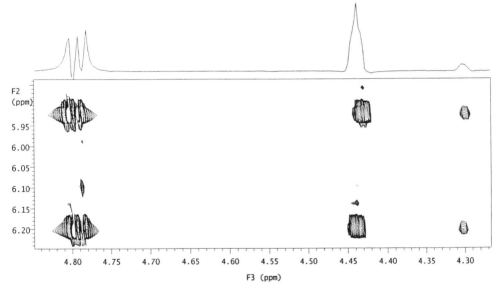

Figure 7–17. A TOCSY plane of a 3D HMQC–TOCSY data set (600 MHz, no ^{13}C-decoupling) obtained on an equimolar mixture of [1′-^{13}C]adenosine, [1′-^{13}C]cytidine, [1′-^{13}C]ribothymidine, and [1′-^{13}C]uridine (10 mM each) in $^{2}H_{2}O$, edited with respect to the C1′ chemical shift of [1′-^{13}C]adenosine. Only the H2′ ($\sim$4.8 ppm), H3′ ($\sim$4.45 ppm), and H4′ ($\sim$4.3 ppm) crosspeaks of [1′-^{13}C]adenosine are observed. As shown in Figure 7-14, the displacement of the paired crosspeaks (split by the large $^{1}J_{CH}$) is used to measure the magnitude and sign of $^{2}J_{C1′,H2′}$, (-3.2 Hz), $^{3}J_{C1′,H3′}$ ($\sim$+5.1 Hz) and $^{3}J_{C1′,H4′}$ (+1.3 Hz).

This shift decreases as the ¹H or ¹³C becomes further removed from the ²H nucleus and the effect has been used to make ¹³C chemical shift assignments (Dais & Perlin, 1987; Lyerla & Levy, 1974). ²H Isotope shifts of ~ 0.3 ppm are common for the α-carbon (i.e., the carbon bonded directly to the ²H). Since the gyromagnetic ratio of ²H is 1/6.5144 that of ¹H, ²H—¹H or ¹³C—²H spin-coupling constants are ~ 1/6 those

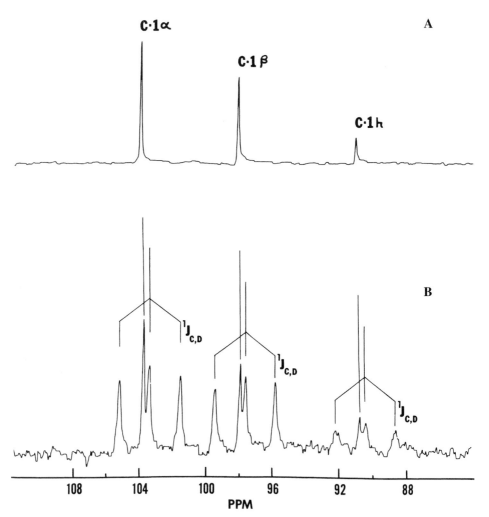

Figure 7–18. ¹³C—²H Spin-coupling and isotope shift in a [¹³C, ²H]-enriched carbohydrate. (**A**) The 15.08 MHz ¹H-decoupled ¹³C NMR spectrum of the enriched region of D-[1-¹³C]threose. The three predominant tautomeric forms in aqueous solution are α- and β-furanose (103.4 and 97.9 ppm, respectively) and acyclic hydrate (91.1 ppm). (**B**) The 15.08 MHz ¹H-decoupled ¹³C NMR spectrum of the enriched region of D-[1-¹³C,²H]threose, showing the splitting of C1 of each form by the directly-bound deuteron. Isotope shift is shown for each species as the difference in the positions between the protonated C1 and the center of gravity of the triplet arising from the deuterated C1. Reprinted from Serianni & Barker, 1979, with permission.

β-D-[1-^{13}C]-
erythrofuranose (**15**)

β-D-[1-^{13}C,^{2}H]-
erythrofuranose (**16**)

of the corresponding ^{1}H—^{1}H or ^{13}C—^{1}H values. For example, $^1J_{C1,H1}$ = 172.3 Hz in β-D-[1-^{13}C]erythrofuranose (**15**) and $^1J_{C1,D1}$ = 26.0 Hz in β-D-[1-^{13}C, ^{2}H]erythrofuranose (**16**) (Serianni & Barker, 1979). The triplet pattern and small magnitudes sometimes make ^{2}H—^{1}H and ^{13}C—^{2}H couplings unresolvable in ^{1}H and ^{13}C NMR spectra, respectively. It should also be mentioned that isotope shifts in ^{13}C resonances upon deuterium exchange at *hydroxyl* protons have been used to facilitate ^{13}C chemical shift assignments in carbohydrates (Pfeffer et al., 1979).

Isotope shifts are also observed for the chemical shifts of ^{13}C nuclei bound to different isotopes of oxygen. Thus, the ^{13}C signal of a ^{13}C—^{18}O fragment will lie slightly upfield of that of the corresponding ^{13}C—^{16}O fragment (Risley & Van Etten, 1979). This effect can be extremely useful in monitoring oxygen exchange reactions and elucidating carbohydrate reaction mechanisms (Mega & Van Etten, 1988).

Nuclear Spin Relaxation

All spectroscopic methods are based on the excitation of the sample by photons followed by one or more relaxation processes. In NMR spectroscopy, atomic nuclei are excited essentially instantaneously by low energy radio-frequency photons, but the manner and rates at which nuclei relax (i.e., dissipate the absorbed energy) depend on various factors. Nuclear spin relaxation can occur *via* several mechanisms and over a wide range of rates, and both must be appreciated in order to use relaxation information in structural studies.

Two fundamental nuclear relaxation processes are possible, nuclear spin–lattice relaxation and nuclear spin–spin relaxation, characterized by the relaxation times T_1 and T_2, respectively. The former may be viewed as an enthalpic process whereby the molecule releases the absorbed energy to the "environment," whereas the latter involves a redistribution of this energy (an entropic process). For small molecules (MW < 1000 Da), $T_1 = T_2$ (this is denoted as the region of extreme motional narrowing), whereas in larger molecules (e.g., polysaccharides, proteins) $T_1 > T_2$. Resonance linewidth in NMR spectroscopy is related to T_2; specifically, $\nu_{1/2} = (\pi T_2)^{-1}$, where $\nu_{1/2}$ = signal linewidth in Hz at 1/2 height of the signal. This relationship is confirmed qualitatively by noting the effect of molecular size on resonance linewidth, namely, larger molecules produce broader NMR signals than smaller molecules, all other factors being equal. Large molecules exhibit smaller T_2 values than smaller molecules, meaning that the spin–spin relaxation rate is greater ($1/T_2$ = spin–spin relaxation rate). A greater rate of relaxation translates into a less well-defined signal (due to the Uncertainty Principle) and thus a broader signal. In the following discussion, we assume the region of extreme motional narrowing applies to the examples cited.

Why does molecular size affect nuclear relaxation rates? Molecular size determines, in part, the motional properties of a molecule in solution, with larger molecules tumbling more slowly than smaller molecules. This molecular tumbling rate is embodied in the so-called rotational correlation time, τ_r, which is directly proportional to $1/T_1$ and $1/T_2$, that is, $1/T_1 = 1/T_2 \propto \tau_r$. Since larger molecules have larger τ_r values (i.e., it takes a longer time for larger molecules to undergo one full molecular rotation), they will exhibit larger $1/T_1$ and $1/T_2$ and thus smaller T_1 and T_2 (Fig. 7-19). Thus, it is apparent that T_1 and T_2 values report on molecular motion in solution.

There are different kinds of motion, however. Considering overall (global) motion first, there are two fundamental types: isotropic and anisotropic. The former describes molecules that tumble as a sphere with no preferred axis of rotation. This type of overall motion is completely described by a single-correlation time. In contrast, anisotropic motion implies that rotation rates about at least two molecular axes (x, y, or z) are different. This situation occurs when the molecular dimensions about these axes are highly dissimilar, as might occur in elongated, extended molecules such as linear unbranched oligosaccharides. It should be appreciated, however, that molecules, especially carbohydrates, probably do not tumble as isolated molecules in aqueous solution, but rather as carbohydrate–water assemblies, and thus the nature of the solvent cage itself may affect tumbling behavior. Anisotropic motion is characterized by distinct correlation times that describe motion about each rotational axis of the molecule.

In addition to global motion, portions of a molecule may experience motion, this being referred to as local (internal) motion. Examples might be rotational reorientation of the exocyclic hydroxymethyl groups of monosaccharides (Scheme 7-8), or motion about the glycosidic linkages of oligosaccharides (Scheme 7-9). This motion will be

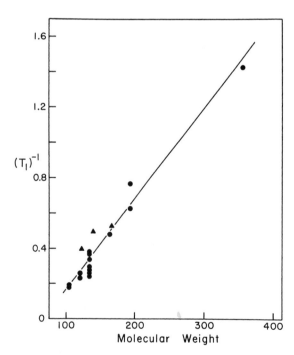

Figure 7–19. Plot of C1 ^{13}C relaxation rates $(T_1)^{-1}$ versus molecular weight (M) for a series of monosaccharides at $30 \pm 1.5°C$ in $^2H_2O:H_2O$ (1:1), 50 mM HOAc/NaOAc (pH 5.0), and 3 mM EDTA. Circles and triangles refer to cyclic and acyclic structures, respectively. Reprinted from Serianni & Barker, 1982, with permission.

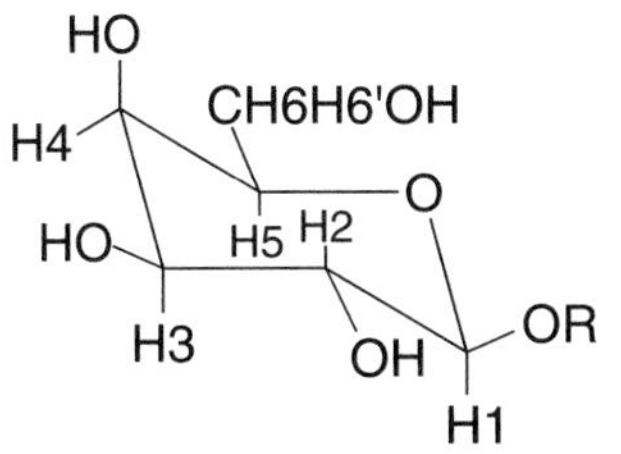

$r_{H1,H2} = 2.78\ \text{Å}$
$r_{H1,H3} = 2.53\ \text{Å}$
$r_{H1,H4} = 3.89\ \text{Å}$
$r_{H1,H5} = 2.29\ \text{Å}$
$r_{H1,H6} = 4.41\ \text{Å}$
$r_{H1,H6'} = 4.52\ \text{Å}$

Scheme 7–13. Approximate ^{1}H—^{1}H internuclear distances involving H1 of a β-D-galactopyranosyl ring determined by X-ray crystallography.

superimposed on the overall motion of the molecule, and thus those nuclei experiencing both kinds of motion will exhibit different relaxation properties than those experiencing only global motion.

Both T_1 and T_2 processes occur *via* several distinct mechanisms. These include dipole–dipole, quadrupolar, spin rotation, scalar coupling, chemical shift anisotropy, and paramagnetic mechanisms. For most carbohydrates in dilute aqueous (^{2}H$_2$O) solution at room temperature with no paramagnetic ions present, the *intramolecular* dipole–dipole mechanism dominates ^{1}H and ^{13}C relaxation (Berry et al., 1979; Dais & Perlin, 1987; Lyerla & Levy, 1974), and thus we focus attention solely on this mechanism in the following discussion. It should be appreciated, however, that relaxation may not always be an *intramolecular* process (i.e., there may be an *intermolecular* contribution), particularly for peripheral nuclei such as ^{1}H which can come into contact with ^{1}H nuclei of other solute molecules (especially at high-solute concentrations) or ^{1}H nuclei in solvent molecules (for example, when studies are conducted in ^{1}H$_2$O solvent).

In addition to motional properties, relaxation times, particularly ^{1}H relaxation times, can be used to determine molecular geometry. To consider why this is so, we examine a specific monosaccharide, methyl β-D-galactopyranoside (**5**) in ^{2}H$_2$O solution. There are five carbon-bound ring protons and two exocyclic carbon-bound protons in **5** (excluding the aglycone CH$_3$ protons); the OH protons in **5** exchange with solvent deuterium and thus can be ignored. Consider the T_1 relaxation of H1. Assuming H1 relaxes exclusively *via* a dipole–dipole mechanism, there are six carbon-bound ^{1}H nuclei within **5** that can potentially promote its relaxation. The efficiency with which a given ^{1}H nucleus **A** promotes the relaxation of H1 is determined by the internuclear distance between it and H1. More specifically, the contribution **A** makes to the overall relaxation rate, $1/T_{1,H1,overall}$, is directly proportional to $1/r_{A,H1}^6$. Thus, the contribution that any particular proton makes to the relaxation of H1 drops off rapidly as the internuclear distance increases. For example, H1 is closest to H3 and H5 in **5** (2.53 Å and 2.29 Å, respectively, according to X-ray analysis [Takagi & Jeffrey 1978]) (Scheme 7-13), and the latter are expected to make a major contribution to the relaxation of H1. In addition, the exocyclic hydroxymethyl protons of **5**, being distant from H1, will make little contribution to H1 relaxation. The overall relationship between $1/T_{1,H1,overall}$ and internuclear distances, assuming only an intramolecular process, is: $1/T_{1,H1,overall} \propto 1/r_{H1,H2}^6 + 1/r_{H1,H3}^6 + 1/r_{H1,H4}^6 + 1/r_{H1,H5}^6 + 1/r_{H1,H6}^6 + 1/r_{H1,H6'}^6$. Using X-ray data to estimate the ^{1}H—^{1}H internuclear distances in the β-D-galactopyranosyl ring in Scheme 7-13, the following % contributions to the overall relaxation rate of H1 are computed: H2, 16%; H3, 28%; H4, 2%; H5, 52%; H6, 1%;

H6', 1%. These results confirm that H3 and H5 are the major relaxation sinks for H1, accounting for 80% of its relaxation.

Since the internuclear ^1H—^1H distance relationships experienced by each proton in **5** differ, the relaxation times of these nuclei will differ even though the molecule tumbles isotropically (i.e., each nucleus experiences the same motion). Thus, for example, using ^1H—^1H distances derived from X-ray crystallography (Takagi & Jeffrey, 1978) to compute the relative magnitudes of T_1 values for the carbon-bound protons of **5**, H2 is predicted to have a significantly larger T_1 (i.e., slower relaxation rate) than the remaining protons (Scheme 7-14) (Kline et al., 1990) since it is relatively spin-isolated, that is, it experiences no 1,3-diaxial interactions with other ring protons and is oriented *trans* with respect to H1 and H3. T_1 Relaxation times measured experimentally by the inversion–recovery method (Vold et al., 1968) confirm this prediction (Scheme 7-14) (Kline et al., 1990). Sometimes differential ^1H T_1s in a molecule can be exploited to resolve overlapping signals in partially relaxed spectra (Garrett & Serianni, 1990; Snyder & Serianni, 1986), thus facilitating spectral assignment and interpretation.

The relaxation of ^{13}C is also exclusively dipole–dipole for carbon atoms bound directly to protons. In this case, the C—H bond length is the prime distance determinant of the relaxation rate. Thus $1/T_{1,C} \propto 1/r_{CH}^6$. In this situation, *assuming that C—H bond lengths are relatively constant in a given molecule*, the T_1 values of carbons bearing the same number of protons should be very similar if the molecule is tumbling isotropically and no internal motions are present. It should also be apparent that, in the above case, ^{13}C relaxation can provide a good estimate of τ_r (isotropic motion), since the only internuclear distance affecting ^{13}C T_1 is relatively constant (i.e., the C—H bond length).

^{13}C relaxation rates of protonated carbons depend on the number of attached protons, that is, $1/T_{1,C} \propto N$, where N = the number of attached protons. For example, C1 of methyl 2-deoxy-β-D-glucopyranoside (**17**) should have a T_1 value twice that of C2. It should be appreciated, however, that this relationship holds only for carbons experiencing the same motions. If this is not the case, then deviations may be observed. For example, the T_1 values for C1, C2, and C3 of methyl α-D-threofuranoside (**10**) in aqueous solution at 30°C average 3.5 sec, whereas that of C4 is 2.5 sec (Serianni & Barker, 1984). Clearly, the latter would be 3.5/2 = 1.8 sec if C4 were experiencing the same motion as C1—C3. The larger than expected T_1 of C4 may reflect the presence of internal motion. The full equation for ^{13}C relaxation (dipole–dipole, isotropic motion, small molecule, nonviscous solvent) is $1/T_{1,C} = (N\gamma_C^2\gamma_H^2\hbar^2\tau_r) r_{CH}^{-6}$, where γ_C and γ_H are the gyromagnetic ratios for ^{13}C and ^1H, respectively, and $\hbar$ = Planck's con-

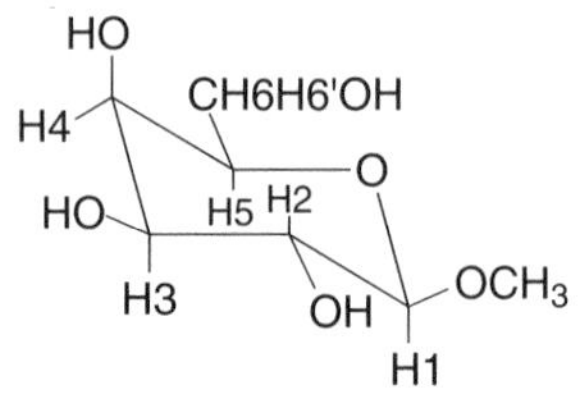

	$T_{1,H1}$ = 1.5 s
	$T_{1,H2}$ = 6.5 s
	$T_{1,H3}$ = 1.9 s
	$T_{1,H4}$ = 1.8 s
	$T_{1,H5}$ = 1.1s
	$T_{1,H6}$ = 1.0 s
	$T_{1,H6'}$ = 0.97 s

Scheme 7–14. ^1H T_1 values for methyl β-D-galactopyranoside (**5**).

methyl 2-deoxy-β-D-glucopyranoside (**17**)
(methyl 2-deoxy-β-D-*arabino*-hexopyranoside)

stant divided by 2π. Thus, τ_r may be calculated from a ^{13}C T_1 of 1.4 sec for a methine carbon as follows: $1/T_{1,C} = 1/1.4$ sec $= [(1) (6728)^2 (26753)^2 (1.054 \times 10^{-27})^2 \tau_r]/$ $(1.07 \times 10^{-8})^{6\cdot}$ where γ_C and γ_H are expressed in radians $s^{-1}gauss^{-1}$, h in erg sec, and r in cm. A value of 2.98×10^{-11} s is obtained for τ_r.

The efficiency of relaxation of carbons bound directly to ^1H can be destroyed by ^2H substitution. Thus, conversion of a ^{13}C—^1H methine group to a ^{13}C—^2H methine group results in a change in the ^{13}C signal multiplicity under broadband ^1H-decoupling conditions (singlet to triplet, since I $= 1$ for ^2H) (Fig. 7-18), a dramatic increase in ^{13}C T_1 (due to loss of the nearby strong ^1H dipole), and a partial or complete loss in nuclear Overhauser enhancement (NOE). All of these factors conspire to make detection of the deuterium-substituted carbon difficult (in principle, some of the sensitivity can be restored by decoupling the ^2H from ^{13}C, which removes the signal multiplicity).

Precise dipole–dipole relaxation equations for T_1 and T_2 processes involving ^1H and ^{13}C have been derived for molecules undergoing either isotropic or anisotropic motion, for molecules tumbling within and outside of the region of extreme motional narrowing, and for molecules experiencing internal motions. We refer the reader to several references for a more comprehensive discussion of these equations (London, 1980; Shimizu, 1962; Woessner, 1962). The key points with respect to ^1H and ^{13}C T_1 relaxation in carbohydrates are: the relaxation mechanisms of both nuclei are normally of the dipole–dipole type; relaxation rates are directly proportional to τ_r (isotropic motion); ^1H relaxation is affected by ^1H—^1H internuclear distances; ^{13}C relaxation (protonated carbons) is affected by C—H bond length.

^1H Relaxation times are, in principle, well suited for use in conformational analysis due to their sensitivity to ^1H—^1H internuclear distances. One approach takes advantage of selective ^2H substitution to perturb the ^1H relaxation times in a molecule and thereby compute ^1H—^1H internuclear distances. This method, first described by Akasaka and co-workers (Akasaka et al., 1975), is known as the DESERT method (DEuterium Substitution Effects on ^1H Relaxation Times). The method gives reasonable estimates of ^1H—^1H internuclear distances in *conformationally-rigid* carbohydrates (Fig. 7-20) (Kline et al., 1990) and nucleosides.

Due to the inherent insensitivity of the ^{13}C nucleus, the measurement of ^{13}C relaxation times is usually not straightforward unless substantial quantities of material are available. While the severity of this problem has been reduced due to the availability of more sensitive high-field spectrometers and microprobes, it remains advantageous to use ^{13}C-labeled compounds or indirect detection *via* higher γ nuclei (e.g., ^1H; see Nicholson et al., 1992) to assist in these relaxation measurements. It should be appreciated, however, that ^{13}C relaxation measurements on ^{13}C-labeled compounds, *specifically those containing contiguous ^{13}C-labeling*, must take into account not only ^{13}C—^1H relaxation pathways in the molecule but also ^{13}C—^{13}C relaxation pathways

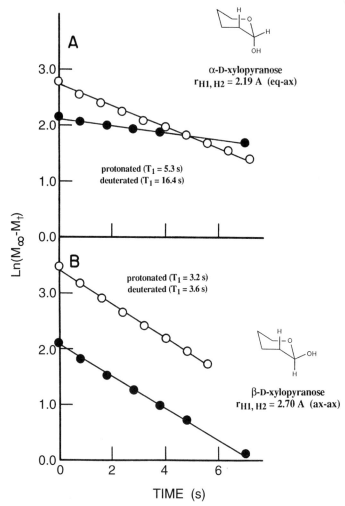

Figure 7–20. Spin-lattice relaxation curves for H1 of α-xylopyranose (5.3 sec) and α-D-[2-^{2}H]-xylopyranose (16.4 sec) (**A**), and β-D-xylopyranose (3.2 sec) and β-D-[2-^{2}H]xylopyranose (3.6 sec) (**B**) at 27°C. Data for natural compounds are shown in open symbols, and data for deuterated compounds are shown in filled symbols. From these data, and τ_r values derived from ^{13}C relaxation measurements, internuclear distances between H1 and H2 in the α- and β-pyranoses were determined (2.19 Å and 2.70 Å, respectively). Reprinted from Kline et al., 1990, with permission.

(London et al., 1975). While the gyromagnetic ratio of ^{13}C is considerably smaller than ^{1}H and thus its dipolar relaxation effect considerably smaller, the contributions that ^{13}C nuclei make to the relaxation of another ^{13}C nucleus directly bonded to them cannot be ignored (since the internuclear C—C distance is only ~ 1.5 Å), and this factor must be taken into account when interpreting the relaxation data.

Conformational Flexibility and the
Interpretation of NMR Parameters

Many molecules, including carbohydrates, do not assume a single conformation in solution. Frequently, one or more portions of the molecule are conformationally flexible, sampling two or more geometries; for example, the C5—C6 bond in aldohexopyranosyl rings is free to rotate, giving rise to different hydroxymethyl (CH_2OH) conformations (Scheme 7-8). When this sampling occurs, the interpretation of NMR parameters can be complicated. This problem has been discussed in detail by Jardetzky (1980), and has resurfaced more recently in conformational studies of oligosaccharides by Carver and co-workers (Cumming & Carver, 1987). Only two limiting conditions are considered here.

Consider a molecule A that assumes two stable conformations in solution, A_1 and A_2. Important in terms of the interpretation of NMR parameters are: (1) What are the relative stabilities of A_1 and A_2? (2) What are the rates of interconversion between A_1 and A_2? (3) Do any intermediates in the interconversion of A_1 and A_2 affect the observed NMR parameters? In the simplest case, A_1 and A_2 interconvert very slowly, and the lifetimes of any intermediates are very short. The terms "very slowly" and "very short" are *relative* times, that is, they are times relative to the frequency of observation (i.e., the NMR observe frequency). A very slow interconversion of A_1 and A_2 means that the lifetimes of A_1 and A_2 will be long relative to the NMR observe frequency, and thus they will be detected as separate species (i.e., distinct signals will be observed for A_1 and A_2 in the spectrum). Furthermore, the intermediates, having very short lifetimes, will not be detected and will therefore not influence the NMR parameters observed for A_1 and A_2. In this system A_1 and A_2 are said to be in *slow exchange*. Since distinct signals are observed for each form, it is straightforward to compute their percentages *via* signal integration, and thus their relative stabilities, in solution.

Now consider what happens when A_1 and A_2 interconvert very rapidly relative to the frequency of observation (fast exchange), again under conditions where intermediates make no contribution to the observed spectrum. In this case, only one averaged species, which we shall call A_{avg}, is observed in the NMR spectrum, and the NMR parameters measured for this species will be determined by those found in the two interconverting forms weighted according to their percentages in solution. For example, a specific ^1H—^1H spin coupling would be averaged as $J_{HH}^{avg} = \rho_{A1}J_{HH}^{A1} + \rho_{A2}J_{HH}^{A2}$, where J_{HH}^{avg} is the observed averaged coupling, ρ_{A1} and ρ_{A2} are the fractional populations of A_1 and A_2, and J_{HH}^{A1} and J_{HH}^{A2} are the couplings expected in the individual forms A_1 and A_2. Now, if in fact we knew that molecule A interconverted only between A_1 and A_2 in solution, then we could interpret the averaged NMR parameter correctly. However, what if we do not possess this knowledge, as is often the case? In this situation, we might interpret the data incorrectly as being consistent with a single conformation rather than a mixture. Furthermore, this "single" conformation would have no physical meaning (i.e., it does not represent what is actually occurring in solution) and is thus referred to as a "virtual" conformation. *It is important to appreciate that the interpretation of NMR data in cases where conformational flexibility or averaging is suspected requires prior knowledge of the number of stable forms present in solution and their structures, and the number and structures of intermedi-*

ates involved in their interconversion and their lifetimes. This kind of information is usually obtained by computational analyses, which may be empirical (e.g., molecular mechanics), semiempirical (e.g., SE2) or *ab initio* (e.g., molecular orbital calculations) in nature. These computations can be used to evaluate the structures and relative stabilities of products and intermediates as well as rates of interconversion (e.g., *via* molecular dynamics simulations). On the experimental side, *studies of systems suspected to be conformationally flexible should attempt to inspect as large a number of diverse NMR parameters as possible in order to test potential conformational models against computational predictions.*

The presence of conformational averaging can complicate the interpretation of NMR parameters that are not subject to a linear response to particular molecular motions (Homans, 1990; Jardetzky, 1980). This effect is illustrated by considering a molecule containing a three-spin system, H_a, H_b and H_c, in which the H_a—H_b distance is fixed at 2.5 Å and the H_a—H_c distance assumes two values, 2.5 Å and 4.0 Å, thereby generating two conformations of the molecule, X and Y, respectively (Scheme 7-15), in solution. We assume that (1) the interconversion between X and Y occurs rapidly on the NMR time scale, (2) the transit time between these forms is rapid relative to their lifetimes, (3) X and Y are present in equal proportions, (4) nuclear relaxation occurs exclusively *via* a dipole–dipole mechanism, and (5) each spin experiences the same isotropic motion with $\tau_c = 2.5 \times 10^{-11}$ sec. Under these conditions, the T_1 of H_a when H_c is 2.5 Å and 4.0 Å away can be computed; these values are 5.72 and 10.80 sec, respectively. Upon substitution of ^2H at H_c, the relaxation of H_a is perturbed; again we can calculate this effect and obtain T_1 values of H_a of 11.02 and 11.47 sec. Since relaxation rates average linearly, we know that $(T_1)^{-1}{}_{Ha,avg} = \rho_X(T_1)^{-1}{}_{Ha,X} + \rho_Y(T_1)^{-1}{}_{Ha,Y}$, where $(T_1)^{-1}{}_{Ha,avg}$ is the observed relaxation rate of H_a, ρ_X and $\rho_Y =$ the fractional populations of X and Y, respectively, where ρ_X and $\rho_Y = 0.5$, and $(T_1)^{-1}{}_{Ha,X}$ and $(T_1)^{-1}{}_{Ha,Y} =$ the relaxation rates of H_a in X and Y, respectively. Using this equation, we find $(T_1)^{-1}{}_{Ha,avg}$ to be 0.134 sec^{-1} and 0.089 sec^{-1} when H_c is ^1H and ^2H, respectively. Using this differential relaxation, the DESERT equations (Kline et al., 1990; Akasaka et al., 1975) can be applied to compute the internuclear distance between H_a and H_c. Doing so yields a value of 2.64 Å. Clearly, this value differs from the *linearly averaged* distance of $(2.5 + 4.0)(0.5) = 3.25$ Å. Despite the 50:50 distribution of X and Y, the H_a—H_c distance determined *via* relaxation measurements *is highly weighted in favor of the form providing the more potent relaxation pathways for H_a (i.e., molecule X).* This skewing effect is due to the r^{-6} dependence of the relaxation rates; likewise, since NOE shows a similar dependence on internuclear dis-

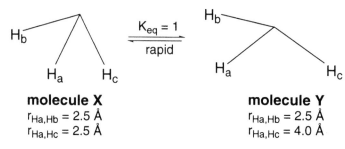

molecule X
$r_{Ha,Hb} = 2.5$ Å
$r_{Ha,Hc} = 2.5$ Å

molecule Y
$r_{Ha,Hb} = 2.5$ Å
$r_{Ha,Hc} = 4.0$ Å

Scheme 7–15.

tance, it will behave similarly. *It should be evident that the accurate interpretation of the experimentally derived value of 2.64 Å depends on the conformational model invoked for the molecule.* If we have knowledge of the 50:50 mixture of X and Y, then we can interpret this distance correctly. On the other hand, if we were to assume a single conformation, this distance would provide a completely erroneous picture of the molecule (a "virtual" conformation). The only way to minimize these effects is to measure as many relaxation times or NOEs possible with the expectation that nonlinear effects will be more readily discerned and appropriately considered.

Specific Applications

In this section, we discuss some examples of the use of ^{13}C-labeled carbohydrates and their derivatives to address structural problems.

Solution Behavior of Reducing Sugars and Glycosides

Reducing sugars exist in solution in various cyclic and acyclic forms (Angyal, 1969, 1991). The common cyclic forms include furanoses and pyranoses, whereas acyclic forms include the carbonyl and hydrate forms. In some cases, oligomeric forms may be present. The relative stabilities of these forms depend in large part on the structure and configuration of the sugar, and the nature of the solvent. Detection, assignment, and quantitation of these forms by NMR are not always straightforward, especially for those present in low abundance; a weak signal arising from a sample impurity can easily be mistaken for signals of a minor form. This problem is ameliorated greatly by use of selective ^{13}C-labeling at the carbonyl carbon(s) of the sugar (i.e., C1 of aldoses, C2 of 2-ketoses, and C1 or C2 of aldos-2-uloses [dicarbonyl sugars]) (Scheme 7-16). ^{13}C-Labeling allows the selective detection of the carbonyl carbon and significantly improves the selectivity and sensitivity of detection, which simplifies the spectral pattern and permits detection of minor forms. It should be appreciated that the detection of weak signals in the presence of strong signals is severely compromised by the dynamic range problem (Martin et al., 1980), such that detection and quantitation of some minor forms using natural compounds is virtually impossible.

Aqueous solutions of short-chain sugars such as glycolaldehyde (**18**) and D-glyceraldehyde (**19**) contain acyclic forms (carbonyl and hydrate) as well as oligomers (Serianni et al., 1979a). The presence of the latter is clearly indicated by the presence

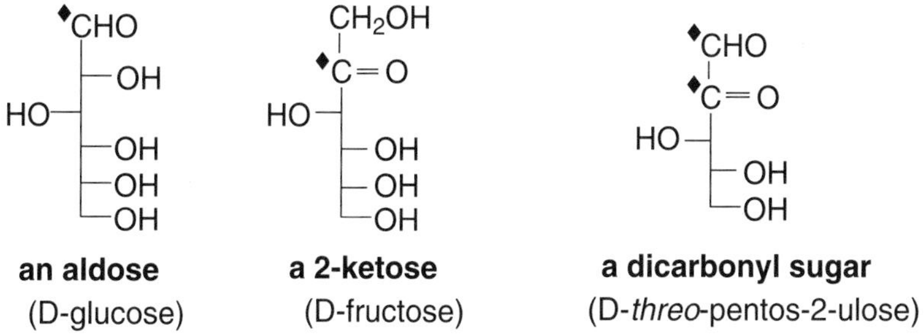

| **an aldose** | **a 2-ketose** | **a dicarbonyl sugar** |
| (D-glucose) | (D-fructose) | (D-*threo*-pentos-2-ulose) |

Scheme 7–16.

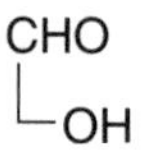

CHO
└─OH

glycolaldehyde (**18**)

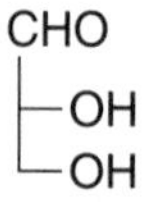

CHO
├─OH
└─OH

D-glyceraldehyde (**19**)

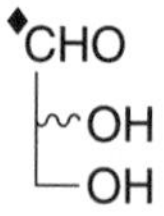

◆CHO
├╌OH
└─OH

DL-[1-^{13}C]glyceraldehyde (**20**)

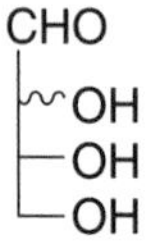

CHO
├╌OH
├─OH
└─OH

D-aldotetroses (**21**)

◆CHO
├─OH
├─OH
├─OH
└─OH

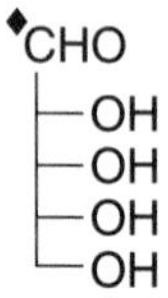

D-[1-^{13}C]ribose (**22**)

of multiple signals clustered between 90 and 100 ppm in the ^{13}C spectrum of DL-[1-^{13}C]glyceraldehyde (**20**) (Fig. 7-21). Extension of **19** by one carbon, generating D-aldotetroses (**21**), permits ring closure to α- and β-aldofuranoses, although measureable amounts of acyclic carbonyl and hydrate forms are detectable in aqueous solutions of these sugars (Serianni et al., 1979a, 1982). Chain extension of **21** to aldopentoses and aldohexoses allows ring closure to both α- and β-furanoses *and* α- and β-pyranoses (Figs. 7-22 and 7-23) (King-Morris & Serianni, 1987; Snyder & Serianni, 1986; Snyder et al., 1989; Wu et al., 1992). In general, the greater thermodynamic stability of the latter forms significantly reduces the amount of acyclic forms in solution, such that for D-ribose, only about 0.05% exists as the acyclic aldehyde (King-Morris & Serianni, 1987), and for D-glucose less than 0.005% (Maple & Allerhand, 1987). Nevertheless, the aldehyde form can be detected with the use of D-[1-^{13}C]ribose (**22**) (Fig. 7-22) (King-Morris & Serianni, 1987) and D-[1-^{13}C]glucose (Maple & Allerhand, 1987).

2-Ketosugars also exist in multiple forms in solution; thus, for example, aqueous solutions of D-[2-^{13}C]*erythro*-2-pentulose (D-ribulose [**23**]) and D-[2-^{13}C]*threo*-2-pentulose (D-xylulose [**24**]) contain α- and β-furanose forms and the acyclic keto form (Fig. 7-24) (Vuorinen & Serianni, 1990b). The keto hydrate form of cyclizable 2-ketoses is usually present in low abundance, but can be detected in noncyclizable 2-ketoses (e.g., D-[2-^{13}C]ribulose 1,5-bisphosphate [**25**]) (Serianni et al., 1979b).

The value of selective ^{13}C-labeling is clearly apparent in studies of the solution composition of dicarbonyl sugars. These compounds exist in multiple acyclic and cyclic forms in solution; for example, D-*erythro*-pentos-2-ulose (**26**) has the potential to exist in 18 monomeric forms in aqueous solution (Scheme 7-17) (Vuorinen & Serianni,

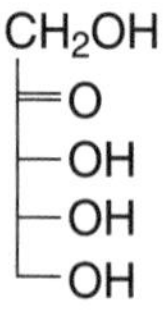

CH$_2$OH
╠═O
├─OH
├─OH
└─OH

D-ribulose (**23**)

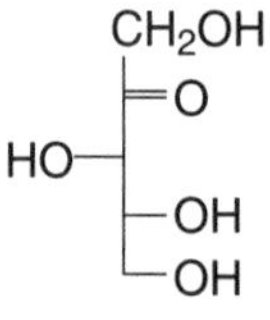

CH$_2$OH
╠═O
HO─┤
├─OH
└─OH

D-xylulose (**24**)

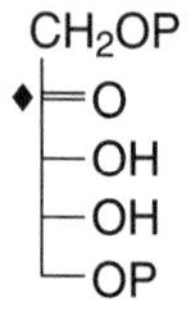

CH$_2$OP
◆╠═O
├─OH
├─OH
└─OP

D-[2-^{13}C]ribulose
1,5-bisphosphate (**25**)

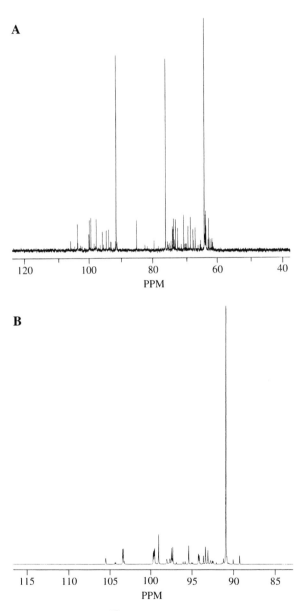

Figure 7–21. The natural abundance ^{13}C NMR spectrum of D-glyceraldehyde (**A**) and the ^{13}C NMR spectrum of DL-[1-^{13}C]glyceraldehyde (**B**) showing only C1 resonances. The intense signal at ~ 91 ppm in **B** is attributed to C1 of the acyclic hydrate (gem-diol) form; the remaining signals arise from C1 resonances of oligomeric forms. Reprinted from Serianni et al., 1979a, with permission.

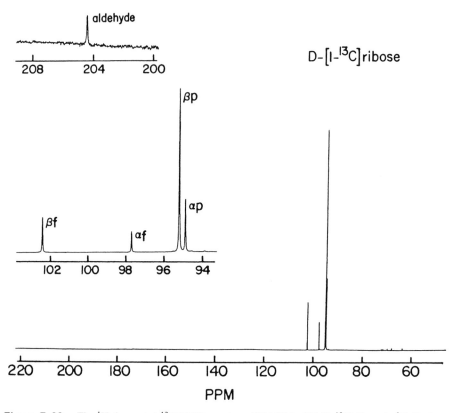

Figure 7–22. The ^{1}H-decoupled ^{13}C NMR spectrum (75 MHz) of D-[1-^{13}C]ribose in ^{2}H$_2$O, showing only the enriched carbons. The intense signals are assigned to α- and β-ribofuranoses and α- and β-ribopyranoses as shown. The weak signal at 204.5 ppm is attributed to the C1 signal of the aldehyde form, present at ~0.05 mol % at 25°C. Reprinted from King-Morris & Serianni, 1987, with permission.

1990a). The introduction of ^{13}C either at C1 or C2 of **26** allows for the detection (and assignment) of these forms (Vuorinen & Serianni, 1990a).

Once detected, anomeric carbon signals can be assigned to specific anomers in different ways. ^{13}C-Chemical shift data alone are insufficient for this purpose, although several rules lead to preliminary assignments. The C1 chemical shift of the acyclic carbonyl forms of aldoses is distinctive ($\sim$ 205 ppm). However, due to its low abundance, the detection of a weak signal in this region of the spectrum is insufficient to make an

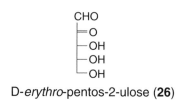

D-*erythro*-pentos-2-ulose (**26**)

assignment. Additional information can be obtained by collecting ^{1}H-coupled ^{13}C data; if the small signal is indeed due to C1 of the acyclic carbonyl form, then a $^1J_{C1,H1}$ value of ~ 180 Hz will be observed (Snyder & Serianni, 1986). C1 Signals arising from acyclic hydrate forms of aldoses have characteristic chemical shifts of 90–92 ppm and $^1J_{C1,H1}$ values of 160–165 Hz.

Chemical shifts and $^1J_{CH}$ values play a role in assigning the C1 signals of furanose and pyranose forms of aldoses (Bock & Pedersen, 1983). For a given aldose, C1 of the furanose forms typically resonates downfield of C1 of the pyranose forms, although this behavior is not always observed and should not be used as the sole criterion for assignment. Assignment of anomeric configuration of aldofuranoses is based partly on chemical shift arguments; the anomer having O1 and O2 *cis* gives a C1 signal *upfield* of that having these atoms *trans* (Scheme 7-18) (Ritchie et al., 1975). $^1J_{C1,C2}$ has also been used to distinguish aldofuranose anomers; aldofuranose rings having O1 and O2 *trans* yield larger $^1J_{C1,C2}$ values (46.2 ± 0.5 Hz) than those having these atoms *cis* (42.7 ± 0.7 Hz) (Scheme 7-18) (Snyder & Serianni, 1987a). In contrast, $^1J_{C1,H1}$ is relatively insensitive to furanose anomeric configuration (Snyder & Serianni, 1987a), ap-

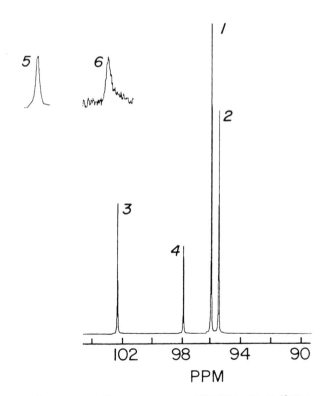

Figure 7–23. The ^{1}H-decoupled ^{13}C NMR spectrum (75 MHz) of D-[1-^{13}C]talose in aqueous solution, showing only signals due to the enriched C1 carbons. Six signals are observed and are assigned as shown to four cyclic (**1–4**) and two acyclic forms (**5,6**) (see Scheme 7-22). The C1 signals of **5** and **6** (inset) are observed at 91.2 and 204.2 ppm, respectively. Reprinted from Snyder et al., 1989, with permission.

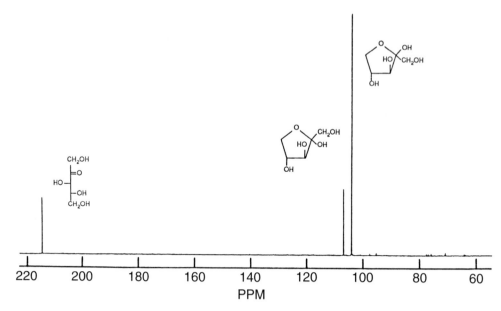

Figure 7–24. The ^{1}H-decoupled ^{13}C NMR spectrum (75 MHz) of D-[2-^{13}C]*threo*-2-pentulose, showing only the enriched carbons. The C2 signals have been assigned to the α-furanose (107.2 ppm), β-furanose (104.4 ppm), and acyclic *keto* (214.4 ppm) forms. Reprinted from Vuorinen & Serianni, 1990b, with permission.

parently because the C1—O1 bond has a propensity to orient quasi-axially in such rings due to the anomeric effect (Serianni & Barker, 1984) (Scheme 7-18). Similar trends have been observed in 2-ketofuranoses with respect to δ_{C2} and $^1J_{C2,C3}$ (Vuorinen & Serianni, 1990b); in addition, $^3J_{C1,H3}$ in these rings is useful to assign anomers, being about 2.3 Hz when C1 and H3 are *cis* and < 0.3 Hz when these atoms are *trans* (Scheme 7-19) (Vuorinen & Serianni, 1990b).

In aldopyranosyl rings, $^1J_{C1,H1}$ is valuable in assigning anomeric configuration. As first reported by Bock and Pedersen (Bock & Pedersen, 1974, 1975; Bock et al., 1973), $^1J_{C1,H1}$ for axial C1—H1 bonds is typically about 160 Hz, whereas $^1J_{C1,H1}$ for equatorial C1—H1 bonds is about 170 Hz. Thus, for example, $^1J_{C1,H1}$ in methyl α-D-mannopyranoside is 171.0 Hz (C1—H1 bond equatorial), whereas $^1J_{C1,H1}$ in methyl β-D-mannopyranoside is 159.5 Hz (C1—H1 bond axial). Intermediate values of $^1J_{C1,H1}$ indicate the presence of conformational heterogeneity (e.g., $^4C_1/^1C_4$ interconversion) and/or the existence of nonchair forms. For example, $^1J_{C1,H1}$ in α-D-idopyranose (**4**) is 165.2 Hz, indicating that forms other than, or in addition to, $^4C_1$ must be present in solution (Snyder & Serianni, 1986). With ^{13}C-labeling at C1 of aldoses, the measurement of $^1J_{C1,H1}$ from ^{13}C spectra is straightforward. Alternatively, ^{13}C signals of protonated anomeric carbons can be assigned *via* selective ^{1}H-decoupling methods or 2D ^{13}C—^{1}H correlation spectroscopy (Fig. 7-25) (Snyder & Serianni, 1986).

In addition to $^1J_{CH}$, ^{13}C—^{1}H spin-couplings across two ($^2J_{CH}$) and three $^3J_{CH}$) bonds are important structural probes in carbohydrates. A number of investigations have examined how these couplings are affected by carbohydrate configuration and confor-

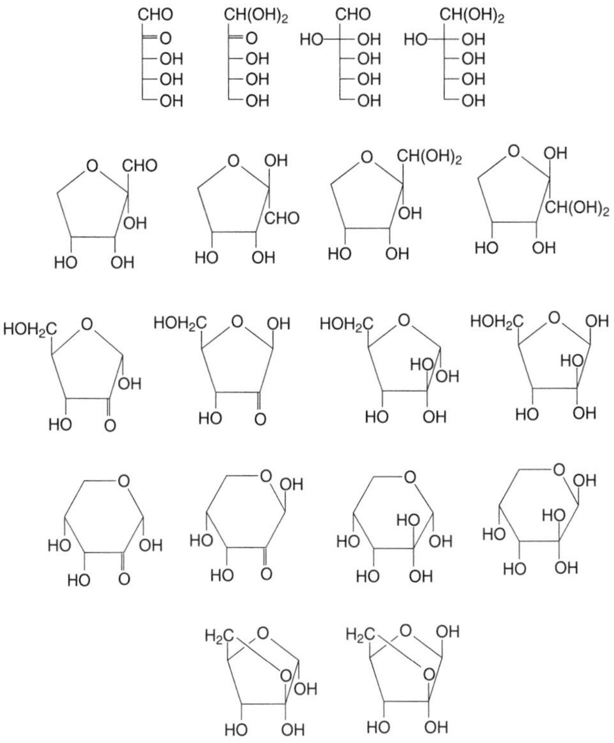

Scheme 7–17. Monomeric forms of D-*erythro*-pentos-2-ulose (**26**) in aqueous solution.

mation (e.g., see Podlasek et al., 1995 and references therein). For $^2J_{CH}$, electronegative substituents gauche to the coupled proton make a *negative* contribution to the coupling, whereas anti substituents make a *positive* contribution (Schwarcz & Perlin, 1972; Schwarcz et al., 1975). A general rule (the projection rule) has been proposed which provides a crude estimation of the magnitude and sign of a given $^2J_{CH}$ within carbohydrates (Bock & Pedersen, 1977). It is important to appreciate that $^2J_{CH}$ can be neg-

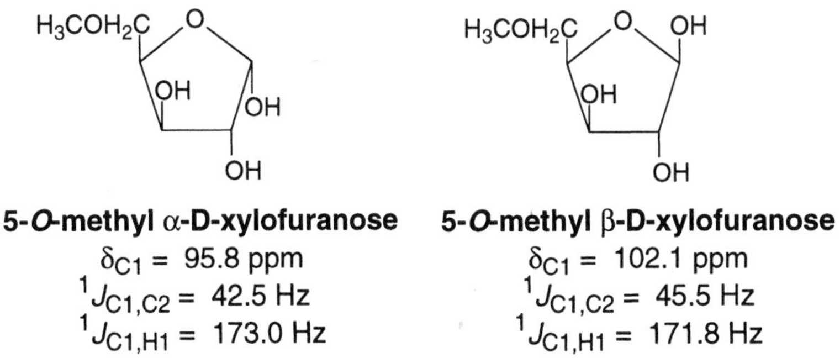

5-*O*-methyl α-D-xylofuranose
δ_{C1} = 95.8 ppm
$^1J_{C1,C2}$ = 42.5 Hz
$^1J_{C1,H1}$ = 173.0 Hz

5-*O*-methyl β-D-xylofuranose
δ_{C1} = 102.1 ppm
$^1J_{C1,C2}$ = 45.5 Hz
$^1J_{C1,H1}$ = 171.8 Hz

Scheme 7–18.

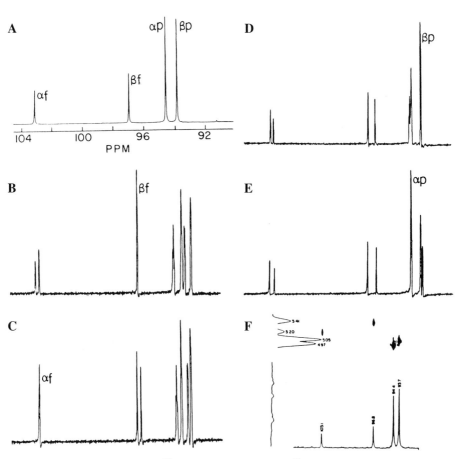

α-D-*threo*-2-pentulose
(α-D-xylulose)
δ_{C2} = 107.2 ppm
$^1J_{C2,C3}$ = 47.0 Hz
$^3J_{C1,H3}$ = 0 Hz

β-D-*threo*-2-pentulose
(β-D-xylulose)
δ_{C2} = 104.4 ppm
$^1J_{C2,C3}$ = 44.7 Hz
$^3J_{C1,H3}$ = 2.3 Hz

Scheme 7–19.

Figure 7–25. (A) The 75 MHz ^{13}C NMR spectrum of D-[1-^{13}C]idose (anomeric carbon region). (B-E) Selective ^{1}H-decoupled ^{13}C NMR spectra of the anomeric carbons of D-[1-^{13}C]idose, showing the effect of specific ^{1}H irradiation of 5.41 ppm (**B**), 5.20 ppm (**C**), 5.05 ppm (**D**), and 4.97 ppm (**E**) on carbon signal multiplicity, allowing the assignment of carbon signals to specific forms. (**F**) The 2D ^{13}C—^{1}H shift correlation contour map of the anomeric ^{13}C and ^{1}H regions of D-idose, confirming the signal assignments based on selective ^{1}H-coupled ^{13}C NMR spectra (**B-E**). Reprinted from Snyder & Serianni, 1986, with permission.

ative or positive in sign in carbohydrates, and that sign information is often essential if $^2J_{CH}$ is to be used in structural studies.

$^3J_{CH}$ in carbohydrates can occur *via* intervening carbon nuclei (e.g., $^3J_{CCCH}$) or oxygen nuclei (e.g., $^3J_{COCH}$). Both types of couplings exhibit a Karplus-like dependency, and several Karplus curves have been proposed for both coupling pathways (Hamer et al., 1978; Mulloy et al., 1988; Schwarcz & Perlin, 1972; Spoormaker & de Bie, 1978; Tvaroska et al., 1989). Due to the sensitivity of $^3J_{CH}$ to substituent effects (van Beuzekom et al., 1990), the dependencies of $^3J_{CCCH}$ and $^3J_{COCH}$ on molecular dihedral angles are not fully understood. Despite this limitation, however, $^3J_{COCH}$ values have been used as probes of *O*-glycoside linkage conformation in oligosaccharides (see below), along with $^3J_{COCC}$ (see below).

The ability to observe the acyclic forms of reducing sugars with the use of selective ^{13}C-labeling permits the application of saturation-transfer NMR methods to determine unidirectional rate constants of anomerization (Serianni et al., 1982). The observation of distinct resonances for each anomer in solution indicates slow exchange on the NMR timescale, suggesting first-order rate constants of chemical exchange $< \sim 20 \; sec^{-1}$. When a saturating rf pulse is applied at the resonance frequency of C1 of the acyclic carbonyl form, the C1 resonances of the cyclic forms decrease in intensity, due to chemical exchange between the cyclic and acyclic forms (Fig. 7-26) (Serianni et al., 1982). This decrease can be treated quantitatively in order to extract unidirectional rate constants of ring-opening (k_{open}) for all cyclic forms present, and of the dehydration of the acyclic hydrate, if present. This experiment permits a detailed assessment of the effects of carbohydrate structure on k_{open} and k_{close} (the latter is obtained indirectly *via* k_{open} and equilibrium constants) under different solution conditions. Using this approach, it has been observed that, in simple aldofuranose anomers, ring-opening is in general more rapid in that anomer having O1 and O2 *cis* as opposed to *trans*; this effect is apparently due to anchimeric assistance of O2 in abstracting OH-1 (**I**, Scheme 7-20) (Serianni et al., 1982; Snyder & Serianni, 1988). Solution pH also affects k_{open}, with the catalytic efficiency of H^+ (Scheme 7-21) dependent on furanose structure (Snyder & Serianni, 1988). In contrast, k_{open} of phosphorylated aldofuranoses (e.g., D-ribose 5-phosphate [**27**]) exhibits a different anomeric dependence (in the monoprotonated species, k_{open} is greater in α-anomers regardless of relative configuration at C1 and C2) and is significantly affected by the ionization state of the phosphate group (Fig. 7-27) (Pierce et al., 1985). The phosphate group is an effective catalyst, apparently assisting in protonation of the ring oxygen (**II**, Scheme 7-20).

Studies of anomerizing systems in which both furanose and pyranose forms are present (e.g., D-talose [**28**]) (Fig. 7-23) reveal larger values of k_{open} and k_{close} for furanoses, indicating that they are more favored kinetically, although the pyranose forms are more favored thermodynamically (Scheme 7-22) (Snyder et al., 1989).

D-ribose 5-phosphate (**27**)
(dianion form)

$R = PO_3^{-2}$

D-talose (**28**)

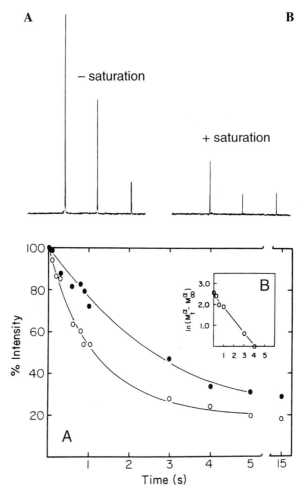

Figure 7–26. ^{13}C Saturation-transfer experiment on D-[1-^{13}C]erythrose (0.1 M) in ^2H_2O (p^2H 5.0) and acetate buffer (50 mM) at 55°C. (**A**) Plot of resonance intensity versus saturation times, showing different rates of decay for the α-furanose (open circles) and β-furanose (closed circles). (**B**) Semilogarithmic plot of the data in A for the α-furanose, from which a ring-opening rate constant of 0.40 sec^{-1} is obtained. Similar treatment for the β-furanose yields a value of 0.19 sec^{-1}. Under these conditions, the effect of saturation on the intensity of the hydrate signal is small, and only an upper limit for the rate of dehydration (< 0.05 sec^{-1}) is obtained. Reprinted from Barker & Serianni, 1986, with permission.

The introduction of alkyl substituents on aldofuranose rings also affects anomerization rate constants; for example, if appended to C1 (generating a 2-ketose), k_{close} is significantly reduced and k_{open} essentially unaffected (Scheme 7-23) (Wu et al., 1990), whereas alkylation at other ring carbons causes k_{close} to be enhanced and k_{open} to be essentially unaffected (Snyder & Serianni, 1987b, 1991). These observations may be viewed as manifestations of Thorpe–Ingold effects in carbohydrate systems. Two-dimensional exchange spectroscopy has also been applied to determine overall rates of anomer exchange in solution (Fig. 7-28) (Snyder et al., 1989).

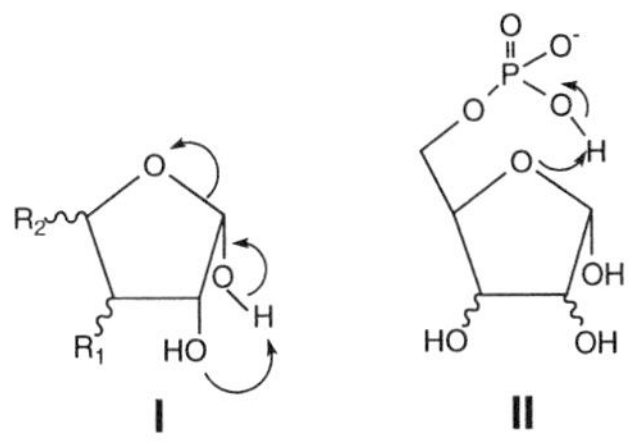

Scheme 7–20.

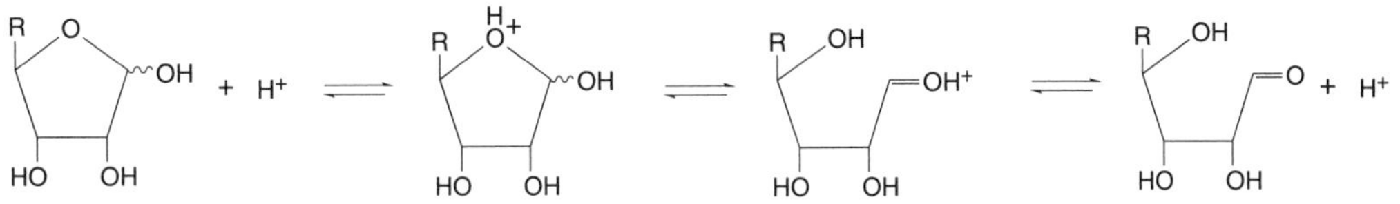

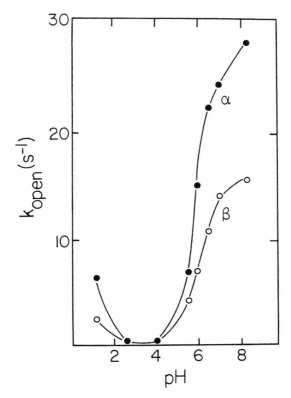

Scheme 7–21. Mechanism of acid-catalyzed furanose anomerization.

Figure 7–27. pH Dependence of the ring-opening rate constants for the α- and β-furanose forms of [1-^{13}C]ribose 5-phosphate (0.3 M sugar in 15% ^{2}H$_2$O at 24°C). Rate constants were obtained from line broadening experiments except for the values at pH 2.3 and 4.0 which were obtained by ^{13}C saturation-transfer experiments. Reprinted from Pierce et al., 1985, with permission.

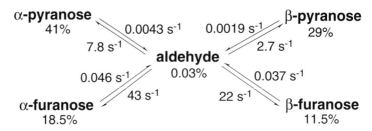

α-pyranose
41%

0.0043 s⁻¹   0.0019 s⁻¹

7.8 s⁻¹

aldehyde
0.03%

2.7 s⁻¹

β-pyranose
29%

0.046 s⁻¹

43 s⁻¹

0.037 s⁻¹

22 s⁻¹

α-furanose
18.5%

β-furanose
11.5%

Scheme 7–22. Talose anomerization (50 mM Na acetate, pH 4.0, 15% v/v ^2H_2O, 28°C).

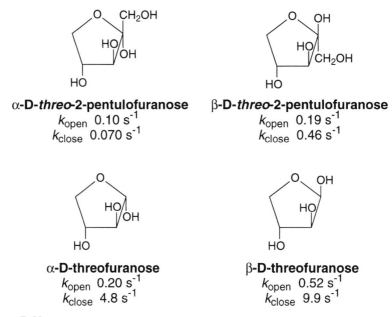

α-D-*threo*-2-pentulofuranose
k_{open} 0.10 s⁻¹
k_{close} 0.070 s⁻¹

β-D-*threo*-2-pentulofuranose
k_{open} 0.19 s⁻¹
k_{close} 0.46 s⁻¹

α-D-threofuranose
k_{open} 0.20 s⁻¹
k_{close} 4.8 s⁻¹

β-D-threofuranose
k_{open} 0.52 s⁻¹
k_{close} 9.9 s⁻¹

Scheme 7–23

Reaction Mechanisms/Enzyme Reactions

The presence of one or more ^{13}C labels in a carbohydrate assists greatly in assessing the course of chemical and biochemical transformations. Examples include: (1) The cyanohydrin reaction applied to D-erythrose was examined with the use of $K^{13}CN$ or ^{13}C-labeled tetrose (Serianni et al., 1980). The time course of the reaction was monitored by ^{13}C NMR, allowing direct detection and characterization of intermediates and products. Key intermediates were identified to be *ribo* and *arabino* imido-1,4-lactones by varying reaction conditions (pH; Fig. 7-29) or directly measuring ^{13}C—^{13}C spin-couplings within doubly-^{13}C-labeled intermediates and comparing them to those observed in structurally related (and chemically stable) aldono-1,4-lactones. (2) ^{13}C NMR studies of the base-catalyzed isomerization of D-[1-^{13}C]mannose showed the presence

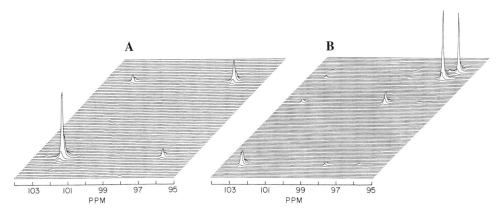

Figure 7–28. 2D ^{13}C chemical exchange maps for D-[1-^{13}C]erythrose (**A**) and D-[1-^{13}C]talose (**B**). The 1D spectrum (C1 signals of cyclic forms) of each aldose is found along the diagonal. Crosspeaks denote sites of chemical exchange and were used to extract overall rate constants of exchange between cyclic forms. Reprinted from Snyder et al., 1989, with permission.

of D-mannose, D-glucose, and D-fructose containing ^{13}C-enrichment at C1 *and* C6 after several days of reaction (Fig. 7-30) (King-Morris & Serianni, 1986). This observation provides indirect evidence of enolization of the carbon chain during the reaction. The observed labeling at the hexose terminal carbons presumably results from the transient generation of a 3,4-enediol **29** which reverts to either C1 or C6 labeled prod-

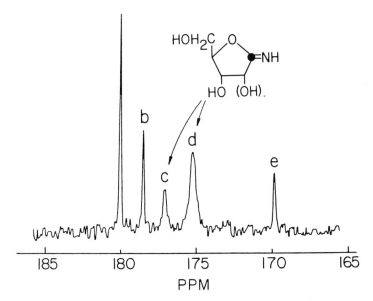

Figure 7–29. Partial ^{1}H-decoupled 15.08 MHz ^{13}C NMR spectrum of [1-^{13}C]-intermediates in the reaction of 0.3 M K^{13}CN and 0.3 M D-erythrose after 11 min at pH 10.5 and 18 ± 1°C, showing imido-1,4-lactone formation. Reprinted from Serianni et al., 1980, with permission.

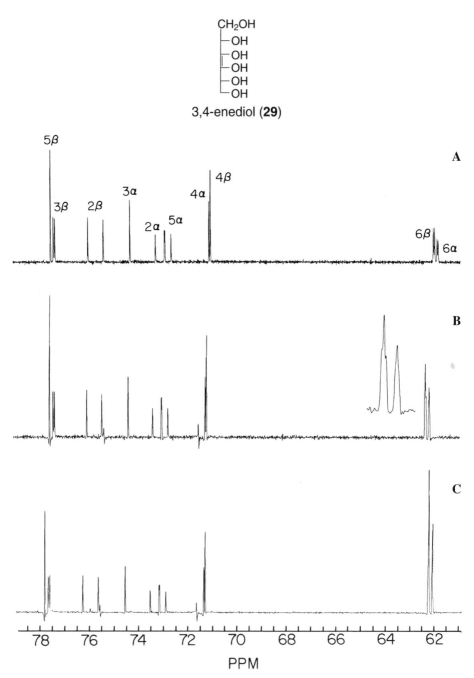

Figure 7–30. (A) ^{1}H-Decoupled ^{13}C NMR spectrum of D-[1-^{13}C]glucose, showing only the natural abundance carbon atoms, with assignments. (**B** and **C**) ^{13}C NMR spectra of D-[1-^{13}C]glucose isolated after 3 and 7 days of reaction, respectively. Only C6 appears to contain [^{13}C]-enrichment (in addition to C1), which increases as the reaction proceeds. The out-of-phase signals are due to "foldover" of the intense, enriched signals caused by quadrature detection. Spectra **A–C** are resolution enhanced. Reprinted from King-Morris & Serianni, 1986, with permission.

ucts *via* back-reactions. (3) The use of specific ^{13}C-labeling revealed an unusual C1—C2 transposition that occurs during molybdate-catalyzed C2 epimerization of al-doses (Hayes et al., 1982a). The C1 (anomeric carbon) signals of a starting [1-^{13}C]-aldose were clearly shifted upfield to the secondary alcohol region as the reaction pro-ceeds, indicating the production of its C2 epimer *labeled at C2* (Fig. 7-31). Subsequent experiments with deuterium and oxygen isotopes showed that C1 and C2 are trans-posed with their attached protons and oxygens intact. A reaction mechanism for this transformation has been proposed (Scheme 7-24) (Hayes et al., 1982a). (4) Enzymic transformations can be monitored *in vitro* and *in vivo* with the use of specific ^{13}C-labeled compounds. For example, sequential treatment of DL-[1-^{13}C]glyceraldehyde with triosephosphate isomerase and fructose 1,6-bisphosphate (FBP) aldolase shows initial (transient) production of D-[3,4-[13]C_2]FBP, followed by the generation of the fi-nal product, L-[3,4-[13]C_2]sorbose 1,6-bisphosphate (Fig. 7-32) (Serianni et al., 1979b). Injection of ^{13}C-labeled D-glucoses into live cecropia pupae allows the *in vivo* biosyn-thesis of labeled sorbitol and fructose to be monitored noninvasively, indicating the presence of aldose reductase (or nonspecific dehydrogenases) and sorbitol dehydroge-nase in this organism (Fig. 7-33) (Podlasek & Serianni, 1994).

Introducing Chirality with [13]C

Achiral molecules can be made chiral with the introduction of the ^{13}C isotope (Garrett & Serianni, 1990). For example, the ^{1}H NMR spectrum of glycerol shows three sets of equivalent nuclei (Fig. 7-34A). The introduction of ^{13}C at either C1 or C3 allows the distinction of two sets of hydroxymethyl protons, since those directly bound to the

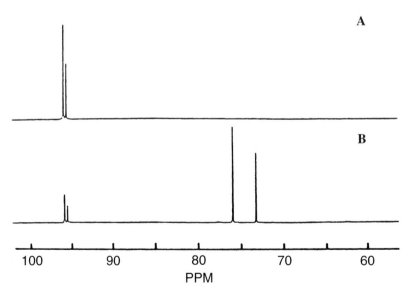

Figure 7–31. 75 MHz ^{1}H-decoupled ^{13}C NMR spectra of D-[1-^{13}C]mannose (**A**) and the reaction mixture resulting after treatment of D-[1-^{13}C]mannose (0.1 M) with molybdic acid (85%, 5 mM) at 90°C for 4 hr (**B**). Only enriched carbons are shown. The new resonances in **B** at 72.9 and 75.6 ppm are due to D-[2-^{13}C]glucose generated during the reaction. Reprinted from Barker & Serianni, 1986, with permission.

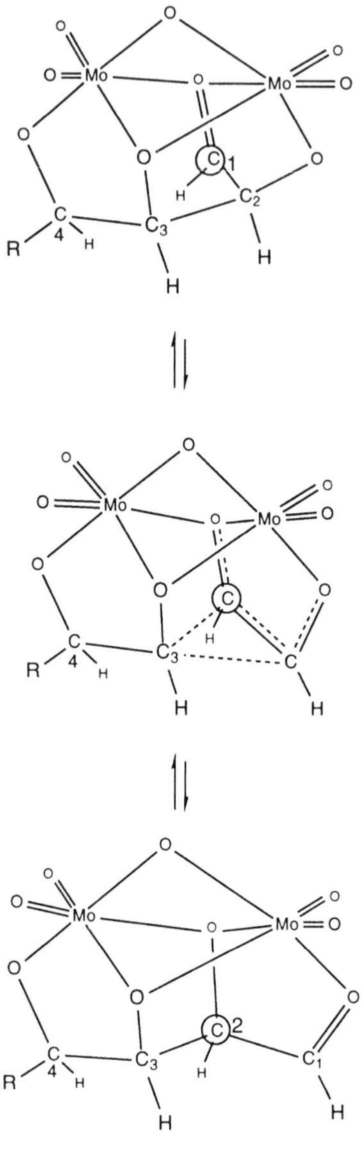

Scheme 7–24.

[13]C exhibit one-bond [13]C—[1]H coupling (Fig. 7-34B). In this instance, it is possible to measure $^3J_{CCCH}$ values between the labeled hydroxymethyl carbon and the second pair of hydroxymethyl protons (2.9 and 3.2 Hz). Similar effects are observed in ^{13}C NMR spectra of symmetric compounds (e.g., [1-^{13}C]xylitol [Garrett & Serianni, 1990]).

Oligosaccharide Conformation

The disaccharide, methyl β-lactoside (**7**), has been labeled with ^{13}C at several positions in the vicinity of the glycosidic linkage, and all of the available [13]C—[13]C and [13]C—[1]H spin-couplings across this linkage have been evaluated (Scheme 7-25) (Hayes

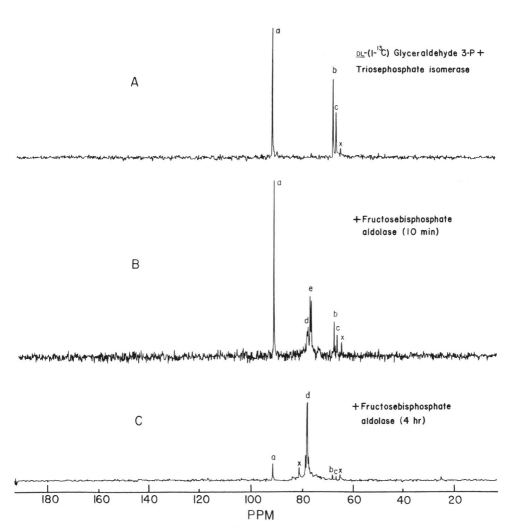

Figure 7–32. The enzymic conversion of DL-[1-^{13}C]glyceraldehyde 3-P to L-[3,4-^{13}C]sorbose 1,6-P$_2$, as followed by ^{13}C NMR. Peaks designated by "x" are unidentified components. Only the resonances of enriched nuclei are shown. (**A**) The ^{13}C NMR spectrum of DL-[1-^{13}C]glyceraldehyde 3-P hydrate (**a**) after the addition of triosephosphate isomerase, producing resonances due to C3 of the *keto* (**b**) and hydrate (**c**) forms of [3-^{13}C]dihydroxyacetone P. (**B**) The addition of D-fructose-1,6-P$_2$ aldolase to **A** causes the appearance after 10 min of resonances due to C3 and C4 of D-[3,4-^{13}C$_2$]fructose 1,6-P$_2$ (**e**), with a smaller amount of L-[3,4-^{13}C$_2$]sorbose 1,6-P$_2$ (**d**). (**C**) The same reaction mixture as analyzed by ^{13}C NMR after 4 hr shows little unreacted [1-^{13}C]glyceraldehyde 3-P and [3-^{13}C]dihydroxyacetone P and the major product, L-[3,4-^{13}C$_2$]sorbose 1,6-P$_2$ (**d**). Reprinted from Serianni et al., 1979b, with permission.

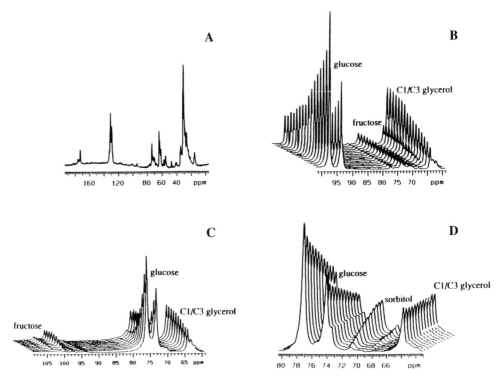

Figure 7–33. (**A**) The natural abundance ^{13}C NMR spectrum of an intact *H. cecropia pupa*. Signals observed at 30, 130, and 180 ppm were attributed to the saturated, unsaturated, and carbonyl carbons of the fatty acids of triglycerides. Two additional resonances at 64.0 and 73.5 ppm arise from the C1/C3 and C2 carbons of glycerol. (**B**) ^{13}C NMR spectra of an intact pupa injected with D-[1-^{13}C]glucose, showing metabolism to D-[1-^{13}C]fructose (65.4 ppm). (**C**) ^{13}C NMR spectra of a pupa injected with D-[2-^{13}C]glucose. Fructose synthesis is apparent from the signals at 99.5, 103.0, and 106.0 ppm, which are attributed to the β-pyranose, β-furanose, and α-furanose forms, respectively. (**D**) ^{13}C NMR spectra of an intact pupa injected with D-[3-^{13}C]glucose. ^{13}C label incorporation was observed into C3 of sorbitol (71.3 ppm), fructose (69.1 ppm), and glycerol (64.0 ppm). Reprinted from Podlasek & Serianni, 1994, with permission.

et al., 1982b). These data point to a *preferred* linkage conformation as depicted in Scheme 7-26.

The computed dependency of $r_{H1\text{-}Gal,H4\text{-}Glc}$ on linkage conformation in **7** (Fig. 7-35) was determined with the use of crystallographic data, and DESERT experiments were performed to determine this distance in solution (Kline et al., 1990). The computed distance, 2.0 Å, is consistent with the linkage geometry deduced from trans-*O*-glycoside ^{13}C—^{1}H and ^{13}C—^{13}C spin-couplings, although this result alone does not constitute definitive proof of a *single* linkage geometry in aqueous solution for reasons discussed above. A similar *preferred* linkage geometry has been predicted in recent computational studies, although these also predict the presence of small percentages of other conformations (Engelsen et al., 1994).

Sucrose (**30**) conformation has been examined using ^{13}C-labeled molecules (Duker

& Serianni, 1993). In this instance, the trans-O-glycoside $^{13}C-^{13}C$ and $^{13}C-^{1}H$ spin coupling data do not appear consistent with a single linkage geometry. Data suggest a relatively constrained ϕ, but significant rotation about ψ. This conformational heterogeneity has also been suggested from other NMR and computational studies (du Penhoat et al., 1991; Poppe & van Halbeek, 1992; Tran & Brady, 1990).

Nucleoside/Oligonucleotide Conformation

Furanose and N-glycoside conformation of nucleosides have been investigated with the use of ^{2}H- and ^{13}C-labeled molecules (Bandyopadhyay et al., 1993; Kline & Serianni, 1988, 1990a,b, 1992; Lemieux et al., 1972). A comparison of $^{13}C-^{1}H$ spin-couplings

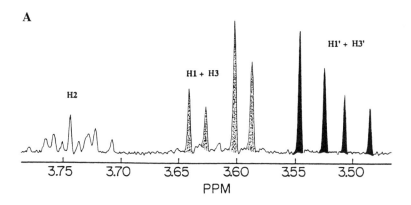

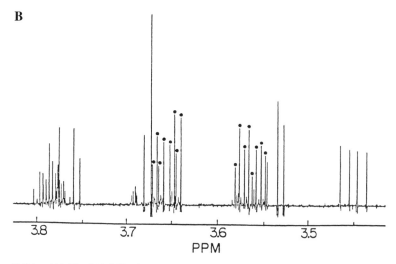

Figure 7–34. (A) The 300 MHz ^{1}H NMR spectrum of glycerol showing signal assignments for H1/H3, H2, and H1'/H3'. (B) The 620 MHz ^{1}H spectrum of [1-^{13}C]glycerol. The signals marked by dots are those of H3 and H3' which are split by ^{13}C at C1 ($^{3}J_{CH}$). The H1 and H1' signals are displaced from those of H3 and H3', respectively, due to their coupling to ^{13}C at C1 ($^{1}J_{CH}$). Reprinted from Garrett & Serianni, 1990, with permission.

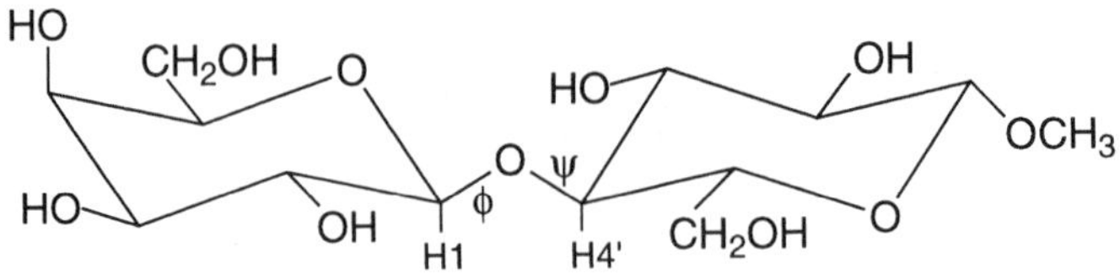

$$^3J_{C4',H1} = 3.8 \text{ Hz}$$
$$^3J_{C2,C4'} = 3.1 \text{ Hz}$$

$$^3J_{C1,H4'} = 4.9 \text{ Hz}$$
$$^3J_{C1,C3'} = 0 \text{ Hz}$$
$$^3J_{C1,C5'} = 1.6 \text{ Hz}$$

Scheme 7–25. Trans-O-glycoside couplings in methyl β-lactoside (7).

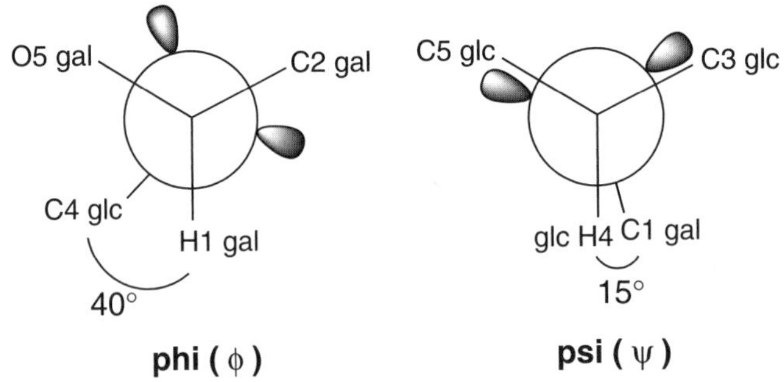

phi (ϕ)   **psi (ψ)**

Scheme 7–26. Glycoside linkage conformation of methyl β-lactoside (7) derived from trans-O-glycoside couplings.

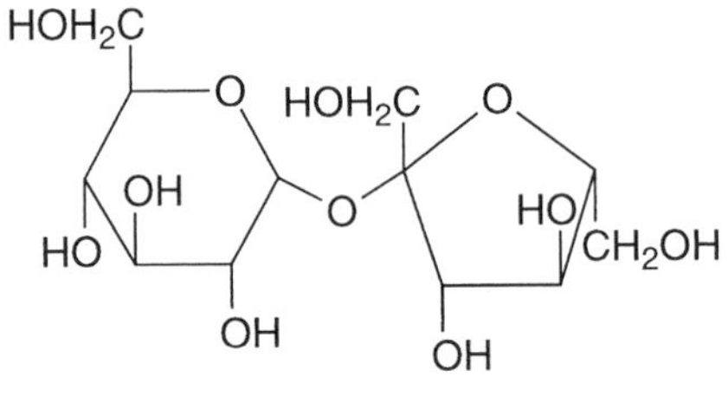

(30)

within the furanose constituents of structurally related nucleosides can be useful in discerning the effect of ring structure on preferred conformation. Consider J_{CH} in [1'-^{13}C]adenosine (31), [1'-^{13}C]uridine (32), [1'-^{13}C]erythroadenosine (33), and [1'-^{13}C]2'-deoxyadenosine (34) (Scheme 7-27). Analysis of $^3J_{HH}$ values in 31 and 32 (Kline & Serianni, 1990a) indicates S/N ratios of 1.9 and 0.82, respectively; here south (S) is defined as ^2E and north (N) as ^3E (see Schemes 7-6 and 7-7). This conclusion is supported by differences in related J_{CH} values in 31 and 32. Converting 31 to 32 causes $^3J_{C1',H3'}$ to decrease, suggesting a greater proportion of N forms in the latter (i.e., the

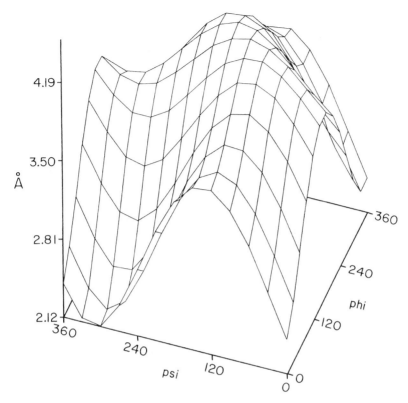

Figure 7–35. Surface-contour plot of H1—H4′ internuclear distance versus O-glycoside dihedral angles, ϕ and ψ, in methyl β-lactoside (**7**). Reprinted from Kline et al., 1990, with permission.

smaller $^3J_{CH}$ in **32** reflects the smaller C1′—H3′ dihedral angle found in N forms) (Kline & Serianni, 1990a). The smaller value of $^3J_{C1',H4'}$ in **32** is also consistent with a greater proportion of N forms (Kline & Serianni, 1990a).

The interpretation of $^2J_{C1',H2'}$ is less straightforward. In **31**, this coupling is -3.2 Hz, whereas in **32** a value of -2.1 Hz is observed. These two-bond ^{13}C—^1H couplings can be modeled by observing related couplings in simple aldohexopyranosides having configurations at C1, C2, and C3 similar to those observed in N and S forms of the β-D-ribofuranosyl ring. Specifically, the α-*manno* configuration is a reasonable mimic of N forms, having an axial–axial–equatorial substitution pattern at C1—C2—C3, whereas the β-*allo* configuration contains an equatorial–equatorial–axial pattern that mimics S forms (Serianni, 1994). Thus, $^2J_{C1,H2}$ in methyl α-D-mannopyranoside (**35**) (-1.2 Hz) may be considered a crude estimate of $^2J_{C1',H2'}$ in N forms, whereas $^2J_{C1,H2}$ in methyl β-D-allopyranoside (**12**) (-6.0 Hz) provides an estimate of $^2J_{C1',H2'}$ in S forms. These model couplings indicate that $^2J_{C1',H2'}$ *becomes less negative as N forms become more favored.* This trend is observed when **31** is converted to **32**.

Removal of the exocyclic hydroxymethyl group at C4′ of **31**, generating **33**, results in a substantial N/S shift in favor of S forms, as indicated by analysis of $^3J_{HH}$ data (Kline & Serianni, 1992). This prediction is consistent with the more negative value of $^2J_{C1',H2'}$ in **33** relative to **31** (see above discussion); interestingly, the shift towards S forms in **33** does not result in a substantial change in $^3J_{C1',H3'}$, for reasons not well understood at present. Furthermore, since the dihedral angle between C1' and H4'R is smaller ($\sim 100°$) than that between C1' and H4'S ($\sim 140°$) in S forms, $^3J_{C1',H4'R}$ should be smaller than $^3J_{C1',H4'S}$ in **33**, as is observed.

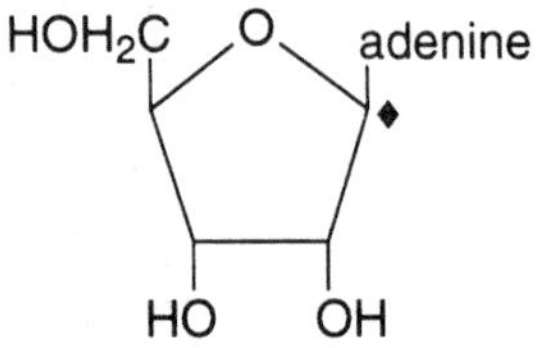

[1'-^{13}C]adenosine (31)

$^1J_{C1',H1'} = 165.6$ Hz
$^2J_{C1',H2'} = -3.2$ Hz
$^3J_{C1',H3'} \approx 5.1$ Hz
$^3J_{C1',H4'} = 1.3$ Hz

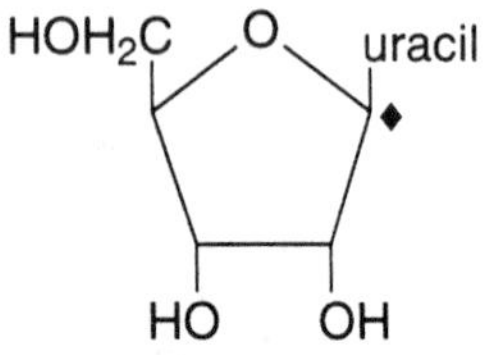

[1'-^{13}C]uridine (32)

$^1J_{C1',H1'} = 170.1$ Hz
$^2J_{C1',H2'} = -2.1$ Hz
$^3J_{C1',H3'} = 3.5$ Hz
$^3J_{C1',H4'} = 0.7$ Hz

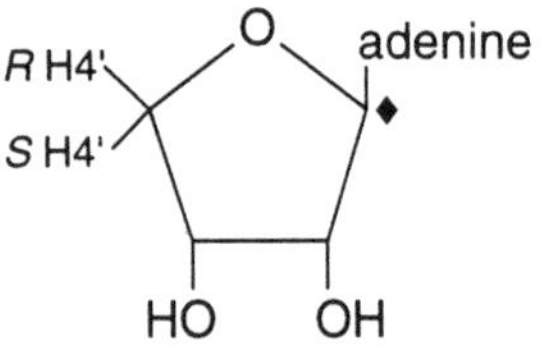

[1'-^{13}C]erythroadenosine (33)

$^1J_{C1',H1'} = 166.0$ Hz
$^2J_{C1',H2'} = -4.1$ Hz
$^3J_{C1',H3'} = 5.2$ Hz
$^3J_{C1',H4'R} = 1.7$ Hz
$^3J_{C1',H4'S} = 3.7$ Hz

[1'-^{13}C]2'-deoxyadenosine (34)

$^1J_{C1',H1'} = 167.4$ Hz
$^2J_{C1',H2'R} \approx 0.4$ Hz
$^2J_{C1',H2'S} \approx -5.7$ Hz
$^3J_{C1',H3'} = 5.3$ Hz
$^3J_{C1',H4'} = 2.9$ Hz

Scheme 7–27.

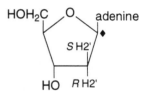

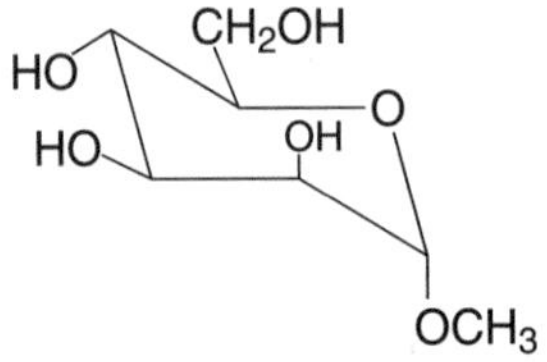

methyl α-D-mannopyranoside (**35**)

Conversion of **31** to **34** also appears to be accompanied by a shift towards S forms based on $^3J_{HH}$ analysis, which is consistent with the larger value of $^3J_{C1',H4'}$ observed in **34** (Bandyopadhyay et al., 1993); changes in $^3J_{C1',H3'}$ cannot be compared since substitution patterns along the coupling pathway differ in **31** and **34**. $^2J_{C1',H2'R}$ and $^2J_{C1',H2'S}$ differ substantially in **34**. Coupling of C1′ to H2′S is large and negative, whereas only a small coupling is observed to H2′R. Recent studies with model deoxyaldopyranosides have shown that this behavior is consistent with S ring forms; in N forms, these geminal ^{13}C—^1H couplings become comparable in absolute magnitude (Bandyopadhyay et al., 1993). This difference in $^2J_{C1',H2'R}$ and $^2J_{C1',H2'S}$ has recently been observed in an intact DNA oligomer (Wu & Serianni, 1994) and was used to confirm the predominance of S forms in specific residues.

This discussion illustrates the potential of J_{CH} values within the furanosyl rings of nucleosides (and, *vide infra*, oligonucleotides) as conformational probes. It is expected that a more detailed understanding of the relationships between J_{CH} and conformation will ultimately provide a more detailed quantitative picture of the solution behavior of these conformationally flexible rings (Serianni, 1994). Indeed, recent work has shown that $^1J_{CH}$ values in β-D-ribofuranose and 2-deoxy-β-D-*erythro*-pentofuranose rings are sufficiently sensitive to ring conformation to be valuable as conformational probes (Bandyopadhyay et al., 1997; Serianni et al., 1995), especially in larger molecules where the measurement of smaller $^2J_{CH}$ and $^3J_{CH}$ values may be precluded by broad signals. Detailed investigations of $^2J_{CH}$ and $^3J_{CH}$ behavior in the β-D-ribofuranosyl ring have shown a number of these coupling constants to be sensitive to ring conformation (Podlasek et al., 1996).

A comparative study has been conducted to assess the effect of C2 deoxygenation on J_{CH} values in the β-D-ribofuranosyl ring (Church et al., 1997); in this work methyl β-D-ribofuranoside and 2-deoxy-β-D-*erythro*-pentofuranoside were prepared singly ^{13}C-labeled at each carbon and a complete set of J_{CH} (and J_{CC}) values were analyzed in each compound. While much has been learned about the behavior of J_{CH} and J_{CC} values in these biologically important furanose rings, considerable effort will be required to quantify structure-coupling relationships sufficiently to permit their application in studies of solution conformation and dynamics. This effort, however, is expected to result in substantial improvements in the understanding of carbohydrate structure, especially in systems where conformational flexibility is likely to be coupled to biological function (e.g., RNA and DNA).

Hydroxymethyl conformation in ribonucleosides and 2′-deoxyribonucleosides has been investigated with compounds stereoselectively deuterated at C5′ (Kline & Serianni, 1988, 1990b). In this manner, the ^1H NMR signals of H5′R and H5′S were assigned unequivocally (in neutral aqueous solution, H5′S resonates downfield of H5′R in both series of compounds) and these assignments were used to estimate the relative populations (i.e., *gg*, *gt*, and *tg*) of C4—C5 rotamers (see, e.g., Scheme 7-8).

A crude Karplus curve defining the dependence of ^{13}C—^{13}C spin-coupling across the N-glycosidic linkage of ribonucleosides has been constructed (Fig. 7-36) and used to assess base–sugar linkage conformation (Kline & Serianni, 1990a). Thus, for example, the very small values of $^3J_{C2',C2}$ and $^3J_{C2',C6}$ observed in [2′-^{13}C]uridine have been interpreted to suggest that the uracil base is oriented approximately orthogonal to the C1′—C2′ bond in both *syn* and *anti* conformers, which presumably interconvert in solution.

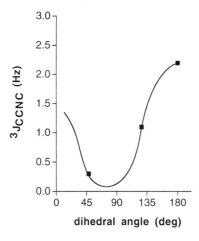

Figure 7–36. $^3J_{CCNC}$ versus torsion angle for the *N*-glycoside bond in ribonucleosides. Standard couplings were determined with conformationally rigid nucleosides. Reprinted from Kline & Serianni, 1990a, with permission.

Summary

Carbohydrates labeled with stable isotopes are likely to play an increasingly important role in structure determinations by NMR spectroscopy in the future, especially as larger molecules (and molecular complexes) attract greater attention. The reasons for this are clear. Compared to proteins, ^{1}H NMR spectra of carbohydrates are considerably more complex, even for simple monosaccharides and disaccharides. Conformational analysis of oligosaccharides containing five or more monosaccharide units is a considerable challenge at the present time (compare this situation with that for proteins, where the current molecular weight limit is 10–15 kDa for unlabeled samples). Thus, any means to resolve these spectra into interpretable components will be essential to further progress in the field.

The advent of multidimensional (MD) methods of data collection, and continued improvement in superconducting magnet technology, will certainly provide greater sensitivity and the much needed enhancement in spectral dispersion. However, heteronuclear MD methods (Fesik & Zuiderweg, 1990), such as HMQC–TOCSY (Wijmenga et al., 1989), HCCH–TOCSY (Fesik et al., 1990) (Fig. 7-37), and related techniques which are arguably the most powerful, are not readily applicable to unlabeled compounds. Indeed, recent progress in nucleic acid structure determination, which suffers from problems similar to those encountered with complex carbohydrates, has depended

---

Figure 7–37. The 2D HCCH–TOCSY spectrum (600 MHz) of methyl α-D-[UL-^{13}C$_6$]glucopyranoside. Correlations are observed between each carbon (chemical shifts in ppm along the F1 axis: C1, 100.7; C2, 72.9; C3, 74.8; C4, 71.3; C5, 73.2; C6, 62.3) and each proton (chemical shifts in ppm along the F2 axis: H1, 4.904; H2, 3.655; H3, 3.761; H4, 3.495; H5, 3.741; H6, 3.965; H6′, 3.852). The C2/C5 signals along F2 are partly resolved, and the H3/H5 signals along F1 are overlapping. In this experiment, the TOCSY correlations are mediated solely by the large $^1J_{CH}$ and $^1J_{CC}$ values in the molecule, and are therefore unaffected by the presence of small $^3J_{HH}$ values that truncate the size of conventional TOCSY signals. Consequently, TOCSY transfer throughout the intact spin system is possible (e.g., the C6 signal at 62.3 ppm shows correlations to H1, H2, H3, H4, H5, H6, and H6′).

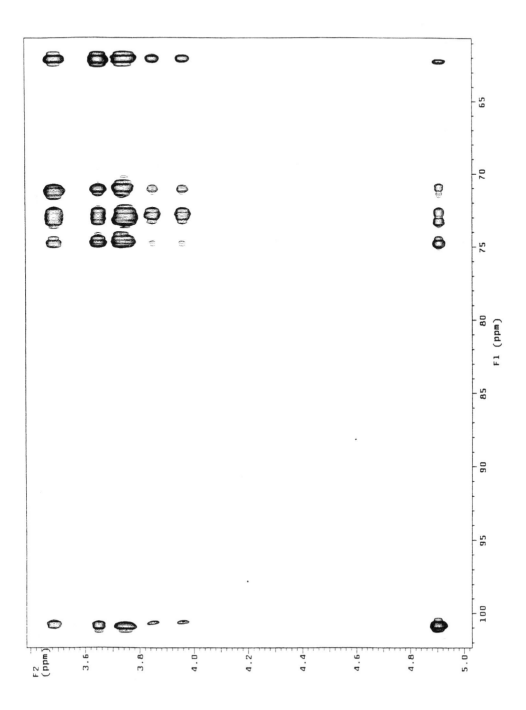

on the use of heteronuclear methods and ^{13}C-labeled material to decipher spectra and permit the measurement of various NMR parameters useful in structure determinations (Kellenbach et al., 1992; Lancelot et al., 1993; Nikonowicz & Pardi, 1992; Pardi & Nikonowicz, 1992; Wu & Serianni, 1994). Similar applications are anticipated in structural studies of carbohydrates.

It should be appreciated that the use of ^{13}C, for example, as an editing nucleus can be extremely valuable in simplifying complex spectra (Bossennec et al., 1990; Wu & Serianni, 1992) or in detecting labeled species in the presence of unlabeled species (e.g., a ^{13}C-labeled compound bound to an unlabeled receptor) (Fesik, 1991). Thus, ^{13}C-labeling enhances not only the *sensitivity*, but also the *selectivity* of NMR measurements. ^{15}N-Labeling of nitrogen-containing carbohydrates (e.g., *N*-acetyl-D-glucosamine), although little exploited to date, may be expected to function in a similar fashion.

The limiting factor in extending carbohydrate structure determination by NMR to larger structures and complexes appears to be access to appropriately labeled materials rather than shortcomings of the NMR method. The combined use of stable isotope labeling and MD NMR at super-high fields (> 800 MHz) is expected to have an enormous impact on the manner in which carbohydrate structure determinations are conducted in the next decade.

8

Cell Adhesion Molecules and Their Cellular Targets

Diane Hollenbaugh, Jürgen Bajorath, and Alejandro Aruffo

The first evidence that white blood cells could leave the bloodstream and move into the surrounding tissue was provided by microscopy studies in the early 1800s by Dutrochet, who examined the flow of blood cells in the transparent tail of tadpoles. These observations were extended in the late 1800s by Cohnheim and Metchnikoff who described the sticking of leukocytes to the vascular cell wall and debated if this mechanism was driven by the vascular endothelial cells (Cohnheim) or the leukocytes (Metchnikoff). The advent of better intravital microscopy techniques in the early part of this century allowed researchers to extend these studies and describe in detail the temporal sequence of events leading to leukocyte extravasation at sites of inflammation (for a review, see Harlan et al., 1992). These studies showed that leukocytes flowing in the vasculature would roll and stick only to the inflamed portion of the vessel.

Leukocyte extravasation at sites of inflammation is a multistep process which is mediated by cell surface molecules on both the leukocyte and endothelial cell wall (Fig. 8-1) (Bevilacqua, 1993). Recruitment is initiated by a subset of adhesion molecules on the leukocyte and the activated vascular endothelial cells, which permit circulating leukocytes to roll on the activated vascular cell wall. This initial interaction allows the leukocyte cell surface to probe the activated vascular endothelial cell surface. If there are appropriate activation stimuli, additional adhesive interactions are induced and the leukocyte sticks to the wall of the activated vessel. This process eventually leads to leukocyte extravasation at the site of inflammation.

Many of the molecular mechanisms utilized for leukocyte recruitment to sites of inflammation are also utilized for leukocyte recirculation during immune surveillance. Leukocytes exit the blood stream into surrounding tissue, are taken up into the lymphatic system and then return to the blood stream. This cycle of recirculation maximizes the possibility of contact with foreign antigens or damaged self cells and allows the amplification of a response when contact is made. For a given leukocyte subset (T vs. B cells, memory vs. naive T cells, etc.) the pattern of leukocyte trafficking is nonrandom, with different leukocyte subsets having distinct migratory patterns. At first, it was proposed that a leukocyte would express a specific type of homing receptor that would dictate its pattern of leukocyte migration. However, more recent data suggest that homing specificity is not likely due to the expression of a single molecule but rather to the expression of a specific combination of a limited number of cell adhesion

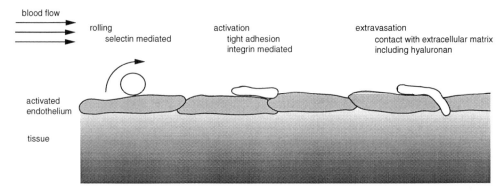

Figure 8–1. Three steps in leukocyte recruitment during inflammation. Schematic of a cross section of a vascular wall lined by endothelial cells.

molecules. In addition, leukocyte trafficking requires that the adhesive properties of leukocytes vary over a large range, from nonadherent to a transiently adherent state. Thus, leukocyte trafficking is a highly regulated and complex molecular event. In recent years, some of the molecules responsible for these adhesive interactions have been identified and their mode of action elucidated.

Extensive studies have shown that protein–carbohydrate interactions are responsible for many of the adhesive events that mediate leukocyte migration. Well studied examples of protein–carbohydrate interactions in cell trafficking include the binding of the selectins to their carbohydrate targets on leukocyte and endothelial cell surfaces (Bevilacqua & Nelson, 1993; Lasky, 1992; Rosen, 1993). The selectins are primarily responsible for leukocyte rolling on the activated vascular endothelium (Doré et al., 1993; Lawrence & Springer, 1991). Another interaction that plays an important role in leukocyte trafficking is the association of extracellular matrix receptors with their glycosaminoglycan targets, an example of which is the binding of CD44 to its ligand hyaluronan (Haynes et al., 1989, 1991; Herrlich et al., 1993; Rosen, 1993). This receptor–ligand pair is thought to play a role in leukocyte migration in tissue. In this chapter, a review of the current understanding of the structure and function of these cell adhesion molecules and their physiological role in the regulation of leukocyte trafficking is provided. Also discussed are experimental approaches used to identify the ligand binding sites in these proteins and to determine which residues within the binding sites are important in mediating the protein–carbohydrate interaction.

Selectins

The selectin family of proteins consists of three members: E-selectin, L-selectin, and P-selectin (Bevilacqua & Nelson, 1993; Lasky, 1992; Rosen, 1993). E-selectin and L-selectin were first defined by monoclonal antibodies which blocked the binding of leukocytes to endothelial cells, while P-selectin was initially identified as a protein that was expressed on activated platelets where it mediated binding to leukocytes. It was later shown that P-selectin is also expressed by activated vascular endothelial cells.

E-selectin (endothelial-selectin, CD62E) was identified initially using cell adhesion assays. Treatment of human vascular endothelial cells with interleukin-1 (IL-1) or tumor necrosis factor (TNF), two cytokines that are expressed during inflammation, resulted in leukocyte–endothelial cell adhesion. This suggested a mechanism whereby vascular endothelial cells at sites of inflammation could bind to leukocytes. Monoclonal antibodies (mAbs) were prepared that bound to activated endothelial cells and blocked the binding of leukocytes to activated vascular endothelial cells . Two of these antibodies (H4/18/7, H4/18) recognized an inducible protein (~115 kDa) on activated vascular endothelial cells and were able to block the adhesion of polymorphonuclear leukocytes to these cells (Bevilacqua et al., 1987). The protein recognized by these two antibodies was named endothelial–leukocyte adhesion molecule 1 (ELAM-1), but has since been renamed E-selectin. Further *in vitro* experiments using cytokine-activated human umbilical vein endothelial cells (HUVEC) and anti-E-selectin monoclonal antibodies suggested that E-selectin expression peaks 4–6 hours postactivation and decays over a period of 24 hours (Bevilacqua et al., 1989). Expression of E-selectin is induced by a variety of stimulators, including IL-1β, TNF, and some bacterially derived lipopolysaccharides (LPS) (Bevilacqua et al., 1987; Pober et al., 1986), and is augmented by interferon-γ (IFN-γ) (Leeuwenberg et al., 1990). Anti-E-selectin mAbs were used together with a mammalian transient expression cloning system to isolate a cDNA clone encoding E-selectin (Bevilacqua et al., 1989). When transfected into COS cells, this cDNA directed the expression of a protein of ~115 kDa which mediated the binding of human peripheral blood neutrophils or the human myeloid cell line HL60 to the COS cells in a calcium-dependent manner. These studies established the role of E-selectin as a cell adhesion molecule. Additional binding studies have shown that E-selectin preferentially mediates the adhesion of neutrophils, monocytes and memory T cells to activated endothelial cells (Bevilacqua et al., 1989; Carlos et al., 1991; Dobrina et al., 1989; Picker et al., 1991a; Shimizu et al., 1991). Aberrant E-selectin expression in tissue sections obtained from both chronic and acute inflamed tissues suggests a physiological contribution of E-selectin to leukocyte adhesion in inflammatory disease (Postigo et al., 1992). These observations have been further corroborated by a number of recent *in vivo* experiments in which anti-E-selectin mAb administration prevented leukocyte-mediated tissue damage in animal models of acute inflammation (Gundel et al., 1991; Ledbetter et al., 1988; Mulligan et al., 1991).

P-selectin (platelet-selectin, also known as PADGEM, GMP-140, or CD62P) was identified initially as a ~140 kDa protein expressed in the alpha granules of resting platelets. The protein is rapidly mobilized to the cell surface following platelet activation (Hattori et al., 1989; Stenberg et al., 1985). P-selectin is also expressed by megakaryocytes and within the Weibel–Palade bodies of endothelial cells (McEver et al., 1989). As in the case of platelets, endothelial cell activation results in the cell surface expression of P-selectin. A cDNA clone encoding P-selectin was isolated by first purifying the protein from platelets. The N-terminal amino acid sequence was determined from the intact molecule and from proteolytic fragments of the molecule. A phage cDNA library, prepared from mRNA isolated from HUVEC, was then screened with oligonucleotide probes designed on the basis of the protein sequence (Johnston et al., 1989). The cDNA sequence revealed that P-selectin had a similar domain structure and significant homology to E-selectin. Binding studies established that P-selectin could mediate platelet–neutrophil and platelet–monocyte cell binding (Larsen et al., 1989, 1990). These results were extended using COS cell transfectants expressing P-selectin and peripheral blood neutrophils and myeloid cell lines (Geng et al., 1990). The experiments showed that E-selectin and P-selectin on activated vascular endothelial cells mediate the binding of leukocytes to the vascular cell wall at sites of inflammation. Unlike E-selectin expression, P-selectin is expressed on the cell surface within minutes of activation because the protein is stored within cells and *de novo* protein synthesis is not required. Inflammatory mediators such as thrombin, histamine and H_2O_2 stimulate P-selectin cell surface expression (Patel et al., 1991; Sugama et al., 1992). *In vitro* studies have shown that P-selectin is reinternalized minutes after expression. This rapid turnover of P-selectin can be prevented *in vitro* by free radicals (Patel et al., 1991), which are known to be produced by activated neutrophils *in vivo*. This suggests that P-selectin expression may be stabilized by the presence of activated neutrophils at sites of inflammation. The physiological contribution of P-selectin to leukocyte recruitment at sites of acute and chronic inflammation has been established by immunohistology (Grober et al., 1993). Like anti-E-selectin mAb, anti-P-selectin mAb, and also a soluble recombinant form of P-selectin, have been shown to block neutrophil mediated tissue damage in *in vivo* models of acute inflammation (Mulligan et al., 1992; Weyrich et al., 1993; Winn et al., 1993). In mice lacking P-selectin, leukocyte infiltration into a site of inflammation is delayed by several hours, consistent with the notion that P-selectin is necessary for the rapid initial recruitment of leukocytes to inflammatory sites (Mayadas et al., 1993).

L-selectin (leukocyte-selectin, also known as MEL-14, Leu-8, LAM-1, or CD62L) was identified initially by the use of an assay developed by Stamper and Woodruff (1976). In this assay, lymphocytes are allowed to adhere to frozen sections of lymphoid tissue, nonadherent cells are removed by washing, and specific adherence to lymph node ultrastructure is assessed. Weissman and colleagues demonstrated the *ex vivo* binding of lymphocytes to the plump endothelial cells lining the lymph node capillaries, known as high endothelial venules (HEV). These investigators then isolated an anti-murine lymphocyte mAb (MEL-14) able to block lymphocyte-HEV binding (Gallatin et al., 1983). Due to the role of the MEL-14 defined antigen in lymphocyte adhesion to HEV, this antigen became known as the "homing receptor." Antibody binding studies showed that MEL-14 recognized a protein of ~90 kDa which was expressed by most leukocytes. When cloned, the three selectins were found to have significant homology and the same domain organization (Lasky et al., 1989; Siegelman et al., 1989). In independent studies, mAbs Leu-8 and TQ1 had been generated and used to define subpopulations of T cells. The antigen defined by these mAbs was cloned independently by several groups and found to be the human homologue of the protein recognized by MEL-14 (Bowen et al., 1989; Camerini et al., 1989; Tedder et al., 1989). Further experiments with both murine and human lymphocytes have shown that the expression of L-selectin is down-regulated following lymphocyte activation. Soluble recombinant L-selectin could specifically bind to the HEV (Fig. 8-2) (Watson et al., 1990b). The physiological role of L-selectin in inflammation has been established by experiments in which the anti-L-selectin mAb and a soluble recombinant form of L-selectin can inhibit neutrophil migration in *in vivo* models of inflammation (Ma et al., 1993; Watson et al., 1991).

The selectins are type-I membrane proteins which means that their N-terminus is located outside the cell. Their extracellular domain is composed of an amino-terminal C-type lectin domain (~120 amino acids) followed by an epidermal growth factor

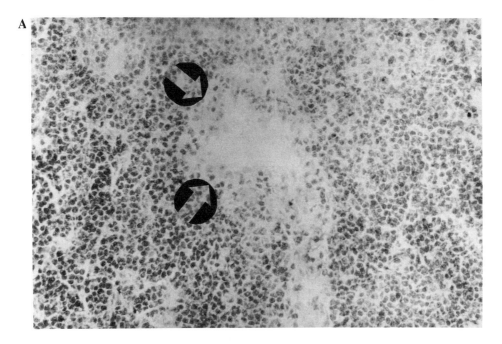

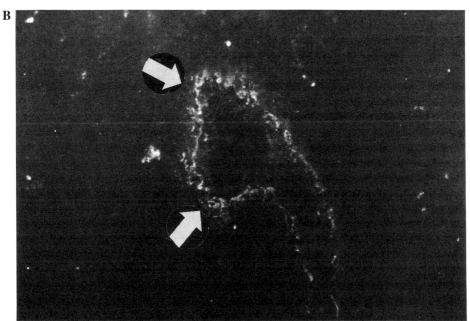

Figure 8–2. L-selectin–Rg binds to peripheral lymph node HEV. Serial frozen tissue sections of murine peripheral node HEV were stained with (**A**) hematoxylin fixed or with (**B**) L-selectin-Rg followed by a fluorescein-conjugated anti-human IgG antibody. Arrows point to the L-selectin–Rg positive HEV.

(EGF, ~30 amino acids) domain and by a variable number of consensus repeat sequences (CR, ~60 amino acids) which have homology to sequences found in complement regulatory proteins. P-selectin, E-selectin and L-selectin contain nine, six and two CR repeats, respectively (Fig. 8-3). In most cases, the selectins are anchored to the cell membrane via a hydrophobic transmembrane domain which is followed by a largely hydrophilic cytoplasmic domain. However, alternatively spliced isoforms of P-selectin and L-selectin which would produce secreted, soluble forms have been reported (Johnston et al., 1989; Schleiffenbaum et al., 1992). Cloning of the genomic fragments encoding the selectins has revealed a similar genomic organization. Each of the functional domains of the selectins (the amino terminal secretory signal sequence, the C-type lectin domain, the EGF-like domain, the complement receptor-like repeats, the transmembrane and cytoplasmic domains) is encoded by a different exon.

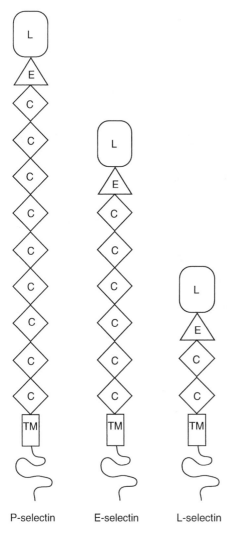

P-selectin   E-selectin   L-selectin

Figure 8–3. Selectins. Line drawing representation of human E-selectin, L-selectin, and P-selectin. The lectin (L), EGF-like (E), complement receptor-like (C) and transmembrane (TM) domains of the selectins are shown. The cytoplasmic domain is represented by a curved line following the transmembrane domain.

Furthermore, all three genes are located within 2×10^6 bases of each other on chromosome 1 (Collins et al., 1991; Ord et al., 1990; Siegelman et al., 1990; Tedder et al., 1989; Watson et al., 1990a). These observations suggest that the selectin genes arose from progenitor-exons *via* exon duplication and shuffling, followed by mutation and functional selection.

The role of the selectins in leukocyte rolling has been addressed experimentally. *In vitro* studies by Lawrence and Springer (1991) showed that purified P-selectin mediates leukocyte rolling under similar shear conditions as those found in the bloodstream. Experiments using intravital microscopy in conjunction with anti-L-selectin mAb or a soluble recombinant form of L-selectin, have also shown that L-selectin can mediate leukocyte rolling (Ley et al., 1991b). Using a flow cell system, Smith and colleagues have shown that E-selectin is able to mediate leukocyte rolling on activated vascular endothelial cells (Briskin et al., 1993). These experiments are consistent with the experiments carried out by Atherton and Born (1972), who showed, almost twenty years before the selectins were cloned and characterized, that leukocyte rolling could be abolished by superfusion with EDTA.

As mentioned above, the role of the selectins in leukocyte adhesion at sites of inflammation has been examined *in vivo* by testing the ability of selectin inhibitors (mAb, soluble recombinant forms of the selectins, or ligand analogues) to block leukocyte-mediated tissue damage at sites of inflammation. Anti-L-selectin mAbs as well as soluble L-selectin were found to block the infiltration of neutrophils into the peritoneal cavity following a thioglycolate injection in rodents (von Andrian et al., 1991; Watson et al., 1991). Anti-L-selectin and anti-P-selectin mAbs have been shown to block the neutrophil-mediated tissue damage in models of reperfusion injury (Ma et al., 1993; Winn et al., 1993; Weyrich et al., 1993) and anti-E-selectin mAb blocks neutrophil-mediated damage in a lung injury model of acute inflammation (Gundel et al., 1991; Mulligan et al., 1991). Infusion of a selectin ligand (sialyl Lewis x [Le^x]) was used to prevent leukocyte-mediated lung injury, a P-selectin dependent process, in rats infused with cobra venom factor (Mulligan et al., 1993).

Many lines of evidence have suggested that selectins mediate their adhesive function via interactions between their lectin domains and their carbohydrate ligands. A very intensive research effort by many groups has led to the identification of the carbohydrate ligands of the selectins, the identification of the proteins which present the carbohydrate ligands to the selectins, and the molecular basis for these protein–carbohydrate interactions. We and others have approached the study of the selectins by preparing soluble forms of the extracellular domains as immunoglobulin fusion proteins (Rg, receptor/recombinant globulins) which are shown schematically in Figure 8-4. These fusion proteins were used, for example, in studies which suggested that the C-type lectin domain represents the ligand binding domain in the selectins (Walz et al., 1990).

Selectin Ligands

The first information about the crucial role of carbohydrates in the adhesion of lymphocytes was obtained before the selectins were identified. Initially, carbohydrates were shown to inhibit lymphocyte adherence to frozen sections of lymphoid tissue in the assay of Stamper and Woodruff (1976). A series of structurally related monosaccharides

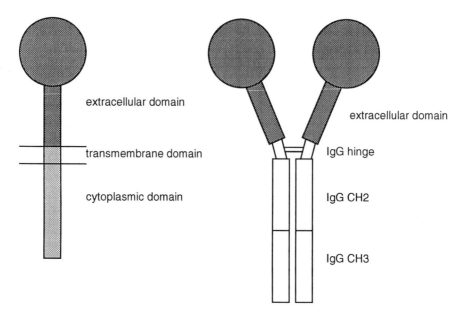

Figure 8–4. Receptor globulin. The receptor immunoglobulin fusion proteins, receptor globulins, consist of the extracellular domain of a type I membrane protein fused onto the hinge, CH2 and CH3 domains of the constant chain of an immunoglobulin. These proteins are expressed as covalently-linked homodimers held together via disulfide bonds in the hinge region of IgG1.

and fucoidin, a polysaccharide, inhibited lymphocyte adherence *in vitro* by competing with a carbohydrate on the endothelium for binding to a proposed lectin on the lymphocyte (Stoolman & Rosen, 1983). The work was extended to demonstrate the inhibitory effect of fucoidin and other polysaccharides on the *in vivo* recirculation of lymphocytes (Braaten et al., 1984; Spangrude et al., 1984; Stoolman et al., 1984). Treatment of tissue sections with neuraminidase, an enzyme which removes sialic acid from complex carbohydrates, destroyed binding of lymphocytes to peripheral node HEV (Rosen et al., 1985). These results, together with the previous observation that leukocyte binding required divalent cations, eventually led to the prediction that the carbohydrate binding activity resided in the amino terminal domain that bears homology to the C-type lectins. This prediction was verified by the demonstration that an antibody that blocked E-selectin binding to its ligand, bound to a chimeric protein containing only the first 90 amino acids of the lectin domain of E-selectin (Walz et al., 1990).

The exact structures of the cellular ligands of the selectins have not yet been elucidated. However, several experimental approaches have led to the identification of carbohydrates to which the selectins bind. The experimental approaches have included the use of blocking antibodies which recognized carbohydrates, the inhibition by, or direct binding to, known carbohydrates or polysaccharides, the expression of glycosyltransferases by transfection, and the enzymatic treatment of ligand-bearing cells or tissues (Corral et al., 1990; Goelz et al., 1990; Imai et al., 1990; Kuijpers et al., 1992; Larsen et al., 1990, 1992; Lowe et al., 1990, 1991; Moore et al., 1991; Stoolman et al., 1987; Yednock et al., 1987a,b; Zhou et al., 1991). As an example, anti-CD15 antibodies, which recognize the trisaccharide 3-fucosyl-N acetyllactosamine (Gal1-4[Fuc1-3]-

GlcNac), Le^x, were found to inhibit the binding of cells expressing the P-selectin ligand to either purified P-selectin or P-selectin expressed by transfected COS cells (Larsen et al., 1990). (A summary of carbohydrate structures discussed here is shown in Fig. 8-5.) This approach, while revealing, suffered from the possibility that the antibody could have cross reactivity with other unknown carbohydrate structures. However, the binding of E-selectin and P-selectin to CD15-type structures has also been demonstrated by other means. The binding of a variety of cell lines to E-selectin-Rg correlated with the expression of sialyl Le^x, but not CD15, and this binding could be inhibited by proteins bearing sialyl Le^x type structures (Tyrrell et al., 1991). The transfection of a plasmid encoding an $\alpha(1,3)$fucosyltransferase activity resulted in the expression of an E-selectin ligand in cells that did not express this ligand originally (Goelz et al., 1990; Lowe et al., 1990). The position of the fucose on the oligosaccharide adjacent to the sialic acid is critical for E-selectin binding as demonstrated by the inability of transfection with a specific fucosyltransferase that generates the VIM2 structure to mediate E-selectin dependent adhesion (Lowe et al., 1991). This notion

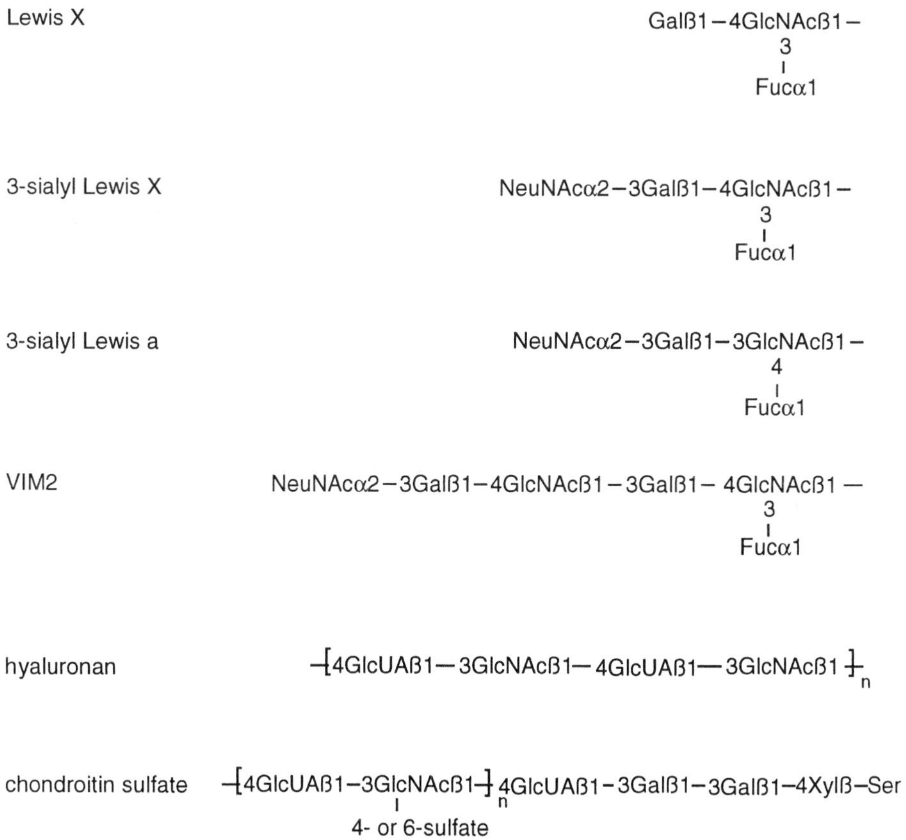

Lewis X
$$\begin{array}{c} \text{Gal}\beta1-4\text{GlcNAc}\beta1- \\ 3 \\ | \\ \text{Fuc}\alpha1 \end{array}$$

3-sialyl Lewis X
$$\begin{array}{c} \text{NeuNAc}\alpha2-3\text{Gal}\beta1-4\text{GlcNAc}\beta1- \\ 3 \\ | \\ \text{Fuc}\alpha1 \end{array}$$

3-sialyl Lewis a
$$\begin{array}{c} \text{NeuNAc}\alpha2-3\text{Gal}\beta1-3\text{GlcNAc}\beta1- \\ 4 \\ | \\ \text{Fuc}\alpha1 \end{array}$$

VIM2
$$\begin{array}{c} \text{NeuNAc}\alpha2-3\text{Gal}\beta1-4\text{GlcNAc}\beta1-3\text{Gal}\beta1-4\text{GlcNAc}\beta1- \\ 3 \\ | \\ \text{Fuc}\alpha1 \end{array}$$

hyaluronan
$$-[4\text{GlcUA}\beta1-3\text{GlcNAc}\beta1-4\text{GlcUA}\beta1-3\text{GlcNAc}\beta1-]_n$$

chondroitin sulfate
$$\begin{array}{c} -[4\text{GlcUA}\beta1-3\text{GlcNAc}\beta1-]_n\,4\text{GlcUA}\beta1-3\text{Gal}\beta1-3\text{Gal}\beta1-4\text{Xyl}\beta-\text{Ser} \\ | \\ \text{4- or 6-sulfate} \end{array}$$

Figure 8–5. Schematic representation of carbohydrate structures. Shown are the structures of carbohydrate ligands implicated in cell adhesion as described in the text. Abbreviations used are Gal, galactose; GlcNAc, N-acetylglucosamine; Fuc, fucose; NeuNAc, N-acetylneuraminic acid, also known as sialic acid; GlcUA, glucuronic acid; Xyl, xylose; Ser, serine.

was supported by the inability of purified VIM2 to support adhesion of E-selectin-expressing COS cells (Tyrrell et al., 1991). The nature of the E-selectin ligand was confirmed directly by the isolation of glycolipids which were able to support adhesion of E-selectin-expressing cells. The glycolipids were subjected to composition, linkage and mass spectrum analyses. It was found that E-selectin bound specifically to terminally sialylated lactosyl ceramides containing at least one fucose in an $\alpha(1,3)$ glycosidic linkage to an internal GlcNAc residue (Tiemeyer et al., 1991). The minimal requirements for E-selectin binding to Le^x and Lewis a (Le^a) type structures were analyzed by examining the ability of these carbohydrates to mediate adhesion of E-selectin-expressing COS cells. In this study, synthetic or purified carbohydrates were used, and it was found that 2,6-sialyl Le^x was not recognized by E-selectin (Tyrrell et al., 1991).

Additional carbohydrate binding studies have shown that some sulfated carbohydrates can also act as selectin ligands. In particular, P-selectin can bind to sulfatide, a sulfated glycolipid (Aruffo et al., 1991). Likewise, sulfated glycans such as heparin, fucoidin, and dextran sulfate can inhibit P-selectin-mediated cell adhesion (Skinner et al., 1991). L-selectin and P-selectin, but not E-selectin, were shown to bind sulfoglucuronyl glycosphigolipids (Needham & Schnaar, 1993). This was consistent with early observations that sulfated polysaccharides inhibited leukocyte rolling *in vivo* (Ley et al., 1991a). More recently it has been shown that sulfated forms of Le^x and Le^a are potent inhibitors of E-selectin-mediated cell adhesion (Yuen et al., 1994). Furthermore, sulfation of GlyCam-1, a mucin which presents carbohydrates to L-selectin, is required for L-selectin binding (Imai et al., 1993).

The proteins that specifically present carbohydrate ligands to selectins have been identified. These proteins include GlyCam-1 (Lasky et al., 1992), CD34 (Baumheter et al., 1993), and MadCam-1 (Berg et al., 1993; Briskin et al., 1993), to which L-selectin binds, and PSGL-1 to which P-selectin binds (Sako et al., 1993). Early reports by Butcher and colleagues showed that both E-selectin and P-selectin could bind a carbohydrate on neutrophil L-selectin (Picker et al., 1991b). These experiments suggest that a discrete number of glycoproteins can efficiently present carbohydrate ligands to the selectins. Thus, selectin-mediated cell adhesion is regulated in multiple ways: first, by the expression of the selectin protein itself, second, by the expression of the appropriate glycosylation enzymes required for the synthesis of the carbohydrate ligand, and third, by the expression of the appropriate protein carriers which can efficiently present the carbohydrate to the selectin. These multiple levels of regulation play a key role in providing the *in vivo* specificity of leukocyte trafficking.

In clinical investigations, two patients with a genetic defect that prevents expression of sialyl Le^x have been identified (Etzioni et al., 1992). This defect is known as leukocyte adhesion deficiency type 2 (LAD-2). The patients show high blood neutrophil counts and recurrent pneumonia and bacterial infections. These manifestations are the result of the inability of leukocytes to efficiently home to sites of infection (von Andrian et al., 1993) and provide the strongest *in vivo* evidence to date that the selectin/carbohydrate interactions play key roles in leukocyte trafficking in humans.

Structure–Function Analyses of the Selectins

With an understanding of the carbohydrate ligands and the domain organization of the selectins, a more detailed analysis of selectin–ligand interactions could be approached. Ideally, the protein in complex with its ligand would be crystallized and subjected to

X-ray crystallography or, alternatively, be studied in solution by nuclear magnetic resonance (NMR) spectroscopy. The experimentally determined structures could then be analyzed, and the residues found to be involved in ligand binding could be subjected to site-specific replacement to assess their importance for binding. It is often not readily possible to produce crystals suitable for X-ray crystallography, and the NMR spectroscopic analysis of macromolecules may be difficult. This has been the case for the selectins and, until recently, protein modeling was the only available method to obtain three-dimensional structural information for the selectins. We will, therefore, provide a brief overview of the theoretical techniques used to generate three-dimensional models of the selectin ligand binding domains.

Theoretically derived protein models are generally less accurate than experimental structures, but a reasonable protein model is often sufficient to guide experiments that are designed to identify the residues or interactions important for biological function. Presently, so-called comparative or knowledge-based protein modeling, pioneered by Greer (1991), produces, in the absence of X-ray crystallographic or NMR spectroscopic data, the most accurate three-dimensional models of proteins. In comparative model building, one or more known experimental structures are used as structural templates to construct a model of a protein with unknown three-dimensional structure. To be successful, the protein with unknown structure must display a three-dimensional fold similar to the selected structural template(s). If it is possible to identify at least one reasonable template structure, three-dimensional constraints can be included in the comparison of the protein sequences and to align the sequence of the protein of unknown structure with that of the structurally defined protein (Greer, 1991). Using the structure-based sequence alignment, structurally conserved regions are assigned and included in the model. These are generally regions of well-defined secondary structure and the hydrophobic core of the protein. Regions with variable structure, often loops, can frequently be modeled using parts of other known structures (Jones & Thirup, 1986) available in the Brookhaven Protein Data Bank (Bernstein et al., 1977) or, alternatively, by application of theoretical methods such as conformational searching (Bruccoleri et al., 1988).

The comparative modeling approach depends critically on an accurate assessment of whether a protein of unknown structure adopts a fold similar to that of a known three-dimensional structure (Bajorath et al., 1993). This is a rather straightforward question if significant sequence similarity (50% or more) exists between proteins of unknown and known three-dimensional structure, since high sequence similarity corresponds to structural similarity (Chothia & Lesk, 1986). However, since three-dimensional structural similarities are generally more conserved than amino acid sequence similarities (Chothia & Lesk, 1986), many proteins display similar structures despite moderate to low sequence identity. In the latter case, the so-called "inverse folding" approach (for reviews see Wodak & Rooman, 1993) is often able to verify structural relationships between proteins even if the sequence similarities are very low. The inverse folding approach focuses on the question of which amino acid sequences are compatible with a given three-dimensional fold. Sequences are computationally "threaded" onto three-dimensional structures, and the respective sequence-structure compatibility is assessed either by analyzing the energetics of pairwise residue interactions (Sippl, 1990; Sippl & Weitckus, 1992) or by analyzing the chemical compatibility of a residue with its specific environment in a three-dimensional structure (Blundell & Johnson, 1993; Bowie & Eisenberg, 1993). If the sequence of a protein with undetermined three-dimensional structure is found to be very compatible with an experimentally determined three-dimensional structure, then reasonable structural similarity can be assumed with some confidence even in the absence of significant sequence similarity.

A major limitation of the comparative modeling method is that a similar and experimentally derived structure is required initially and that the structural similarity must be recognized. This means that only a fraction of novel protein sequences can be modeled using this approach. The search for a reasonable structural template for model building of the ligand binding domain of the selectins had long been without success. It was the X-ray crystallographic analysis of the C-type lectin domain of the mannose binding protein (MBP) from rat by Hendrickson and colleagues (Weis et al., 1991) that provided the first view of the structural features of the mammalian C-type lectin family including the selectins. The structure of MBP displayed a previously unobserved protein fold. Sequence comparison with the C-type lectins, taking the three-dimensional constraints provided by the MBP structure into account, suggested that MBP provided the "prototype" protein fold of this protein family (Weis et al., 1991). Since the sequence identity between MBP and the selectin lectin-like ligand binding domains is only approximately 25%, such sequence-structure comparisons and alignments have significantly aided in the development of selectin models. The structural similarity of the selectins and MBP was further supported by inverse folding analysis, which suggested that the sequence of P-selectin was fully compatible with the C-type lectin fold (Hollenbaugh et al., 1993). This information provided the basis for comparative modeling and structure-function analysis of the selectins.

Three-dimensional models for the lectin-like domains in E- and P-selectin have been generated and analyzed by us (Bajorath et al., 1993; Hollenbaugh et al., 1993) and others (Erbe et al., 1992, 1993). Figure 8-6a shows a comparison of the P-selectin model and MBP, its crystallographic "parent" structure, and in Figure 8-6b our E- and P-selectin models are compared. Both models were generated using an equivalent modeling protocol (Bajorath et al., 1993). The following analysis focuses on P-selectin. One of the calcium binding sites of MBP was thought to be conserved in the selectins. Based on the calcium requirement for ligand binding to the selectins, and based on computer graphic analysis of our model, we selected a shallow groove proximal to the hypothesized calcium as the potential ligand binding site in P-selectin. Residues in this region, distant in sequence but predicted to be spatially close, were selected and subjected to site-specific mutagenesis experiments. In total, fourteen P-selectin mutants were generated. The binding of the mutant proteins to HL-60 cells, which carry the cellular P-selectin ligand, was assessed using two different cell adhesion assays (Hollenbaugh et al., 1993). We found that several residues in the shallow groove of P-selectin were crucial for the binding to HL-60 cells (Fig. 8-7). These residues include two tyrosines (Tyr48 and Tyr94) and a lysine (Lys113) whose substitution completely eliminates binding. These residues are conserved in E- and P-selectin. The results suggest that the identified groove is the binding site in P-selectin for its carbohydrate ligand. In an independent study on E-selectin, similar results were obtained (Erbe et al., 1992).

Once the binding region and critical residues in P-selectin were identified, the orientation of the carbohydrate in the P-selectin ligand binding site could be addressed. However, any theoretically derived models of protein–carbohydrate interactions need to be considered with caution. Current modeling methods are, in general, insufficient

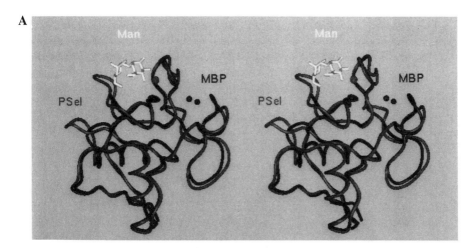

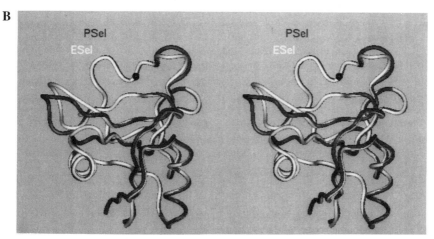

Figure 8–6. Comparison of the MBP, P-selectin and E-selectin. (**A**) Superposition of MBP and P-selectin. The stereo view shows the crystal structure of the C-type lectin domain of the rat mannose binding protein (MBP, darker gray) in its crystallographic complex with mannose (Man, white) and the model structure of the P-selectin lectin-like domain (PSel, lighter gray). The protein backbones are shown in solid ribbon representation. Conserved secondary structure elements in MBP and PSel were used to generate the superposition. The three calcium ions found in the complex crystal structure of MBP and oligomannose are shown as balls. The smaller dark gray balls represent calcium ions unique to the MBP complex. The larger lighter gray ball represents the calcium which is conserved in MBP and in the selectins and which participates directly in carbohydrate binding to MBP. Only the two terminal units of the oligosaccharide are shown since they interact with MBP. The figure illustrates the previously unobserved protein fold of the C-type lectin domain and the calcium-dependent carbohydrate binding to the mammalian lectin. Furthermore, the comparison illustrates the main premise of comparative model building: the protein of unknown structure (PSel) displays a core protein fold very similar to the selected structural template (MBP). (**B**) Comparison of P- and E-selectin. The model structures of the P- (gray) and E-selectin (white) ligand binding domains generated in our laboratory are shown analogous to Figure 8-6A. The calcium which is conserved in MBP and the selectins is shown in black.

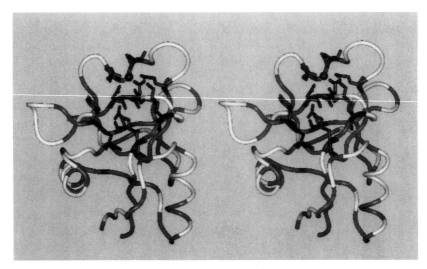

Figure 8–7. The predicted structure of the ligand binding site in P-selectin. The stereo view is obtained from Figure 8-6 by approximately 90° rotation around the vertical axis. The side chains of the residues which participate in the conserved calcium coordinate sphere and of the residues which were shown to be critical in the binding of P-selectin to its cellular ligand are shown in dark gray. The residues which form the ligand binding site in P-selectin are conserved in E-selectin. Gray residues are conserved in P- and E-selectin, and residues shown in white are not conserved. The conserved binding site in the selectins is surrounded by loop regions which are variable in sequence. This structural arrangement of a conserved central scaffold with a different environment may give rise to the overlapping, but not identical, ligand specificities found for E- and P-selectin.

to realistically predict details of protein–carbohydrate interactions (Bundle & Young, 1992). The great flexibility of carbohydrates, the possibility of conformational changes in protein and carbohydrate upon binding, and the important role of defined water molecules in mediating protein–carbohydrate interactions (Bourne & Cambillau, 1993) represent major problems for current predictive methods. In addition, the limited accuracy of some parts of the model, for example, side chain conformations, has to be taken into account when analyzing protein models. The detailed structural analysis of protein–carbohydrate interactions requires the availability of X-ray crystallographic protein–carbohydrate complexes refined to high resolution. We have, therefore, limited our efforts to the generation of simple and rather schematic models of P-selectin–sialyl Le^x interactions.

The first insight into carbohydrate binding to C-type lectins was obtained from the crystal structure of MBP in complex with oligomannose (Weis et al., 1992). This structure revealed that the calcium conserved in the MBP and selectins structures participates directly in ligand binding. The terminal mannose unit of the oligomannose ligand binds directly to the calcium by participating in its coordination sphere with two equatorial hydroxyl ligands (Weis et al., 1992). This previously unobserved carbohydrate–protein interaction provided a molecular rationale for the calcium-dependence of carbohydrate binding by the C-type mammalian lectins. Using site-specific mutagenesis, it has been shown that amino acid substitution of two residues of the calcium coordination sphere is sufficient to change the primary carbohydrate specificity of MBP from mannose to galactose (Drickamer, 1992).

Approximate modeling of the P-selectin–sialyl Le^x interaction began with the placement of the fucose of sialyl Le^x in contact with the conserved calcium, analogous to the MBP–mannose structure. The calcium–fucose interaction limits the number of possible orientations that the conformationally restricted sialyl Le^x ligand can have in the modeled P-selectin ligand binding site. The simple hypothetical model depicted in Figure 8-8 has helped us to better understand some features of this binding site. Most importantly, the negative charge of the sialic acid moiety in sialyl Le^x is at an appropriate distance for an ionic interaction with P-selectin residues Lys111 or Lys113. Lys113 was found by us to be particularly critical for the function of P-selectin. The side chain of this residue was predicted to be ~10 Å distant from the conserved calcium. A similar modeling approach suggested that sulfatide bound to the same site. Analysis of the binding of mutants confirmed that the binding sites in P-selectin for its cellular ligand and for negatively charged sulfatide are overlapping.

The architecture of the ligand binding site in P-selectin has several implications for carbohydrate binding to the selectins, which are, in part, illustrated by the schematic model of sialyl Le^x binding to P-selectin. Proteins often bind carbohydrate ligands in cavities or pockets where the carbohydrate is involved in a network of hydrogen bonds and van der Waals interactions, and largely shielded from the solvent environment (Quiocho, 1989; Vyas, 1991). As mentioned above, the binding of mannose to MBP revealed a previously unobserved mode of carbohydrate binding (Bundle & Young, 1992) which involved direct participation in a calcium coordination sphere. Structural models of the selectins in combination with mutagenesis data suggest that carbohydrate binding to the selectins also involves a calcium binding site in addition to other

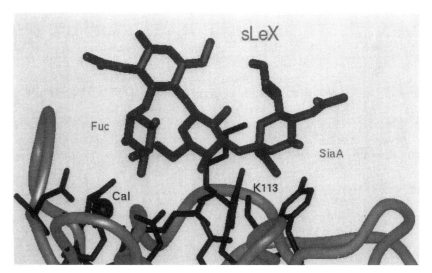

Figure 8–8. A schematic model of P-selectin–ligand interactions. Sialyl Le^x (sLe^x) is shown in dark gray. This close-up view is obtained from Figure 8-6 by an approximately 180° rotation around the vertical axis. The side chains of P-selectin binding site residues are shown according to Figure 8-6. The fucose (Fuc) and sialic acid (SiaA) moieties of sLe^x are proposed to interact with the calcium coordination sphere (Cal) and with Lys 113 (K113), respectively.

residues on the surface. Sialyl Le^x-like structures are therefore suggested to bind to a region of the selectin–protein surface rather than in a crevice or pocket. It follows that only one side of the ligand would participate in interactions with the protein (Tyrrell et al., 1991) whereas the other side remains essentially exposed to solvent. This binding on the selectin surface would be consistent with low affinity interactions between single sialyl Le^x molecules and P-selectin (Foxall et al., 1992).

Lasky and co-workers (Erbe et al., 1992) used a different approach to identify the ligand binding site in E-selectin. They generated a number of anti-E-selectin antibodies which mapped to the lectin domain and blocked E-selectin binding to sialyl Le^x. They then generated a number of mutants in the E-selectin lectin domain and examined their ability to bind to their mAbs and to sialyl Le^x. Using this approach, E-selectin mutants were characterized that were not recognized by blocking mAbs and that did not bind to sialyl Le^x. These mutations were then superimposed on a three-dimensional model of E-selectin to map the region of E-selectin involved in ligand binding. Based on these observations, analogous P-selectin mutants were prepared and it was shown that the same region in P-selectin and E-selectin are involved in ligand binding (Erbe et al., 1993).

The crystal structure of the E-selectin lectin and epidermal growth factor (EGF) domains has now been reported (Graves et al., 1994). This structure confirmed the proposed structural similarity between MBP and the selectins and the presence of the functional calcium binding in the selectins. Mutational experiments based on the E-selectin crystal structure resulted in the identification of the same residues previously found to be important for ligand binding.

CD44

Monoclonal antibodies against the lymphocyte cell surface protein CD44 have been developed by several groups. Butcher and colleagues have obtained data that suggest an important role for CD44 in the adhesion of lymphocytes to certain endothelial cells that line the capillaries in Payer's patches (Jalkanen et al., 1987). Payer's patches are lymphatic organs found in the gut. Other groups have used anti-CD44 mAbs to show that CD44 can act as a co-stimulatory molecule in T-cell activation (Huet et al., 1989; Shimizu et al., 1989). Independently, a hyaluronan receptor expressed by hamster fibroblast was identified with mAb K-3 (Tarone et al., 1984). This antibody was subsequently shown to recognize CD44 (Aruffo et al., 1990).

Two groups independently isolated cDNAs encoding a 90 kDa isoform of CD44. Anti-CD44 mAb was used to purify CD44 and a partial amino acid sequence was obtained. This information was used to design oligonucleotide probes to screen a cDNA library prepared with mRNA from lymphocytes and isolate a cDNA clone encoding CD44 (Goldstein et al., 1989). A cDNA clone encoding CD44 was also obtained using a mammalian cell-based transient expression cloning system in conjunction with a number of anti-CD44 mAbs (Stamenkovic et al., 1989). Analysis of the deduced amino acid sequence of CD44 showed that these cDNAs encoded a type I membrane protein with an amino terminal extracellular domain (248 amino acids), followed by a hydrophobic transmembrane domain (21 amino acids) and a 72-amino acid cytoplasmic domain. Comparison of the amino acid sequence of CD44 with that of other known proteins showed that CD44 contains at its amino terminus, a ~130 amino acid domain with significant sequence homology (~30%) to cartilage link proteins (CLPs). CLP binding sites have been found in other proteins including aggrecan, versican, and the tumor necrosis factor-inducible protein TSG-6 (Yang et al., 1994). Like CD44, these proteins are involved in extracellular matrix binding.

After the isolation of the 90 kDa isoform of CD44 (also known as CD44H due to its expression by cells of hemopoietic origin), a cDNA encoding a 150 kDa isoform expressed by epithelial cells (CD44E) was isolated (Stamenkovic et al., 1991). Comparison of the predicted amino acid sequence of this protein with that of CD44H revealed the presence of an additional 135 residues between the extracellular and transmembrane domains of CD44H. The amino acid sequence of this new CD44 domain had no obvious homology to known proteins. The analysis of carbohydrate moieties which are added to CD44, showed that this protein is posttranslationally modified by the addition of heparan, chondroitin and keratan sulfate (Carter & Wayner, 1988; Stamenkovic et al., 1991); thus, CD44 is a proteoglycan.

Following the isolation of the cDNA encoding CD44E, cDNAs encoding additional isoforms of CD44 were isolated and characterized (Fig. 8-9) (Brown et al., 1991; Dougherty et al., 1991; Günthert et al., 1991; Jackson et al., 1992). To date, at least 30 different isoforms of this protein are known. Most cells express one or another isoform of CD44; however, expression of more than one isoform at a time has also been reported (Screaton et al., 1992; Tölg et al., 1993). The isolation and characterization of a genomic clone encoding CD44 has revealed that this protein is encoded by 19 exons (Screaton et al., 1992). The various CD44 isoforms result predominantly from the variable splicing of 10 exons (exons 6–14). The use of these exons results in the ad-

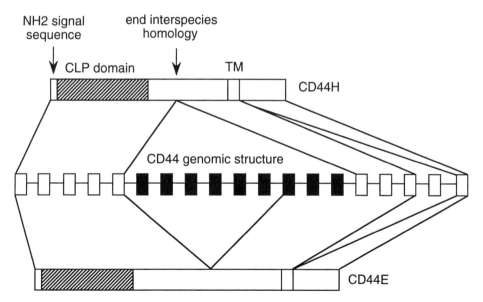

Figure 8–9. CD44. Line drawing representation of the CD44 genomic structure and the use of exons which leads to the expression of CD44H or CD44E. The CLP domain is represented by a striped rectangle, the transmembrane domain is depicted by a lightly shaded box. The variably spliced membrane proximal exons are represented by small dark rectangles.

dition of amino acid sequences between the extracellular and transmembrane domain of CD44. For example, CD44E is encoded by exons 1–5 and 12–19 (Screaton et al., 1992). In addition to isoforms with variably spliced exons 6–14, other CD44 splice variants have been reported. These include variants with different cytoplasmic domains and soluble isoforms (Haynes et al., 1989). All CD44 isoforms identified to date contain exons 1–5, which code for the region of the protein required for hyaluronan binding. Interestingly, not all CD44 isoforms bind hyaluronan *in vitro* (Stamenkovic et al., 1991). Based on this observation, it has been proposed that the alternatively spliced exons of CD44 encode domains that are responsible for modulating the hyaluronan binding activity of CD44 (Screaton et al., 1992). But it is also possible that the variably spliced exons of CD44 code for additional functions. It is known, for example, that a number of cytokines and chemokines can bind to glycosaminoglycans and proteoglycans. Shaw and colleagues have proposed that the proteoglycan isoform of CD44 may be involved in the presentation of chemokines to leukocytes at sites of inflammation (Webb et al., 1990). Furthermore, it has been shown that CD44 can bind chondroitin sulfate (Aruffo et al., 1990), fibronectin (Jalkanen & Jalkanen, 1992) and collagen (Carter & Wayner, 1988). It is possible that alternatively spliced exons of CD44 are involved in binding to other glycosaminoglycans or these extracellular matrix proteins.

Using transfected cell lines expressing high levels of CD44, it was found that CD44 can trigger homotypic cell aggregation (St. John et al., 1990). Experiments with stably transfected cell lines expressing CD44H with and without a cytoplasmic domain have shown that intact CD44H can mediate cell motility on plates coated with hyaluronan (Thomas et al., 1992). Such experiments have established that CD44 can also play a role in cell–cell adhesion and cell locomotion. In addition to a normal physiological role in cell–cell and cell–extracellular matrix adhesion, cell locomotion and cell activation by CD44 have been found to play a role in tumor growth and metastasis. In a rat model of tumor metastasis, it was found that expression of a variably spliced CD44 isoform rendered a nonmetastatic tumor line highly metastatic (Günthert et al., 1991). Overexpression of CD44H in human tumor lines allowed them to grow readily in nude mice, while the parent cell line was not very aggressive *in vivo* (Sy et al., 1991, 1992). Although the exact role of CD44 in cancer remains to be clarified, these and other studies indicate a possible role for CD44 in tumor growth and metastasis.

The CD44 Ligand

Various approaches led to the identification of CD44 as a hyaluronan receptor. An immunoglobulin fusion protein of the extracellular domain of CD44H was used in immunohistological studies and in *in vitro* binding assays to demonstrate that CD44 bound hyaluronan (Fig. 8-10) (Aruffo et al., 1990). These studies also showed that CD44 is capable of binding chondroitin sulfate. (Carbohydrate structures are shown in Fig. 8-5.) The binding of CD44 to hyaluronan could be abolished by hyaluronidase. Hyaluronan binding to CD44 was further established when it was shown that the hamster protein recognized by mAb K-3, which was known to bind hyaluronan, was the homologue of CD44H (Aruffo et al., 1990). The binding of CD44 to hyaluronan was also demonstrated using a binding assay in which the adhesion of cells expressing CD44H to plates coated with hyaluronan and other glycosaminoglycans was examined

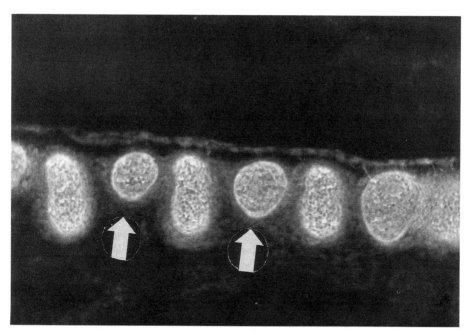

Figure 8–10. CD44H-Rg binds to hyaluronan in the developing vertebral column. Frozen tissue sections of the vertebral column of a 13-day-old murine embryo were stained with CD44H-Rg followed by a fluorescein conjugated anti-human IgG antibody. Arrows point to the vertebrae.

(Miyake et al., 1990). The binding of CD44 to plastic-coated with hyaluronan could be blocked by pretreatment with hyaluronidase or with anti-CD44 antibodies. Carter and Wayner (1988) showed that CD44 interacted with collagen, and others have shown that CD44 could bind to the heparin-binding domain of fibronectin (Jalkanen & Jalkanen, 1992). The binding of CD44 to fibronectin appears to be mediated by chondroitin sulfate-modified CD44 (Jalkanen & Jalkanen, 1992).

Structure–Function Analysis of CD44

We have previously outlined the use of structural models in combination with site-directed mutagenesis to identify the carbohydrate binding site in the selectins and to identify residues within this site that are crucial for selectin function. This approach was feasible because a reasonable structural template could be identified for the selectins and a reliable three-dimensional model of the selectin–lectin domain generated. A different approach was needed to gain insight into the possible location of the hyaluronan binding site in CD44 and to identify residues critical for the ability of CD44 to bind to hyaluronan.

First, the minimal region of the extracellular domain of CD44H required for hyaluronan binding was defined by preparing CD44–Rg fusion proteins which had progressively larger truncations in the carboxy-terminal region of the extracellular domain of CD44 (Peach et al., 1993). Four different CD44–Rg fusion proteins were prepared and tested for their abilities to bind hyaluronan. The first construct coded for the first 131 amino acids of the extracellular domain of CD44, known as the CLP homologous domain. The second, third, and fourth constructs encoded increasingly larger portions of the extracellular domain of CD44 (145, 186, and 210 residues, respectively). Only the fusion proteins containing the terminal 186 and 210 amino acids of the extracellular domain of CD44 were capable of binding to hyaluronan. Interspecies extracellular domains homology of CD44 (human, baboon, bovine, and mouse) end at amino acid 195 and variably spliced exons are inserted at amino acid 200. Taken together, this suggests that sequences C-terminal from the CLP homologous domain of CD44, which are conserved among different species, are required for hyaluronan binding and that sequences which are membrane proximal, as well as the variably spliced exons, are not.

The alteration of a protein sequence by deletion or point mutations may result in incorrectly folded proteins. Thus, lack of activity might be attributed to a disruption of tertiary structure rather than to the loss of a functional domain or the change of an important contact residue. A convenient, albeit indirect and somewhat limited, method to assess the structural integrity of proteins altered by mutagenesis is to test the ability of such mutants to bind mAbs which recognize the folded protein as opposed to a linear sequence. Antibodies which recognize such conformational epitopes usually do not bind to denatured forms of the protein and loss of binding indicates misfolding in the region of the epitope. The ability of the four CD44H truncation mutants to bind to three anti-CD44 mAbs was tested. All three antibodies bound to the CD44H mutants with one exception. Antibody IM7 did not bind to the fusion protein consisting of only the CLP homologous domain. This suggested that either this protein was incorrectly folded or that the CD44 epitope recognized by this antibody was not located within the CLP domain.

Inspection of the amino acid sequence of the smallest hyaluronan binding domain of CD44H showed two clusters of positively charged residues in the linear sequence (Arg29, Lys38, Arg41 and Arg150, Arg154, Lys158, and Arg162). It was already well known that the binding sites of other glycosaminoglycan binding proteins are composed of two adjacent positively charged regions in the protein that presumably fold together to create a positively charged ligand binding pocket suitable to host the negatively charged carbohydrate ligand. The proposed interaction between basic residues in the protein binding site and negatively charged groups (sulfates and carboxyl groups) on the glycosaminoglycan ligands was not without precedent. For example, the heparin binding domains of antithrombin II and heparin cofactor II, and the hyaluronan binding domains of the Receptor for Hyaluronan-Mediated Motility (RHAMM) and of CLP are known to be composed of clusters of basic amino acids (Goetinck et al., 1987; Yang et al., 1993, 1994). To investigate the contribution of individual basic residues to the CD44–hyaluronan interaction, each basic residue in these clusters was individually replaced with an alanine residue. Each point mutant was tested for its ability to bind hyaluronan. The structural integrity of each of the CD44H point mutants was examined with five different anti-CD44 mAbs. It was found that the ability of all CD44H point mutants to bind to hyaluronan was affected. The mutation of Arg41 to Ala completely abolished the hyaluronan binding activity of CD44H. The simultaneous change of Lys58 and Arg162 to Ala resulted in a double mutant that bound to hyaluronan very poorly but still above background levels. The mutagenesis experiments suggested that all basic residues within the putative hyaluronan binding domains of CD44 contribute,

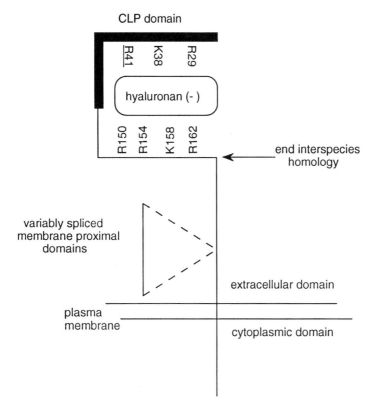

Figure 8–11. CD44 hyaluronan binding site. Line drawing representation of CD44 binding to hyaluronan. Clusters of basic residues found to contribute to CD44-hyaluronan binding are shown. Arg41 (R41), which was found to be critical for the CD44–hyaluronan interaction, is shown underlined.

directly or indirectly, to hyaluronan binding, but that the presence of Arg41 is crucial (Fig. 8-11).

Summary

We have outlined important mechanisms underlying cellular adhesion and have introduced different proteins which are, in part, responsible for these effects. Carbohydrate structures presented on cell surfaces and specific protein–carbohydrate interactions play a major role in mediating cellular adhesion. The identification of such adhesion proteins and of their carbohydrate ligands, and the analysis of the biological effects of their interactions, have been described in some detail. Furthermore, we have discussed experimental approaches to identify, at the level of individual residues, the ligand binding sites of two proteins involved in cell adhesion. In the case of P-selectin, the availability of the crystal structure of a related mammalian lectin allowed the derivation of a three-dimensional model of the P-selectin ligand binding domain. The analysis of this model, in conjunction with site-directed mutagenesis, was used to identify the ligand binding site and individual residues that are critical for ligand binding. The results of this study were recently confirmed by X-ray crystallographic analysis and mutagenesis of E-selectin. In the case of CD44, the location of the ligand binding site in this protein was identified by generating the smallest CD44 fragment capable of binding ligand. Comparison of the amino acid sequence of this CD44 fragment with that of other proteins which bind glycosaminoglycans showed that, like these proteins, the CD44 fragment contained two clusters of basic residues. Site-directed mutagenesis of these residues was for the interaction of CD44 with its carbohydrate ligand.

Acknowledgments

We thank Dr. Peter Senter for critical review of the manuscript, Dr. Ivan Stamenkovic and Dr. Bruce Fenderson for the L-selectin–Rg and CD44–Rg immunohistology, respectively, and Debby Baxter for help in the preparation of this manuscript.

9

Models of Cell Surface Carbohydrates and Biological Adhesion

François Tropper and Mark Bednarski

The events that take place at the surface of a cell are responsible for the survival and growth of an organism. Cells are able to communicate with each other to form tissues and initiate growth and repair. Cell surface chemistry of carbohydrate molecules plays an important role in this phenomenon and also in most major diseases of tissues ranging from infectious agents to autoimmune disorders, cancer, and stroke (Sharon & Lis, 1993; Varki, 1993). Infectious disease agents such as bacteria, viruses and parasites use carbohyrate molecules on the surface of cells as a point of attachment during the process of infection. The influenza virus, for example, uses the cell surface carbohydrate sialic acid to gain entry to respiratory epithelial cells (Wiley & Skehel, 1987). The bacterium *Helicobacter pylori* is now believed to be one of the primary causative agents for ulcers; the bacterium binds to the gastric mucosa via a glycolipid ligand as a point of attachment for colonization (Boren et al., 1993). In cancer, the changes in cell surface carbohyrate structures is involved in controlling cell growth and migration (metastatic disease). In ischemic diseases, such as those caused by heart attack and stroke, the surface of endothelial cells that line the blood vessels express receptors that bind to carbohydrate molecules on circulating leukocytes that mediate the inflammatory response which kills the tissue. Clearly, therefore, the need to study cell surface carbohydrates and their conjugates at the molecular level has led to the formation of an area of active research in bioorganic chemistry now called glycobiology.

Model systems to study cell surface carbohydrates and biological adhesion have led to unique insights into cell adhesion molecules, antibody–antigen interaction, viral adhesion, baterial adhesion, entry of toxins into cells, cell differentiation, cancer metastasis, cell growth, and fertilization. An important goal of bioorganic chemistry and glycobiology is to gain a clear understanding of these processes in order to exploit them for therapeutic applications. This type of understanding is only possible through the development of structurally well defined molecules and model systems that can be used to study these processes. Unlike enzymes that usually operate on soluble substrates and individual molecules, the carbohydrate-based adhesion processes operate on surfaces that are multivalent in nature. This makes the determination of the kinetics and thermodynamics of interaction difficult because they cannot be calculated using a pure protein and soluble substrate.

The task of developing model systems to study carbohydrate adhesion depends critically on the use of methods in organic chemistry that can handle macromolecular assemblies. These model systems have to be synthesized, purified, characterized and tested for their biological activity. This type of science has recently been the focus of research in organic chemistry as the targets for synthesis have gone from small molecular species with well defined structures, to macromolecules, particles and materials. This chapter will focus on the construction of macromolecular systems that are used to study biological adhesion mediated primarily by carbohydrate-binding interactions.

We have focused on three model systems used to study carbohydrate-mediated biological adhesion: polymers, liposomes and films. The common ground for these systems is that they all try to deal with the concept of polyvalency of ligands that are interacting with receptors or enzymes at an interface. The effect of multivalency on affinity (avidity) is particularly important when considering cell surface carbohydrate binding interactions. For example, a single monovalent binding event between a cell surface receptor protein, such as a lectin, and its carbohydrate ligand is typically weak with binding constants in the low mM or high μM range. The force of adhesion between two complementary multivalent surfaces, however, is greatly reinforced through a multiplicative effect. The resulting strength of adhesion becomes considerably larger than the sum of all the single binding events.

The most difficult problem in using glycoconjugate materials to study biological adhesion lies in the construction and detailed structural analysis of these materials. Polymers have the advantage that relatively straightforward spectroscopic and chromatographic techniques can be used in their analysis. They have the disadvantage that alone, as soluble species, they are poor mimics of cell surfaces. Liposomes represent the best mimic of the cell surface but their analysis is very difficult. Films represent a compromise, as they serve as a well defined model surface for a liposome that can be probed structurally by modern surface science analytical techniques.

Glycopolymers

Of the various types of neoglycoconjugates (i.e., synthetic or semisynthetic materials containing attached carbohydrate residues) developed for studying cell surface carbohydrate receptors, a class of conjugates generally referred to as glycopolymers has attracted much interest in recent years. An excellent review dedicated to neoglycoconjugates has recently been published (Lee & Lee, 1994b). The term "glycopolymer" (Roy et al., 1992b) has been introduced as a replacement for "pseudopolysaccharide" (Kraska & Mester, 1978); it is analogous to the general names, glycoprotein and glycolipid, that are common terms in the literature. The term glycopolymer is used to describe multivalent, synthetic water soluble polymers bearing covalently bound carbohydrate residues. It should also be equally applicable to insoluble crosslinked polymer systems such as those used in affinity chromatography. This section focuses mainly on water soluble glycopolymer systems and describes existing strategies for their synthesis, their use in studying carbohydrate adhesion, and their advantages and disadvantages over other systems.

In contrast to glycopolymers, neoglycoproteins (semisynthetic glycoproteins) have long served biochemists and immunochemists as tools to study complementary carbohydrate–receptor binding interactions. Neoglycoproteins are synthesized by the random coupling of carbohydrate molecules to a protein, or by the modification or partial degradation of glycoproteins. Many approaches to their preparation exist and have been described in detail elsewhere (Kennedy, 1988). For example, reducing sugars can be attached by reductive amination to ε-amino groups of lysine residues on the protein, or natural glycoproteins can be desialylated by treatment with neuraminidase (Fig. 9-1). As models for studying the fundamental aspects of adhesion, neoglycoproteins are not without certain drawbacks. The very nature of the protein carrier, which is typically comprised of 20 common different amino acids arranged in various sequences, makes the absolute structural characterization of neoglycoproteins very difficult. Neoglyco-proteins are also sensitive and biodegradable compounds often having only relatively short shelf lives and, therefore, must be handled with care. In addition, sugar densities on neoglycoproteins can vary from batch to batch and are not easily controlled. In addition, carbohydrate loading densities are limited to the number of amino acids bearing the required functionality for the carbohydrate conjugation reaction. They can also be relatively costly or difficult to prepare, handle and purify on large scale. One other very important consideration is that neoglycoproteins can give rise to non-carbohydrate-specific interactions directed at the protein carrier which can interfere with investigations during immunochemical or serological assays. As an alternative, synthetic glycopolymers can alleviate many of these drawbacks.

Glycopolymers offer many advantages over neoglycoproteins. They are cheap and easy to prepare in either small or large quantities; milligram to multigram preparations have been achieved. Their purification and isolation is usually done by simple dialysis in water followed by lyophilization or by precipitation from the aqueous reaction solution with organic solvents. They are much more resistant to degradation than their neoglycoprotein counterparts, thermally, chemically and biologically. For example, lyophilized copolyacrylamide glycopolymers have been shown to be stable without any loss of binding activity after being stored at room temperature for many years. Their simple chemical structures allow for easier and more accurate composition and structural characterization. The molecular weights of glycopolymers can be controlled as needed and determined by standard methods such as gel permeation chromatography, viscometry or low angle laser light scattering. Carbohydrate densities are easily determined by NMR, colorimetric chemical assays or elemental analysis. The artificial and hydrophilic nature of the carrier backbone (usually polyacrylamide) makes nonspecific binding interaction between it and the receptor proteins of interest, or other proteins that may be present, essentially undetectable. This is a quality exploited in acrylamide gel electrophoresis techniques. Unlike neoglycoproteins, the artificial nature of the acrylamide backbone also does not interfere with quantitative immunochemical assays such

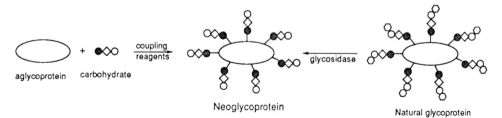

Figure 9–1. Neoglycoprotein synthesis by carbohydrate conjugation to aglycoproteins or by enzymatic degradation of natural glycoproteins.

as immunoprecipitation that depend on colorimetric protein assays to measure protein content in ligand–receptor complexes. This renders such assays directly and unambiguously quantitative. Multivalent glycopolymer antigens also appear to be much more sensitive and give less background interference during serological screening assays. For example, subnanogram quantities of monosaccharide-derived glycopolymers have been used effectively in solid phase enzyme linked immunoassays with lectins. Their quick and efficient synthesis is versatile in the sense that the incorporation of carbohydrate ligands and molecular weights of the resulting conjugates can be controlled in a reproducible manner. In addition, the use of different co-monomers or the addition of extra co-monomers can give glycopolymers having enhanced solution or absorptive properties. Glycopolymers having multiple specificities and utilities can also be synthesized by incorporating controlled amounts of polymerizable spectroscopic, immuno- or radiochemical probes. They do have the disadvantage that alone, as soluble species, they are poor mimics of cell surfaces. However, they can represent good mimics of glycosylated cell surfaces when they are adsorbed onto solid surfaces such as during solid phase enzyme-linked immunosorbant (ELISA) assays, or when they are synthesized as insoluble crosslinked polymer supports or coatings.

Synthetic Strategies

Two types of strategies exist for the synthesis of glycopolymers, as depicted in Figure 9-2. Pendant carbohydrate residues can be introduced onto a polymer by either copolymerization of suitably functionalized glycosides with complementary monomers or by grafting carbohydrates to synthetic polymer chains. Both strategies have been explored for preparing water soluble conjugates and insoluble crosslinked hydrogels. The radical copolymerization of vinylic monomers in water is more commonly used for the preparation of water soluble antigenic glycopolymers.

Copolymerization strategies have involved mainly the radical copolymerization of alkene-terminated glycosides with acrylate-type monomers. Acrylamide has been the comonomer of choice for this strategy. The choice of acrylamide as comonomer is well

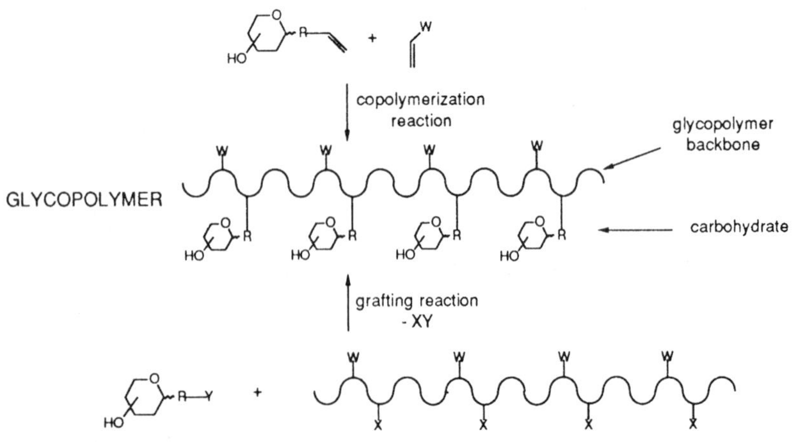

Figure 9–2. Synthetic strategies used for preparing glycopolymers.

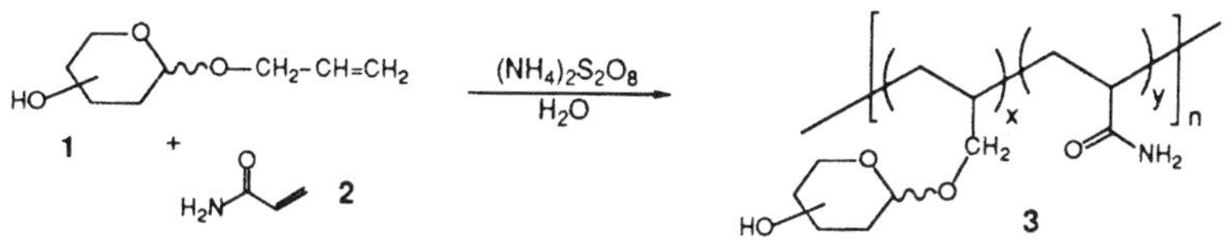

Figure 9–3. Allyl glycoside–coacrylamide glycopolymer synthesis.

suited for glycopolymer synthesis. The resulting polymers are stable, nonionic and extremely soluble in water even at high electrolyte concentrations. In addition, because polyacrylamide gels are used extensively by biochemists and immunochemists for electrophoresis work, experience, as well as the materials required for preparing such glycopolymers, is already present in laboratories where they are probably most often required. It is worth mentioning that acrylates and acrylamide monomers are very toxic and care should be exercised when working with them.

The copolymerizations are usually carried out in deoxygenated water using deprotected glycosides with ammonium persulfate ($[NH_4]_2S_2O_8$) as the initiator, in combination with heat or N,N,N',N'-tetramethylethylenediamine (TEMED) at room temperature to generate the radical species. Hydrogen peroxide or a variety of other aqueous initiator systems can also be used to initiate polymerization. The addition of a small amount of crosslinking agent, such as N, N'-methylenebisacrylamide, to the polymerization reaction, results in the formation of three-dimensionally crosslinked gels, useful in affinity purifications or as cell culture surfaces.

Glycopolymers from Allyl Glycosides

More than 15 years have passed since Horejsi, Smoleck, and Kocourek first described their approach of making soluble, artificial, multivalent carbohydrate antigens for studying lectin binding (Horejsi et al., 1978). In this approach, allyl glycosides (**1**) were copolymerized with acrylamide (**2**) using a radical initiator in water to yield glycopolymers (**3**) of various monosaccharides and a disaccharide (lactose) (Fig. 9-3). This general method continues to be popular with many researchers as allyl glycosides are often employed in block oligosaccharide syntheses to protect the reducing terminus. The resulting multivalent carbohydrate antigens have proven effective in studying both lectins and antibodies directed to the carbohydrate determinants by standard methods such as double radial diffusion/precipitation assays in agarose gels, quantitative agglutination or immunoprecipitation assays, and solid phase enzyme immunoassays (EIA/ELISA).

Molecular dynamics studies of soluble acrylamide/allyl D-glucosaminide glycopolymers have been performed to facilitate a clearer understanding of the solution behavior of glycopolymers (Roy et al., 1993b). ^{13}C NMR relaxation data, viscometry and molecular mechanics calculations have shown that these neoglycoconjugates behave as dynamic random coils in aqueous solution and that the pendant carbohydrate units retain a good range of motional freedom. In this case the β-anomerically linked residues are more mobile than α-linked residues. This behavior is different than typical neoglycoprotein conjugates that exist mainly as tight globular species. One question that still needs to be addressed, however, concerns the behavior of glycopolymers when they are adsorbed onto solid surfaces such as the coating antigens used in ELISA type immunoassays. It is unclear if the glycopolymers remain as coiled entities or if they become fully or even partially uncoiled and extended across the solid surface, thus exposing the carbohydrate residues to the aqueous environment.

Although popular, the preparation of glycopolymers by the copolymerization of allyl glycosides with acrylamide has serious drawbacks. Because the polymers are formed from two distinctly different types of alkenes that have different reactivities, this leads to the formation of a less desirable "block" polymer where carbohydrate monomers are concentrated in specific areas rather than being randomly distributed along the polymer backbone. In addition, glycoside incorporation does not directly reflect the ratio of monomers present prior to polymerization. This makes the carbohydrate density in the resulting glycopolymer difficult to predict and control. Also, only a short methylene bridge exists as a spacer between the polymer backbone and the carbohydrate ligands. Glycopolymers with elongated spacers have proven more effective in lectin and antibody binding studies when compared with their allyl glycoside-derived counterparts (Gamian et al., 1991).

Apart from allyl glycosides, relatively few unconjugated alkene-based glycoside monomers have been used to prepare water soluble antigenic glycopolymers. Coacrylamide glycopolymers prepared with simple alkenyl-type glycosides, such as n-pentenyl or n-undecenyl glycosides, have provided improved spacer lengths but also result in block copolymers where glycoside incorporation can be difficult to predict and control (Nishimura et al. 1990). There have also been attempts at preparing polystyrene glycopolymer systems, some as early as 1952 (Helferich & Hoffman, 1952).

Polystyrene-based glycopolymers prepared from vinylphenyl glycosides and vinylbenzyl glyconamides have provided useful antigenic materials but have not been commonly adopted. Kobayashi and co-workers have studied such polystyrene glycopolymer systems to immobilize and grow hepatocyte cells in an attempt to develop a substrate on which an artificial liver could be constructed (Kobayashi et al., 1994). Their success is based on targeting the asialyl glycoprotein receptor on the hepatocyte cell surface using galactose-functionalized glycopolymers. In this case, the glycopolymers were prepared as homopolymers and adopted a tightly coiled structure in water due to the amphiphilic nature of the polymer, consisting of water-soluble carbohydrate structures attached to a very hydrophobic polystyrene backbone. While the carbohydrate residues provide points of adhesion for the cell receptors, the hydrophobic backbone provides good adsorption of the glycopolymer to hydrophobic surfaces of ELISA plates and latex microparticles. One advantage to the polyvinylbenzyl glyconamide system is that the glycopolymers can be prepared from unprotected carbohydrates in high yields with relative ease. The method does suffer from the fact that homopolymers must be synthesized to achieve acceptable solubility in water. In addition, the reducing terminus of carbohydrate structure is sacrificed to provide an acyclic spacer (Fig. 9-4).

Improvements to Glycomonomer and Glycopolymer Preparation

Improvements in the original coacrylamide glycopolymer synthesis strategy have included increasing the aglycone length of glycosides and changing the polymerizable function on the glycoside. Increasing the aglycone length provides a better spacer between the antigenic determinant and the polymer backbone, thus providing a decrease in steric hindrance during receptor binding. By changing the polymerizable group on the glycosides from simple alkenes to more reactive N-substituted acrylamides, a more

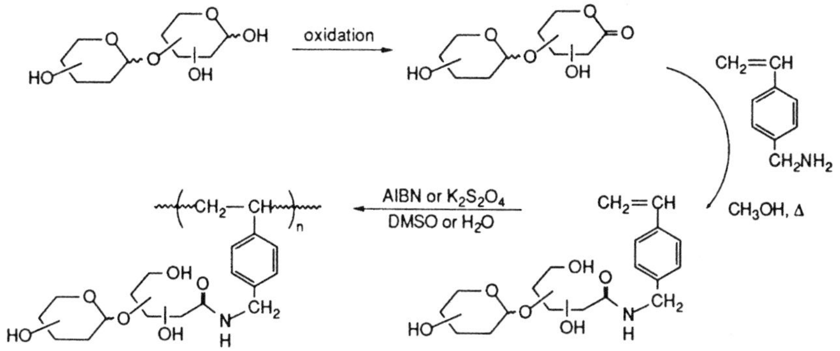

Figure 9–4. Synthesis of polystyrene-based glyconamide glycopolymers.

desirable "randomly alternating" copolymer (**6**) can be produced with acrylamide. An additional benefit from the latter change is that the incorporated monomer ratio within the glycopolymer becomes identical to the ratio of monomers added to the reaction vessel. The preparation of acrylamide-type glycomonomers called *N*-glycosidylacrylamides (**5**) also offers the advantage that these materials may be used for the preparation of neoglycoproteins by conjugate addition (Romanowska et al., 1994; Roy et al., 1991). Since the carrier is now the only variable, a strategy is created by which antibodies directed solely at the carbohydrate hapten can be identified. For example, after eliciting antibodies in vivo against the neoglycoprotein **7**, only those binding to carbohydrate determinants will bind to glycopolymer **6** (Fig. 9-5). This approach allows

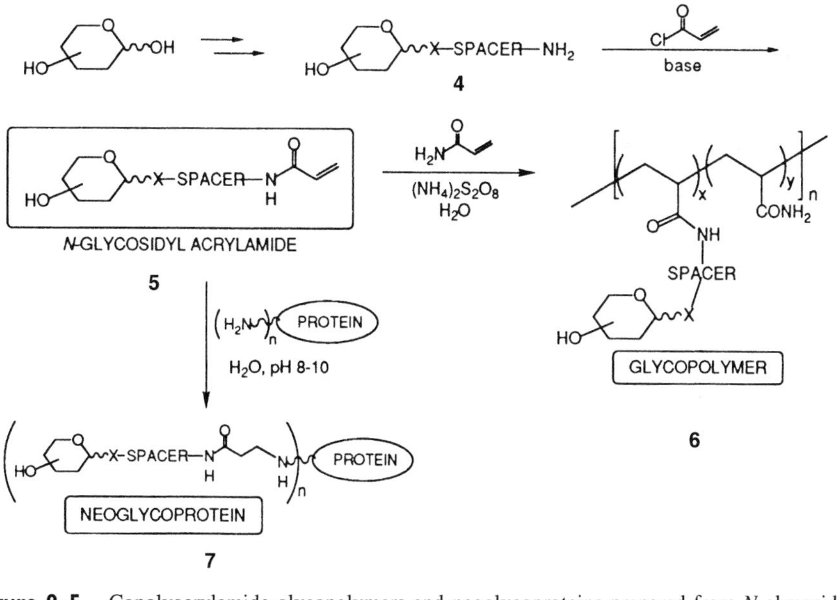

Figure 9–5. Copolyacrylamide glycopolymers and neoglycoproteins prepared from *N*-glycosidyl-substituted acrylamides.

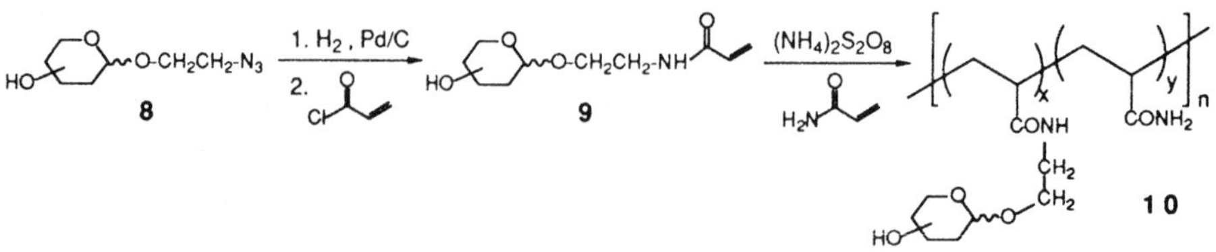

Figure 9–6. Synthesis of coacrylamide glycopolymers from 2-azidoethyl glycosides (8).

one to obtain antibodies that are specific for the carbohydrate groups and not against peptide fragments. A brief summary of some methods currently used to prepare such N-glycosidylacrylamide glycomonomers follows.

N-glycosidylacrylamides have been prepared in a variety of fashions by different research groups resulting in aglycon lengths of varying sizes and complexities. In almost all cases, an amino-terminated glycoside **4** is synthesized and acryloylated with acryloyl chloride in a final step. Alternatively, glycosylation of an acrylamide-terminated alcohol or thiol followed by deprotection of the carbohydrate can be used. However, the latter approach often leads to extensive degradation or unwanted early polymerization of the valuable glycoside. Some researchers have adopted the use of 2-acrylamidoethylglycosides (**9**) to prepare a large number of coacrylamide glycopolymers (**10**). These materials are important models of bacterial capsular antigen repeating units used to identify and study the specificity of antibodies (Chernyak et al., 1991). In this strategy, 2-azidoethyl glycosides (**8**) are synthesized, the amine is then liberated by hydrogenation and acryloylated, then finally copolymerized (Fig. 9-6).

Lee and co-workers have used 6-acrylamidohexyl glycosides (similar to **9**) and a variety of 2-([N-acrylamidoalkyl]aminocarbonyl)ethyl thioglycosides (**13**) in their efforts to provide glycopolymeric gels for studying animal hepatocyte binding (Schnaar et al., 1978). In addition, these researchers have also provided an interesting method for preparing thioglycoside monomers. Their method involves the conjugate addition of 1-thio sugars (**11**) to symmetric N,N'-alkyl bisacrylamides (**12**) (Fig. 9-7) (Lee & Lee, 1982). Thioglycoside neoglycoconjugates are more resistant to enzymatic hydrolysis and thus represent an important type of useful glycomaterial.

Alternatively, Roy and co-workers have adopted two different strategies resulting in either thioalkyl or aromatic spacers. Each strategy takes advantage of commonly encountered substrates, allyl and nitrophenyl glycosides. Allyl glycosides (**1**) are effectively converted to polymerizable 3-(2-(2-propenamido)ethylthio)propyl glycosides (**14**) by the aqueous, light-catalyzed addition of 2-aminoethanethiol followed by N-acryloylation and copolymerization (Fig. 9-8) (Roy & Tropper, 1988). This technique has since been applied by others such as Paulsen and Höffgen to prepare antigenic glycopolymers of the gram negative bacteria lipid A core region trisaccharide unit L-α-

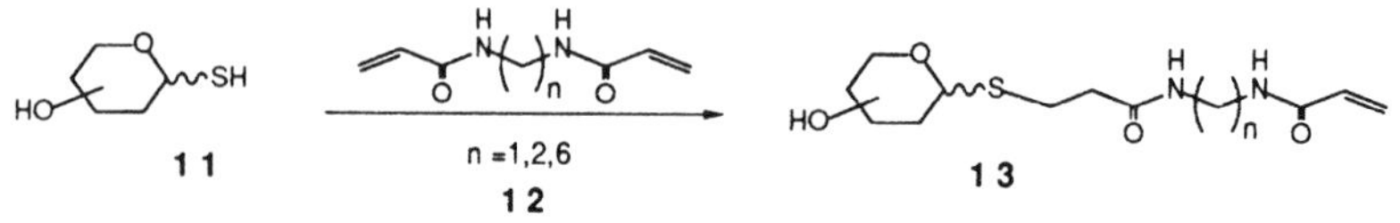

Figure 9–7. Synthesis of N-thioglycosidyl acrylamide glycomonomers from thio sugars.

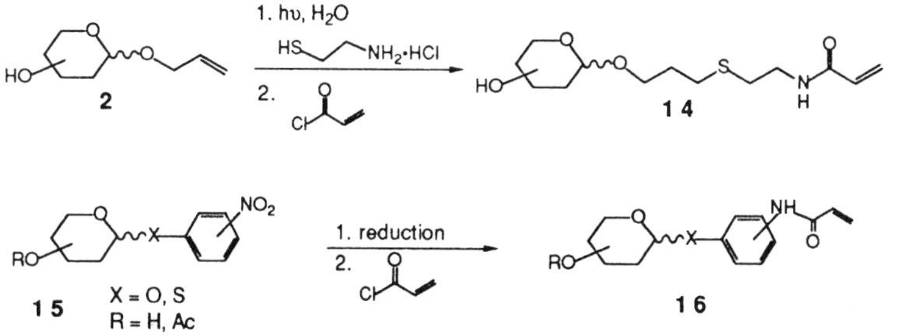

Figure 9–8. Synthesis of *N*-glycosidyl acrylamide glycomonomers from allyl and nitrophenyl glycosides.

D-Hep(1-3)-L-α-D-Hep(1-5)-α-Kdo (Paulsen & Höffgen, 1991). Conversely, propenamidophenyl glycosides (**16**) can be prepared quickly and efficiently from nitrophenyl glycosides (**15**) that are commonly used as glycosidase substrates or as precursors to aminophenyl glycosides (Roy et al., 1992b). These aminophenyl glycosides are also commonly used to prepare neoglycoproteins via diazotization or isothiocyanate formation. This method of preparing polymerizable glycomonmers is attractive in view of the fact that many nitrophenyl glycosides are already commercially available and it only requires the simple reduction of the nitro group to the corresponding amine followed by *N*-acryloylation. Glycopolymers derived from these monomers have been used to study both lectins and antibodies. In addition, some sialic acid glycopolymers prepared from these types of monomers were demonstrated to be potent inhibitors of the influenza virus (Roy et al., 1992a). The influenza virus requires a sialylated cell surface for attachment prior to infection. Other sialic acid coacrylamide glycopolymers have been prepared by other groups (Byramova et al., 1991; Matrosovich et al., 1990; Sigal et al., 1996; Sparks et al., 1993); poly-L-lysine-based sialic acid glycopolymers have also been shown to inhibit viral adhesion (Roy et al., 1993a).

The use of a terminal nitro- or aminophenyl group as precursor to a polymerizable acrylamide function has also been exploited by Andersson and co-workers (Andersson & Oscarson, 1992). In this case, an extended O–N linked aglycon was formed by a ring opening amidation reaction performed on oxysuccinimide glycosides (**17**) with 2-(4-nitro- or aminophenyl)ethylamines. The final glycosides were then converted to their corresponding acrylamides (**18**), the latter of which were then copolymerized with acrylamide (Fig. 9-9). These glycopolymers were used to study the previously unexplored chemical stability of rare O–N linked glycosides.

Methods have also been described for the derivatization of reducing oligosaccharides into glycosylamines or aminoalditols which can then be acryloylated and copoly-

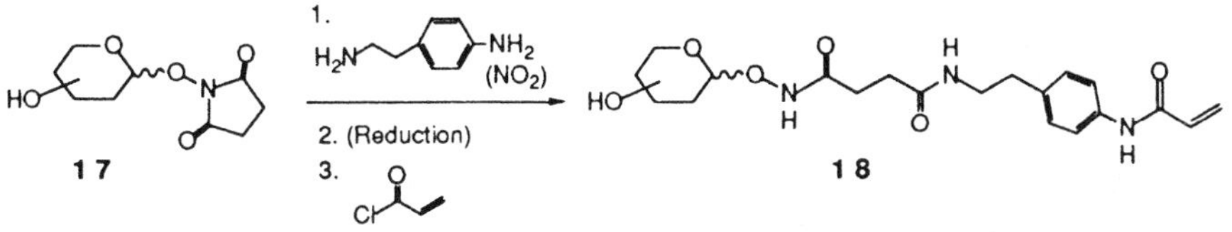

Figure 9–9. Synthesis of *N*-glycosidyl acrylamide glycomonomers from oxysuccinimide glycosides.

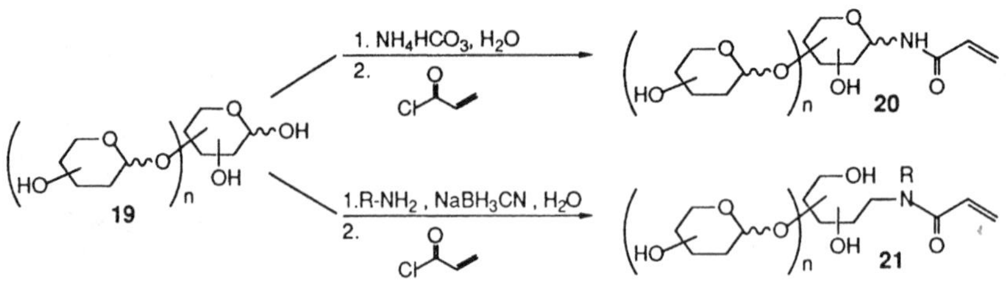

Figure 9–10. Synthesis of N-glycosidyl acrylamide glycomonomers from glycosylamines or aminoalditols.

merized with acrylamide (Fig. 9-10) (Kallin, 1994). Glycosylamines can be prepared by reacting ammonia or amines with reducing sugars (**19**) in various solvent systems. A particularly useful strategy involves reacting a reducing sugar with ammonium carbonate in water, a solvent in which most carbohydrates are soluble. After acryloylation, an N-glycosylacrylamide monomer (**20**) is obtained. To prepare aminoalditols, methods of reductive amination are used. This typically involves mixing an amine and reducing sugar in water in the presence of a reducing agent such as NaCNBH$_3$. After acryloylation, an acrylamidoalditol monomer (**21**) is obtained. Both methods have the advantage of allowing the preparation of glycopolymers from complex oligosaccharides which might be isolated in small quantities from natural sources, without any need for intricate protective derivatizations and glycosidation steps. In the case of acryloylating aminoalditols, this strategy suffers (or benefits) from the fact that the structural integrity of the reducing saccharide unit is lost by being converted to an acyclic spacer. The structural loss of the cyclic reducing sugar unit, however, does not present a problem if one is dealing with an extended and repetitive oligosaccharide sequence.

Terpolymeric Systems

Further improvements in glycopolymer systems have been attempted by incorporating various effectors into the the the final conjugate by terpolymerization (the copolymerization of three comonomers). A variety of interesting terpolymeric glycopolymers have been synthesized and studied in the hope of imparting enhanced solid phase adsorption properties for use as coating antigens in ELISA-type assays. The incorporation of useful spectroscopic, immuno- and radiochemical probes in the glycopolymers has also been achieved by terpolymerization. In addition, glycopolymers bearing two different carbohydrates have been engineered to be specific toward different lectins or antibodies. For example, Roy and co-workers have attached biotin to poly-L-lysine glycopolymers of colominic acid (**23**) (Roy et al., 1993a) and to copolyacrylamide glycopolymers of β-lactose, β-N-acetylglucosamine and lacto-N-teraose (**22**) (Fig. 9-11) (Roy et al., 1992c; Tropper et al., 1994). These synthetic antigens were shown to be specific towards both avidin (or streptavidin) and their complementary lectins or antibodies. A bispecific antigen (**24**) was also created by copolymerizing a combination of α-L-rhamnosyl and β-D-N-acetylglucosamine acrylamide glycomonomers with acrylamide. A spectroscopically labeled glycopolymer was prepared by grafting fluorescein isothiocyanate (FITC) to free amine groups on a terpolymer made of acrylamide, allylamine and allyl β-D-galactoside (Ticha & Kocourek, 1991). Also, a glycoterpolymer capable of being radiolabelled by iodination was prepared by acryloylating 2-(4-hydroxyphenyl)ethylamine (tyrosinamine) and copolymerizing with acrylamide and 3-(2-[2-propenamido]ethylthio)propyl-β-D-N-acetylglucosamine (**27**) (Fig. 9-12).

In an attempt to prepare glycopolymers having improved adherence to polystyrene microtiter plate surfaces, a glycoterpolymer having pendant strearyl chains (**28**) was synthesized. This material was then compared to simple coacrylamide (**25**) and comethacrylamide (**26**) glycopolymers, as well as a tyrosinylated glycoterpolymer (**27**). All of these glycopolymers had identical sugar contents and were assayed by a solid phase enzyme-linked lectin assay. The results indicated that adding hydrophobic groups to a glycopolymer does increase its effectiveness as a coating antigen in ELISA type

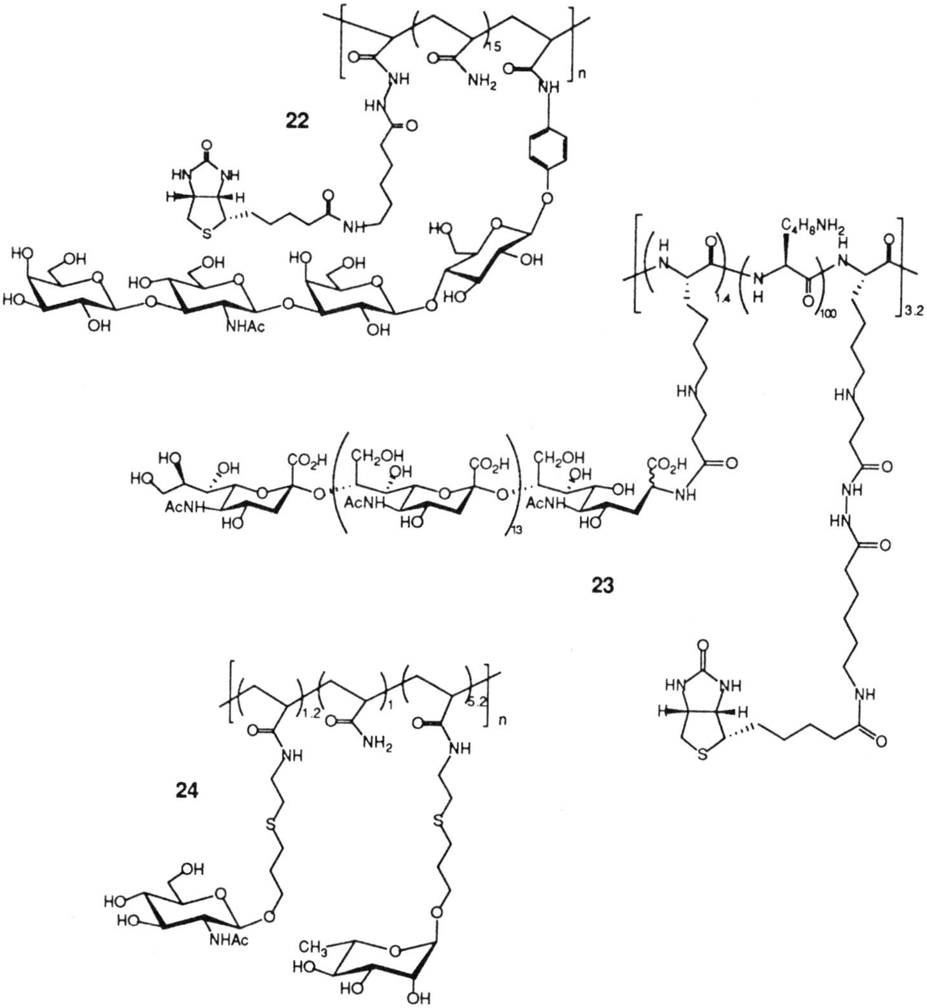

Figure 9–11. Examples of different glycoterpolymers.

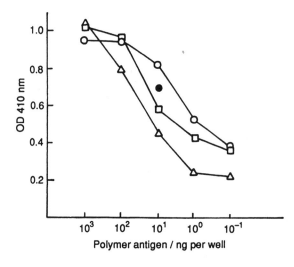

	R¹	R²	X
25	H	H	0
26	CH3	H	0
27	H	$CH_2CH_2C_6H_4OH$	0.1
28	H	$n\text{-}C_{17}H_{36}$	0.025

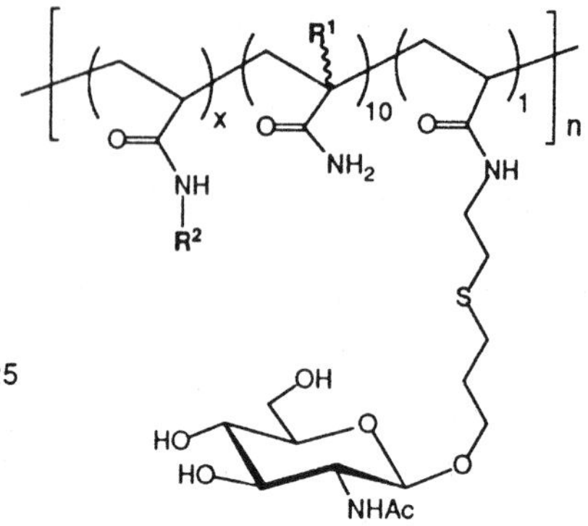

Figure 9–12. Glycopolymers of varying hydrophobic character.

assays when performed in polystyrene microtiter plates. The strearylated glycopolymer (**28**) outperformed the methacrylamide glycopolymer (**26**), which in turn was superior to the tyrosinylated glycopolymer (**27**) that surpassed the more traditional copolyacrylamide glycopolymer (**25**) (Fig. 9-13).

Dendritic Glycooligomers

Apart from the coacrylamide types of glycopolymers described above, a new type of polymeric, or rather oligomeric, glycopolymer has been synthesized. Roy and coworkers have recently developed a strategy for preparing dendritic glycopeptides by solid phase synthesis (Roy et al., 1993c). After attachment and deprotection of the first lysine unit on a solid phase support, additional lysine units are attached to both amine groups of the previous lysine unit and the process repeated a total of n times to give a hyperbranched fractal structure having 2^n terminal amine groups after final deprotection. Once the desired valency is achieved, the terminal amines are capped with *N*-

Figure 9–13. Enzyme-linked lectin assay (ELLA) of the terpolymers **25–28** when used as coating antigens with horseradish peroxidase-labeled wheat germ agglutinin ($1\,\mu g/100\,\mu L$) and 2,2′-azinobis (3-ethylbenzothiazoline-6-sulfonic acid)/H_2O_2 as enzyme substrate; **25** (△), **26** (●, one point), **27** (□), **28** (○).

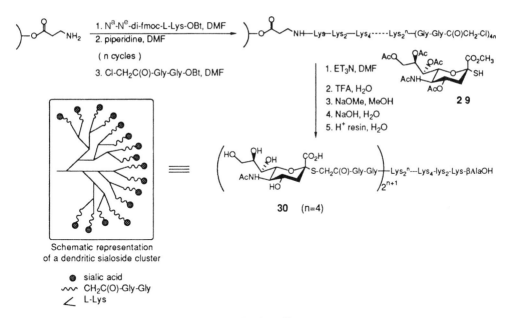

Schematic representation
of a dendritic sialoside cluster

⊗ sialic acid
⌇ CH₂C(O)-Gly-Gly
∠ L-Lys

Figure 9–14. Synthesis of a dendritic sialoside glycooligomer.

chloroacetylglycylglycine which acts as a spacer to which carbohydrates can be attached. For example, 2-thiosialic acid **29** was attached to the dendritic structure by nucleophilic displacement of chloride. The dendrimers with 2^{n+1} valency were then cleaved from the resin and deprotected to give the glycooligomer **30**. The hexadecameric sialic acid conjugate **30** contains thioglycoside linkages and is thus resistant to neuraminidase (an enzyme that cleaves sialic acid O-glycosides). These materials were used effectively as multivalent inhibitors of influenza virus or as coating antigens in solid phase enzyme-linked assays (Fig. 9-14).

These materials are conceptually similar to glycoside clusters pioneered by Lee and co-workers. Glycoside clusters proved very useful in evaluating the optimal intersugar distances for binding to hepatic or other lectins. The effect of clustering carbohydrate ligands was shown to increase binding affinity dramatically in some cases, while diminishing it in others (Lee & Lee, 1994a). Similarly, Hakomori and co-workers have also developed L-lysyl-L-lysine-based triclusters of β-D-lactosides and investigated their effectiveness as tumor metasthesis inhibitors (Dean et al., 1993). In this case, coinjection of monomeric lactoside clusters and clusters conjugated to a poly-L-lysine carrier, with metastatic B16 murine melanoma cells, inhibited the formation of colonies in mice. The length of the cluster spacer arms proved to be important for inhibitory activity.

Crosslinked Glycopolymers

The first glycopolymeric gels for use in affinity chromatography (Horejsi & Kocourek, 1974; Horejsi et al., 1978) and affinity gel electrophoresis (Horejsi et al., 1977a,b) were made by copolymerizing allyl glycosides and acrylamide in the presence of N,N'-methylenebisacrylamide. The affinity electrophoresis gels were found to be effective in determining the binding constants of a wide range of plant and animal lectins. Glycopolymeric gels having elongated spacers between the carbohydrates and the gel matrix have been made by copolymerization and grafting approaches (Schnaar et al., 1978, 1982; Schnaar & Lee, 1975; Weigel et al., 1978a,b; Weisz & Schnaar, 1991). These gels provided materials for cell affinity columns and culture surfaces for the study of whole cell animal hepatocyte adhesion mediated by cell surface galactose and N-acetylgalactosamine lectins. The affinity gels were prepared by the addition of 6-aminohexyl- and similar glycosides to N-hydroxysuccinimide ester groups in acrylamide-N-succinimidyl acrylate copolymers gels (Fig. 9-15). This grafting approach has the drawback that the N-hydroxysuccinimide esters are susceptible to hydrolysis prior to amide formation, thereby forming an ionic matrix. Alternatively, the terminal amines of O- and S- glycosides were converted to N-substituted glycosidyl acrylamides and then copolymerized directly with acrylamide and N,N'-methylenebisacrylamide to provide essentially identical glycopolymeric gels.

Grafting carbohydrates to synthetic polymer chains is a versatile method where a variety of existing conjugation chemistries can be utilized depending on the polymer backbone chosen. Additional approaches to provide soluble glycopolymers by this strategy include the conversion of commercial polyacrylic acid to substituted polyacrylamide by 1-(3-dimethylaminopropyl)-3-ethylcarbodiimide hydrochloride (EDC)-mediated amidation with glycosides having terminal amine groups and ammonium chloride to complete the capping. Other methods include, for example, Michael addition of acrylamide terminated glycosides to poly-L-lysine (Roy et al., 1993a), the reaction of epoxypropyl galactosides with polyvinyl alcohol (Kraska & Mester, 1978), and the formation of malto-oligosaccharide hydrazones with polyacrylic acid hydrazide (Andresz et al., 1978).

Homogeneous and structurally well-defined carbohydrate antigens are required to study protein–carbohydrate binding interactions properly. The fundamental principles associated with the multivalent nature of these interactions and associated cell surfaces can more accurately be addressed with structurally well defined systems. Glycopolymers such as those presented here constitute such materials. Aside from their stability and homogeneous structure, a great advantage to glycopolymer antigens is the ability to prepare them in a controllable manner. In essence, they can be tailor made to suit any requirement. Glycopolymers can be adapted to many uses. They can be useful in

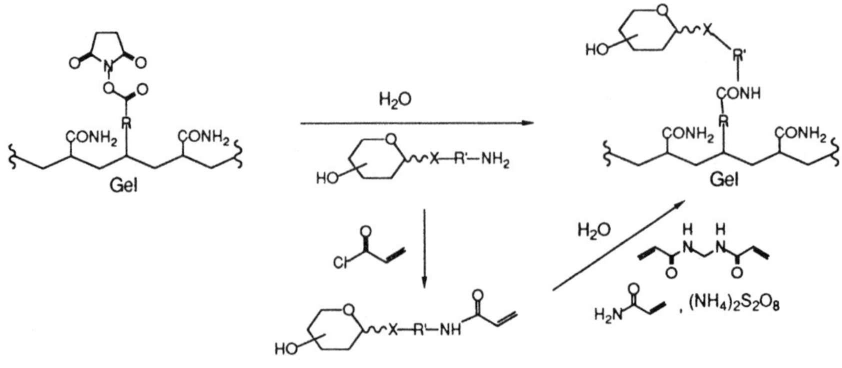

Figure 9–15. General synthetic strategies for crosslinked glycopolymer gels.

diagnostic or purification applications as well as in therapeutic applications as multi-valent inhibitors. Glycopolymers will no doubt continue to fulfill an important role in the study of carbohydrate receptors in years to come.

Langmuir Films

This section discusses two model cell surface membrane systems: films and liposomes. These model membrane systems allow the study of biological membrane phenomena without the difficulties found in natural cellular models. Results from these model systems can then be compared with the behavior of cells in biological assays in order to evaluate and refine the model. The lipid molecules needed for the formation of model membrane systems can be prepared and the surface functionality of the liposome or monolayer can be controlled and analyzed for its biological and chemical behavior. In addition, unnatural molecules can be constructed to be both chemically and enzymatically stable in order to evaluate these systems *in vivo*. This section will specifically explore ways of mimicking biological cell surface carbohydrate interactions on monolayers and liposomes. We will first discuss how liposomes and monolayers are constructed and characterized. Investigations of the binding of pathogens and cells to these surfaces will next be presented. Finally, we will give some of the most recent uses of these systems for the study of biological phenomenon at these well characterized interfaces.

Self-Assembly

The process of self-assembly necessary for the formation of liposomes and monolayers is driven by the hydrophobic effect and its influence on amphiphilic molecules (Tanford, 1980). An amphiphilic molecule is one that has both hydrophobic and hydrophilic parts. The hydrophobic constituent is usually a long chain hydrocarbon group known as the tail. The hydrophilic part is usually a charged or polar group and is referred to as the head group. A typical amphiphile contains one or two hydrocarbon tails that may be branched or possess an unsaturated or aromatic functional group. When placed in water the tail group of the amphiphile aggregates to exclude water and form vesicles that have exposed hydrophilic head groups. This ordering or self-assembly process is due to an entropic effect. The increase in entropy is caused by the exclusion of water when the hydrocarbon tails aggregate, and is the driving force in free energy for the process to occur spontaneously. The two best studied systems that self-assemble are monolayers formed at the air–water interface and liposomes generated in aqueous solution.

Monolayers at the Air–Water Interface

The study of self-assembled monomolecular films at the air–water interface was first described by Pockels over 100 years ago (Gaines, 1966; Pockels, 1893). In 1917, Langmuir laid the theoretical groundwork for the interpretation of Pockels' initial experiments. This analysis has survived through the intervening decades and is the basis for the modern interpetation of monolayer properties (Langmuir, 1939). Langmuir found that amphiphilic molecules become ordered at the air–water interface with their polar head groups in the water surface and their hydrophobic tail chains directed vertically away from the water surface. The films were also described as being only one molecule thick, thus forming a monolayer. It has been shown that the cross-sectional area occupied by one amphiphile (e.g., fatty acid) in a monolayer is approximately 20 Å^2. This value agrees with the van der Waals cross sectional area of an alkyl chain when viewed down the long axis of the molecule. These films have become known as Langmuir films and their study continues today using equipment and techniques similar to those first described by these researchers. Figure 9-16 depicts a modern Langmuir trough. The modern Langmuir trough not only provides a dependable means to form monomolecular films but offers the means to measure basic properties of the film.

Organized molecular assemblies can be created by the Langmuir–Blodgett (LB) technique, first reported by Katherine Blodgett in 1935. Blodgett further advanced the Langmuir technique by refining the analysis of films and describing methods to transfer the films from the air–water interface to other surfaces. Thus, these materials are now referred to as Langmuir–Blodgett (LB) films. LB films are formed by spreading a solution of the amphiphiles onto a clean water surface contained in a Langmuir trough (Fig. 9-16). Upon evaporation of the spreading solvent (usually chloroform or hexane), the insoluble film on the water surface can be compressed by a moveable barrier. The position of the barrier determines the surface concentration (molecules/area) of the lipid, that generally determines the molecular phase. Depending on the chemical nature of the amphiphile, it is possible to observe gaseous, liquid, or solid phases of the monolayer. Using a surface balance, simultaneous recording of the surface pressure during compression results in the surface–pressure-area isotherm two-dimensional [2D] analogues of pressure-volume isotherms. The monolayer phases can generally be inferred from the slopes of the surface–pressure-area isotherm. In the condensed or oriented phase the hydrophilic head group is associated with the water, and the hydrophobic tail is oriented towards the gas (typically air). These structures, self-organized in two dimensions, formally correspond to one half of the lipid bilayer of a cell membrane.

Dipping of a substrate perpendicularly through the condensed film transfers the monolayer from the air–water interface to the solid–air interface. This allows the film to be transfered to a solid support in order to analyze its structure. Multiple dips through the film result in buildup of molecular assemblies. Depending on whether deposition occurs on the downstroke (hydrophobic substrate), or upstroke (hydrophilic substrate), different molecular orientations are possible. The degree of organization depicted in the cartoons of the floating, compressed monolayer on water, and the supported multilayer on the substrate is confirmed by numerous film characterization studies that include X-ray and electron diffraction, scanning and transmission electron microscopy, scanning tunneling and atomic force microscopy, ellipsometry, polarized infrared spectroscopy, and X-ray photoelectron spectroscopy. These techniques are discussed in a later section. The main results from such characterization is that LB films generally possess a highly crystalline lamellar structure, where crystal domains are on the order of 0.1–100 μm. The alkyl chains can vary from being either slightly tilted or perpendicular to the plane of the substrate.

Polymerizable amphiphilic monomers that are in the condensed phase at the air–water interface can undergo a solid state reaction following exposure to radiation (e.g. UV, gamma, electron beam) or chemical initiating agents to yield a polymerized LB film (Cemel et al., 1972; Day & Ringsdorf, 1985). This process can lead to high molecular weight LB films containing isolated or conjugated double or triple bonds with

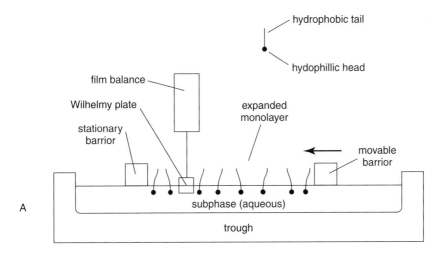

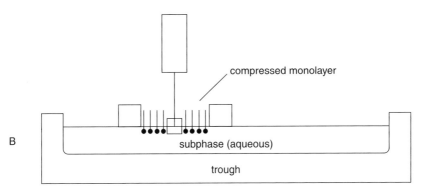

Figure 9–16. A typical Langmuir trough for the study of monolayers at the air-water interface. The trough is composed of a teflon well that is filled with the aqueous subphase. (**A**) The molecules to be investigated are disolved in a volatile organic solvent and spread at the surface. (**B**) The molecules are compressed by decreasing the area of the trough with a moveable barrier. A Wilhelmy film balance records the surface pressure. A plot of the surface pressure versus the area per molecule is called an isotherm.

superior stability to solvent, pH and temperature compared to monolayers of the precursor monomer. These materials are, therefore, potentially useful for applications in severe environments. The simplest examples include photochemical polymerization of monomers containing vinyl groups (Laschewsky et al., 1988) or diacetylenes (Hub et al., 1980; Tieke, 1985).

   The most common way to characterize an amphiphile at the air–water interface is to measure the surface pressure as a function of the relative area occupied per molecule in the film. The area occupied per molecule is easily calculated by knowing the surface area of the Langmuir trough and the amount of material (i.e., the number of molecules or moles) that was applied to the surface during the experiment. A plot of the surface pressure versus the area occupied by each molecule is known as an isotherm since the experiment is performed at a constant temperature. In a typical experiment, the isotherm is obtained by adding the amphiphile being investigated to a clean water surface as a solution in a volatile organic solvent. The surface pressure is then monitored as the area of the trough is reduced by a movable barrier. The behavior of 2D monolayers at the air–water interface is analogous to three-dimensional (3D) phase diagrams. This can be best illustrated by an ideal isotherm diagrammed in Figure 9-17.

Molecules can exist at the air–water interface in states similar to normal solids, liquids and gases. The state in which a monolayer exists can be obtained from the pressure-area isotherm (abbreviated the P-A isotherm). At large areas without compressing the amphiphile, the molecules behave as a 2D gas. They exert little force on each other and the hydrocarbon tails are believed to make contact with the water layer. As one compresses the monolayer, the pressure rises asymptotically until a plateau is reached. This rise in pressure is called the liquid-analogous phase and with continued compression directly forms the liquid–solid coexistence plateau. The continued compression of this phase gives rise to the another phase transition due to the formation of a solid-analogous phase. The phase transition from liquid-analagous to solid-analagous phase (*via* the liquid–solid coexistence region) is the main phase transition for amphiphiles such as phospholipids, the molecules that are found in most naturally occurring cell membranes. This transition is often measured by differential scanning calorimetry (DSC) and is defined as the T_c for the lipid amphiphile. The physical phenomenon that gives rise to this phase transition is believed to be the ordering of the alkyl side chains to the all-*trans* (gauche) conformations. When the lipid reaches the solid-analogous phase, very low compressibility is observed. Extrapolation of the slope of the solid-analogous phase to zero surface pressure gives the limiting area (A_L) of the lipid amphiphiles in the film. The values of A_L correspond well with values from X-ray data for the packing of the alkyl chains.

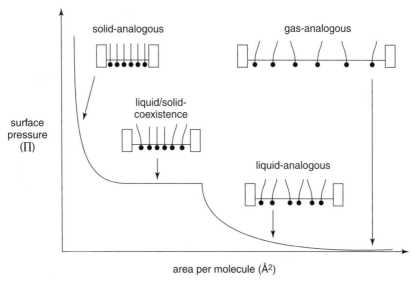

Figure 9–17. Pressure-area diagram (isotherm) of an idealized monolayer. The phases are indicated as gas-, liquid-, and solid-analogous.

The pressure-area isotherms are a useful technique to characterize mixed monolayers that contain more than one lipid amphiphilic molecule. Comparison of the P-A data from single component lipids with those from mixed monolayers give information on the miscibility of the lipids used in the study. These physical properties are often measured when constructing LB films for adhesion studies. With this background we will now turn our attention to specific examples of the use of Langmuir films to study biological membrane chemistry.

Biological Interactions on Langmuir Films

Ringsdorf and co-workers have used Langmuir monolayers to investigate biomembrane adhesion processes involving carbohydrate ligands (Bader et al., 1984). In their early studies, they have investigated the carbohydrate binding protein concanavalin A (con A) using polymerized monolayers containing glycoside-coated surfaces. These films were formed from synthetic lipid molecules that contain a diacetylene group that is easily polymerized by UV irradiation. They demonstrated the capability of the lectin con A to bind specifically to α-D glucopyranoside surfaces and not to surfaces derivatized with similar carbohydrates such as galactopyranosides.

These investigators also studied mixed monolayers expressing biotin at the surface in order to crystallize the protein streptavidin and to form multiple protein layers using a carbohydrate–lectin interaction as a key component of these films (Tieke, 1985). Avidin and streptavidin are two homologous water soluble proteins, each containing four specific binding sites for biotin (vitamin H) (Bater & Wilcheck, 1989; Buckland, 1986; Green, 1975). Even though the interaction involves no covalent bonds, the high binding constant of avidin or streptavidin to biotin (approx. 10^{15} M^{-1}) makes the binding virtually irreversible. A crystalline biotin–streptavidin layer was used to anchor a bifunctional molecule containing biotin and mannose. The concanavalin A lectin was then attached to the new mannose surface. This process was repeated to build up multiple protein layers (Fig 9-18).

Glycosylated monolayer films coupled to a quartz crystal microbalance (QCM) have recently been used to measure important physical properties of a carbohydrate–lectin binding interaction (Ebara & Okahata, 1994). Ebara and Okahata imbedded glyconamide-terminated phospholipids in a phosphoethanolamine monolayer formed at an air–water interface and exposed it to con A. The effects of con A binding to the glucose-terminated maltonamide ligands were sensed by an ultra sensitive QCM placed horizontally on the surface of the monolayer. During the binding event, the QCM undergoes a change in resonance frequency that is related to a change in mass of the monolayer. The QCM is sensitive to changes in mass at the nanogram level. Using this technique, the researchers were able to determine the amount of con A binding to the monolayer as a function of time (Fig. 9-19). The results indicated that the maximum amount of con A that bound to the monolayer was independent of the ligand density while the rates of binding (k_1) and dissociation (k_{-1}) were largely affected by ligand density. The lectin association constant ($K_a = k_1/k_{-1}$) determined using this technique ($1–3 \times 10^6$ M^{-1}) was in agreement with the value determined by the agglutination of erythrocyte or lymphocyte cells ($1–5 \times 10^6$ M^{-1}).

Ribi and co-workers have used Langmuir films to study the binding of cholera toxin to gangliosides and to elucidate the toxin structure by electron microscopy. The cholera toxin binds to the carbohydrate portion of GM$_3$ as a ligand to support the entry of the

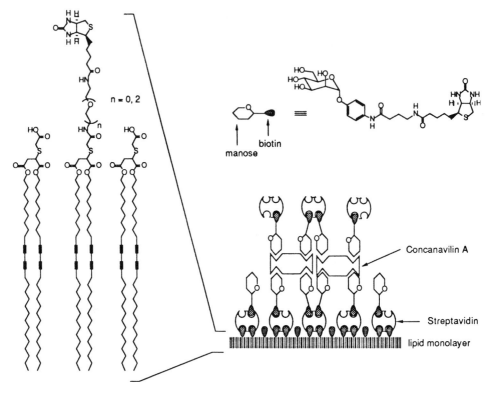

Figure 9–18. Streptavidin–con A monolayer sandwich.

toxin into the colonic epithelium. The investigators incorporated GM3 ganglioside (a sialylated complex glycolipid) into a film which was lifted onto an electron microscopy grid and then treated with the cholera toxin protein (Ribi et al., 1988). Using this approach, cholera toxin was then crystallized at the interface and evaluated by electron microscopy.

Charych and co-workers have investigated the binding of the influenza virus to sialic acid at a polymerized monolayer film interface with the goal of constructing a diag-

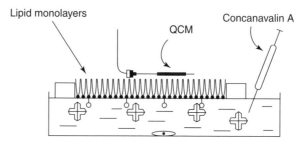

Figure 9–19. Schematic illustration of the experimental setup used to measure carbohydrate–lectin binding on a monolayer.

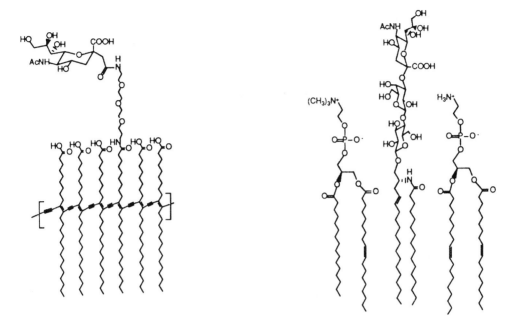

Figure 9–20. Representative polymerized and unpolymerized glyco-monolayers.

nostic sensor for biological adhesion (Fig. 9-20) (Charych et al., 1993). The bilayer is composed of a self-assembled monolayer of octadecylsilane and a Langmuir–Blodgett monolayer of polydiacetylene lifted onto the modified glass substrate. The polydiacetylene layer is functionalized with an analogue of sialic acid that binds to the influenza virus hemagglutinin. The sialic acid ligand serves as a molecular recognition element while the conjugated polymer backbone signals binding at the surface by a chromatic transition. The color transition is readily visible to the naked eye as a blue to red color change and can be quantified by visible absorption spectroscopy.

The bilayer assembly described above contains a carbohydrate ligand for receptor binding and the capability to signal the binding event. This mechanism has been called affinitychromism. Since ligands other than sialic acid could be incorporated into the film, affinitychromism offers the possibility of a general method for the direct detection of receptor–ligand interactions. These films could potentially be used for screening new drug candidates by inhibition of the colorimetric response.

Self-Assembled Molecular Films

Certain molecules when placed in contact with a surface can bind to form well-ordered two dimensional arrays (Bigelow et al., 1946). This process is called molecular self-assembly, and it is emerging as an important method to control the interactions between a surface and its environment (interfacial properties). Molecular self-assembly uses the fundamental forces between molecules (van der Waals interactions, hydrophobic effects and hydrogen bonding) to form highly ordered macromolecular systems. This process has been used to synthesize molecular films on a variety of surfaces including glass, gold, alumina, platinum and silicon (Bain & Whitesides, 1989; Swalen et al., 1987; Whitesides & Ferguson, 1988; Whitesides & Laibinis, 1990). The objective of this section is to describe the application of molecular self-assembled films to the study of carbohydrate binding systems. Specifically, we will survey methods to synthesize and characterize surfaces that contain carbohydrate molecules that bind to proteins, bacteria and viruses. We will also evaluate the use of surface science techniques to characterize the physical and chemical properties of self-assembled films, and we will attempt to correlate these properties with the adhesion of biological molecules. Finally, we will outline some present and future applications of these surfaces.

Organic molecules can be synthesized with functional groups that covalently attach to surface atoms. This covalent attachment coupled with intermolecular forces drives the formation of a uniform molecular film (Fig. 9-21). Organic self-assembled films allow one to vary the properties of a surface by changing the individual organic monomers. The chemical properties of these monomers, therefore, determine the optical, electronic and adhesive properties of the resultant surface. Molecular films have been studied extensively on gold and silicon. These films have been used to control the properties of wetting (adhesion) and molecular recognition.

The process of forming self-assembled monolayers (SAMs) involves exposing a substrate to a solution of the assembling molecules which contain reactive functionalities. Most typically for silicon surfaces, the self-assembling functionality is a trichloro or triethoxysilane that undergoes a subsequent condensation with surface hydroxyl groups. The assembly, therefore, consists of a crosslinked network of siloxane (Si—O—Si) bonds near the solid surface, and some exposed functionality at the other end of the molecule, expressed at the film/ambient interfacial region. The exposed surfaces of silicon and glass were modified with a variety of functional groups including

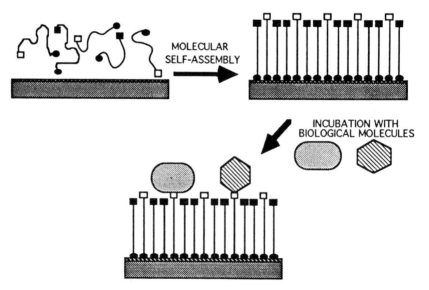

Figure 9–21. Schematic representation of molecular self-assembled monolayers and their interaction with biological molecules and cells.

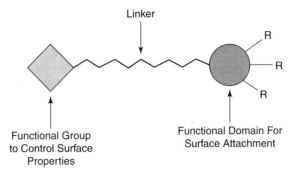

Linker

R

R

R

Functional Group
to Control Surface
Properties

Functional Domain For
Surface Attachment

Figure 9–22. Schematic representation of molecules used to form self-assembled monolayers. These bifunctional molecules contain groups both for surface attachment and for controlling surface properties.

alkyl chains (CH_3), amino groups, thiols, alcohols, esters, amides and heterocyclic compounds and, most importantly, carbohydrates using appropriate tailoring of the surface groups in alkylsilanes (Fig. 9-22).

The other common self-assembling functionality is thiol or disulfide which reacts with surfaces of gold, silver, and platinum. Once the surface reaction is complete, the functional groups at the other end of the molecule are addressable at the film/ambient interface. The wettability and surface structure of these thin film materials has been studied extensively.

Film Characterization

A variety of surface science techniques may be used to characterize self-assembled organic molecular films. These techniques have been previously reviewed in detail (Whitesides & Ferguson, 1988); several that are commonly used in many laboratories to characterize these films are described below. These include XPS (X-ray photoelectron spectroscopy), contact angle measurements, ellipsometry, and the scanning probe microscopies of STM (scanning tunneling microscopy) and AFM (atomic force microscopy).

X-ray Photoelectron Spectroscopy

The XPS technique uses X-rays to eject core electrons from atoms at the surface. The energy required to eject an electron from an atom is called the binding energy (BE) and is characteristic of each element present at the surface. The XPS spectrometer irradiates the surface with high-energy X-ray photons which collide with surface atoms and eject electrons from the core energy levels. The analyzer subtracts the kinetic energy (KE) of the emitted electrons from the energy of incident X-ray photons ($h\nu$) to give the binding energy (BE).

$$BE = h\nu - KE \qquad (9-1)$$

XPS can provide an unambiguous chemical signature of a sample. The peak heights, widths and positions in an XPS spectrum provide more information than simple elemental identification. The shifts in binding energy reflect the oxidation states of sur-

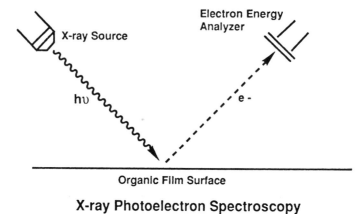

X-ray Photoelectron Spectroscopy

Figure 9–23. Diagram for an experimental setup for an X-ray photoelectron spectroscopy (XPS) experiment. XPS gives information about the types of functional groups at an interface.

face atoms and allow one to determine chemical groups such as methylene carbons versus the carbon of a carbonyl group. The experimental set-up for XPS is diagrammed schematically in Figure 9-23. The data from XPS are plotted as binding energy versus number of electrons.

The experimental set-up of XPS may be modified to yield information relating the change in chemical composition as a function of depth (i.e., XPS depth profiling). The angle between the sample normal and the analyzer axis can be varied to change the amount of material through which the photoelectrons pass (Fig. 9-24). Increasing the angle to a maximum of 90 degrees increases the contributions from surface atoms and attentuates electrons from the bulk. This method of analysis, termed angle-resolved XPS, provides a sensitive depth profile characterization of the physical and electronic properties at the interface.

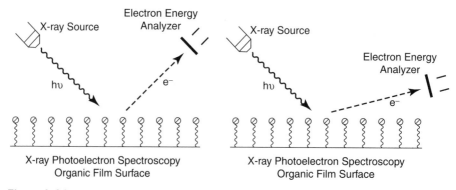

Figure 9–24. Experimental diagram for XPS depth profiling experiments.

Contact Angle Measurements

The angle between the surface and a line tangent to a drop of liquid placed on the surface is defined as the contact angle q (Fig. 9-25). Contact angles are a measure of the macroscopic surface property of wettability (DeGennes, 1985; Fowkes, 1964; Israelachvili, 1985). Wettability is sensitive to surface free energy and is quantified by measuring the contact angle at the solid–liquid–vapor interface. Various techniques for measuring contact angles have been reviewed in detail (DeGennes, 1985). The most common method is to measure q directly for a drop of liquid resting on a flat surface using a contact angle telegoniometer.

There are three forces that control the spreading of a drop of liquid on a surface. Consider a drop of water at the Si/SiO_2 interface. The water interacts with itself, the surface and the environment. The attraction of a drop of water for the surface may outweigh the attractions of the water for itself and for the environment. In the case of SiO_2, the attraction between water and the surface is strong, the drop spreads out, and the contact angle is measured to be a small number, typically 5–10 degrees. However, when SiO_2 is modified with a hydrocarbon-terminated surface the water contact angle rises to 110 degrees. The interaction of water with itself and the environment now outweighs the attraction of the water for the surface. The long chain hydrocarbon, therefore, causes the water to bead up on the surface. We say that the clean SiO_2 exhibits a larger surface free energy of interaction with water than the hydrocarbon-terminated surface. Contact angle measurements are, therefore, a useful tool to study self-assembled organic films. The measurement is a sensitive determination of the structure and properties of the topmost atomic layers of a surface. The measurement of contact angles can be performed rapidly and accurately, giving a quick, qualitative analysis of the surface.

The main disadvantage to contact angle measurements, however, is that the actual interpretation of the data, in terms of microscopic properties, can be very difficult. The relationship between the wettability of a surface (measured by q or cosq) and the microscopic structure and properties of that surface was first discussed by Young (1805). Consider an ideal case where the liquid, vapor, and solid phases are in thermodynamic equilibrium. Young derived an expression for the contact angle in terms of interfacial free energies g.

$$gl_v cosq = g_{sv} - g_{sl} \text{ (Young's Law)} \tag{9-2}$$

CONTACT ANGLE

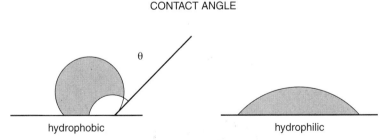

Figure 9–25.   Diagram of a water droplet on a surface demonstrating a contact angle measurement. A hydrophobic surface gives rise to a high contact angle (θ) while a hydrophilic surface gives a small contact angle.

$$gl_v = \text{liquid–vapor}$$
$$g_{sv} = \text{solid–vapor}$$
$$g_{sl} = \text{solid–liquid}$$

However, real surfaces are seldom ideal, often having roughness and inhomogeneity. Therefore, the utility of contact angle measurements is questionable. Despite the limitations of contact angle measurements, however, their simplicity and generality make them useful in the study of organic films.

Ellipsometry

Light experiences a change in velocity which is related to the material through which it is travelling. This change is measured as the refractive index of the material and is expressed as:

$$n = c/v \tag{9–3}$$

where n is the refractive index, c is the speed of light in vacuum, and v is the speed of light in a material (Eq. 9-3). Changes in refractive index of materials at an interface are measured with the technique of ellipsometry. The reader is referred to detailed discussions of ellipsometry for a detailed treatment of the topic (Azzam & Bashara, 1977).

In an ellipsometry measurement, plane polarized light (with a characteristic amplitude and direction) hits the surface and is reflected (Fig. 9-26). The reflected light typically has both a different amplitude and direction. These changes are related to the amount of material (for instance, an organic monolayer on silicon) that the light must pass through. The thicker the material, the more the change in the amplitude and direction of the reflected light. One cannot, however, compare the thicknesses for two different materials without considering their respective indices of refraction. When the indices of refraction are included in the calculations, then the relative thicknesses of a variety of different materials can be evaluated. Ellipsometry data is reported directly as a thickness of material (in Å) on the surface.

Ellipsometry

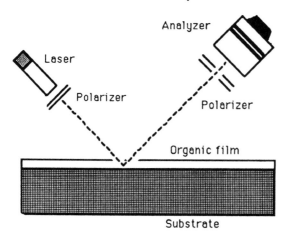

Figure 9–26. Diagram of an experimental set-up for an ellipsometry experiment. Ellipsometry measures the depth of a monolayer.

One of the major drawbacks of ellipsometry is that one must estimate the refractive index of organic materials on the surface. Additionally, the spatial resolution of the technique is no better than the spot size of the incident laser light. Furthermore, the data from three component systems such as Si/SiO_2/organic film is more difficult to interpret because light is reflected from an additional interface. However, the ellipsometric technique is nondestructive, quick and simple and provides a relatively accurate measurement of film thickness.

Scanning Probe Microscopies

Two potentially powerful techniques to investigate organic surfaces at the molecular level are STM and AFM. These two techniques provide real-space, real-time, atomic resolution images of surfaces on materials ranging from insulators (glass and mica) to semiconductors (Si and Ge) to metals (Ag, Au, Pt). Additionally, these techniques do not have the same ultrahigh vacuum limitations of many standard surface science techniques. STM and AFM may be performed under solvents and electrolytes. This allows the surface to be studied in an environment similar to that which will be encountered in a device application. A review of the state of the art in scanned microscopies is beyond the scope of this chapter; the reader is referred to several excellent review articles (Behm & Hosler, 1986; Golocchenko, 1986; Hansma & Tersoff, 1987; Quate, 1973). However, illustrated below are data pertinent to the potential use of these techniques for the study of self-assembled films at molecular resolution.

AFM does not require electrical conductivity. The AFM rasters a stylus across the sample and records the attractive or repulsive forces between tip and sample. These forces are measured as the deflection of a cantilever to which the tip is mounted. AFM has not yet achieved the resolution of STM, but it can be applied to nonconductive surfaces, which are typical for organic film applications. AFM has imaged the surfaces of glass (Marti et al., 1988), insulating materials (Albrecht & Quate, 1987) and organic polymers (Mate et al., 1989).

While these two techniques have only recently been applied to self-assembled molecular films, we have confidence that this exciting field may soon merge with organic films and monolayers to provide the first detailed, direct studies of organic materials at nanometer resolution. The analysis of biological surface models holds the key to the future of developing a greater understanding of biological surfaces in order to control their properties.

Biological Adhesion on SAMs

In comparison to most of the subfields outlined in previous sections that employ molecular assemblies, the area of biological adhesion and carbohydrate binding is probably the least studied with SAMs.

Probably the most popular model system investigated to date for molecular recognition studies is based on the biotin–avidin interaction (Ahlers et al., 1990; Helm et al., 1991; Samuelson et al., 1992; Schmitt et al., 1991). Biotin derivatives of alkyl thiols have been synthesized and self-assembled onto a gold surface. Recognition and subsequent binding of streptavidin was detected by surface plasmon spectroscopy whereby adsorption to the surface is measured by a change in the resonance angle (Haussling et al., 1991). The resonance angle in the reflected laser light occurs at a minimum due to resonantly exciting surface plasmons. Using this technique, saturation coverage of avidin onto the biotinylated surface occurred in roughly five minutes. However, tightly packed biotin-monolayers could not bind as much avidin as loosely packed films which incorporate a hydrophilic linker, presumably due to steric hindrance. The streptavidin, once bound to the monolayer, could not be rinsed away.

An exciting, yet relatively underexplored, use of SAMs is in derivatization of implant materials with carbohydrates, peptides or glycoproteins to manipulate responses from animal cells. The design of biomaterials poses several challenges related to modification of the implant surface. While adhesion and spreading of cells to the implant material is desirable, adhesion of bacteria is presumed important in the mechanism of infection (Merritt, 1984). In fact, a major problem associated with the use of inplants is the formation of bacterial biofilms which reduce the number of organisms required for the onset of infection (Barth et al., 1985; Gristina, 1987).

Sukenik and co-workers modified the surfaces of titanium, an implant material in pacemaker casings, orthopedic devices, and dental apparatus with self-assembled monolayers of silanes. The exposed functionalities on the surface included bromine, alcohols, and alkenes (Sukenik et al., 1990). For comparison, similar surfaces on glass substrates were prepared. Fibronectin, an extracellular glycoprotein that mediates cell adhesion, was adsorbed to the derivatized glass and titanium surfaces. It was believed that the conformation of fibronectin would depend on its interaction with the functionalized surfaces, which in turn would modify the response of animal cells (in this case neuroblastoma cells) binding to the surfaces. Indeed, these investigators found that while cell attachment was comparable for all surfaces, cell spreading and cellular responses were different. Underivatized titanium or glass adsorbed with fibronectin produced bipolar cells with short neurite-like processes, while surfaces containing exposed diol functionalities with fibronectin produced long thin neurites, extending as much as 100 mm from the cell body. Bromine-terminated surfaces produced responses similar to that of underivatized surfaces, while the alkene functionalities produced pseudopodial processes, more like fibroblasts in appearance rather than nerve cells. These results are intriguing in that the chemical functional groups interacting with fibronectin molecules alter its conformation in an unknown fashion which, in turn, forces the neuronal cells to use different receptors to interact with the various exposed binding domains of fibronectin. In functionalized glass or titanium containing no adsorbed fibronectin, cells attached but did not spread, and eventually died. This emphasizes the importance of the adhesion-promoting glycoproteins. No differences were found between titanium and glass substrates, suggesting that the monolayer films isolate the substrate from the environment, leaving only the chemical functionality–fibronectin interaction as the important one determining cell adhesion and spreading. This may enable other bulk substrates to be utilized as biomaterials if they can be functionalized suitably. Bacterial adhesion to underivatized surfaces and methyl-terminated SAMs were difficult to determine since no trends were evident. However, the methyl-terminated titanium surface showed the least bacterial adhesion. Although these results are not very clear, the authors point out that by modifying the surface, the responses of mammalian cell adherance could be controlled without increasing bacterial adherance.

Bednarski and co-workers synthesized SAMs of carbohydrate-terminated alkyl silanes on smooth silicon surfaces in their attempts to develop sensors with semiconductors (Mastendrea & Bednarski, 1990). Different glycosylated materials were prepared and compared for adhesion to various cell preparations. As expected, cells of the bacterium *Salmonella* were shown to adhere specifically to mannosylated surfaces and not to other carbohydrate derived surfaces. These researchers also demonstrated the ability to modify glycosylated silicon surfaces by enzymatically extending some carbohydrate sequences and phosphorylating others.

Liposomes

A liposome is a spherical bilayer assembly that encloses an aqueous interior and is equivalent to the lipid portion of a cell membrane that is devoid of the proteins, cytoskeleton and other membrane associated components found in living cells (Gregoriadis, 1984; Knight, 1981; Lelkes, 1984; New, 1990; Ostro, 1983, 1987). Lipo-somes have been used extensively as model membranes to study biological interactions of membrane bound proteins (Bretscher, 1985; Luna & Hitt, 1992). Liposomes have also been used as carriers for entrapped molecules (i.e., drug delivery vehicles). This section describes how liposomes are formed and characterized, and illustrates some recent examples of their use in studying carbohydrate adhesion events at interfaces.

Bangham, Papahadjopoulus and Watkins postulated during their studies on the hydration of purified phospholipids that these molecules could form closed spherical membrane structures (Bangham et al., 1965a,b; Roy et al., 1993a,b). Since their discovery, liposomes have been formed using a wide variety of molecules besides natural phospholipids. The relative hydrophobicity and hydropholicity of the amphiphilic molecules, as well as their chemical structures, determine what types of aggregates are formed when these molecules are hydrated.

The most common way in which liposomes are characterized relies on a description of their size (large versus small) and the number of bilayers (lamellarity) of the particle (Ostro, 1983). A liposome that has only one bilayer is similar to a cell and is referred to as being unilamellar. Liposomes that have more than one bilayer are called multilamellar. The bilayers represent a series of concentric spheres, each with an aqueous volume trapped between the lipid layers, similar to onion skin. Figure 9-27 shows a schematic representation of the common liposome structures that have been described in the literature. It is important to note that most liposome preparations give rise to a variety of sizes and lamellarities analagous to the molecular weight distributions found in polymer systems.

Multilamellar vesicles (MLVs) are typically large in diameter (100–1000 nm) and have at least five bilayer assemblies for each liposome. Large unilamellar vesicles (LUVs) have diameters similar to the MLVs (1000 nm) but consist of a single bilayer. Small unilamellar vesicles (SUVs) have sizes that range from 15 to 100 nm and are the minimum size for most phospholipid vesicles. SUVs also have a more uniform size distribution and, like LUVs, they have only one lipid bilayer associated with them. The method of preparation and the structure of the lipid molecule employed in the process are important for determining the type of liposome structures that are obtained (Hope et al., 1986).

Liposome Preparation

Many methods for the preparation of liposomes have appeared in the literature since the first report in 1965 by Bangham for the construction of multilamellar liposomes. Although these methods are too numerous to describe individually, the most common

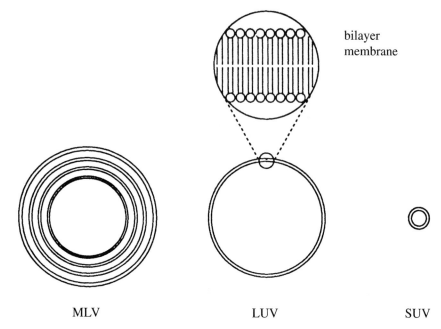

bilayer
membrane

MLV   LUV   SUV

Figure 9–27. General types of spherical liposomes. The multilamellar vesicle (MLV) is composed of several bilayers with a total diameter of 100–1000 nm. The large, unilamellar vesicle (LUV) has a single bilayer and a diameter of approximately 1000 nm. The small unilamellar (SUV) is composed of a single bilayer and is about 15–100 nm in diameter.

are detergent solubilization solvent dispersion (Deamer, 1984a,b) and mechanical dispersion. Mechanical dispersion is by far the simplest and most widely used method of liposome formation and a brief overview on the use of this method is given below.

Methods using mechanical dispersion for liposome formation range from simply shaking an aqueous suspension of lipid by hand to high energy agitation methods such as sonication. In these methods, the dispersed lipids form kinetic structures that are trapped in the form of lipid bilayer assemblies. The method of extrusion is perhaps the most reproducible for the preparation of liposomes with uniform size distribution. The extrusion method was first described by Olson and relies on pushing MLVs through a polycarbonate filter having a defined pore size (Olson et al., 1979). Filters with narrow pore sizes are available that range from 30 to 600 nm. Electron microscopy indicates that the liposomes produced using this technique have a similar distribution of sizes that is centered around the pore size (Mayer et al., 1986). MLVs are usually prepared by light shaking or vortexing a lipid mixture in water. The extrusion involves forcing the MLVs through a filter at high pressure and at a temperature above the T_m. This process is repeated many times to maximize uniformity. The sample can be frozen and thawed between the extrusion cycles. The freeze–thaw cycles are used for entrapping materials in the interior of the liposome and seem to improve the efficiency of this process (MacDonald et al., 1991; Mayer et al., 1985).

A frequently used mechanical dispersion method is sonication (Kirby & Gregoriadis, 1984; Szoka & Papahadjopoulos, 1980). Two types of sonication methods are use to form liposomes: bath and probe sonication. The amount of energy transferred in bath sonication is usually not adequate to reach the smaller size limits for most lipids, though the technique is very mild. Probe sonication is a high energy process which also generates heat that could potentially destroy sensitive lipid molecules. This method also is not effective below the transition temperature (T_c) of the lipid molecules; the heating is, therefore, an integral part of this procedure. Despite its simplicity and effectiveness, probe sonication is difficult to control from batch to batch. For this reason, the procedure is usually limited to the preparation of "over-sonicated" liposomes that results primarily in the formation of SUVs.

Liposome Characterization

The most common methods for the determination of the size distribution of liposomes is transmission electron microscopy (TEM) and dynamic (quasi-elastic) laser light scattering (DLS). By far the most accurate is negative staining electron microscopy. The staining method relies on the use of heavy metal salts (negative stain) that deflect the electron beam in the microscope. Since the metal salt cannot cross the intact lipid bilayer membrane, liposomes appear as bright transparent spots on a dark background. The stains most commonly used for liposomes are phosphotungstanate and uranyl acetate. The image of the liposome can be directly measured and the size determined. However, this procedure is time consuming and can only be performed under high vacuum.

Another common technique is the use of dynamic laser light scattering. This method relies on the fact that small particles (liposomes) can deflect or scatter light. Since liposomes are in motion, there is a time-dependent fluctuation in the intensity of scattered light in the sample. This fluctuation is related to the rate of diffusion of the particles and, therefore, on the particle size. The translational diffusion rate is related to the hydrodynamic radius by the Stokes–Enstein equation. One advantage of DLS is that the sample is observed in solution and in real time. However, for liposomes, a number of assumptions must be made for data analysis that make this method somewhat unreliable.

The use of atomic force microscopy for the analysis of liposomes has recently been investigated. Figure 9-28 compares the structure of liposomes prepared by the probe sonication method using both TEM and AFM. The main advantage of the AFM is that it is quick and can be performed under water. The disadvantage of AFM is that the liposome must be strongly attached to a smooth surface (for example mica) to be imaged properly.

Biological Models Using Liposomes

The incorporation of membrane bound proteins into liposomes has been studied extensively and is one of the only methods available to study cell surface glycoproteins and glycolipids with complementary receptors (Gosh & Bachawat, 1980; Juliano, 1983; Juliano & Stamp, 1976; Ketis et al., 1980; Redwood et al., 1975; Sharom et al., 1977). These studies exemplify the use of liposomes to derive a basic understanding of multivalent carbohydrate receptor–ligand interactions at a biologically significant interface. A specific example is the agglutination of mannose-terminated glycolipids incorpo-

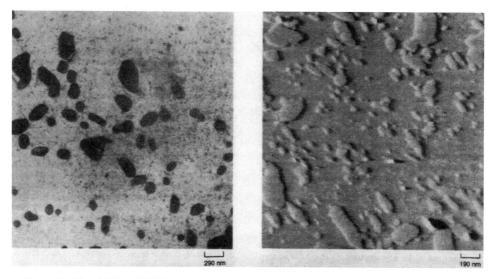

290 nm   190 nm

Figure 9–28.   AFM and TEM of polymerized liposomes on cleaved mica.

rated into liposomes with concanavalin A. This study revealed that a 5% surface density of the carbohydrate ligand was needed before agglutination occurred (Rando & Bangerter, 1979). The rate of agglutination is also proportionally sensitive to the surface density of the ligand (Surolia et al., 1975). Other investigators examined the importance of the distance between the carbohydrate and the lipid bilayer using variable hydrophobic spacers (Slama & Rando, 1980). A similar spacer effect was observed when comparing the agglutination behavior of liposomes containing lactosyl ceramide or galactosyl ceramide with *Ricinus communis* agglutinin (Curatolo et al., 1978). Although this lectin is capable of recognizing terminal galactose residues, only the lactosyl ceramide liposomes agglutinated. The authors felt that that the glucose portion of the lactose molecule was serving as a spacer group, allowing the lectin access to the galactose ligand. Other experiments have shown that the surface charge of the liposome is a major determinant of lectin binding capability. Concanavalin A was observed to bind to negatively charged liposomes bearing a glucose-derived ligand, while no binding was observed in uncharged liposomes (Hampton et al., 1980).

Experiments aimed at determining the stability of ligand-containing liposomes toward the binding of lectins have shown that the liposome structures are usually intact after agglutination. For example, lectin binding does not increase the release of encapsulated compounds inside the liposome (Curatolo et al., 1978; Juliano & Stamp, 1976). The fluidity of the liposome bilayer was also shown to be unaffected by the receptor–ligand interactions based on fluorescence polarization studies (Roche et al., 1977). The curvature of the liposome has also been implicated in affecting the rate of carbohydrate receptor–ligand interactions and the agglutination of ganglioside-bearing liposomes.

It has been shown recently that liposomes can be effective agents for blocking the binding of influenza virus to cells. A hydrolytically stable sialic acid glycolipid was incorporated into a polymerized liposome. Approximately 1% of surface coverage of sialic acid on the liposome was needed to inhibit the attachment of the virus to red blood cells (Spevak et al., 1993). In addition, these studies also demonstrated that the liposome particles could block virus infectivity in *in vitro* cell culture assays (Fig. 9-29).

Paulson and coworkers demonstrated the importance of multivalency on carbohydrate-mediated cell surface adhesion processes involved in inflammation. These researchers made liposomes containing sialyl Lewis[x] (sLe[x]) on their surface. SLe[x] is a carbohydrate ligand thought to be required for neutrophil adhesion to endothelial cells. In *in vitro* studies, the sLe[x] liposomes were shown to have a much higher affinity than monomeric ligands (Phillips et al., 1990).

Hakomori and co-workers have used various liposome systems to investigate carbohydrate-mediated cell adhesion. In one study, phosphatidylcholine liposomes containing the GM$_3$ ganglioside and $\alpha_5\beta_1$ integrin were shown to bind to fibronectin (Zheng et al., 1992). The binding was shown to be sensitive to the amount of GM$_3$ in the liposome. This study suggests that the glycolipid GM$_3$ regulates integrin receptor function that is essential for cell adhesion to fibronectin.

In another interesting model for studying carbohydrate-mediated cell surface adhesion, Hakomori used a parallel plate laminar flow system that could mimic *in vivo* flow conditions encountered on vascular endothelium (Kojima et al., 1992). Plastic surfaces were coated with E-selectin, an important lectin involved in neutrophil adhesion during inflammatory processes. These surfaces were then placed in the shear flow system. HL60 cells were treated with sialidase to cleave sialic acid units as well as different carbohydrate-binding antibodies to expose specific carbohydrate squences on the cell surface. These cells were then passed through the sytem and cell adhesion to the E-selectin surface visualized with a microscope (Fig. 9-30). The study concluded that a synergistic combination of sLe[x] and Le[x] carbohydrate epitopes was required for max-

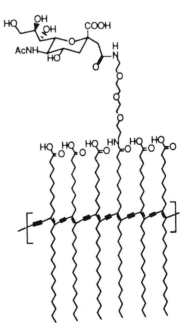

Figure 9–29. Schematic representation of the surface of the polymerized NeuAc liposomes.

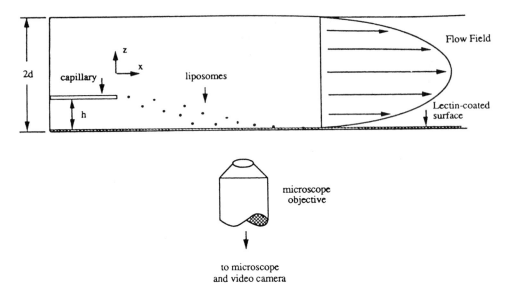

Figure 9–30. Cross section diagram of a laminar flow cell experimental set-up used to visualize liposome binding to complementary surfaces.

imal binding. In a similar type of experiment, the binding of liposomes coated with a glycoprotein (glycophorin) to surfaces coated with a lectin (wheat germ agglutinin) was investigated (Wattenbarger et al., 1990). This study demonstrated the importance of particle size, receptor density, and flow rate on adhesion. Similar studies conducted with cell systems have also demonstrated the importance of such common sense parameters.

The success of liposomes as models of biological membranes has led to an increased focus on the use of these agents for drug delivery (Huang, 1993). Fusion (Weinstein et al., 1977), adsorption (Szoka et al., 1980), endocytosis (Raz et al., 1981; Wu et al., 1981) or induced leakage are all well documented interactions that occur between synthetic liposomes and natural cells. Drugs encapsulated in the interior of the carbohydrate-derivatized liposomes can be delivered to cells by the mechanisms described above. For example, galactose-coated liposomes are being investigated *in vivo* to target the asialoglycoprotein receptor for drug delivery to the liver (Reimer et al., 1992). In addition, the use of liposomes for drug delivery has recently received attention in the context of carriers having antibodies attached to the surface of the liposome (Alving et al., 1983).

Finally, liposomes have also served as models to study immune response. McConnell and co-workers incorporated dinitrophenol (DNP)-containing lipids into liposomes and studied the binding of anti-DNP antibodies to the surface of the liposome (Humphries & McConnell, 1977; Parce et al., 1978). The antibody–liposome complex was then treated with complement factors. Complement is a group of proteins that can produce cell lysis when bound to the Fc portion of an antibody. These investigators used this system to lyse the DNP-liposomes. These experiments set the precedent for the development of immunoadhesins. The mannose receptor on the surface of a pathogenic strain of *E. coli* was exploited to attach antibodies to the surface of the bacterium.

These antibodies were shown to be recognized by complement factors and macrophages that were able to destroy the organism (Bertozzi & Bednarski, 1992).

Oligosaccharide ligands on the surface of liposomes have also been shown to activate natural killer cells. For example, GM2 was incorporated into a liposome which was then used to treat tumor cells. Upon addition of natural killer (NK) cells, the tumor cells could be specifically lysed. Under the same conditions, liposomes containing lactosyl ceramide ligands did not result in lysis of the tumor cells. These experiments demonstrate that glycoconjugates are important in targeting the immune response of NK cells.

Liposomes constitute a very interesting and important class of glycoparticles because of their size, construction, and numerous potential medical applications. Besides these "glycosomes," other types of glycoparticles have been prepared for various studies. These materials are typically less homogeneous and less well characterized. Some examples include glycosylated latex microspheres, dextrans, gelatins or silica particles. Their uses have ranged from affinity chromatography materials (Mo et al., 1994) to agglutination-based diagnostic devices for the detection of antibodies or lectins (Roy & Tropper, 1988). Other applications include supports for the solid phase synthesis of oligosaccharides (Halcomb et al., 1994) and immunoadsorbants for the selective removal of antibodies or microorganisms from blood (Lemieux et al., 1980).

Summary

Carbohydrates represent interfacial structures that form the basis for the interactions of cells with their environment. It is this property that has brought this class of molecules to play a role in bioorganic chemistry as important as those of nucleic acids and proteins. It is also apparent that the development of clear and well defined model systems to study carbohydrate interactions with surfaces represents an important and challenging area of research in bioorganic chemistry. The difficulty in developing these models derives from the nature of the processes by which carbohydrate molecules interact with their environment. The binding is often multivalent and requires the construction of model surfaces that can mimic this phenomenon. Advances in developing these models entails the practical construction of complex organic molecules such as glycolipids and glycoproteins, and the development of new analytical equipment for permitting quantitative analysis of these interactions. Well defined surfaces such as Langmuir films, self-assembled monolayers and liposomes serve as the best models to study these interactions. The construction of these systems is itself an ambitious task and their analysis can be even more difficult. However, the rewards from understanding cell surface carbohydrate interactions and biological adhesion can range from the development of new materials to a deeper understanding of many human diseases. It is such applications, in addition to the intellectual challenge to understand how cells interact, that will set the goals of bioorganic chemistry in the next century.

10

Recognition of Carbohydrate Antigens by Antibody Binding Sites

David R. Bundle

The interaction of carbohydrates with protein receptors is a molecular recognition event that is essential for vital biological functions within the cell and at its surface where cells communicate with other cells. The ability of receptors to interact specifically with a heavily solvated, hydrophilic oligosaccharide at low concentrations in 55 M water is a truly remarkable process about which much remains to be learned. It permits the information encoded in oligosaccharide three-dimensional structure to be translated into signals that can determine the destination of cells and glycoproteins (Varki, 1993).

In the last decade a great deal of structural information has been collected for ligand complexes with carbohydrate transport proteins (Quiocho, 1986, 1988, 1989), enzymes (Johnson et al., 1988), lectins (Bourne et al., 1993; Rini, 1995) and antibodies (Bundle, 1997). Since they have played such a central role in the development of immunochemistry (Kabat, 1976, 1983, 1988; Landsteiner, 1945) it is surprising to realize that the first crystal structure of an oligosaccharide–antibody complex was reported only recently (Cygler et al., 1991). In this chapter, recent observations that have established the structure and energetics of carbohydrate–antibody interactions are described, and are related to developments in vaccine research and efforts to develop carbohydrate-based therapeutic agents (Jennings, 1983; Karlsson, 1991; Mulligan et al., 1993; van Boeckel & Petitou, 1993).

The results of recent research with antibodies (Bundle et al., 1994a,b; Cygler et al., 1991) and lectins (Bourne et al., 1993; Lemieux, 1989) suggest that the size of the oligosaccharide recognition element that actually makes contact with protein receptors is restricted to a disaccharide or trisaccharide unit. Such epitopes are often presented as part of a larger antigenic determinant and the recognition can involve small portions of the flanking residues. For comparably sized receptor sites involved in cell–cell interaction this is exciting news, since trisaccharide-sized carbohydrate drugs are accessible by current synthetic methodologies. Even for a large molecule such as heparin (Grootenhuis & van Boeckel, 1991; van Boeckel & Petitou, 1993), it is now possible to envisage synthetically accessible, low molecular weight carbohydrate-based therapeutics. Recent examples include the antiinflammatory sialyl Lewis x (sLe^x) tetrasaccharide (Mulligan et al., 1993) and therapeutics for influenza treatment (von Itzstein et al., 1993).

For antibodies, advances leading to new insights have come from multidisciplinary research that has employed hybridoma techniques combined with synthetic organic chemistry as a source of unnatural ligands that are not available by degradation of the native antigen. The combination of advances in the chemistry of carbohydrates with multidisciplinary techniques such as protein engineering, protein crystallography, and biophysical measurements such as NMR spectroscopy and microcalorimetry provides an example of the progress that is now made possible by application of multifaceted approaches in the biomedical sciences (Bundle, 1997).

Carbohydrate Antigens and Antibodies

Carbohydrate antigens at the surfaces of bacteria, red blood cells, and tumor cells may provoke a powerful immune response that can have beneficial effects, for example, in vaccines (Jennings, 1983) or antitumor treatment (Hellström et al., 1990; Honsik et al., 1986; Trail et al., 1993; Vadhan et al., 1988). In addition, deleterious effects such as transfusion reactions and transplant rejection can also result from responses to carbohydrate antigens (Nelson et al., 1992; Oriol et al., 1980, 1993). Conversely, adverse effects can also be treated by carbohydrate-based therapies, such as removal of antibodies by carbohydrate-based affinity matrixes or soluble inhibitors (Bannett et al., 1987; Bensinger et al., 1981, 1987; Cooper et al., 1993; Romano et al., 1994; Takahashi et al., 1993; Wilbrandt et al., 1969). In these immune recognition events the active molecules of the host response against the foreign substance are antibody molecules with combining sites that have shapes complementary to the carbohydrate antigens.

The relatively small area of an oligosaccharide to which the binding site is complementary is referred to as the antigenic determinant or epitope (Kabat, 1976). In some instances, as for example the human blood group antigens and human antibodies that bind them, the combining sites of the blood group A and B antibodies are able to distinguish trisaccharide epitopes that differ by the presence of either an -OH or an -NHAc group at a single carbon atom of one pyranose ring (Fig. 10-1). Incompatible transfu-

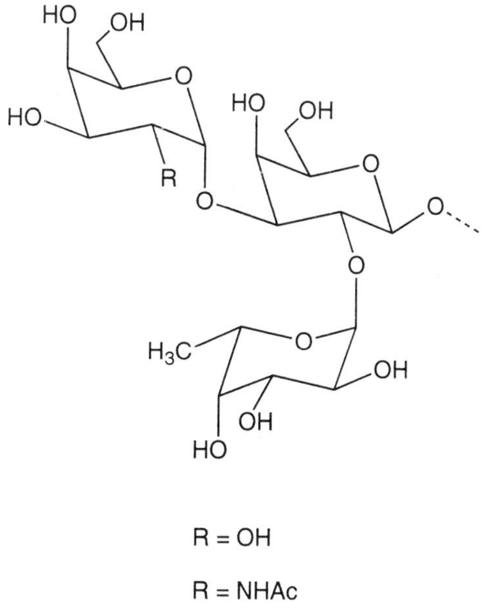

Figure 10–1. The trisaccharide antigenic determinants of blood group A (R = NHAc) and B (R = OH) differ only in the substituent on the C2 position of the terminal α-galactopyranosyl residue. Despite the otherwise identical structural features, blood group antibodies in the serum of a blood group B individual only bind the A antigen, while serum from a group A individual contains antibodies specific for the B antigen.

R = OH

R = NHAc

sion of an individual of blood group A with blood type B red cells results in a severe transfusion reaction as the B antigen-specific antibodies of an A individual destroy the transfused red cells (Mollison, 1983). High-fidelity binding such as this contrasts with cross reactivity (generally but not always recognition of common structural features), between antibodies evoked by distinct bacterial polysaccharide antigens (Heidelberger et al., 1942; Jennings 1983; Jones & Perry, 1957), sometimes with no obvious structural correlation (Bax et al., 1988). Both high- and low-fidelity profiles are observed among antibodies that bind oligosaccharide epitopes. Well-defined antibodies that exhibit such fine discrimination find uses in diagnostic (Bundle et al., 1982, 1989; Carlin et al., 1987a,b; Luk & Lindberg 1991; Perry & Bundle 1990; Perry et al., 1988; Pinto & Bundle, 1984) or therapeutic applications (Honsik et al., 1986; Trail et al., 1993; Vadhan et al., 1988). The ability of an antigen to induce antibodies with a cross-reactive profile may be a desirable outcome in the application of polysaccharide vaccines (Jennings, 1983). By understanding the molecular interactions that occur between antibodies and carbohydrate antigens one may hope to acquire the basic knowledge necessary to tailor carbohydrate-specific antibodies to diagnostic and therapeutic applications and more generally to design carbohydrate-based therapeutics (Karlsson, 1991; Mulligan et al., 1993; van Boeckel & Petitou, 1993; von Itzstein et al., 1993). This chapter describes progress toward these objectives through the combination of advances in the chemical synthesis and conformational analysis of oligosaccharides with contemporary techniques of immunology and molecular biology.

The Immune Response

In order to embark upon structural studies of oligosaccharide–antibody interactions, a plentiful supply of antibody is required. Although special techniques such as the hybrid myeloma method (hybridoma antibodies) can provide gram quantities of uniform antibody molecules (Goding, 1983; Köhler & Milstein, 1975a,b), the mammalian immune response to carbohydrate antigens influences the type and supply of pure antibodies (Milstein, 1986). Prior to the hybrid myeloma method, classical studies of oligosaccharide–antibody interactions were severely limited by the supply of antibody. Serum (the clear liquid remaining after red-cell clotting), the most common source of antibody, contains heterogeneous populations of antibody molecules (Kabat & Mayer, 1961). Only a fraction of serum antibodies from a hyperimmunized animal are able to bind the immunizing antigen and within the serum of a single animal, many sets of antibodies bind antigen in different ways (Kabat, 1976; Kabat & Mayer, 1961).

The immune response to an antigen is composed of humoral and cellular components (Klein, 1982; Roitt et al., 1985). The humoral response refers to the stimulation of a class of lymphocytes called B cells, that under the influence of antigen differentiate to memory B cells and plasma cells that produce and secrete antigen-specific antibodies (Ahmed & Gray, 1996; Klein 1982; Nossal & Ada 1971; Paul, 1989; Roitt et al., 1985). The cellular response is composed of antigen-primed T cells (also lymphocytes), some of which become resting memory T cells, while others develop the capacity to attack invading cells. Other T cells may release lymphokines that modulate the B-cell or antibody response. The structural features of the antigen, its antigenic determinants, that are recognized by the T-cell and B-cell receptors need not be identical but must be part of the same molecule.

During a typical immune response when a host first encounters the foreign substance (antigen), a small number of the total 1 to 100 million B cells recognize and bind antigenic determinants on the foreign molecule *via* a membrane bound form of the immunoglobulin molecule located on their own cell surface. Each cell is capable of making only one type of antibody with a binding site similar to that of its membrane bound molecule (Klein, 1982; Milstein, 1986; Nisonoff, 1982; Roitt et al., 1985). Antibody diversity, that is, the ability of an individual to respond to an almost unlimited number of antigens, is determined by the number of inherited V genes (germline genes) that code for the binding site domain of an antibody. However, this pool of genes does not represent the limit of antibody diversity. Gene recombination generates additional binding-site diversity, when the separate gene segments coding for the antibody V, D, and J regions are brought together to create a single antibody binding site (Milstein, 1986; Tonegawa, 1983). The concept of one cell one antibody still applies but the originally stimulated B cell has given rise to hundreds of progeny with modified binding sites. Under selective pressure from persistent antigen, B cells that express surface Ig with increased affinity for antigen are subjected to further rounds of clonal expansion and somatic mutation (Berek et al., 1985; Milstein, 1986). These events combine to expand the pool of functional antibodies, and especially those with increased affinity. The antibodies in the serum of an immunized animal are a reflection of these cellular processes.

Following the first exposure to antigen, the primary immune response produces maximum antibody levels after approximately 10–14 days, and the initial antibody response belongs predominantly to the IgM class (Klein, 1982; Roitt et al., 1985). This molecule possesses ten binding sites and has a molecular weight of ~900,000 Daltons. Generally the intrinsic affinity of individual sites in this molecule is low, but because some antigens (especially carbohydrates) possess a repeating unit motif, multivalency of the antigen and antibody, involving several of the 10 IgM binding sites, ensures high-functional affinity (avidity) by amplification of the intrinsic affinity of each site (Metzger, 1970).

At the secondary challenge by antigen, the immune system responds more rapidly, *via* memory B cells. Antibody production reaches a peak after only 7 days. Under the influence of antigen-primed T cells and cytokines released by them, the B-cell clones bearing receptor Ig molecules that bind antigen are stimulated to produce immunoglobulin molecules of the IgG rather than the IgM class, and the intrinsic affinity of binding sites increases with prolonged exposure to antigen. This class switch and affinity maturation are key features of the secondary immune response. As the process of clonal selection continues, antibodies of progressively higher affinity evolve by a process of gene recombination, antigen selection, and somatic mutation in those portions of the antibody genes that code for the amino acids that contact antigen (Berek et al., 1985; Milstein, 1986), the hypervariable loops, or complementarity determining regions (CDRs) (Kabat et al., 1991).

The B-cell or antibody response is critically dependent upon a concurrent T-cell response. These lymphocytes are so named because they differentiate in the thymus rather than the bone marrow. Antigens may be divided into two classes: T-dependent and T-independent. For T-dependent antigens (most proteins), processing by accessory cells such as macrophages produces peptide fragments that are bound by a major histocompatibility complex (MHC) molecule and presented to the T-cell receptor, which recognizes the bound peptide plus parts of the MHC (Bjorkman et al., 1987a,b).

Activated T cells then release lymphokines that stimulate antigen-activated B cells and program the secondary response that includes an immunoglobulin class switch and shift toward higher affinity antigen recognition (Ahmed & Gray, 1996; Paul, 1989).

Pure carbohydrate antigens generally fail to evoke a cellular response and belong to the class of T-independent antigens. Typical immune responses to such antigens are characterized by IgM antibody, poor immunological memory, and absence of a class switch following secondary antigen challenge (Perlmutter et al., 1978). Thus, many mammalian tumor-associated antigens are frequently carbohydrates, either glycolipids or glycoproteins, and the majority of tumor-specific monoclonal antibodies to these antigens are IgM antibodies (Hakomori, 1989; Thurin, 1988). Conjugation of carbohydrates to protein carrier molecules results in a T-dependent antigen and these antigens produce immunological memory and a typical IgM-to-IgG class switch (Jennings, 1983; Mäkelä et al., 1984, 1985; Moreno et al., 1979; Seppälä & Mäkelä, 1989).

The Antibody Molecule

Structural Principles

The basic structure of all immunoglobulin (Ig) molecules is a symmetrical monomer composed of two identical light (L) chains and two identical heavy (H) chains, held together by disulfide bonds (Nisonoff, 1982) (Fig. 10-2). There are two classes or isotypes of light chains, λ and κ, while there are five different heavy chain isotypes (μ, γ, α, ε, and δ). These five isotypes define the different functional classes of immunoglobulin, that is, IgM, IgG, IgA, IgE, and IgD. The major type of immunoglobulin in normal human or mouse serum is IgG, with lesser amounts of IgM. The latter possesses a pentameric structure with an IgG-like monomeric unit (Davies & Metzger, 1983). Each immunoglobulin chain is composed of domains containing approximately 110 amino acids (Nisonoff, 1982). Light chains have two domains and heavy chains have four, with the exception of the μ chain (IgM) and ε chain (IgE) which have five (Davies & Chacko, 1993; Klein, 1982; Metzger, 1970; Nisonoff, 1982; Roitt et al., 1985).

Amino acid sequence data reveals that the N terminal domains V_L and V_H possess highly variable sequences, while the other domains have constant sequences for each class (Kabat et al., 1991). These constant domains are designated C_L for light chain and C_{H1}, C_{H2}, and C_{H3} for the IgG heavy chain.

Analysis of V_L and V_H sequences reveals four segments of relatively high homology, the so-called framework regions. These are responsible for the stability of the domain and domain–domain association in intact IgG. The three smaller regions of variable sequence, hypervariable regions, or complementarity determining regions, CDR1–CDR3, vary in sequence and length. CDRs correspond to the loops that come together in space to define the size and shape of the binding site. The constant domains, as the name implies, show nearly identical sequence for each isotype or antibody class (Kabat et al., 1991).

Three-Dimensional Structure

All of the domains, from both H and L chains, have a similar overall three-dimensional structure referred to as the immunoglobulin fold (Branden & Tooze, 1991). This consists of two antiparallel β sheets packed tightly against each other. All con-

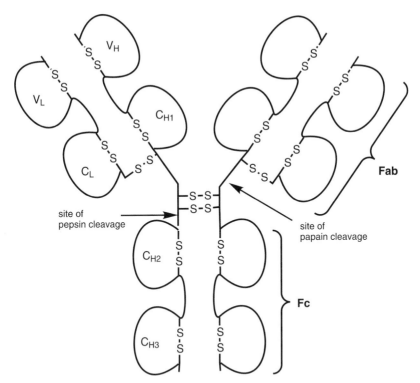

Figure 10–2. The immunoglobulin molecule IgG is built from four polypeptide chains. Two copies of the light chain and two copies of the heavy chain are connected by intrachain disulfide bonds in a symmetrical arrangement. Each chain folds into domains of approximately 110 amino acids (L chain, 2 domains and H chain, 4 domains), and intrachain disulfide bonds help to stabilize the domains. Two antigen binding sites are located at the N terminus of each L and H chain pair. The binding site is formed by six hypervariable loops, three from each of the heavy and light chain variable domains, V_H and V_L. The hinge region is an exposed peptide loop segment that is susceptible to proteolytic cleavage by enzymes such as papain or pepsin. The fragment formed from $V_L C_L$ and $V_H C_{H1}$ domains is called a Fab, and its dimeric counterpart a $F(ab)_2$.

stant domains are built from seven β strands, four in one and three in the second sheet (Fig. 10-3a). Viewed from above and to the side, this arrangement resembles a barrel where the β-strands are the staves (Fig. 10-3b). Variable domains are similar but possess an additional two β-strands, 3b and 3c, inserted into the one sheet between strands 3 and 4 (Fig. 10-4). Hairpin loops connect the strands of the β-sheets and the three loops connecting strand 2 to 3, strand 3b to 3c, and strand 6 to 7 constitute the respective hypervariable loops CDR1–CDR3 (also called L1, L2, and L3 for the light chain and H1, H2, and H3 for the heavy chain). All of the strands of the inner five-strand sheet contribute to a CDR, whereas only strand 2 from the outer four-strand sheet participates as a cross-over loop between the sheets (Fig. 10-5a).

The binding site of the antibody is created by the spatial proximity of the six CDRs (Novotny et al., 1983). There are two binding sites per IgG molecule, each one located in the Fab arm of the antibody molecule at the site created where V_H and V_L associate, V_H with V_L and C_{H1} with C_L (Fig. 10-2). For constant domains, the four-strand sheets associate *via* conserved residues that form a hydrophobic core. The variable domains V_H and V_L associate in a different manner *via* the five-strand β sheet, but again form an eight-stranded β-barrel. Each domain completes half of

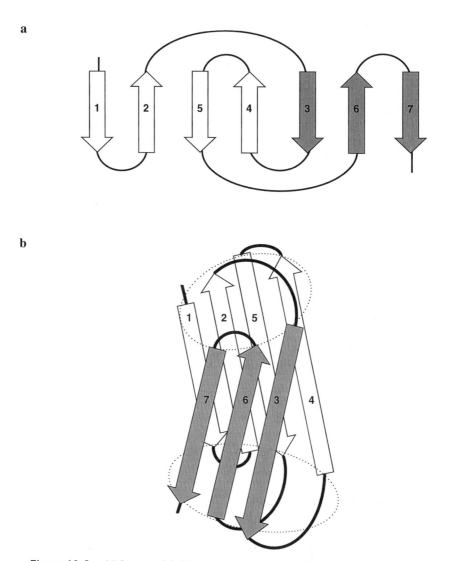

Figure 10–3. (a) Immunoglobulin constant domains are built from seven β-strands connected by hairpin loops. The amino acid sequence of each constant domain is highly conserved. The seven strands are arranged into three-strand and four-strand β sheets (open and shaded strands). (b) The sheets pack against each other in a compressed antiparallel β-barrel. The amino acid residues of the β-sheet come mainly from framework regions. Association between constant domains within a single immunoglobulin molecule occurs *via* interaction between amino acids of the four strand sheets.

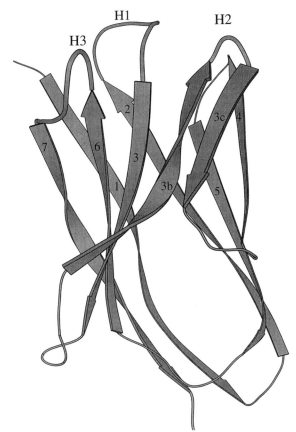

Figure 10–4. The V_H domain of a monoclonal antibody (Se155.4) is displayed in schematic form to show the nine β-strands typical of all variable domains. Compared to a constant domain, variable domains possess two extra β-strands, 3b and 3c. The two β-sheets are constructed from four strands 1, 2, 5, and 4, and five strands 7, 6, 3, 3b, and 3c. The hairpin loops H1, H2, and H3 are the regions of the V_H domain that contact antigen and generate the surface of the binding site. The six hypervariable loops of the V_H and V_L domains generate the antigen binding surface of an antibody.

the barrel with the 3c strand positioned outside the barrel (Fig. 10-5a and 5b). The CDRs are all positioned at the same end of this barrel and can create either a flat extended binding surface, frequently used to bind protein antigens, or deep cavities or grooves that are employed to bind small haptens and oligosaccharide epitopes (Novotny et al., 1983). Typical dimensions for the large interacting area employed to bind protein antigens are 20 × 30 Å, while for oligosaccharides, grooves 10 Å deep by 10 Å wide and approximately 20 Å long have been observed (Rose et al., 1993; Suh et al., 1986).

Carbohydrate Antigen–Antibody Interactions

Polyclonal Antibodies and Antibodies with Restricted Heterogeneity

The foundations of modern immunochemistry were laid by Michael Heidelberger and Elvin Kabat, investigators whose principal interests were the chemical structures of carbohydrate antigens and the properties of their antibodies (Kabat 1983, 1988). Kabat's

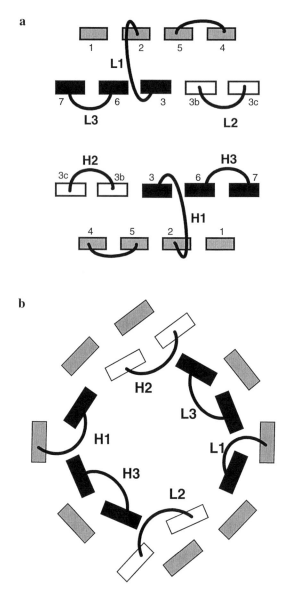

Figure 10.5. (a) Schematic of the association of the V_L and V_H domains to form the F_v portion of an antibody. The five-stranded β-sheets of each variable domain pack against each other to make the domain–domain interactions. The labeled loops H1–H3 and L1 and L3 are the hypervariable regions. The two unlabeled loops are not involved in antigen binding but like the other loops shown they join β-sheets *via* loops across the top of the variable domain. (b) The four-stranded β-sheets generate the outer surface of the β-barrel. The H2 and L2 loops with their associated 3c β-strands are positioned at the edge and are offset at the periphery of the β-barrel. It is often observed that small ligands fail to make significant contacts with one or both of these loops. (c) The β-barrel of the carbohydrate binding Mab Se155.4, that is discussed later in the chapter is displayed without the outer four-strand sheets of the V_L and V_H domains. The six hypervariable loops L1–L3 and H1–H3 are clustered at one end of the β-barrel to make contact with the antigen. Only one hairpin loop, L1 and H1, in each of the two domains links the outer four-stranded β-sheet to the inner five-strand sheet. A disulfide bond between strands 2 and 6 also secures the outer to the inner sheet in each domain.

work in particular established estimates of the size of the antibody binding site (Kabat, 1962, 1966, 1976), the structure of the human blood group antigens (Kabat, 1973), and with Heidelberger laid the foundations for the use of polysaccharide vaccines (Kabat, 1983, 1988). The early work established the importance of accurate quantitative analysis of antibody–antigen complexes (Kabat & Mayer, 1961). In particular, the use of quantitative immunoprecipitation and its related inhibition assay permitted early application of the epitope mapping concept to identify the antigenic determinants of blood

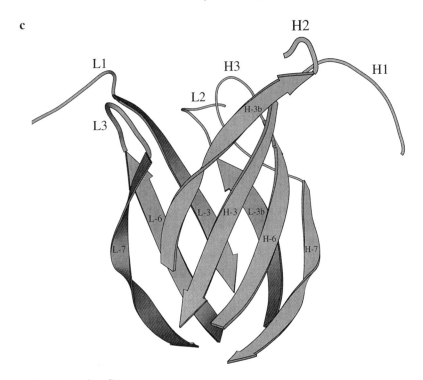

Figure 10.5. (continued)

group antigens and bacterial polysaccharides (Hakomori, 1981; Kabat, 1962, 1973, 1976; Lloyd et al., 1968; Watkins, 1980). The work led naturally to studies of the size of antigenic determinants and their complementary binding sites (Kabat, 1962, 1966, 1976; Kabat & Mayer, 1961).

In these studies, immunization of rabbits or large animals (goats, sheep, and horses) provided copious amounts of serum with an average antibody concentration in the range of 1–3 mg/mL (Kabat & Mayer, 1961). Pooled sera reflect polyclonal responses and represent average properties of a population of antibody molecules, each cocktail being unique to a particular batch of serum. Obviously nonuniform material of this type was not suitable for amino acid sequencing or crystallography, which require homogeneous protein preparations (Bott et al., 1982).

Studies of individual or pooled polyclonal sera provided informative snapshots of the average immune responses to a particular antigen as assessed by hapten inhibition (Kabat & Mayer, 1961), an early form of epitope mapping. This technique required a polyvalent antigen capable of producing immunoprecipitates to be mixed with specific antibody in a fixed volume of serum to which increasing amounts of mono-, di-, and trisaccharides had been added. The reduced amount of precipitated antibody in the presence of oligosaccharide inhibitor provided a quantitative measure of inhibition. Of course early studies (prior to 1970) were limited by the availability of well-defined oligosaccharide structures. Synthetic methods generally could only conveniently provide disaccharides, and larger structures would have to be obtained by a very limited number of clean degradative methods. However, simple monosaccharides or their glycosides were used widely to identify the monosaccharide contributing most to overall binding interactions (Kabat, 1962, 1966, 1976). This element of the antigenic determinant came to be known as the immunodominant sugar (Lüderitz et al., 1966). The general conclusions of such studies were that antibody sites could be directed toward epitopes as small as a disaccharide or as large as an octasaccharide (Arakatsu et al., 1966; Kabat, 1962, 1966). Although a trisaccharide might provide 90% of the binding energy for a hexasaccharide–antibody interaction, the remaining 10% could be contributed by all of the additional three residues that made up a hexasaccharide. In the immediate period following the publication of this data, the conclusion that a binding site was on average, sized to accommodate a hexasaccharide became a virtual dogma. However, this oversimplification ignored the subtleties of Kabat's findings, that had clearly pointed out the existence of a spectrum of binding site sizes and, furthermore, clearly indicated that by far the largest proportion of the binding energy could be accounted for in the binding of the first 2–4 saccharide residues (Kabat, 1966). Subsequent crystallographic data have validated the concept of a smaller-sized epitope (Bundle, 1997).

Kabat's work was performed with dextrans and blood group oligosaccharides from the human blood group A and B antigens, whose structures were only then becoming known (Kabat, 1973; Lloyd et al., 1968). It is interesting to observe that in the quest for larger amounts of antibodies Professor Kabat immunized himself with a variety of antigens and then donated 7.5 liters of his own serum by 15 successive plasmaphoreses (Kabat, 1980, 1983). Despite this endurance, the quantity of antibody and its heterogeneity precluded any attempts to determine amino-acid sequences of even refined subpopulations. The solution to the problem of antibody supply and uniformity marks a convenient transition point between the classical and modern approaches in immunochemistry.

Myeloma Proteins and Monoclonal Antibodies

Sequence and crystallography studies demand homogeneous populations of antibodies, and myeloma proteins (Nisonoff, 1982) provided both an immediate and eventually a permanent solution to the availability of large quantities of monoclonal antibody. These proteins are secreted as the result of a pathological condition in which one or several B-cell clones proliferate and secrete significant amounts of immunoglobulin, the product of each B-cell clone being in fact a monoclonal immunoglobulin molecule. A wide range of well-characterized myeloma cell lines were established in inbred mouse strains by Michael Potter at the National Institutes of Health (NIH) (Potter, 1972) and these antibodies could be used for sequence studies (Kabat et al., 1991) and crystallography (Davies & Metzger, 1983). Since the cell lines arose by spontaneous and random transformation of B cells, their antigen specificity was unknown and could not be preordained. To establish the probable identity of the complementary antigen necessitated screening for high-affinity antigens or ligands and was essential prior to studies of the binding site. Even though the original antigen specificity of myeloma proteins could never be determined with certainty, valuable insight into antigen binding could be obtained. A preoccupation at this time was the question of specific antigen versus cross-reacting antigen. Put another way, were myeloma proteins actually typical of the antibodies that the immune response would produce in a natural response against antigens? Subsequent experiments satisfied these concerns when it was demonstrated that antibodies identical to myelomas that bound β-galactan epitopes were generated in a normal immune response to antigens bearing this epitope (Potter et al., 1979).

Mouse and human myeloma proteins proved to be an invaluable source of aminoacid sequence data (Hilschmann & Craig, 1965; Wu & Kabat, 1970) and some crystal structure data (Amzel et al., 1974; Edmunson et al., 1974; Padlan et al., 1973, Poljak, 1978). Among the myeloma proteins, a large proportion were found to bind carbohydrates with association constants between 10^4 and 10^6 M^{-1}. The group of Neil Glaudemans at the NIH carried out pioneering work in the study of a β-galactan that bound the J539 myeloma. It was also later shown that J539 clonotypes were present when mice were immunized with β-galactan type antigens, thereby correlating the origin of the J539 myeloma with this antigen (Glaudemans, 1991; Potter et al., 1979).

Another important source of antibodies with more uniform properties arose from an observation that hyperimmunization of high-responder rabbits with bacterial vaccines was a practical route whereby an otherwise polyclonal response could be constrained to a highly restricted response and one in which the level of serum antibodies ranged from 2–20 mg/mL and in a few cases exceeded 40 mg/mL (Kimball et al., 1971). The two bacteria that were studied by this approach were *Streptococcus pneumonia* (Haber, 1970; Pincus et al., 1970a,b) and *Streptococcus heamolytica* (Krause, 1970). Despite the restricted clonal response complete sequencing of such antibodies was a considerable challenge (Braun et al., 1969, 1973; Jaton & Braun, 1972). These antibodies were used for detailed studies of binding kinetics (Jaton et al., 1975; Mäda et al., 1977).

Hybridoma Antibodies

Although the myeloma proteins represent a crucial stage in the development of our understanding of antibody structure, it was not until the development of hybridoma antibodies that a limitless supply of antibody with predefined specificity could be contemplated.

Mouse myelomas are tumor cells that can be propagated indefinitely *in vivo* or in tissue culture. Consequently myeloma cell lines were used for *in vitro* study of immunoglobulin biosynthesis (Cotton et al., 1973). During such studies, Cesar Milstein and George Köhler at the MRC laboratory in Cambridge succeeded in creating stable myeloma cell lines with predetermined antigen-binding activity (Köhler & Milstein, 1975a,b). Since the cells were tumorigenic, they could be propagated at will and, therefore, represented an indefinite source of antibody. The discovery, termed the hybridoma technique for which Milstein and Köhler were awarded the 1984 Nobel Prize in Physiology and Medicine, provided unlimited amounts of monoclonal antibody ideally suited to detailed structural studies of antigen-binding sites. The majority of solved antibody crystal structures now derive from antibodies of this type (Davies & Chacko, 1993).

In principle, the hybridoma technique does not require highly refined antigens to generate monoclonal antibodies, since the ability to clone and reclone tissue culture cell lines to monoclonality ensures that a single, monospecific antibody is derived (Goding, 1983). Purified antigens may also be a disadvantage since antigens presented to the immune system as components of whole cells or bacterial vaccines often induce a stronger immune response than the purified antigen. One example of this phenomenon is the observation that monoclonal antibodies that bind carbohydrate epitopes of glycolipids and glycoproteins were frequently observed when mice were immunized with mammalian tumor cells (Thurin, 1988), whereas glycoconjugates are poor antigens (Bundle et al., 1982). Nevertheless, pure antigen is highly desirable since it may be used in a solid phase screening assay which identifies precise antigen binding specificity early in the hybridoma protocol, thereby simplifying the experimental procedure and increasing the likelihood or recovery of monoclonal cell lines secreting the desired antibody specificity. Synthetic chemistry may make an important impact at this point and its exploitation in recent work on carbohydrate antigens will be discussed (Bundle, 1990; Bundle et al., 1982, 1994a,c).

Synthetic Carbohydrate Antigens and Oligosaccharide Inhibitors

The synthesis of glycoconjugates (sugars covalently attached to proteins) has been a long-standing approach to the production of hapten-specific antibody (Avery & Goebel 1931; Goebel & Avery, 1931) but interest was rekindled during the early 1970s as diverse strategies were developed for the covalent attachment of oligosaccharides to protein (Ashwell, 1972; Himmelspach et al., 1971; Lemieux et al., 1975a,b; Lönngren et al., 1976; McBroom et al., 1972; Schwartz & Gray, 1977). At the same time methods for the rational synthesis of oligosaccharides matured and became more widely applied (Lemieux & Driguez, 1975a,b; Lemieux, 1978; Paulsen, 1982). The techniques were applied with particular effect to bacterial epitopes of *Salmonella* by the groups of Garegg and Lindberg in Stockholm (Borén et al., 1972; Ekborg et al., 1975,1977; Jörback et al., 1984; Lindberg et al., 1983), and for the blood group antigens by Lemieux's group in Alberta (Lemieux et al., 1975a,b, 1977). As the hybridoma technique was adopted, it was clear that oligosaccharide haptens coupled to soluble protein carriers or to cells could be used to immunize mice for the development of hybridomas (Bundle et al., 1982). It was also equally clear that the same soluble glycoconjugates were excellent antigens for the solid phase assays used to screen fusion experiments conducted with cells from mice immunized by crude antigens (Carlin et al., 1986, 1987a,b).

Since the development of quantitative inhibition assays, conventional epitope mapping had been limited by the availability of well-defined oligosaccharides. Unless these were commercially available or accessible by degradation of natural antigens as part of concomitant structural studies, as was the case for blood group antigens, it was difficult to complete a comprehensive characterization of an antibody binding profile. This situation changed as chemical synthesis of oligosaccharides as large as hexasaccharides became easier (Bundle et al., 1983, 1994a; Glaudemans, 1991; Lemieux, 1989; Lemieux et al., 1984; Norberg et al., 1985). In addition strategies began to be employed not only to map out the extent of an epitope, but also to identify the functional groups that interacted with the binding site amino acids (Glaudemans, 1991; Nikrad et al., 1992; Lemieux, 1989, 1993, 1994; Lemieux et al., 1988a,b; Spohr et al., 1985a,b). Since conformational analysis for oligosaccharides in aqueous solution was developing rapidly it was also possible to infer bound orientations by making simple assumptions about the conformational preferences of the glycosidic torsional angles (Lemieux et al., 1980; Thørgersen et al., 1982). For example, if the bound and free solution conformers were similar or at least not grossly dissimilar, those functional groups residing on the same side of the molecule could be predicted. Replacement of functional groups one at a time, coupled with activity measurements, could then identify those areas of the epitope in contact with protein or bulk solvent (Lemieux, 1989, 1993, 1994; Nikrad et al., 1992). This type of strategy was adopted by several groups and developed to an advanced stage (Bock, 1983; Bundle, 1989, Bundle et al., 1994a; Lemieux, 1989, 1993; Lemieux et al., 1984, 1985; Rivera-Sagredo et al., 1991; Sierks et al., 1992).

Quantifying Oligosaccharide–Antibody Interactions

Solid Phase Assays

Enzyme and radioimmunoassays (EIA and RIA) are now most frequently performed as solid phase assays. This format provides a convenient and rapid assay of antibody binding and, when performed as an inhibition assay, it allows large numbers of inhibitors to be compared in a convenient manner in 96-well EIA plates. Several assay formats are possible. Indirect and direct detection are two of the simplest and most readily used binding assays that are also readily adapted for inhibition measurements. Indirect assays employ antigen immobilized on the solid phase and antibody binding to this antigen is detected by a protein that binds to the antibody recognizing the antigen. This protein may be a anti-Ig antibody raised in a different species to the antibody under investigation (e.g., for a mouse antibody a goat antimouse IgG could be employed). The recognition could also be based on a protein such as protein A that binds immunoglobulin molecules (Fig. 10-6). The detecting protein is covalently linked to an enzyme reporter molecule. The concentration of protein or antigen used to coat the wells of plates falls in the 1 μg/mL range while oligosaccharides are used at concentrations in the millimolar to micromolar range. Consequently the method is highly economical with respect to valuable reagents. Multiple washing steps are included between each of the following assay steps of an indirect assay:

- sensitize microtiter plate
- wash
- add test antibody
- wash
- add enzyme–protein A conjugate
- wash
- add chromogen
- read absorbance

The complex formed between antigen-bound specific antibody and protein A-enzyme conjugate is able to generate a colored product when the enzyme reacts with a colorless substrate.

a
indirect detection

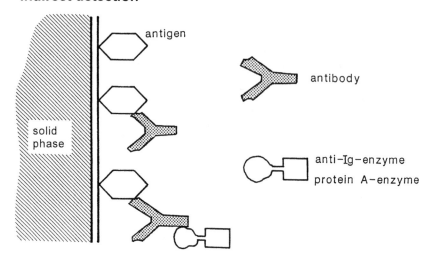

antigen

antibody

anti-Ig-enzyme

protein A-enzyme

solid phase

b
direct detection

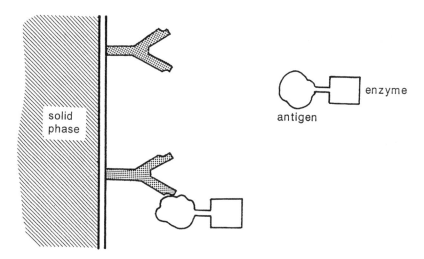

enzyme

antigen

solid phase

Figure 10–6. (a) Solid phase enzyme immunoassays (EIA) conducted in the indirect format employ antigen absorbed to the surface of plastic 96-well microtiter plates. Antibody binds to the antigen and bound antibody is detected by a universal detection system that recognizes mouse antibody. This may be a goat antimouse antibody for instance, or a molecule such as Protein A that recognizes IgG molecules. The detecting protein is conjugated to a reporter protein, usually an enzyme such as alkaline phosphatase or horseradish peroxidase. When the substrate for the enzyme is added a chromogen is produced. Assay conditions are established such that the color produced (optical density) is proportional to the concentration of bound antibody. (b) The direct detection format uses purified antibody as the immobilized phase, and the antigen is enzyme or biotin-labeled. Specific binding is detected in a similar fashion to that described for the indirect EIA. In the case of a biotin-labeled antigen, enzyme-conjugated streptavidin is used to detect bound antigen. This format is very well suited to inhibition assays.

The direct EIA employs immobilized antibody as the solid phase and this binds an enzyme-labeled antigen (Fig. 10-6b). This format is well suited to clinical assays for low molecular weight analytes and is also the preferred format for inhibition assays designed to compare relative binding strengths of structural analogues (Meikle et al., 1990; Vorberg & Bundle, 1990). Inhibition assays of this type are able to provide a fairly good estimate of the absolute association constant (K_A), and hence the free energy ΔG. Even more accurate comparisons of relative binding energies, $\Delta(\Delta G)$, are obtained when the inhibitory powers of analogues are compared to the inhibition obtained with a standard reference compound, generally the native ligand.

Isothermal Titration Calorimetry

The most general and direct method to determine the thermodynamics of chemical reactions is calorimetry. Recently, very sensitive titration microcalorimeters have become commercially available (Brandts et al., 1990; Wiseman et al., 1989). These have made possible small-scale calorimetric measurements of antibody–antigen binding, yielding information on both enthalpy and binding constant in one experiment. Previous calorimetric studies of antigen–antibody interactions only determined enthalpy or changes in heat capacities (ΔCp) and had to rely on independent determinations of binding constants to achieve a complete thermodynamic description (see, e.g. Zidovetzki et al., 1988). Titration microcalorimetry is ideal for monitoring antibody–antigen binding in solution (Bundle & Sigurskjold, 1994; Sigurskjold et al., 1991).

In a titration calorimetric experiment, small volumes of a ligand solution are added to a solution ($\sim$1 mL) of protein. Each addition gives rise to the evolution or absorption of a certain amount of heat, the magnitude of which depends on the reaction volume, concentrations, molar enthalpy, binding constant, heat of dilution, stoichiometry, and the amount of ligand added previously. The heat evolved during titration of the ligand becomes progressively smaller as the binding sites become filled by ligand. Typical raw data in the form of a thermogram and the deduced binding isotherm for the titration of a oligosaccharide ligand against a monoclonal antibody are shown in Figure 10-7. The total area of the thermogram yields $\Delta H°$ and K_A may be obtained by nonlinear regression analysis of the binding isotherm. This allows the calculation of $\Delta G°$ and $\Delta S°$.

Fluorescence Quenching and UV Difference Spectroscopy

Often antibodies show a change in fluorescence quenching and, to a lesser extent, changes in UV absorption following addition of ligand. The degree of change is proportion to the extent of ligand saturation and a plot of the change versus ligand concentration can allow accurate estimates of association constants as low as 10^2 M^{-1} (Glaudemans, 1991).

Surface Plasmon Resonance

When a protein in a miniaturized flow cell binds to a ligand immobilized in a dextran matrix that is covalently attached to a gold film coated on a glass surface (sensor chip), the refractive index close to the gold film changes. Light incident on the side opposite to the gold film is reflected and surface plasmon resonance occurring at a certain angle of reflected light is seen as a minimum in the intensity of reflected light. A sensorgram of the change allows the determination of both the kinetics of association and

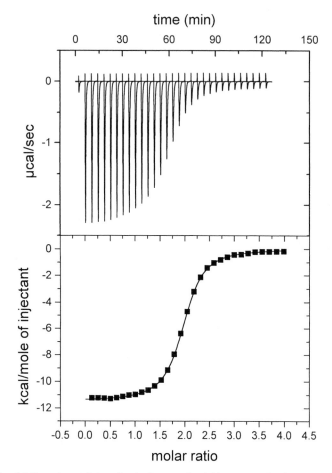

Figure 10–7. (**a**) Experimental data for the heat evolved (thermogram) when a twofold excess of saccharide ligand is titrated against an antibody solution (trace 1). As incremental amounts of ligand are added, the concentration of free binding sites drops until the antibody is saturated. At this point, the heat evolved for each ligand injection does not exceed the heat of mixing for a blank. (**b**) Regression analysis of the thermogram peak areas yields an isotherm that clearly shows saturation of the binding site by ligand. The number of protein binding sites may be calculated together with the equilibrium association constant K_a and the enthalpy $\Delta H°$.

disassociation and hence the equilibrium association constant K_a (Karlsson et al., 1991). The technique requires very little material to obtain parameters that are normally difficult to measure. However, sample preparation and data analysis require considerable care and experience if ambiguous results are to be avoided (Deng et al., 1994).

Mapping Carbohydrate Epitopes

The relative strength with which mono- and oligosaccharides (up to penta-/hexasaccharides) inhibit antibody–antigen binding establishes an estimate of the size of an antigenic determinant (Kabat, 1976). Systematic variation of the structure of each pyranose ring when it forms part of the antigenic determinant provides an even more de-tailed picture of its contribution to the free energy of association, even down to the im-portance of each hydroxyl group within a pyranose residue. Since all but the simplest oligosaccharides are unavailable commercially, estimating the size of an antigenic de-terminant requires a commitment to a program of oligosaccharide synthesis. When it is desired to complement knowledge of epitope size with an appreciation of the func-tional groups that are essential for recognition, then the synthetic undertaking involves the chemical synthesis of an extensive panel of oligosaccharide analogues (congeners). An example of the scope of this undertaking can be appreciated by the more than 60 discrete tetrasaccharide analogues of the blood group Lewis b (Le^b) antigenic deter-minant synthesized and assayed by the group of Lemieux during studies of the *Griffonia simplicifolia* lectin IV (Lemieux et al., 1988a, 1990; Lemieux, 1989, 1993; Spohr et al., 1985a; Spohr & Lemieux, 1988).

Functional Group Replacement

Several strategies have been devised to identify the key functional groups that are essential for bioactivity (Fig. 10-8). Lemieux's group developed the use of three-dimensional mapping to a highly refined level through the use of monodeoxy ana-logues in combination with conformational analysis; he was able to predict the bind-ing topography of oligosaccharide epitopes (Lemieux, 1989). Later he used mono-methyl congeners as a supplementary technique to identify solvent exposed hydroxyl groups and those located at the periphery of the complex. While monodeoxy ana-logues are most likely to show virtually unchanged binding potency in either situa-tion, mono-methylation discriminates hydroxyl groups that are located at the pe-riphery of the sugar–protein complex, since these show substantially reduced binding, due to the steric bulk of the methyl group (Fig. 10-8) (Nikrad et al., 1992). Hydroxyl groups exposed to bulk solvent exhibited unchanged binding activity when methy-lated.

Lemieux successfully integrated inhibition data for these types of congeners with an inferred model for the topography of the Le^b that contacts the lectin, by making the assumption that the bound oligosaccharide conformation closely resembles low-energy conformers observed in solution by NMR methods. It could be further inferred from this modeling that both the blood group Le^b and Lewis Y (Le^Y) tetrasaccharides ex-hibit very similar topography (Spohr et al., 1985a; cf. Lemieux et al., 1979b), at least for the position and orientation of the galactose and both fucose residues (Fig. 10-9). This three-dimensional model of the oligosaccharide was in good agreement with that subsequently identified for the bound form present in a crystal structure of the tetrasac-charide-*Griffonia simplicifolia* (GS-IV) lectin complex (Delbaere et al., 1990, 1993; Vandonselaar et al., 1987).

Whereas replacement of a hydroxyl group by hydrogen abolishes both the hydro-gen bond donor and acceptor function, crystal structure data for small molecules sug-gests that substitution of a hydroxyl group by fluorine can in principle preserve a weak hydrogen bond acceptor role at a given site (Murray-Rust et al., 1983) (Fig. 10-8). Fluorine has been employed to great effect in studies of glycosyl hydrolase enzymes and in studies of the binding sites of myeloma proteins (Glaudemans, 1991, Glaudemans et al., 1994; Street et al., 1986). In order to confirm the hydrogen bond acceptor role of a monodeoxyfluoro analogue, the activity is compared with that of the correspond-ing monodeoxy congener (Street et al., 1986; Glaudemans, 1991). The fluorine strat-

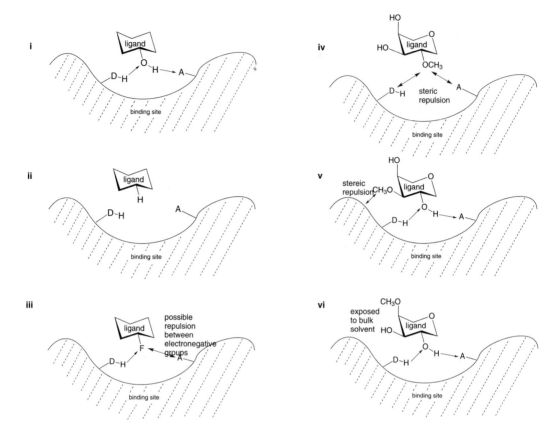

Figure 10–8. Strategies to chemically map the role of oligosaccharide hydroxyl groups in protein complexes use three different chemical modifications. (**i**) When a saccharide hydroxyl groups forms a hydrogen bond to protein it may do so as a donor or acceptor. In fact buried hydroxyl groups always donate their hydrogen atoms to an acceptor, if they accept a hydrogen bond from the protein. At more solvent-exposed areas of the binding site, water molecules often function as the acceptor. (**ii**) If the hydroxyl group is replaced by a deoxy function the role of hydrogen acceptor and donor is abolished. (**iii**) However, if the hydroxyl group is replaced by a fluorine atom, this atom may function as a hydrogen bond acceptor. (**iv**) The use of O-methyl ethers can distinguish between hydroxyl groups that are solvent exposed or located at the periphery of the binding site. Hydroxyl group that are buried become inactive when they are methylated. (**v**) Hydroxyl groups that are involved in protein interactions yet remain partially solvent exposed, experience a large decrease in activity. (**vi**) Solvent-exposed hydroxyl groups show no decrease in binding strength after methylation.

egy suffers from one drawback which is the extreme electronegativity of the fluorine atom. One crystal structure of a complex between the arabinose-binding protein and a monosaccharide has shown that conformational changes are induced in the C5–C6 rotamers of a 6-deoxy-6-fluoro-D-galactose. Repulsion between the fluorine group of the ligand and the electronegative amino acid side chain of the protein that formerly served as proton acceptor was identified as the cause of the conformational change (Vermersch et al., 1992).

Successful Epitope Mapping of the Le^b Blood Group Antigen

Lemieux's group used the chemical modification approach described above to map thoroughly the topology of Le^b, Lewis a (Le^a), H-type 2, and B blood group antigenic determinants that are bound by hybridoma antibodies and several lectins. The key polar contact groups were identified (Cromer et al., 1992; Du et al., 1994; Hindsgaul et al., 1985; Lemieux et al., 1985, 1988a,b, 1994; Spohr et al., 1985a,b, 1992a,b; Young et al., 1983), and several general conclusions emerged from these studies. A limited number of hydroxyl groups were essential to binding (2–4), and as a rule these were clustered to one side of the molecule adjacent to a less hydrophilic part of the oligosaccharide that might participate in important nonpolar interactions. The results of the extensive studies performed by Lemieux's group on GS-IV, as well as other lectins and antibodies have been summarized in reviews (Du et al., 1994; Lemieux, 1989, 1993, 1994), and their approach has been widely adopted by many groups (Bock, 1987; Bundle et al., 1994c; Kihlberg et al., 1989). Further, the data arising from the study of several recognition systems has lead to provocative ideas about the role of water in oligosaccharide-protein complexation (Beierbeck & Lemieux, 1990; Beierbeck et al., 1994; Lemieux, 1996). While Lemieux has proposed that favorable enthalpic interactions may arise from reordering of water, Carver has favored the interpretation of an entropic penalty for binding flexible saccharide ligands (Carver et al., 1991).

The Role of Conformational Analysis

An example of the insights that arise by considering the stereochemistry and conformational preference of oligosaccharide epitopes is illustrated well by the Le^b and Le^Y tetrasaccharides. These structures are built on the type 1, β-D-Gal(1 → 3)β-D-GlcNAc, and type 2, β-D-Gal(1 → 3)β-D-GlcNAc, disaccharides that are the precursors of the Le^a and Lewis x (Le^x) antigens (Fig. 10-10). Although the core Gal-GlcNAc element

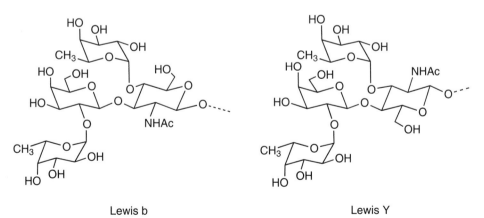

Lewis b   Lewis Y

Figure 10-9. The human blood group Le^b and Le^Y tetrasaccharides. Although the glycosidic linkages between the residues differ, these structures are almost isosteric with respect to the fucose and galactose residues.

differs in the linkage positions of β-D-Gal and the α-L-Fuc to the GlcNAc unit, there remains a virtual identity in the topology of the β-D-Gal and the two α-L-Fuc residues of the Le^b and Le^Y antigens (Hindsgaul et al., 1985; Spohr et al., 1985a). Consequently, the lectin GS-IV binds the Le^Y almost as well as Le^b tetrasaccharide (Shibata et al., 1982; Spohr & Lemieux, 1988; Spohr et al., 1985a). The crystal structure of the Le^b epitope bound by the GS IV lectin shows that the Gal residue is buried in a shallow binding site of the lectin flanked by both fucose residues each making contacts to the protein (Fig. 10-11), while the GlcNAc residue remains in a solvent exposed position (Delbaere et al., 1990, 1993). The bound conformation of the ligand is virtually superimposable upon that proposed for the solution conformation of the Le^b tetrasaccharide (Delbaere et al., 1993; Spohr & Lemieux, 1988) and the protein–sugar contacts are consistent with the Le^b epitope mapping studies, the observed activity for the

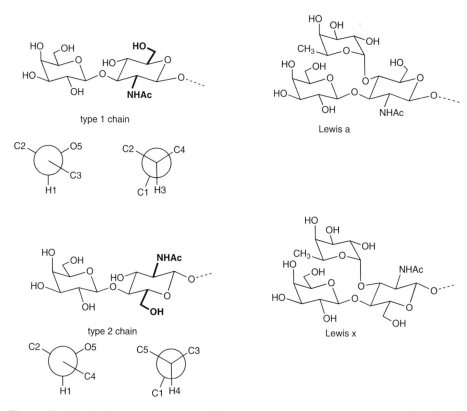

Figure 10–10. The Le^a and Le^x trisaccharide epitopes are, respectively, precursors of the Le^b and Le^Y antigens. The type 1 and type 2 disaccharides on which these structures are built exhibit interesting conformational properties. The glycosidic torsional angles that determine conformation are shown as Newman projections for the mostly heavily populated conformers (Lemieux et al., 1979a). Thus, although the position of the GlcNAc ring is rotated 180 degrees the positions of the galactose and fucose residues are virtually superimposable.

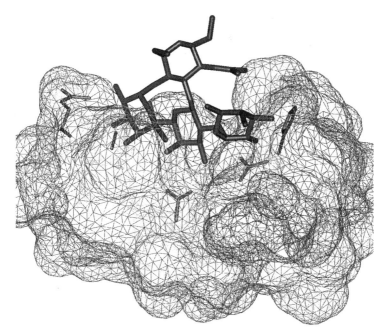

Figure 10–11. The crystal structure of the lectin *Griffonia simplicifolia* IV binding site filled by the human blood group Lewis b tetrasaccharide. The GlcNAc residue is almost completely solvent exposed and holds the galactose and fucose towards the appropriate areas of the binding site. The galactose residue is the central residue and is the most buried.

Le^Y antigen, and the conformational model for both tetrasaccharides (Lemieux, 1989) (Figs. 10-9 and 10-10).

Simple binding assays provide a broad picture of ligand activity and permit an estimate of $\Delta(\Delta G)$ for each structural modification, but more detailed thermodynamic analysis of the binding process can provide additional insights, and an appropriate return on the time invested in the synthesis of a large panel of modified oligosaccharides. The most attractive technique for this purpose is titration microcalorimetry (Raffa & Porreca 1989; Wiseman et al., 1989) which yields a full thermodynamic description of the interaction (Chervenak & Toone, 1994; Sigurskjold et al., 1991). Other methods such as changes in UV absorbance on binding (Neurohr et al., 1982b), fluorescence quenching (Glaudemans, 1991), NMR line broadening, or chemical shift changes (Kronis & Carver 1985; Neurohr et al., 1982a) allow the measurement of K_a as a function of temperature and hence the estimation of ΔH and ΔS by employing van't Hoff plots.

Physical Methods for the Study of Binding Site Structure

Thermodynamic measurements provide no information about the structure of the antibody oligosaccharide complex, although a binding model and antigen topography may be inferred from a set of thermodynamic data for oligosaccharide congeners. In principle, NMR methods might be used to determine the structure of the antibody site for small fragments such as V_H:V_L pairs (Fv) or single chain antibodies, although with a molecular weight of 25,000 Daltons such proteins are close to the limit of current NMR methodology (Kay, 1995; Levy et al., 1989; Odaka et al., 1992; Riechmann & Weill, 1993; Scherf et al., 1992; Takahashi et al., 1992). Transferred NOE experiments (Clore & Gronenborn, 1982, 1983) can provide important information on the bound conformer if the fast exchange conditions are fulfilled and relayed NOE via protein hydrogen atoms are rigorously identified (Arepali et al., 1995; Bundle et al., 1994b; Glaudemans et al., 1990; Weimar et al., 1995). At present none of these methods rivals crystallography for revealing detailed interactions at the level of atomic resolution.

Protein Crystallography

Provided a ligand can be cocrystallized with the antibody Fab, the method of choice for detailed three-dimensional structure determination is protein crystallography (Davies & Chacko, 1993; Davies et al., 1990). Numerous Fab samples have been crystallized without antigen but, unfortunately, such structures lack the desired information on antigen–antibody interactions (Rose et al., 1993). It is possible to model ligands into a site whose structure has been solved to high resolution, but considerable uncertainties exist for the positioning of the ligand (Evans et al., 1995; Glaudemans, 1991; Oomen et al., 1991; Padlan & Kabat, 1988). Docking carbohydrate ligands with protein receptors appears to be particularly difficult and at the limits of current computational methods, since a universally accepted force field has yet to be reported for carbohydrates alone or in combination with protein (Ha et al., 1988; Homans, 1990; Imberty & Perez, 1994; Imberty et al., 1990; Stuike-Prill & Meyer, 1990; Weiner et al., 1986). Therefore, the emphasis of experimental work is focused on solving the structure of ligand–protein complexes at a resolution high enough to distinguish as much atomic detail as possible, including hydrogen bonding and structural data for organized water molecules (Davies & Chacko 1993; Quiocho, 1988, 1989).

Structures of Salmonella Antigen–Antibody Fab Complexes

Hybridoma antibodies generated against specific carbohydrate antigens provide the ideal opportunity to study carbohydrate–antibody interactions. However, to date only three structures of antibody–oligosaccharide complexes have been solved and reported in the literature (Cygler et al., 1991; Jeffrey et al., 1995; Vyas et al., 1993). The general inability of carbohydrate antigens to evoke immunological memory and the concomitant class switch, no doubt accounts for the low number of carbohydrate-specific IgG antibodies (the class that gives good yields of Fab). In the following sections two discrete bacterial antigen antibody systems that have reached this detailed structural level are discussed.

Well-defined three-dimensional structures offer a valuable opportunity to assess the success of epitope mapping studies and molecular modeling approaches such as those described for the Lewis b antigen. The crystal structure of this antigen with the lectin GS IV provides an ideal opportunity to compare modeling studies of a lectin site with crystal structure data. In analogous fashion solved crystal structures of the antibacterial polysaccharide monoclonal antibodies Se155.4 and SYAJ6 in complexes with a variety of oligosaccharides, permit comparisons with modeling studies, not only for the key polar contacts and epitope topology, but also concerning the general question of bound oligosaccharide conformation. Comparisons of this type are important since only a fraction of the available sugar binding sites of proteins are likely to be successfully described by protein crystallography. Since it seems likely that structural inferences from chemical probing of protein sites will provide the largest body of data, it is important to compare the results of epitope mapping with well-defined crystal structures.

In the following sections, the crystal structures of two bacterial lipopolysaccharide antigens of *Salmonella* and *Shigella* are described in complexes with their respective monoclonal antibodies. Chemical synthesis of the native antigen repeating units are described as well as the synthesis of a range of analogues designed to map the binding sites. The thermodynamics of the interaction of IgG with the native epitope and its congeners are rationalized with the detailed three-dimensional structures of the binding site.

Salmonella Polysaccharide Antigens: Structure and Immunochemistry

Antibodies that bind certain bacterial antigens exhibit an ability to discriminate very subtle stereochemical differences; intrinsic interest in this type of recognition provides the impetus for the structure–activity studies described in the following section (Fig. 10-12).

In a 1960 experiment that echoed the 1930 discovery that artificial carbohydrate antigens could be used to immunize rabbits and induce immunological protection to otherwise lethal challenge with live bacteria (Goebel, 1939, 1940; Goebel & Avery, 1931), Lüderitz, Westphal, and Staub demonstrated that glycoconjugates of 3,6-dideoxy-L-galactose and other 3,6-dideoxyhexoses could induce bactericidal (bacteria-killing) antibodies in a goat (Lüderitz et al., 1960). The sera could also differentiate bacteria whose lipopolysaccharides (LPS) contained related but stereochemically distinct 3,6-dideoxyhexoses. Four of these sugars were identified in lipopolysaccharides from gram negative bacteria and each was given a trivial name, both D and L isomers of 3,6-dideoxygalactose (D isomer, abequose and L isomer, colitose), 3,6-dideoxy-D-mannose (tyvelose) and 3,6-dideoxy-D-glucose (paratose). Between 1950 and 1960 five of eight potential members of this new class of hexoses were isolated and identified as important constituents of the O-antigen component of bacterial LPS (Westphal, 1952; Westphal & Lüderitz, 1960). Lipopolysaccharides are found in Gram-negative bacteria and their chemical structure displays three broad structural domains, the Lipid A (responsible for endotoxin activity), the core domain bridging Lipid A and the O-antigen, the exposed polysaccharide that protrudes from the bacterial outer membrane. It became clear that the rabbit antibodies used to serotype such Gram-negative bacteria possessed binding sites that principally recognized the dideoxyhexose and the term immunodominant sugar was coined to describe this property (Lüderitz et al., 1966). Initially the inferences about the structures in which the 3,6-dideoxyhexoses occurred were based on partial chemical structures and immunochemical data (Lüderitz et al., 1966, 1971; Stirm et al., 1966) but in the late 1960s new structural methods in polysaccharide analysis (i.e., combined gas chromatography–mass spectrometry (Björndal et al., 1970)) began to yield full structural information on the LPS of *Salmonella*, leading eventually to a complete description of their O-antigen structures (Hellerqvist et al., 1968, 1969a,b, 1970a,b, 1971).

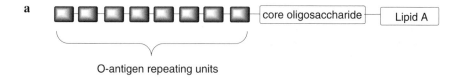

a

O-antigen repeating units

b

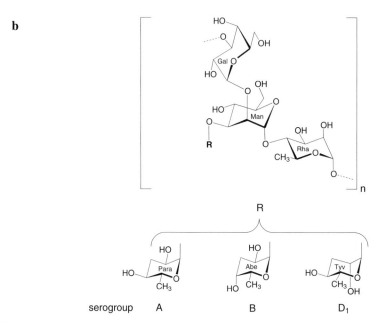

serogroup   A   B D_1

Figure 10–12. (a) Schematic representation of lipopolysaccharide structure that shows the three principal domains, Lipid A (the hydrophobic outer membrane anchor), the core oligosaccharide, and the O-polysaccharide that consists of O-antigen repeating units that may range in size from monosaccharide to hexasaccharide. (b) The chemical structures of the O-antigenic polysaccharides that are found on the outer membrane of *Salmonella* serogroup A, B, and D_1. The three serogroups are determined by antibodies that recognize the otherwise structurally uniform repeating units, that are made unique by the stereochemistry of the 3,6-dideoxyhexoses, 3,6-dideoxy-α-D-*ribo*-hexopyranose (paratose), 3,6-dideoxy-α-D-*xylo*-hexopyranose (abequose) and 3,6-dideoxy-α-D-*arabino*-hexopyranose (tyvelose).

The chemical tetrasaccharide repeating unit of the *Salmonella* group B O-antigen is based on a main chain trisaccharide α-D-Man(1 → 4)α-L-Rha(1 → 3)α-D-Gal that is common to *Salmonella* of groups A, B, and D_1 (Fig. 10-12). In these antigens a 3,6-α-D-dideoxyhexose residue forms a branch point to the α-D-Man residue, and is an immunodominant monosaccharide, that is, antibodies specific for the three serogroups A, B, and D_1 use the dideoxyhexose as the principal site for binding (Lüderitz et al., 1966). The stereochemistry of the 3,6-dideoxy-α-D-hexose (abequose in serogroup B) is the sole variable in this family of structurally related antigens, hence the stereochemistry of single hydroxyl groups on this moiety determine antibody binding (Fig. 10-12). Conformational analysis suggested that accessibility of the 3,6dd Hex(1 → 3)Man element in the serogroup A, B, and D_1 determinants as well as the constrained flexibility of this glycosidic linkage account for this property (Bock et al., 1984a,b). In order to carry out such conformational studies and to produce the defined monoclonal antibodies that eventually led to a crystal structure, synthetic oligosaccharides and glycoconjugates derived from them were synthesized chemically.

Chemical Synthesis of Glycoconjugate Antigens and Oligosaccharide Inhibitors

Glycoconjugates were synthesized using the 9-carbon linking arm, 8-methoxycarbonyl-1-octanol (**1**) (Lemieux et al., 1975a,b). In the initial synthetic strategy it was concluded that the lower reactivity of the *O*-2 position of mannose, relative to the *O*-3 position, made it advisable to introduce the galactose residue prior to the 3,6-dideoxy-hexose, abequose. The approach also allowed the synthesis of all three trisaccharide epitopes of the *Salmonella* serogroups A, B, and D_1 from a single disaccharide intermediate (**9**) (Iversen & Bundle, 1982; Pinto & Bundle, 1983). The mannose glycoside **3** was synthesized by a mercuric cyanide-assisted reaction of the mannosyl bromide **2** with alcohol **1**. After deacylation, **4** was converted to the 4,6-cyclohexylidene acetal **5** under kinetic control. Selective benzoylation gave the 3-*O*-benzoyl derivative **6**. Silver trifluoromethanesulfonate-assisted glycosylation of **6** by tetra-O-benzyl-α-D-galac-topyranosyl chloride (**7**) afforded the protected disaccharide **8**; following transesterification the selectively protected disaccharide alcohol **9** was obtained (Fig. 10-13a). This alcohol served as an acceptor for glycosylation by activated derivatives of each of the 3,6-dideoxyhexoses found in *Salmonella* O-antigens. Reaction with the glycosyl chloride donors **10–12** led to the fully protected trisaccharides **13–15** (Fig. 10-13b). Removal of protecting groups by hydrogenation, transesterification, and acetal hydrolysis gave the deprotected trisaccharide glycosides **16–18**. These were converted to the corresponding acyl hydrazides **19–21**; following reaction with dinitrogen tetroxide, the acyl azides were each treated (without isolation) with BSA (Fig. 10-14) to yield glycoconjugates **22–24** (Fig. 10-14). A related strategy was employed by Bock and Meldal (1984a,b) for the synthesis of the tetrasaccharide ligands **25–29** (Fig. 10-15) and ultimately the corresponding glycoconjugates. These tetrasaccharides as well as a trideoxy analogue were coupled to BSA to yield glycoconjugates **30–34** (Fig. 10-16). Together, the glycoconjugates **22–24** and **30–34** were used to screen hybridoma experiments for monoclonal antibodies and to probe their fine specificity.

In order to synthesize native trisaccharide **48**, modified congeners **49–56** and monodeoxy trisaccharide glycosides **57–60** (Fig. 10-17), the initial synthetic strategy was modified to avoid acetal functions that required moderately strong acid hydrolysis for deprotection. By employing acetylated **41–43** or benzylated glycosyl donors **36, 39**, and **40** with benzylated acceptor molecules **35, 38, 46**, and **47**, trisaccharide congeners of the *Salmonella* serogroup B epitope were obtained that required only a hy-drogenolysis step for deprotection, or else a transesterification and hydrogenolysis. The approach is illustrated for a variety of ligands **48–52** (Fig. 10-18a) and **57, 58** (Fig. 10-18b). In these schemes the abequose residue was introduced onto the mannose acceptor **35** prior to the galactose unit. A common disaccharide **38** was employed for synthesis of all the trisaccharides containing either modified abequose or galactose residues,

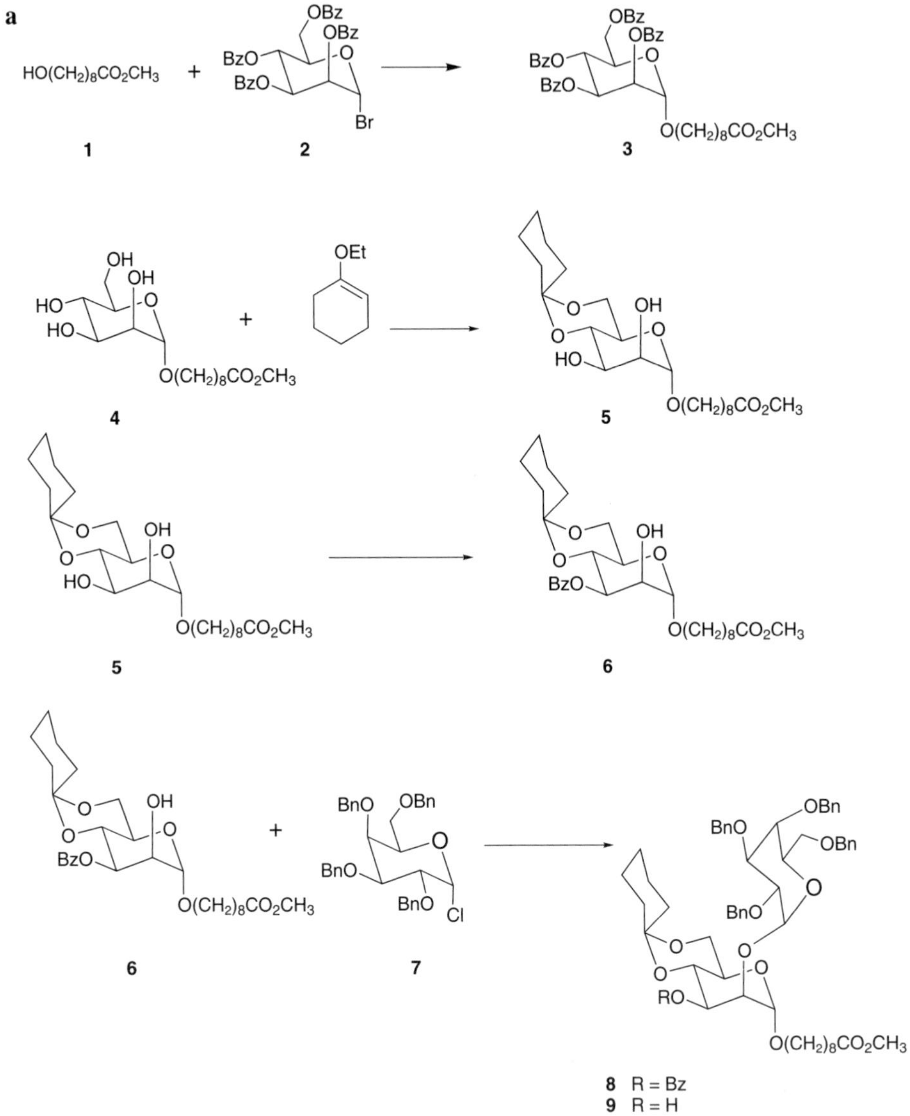

a

Figure 10–13. Synthetic strategy for obtaining all three *Salmonella* O-antigen epitopes from a common disaccharide glycosyl acceptor.

whereas trisaccharides with modified sites within the mannose unit were each synthesized from the respective 4- or 6-monodeoxy mannose glycosides **44**, **45** (Fig. 10-18b). Disaccharide epitopes **54–56** and monosaccharide derivatives **61–63** were synthesized by chemistry that had formed the basis for the oligosaccharide syntheses. Together with the trisaccharide congeners, an extensive panel of ligands was available to characterize monoclonal antibodies.

b

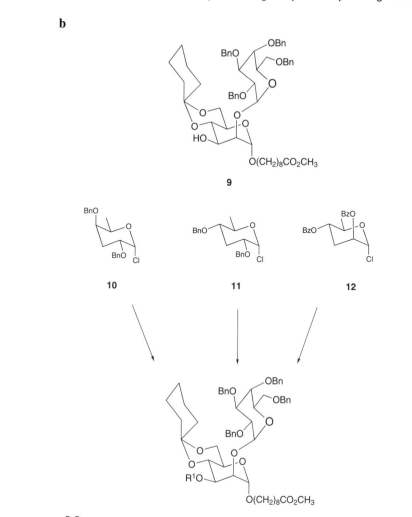

Figure 10–13. (continued)

Glycoconjugate Selection of Salmonella Monoclonal Antibody Se155.4

Synthetic ligands and glycoconjugates **22–24**, **30–34** provided a variety of opportunities to augment the intrinsic power of the hybridoma technique to provide uniform antibody molecules. Covalent attachment of a ligand to protein or other carriers provides a highly specific immunizing antigen, an approach used earlier by us to generate a blood group B antibody (Bundle et al., 1982). An alternative approach, the one used

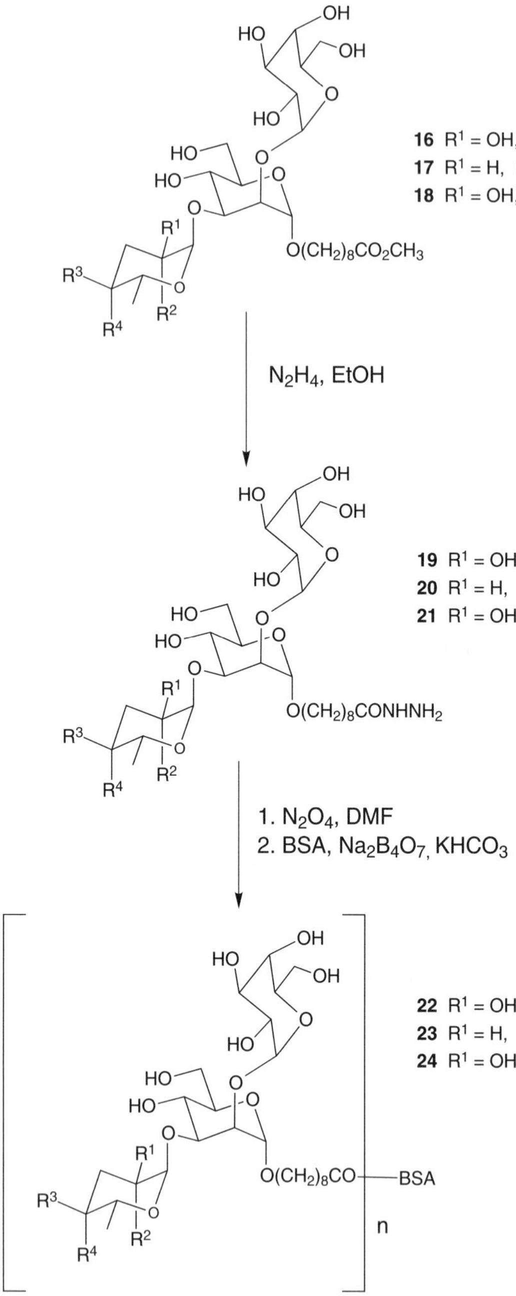

16 R¹ = OH, R² = H, R³ = H, R⁴ = OH
17 R¹ = H, R² = OH, R³ =OH, R⁴ = H
18 R¹ = OH, R² = H, R³ = OH, R⁴ = H

N₂H₄, EtOH

19 R¹ = OH, R² = H, R³ = H, R⁴ = OH
20 R¹ = H, R² = OH, R³ =OH, R⁴ = H
21 R¹ = OH, R² = H, R³ = OH, R⁴ = H

1. N₂O₄, DMF
2. BSA, Na₂B₄O₇, KHCO₃

22 R¹ = OH, R² = H, R³ = H, R⁴ = OH
23 R¹ = H, R² = OH, R³ =OH, R⁴ = H
24 R¹ = OH, R² = H, R³ = OH, R⁴ = H

Figure 10–14. Illustration of the synthetic transformations employed to couple a deprotected oligosaccharide to protein. Nitrous acid generated *in situ* or dinitrogen tetroxide produce an acyl azide which reacts with lysine on *N*-terminal amine groups.

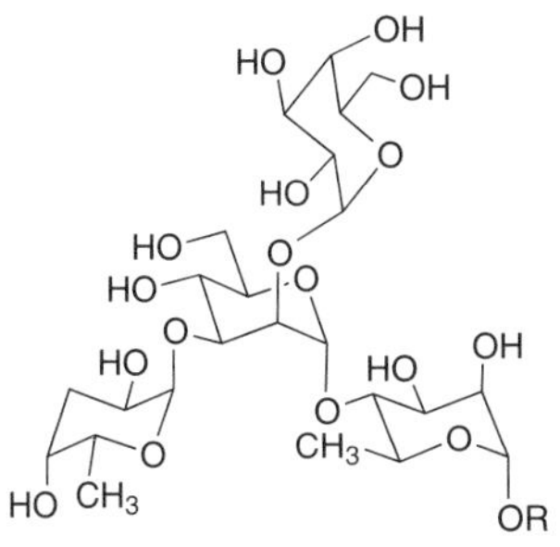

25

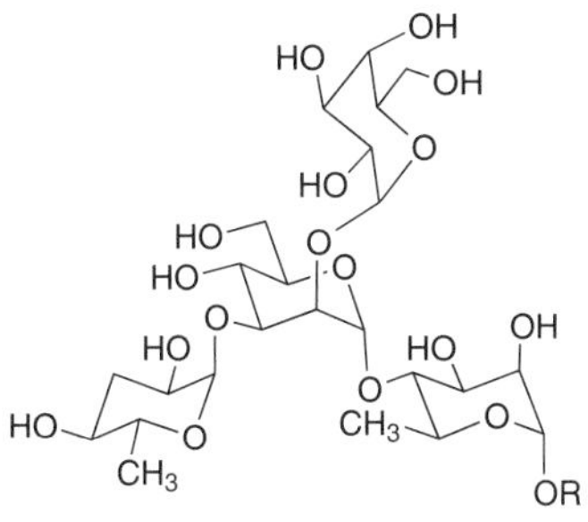

26

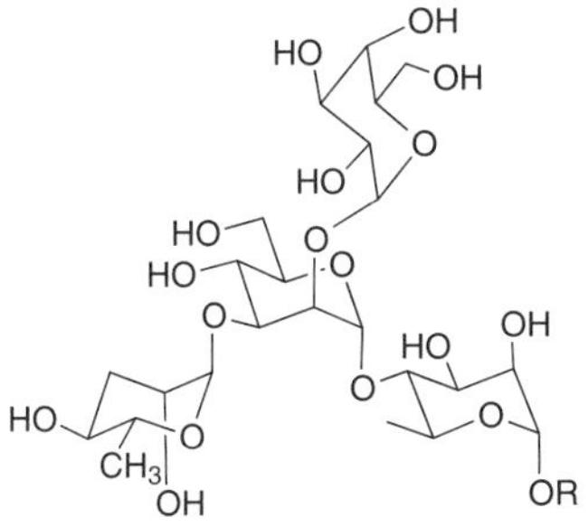

27

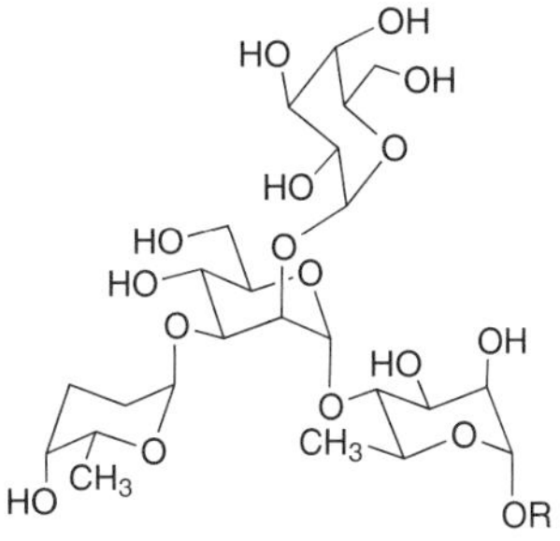

28

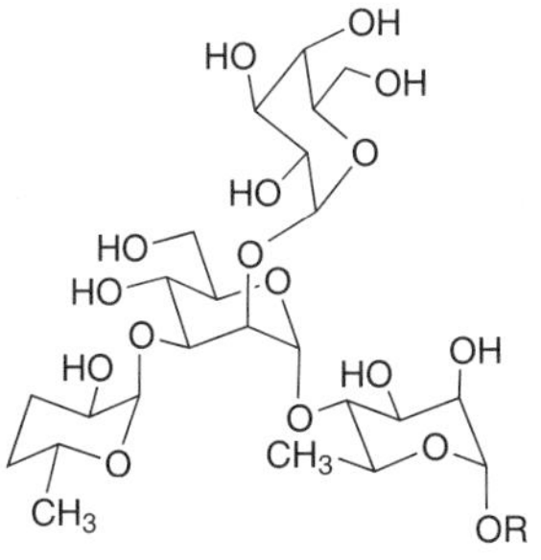

29

$R = CH_2(CH_2)_7CO_2CH_3$

Figure 10–15. Synthetic tetrasaccharide inhibitors functionalized for glycoconjugate synthesis.

here, employed the natural antigen presented on the bacterial cell wall to induce a strong immune response and the glycoconjugates were used to select cell lines that secreted antibody with defined binding characteristics. Mouse hybridomas were generated from spleen cells of mice hyperimmunized (eight injections) with killed, whole cell vaccines of *Salmonella essen* bacteria (Fig. 10-19).

Successful crystallography depends upon the crystallization of the smallest antigen-

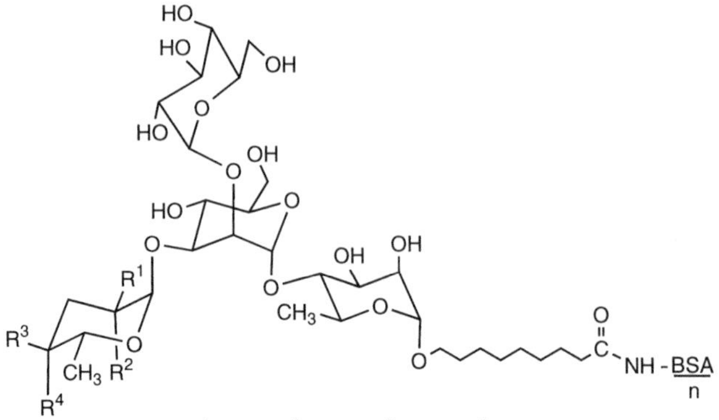

30 R^1 = OH, R^2 = H, R^3 = H, R^4 = OH
31 R^1 = H, R^2 = OH, R^3 = OH, R^4 = H
32 R^1 = OH, R^2 = H, R^3= OH, R^4 = H
33 R^1 = H, R^2 = H, R^3 = H, R^4 = OH
34 R^1 = OH, R^2 = H, R^3 = H, R^4 = H

Figure 10–16. Structures of the tetrasaccharide glycoconjugates **30–34** used with trisaccharide glycoconjugates **22–24** to select and characterize monoclonal antibody specific for the serogroup B *Salmonella*.

binding fragment, generally a Fab. In order to obtain this 50,000-Dalton fragment in acceptable yield by proteolysis of the immunoglobulin molecule, it is essential to employ antibodies of the IgG subclass. Therefore, it was necessary to adopt a screening protocol that not only detected hybridoma antibodies with the desired binding profile, but also identified monoclonal antibodies belonging to the IgG class. This facilitated a good yield of the antigen binding, Fab fragments, the starting point for crystallization trials. Class selection (detection of an IgG) could be accomplished by using a heavy chain specific, second antibody in the revealing step of solid phase assays where mouse monoclonal antibody has been captured by immobilized glycoconjugates **22** and **30** (Figs. 10-14 and 10-16). The actual screening protocol employed natural LPS antigens and glycoconjugates in parallel solid phase assays (Fig. 10-6a) to identify the binding profile of monoclonal antibodies that bound to the serogroup B antigen but not the group A or D$_1$ antigens (Fig. 10-20). The selected clone Se155.4 was one of five specific IgG antibodies that exhibited good EIA titers ($> 10^3$) against the two conjugates **22** and **30** (Fig. 10-20). It also precipitated these multivalent glycoconjugates **22** and **30**, as well as the homologous, alkali-treated LPS. Since crosslinking of polysaccharide chains is required for immunoprecipitation this observation implied binding to an epitope that was repeated along the polysaccharide chain.

Epitope Mapping of the Monoclonal Antibody Se155.4 Binding Site

The Se155.4 binding site was probed first by the synthetic glycoconjugates **22–24**, **30–34**. Weak titers were observed for Se155.4 with the LPS from the serogroup D$_1$ and serogroup A LPSs; significantly, the tyvelose or paratose containing trisaccharide glycoconjugates **23**, **24**, and the corresponding tetrasaccharide glycoconjugates **31**, **32** were not bound. The antibody did not bind to glycoconjugates **33** and **34** modified by deoxygenation at *C*-2 and *C*-4 of the abequose residue (Fig. 10-16). These experiments established that the Se155.4 antibody binding site possessed an absolute requirement for abequose, since the glycoconjugates bearing the other dideoxyhexose sugars failed to bind the Se155.4 antibody.

A detailed picture of the topography of the oligosaccharide epitope was obtained by analogue (**48–63**) (Fig. 10-17) inhibition of biotin-labeled polysaccharide antigen binding to solid phase-immobilized Se155.4 antibody (cf. Fig. 10-6b). Microcalorimetry was employed to determine the thermodynamics of interaction with Se 155.4 in solution (Table 10-1). Some of these data are summarized in terms of the free energy difference for specific functional group replacements (Fig. 10-21). Both hydroxyl groups of the abequose residue were judged essential for binding since deoxygenation at either *C*-2 or *C*-4 produced inactive glycoconjugates **33** and **34**. The mannose *C*-4 hydroxyl group also appears to be an important polar contact since its conversion to a deoxy function (**57**) caused a loss in ΔG of 0.9 kcal mol^{-1}. Thus three hydroxyl groups appear to be the key polar contacts. Since deoxygenation of the hydroxyl groups at Man *O*-6 (**58**), Gal *O*-6 (**60**), and *O*-3 (**59**) failed to cause a large change in binding

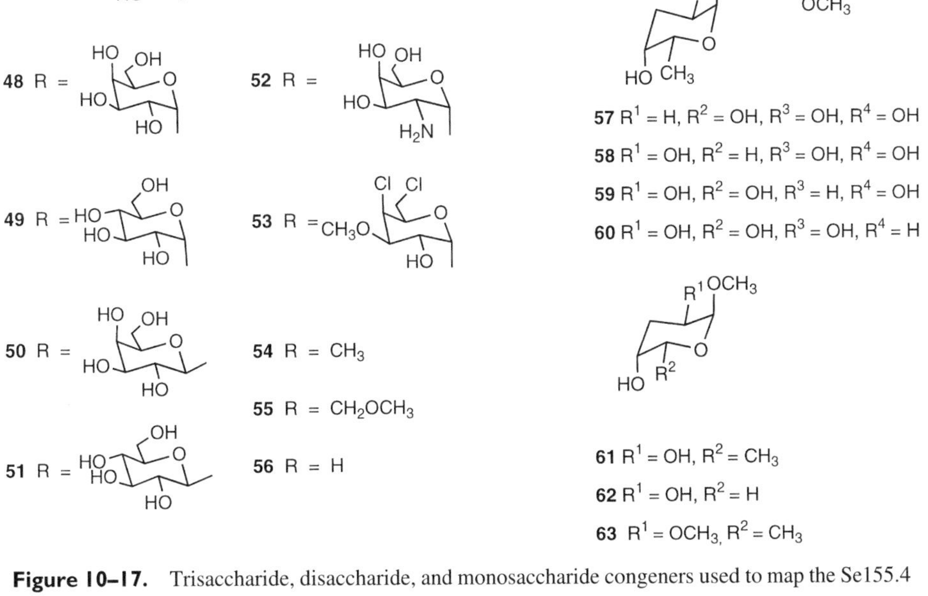

Figure 10–17. Trisaccharide, disaccharide, and monosaccharide congeners used to map the Se155.4 binding site.

a

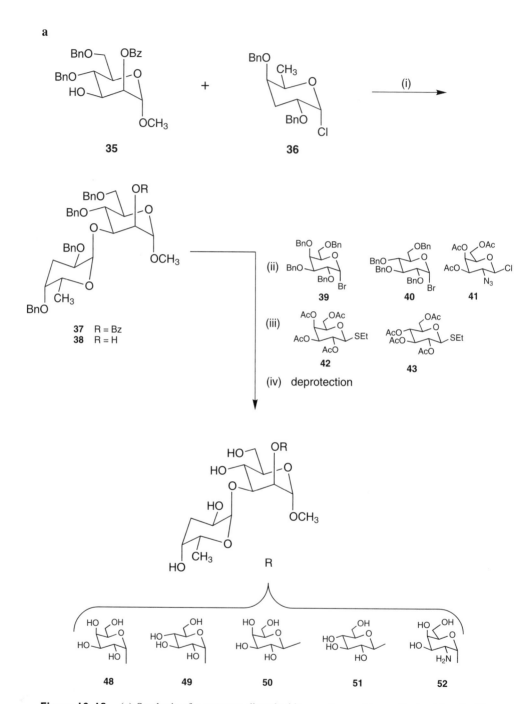

Figure 10–18. (a) Synthesis of a common disaccharide acceptor used to synthesize trisaccharide congeners **48–52**. (b) Synthesis of the 4-deoxymannose and 6-deoxymannose trisaccharide congeners **57** and **58**.

b

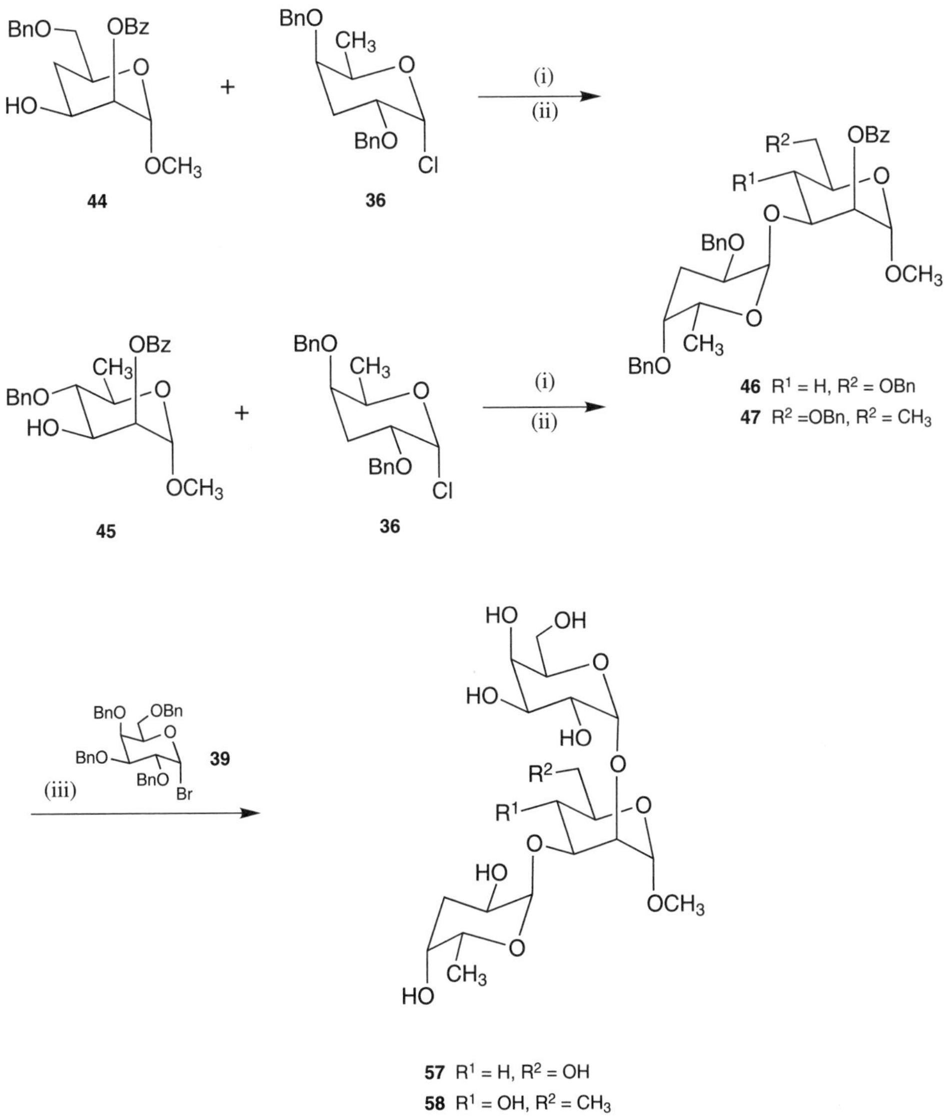

Figure 10–18.   (continued)

energy, these positions are likely exposed to bulk solvent in the complex. Both hydroxyl groups of the immunodominant sugar, abequose, are vital for activity and presumably make key polar contacts with the protein. Consequently, it is inferred that the dideoxyhexose is in intimate contact with the binding site. The galactose residue appears to be the most solvent exposed of the three residues since it may be replaced by

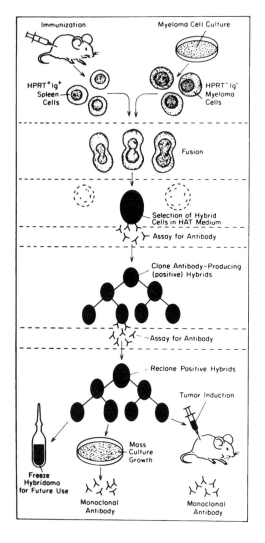

BALB/c mice immunized with
Salmonella essen 10^8 phenol - killed cells.
Three injections over 4 weeks.

fusion of 10^8 spleen cells with
10^7 plasmacytoma cells (Sp2/0).

In polyethylene glycol

Indirect ELISA (*S.essen* LPS)

Select interesting hybrids.
Clone twice, semi-solid agar.
Mouse spleen cells as feeders.

Pristane primed mice,
injected 10^6 hybridoma cells
Tap at 7 - 10 days

Whole cell agglutination
Immunoprecipitation
ELISA titration

Figure 10–19. Hybridoma protocol employed to derive the monoclonal antibody Se155.4.

a methyl group (**54**) with only a minor change in binding energy $\Delta(\Delta G)0.6$ kcal mol^{-1}. Indeed the sequential build up of the trisaccharide from abequose to the trisaccharide is seen to proceed to higher free energy not by increased enthalpy but an increase in the entropy of binding (cf. **51**, **56**, and **48**). Simple $\Delta(\Delta G)$ when comparing trisaccharide **48**, disaccharide **56**, and monosaccharide **61** do not yield the free energy contributions of the component monosaccharides because interaction energies between monosaccharide contribute toward the total free energy change (Jencks, 1981). However, the

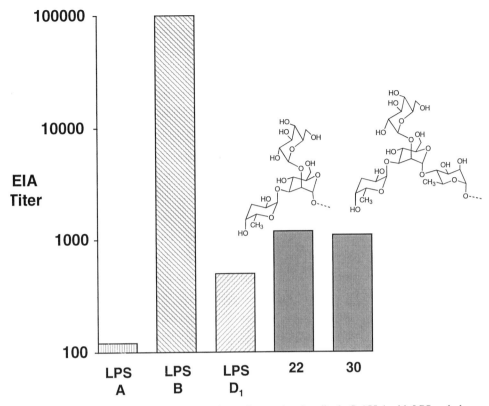

Figure 10–20. EIA binding profile and titers of monoclonal antibody Se155.4 with LPS and gly-coconjugate antigens **22** and **30**.

Table 10–1 Thermodynamics of the Binding of *Salmonella* Serogroup B O-Antigen Analogues by Se155-4

Methyl α Glycoside Ligands	K_A (M^{-1})	ΔG	ΔH kcal mol^{-1}	$-T\Delta S°$
Gal[Abe]Man (**48**)	$2.1 \pm 0.2^a \times 10^5$	-7.3 ± 0.1	-4.9 ± 0.1	-2.4 ± 0.2
Glc[Abe]Man (**49**)	$1.3 \pm 0.2 \times 10^5$	-7.0 ± 0.1	2.0 ± 0.3	-9.0 ± 0.3
Gal[Abe](4-deoxy-Man) (**57**)	$5.1 \pm 1.7 \times 10^4$	-6.4 ± 0.2	-6.6 ± 0.9	0.2 ± 0.9
Gal[Abe](6-deoxy-Man) (**58**)	$8.6 \pm 1.2 \times 10^4$	-6.8 ± 0.1	-6.3 ± 0.3	0.4 ± 0.3
(3-Deoxy-Gal)[Abe]Man (**59**)	$8.8 \pm 1.4 \times 10^4$	-6.8 ± 0.1	-7.9 ± 0.5	1.1 ± 0.5
(6-Deoxy-Gal)[Abe]Man (**60**)	$9.0 \pm 2.3 \times 10^4$	-6.8 ± 0.1	-4.9 ± 0.4	-1.9 ± 0.4
(2-Amino-2-deoxy-αGal)-[Abe]Man (**52**)	$1.3 \pm 0.3 \times 10^5$	-7.0 ± 0.1	-6.4 ± 0.3	-0.6 ± 0.3
(3-O-CH$_3$-4,6-dichloro-αGal)-[Abe]Man (**53**)	$3.0 \pm 0.8 \times 10^5$	-7.5 ± 0.1	-8.1 ± 0.6	0.7 ± 0.6
Abe(2-O-CH$_3$-Man) (**54**)	$1.2 \pm 0.2 \times 10^5$	-6.9 ± 0.1	-7.7 ± 0.2	0.7 ± 0.2
Abe(2-O-CH$_2$O CH$_3$-Man) (**55**)	$6.9 \pm 0.6 \times 10^4$	-6.6 ± 0.1	-6.4 ± 0.1	-0.2 ± 0.2
AbeMan (**56**)	$2.4 \pm 0.2 \times 10^4$	-6.0 ± 0.1	-6.9 ± 0.2	0.9 ± 0.2
Abe (**61**)	$1.5 \pm 0.4 \times 10^3$	-4.3 ± 0.1	-7.9 ± 1.1	3.6 ± 1.2

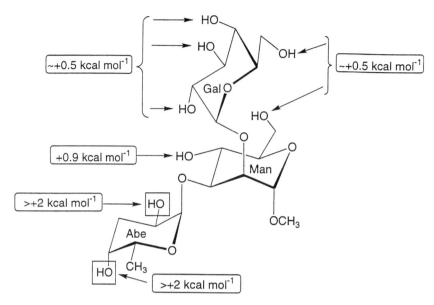

Figure 10–21. The free energy changes associated with monodeoxygenation at selected sites. Deoxygenation of OH groups in boxes yields inactive trisaccharides. The free energy change Δ (ΔG) is related to the native trisaccharide **48** (ΔG = 0). Positive numbers indicate inhibitors less active than the native trisaccharide.

trend shows entropic gains as larger saccharide segments fill the binding site and suggests the release of water from the binding site. The activity of the dichlorodideoxy analogue **53** which is unable to make strong hydrogen bonds between the protein and the galactose residue suggests that hydrogen bonds between O-3, O-4, and O-6 of the galactose residue of the native ligand and the protein binding site are weak or nonexistent (cf. Watson et al., 1994).

Antibody and Fab Purification for Crystallography

Crystallization trials require 50–100 mg of purified IgG, from which the lower molecular weight Fab fragment is derived in variable yield by proteolysis. Consequently, it was necessary to devise a simple purification procedure to secure this amount of material from ascites fluid, the antibody-rich fluid that accumulates in the peritoneal cavity of a mouse injected with hybridoma cells. An affinity matrix was prepared from bacterial O-polysaccharide, selectively aminated to allow covalent attachment to activated Sepharose gel (Altman & Bundle, 1994). The preferred approach for coupling polysaccharide to the affinity matrix was direct conjugation of a small tether, 1,3-diaminopropane, to the terminal 3-deoxyketooctulosonic acid (dOclA) residue (Lüderitz, 1970; Rietschel et al., 1992) of the polysaccharide. This eight-carbon ketosugar terminates the inner core of polysaccharide chains that have been liberated from the Lipid A by mild acid hydrolysis of LPS, and provides a site for attachment of the tether by reductive amination. Covalent attachment of the mono-aminated O-polysaccharides to either epoxy or succinimide ester-activated Sepharose 4B (Altman & Bundle, 1994) gave an affinity column with a capacity of at least 5.8 mg of antibody/mL of gel. A pH profile obtained by enzyme immunoassay (EIA) showed that the Se155.4 antibody lost activity as the pH approached 5.0. It could, therefore, be readily eluted from the affinity column with sodium acetate buffer, pH 4.5 (Fig. 10-22). The pH sensitivity of binding of the antibody is consistent with the presence of histidine residues in the binding site (see below). Antibody recovered from the affinity matrix by this very mild dissociation procedure was shown to retain full activity; samples purified in this way were used for amino acid sequencing, proteolytic cleavage to yield Fab, EIA competitive binding assays, and microcalorimetry.

Conditions were established in preliminary experiments that gave a 50% yield of Se155.4 Fab by papain digestion of mildly reduced IgG$_1$. The Fab and Fc products were readily separated on the affinity column. If the protein was not alkylated after the digestion, the H—L interchain disulfide bond was reformed by dissolved oxygen oxidation during the affinity chromatography, giving a stable species for crystallization (Rose et al., 1990).

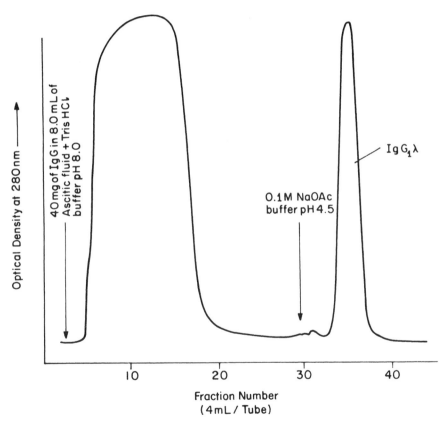

Figure 10–22. Affinity purification of the Se155.4 antibody on the polysaccharide affinity matrix. Affinity purification of the papain digest of the purified IgG$_1$ antibody can be achieved in the same manner. Antibody Fc is recovered from the flow through and Fab is eluted by lowering the pH to 4.5. Dissolved oxygen in the buffer reoxidized interchain disulfide bonds that were reduced during papain proteolysis.

Crystal Structures of Se155.4 Oligosaccharide Complexes

Several crystal structures of the *Salmonella* monoclonal antibody Se155.4 Fab or its single-chain F_v version complexed with a variety of ligands have been solved. In these, the bound trisaccharide ligand is observed to adopt either of two bound conformations. The conformational change is accompanied by changes in the number and positions of bound water molecules; this highlights the important role of ordered water molecules in the hydrogen bonding scheme and its effect on bound conformation as well as the thermodynamics of binding. The following section describes the observations for the crystal structures and solution studies of bound antigen conformation by NMR.

A Dodecasaccharide Complex

The first solved X-ray structure was obtained with data collected from crystals of a Fab–dodecasaccharide complex (Cygler et al., 1991). The dodecassaccharide consisting of three antigen repeating units was obtained by phage enzyme hydrolysis of the O-polysaccharide at the Rha–Gal linkage to give a mixture of oligomers, $n = 2–5$. The dodecasaccharide–Se 155.4 Fab complex (Cygler et al., 1991), revealed electron density that defined the positions of the abequose, mannose, and galactose residues of one repeating unit but showed no well-defined density for the remaining nine hexose residues; their position was undetermined. The crystal structure of a Fab–trisaccharide methyl glycoside complex showed a virtually identical coordination for the amino acid and carbohydrate atoms (Fig. 10-23) (Bundle et al., 1994b).

In these two structures (Table 10-2, entries 1 and 2), the antibody interacts with antigen through contacts with both the V_H and the V_L domains. The protein surface is relatively flat with the exception of a pocket approximately 8Å deep and 7Å wide formed by the CDRs near the pseudo-twofold axis of the variable domains (close to the middle of the V_H:V_L interface). The abequose residue fills and is completely buried in this depression (Fig. 10-24) while the Gal and Man residues lie across the surface of the V_H and the V_L domains and are partially exposed to solvent (Fig. 10-25). The direction of the polysaccharide chain, inferred from the positions of Gal and Man residues, is nearly perpendicular to the V_H:V_L interface. In the dodecasaccharide–Fab complex, the position of the remaining nine hexose residues could not be established

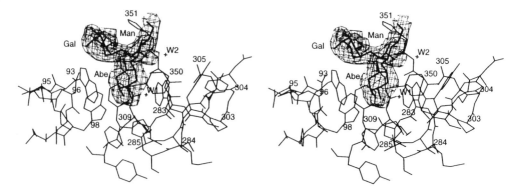

Figure 10–23. Stereo display of the Se155.4 Fab binding site with cocrystallized trisaccharide (van der Waals surface for the trisaccharide shown as wire cage).

Table 10–2 *Salmonella* Se155.4 Structures with Bond Saccharide Ligand

Complex	Electron Density for Pyranose Residues	Conformation of Bound Ligand	Intramolecular Hydrogen Bond	Resolution of Solved Structure	Reference
Fab + (Gal[Abe]ManRha)₃	3	I	Abe → Gal	2.05 Å	Cygler et al., 1991
Fab + Gal[Abe]Man → OMe (48)	3	I	Abe → Gal	2.1 Å	Bundle et al., 1994a,b,c
Fab + GalManRhaGal-[Abe]ManRha (64)	7	II	Abe → Gal Bridged by water	2.1 Å	Bundle et al.
scF$_v$ + Gal[Abe]Man → OMe (48)	3	II	Abe → Gal Bridged by water	1.7 Å	Zdanov et al., 1994

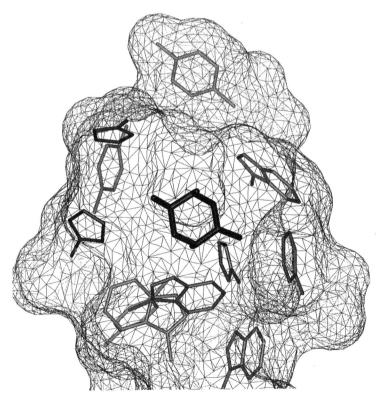

Figure 10–24. The Conelly surface (Conelly, 1983) of the Se155.4 Fab binding site cut away to illustrate the buried abequose residue viewed from above.

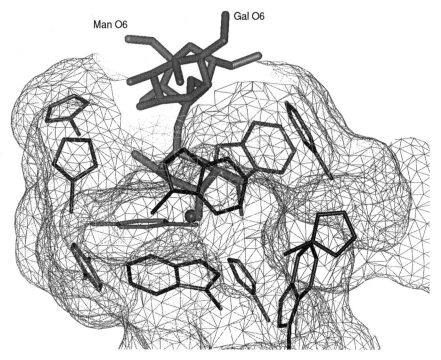

Figure 10–25. Side view of the trisaccharide **48** in the binding site. The protein is displayed as a Conelly surface (Conelly, 1983). The abequose residue is completely buried, while the upper surfaces of the Gal and Man residues are solvent exposed.

and it is impossible to be sure which one of the three repeating units occupied the site, and if there was in fact a preference. In principle, all three repeating units may bind but calorimetry data discussed later suggest that the repeating unit at the reducing end of the oligomer should have the greatest affinity for the site.

All of the amino-acid residues that interact with antigen come from the hypervariable loops (CDRs) but not all of the CDRs contact antigen. Loop L2 (light chain residues 50–56) is quite short and does not contact antigen. The other five CDRs have residues within 4.2 Å of the antigen (Table 10-3). The shape of the oligosaccharide and the an-

Table 10–3 Fab Amino Acid Residues in Contact with the Trisaccharide Epitope

Hypervariable Regions	Amino Acid Sequence
CDRL 1	R S S T G T V T S **G** N **H** A N
CDRL 2	D T N N R A P
CDRL 3	A L **W** C N N H W I
CDRH 1	N Y **W** M **H**
CDRH 2	A I Y P G N S **A** T **F** Y N H K F R A
CDRH 3	G **G** H G **Y** Y G D Y

Bold = residues contributing to Fab surface buried by trisaccharide hapten.

<u>Bold</u> = residues within 4.0 Å of the trisaccharide hapten.

tibody surface are highly complementary. A water molecule (Wat1) positioned at the base of the pocket also helps to contour the binding pocket, in addition to its important role in coordinating multiple hydrogen bonds (Fig. 10-26). The buried antibody surface area calculated with a 1.7 Å probe radius is 304 Å² while that for the antigen is 255 Å². Half this area (121 Å²) comes from protein–abequose contacts with approximately equal areas for Gal and Man. The total contact area is similar to that seen for other ligand antibody complexes (~300 Å²) but less than half the buried surface area seen for typical protein–Fab complexes (Davies et al., 1990).

Hydrogen bonds and nonpolar interactions appear to be important for binding. Based on interatomic distances (Table 10-4) and donor acceptor geometry it was concluded

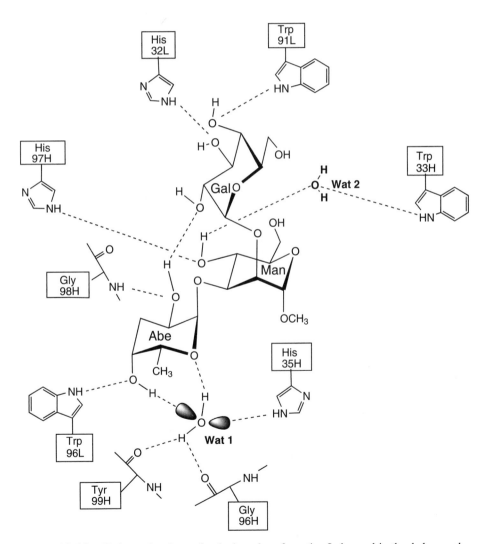

Figure 10–26. Hydrogen bond map for the bound conformation I observed in the dodecasaccharide–Fab complex (Cygler et al., 1991).

Table 10–4 Fab-Oligosaccharide Hydrogen Bonds

		Heavy Atoms of Hydrogen Bond Donor-Acceptors		Interatomic Distance (Å)
Abequose	O2	Galactose	O2	2.91
		Gly102H	NH	2.69
	O4	Trp98L	NE1	2.88
		Water #1	O	2.97
	O5	Water #1	O	3.02
Galactose	O2	Mannose	O2	2.83
			O3	2.88
	O3	His34L	NE2	2.94
	O4	Trp93L	NE1	3.19
		Water #2	O	3.28
Mannose	O4	His101H	ND1	3.00
Water #1	O	His35H	NE2	2.84
		Tyr103H	O	3.20
		Aberquose	O4	2.97
			O5	3.02

that all the hydroxyl groups with the exception of those at Gal C-6 and Man C-6 could form hydrogen bonds. In agreement with the epitope mapping studies, the critical hydrogen bonds obviously involve the buried abequose residue. Abequose O-4 accepts a hydrogen bond from Trp96L of CDR L3 and the proton of the hydroxyl group is donated to the bound water molecule Wat1, which is itself hydrogen bonded to one or both backbone carbonyl groups of CDR H3 residues Tyr99H and Gly96H (Fig. 10-26). In general it is unusual for hydroxyl groups of carbohydrates to act only as acceptors. In fact, considerable energy penalties can arise if a buried OH group fails to form appropriate hydrogen bonds (Quiocho, 1989). Consequently as first noted by Quiocho, bound carbohydrates are frequently involved in networks of hydrogen bonds of the type NH $\rightarrow$ O—H $\rightarrow$ O. In the Se155.4 structure, the water molecule forms part of such a network with additional hydrogen bonds from His35H of CDR H1 and donation from water to the Abe O-5. Gly98H of CDR H3 residue, donates a hydrogen bond to Abe O-2; its proton is positioned to form a hydrogen bond to the closest acceptor atom, which happens to be another sugar residue, Gal O-2 (Table 10-4). In order to bring Gal O-2 sufficiently close to Abe O-2, the Gal $\rightarrow$ Man torsional angles ϕ, ψ must adopt values that are shifted (by some 40° and 20°, respectively) from those of the minimum energy conformation (Table 10-5). Without considering the energy gained by formation of the intersaccharide hydrogen bond, the potential energy of this conformation lies some 3–4 kcal mol^{-1} above that of the trisaccharide minimum energy conformation. Mannose O-4 is hydrogen bonded to the CDR H3 residue His97H, and a bound water molecule presumably acts as the acceptor of the OH proton. The light chain CDR L3 residue Trp91L is close enough for a hydrogen bond with O-4 of Gal, while another L chain residue His32L, (CDR L1) may be a hydrogen bond donor to the Gal O-3. The tryptophans 91L and 33H provide important nonpolar interactions between the corresponding hydrophobic surfaces (Cygler et al., 1991).

This Fab binding site is unusual among carbohydrate binding proteins in that only aromatic side chains contact the oligosaccharide (Tables 10-3 and 10-4). Apart from the backbone amide of Gly102H and the water molecule Wat1, tryptophan or histidine residues are the sole hydrogen-bonding partners with the oligosaccharide oxygen atoms. No charged or polar, planar side chains are to be seen among the hydrogen-bonding partners, in contrast to the active sites of lectins, transport proteins and enzymes (Johnson et al., 1988; Quiocho 1986, 1988, 1989). This difference is manifest in thermodynamic studies where hydrophobic contributions to the binding have been noted (Sigurskjold & Bundle, 1992). The gain in entropy as larger saccharides fill the binding site (Table 10-2) is also consistent with these observations.

Conformational Changes on Binding

In published antibody structures the extent of antigen and antibody movement upon complex formation has been shown to span a range of behavior. At one extreme the structure of a protein antigen and antibody binding site showed no change on complex formation (Bhat et al., 1990; Davies & Chacko 1993; Rini et al., 1992). While mutual antigen–antibody adaptations have been observed in several crystal structures (Colman, 1988; Davies et al., 1990), ligand conformational shifts (Glaudemans et al., 1990; Scherf et al., 1992), or shifts in antibody hypervariable loop structure (Arevalo et al., 1993; Rini et al., 1992; Stanfield et al., 1990; Wilson & Stanfield, 1994) have also been noted. Comparison of the various complexed and uncomplexed structures of Se155.4 Fab shows that despite differences in intermolecular contacts, the conformational changes of the antibody are very small, especially in the area of the binding site. Most of the small changes involve movements of histidine side chains. The binding pocket retains its size, shape, and side chain orientation in the presence or absence of abequose. The water at the base of the binding site does not change in any structure so far solved. When the sugar epitope is absent, four water molecules enter the cavity and two are found near the positions formerly occupied by sugar hydroxyl groups.

In the trisaccharide–Fab complex it appears that the antibody induces a conformational change on the most exposed portion of the trisaccharide epitope, the galactose residue, the movement of which occurs most readily *via* adjustments in the Gal → Man glycosidic torsional angles (Table 10-5). This raises the possibility that the trisaccharide intramolecular hydrogen bond observed in the crystal structure (Fig. 10-26) corresponds to a significantly populated solution conformer. In this case the ligand con-

Table 10–5 Glycosidic Torsional Angles that Define Bound Conformations I and II of the Trisaccharide Epitope

	Glycosidic Torsional Angles	
Bound Conformation	Abe1 → 3Man ϕ,ψ	Gal 1 → 2Man ϕ,ψ
Conformation I (Fab + trisaccharide)	75°, 97°	107°, 85°
Conformation II (scFv + trisaccharide)	72°, 102°	76°, 145°
Global Minimum Energy	65°, 105°	70°, 105°
NMR transferred NOE (solution)	50°–70°, 80°–100°	90°–110°, 90°–110°
		60°–70°, 90°–110°

formational shift on binding would be minimal. On the other hand, there seems to be no reason for intramolecular saccharide–saccharide hydrogen bonds to form in the high dielectric medium of aqueous solution. Solution NMR studies not only provide a tool to investigate the solution conformation and possible hydrogen bonding of the free ligand but are especially useful under conditions of fast exchange, when the transferred NOE experiment can report on the conformation of bound oligosaccharide (Clore & Gronenborn, 1982, 1983).

Hydrogen Bonding in the Unbound Oligosaccharide Epitope

The crystal structure of the trisaccharide–Fab complex showed an Abe O-2 to Gal O-2 interatomic distance of 2.9 Å (Table 10-4). Is the hydrogen bond present in solution when the trisaccharide is not bound? If so it should be detectable by NMR methods such as those recently applied to sucrose and sialyl lactose (Poppe & van Halbeek, 1991, 1992; Poppe et al., 1992). Generally it is difficult to observe intramolecular hydrogen bonds in aqueous solutions due to chemical exchange phenomena but presteady-state NOE experiments (Poppe & van Halbeek, 1991) and isotope effects of O^1H/O^2H hydroxyl groups on ^{13}C chemical shifts (Christofides et al., 1986) have been used to detect intramolecular hydrogen bonds for oligosaccharides dissolved in H_2O solutions (Poppe et al., 1992). Neither of these techniques revealed the existence of a persistent hydrogen bond in aqueous solution.

Observation of isotope shifts in dimethyl sulfoxide (DMSO) solution may detect intramolecular hydrogen bonds in oligosaccharides but in this solvent the conformational behavior may differ from that in water. The anhydrous trisaccharide was dissolved in dry DMSO and the ^1H NMR spectra were observed as increasing amounts of D_2O were titrated into the trisaccharide solution. Hydrogen-bonded OH resonances, that in the simplest case appear as doublets (Fig. 10-27a), experience isotope shifts and appear as multiplets as D_2O is added, allowing the identification of the hydrogen bond donor and acceptor pairs (Bock & Lemieux, 1982). Thus a hydrogen-bonded hydroxyl group will exhibit a deuterium isotope shift as the hydrogen atom attached to an acceptor oxygen atom is exchanged for deuterium (Fig. 10-27b). In DMSO no hydrogen bond could be observed between O-2 Abe and O-2 Gal but one was detected between Abe O-2 and Man O-4 (Fig. 10-27a, middle and lower spectra) (Bundle et al., 1994b). In water, the intramolecular saccharide–saccharide hydrogen bonds are not sufficiently strong to stabilize a preferred conformation, and, if formed, they are only transient as various families of conformations are sampled. In DMSO, a solvent-dependent conformational shift occurs. If an Abe O-2 to Man O-4 distance of 2.8 ± 0.25 Å is assumed for a hydrogen bond, the Abe–Man torsional angles must change to satisfy this condition and these glycosidic torsional angles are substantially shifted from the global minimum energy conformation (Table 10-6). DMSO-dependent conformational shifts have been reported for a galabiose structure, α-D-Galp$(1 \rightarrow 4)$ β-D-Galp (Bock et al., 1988).

Solution Conformation of the Free Oligosaccharide

The conformational preference of oligosaccharides are dominated by stereo-electronic effects (exo-anomeric effect, [Lemieux 1978; Lemieux et al., 1979a]), and nonbonded interactions. If allowance is made for motion solution, structures may be investigated by quantitative NOE measurements. These experiments provide a time-averaged distance between the anomeric proton and the proton attached to the glycosylated carbon

a

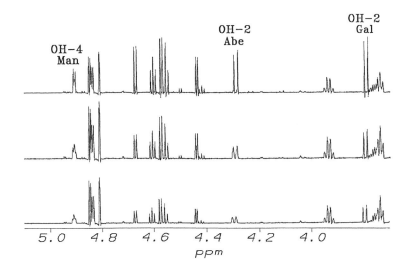

b

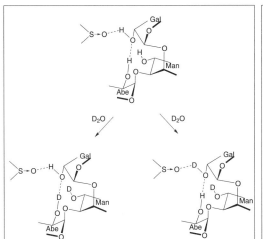
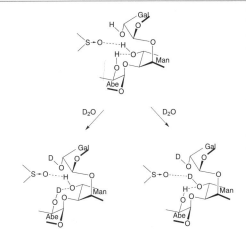

Figure 10–27. (a) ^{1}H NMR spectra of trisaccharide **48** in DMSO d$_6$ solution. The top trace is the spectrum prior to titration with D$_2$O. The subsequent 2 traces correspond to the spectrum with increasing amounts of D$_2$O. The Gal 2-*OH* resonance is unchanged but those of Abe 2-*OH* and Man 4-*OH* become more complex multiplets as the protons experience deuterium isotope effects. (b) The two possible intramolecular hydrogen bonds between Abe—Gal or Abe—Man are illustrated together with the isotopic substitution that occurs when 0.5 molar equivalent of D$_2$O is added.

Table 10–6 Potential Energy Minima for Trisaccharide **48**

Potential Energy Minima	Gal ϕ,ψ	Abe ϕ,ψ	Relative Energy (kcal mol^{-1})	Hydrogen Bond Formation	Gain in Relative Free Energy Due to Hydrogen Bond (kcal mol^{-1})
M1	70°, 105°	65°, 105°	0	Abe O—2 . . . Gal O—2	−0.8
M2	80°, 145°	70°, 125°	−0.5	Abe O—2 . . . Gal O—2	−0.3
M3	95°, 95°	60°, 95°	−2.4	Abe O—2 . . . Gal O—2	
M4	125°, 125°	135°, 145°	−4	Abe O—2 . . . Man O—4	−0.8
M5	80°, 120°	100°, 100°	−3.1	Abe O—2 . . . Man O—4	−0.5

$\phi = $ O5—C1—O1—C$'_x$, $\psi = $ C1—O1—C$'_x$—C$'_{x+1}$.

atom of the adjacent ring. Several inter-ring distance constraints may be used to define a range of ϕ,ψ angles which are consistent with the observed NOEs. However, there are many caveats to the quantitative interpretation of NOE data and reference to these are to be found in reviews of the recent literature (Homans, 1990; Peters & Pinto, 1996; Rutherford et al., 1993). Flexibility and the time scale of motions about the glycosidic torsional angles are significant causes of divergence between experimentally observed and calculated NOEs. Approximately 12 years ago we carried out a detailed analysis for the *Salmonella* antigenic determinants and a model was proposed for the solution conformation (Bock et al., 1984a,b). In a more recent work, similar data were collected and re-examined using methods that allow a more realistic treatment of motional averaging and weighting of the quantitative NOEs across the potential energy surface of the glycosidic linkages (Cumming & Carver, 1987). This work demonstrated that considerable conformational averaging occurred; no single conformer corresponding to global or local minima or even a Boltzman distribution over all conformational microstates could account for the experimental NOE data (Bundle et al., 1994b). The underlying causes for the inability to reconcile the experimental data with a conformational model for the oligosaccharide were attributed to anisotropic motion (which requires special treatment of the NOEs arising from a preferred but undefined axis of rotation for the oligosaccharide) and an unfavorable time scale for the internal motions about the glycosidic linkages (Hricovini et al., 1992).

Although a precise conformational model for free trisaccharide eluded the detailed analysis, it was clear that the NOEs to be expected from any of five local potential energy minima were sensitive to ϕ,ψ changes and the observed NOEs were quite distinct from those seen for the bound state. These data implied that a conformation change occurred when the trisaccharide was bound by antibody and this could be substantiated by a detailed analysis of transferred NOE experiments (TRNOE).

Bound Trisaccharide Conformation

Transferred NOE measurements showed that the problems of interconverting oligosaccharide conformations detected by NOE difference spectroscopy for free ligand were simplified considerably when oligosaccharide was bound by Fab. When the trisaccharide was added to Fab, one-dimensional NOE experiments showed that previously positive NOEs for the oligosaccharide became negative, indicating that the experiment

Table 10–7 NOE Changes on Binding of Trisaccharide **48** to Se 155.4 Fab

Inter-residue NOEs		Solution	Bound State
Gal H1	Abe H3ax	w	s
	Abe H5	m	w
Gal H2	Abe H3ax	—	w
	Abe H5	—	—
Gal H5	Man H1	m	m
Abe H1	Man H3	s	s

w, weak NOE; m, medium NOE; s, stray NOE.

was sampling the bound conformer (Clore & Gronenborn, 1982, 1983). Long-range NOEs between two pyranose residues that are not covalently linked provide a very sensitive probe of the oligosaccharide conformation since small changes in the ϕ, ψ angles lead to large changes in interproton distances. Thus the diminution of the NOE Gal H-1 to Abe H-5 and the appearance of a new NOE Gal H-1 to Abe H3ax (Table 10-7) were the most notable features of the TRNOEs (Fig. 10-28). These changes when the ligand is bound are consistent with an induced fit of the trisaccharide epitope.

A series of phase-sensitive NOESY experiments were recorded for seven discrete mixing times in the range 50–300 ms. The number of interproton distance constraints obtained from the TRNOE experiment was insufficient to define a unique bound conformation. Consequently, self-consistent solutions that satisfied the set of interproton distances were compared with energetically accessible conformers.

Conformations of bound trisaccharide that satisfied the distance constraints between the noncovalently linked Abe and Gal residues (Gal H-1/ Abe H-3ax and Gal H-1/ Abe H-5) fell within the ~5 kcal mol^{-1} potential energy contour of the trisaccharide potential energy map. This holds for both the Abe–Man and Gal–Man glycosidic linkages, whether the contour is calculated without or with a hydrogen bonding potential for the Abe O2–Gal O2 interaction. The Abe–Man glycosidic linkage is unaffected by

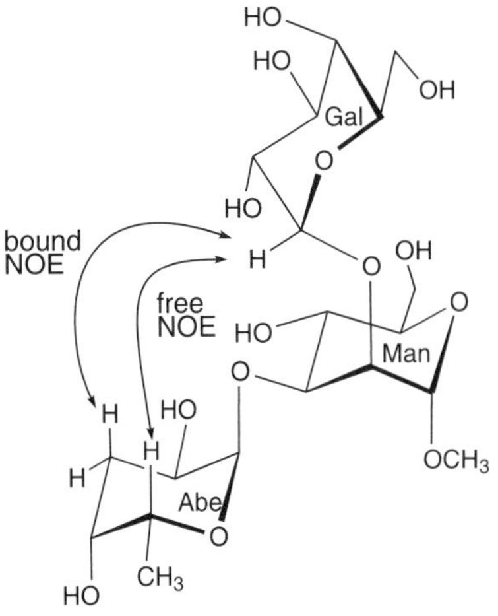

Figure 10–28. Changes in observed NOEs between the free and bound states of the trisaccharide **48**. In the bound state the NOE from Gal H1 to H3ax is observed, while in solution this NOE is not observed. An NOE is seen from Gal H1 to H5 Abe.

the energy constraints, since all distance constraints place acceptable ϕ,ψ values for this linkage within a 1–5 kcal mol^{-1} contour. However, the situation is distinct for the Gal–Man linkage, where the inclusion of a hydrogen bonding term causes a segregation of solutions at the 1 kcal mol^{-1} level. Consequently, there are two Gal–Man conformations at $\phi,\psi \sim 70°/100°$ and $\sim 100°/100°$ (Table 10-5), that are consistent with the combined TRNOE and energy constraints. When error bounds are considered, the second conformer is essentially indistinguishable from the conformer I found in the crystal structure of the Fab–dodecasaccharide complex, while the former is closer to the global potential energy minimum. The solutions are separated by an estimated 3–4 kcal mol^{-1} barrier. Thus the NMR data suggests the possibility of two distinct bound conformations or averaging between them. Subsequently this conclusion was supported by the observation of a second and different bound saccharide conformation in two additional crystal structures (Table 10-2).

Comparison of Oligosaccharide Solution NMR Conformation and Bound Conformations in Crystal Structures

The solution conformations may be compared with those determined for the crystalline state, but a difficulty does arise. Crystallography accurately determines interatomic distances from heavy atom coordinates and, consequently, the conversion of these into a set of torsional angles with realistic error bounds presents certain difficulties. However, within $\pm 10°$ error bounds, the ϕ,ψ angles for the crystal structure were 104°, 86° for the Gal–Man linkage and 78°, 95° for the Abe–Man linkage. As mentioned above, the NMR-determined Gal–Man torsional angles $\phi,\psi = 100°/100°$ are within experimental error of the $\phi,\psi = 104°/86°$ intramolecular hydrogen-bonded crystal structure conformation.

A second ligand conformation II with altered hydrogen bounds and protein–galactose distances was observed in a recent 1.7 Å resolution structure of a trisaccharide complexed with the single-chain variant of the Se155.4 antibody. The Abe–Man torsional angles were unchanged but the Gal–Man torsional angles were estimated at 77°, 144° a conformer that is closer to the minimum energy conformer and the second NMR-determined conformation of $\phi,\psi = 70°/100°$ (Fig. 10-29). The new conforma-

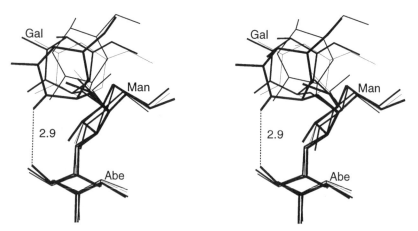

Figure 10–29. In order of decreasing thickness, stereoplots of bound conformation I, the minimum energy molecule and the two conformers inferred by TRNOE NMR experiments.

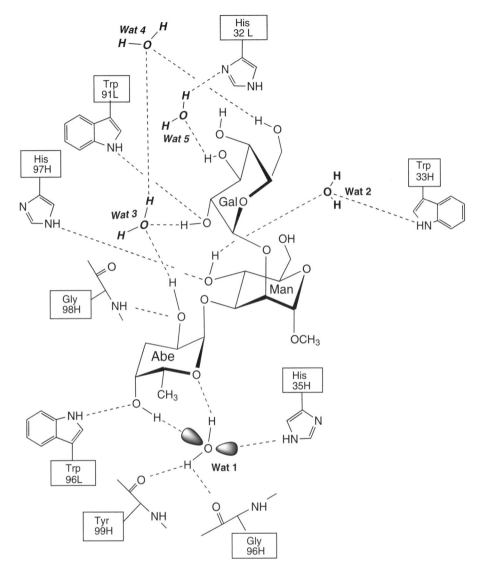

Figure 10–30. Hydrogen bond map for the bound conformation II observed in the heptasaccharide–Fab complex and the trisaccharide **48**–scFv complex (Zdanov et al., 1994).

tion of the galactose residue results from the presence of a new ordered water molecule, Wat2, that bridges the Abe O-2—Gal O-2 hydrogen bond (Fig. 10-30). The hydrogen bonding pattern involving the Gal residue of this structure now more closely matches that deduced by epitope mapping.

It is of interest that the two bound conformations fall within a narrow range of conformers, all of which lie close to minima predicted by potential energy calculations. In the sense that it is possible to predict the approximate relative orientations of adjacent pyranose rings and hence a model of antigen topography, these studies justify the cautious use of such models to rationalize the thermodynamic changes that accompany functional group replacements (Bundle et al., 1994a).

Thermodynamics of Oligosaccharide-Antibody Interactions

Resolution of the Gibbs free energy $\Delta G°$ into enthalpic ($\Delta H°$) and entropic ($\Delta S°$) contributions provides an estimate of the relative contributions of attractive forces, hydrogen bonding and van der Waal interactions, and of changes in the molecular order, particularly solvent reorganization, when an oligosaccharide–protein complex is formed. Interpretation of the $\Delta H°$ and $T\Delta S°$ at the molecular level is fraught with problems since multiple and simultaneous events generate compensating effects. However, without accurate values of $\Delta H°$ and heat capacity, ΔCp, even a modest level of comprehension of intermolecular forces is impossible. In studies involving functional group replacement, comparison of $\Delta(\Delta H)$ and $\Delta(\Delta S)$ values are simpler to correlate with structural variation than comparison of individual $\Delta H°$ and $\Delta S°$ quantities. Calorimetry is one of the few and certainly the most reliable technique for determining $\Delta G°$, $\Delta H°$, $\Delta S°$, and ΔCp, and improvements in calorimeter design now allow these parameters to be determined with impressive accuracy (Wiseman et al., 1989).

Strong enthalpic interactions originating with a single monosaccharide dominate the oligosaccharide–Se 155.4 antibody interactions, even though the antibody binding site is complementary to a trisaccharide-sized epitope. Thermodynamic data for binding of trisaccharide **48**, the monosaccharides **61–63** and disaccharide derivatives **54–57** show in striking fashion that the abequose residue provides the enthalpy of binding, which is consistent with the hydrogen bonds and tight fit of the abequose residue at the base of the binding site (Figs. 10-24–26). In contrast, galactose and mannose contribute only modestly to free energy. Furthermore, the latter contributions arise mainly *via* favorable entropy, suggesting that an important driving force for binding of the mannose and galactose components is solvent displacement, since this is the only source of favorable entropy in such association processes (Finkelstein & Janin, 1989). Several ordered water molecules are observed in the crystal structure of the binding cavity of the native, ligand free Fab and recent estimates suggest that their release to bulk solvent could provide several kcal mol^{-1} of free energy (Dunitz, 1994).

As with all other carbohydrate–protein complexes, hydrogen bonds play a crucial role in complex stability and specificity. Epitope mapping studies correctly identified the OH groups that participate in three antibody–oligosaccharide hydrogen bonds described in Figures 10-26 and 10-30: Abe O-4, Abe O-2, and Man O-4. From the relative magnitude of the (ΔG) values for **57** and **48** (Table 10-1), the hydrogen bond from Man O-4 to His101H is inferred to be a neutral donor–acceptor pair (Street et al., 1986) and this conclusion is supported by calorimetry measurements in two different buffer systems (Sigurskjold & Bundle, 1992). Attempts to replace His101H by Asp, Glu, Asn, and Gln were made during mutagenesis studies in order to introduce a charged or neutral amino acid side chain that was capable of providing bidentate hydrogen bonding to Man O-4 and O-6 (cf. Quiocho, 1986). All of these mutations were less active than the native Fab (Brummell et al., 1993), suggesting that steric and geometric demands for bidentate bonds could not be satisfied. Both hydroxyl groups of abequose are involved in conjugated networks of hydrogen bonds at the base of the binding pocket involving a bound water molecule and Trp98L. The coordination of multiple hydrogen bonds between abequose and protein accounts for the absolute stereochemical requirements of the binding site and explains how the weak hydrogen bond strength of a neutral donor-acceptor pair (Street et al., 1986) has been amplified *via* a water molecule (Wat1, Fig. 10-26) and the formation of conjugated hydrogen bonds (Quiocho, 1988, 1989).

The importance of complementarity and good fit are emphasized by comparison of the methyl glycosides of abequose **61** and 3-deoxyarabinose (**62**). Removal of the 6-CH_3 group reduces ΔG by a surprising 1.5 kcal mol^{-1}, with significantly less enthalpy reflecting lost van der Waals interactions, while a relatively more favorable entropy term—$T\Delta S$ indicates the residues forming the binding pocket have gained motional freedom (cf. Figs. 10-24–10-26).

Structural and stereochemical changes to the α-Gal residue are accompanied by large swings in the enthalpy and entropy of binding (Table 10-1) but even major changes in this residue provide active derivatives. The range of these changes lead to the conclusion that although hydrogen bonds to the galactose residue are postulated based on the crystal structure, the net gain in energy from such hydrogen bonds must be very small. For example the α-Glc analogue **49** with an equatorial hydroxyl group is unlikely to form a hydrogen bond to Trp91L, yet this loss of a hydrogen bond causes a free energy change of only 0.3 kcal mol^{-1}. Despite the small energy changes when Gal was replaced by either methyl or methoxymethyl groups, the hydrogen bonding map (Fig. 10-26) indicates 2 or 3 hydrogen bonds between Gal and the protein. The relative strength of the intersaccharide hydrogen bond between Abe O-2 and Gal O-2 is particularly interesting, since its formation is made possible by a conformational shift that occurs when the ligand is bound (Bundle et al., 1994b). Compound **52**, in which the Gal 2-OH has been replaced by an amino group, retains its activity, which suggests that, since the amino group is partly uncharged at pH 8.0, this group may still participate in the postulated hydrogen bonds. The activities of the deoxy compounds **59** and **60** suggest for the native antigen that neither the glycosylated Gal O-3 nor the Gal 6-OH groups forms productive hydrogen bonds with the antibody. This conclusion, as well as that predicting weak or no hydrogen bond to the Gal 4-OH, are supported by the activity of the dichloro compound **53**. A methyl group at O-3 of Gal mimics the anomeric carbon of the preceding rhamnose unit in the polysaccharide antigen and the two chlorine atoms, which are largely isosteric with hydroxyl groups, replace the 4-OH and 6-OH of Gal. This is the only compound that binds as strongly as, if not slightly stronger than, the natural epitope **48**. Thus, it seems that the hydroxyls at positions 4 and 6 of Gal are mainly involved in van der Waals interactions with the antibody. The increase in enthalpic interactions for compound **53** relative to **48** is consistent with this conclusion, since chlorine produces stronger van der Waals interactions than OH does.

The galactose residue occupies a position at the periphery of the binding site with some freedom of movement and the thermodynamic parameters observed for functional group replacements **48–56** are typical of those reported for lectin–oligosaccharide interactions (Lemieux et al., 1991; Spohr et al., 1992a,b). The enthalpy–entropy compensation seen by Lemieux's group was ascribed to changes in water structure at the water-complex interface (Lemieux, 1996).

Interactions Beyond the Immediate Binding Site

If the trisaccharide epitope is extended by addition of appropriate hexose residues that adopt ϕ,ψ glycosidic torsional angles corresponding to low energy conformations, a "picture" of the polysaccharide–Fab complex can be simulated. When this is done an immediate problem becomes apparent. The polysaccharide chain makes severe steric contact with the antibody surface and to avoid the clash of sugar atoms with those of amino acids, a substantial reorientation of the polysaccharide chain must occur. The glycosidic linkages involved are those between the mannose that resides in the binding site and the rhamnose, and between this rhamnose and the adjacent galactose residue (Fig. 10-31). The polysaccharide on the other side of the binding site experiences no such bad contacts in low-energy conformations. In order to examine this problem, two heptasaccharides were prepared by partial acid hydrolysis of an octasaccharide derived from the polysaccharide antigen by a phage enzyme hydrolysis (Baumann et al., 1993) (Fig. 10-32). Each heptasaccharide is forced to bind in monogamous fashion since only one abequose epitope is present. Thus the interactions on either side of the binding site may be probed.

Heptasaccharides **64** and **65** bind with identical enthalpies, but a ~2.5 kcal mol^{-1} difference in entropy (TΔS) accounts for the 50-fold weaker binding of **65** (Table 10-8). Molecular modeling of heptasaccharides in conformations that are close to the global potential energy minima show that while **64** may be accepted in the binding pocket without serious steric clashes, the terminal reducing galactose (including subsequent residues in larger structures) would collide with the protein surface in such conformations. These clashes could be avoided for **65** by changes to some or all of the glycosidic torsion angles between Man–Rha or Rha–Gal (Fig. 10-33). The smaller favorable entropy when heptasaccharide **65** binds results from the loss of conformational entropy corresponding to the forbidden conformers about these two glycosidic linkages. A crys-

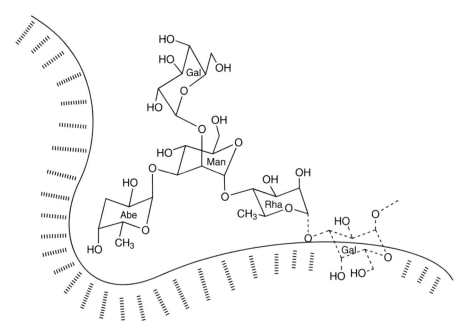

Figure 10-31. Schematic presentation of the steric clash that occurs if polysaccharide chains adopt low energy glycosidic torsional angles. Rha and Gal residues which flank the binding site clash with the protein H2 loop unless a change of conformation occurs on binding.

α-**D**-Gal(1→2)α-**D**-Man(1→4)α-**L**-Rha(1→3)α-**D**-Gal(1→2)α-**D**-Man(1→4)α-**L**-Rha

| 1,3

α-**D**-Abe

| 1,3

α-**D**-Abe

mild acid hydrolysis

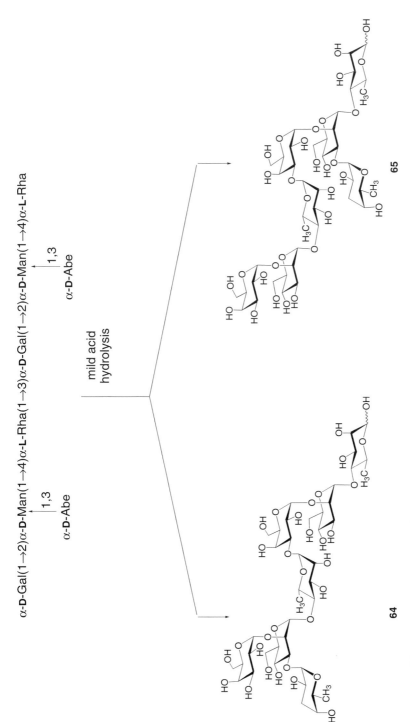

64

65

Figure 10-32. Two heptasaccharides (**64**, **65**) each containing a single abequose residue are obtained from an octasaccharide by mild acid hydrolysis. The octasaccharide was released from the polysaccharide antigen by a phage rhamnosidase enzyme. Each heptasaccharide can bind in only one orientation with Se155.4, thereby enabling the interactions to be probed at either side of the binding site.

Table 10–8 Entropy Effects for the Binding of Heptasaccharides **64** and **65**

Methyl α Glycoside Ligands	$\Delta G°$	$\Delta H°$ (kcal mol^{-1})	$-T\Delta S°$
Gal[Abe]Man (**48**)	−7.3	−4.9	−2.4
GalManRhaGal[Abe]ManRha (**64**)	−6.3	−9.0	+2.7
GalManRhaGal[Abe]ManRha (**65**)	−8.7	−8.9	+0.2

tal structure for the heptasaccharide **64** with Fab has been solved and the position of all seven monosaccharides residues are well defined (Cygler et al., unpublished results).

Role of Bound Water Molecules

The several solved crystal structures for the Se155.4 binding site reveal multiple roles for water. The immobilized Wat1 not only maximizes the number of hydrogen bonds that secure it and abequose to the antibody, it also contours the base of the site to provide improved van der Waals contacts. This water molecule and Wat2 are common features of both bound ligand conformations. The three additional water molecules observed with the bound conformation II in the crystal structure of trisaccharide **2** bound

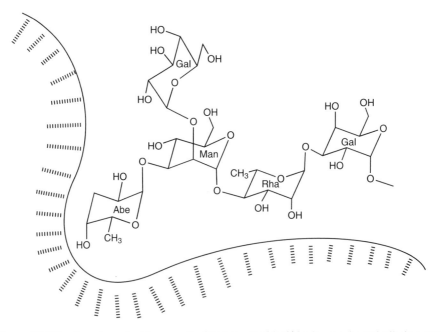

Figure 10–33. A pentasaccharide segment of heptasaccharide **64** is shown schematically in a conformation (Rha $\psi \sim 180°$) that avoids the steric clash shown in Fig. 10-31.

to single chain Fv are particularly interesting. As the Gal–Man torsional opens up to give a lower energy conformation, the hydrogen bonds located in the solvent-exposed regions of conformation I can no longer be maintained and Wat3 and Wat5 mediate these hydrogen bonds *via* an extended network of conjugated hydrogen bonds. Wat3 becomes hydrogen bonded to Wat4. This rearrangement of ordered water molecules in the immediate vicinity of the galactose residue may be the trigger that accounts for the altered conformation about the Gal–Man glycosidic linkage. Extended networks of conjugated hydrogen bonds, often involving ordered water molecules, are common features both in the crystal structure of low molecular weight saccharides (Jeffery & Sänger, 1991) and in a variety of oligosaccharide–protein complexes (Bourne et al., 1992; Quiocho 1986, 1989).

Finally, analysis of native Fab structure shows that five water molecules occupy the base of the empty binding pocket. While Wat1 has an occupancy of 1, the four other water molecules have an occupancy less than 1 per Fab molecule. It seems significant that these water molecules are located at sites that correspond to the position of Abe O2, O4 and O5, and C6. Based on recent estimates of the energy required to position ordered water molecules, the displacement of these molecules to bulk solvent can be expected to provide a significant contribution to binding energy (Dunitz, 1994).

The X-ray crystal structures (Table 10-8), as well as NMR and calorimetry experiments for the *Salmonella* system provide a self-consistent picture of the oligosaccharide–protein complex and binding site. In cases where a crystal structure is lacking it should be possible to build a model identifying the principal features of a binding site using functional group replacement and tandem microcalorimetry studies, possibly supported by NMR studies. However, unlike X-ray crystallography, these approaches will not identify the role of water molecules that are often essential for complex formation.

Predicting Binding Site Topography Prior to X-Ray Crystallography

A *Shigella flexneri*-specific monoclonal antibody SYAJ6 yielded a crystal structure (Vyas et al., 1993) after modeling by an extensive panel of congeners (Auzanneau & Bundle, 1993a,b; Auzanneau et al., 1993; Bock et al., 1982; Bundle, 1989; Bundle et al., 1983, 1994a,b,c; Bundle & Josephson, 1979a,b, 1980; Hanna & Bundle 1993; Josephson & Bundle, 1979, 1980; Pinto et al., 1989, 1990a,b; Wessel & Bundle, 1983). This section describes the attempts to deduce the structural features and topography of the antigen–antibody complex by modeling studies that employed conformational analysis and extensive epitope mapping. Since a solved structure of ligand–Fab complexes subsequently became available, the accuracy of the model may be compared with the crystal structure.

Preparation of Monoclonal Antibody Against an Unbranched Shigella Flexneri Antigen

The antigen is a *Shigella flexneri* lipopolysaccharide but unlike the branched *Salmonella* O-antigen this polysaccharide has a linear, unbranched structure (Fig. 10-34):

$$[\rightarrow 2)\alpha\text{-L-Rha}p(1 \rightarrow 2)\alpha\text{-L-Rha}p(1 \rightarrow 3)\alpha\text{-L-Rha}p(1 \rightarrow 3)\text{-}\beta\text{-D-GlcNAc}p(1\text{-}]$$

$$\quad\;\; \mathbf{A} \qquad\qquad\quad \mathbf{B} \qquad\qquad\quad\;\; \mathbf{C} \qquad\qquad\quad\;\; \mathbf{D}$$

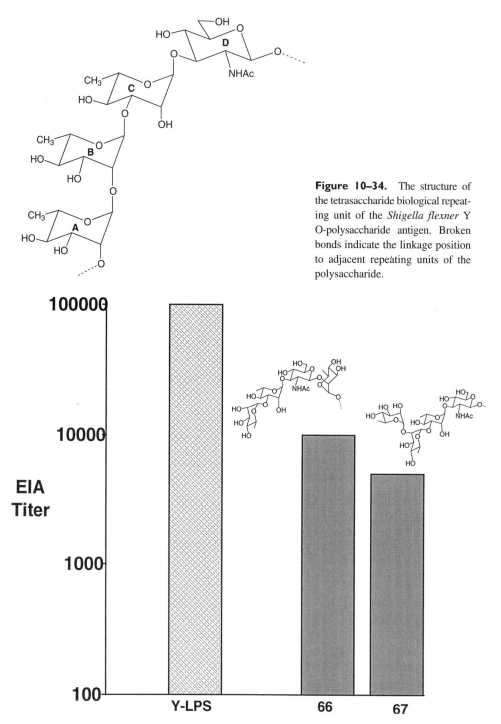

Figure 10–34. The structure of the tetrasaccharide biological repeating unit of the *Shigella flexner* Y O-polysaccharide antigen. Broken bonds indicate the linkage position to adjacent repeating units of the polysaccharide.

Figure 10–35. EIA binding profile and titres of monoclonal antibody SYAJ6 with homologous Y-LPS and glycoconjugate antigens **66** and **67**.

The biological repeating unit of the polysaccharide was established as the sequence ABCD (Carlin et al., 1984). Several monoclonal antibodies to this structure have been developed and studied (Bundle, 1986, 1989; Carlin et al., 1986, 1987a) but only the IgG$_3$ antibody, SYAJ6, was amenable to proteolytic digestion and the formation of Fab crystals. The antibody reacted with purified LPS antigen and with tetrasaccharide glycoconjugates that possessed the residue sequence ABCD, and BCDA (Figs. 10-35 and 10-36). It precipitated polymeric epitopes such as the glycoconjugates and the antibody–polysaccharide complex formed a precipitation lattice. It appeared most likely that the antibody bound to a reoccurring internal epitope encompassing the residues BCDA and spanning the junction of the biological repeating units of the O-polysaccharide.

Chemical mapping of the binding site was far advanced before X-ray crystal structure studies became a reality; therefore, the order of the presentation of results follows closely the chronology of laboratory experiments. Specifically, the binding mode of the oligosaccharide was inferred before the structure of the complex was known (Bundle, 1989) and later refined (Bundle et al., 1994c).

Establishing the Size of the Antibody Site

The approximate size and location of the epitope along the polysaccharide chain could be gauged from glycoconjugate binding experiments (Fig. 10-35); however, the multivalent nature of the glycoconjugates enhances avidity and the fine specificity of the antibody binding site is masked by this effect. For example, antibody binds strongly to the glycoconjugate BCDA (**66**) and almost as well to the glycoconjugate ABCD

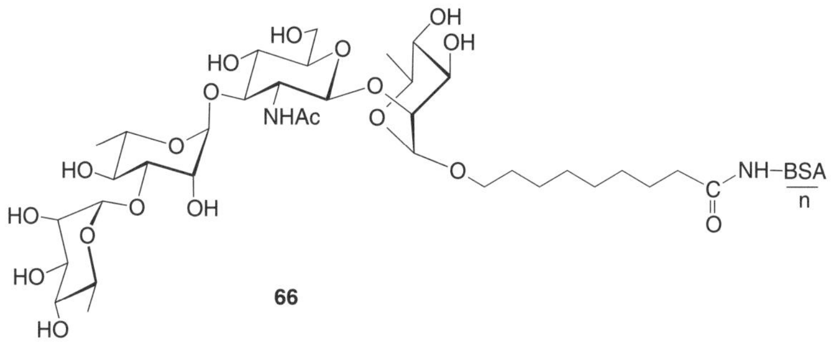

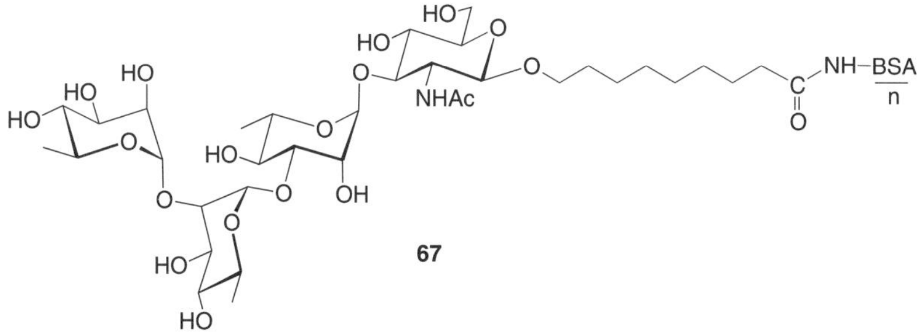

Figure 10–36. The structures of frame shifted glycoconjugate antigens **66** and **67**.

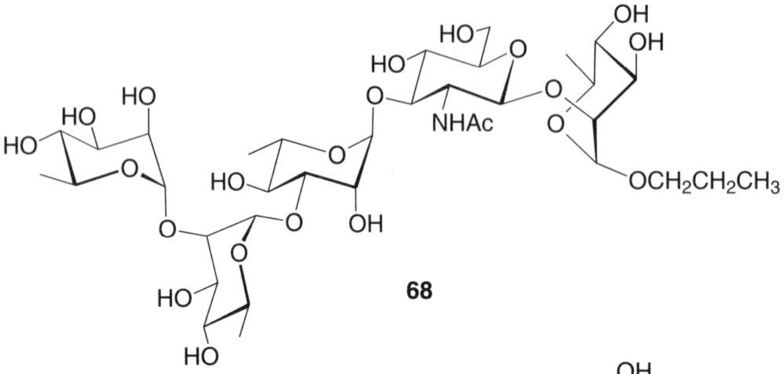

68

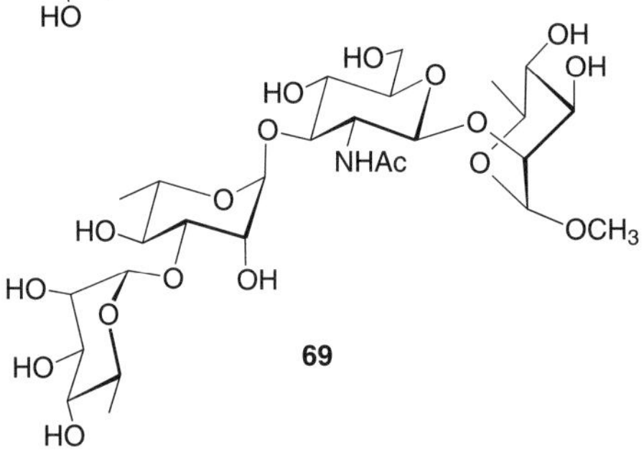

69

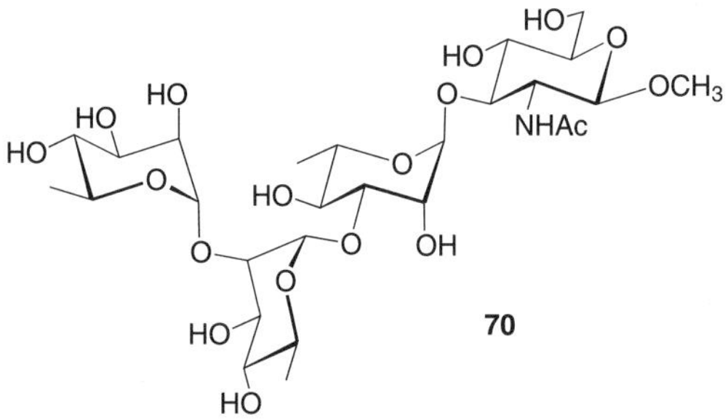

70

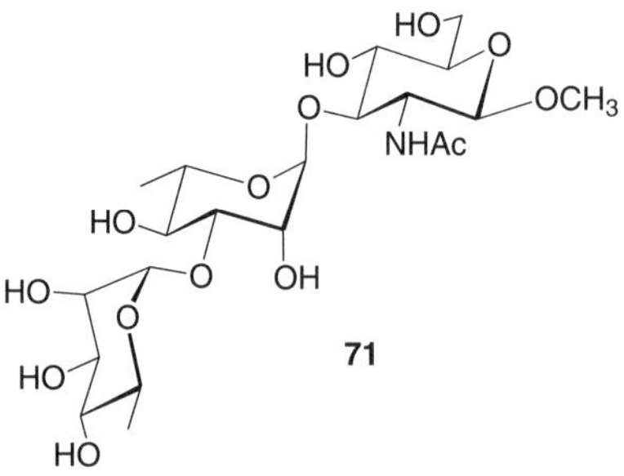

71

Figure 10–37. Structures of *Shigella flexneri* oligosaccharides including trisaccharide **71**, tetrasaccharides **69** and **70**, and pentasaccharide **68**.

(**67**) (Bundle, 1986) (Fig. 10-35). The loss of fine specificity renders it impossible to appreciate whether antibody SYAJ6 can discriminate between glycoconjugates **66** and **67** (Fig. 10-36). In order to resolve this point, the size of the SYAJ6 binding site was probed by a solid phase EIA in which purified antibody was coated on an assay plate and inhibitors were ranked for their ability to compete for antibody with an enzyme-labeled polysaccharide antigen (cf. Fig. 10-6). The assay affords a reasonable estimate of association constants (Meikle et al., 1990; Vorberg & Bundle 1990). The inhibitors included a panel of frame-shifted epitopes, such as pentasaccharide **68**, tetrasaccharides **69** and **70**, and modified derivatives such as trisaccharide **71**, and tetrasaccharides **72** and **73** (Figs. 10-37–10-39). Hexasaccharide ABCDA′B′ and heptasaccharide ABCDA′B′C′ were tested but found to exhibit lower molar activities than the penta- and tetrasaccharides. The relative free energy of binding for the series of di-, tri- to heptasaccharides established that tetrasaccharide BCDA (**69**) was only slightly less active than the pentasaccharide ABCDA′ (**68**); since the trisaccharide ABC was inactive and the trisaccharide CDA was 1.8 kcal less active than tetrasaccharide **69** (Auzaneau et al., 1993), it seemed initially that the tetrasaccharide BCDA represented the minimal epitope size. The minimal epitope size was subsequently revised to the trisaccharide BCD **71** in the light of calorimetry data (Bundle et al., 1994c), but attempts to establish the orientation of the epitope toward the protein were performed on the tetrasaccharide BCDA.

Topography of the Epitope Surface

Solution NMR studies conducted on di-, tri-, and tetrasaccharide ligands had established a model for the conformation of the O-polysaccharide (Bock et al., 1982). This model (Fig. 10-40) was then used as an essential reference point for the interpretation of inhibition experiments that were designed to elucidate the binding surface of the oligosaccharide epitope. The task of defining the oligosaccharide surface that contacts the binding site was tackled in three stages. Once the size and extent of the epitope was correlated with the tetrasaccharide BCDA, its orientation in the binding site was probed by replacing the rhamnose residues B and C sequentially by the more sterically demanding L-mannose residue in tetrasaccharides MCDA (**72**) and BMDA (**73**) (Fig. 10-38) (Wessel & Bundle, unpublished). L-Mannose possesses a hydroxymethyl group rather than the 6-methyl group of rhamnose. The tetrasaccharide BCDA (**72**) showed

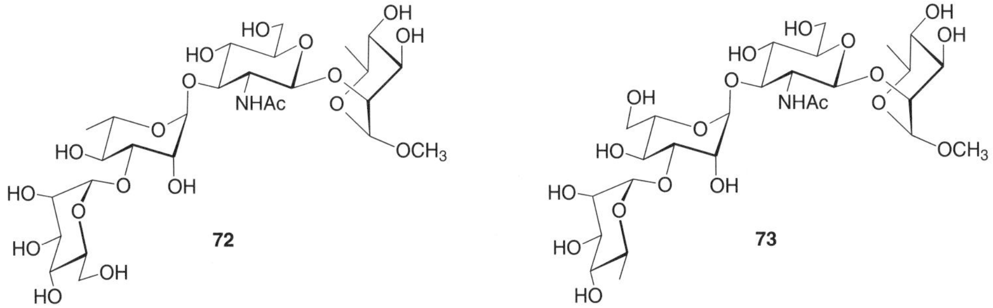

Figure 10–38. Congeners containing L-mannose at positions B and C in the Rha-Rha-GlcNAc-Rha tetrasaccharides **72** and **73**.

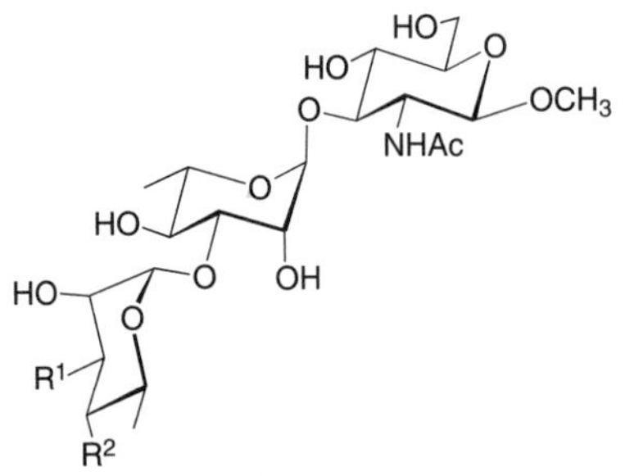

74 $R^1 = H, R^2 = OH$
75 $R^1 = OH, R^2 = H$

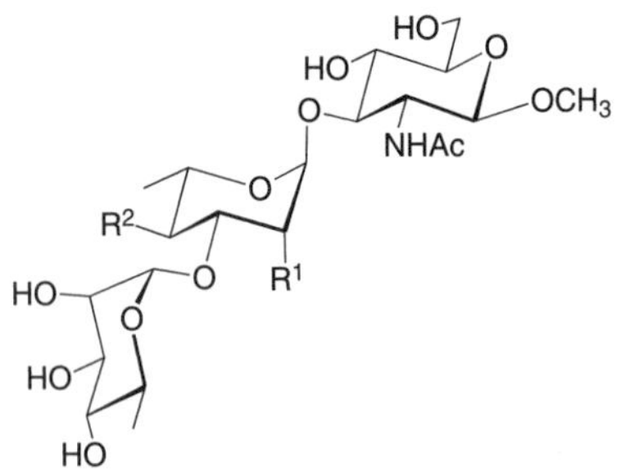

76 $R^1 = H, R^2 = OH$
77 $R^1 = OH, R^2 = H$

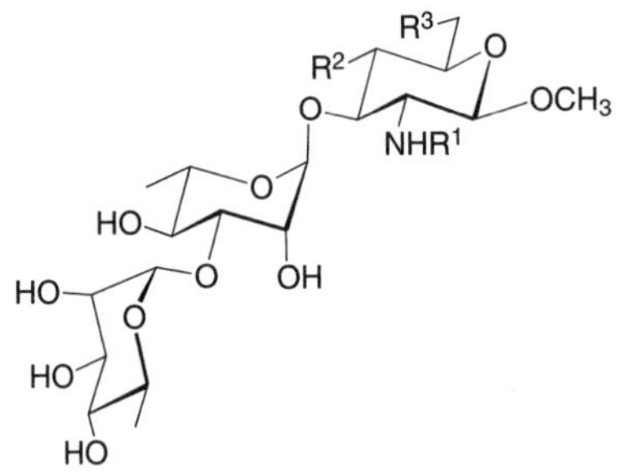

78 $R^1 = CH_3CO, R^2 = H, R^3 = OH$
79 $R^1 = CH_3CO, R^2 = OH, R^3 = H$
80 $R^1 = CF_3CO, R^2 = R^3 = OH$
81 $R^1 = C_2H_5CO, R^2 = R^3 = OH$

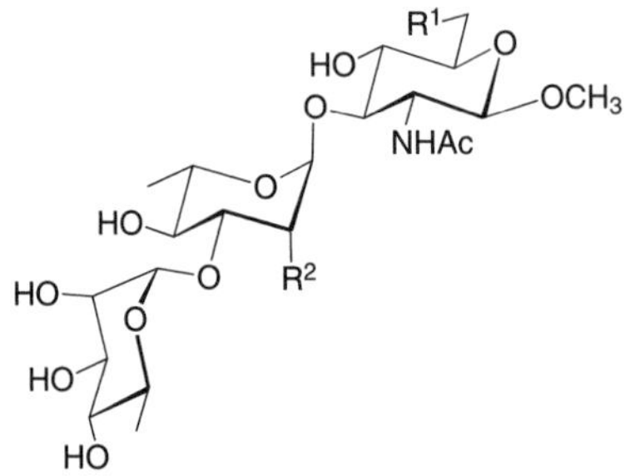

82 $R^1 = Cl, R^2 = OH$
83 $R^1 = OH, R^2 = Cl$

Figure 10–39. Monodeoxy congeners and *N*-acyl analogues **74–81**, and the chlorodeoxy analogues **82** and **83**.

a small change in binding energy but BCDA (**73**) exhibited a ~1.0 kcal mol^{-1} loss of binding energy (Bundle, 1989). Insertion of a hydroxymethyl group within the C residue was concluded to result in unfavorable contacts with the protein surface but these were minimal when it replaced residue B. Consistent with this observation, there was essentially no change in binding energy when Rha was replaced by the pentose L-lyxose (H for CH_3). Substitution of the β-D-GlcNAc residue by a β-D-Glc unit in the disaccharide Rha1 → 3Glc reduced binding by ~1 kcal mol^{-1}, suggesting that at least the acyl residue and possibly the amide group contacts protein, since the amino derivative of trisaccharide BCD was also inactive (Auzanneau & Bundle, 1993a,b). This suggests that this group may be one of the antibody–saccharide polar contacts; the remaining contacts of this type were identified by subsequent calorimetry studies with a set of monodeoxy congeners (Bundle et al., 1994c; Chervenak & Bundle, unpublished). These preliminary EIA-based data suggested that the protein surface contacts the antigen at the 6-deoxy group of residue C, but that the 6-deoxy group of rhamnose B is solvent exposed. In the ball-and-stick projection of the tetrasaccharide BCDA (Fig. 10-

40) the protein surface would be positioned at the bottom and solvent-exposed residues at the top.

The calorimetry data (Table 10-9) deals mainly with trisaccharide BCD (**71**) (Fig. 10-37), its monodeoxy (**74–79**), and *N*-acyl (**80, 81**) analogues (Fig. 10-39). The activity of these congeners sheds light on the hydrogen bonding pattern. For the purposes of correlating these data it is meaningful to note that certain generalizations have emerged from mapping the energetics of ligand–protein and especially carbohydrate–protein interactions (Bock, 1983; Bundle et al., 1994a; Fersht, 1987; Glaudemans, 1991; Lemieux, 1993, 1994; Lemieux, et al., 1991; Street et al., 1986). These may be summarized as follows: (1) for replacement of a hydroxyl group by hydrogen (monodeoxygenation) changes of binding energy $\Delta(\Delta G) \geq 1$ kcal mol^{-1} are often associated with hydrogen bonds between charged donor or acceptor pairs, (2) $\Delta(\Delta G) \geq$ 0.5–1.0 kcal mol^{-1} suggests the possibility of neutral donor/acceptor hydrogen bonds, and (3) $\Delta(\Delta G) \sim$0.25–0.5 kcal mol^{-1} are assigned to weaker polar contacts or solvent-exposed hydrogen bonds (Nikrad et al., 1992). Comparing these criteria with the data for monodeoxy congeners (Table 10-9) it is dramatically evident that only one monodeoxygenation produces an inactive inhibitor. A $\Delta(\Delta G)$ of at least 2–3 kcal mol^{-1} (**78** vs. **71**) suggested that the hydrogen bonding partner of GlcNAc 4-OH may be a charged amino acid side chain (Bundle et al., 1994a,b,c). Both GlcNAc 6-OH (**79**) and Rha B 4-OH (**75**) ($\Delta[\Delta G] \sim$1 kcal mol^{-1}) appear to be bound *via* intermediate-strength hydrogen bonds, most likely involving neutral acceptor/donors. Hydrogen bonds of a

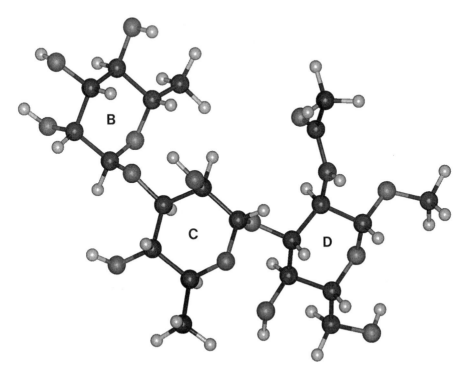

Figure 10–40. Ball-and-stick model of the low energy conformation of trisaccharide BCD (**71**) that is used as the basis for the interpretation of inhibition and thermodynamic data.

Table 10–9 Thermodynamic Parameters for Binding of Native and Modified Trisaccharide Ligands to *Shigella flexneri* Antibody SYA/J6 at 25°C[a]

Ligand	K_A (M^{-1})	$\Delta G°$	$\Delta H°$ (kcal mol^{-1})	$-T\Delta S°$
Rha-Rha-GlcNAc-OMe (**71**)	9.5×10^4	-6.8	-4.3	-2.5
Rha-Rha-GlcNAc-Rha-OMe (**69**)	2.0×10^5	-7.2	-3.4	-3.8
Rha-Rha-Rha-GlcNAc-Rha-OMe (**68**)	2.5×10^5	-7.4	-1.5	-5.9
3-deoxyRha-Rha-GlcNAc-OMe (**74**)	3.8×10^4	-6.3	-7.5	$+1.2$
4-deoxyRha-Rha-GlcNAc-OMe (**75**)	3.3×10^4	-6.2	-2.9	$+3.3$
Rha-2-deoxyRha-GlcNAc-OMe (**76**)	1.7×10^6	-8.4	-10.1	$+1.6$
Rha-4-deoxyRha-GlcNAc-OMe (**77**)[a]	4.7×10^4	-6.4	-2.5	-3.9
Rha-Rha-4-deoxyGlcNAc-OMe (**78**)[a]	Inactive			
Rha-Rha-6-deoxyGlcNAc-OMe (**79**)[a]	2.9×10^4	-6.1	-4.5	-1.6
Rha-Rha-GlcNCOCF$_3$-OMe (**80**)	2.8×10^4	-6.1	-1.8	-4.2
Rha-Rha-GlcNCOEt-OMe (**81**)	9.1×10^4	-6.8	-2.5	-4.3
Rha-2-chloro-2-deoxyRha-GlcNAc-OMe (**82**)	3.2×10^6	-8.9	-6.6	-2.3
Rha-Rha-6-chloro-6-deoxyGlcNAc-OMe (**83**)	1.2×10^6	-8.3	-5.8	-2.5

[a]Data from Bundle et al., 1994c.

similar nature but of slightly weaker strength are inferred for the interactions between protein and Rha B 3-OH (**74**) and Rha C 4-OH (**77**). The binding site can accommodate an extension of the *N*-acyl group by 1 carbon atom (**81**) with only a small loss of binding, but if the polarity of this acyl group is altered (**80**; COCF$_3$ vs. COCH$_3$) the binding is reduced by 0.5 kcal mol^{-1}. A surprising observation was the dramatic gain in binding energy accompanying the conversion of Rha C to a 2-deoxy group (**76**) or a 2-chloro-2-deoxy congener (**82**). This effect may be explained by the not so well appreciated energy cost that can occur if a ligand hydroxyl group is placed in the buried environment of the protein binding site, where it cannot be accommodated through hydrogen bond formation (Quiocho, 1989). Here the gain in binding energy on converting -OH to -H was 2.5 kcal mol^{-1}. From the data accumulated thus far it is now possible to deduce the surface of the oligosaccharide that interacts with antibody.

The conformational model of the polysaccharide inferred from NMR and model building studies shows that, in its most ordered form, the α-L-Rha*p*$(1 \rightarrow 3)$–β-D-GlcNAc link introduces a kink between the array of rhamnose residues (Fig. 10-40) (Bock et al., 1982). A detailed examination of low-energy conformers of the tetrasaccharide sequence BCDA (**69**) shows that the GlcNAc 4-OH and 6-OH groups may contact protein at the same time the Rha C 2-OH group is buried and the C-6 of Rha B is solvent exposed at the opposite face. Thus it is possible to position these key functionalities and those involved in potential hydrogen bonds either in buried or partially buried environments. In this model the acetamido group is positioned so that its methyl group lies at the lip of the binding surface, allowing the potential for the amide nitrogen to hydrogen bond with protein. This is a likely event given the requirement for the amide group in binding of disaccharide analogues (Auzanneau et al., 1993). Using this approach a continuous contact surface for the oligosaccharide has, therefore, been identified (Bundle, 1989; Chervenak & Bundle, unpublished).

Crystal Structures of the Oligosaccharide-SYAJ6 Fab Complexes

The binding site exists as a groove that lies along the interface of the V_H and V_L (Fig. 10-41). Several water molecules are located in the site, which appears to be filled by the trisaccharide element (**71**). The antigen topography inferred from modeling studies agrees well with the X-ray crystal structure data for the antibody–oligosaccharide complexes, a tight binding trisaccharide BC'D (**76**), and the native pentasaccharide (**68**) (Vyas et al., 1993). The solved crystal structure for the antibody complexed with the 2'-deoxy congener **76** unambiguously identified the position of all three pyranose residues, while in the complex with the native pentasaccharide **68**, ABCDA', the monosaccharide residues Rha C and GlcNAc D were well defined. Electron density for the exposed residue Rha B did not readily distinguish between alternate chair conformations of the pyranose ring, suggesting that the B ring may adopt a distorted conformation in the bound state that differs from the $^1C_4$ chair observed in solution. At this point it is not clear why the B residue can be positioned in the deoxytrisaccharide–Fab complex while the same residue cannot be clearly identified in the crystal structure of the native pentasaccharide–Fab complex. Accommodation of the Rha C 2-OH group, as well as the requirement to position the flanking A and A' residues away from contact with the protein, may cause disorder at the periphery of the site. Nevertheless, the crystal structure of the trisaccharide–Fab complex shows clearly that the hydroxymethyl group of GlcNAc and the C-methyl group of Rha C are buried, and the acetamido group points toward the rim of the binding site with the amide nitrogen able to participate in a protein–

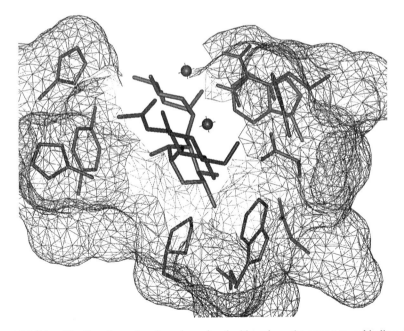

Figure 10–41. The Conelly surface in end-on view looking along the groove type binding site of antibody SYAJ6. The trisaccharide ligand **71** is shown in the binding site, together with two ordered water molecules.

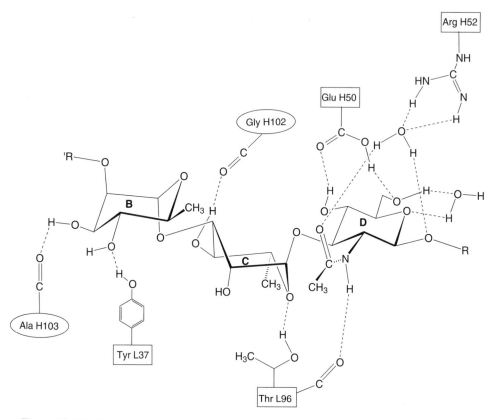

Figure 10–42. The hydrogen bond map for the complex between the trisaccharide **71** and the SYAJ6 antibody.

saccharide hydrogen bond (Figs. 10-41 and 10-42). The general agreement between the proposed and observed binding modes validates the simplified method of model building when it is coupled with an extensive panel of natural and modified oligosaccharide ligands. This success, however, leaves unanswered the more taxing problem of developing a reliable model for the position of a ligand in the binding site in the absence of crystal structure data. The problem is illustrated later for monoclonal antibody that binds the *Brucella abortus* antigen (Bundle, 1989).

Conformation of the Bound Oligosaccharide

The glycosidic torsional angles that define the conformations of the bound pentasaccharide **68** and the bound trisaccharide **76** were extracted from the crystal structures of the two solved complexes and found to lie close to the low energy conformations predicted by potential energy calculations and inferred on the basis of solution NMR experiments (Bock et al., 1982). Upon complex formation, the ligand is not required to undergo any significant conformational change.

Contact Amino Acids and Hydrogen Bonds

Polar and nonpolar residues make up the amino acids that contact antigen (Table 10-10). This feature is also notable in the amino acids that form hydrogen bonds to the antigen (Fig. 10-42). It is also of note that no residues from the L2 CDR contact the antigen, a common feature of hapten-specific antibodies (Novotny et al., 1983). A charged amino acid residue, Glu H50, forms bidentate hydrogen bonds to GlcNAc *O*-4 and *O*-6 (Fig. 10-42). It may also be seen that the remaining hydrogen bonds are of the neutral donor-acceptor type (Table 10-11). The Rha C2-*OH* points into an area of the binding site where there is no available amino acid hydrogen bond acceptor. Intramolecular hydrogen bonding to neighboring saccharide ether oxygen atoms or hydroxyl groups may be possible, but the structure of the complex with pentasaccharide **68** cannot resolve this point. However, it is clear that removal of this hydroxyl group to form a 2-deoxy congener (**76**) relieves an unfavorable situation in this area of the binding site. In sharp contrast to the Se155.4 binding site, that of SYAJ6 shows a much more representative range of contact amino acids that include glycine, alanine, and charged residues such as glutamic acid and arginine (Table 10-10).

Comparison of the native Fab and Fab–oligosaccharide complexes reveals that the side chain of arginine 52H moves almost 3.0 Å upon complex formation. This residue utilizes a water molecule to form hydrogen bonds to the GlcNAc residue.

Role of Water in the Shigella Complex

Two water molecules mediate hydrogen bonds in the trisaccharide (**76**)–Fab complex. Arg 52H does not contact the protein directly but does so through a water molecule, thereby forming hydrogen bonds to the carbonyl group of the acetamido function and the glycosidic oxygen atom of GlcNAc. The second water molecule acts as a hydrogen bond acceptor to GlcNAc 6-*OH* and as a hydrogen bond donor to the GlcNAc ring oxygen atom. The extent of bound trisaccharide exposure to a water probe suggests that rhamnose C is completely buried, while rhamnose B is 75% exposed to solvent and the GlcNAc residue D is 25% water exposed. It is of interest to note that the most energetically favorable hydrogen bonds involve the GlcNAc residue while the buried

Table 10–10 Fab Hypervariable Loops and Amino Acid Residues in Contact with Hapten

Hypervariable Regions	Amino Acids
CDRL 1	R S S Q S L L **H** S D G N T **Y** L H
CDRL 2	No Contacts
CDRL 3	S Q **T T H** V P T F
CDRH 1	N **Y** W M **E**
FR2	W(H47)
CDRH 2	**E** I R L K S N N Y A T H Y A E S V A G
CDRH 3	G **G A** V G A M D Y

Bold = residues contributing to Fab surface buried by and within 4.0 Å of trisaccharide hapten.

<u>Bold</u> = residues within 3.5 Å of the trisaccharide hapten.

Table 10–11 Hydrogen-Bond Contacts and the Contact Surface Area Between the Oligosaccharide and SYAJ6 Fab-Ordered Water Molecules

Antigen Atom	Distance (Å)	Contacting Atom
Rha B		
O3	2.9	O (Ala-103H)
O3	2.9	O-Wat 11
O4	3.4	HO Tyr (37L)
Rha C		
O4	3.1	O (Gly-102H)
O5	3.5	HO (Thr 96L)
GlcNAc D		
O1	3.4	O (Wat 102)
N2	2.8	O (Thr 96L)
O4	3.1	Oε2 (Glu 50H)
O5	2.9	O Wat 6
O6	3.1	Oε1 (Glu 50H)
O6	3.0	O Wat 6
O7	3.2	O Wat 102

rhamnose residue C appears to contribute to the binding *via* van der Waals interactions and relatively lower-energy hydrogen bonds (Fig. 10-42).

Docking Oligosaccharides in Fab Binding Sites

Brucella Abortus O-Antigen Monoclonal Antibody

The structure determination of Fabs without a complexed ligands is only of limited interest since the structural details of Fab are well appreciated. The *Brucella* A O-antigen specific antibody, YsT9.1 (Bundle et al., 1984), has been crystallized and its structure without antigen solved (Rose et al., 1993). The binding site has been mapped by synthetic oligosaccharides (Bundle, 1989) and attempts were made to use that data to guide ligand docking. This was combined with an attempt at homology modeling of the binding site since closely related variable domains V_H and V_L had been separately observed in published crystal structures of unrelated antigens (Oomen et al., 1991).

The value of the antibody structures without bound sugar is dependent upon an ability to reliably model and dock the appropriate oligosaccharide ligand into the binding site. Given the difficulties of modeling flexible polysaccharide chains, and the problems of developing a suitable force field that can treat the interaction of protein and carbohydrate, this is a daunting task. Using binding data for oligosaccharide homologues of the *Brucella* O-antigen, a linear 1,2-linked homopolysaccharide of 4,6-dideoxy-4-formamido-α-D-mannopyranosyl residues (Bundle, 1989; Caroff et al., 1984), attempts were made to dock by inspection, a low energy conformer of an octasaccharide (Oomen et al., 1991). Given the somewhat low association constant (Bundle et al., 1989) and complete absence of information on possible contact residues, the validity of the model is difficult to evaluate. The homology modeling showed a groove type binding site and this groove was complementary to a tetrasaccharide (Bundle et al., 1989; Kihlberg & Bundle, 1991; Kihlberg et al., 1991; Oomen et al., 1991; Peters et al., 1990). However, it is impossible to know in which direction the oligosaccharides lies in the binding site.

Meningococcus Group B Capsular Polysaccharide Specific Antibody

Modeling of the helical epitope of α-2,8 polysialic acid polysaccharide into the site of the mAb735 antibody (Frosch et al., 1985) was more satisfying (Evans et al., 1995). Here it was possible to match the complementary shape of an extended antibody depression against the postulated helical conformation of the group B meningococcal polysaccharide antigen (Brisson et al., 1992). Specifically, patches of antigen and antibody bearing opposite charges could be juxtaposed, with the carboxyl groups of the antigen approaching lysine residues. The model conveniently explained several of the immunochemical characteristics of this important antigen, that not only occurs in capsules of pathogenic bacteria but also in neonatal tissue and as a tumor-associated antigen (Cho & Troy, 1994; Michon et al., 1987). One of the most interesting features of the Fab structure was the extended groove type site, that essentially carved out half the circumference of the Fab molecule, along a line that closely followed the V_H and V_L interface. The size of this site could accommodate about 8–10 Neu5Ac residues, and as such represents the upper-size limit of an antibody site for polysaccharide antigens. The antibody is a rare clone that was isolated from a large number of hybridoma experiments (Frosch et al., 1985). It is probably atypical with respect to binding site dimensions and may represent a compromise situation, whereby the mouse immune system could respond to a potential self antigen without the possibility of small ($n < 5$) oligo-2,8-linked Neu5Ac$_n$ sequences, known to be components of gangliosides and glycoproteins, binding to the antibody (Brisson et al., 1992; Michon et al., 1987).

Myelomas Proteins That Bind Oligosaccharides

Among other interesting examples, myeloma protein J539 is perhaps the best known Fab that has been crystallized without ligand. This has been the subject of extensive modeling studies with synthetic ligands (Glaudemans, 1991; Gluademans & Kovac, 1985, 1988). Although they lack the detailed picture provided by a solved ligand–protein complex, studies of myelomas and related hybridomas that bind similar oligosaccharide epitopes exemplify several aspects of contemporary approaches to binding site mapping. Indeed this system was one of the first carbohydrate–antibody systems to be investigated in depth (Glaudemans, 1991).

The monoclonal immunoglobulin A (IgA) J539 is a mouse myeloma protein that binds a $(1 \rightarrow 6)$-β-D-galactopyranose homopolysaccharide (Sher & Tarikas, 1971). The crystal structure of its Fab was originally solved in the absence of ligand (Suh et al., 1986) and the structure of the ligand-complex has yet to appear in the literature. This antigen–antibody system and the myeloma X-24 with a similar binding profile are the most extensively studied of the myeloma proteins to have been investigated by the NIH group of Glaudemans, beginning with the initial reports in 1973 (Jolley et al., 1973, 1974). This protein was an attractive candidate for extensive binding studies because it had a solved crystal structure, bound an antigen containing a linkage type commonly encountered in oligosaccharide epitopes, and, perhaps most conveniently, exhibited ligand-induced fluorescence changes. This property allowed a sensitive and convenient assay to yield accurate estimates of association constants as low as 10^1–10^2 M^{-1} (Glaudemans, 1991; Glaudemans et al., 1994). The linear and homoglucan antigen structures are a particular disadvantage in studies of this antigenic determinant, since possible frame shifting makes it difficult to distinguish the contributions of neighboring pyranose residues. It also becomes very difficult to examine the incremental contributions to binding of a linear structure because of interaction free energies between homologues that vary by one hexose residue (Jencks, 1981).

Early results suggested that a tri- or tetrasaccharide ligand fills the antibody binding site, since ΔG begins to plateau for the tri- and tetrasaccharide and does not increase thereafter (Jolley et al., 1974; Zeigler et al., 1990). These data were interpreted in terms of a binding site model involving different affinity subsites. Despite the size of the antigenic site, methyl β-D-galactopyranoside contributed at least 50% of the total binding energy. The 2-fluoro and 3-fluoro analogues did not bind, whereas binding of the 4-fluoro compound was essentially unchanged. The presence of a 2-fluoro substituent on the pyranose ring makes synthesis of the series of β1,3-linked oligosaccharides difficult because the nonparticipating group at C-2 would favor α- rather than β-glycoside formation. Consequently, the nonbinding 3-fluoro analogue was used as a building block in chemical synthesis of inhibitors. Introduction of a pyranose residue, which by itself was not active as an inhibitor, provided a means to disentangle the problem of frame shifting. It was reasoned that if the nonbinding residue of an oligosaccharide would be competed from the highest affinity subsite by unmodified galactose residues also present in a synthetic series of oligosaccharides, the relative affinities of the subsites and their sequential ordering might be determined. Since the disaccharide 6-O-(β-D-galactopyranosyl)-β-D-galactopyranoside contributes 80% of the total binding energy, it was reasoned that the two highest affinity subsites were adjacent. Trisaccharides alternatively fluorinated in each of the galactose residues then showed that the two structures offering contiguous unsubstituted disaccharide elements bind as well as the trisaccharide. A difluorodideoxy trisaccharide with fluorine residues at each of the terminal galactose residues had affinity comparable to the native trisaccharide, while a monofluorodeoxy trisaccharide with only one central fluorinated residue bound with the affinity of a disaccharide. The former derivative establishes that the A subsite must be positioned between subsites B and C. By invoking the involvement of two tryptophan residues for maximum ligand-induced fluorescence, the fourth subsite D was inferred to be adjacent to subsite B. The conclusions have been shown to be consistent with the observed binding strengths of a series of monodeoxy analogues (Glaudemans, 1991; Glaudemans et al., 1994).

Dextran-Binding Myelomas

Another myeloma protein that binds saccharides with a fluorescence change is the IgA W3129. However, unlike the J539 myeloma, this antibody binds an α1,6-linked dextran and does so *via* the terminal residue of the chain rather than at internal repeating sequences. Using a similar approach to that employed for the J539 antibody, native and deoxyfluoro ligands were used to demonstrate the existence of four subsites (Nashed et al., 1990). These possessed sequentially decreasing affinities, with the disaccharide sequence providing the major portion of the binding energy. In this work it was possible to construct a model of the antibody binding site since the V_H and V_L domains possessed significant homology with those of two X-ray crystal structures (Padlan & Kabat, 1988). However, without more evidence of precise saccharide–protein interactions than was available from this study, the docking of a ligand must be regarded as tentative.

Summary

In the detailed modeling and crystal structure studies of four antibodies presented and discussed here, the size of the antibody binding site is seen to be complementary to an epitope as small as a trisaccharide and typically within the range of 3 to 4 hexose residues (Table 10-12). The complex between the Le^Y antigen and its Fab (not discussed here) also shows a binding site accommodating four hexose residues (Jeffrey et al., 1995). A single exception is noted but in the case of polysialic acid the immune system may be conditioned to respond with antibodies that do not recognize small epitopes that can be part of "self" glycoproteins or glycolipids. Groove-shaped sites appear to be most typical for oligosaccharide antigens, although in the case of branched oligosaccharides, cavity or pocket-shaped sites may readily accommodate the spatially more exposed branching sugar.

In all complexes studied by functional group replacement, the most striking observations has been the relatively small number (from 1 to 3) of hydroxyl groups that form essential hydrogen bonds. Without these polar contacts the complex cannot form, yet all other hydroxyl groups of the epitope maybe replaced in turn, generally with small energy penalties. It is also the case that the essential hydroxyl groups are clustered together on the epitope surface. Water molecules are frequently an integral structural feature of the oligosaccharide–Fab complex and participate in networks of hydrogen bonds between the sugar and protein. It is observed that favorable entropy contributions occur when the optimum epitope fills the binding site; this almost certainly occurs through the displacement of water molecules from the binding site.

It is noted in one case that protein–saccharide contacts adjacent to the binding site can modulate the affinity of binding in either a positive or negative sense. Fine tuning of affinity *via* contacts with flanking residues may be one way that oligosaccharides epitopes appear to be larger than the number of residues that actually fill the binding site.

Table 10–12   Size and Shape of Antibody Binding Sites

	Binding Site	
Antigen	Size	Shape
Branched epitope, LPS (*Salmonella*)	3 Hexose residues	Cavity
Linear epitope, LPS (*Shigella*)	3 Hexose residues	Groove
Linear epitope, LPS (*Brucella*)	3-4 Hexose residues	Groove
Linear epitope, Capsular polysaccharide (*Meningococcus*)	8–10 Nonulosonic acid residues	Groove

A resurgence in the use of vaccines composed of a relatively small number of saccharide residues covalently linked to a carrier protein (conjugate vaccines) can benefit from the general structural principles that are emerging for oligosaccharide–antibody complexes. Vaccines designed to prevent bacterial disease seem to require intermediate-sized oligosaccharides (Mäkelä et al., 1984, 1985; Seppälä & Mähelä, 1989), while potential cancer vaccines designed to induce antibodies against tumor-associated antigens are much smaller (Ragupathi et al., 1997). With advances in the techniques of protein engineering it may even be possible to envisage cost-effective therapeutic applications of carbohydrate-specific antibody molecules or their fragments in acute situations (Wilson & Stanfield, 1995). In such circumstances appreciation of the subtleties of sugar–protein interactions will be of central importance.

Acknowledgments

The financial support from the National Science and Engineering Research Council of Canada and the Canadian Network Centres of Excellence for Bacterial Dieseases and Protein Engineering is gratefully acknowledged. The author expresses his gratitude to his collaborators and co-authors, whose work forms the basis of this chapter.

11

Atomic Interactions Between Proteins/Enzymes and Carbohydrates

Florante A. Quiocho and Nand K. Vyas

"Stuctural glycobiology," a term describing a subdiscipline of the broad field of structural biology, has been coined recently in recognition of the recent upsurge in the determination by X-ray crystallography of three-dimensional structures of enzymes and proteins that bind carbohydrates (Quiocho, 1993). While structural glycobiology has its underpinning in these tertiary structures, it also encompasses studies employing other biophysical techniques (e.g., nuclear magnetic resonance spectroscopy, calorimetry, mass spectrometry, theoretical analysis, etc.). These techniques also have been used to study the conformations and functions of carbohydrates, especially oligosaccharides, as structural entities.

The first major review pertinent to this chapter, which appeared more than 10 years ago, was based on only a few X-ray structures of protein–carbohydrate complexes (Quiocho, 1986). Since then the number of refined structures of these types of complexes has increased considerably (Table 11-1). This increase nearly parallels the exponential growth in the total number of protein structures that have been determined.

This chapter is focused mainly on the major features of the atomic interactions between proteins/enzymes and carbohydrates as observed in refined X-ray structures. It is an amalgamation and update of two recent reviews by each of the authors (Quiocho, 1993; Vyas, 1991). As in the previous reviews, we consider only complexes of proteins or enzymes with uncharged monosaccharides or oligosaccharides with sugar units in the $^4C_1$ or $^1C_4$ pyranose form.

Atomic Features of Protein–Carbohydrate Interactions

Although the structures of a variety of proteins that bind carbohydrates are now known (Table 11-1), it is possible, as originally proposed (Quiocho, 1989, Quiocho et al., 1991; Vyas, 1991), to broadly divide or classify these structures into two major groups according to the following characteristic features. Proteins belonging to Group I have sites able to sequester bound sugars from the bulk solvent. Representatives of this group are periplasmic sugar receptors for active transport and chemotaxis and glycogen phosphorylase (catalytic site), and hexokinase (Table 11-1). These proteins generally have a two-domain structure with a cleft between the domains wherein the sugar is bound.

Table 11-1 Refined Structures of Carbohydrate-Binding Proteins[a]

Proteins	Complex	(Å) Resolution	R-factor	References
α Amylase	—	2.9	—	Buisson et al., 1987
β Amylase	α-Cyclodextrin	2.0	0.18	Mikami et al., 1993
Taka-amylase	—	3.0	—	Matsuura et al., 1984
Endoglucanase CelD (*C. thermocellum*)	β-Cellobioside	2.3	0.19	Juy et al., 1992
Glucoamylase-II (*A. awamori var. X100*)	—	2.2	0.23	Aleshin et al., 1992
Cyclodextrin Glucosyltransferase	—	2.0	0.18	Klein & Schulz, 1991
Lysozyme	Trisaccharide	1.5	0.17	Strynadka & James, 1991
Hexokinase A isozyme	Glucose	2.1	—	Bennet & Ssteitz, 1980
Glycogen Phosphorylase a	Glucose	2.1	0.30	Sprang et al., 1982
	Maltoheptose	2.5	—	Goldsmith et al., 1982
	Acarbose	2.5	0.13	Goldsmith et al., 1987
Glycogen Phosphorylase b	Glucose	2.3	0.19	Martin et al., 1990
	Heptulose-2-phosphate Oligosaccharide	2.9	0.20	Johnson et al., 1990
	Glucose-6-phosphate	2.3	0.20	Johnson et al., 1993
Neuraminidase	Inhibitor	2.8	0.18	Bossart-Whitaker et al., 1993
	Sialic acid	2.2	0.15	Burmeister et al., 1992
Xylose isomerase	Xylose	1.6	0.14	Whitlow et al., 1991
UDP-Galactose 4-epimerase	—	2.5	0.18	Bauer et al., 1992
Galactose Oxidase (*D. dendroides*)	—	1.7	0.18	Ito et al., 1991
Galactose Oxidase (*A. niger*)	—	2.3	0.18	Hecht et al., 1993
Glucose Permease, Domain IIA (*B. subtilis*)	—	2.2	0.20	Liao et al., 1991
Phosphoglucomutase	—	2.7	0.20	Dai et al., 1992
Cellobiohydrolase II	Glucose-cellobiose	2.0	0.16	Rouvinen et al., 1990
Arabinose-binding protein	L-Arabinose	1.7	0.14	Quiocho & Vyas, 1984
	Galactose	1.8	0.13	Quiocho et al., 1989
	Fucose	1.9	0.13	
Galactose/Glucose-binding protein	Glucose	1.9	0.15	Vyas et al., 1988
(*E. coli*)	Galactose	2.0	0.17	Vyas et al., 1994

Table II–I (continued)

Proteins	Complex	(Å) Resolution	R-factor	References
Galactose/Glucose-binding protein	Glucose	2.4	0.16	Mowbray et al., 1990
(*S. typhimurium*)	Galactose	1.7	0.19	Zou et al., 1993
Maltose-binding	Maltose	2.3	0.25	Spurlino et al., 1991
protein	Maltotriose	1.7	0.20	Spurlino & Quiocho, unpublished
	Maltotetraose	1.7	0.20	Rodseth & Quiocho, unpublished
	β-Maltodextrin	1.8	0.21	Sharff et al., 1993
Ribose-binding protein	Ribose	1.7	0.19	Mowbray & Cole, 1992
Hemmagglutinin	Sialic acid	2.9	0.22	Weis et al., 1988
Wheat germ agglutinin	NeuLac	2.2	0.17	Wright, 1990
Concavalin A	Mannose	2.9	0.19	Derewenda et al., 1989
Favin (*Vicia fabia*)	Glucose	2.8	0.38	Becker & Reeke, 1991
Pea Lectin	Trimannoside	2.6	0.18	Rini et al., 1993
Lentil Lectin	—	2.3	0.16	Loris et al., 1993
Ricin (heterodimer)	Lactose	2.5	0.22	Rutenber et al., 1991
Ricin A chain	—	2.5	0.22	Katzin et al., 1991
Ricin B chain	Lactose	2.5	0.22	Rutenber et al., 1991
Lectin *E. corallodendron*	Lactose	2.0	0.19	Shaanan et al., 1991
Lectin IV of *G. simplicifolia*	Tetrasaccharide	2.8	0.21	Delbaere et al., 1990
Lectin (*Lathyrus ochrus*)	Glucose	2.2	0.18	Bourne et al., 1990a
	Mannose	2.0	0.18	
	Trisaccharide	2.1	0.18	Bourne et al., 1990b
	Octasaccharide	2.3	0.19	Bourne et al., 1992
S-lectin (bovine spleen)	Disaccharide	1.9	0.17	Liao et al., 1994
C-type animal lectin	Glycopeptide	1.7	0.17	Weis et al., 1992
Fab (Shigella)	Trisaccharide	2.5	0.19	Vyas et al., unpublished
	Pentasaccharide	3.0	0.25	Vyas et al., unpublished
Fab (Brucella)	Oligosaccharide	2.5	0.22	Rose et al., 1993
Fab (*E. coli*)	Dodecasaccharide	2.1	0.19	Cygler et al., 1991
Fc fragment (human)	Glycosylated	2.8	0.22	Deisenhofer, 1981
Enterotoxin (*E. coli*)	Lactose	2.3	—	Sixma et al., 1992
Verotoxin (*E. coli*)	—	2.2	0.18	Stein et al., 1992

[a]As of 1994

A hinge-bending motion between the two domains modulates access to and from the cleft. In contrast, Group II proteins, represented by lectins and enzymes such as lysozyme, phosphorylase (glycogen storage site), have carbohydrate-binding sites that are shallow and close to the protein surface. Antibodies raised against oligosaccharides tend to belong to Group II proteins. Other features further distinguish the two groups.

Group I proteins have high affinity carbohydrate-binding sites (dissociation constant or K_d values $\leq 10^{-6}$ M), whereas Group II proteins tend to have lower affinity sites (K_ds in the millimolar range) (Quiocho, 1990; Quiocho et al., 1991; Vyas, 1991). The binding sites of Group I proteins are genrally designed for monosaccharides whereas those of Group II proteins are designed for oligosaccharides. Indeed, it is noteworthy that members of Group II proteins such as lectins and antibodies are multivalent. In the carbohydrate-bound structures, there are fewer ordered water molecules bound in Group I protein binding sites than in Group II protein binding sites. Finally, atomic thermal B-factors of binding site residues and bound sugars in Group II proteins are significantly higher than those bound in Group I proteins.

Features of proein–carbohydrate interactions discussed are related to (1) hydrogen bonds, (2) van der Waals forces, (3) stacking interactions, (4) the role of water molecules, and (5) the role of metal ions. With the exception of the role of metal ions, which is pertinent only to lectins, all the above features are observed in almost all protein–carbohydrate complexes. Since many of these features are found in the well-refined, very high resolution structures of the complexes of periplasmic sugar receptors or binding proteins with carbohydrates, we have used these structures extensively to illustrate many of these features. These structures include the L-arabinose-binding protein (ABP), D-glucose/D-galactose-binding protein (GGBP) and maltodextrin-binding protein (MBP) (Figs. 11-1–11-6) (Quiocho, 1989; Quiocho et al., 1991; Quiocho & Vyas, 1984; Spurlino et al., 1991; Spurlino & Quiocho, unpublished data; Vyas et al., 1988). MBP binds not only linear but also cyclic oligosaccharides (Miller et al., 1983). The structures of MBP complexed with maltose, maltotriose, maltotetraose, or β-cyclodextrin have been determined and refined at very high resolution (Rodseth et al., unpublished data; Sharff et al., 1993).

Hydrogen Bonds

Hydrogen bonds, which are highly directional, are heavily involved in conferring specificity to protein–carbohydrate interactions, as well as contributing to the affinity. Hydroxyl groups are the sole polar and highly exposed groups of carbohydrates; hence, they are always the ones involved in hydrogen-bonding interactions. (In oligosaccharides, sugar-ring oxygens are rarely involved in hydrogen-bonding interactions.) A hydroxyl group has the capacity, especially when it is buried in the protein surface and not accessible to the bulk solvent, to serve as a hydrogen bond donor and an acceptor by way of the oxygen lone pairs. Ideally, a hydroxyl group can donate one and accept two hydrogen bonds. In complexes of sugars with Group I proteins (e.g., Figs. 11-1, 11-2, and 11-3), individual sugar hydroxyl groups frequently participate simultaneously as a donor and an acceptor.

The classes of hydrogen bonds present in protein–carbohydrate interactions are: cooperative hydrogen bonds, bidentate hydrogen bonds, and networked hydrogen bonds (Quiocho, 1986, 1989; Quiocho et al., 1991; Quiocho & Vyas, 1984; Vyas, 1991; Vyas et al., 1988). Cooperative hydrogen bonds result from the simultaneous participation of a sugar hydroxyl group as a hydrogen bond donor and acceptor group (Figs. 11-1, 11-2, and 11-3). With few exceptions, the cooperative hydrogen bonds follow the simple scheme (Quiocho, 1986)

$$(NH)_n \rightarrow OH_c \rightarrow O$$

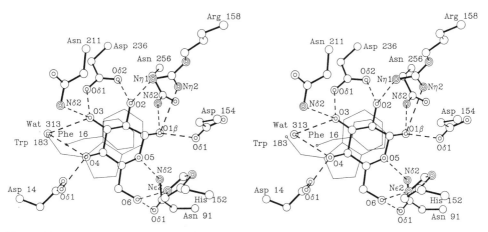

Figure 11-1. Stereoview of the hydrogen-bonding and stacking interactions between β-D-glucose and polar and aromatic residues of the D-glucose/D-galactose-binding protein that serves as initial receptor of active transport and chemotaxis in bacterial cell (Vyas et al., 1988). The sugar is sandwiched between Phe16 and Trp183. Phe16 and Trp183 stack against the B-face and A-face of the sugar, respectively.

where OH_c represents a carbohydrate hydroxyl group and $(NH)_{n = 1 \, or \, 2}$ and O correspond to donor and acceptor groups of the protein, respectively. According to this scheme, there is a great propensity for a sugar hydroxyl group to accept a hydrogen bond from NH groups and to donate a hydrogen bond to oxygen atoms. The NH donor and O acceptor groups originate mostly from protein planar polar side chains, most frequently from Asn, Asp, Gln, Glu, and Arg residues. In a few cases hydroxyl side chains have been found to serve as hydrogen bond donors, for example, in the binding of an inhibitor in the catalyitc site of phosphorylase (Johnson et al., 1990) and a carbohydrate substrate to wheat germ agglutinin (Wright, 1990). In a few instances, ordered water molecules are involved as hydrogen bond donors and acceptors (see below). It has been observed in refined, very high resolution structures that hydrogen bonds in which sugar hydroxyls serve as donors are generally slightly shorter, implying stronger bonds, than those in which they serve as acceptor groups (Bourne et al., 1990a; Derewenda et al., 1989; Quiocho, 1989; Quiocho & Vyas, 1984; Vermersch et al., 1990; Vyas et al., unpublished). Cooperative hydrogen bonds are more frequently used in binding of carbohydrates by Group I proteins than by Group II proteins. They enhance charge delocalization and thus effectively strengthen hydrogen bonds.

Bidentate hydrogen bonds are those in which a pair of adjacent hydroxyl groups of the pyranoside interacts with a pair of atoms of the same side chains of planar polar residues such as Asn, Asp, Gln, Glu, and Arg residues (e.g., see Figs. 11-1, 11-2, and 11-3). Although NH and CO groups of consecutive peptide bonds could similarly serve as hydrogen bond donors and acceptors, this has yet to be observed.

Bidentate hydrogen bonds require specific geometry (Figs. 11-1, 11-2, and 11-3). Those observed thus far involve adjacent pairs of hydroxyl groups that are both equatorial or in which one is equatorial and the other axial (Fig. 11-1) (Vyas et al., 1988).

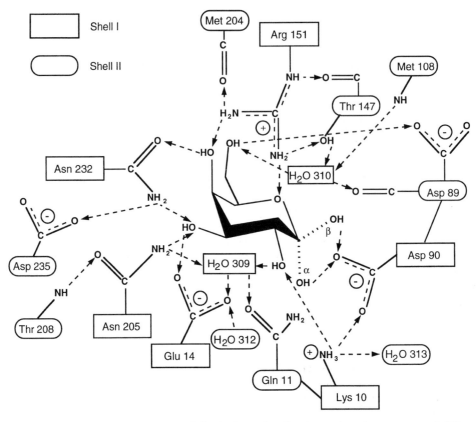

Figure 11–2. Schematic diagram of the hydrogen-bonding interactions between α- and β-D-galactose and the arabinose-binding protein (Quiocho et al., 1989; Vyas et al., 1991).

Occasionally, the ring O5 oxygen in hexoses, together with OH4 or OH6, also participate in bidentate H bonds (Quiocho & Vyas, 1984; Vyas et al., 1988). Bidentate hydrogen bonds involving the pair O5 and OH1 have also been observed (Mowbray & Cole, 1992). Freedom of torsional rotations of C5—C6 and C6—O6 bonds allows the OH6 of the hexose sugar to be paired with either O5 or OH4 (axial or equatorial) to form bidentate bonds (Bourne et al., 1990a; Delbaere et al., 1990; Derewenda et al., 1989; Quiocho et al., 1989; Vyas et al., 1988). For example, in the pairing of OH6 and O5 in the bidentate H bonds observed in the complex GGBP with Glc, the $\chi5$ torsion angle defined by C4—C5—C6—O6 of the bound D-glucose is 177° (Fig. 11-1) (Vyas et al., 1988). In addition, in the pairing of OH4 (equatorial) with OH6 of the mannoside/glucoside in bidentate bonds observed in the monosaccharide complexes with lectin LOL I, the $\chi5$ torsional angle is approximately 90° (see Fig. 6 in Bourne et al., 1990a).

In all cases the distance between a pair of sugar oxygen atoms is about 2.8 Å, ideal for bidentate hydrogen bonding with planar polar side chains. Bidentate bonds involving an adjacent pair of axial hydroxyls of a pyranoside has never been observed, principally due to the fact that both hydroxyls are much farther apart (3.7 Å) and point in opposite directions. Whereas the interatomic distance of 2.8 Å between the ring O5 oxygen and axial OH4 is favorable for the formation of bidentate hydrogen bonds, the distance of about 3.7 Å between the O5 and equatorial OH4 group disfavors the formation of bidentate hydrogen bonds. Group II proteins contain far fewer bidentate hydrogen bonds than Group I proteins.

Networks of hydrogen bonds in protein–carbohydrate interactions are due to the further involvement of polar side chains, which are engaged in holding the sugar in place, in hydrogen bonding interactions with other polar groups (Figs. 11-2 and 11-3).

Generally, carbohydrate binding to Group I proteins makes more use of cooperative, bidentate, and networked hydrogen bonds than binding to Group II proteins. Conditions that would cause variations of the different types of hydrogen bonds in protein–carbohydrate complexes have been discussed (Quiocho, 1989). The key feature is that hydroxyl groups of a buried saccharide need to be paired with hydrogen-bonding residues or bound ordered water molecules. Solvent-exposed hydroxyl groups do not necessarily have to be paired.

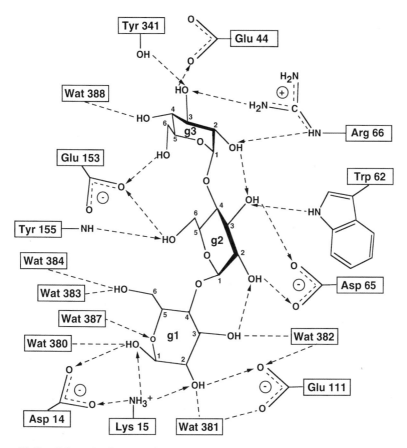

Figure 11–3. Schematic diagram of the hydrogen-bonding interactions between maltotriose and maltodextrin-binding protein (Spurlino & Quiocho, unpublished data).

van der Waals Forces

van der Waals or dispersion forces also contribute significantly to the stability of protein–carbohydrate interactions. There are two sources of these forces. One source results from hydrogen bonding interactions that enable atoms of the residues and sugars to come within van der Waals distance of each other.

Stacking Interactions Between Sugars and Aromatic Residues

The other source of van der Waals forces comes from the stacking of aromatic residues against the faces of sugars. This stacking interaction has emerged as a completely unexpected common feature of protein–carbohydrate interactions (Quiocho, 1986; Vyas, 1991). First observed in the structure of lysozyme (also the first structure of an enzyme) complexed with an oligosaccharide (Phillips, 1966), stacking interactions are present in all but one or possibly two structures of protein–carbohydrate complexes determined thus far (Table 11-2). Stacking interactions contribute not only stability but also specificity to complexes. Interestingly, the structure of a glycosylated human Fc fragment also shows stacking interactions between sugar rings and aromatic residues (Deisenhofer, 1981).

The CH groups exposed from both faces of the pyranose ring, together with the $C(6)H_2$ group, aggregate as nonpolar patches or clusters. (The faces of the sugar rings and asymmetric cyclic protein side chains such as those of Trp and His residues are identified as A-face and B-face according to the numbering of the ring atoms, which increase in clockwise and counterclockwise directions, respectively.) These patches are involved principally in the stacking interactions.

Monosaccharides have about 75% polar and 25% nonpolar surfaces (Vyas et al., 1988). In polysaccharides, the ratio of polar to nonpolar surfaces decreases as the number of monosaccharide units in the polymer increases. For example, the disaccharide maltose has 55% polar and 45% nonpolar surfaces (Spurlino et al., 1991). The reduction of polar surface in polysaccharides is due to the formation of glycosidic bonds and intersugar hydrogen bonds and van der Waals contacts. Therefore, oligosaccharide binding sub-sites, especially those belonging to Group I proteins, have a considerable number of nonpolar or aromatic residues interacting with the nonpolar surfaces of the oligosaccharide (Quiocho et al., 1991; Spurlino et al., 1991; Vyas et al., 1991).

Three examples from our studies of the periplasmic binding proteins (ABP, GGBP, and MBP; Figs. 11-4, 11-5, and 11-6) illustrate stacking interactions in carbohydrate binding beautifully (Table 11-2). In the complex of ABP with L-arabinose, the B-face of the sugar, which has a hydrophobic patch larger than that on the A-face, is stacked against the B-face of Trp16 (Fig. 11-4) (Quiocho & Vyas, 1984; Vyas et al., 1991). Interestingly, although the hydrophobic patch from C2 on the A-face is smaller, it nevertheless makes contact with the terminal methyl group of Met204 side chain.

The second example is found in the complex of GGBP with D-glucose or D-galactose (Vyas et al., 1988). As shown in Figure 11-5, the bound D-glucose is sandwiched between phenylanine and tryptophan residues. Very similar stacking interactions involving two aromatic residues are also present in the complexes of the endoglucanase with a sugar inhibitor (Juy et al., 1992) and the D-ribose-binding protein with D-ribose (Mowbray & Cole, 1992) (Table 11-2).

Table 11–2. Stacking Interactions Between Sugars and Aromatic Residues of Proteins

Protein (Sugar Bound)	Sugar-Site Face[a]	Sugar Residues	Aromatic	References
Lysozyme	GlcNAc-B	B	Trp62	Johnson et al., 1988
Glycogen Phosphorylase a	Glc S5	B	Tyr404	Goldsmith et al., 1991
Cellobiohydroase II	Glc	A	Trp367	Rouvinen et al., 1990
Xylose Isomerase	Xyl	B	Trp16	Whitlow et al., 1991
Endoglucanase (*C. thermo.*)	Glc	A	Phe206	Juy et al., 1992
	Glc	B	Tyr551	
β-Amylase (Soybean)	Glc-4	A	Phe200	Mikami et al., 1993
Neuraminidase	Sialic acid	—	—	Burmeister et al., 1992
L-Ara-binding protein	L-Ara/Gal/Fuc	B	Trp17	Quiocho & Vyas 1984; Quiocho et al., 1989
Gal/Glc-binding protein	Gal/Glc	A	Phe16	Vyas et al., 1988
(*E. coli/Salmonella T.*)	Gal/Glc	B	Trp183	Zou et al., 1993
Maltose-binding protein	Glc-1	A	Tyr155	Spurlino et al., 1991
	Glc-1	B	Trp62	
	Glc-1	A	Trp230	
	Glc-2	A	Trp340	
	Glc-2	B	Trp62	
	Glc-3	B	Trp62	Spurlino & Quiocho, unpublished
	Glc-3	A	Try341	
Ribose-binding protein	Rib	A	Phe13	Mowbray & Cole, 1992
		B	Phe164	
Ricin B-chain	Gal-domain1	B	Trp37	Rutenber & Robertus, 1991
	Gal-domain2	B	Try248	
Lectin (WGA I)	NeuNAc	B	Tyr66	Wright, 1990
	NeuNAc	B	Phe109′	Wright, 1992
	NeuNAc	B	Tyr66	
	NeuNAc	B	Tyr23	
Lectin (WGA II)	NeuNAc	B	His66	Wright, 1990
Lectin (GS4)	Gal-B	B	Trp133	Delbaere et al., 1990
	L-Fuc-D	B	Phe108	
Concanavalin A	Man/Glc	B	Tyr12	Derewenda et al., 1989
Lectin (LOL I)	Man-3	B	Phe123	Bourne et al., 1990b
Lectin-octasaccharide	Man-4	B	Phe123	Bourne et al., 1992
	Man-5′	B	Phe123	
Pea Lectin	Man	B	Phe123	Rini et al., 1993
Favin	Glc	B	Phe124	Becker & Reeke, 1991
Lectin (*E. corallo.*)	Gal	B	Phe131	Shaanan et al., 1991
Lectin (calcium-dependent)	Man-8	A	His1189	Weis et al., 1992
	Man-9	A	His2189	
S-Lectin (bovine spleen)	Gal	A	Trp68	Liao et al., 1994
Hemagglutinin	NeuAc	A	Trp153	Weis et al., 1988
Anti-Oligosaccharide Fab	Gal	B	Trp58L	Cygler et al., 1991
(*Salmonella T.*)	Abe	B	Trp58L	
Anti-Oligosaccharide Fab	Perosamine	A	Tyr103H	Rose et al., 1993
(*Brucella A*				
Anti-Oligosaccharide Fab	Rha-3	A	Trp33H	Vyas et al.,
(*Shigella*)				(unpublished
Enterotoxin (*E. coli*)	Gal	B	Trp88	Sixma et al., 1992

[a]Sugar faces A and B are defined as in the text.

The content of hydrophobic residues in the deep oligosaccharide-binding groove of the MBP and the extent of participation of these residues in carbohydrate binding are unsurpassed. At least nine aromatic residues line the groove, including those shown in Figure 11-6 (Spurlino et al., 1991). The involvement of these residues in ligand binding provide the best illustration not only of the extensive use of stacking interactions but also the role of these residues in molecular recognition. Five aromatic residues (Trp62, Tyr155, Trp230, Trp340, Tyr341) are intimately associated in the binding of maltooligosaccharides (Fig. 11-6). Trp230, Tyr155, and Trp230, arranged like a cradle, appose the A-faces of the two glucosyl units of maltose (Fig. 11-6).

Meanwhile, Trp62 is coplanar with the B-face of the reducing sugar and perpendicular to the B-face of the second or nonreducing sugar unit. These four aromatic residues are also similarly associated in the binding of maltotriose. Furthermore, although Tyr341 does not participate in binding of the maltose, it undergoes a conformational change upon binding of maltotriose in order to make contact with the A-face of the third glucosyl unit (Fig. 11-6) (Spurlino & Quiocho, unpublished data). In order to completely surround the faces of the maltotriose, part of the B-face of the third glucosyl unit is also stacked against Trp62.

Thus the hydrophobic patches of sugars bound to the three periplasmic binding proteins are shielded by aromatic residues. In addition, the polar hydroxyl groups, which are located approximately in the plane of the pyranose ring and form a peripheral polar surface, are involved in hydrogen-bonding interactions (e.g., Figs. 11-1, 11-2, and 11-3).

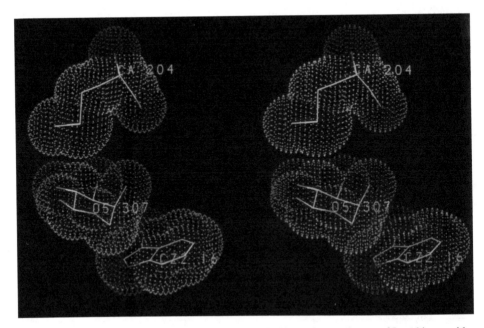

Figure 11–4. Stereoview of the interaction of hydrophobic patches or clusters of L-arabinose with hydrophobic residues of the arabinose-binding protein. Atoms are represented as van der Waals dot surface. Whereas the larger patch on the B-face is stacked against the A-face of Trp16, the smaller patch on the A-face is close to the S-methyl group of Met204.

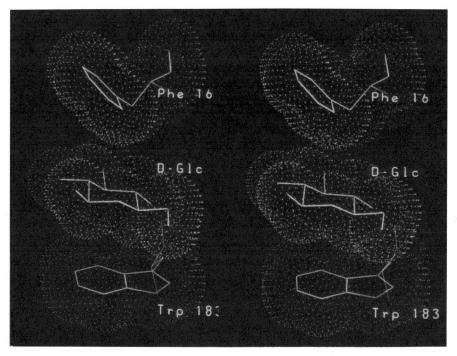

Figure 11–5. Stereoview of the stacking interaction of D-glucose with two aromatic residues, Phe16 and Trp183, of D-glucose/D-galactose-binding protein (Vyas et al., 1988).

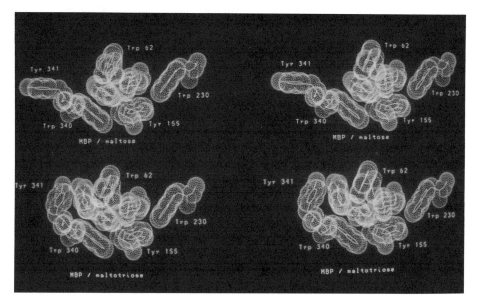

Figure 11–6. Stereoview of the interactions of aromatic residues of the maltodextrin-binding protein with maltose **(top)** (Spurlino et al., 1991) and with maltotriose **(bottom)** (Spurlino & Quiocho, unpublished data). Note that Tyr341 undergoes a conformational change upon binding of maltotriose.

A nearly complete survey of stacking interactions in structures of protein–carbohydrate complexes reveals two predominant features (Table 11-2). Tryptophan residues and the B-faces of the pyranoses are most frequently associated with these interactions.

The combining site of the antibody Se155-4 (IgG$_1$, λ, mouse) raised against the serogroup B oligosaccharide O-antigen of *Salmonella* is believed to be unusual in that it apparently contains more aromatic residues (mostly His and Trp residues) than polar planar residues (Cygler et al., 1991). Polar residues are more abundant in the carbohydrate-binding sites of all other proteins (e.g., Figs. 11-1 and 11-2). As the oligosaccharide O-antigen is composed of sugar units lacking a significant number of hydroxyl groups and thus having unusually extensive hydrophobic clusters, it is not surprising that the combining site contains many more aromatic residues by way of somatic mutations. Nevertheless, surprisingly, it appears that only one of these hydrophobic residues (Trp58L) is face-to-face with a sugar unit of the bound oligosaccharide (Table 11-2). All the hydroxyl groups of the oligosaccharides are involved in hydrogen bonding interactions, several with NH groups of the Trp and His side chains in the combining site. Many of these interactions follow the scheme indicated above for cooperative hydrogen bonds. The involvement of the NH groups of the His and Trp side chains in hydrogen bonding to sugars are typified by His152 of GGBP (Fig. 11-1) and Trp62 of the maltodextrin-binding protein (Fig. 11-3). This involvement makes it less likely that these residues also participate in stacking interactions.

In contrast to the structure of Fab of monoclonal antibody Se155-4 with *Salmonella* cell surface polysaccharide, structures of the Fab fragments of the anti-carbohydrate antibodies *Brucella* YsT9.1 (IgG$_{2b}$, κ) and *Shigella flexeneri* SYA/J6 (IgG$_3$, κ) with cell surface oligosaccharides show combining sites with features typical of those seen in all other proteins that bind oligosaccharides (Rose et al., 1993; Vyas et al., unpublished results).

The stacking interaction between aromatic residues and sugar faces has dual functions. First, it contributes to the stability of protein–carbohydrate complexes by way of van der Waals forces and hydrophobic effects. The structure of the complex of MBP with maltose has revealed another means whereby an aromatic residue can contribute to stability. The β-anomeric hydroxyl group of maltose is pointed toward the center of the six-membered ring (on the A-face side) of the indole side of Trp230 (Spurlino et al., 1991) in a manner consistent with an aromatic or π hydrogen-bonding with the hydroxyl group, as previously suggested by Levitt and Perutz (1988).

The second role of stacking interactions is related to ligand specificity. The placement of aromatic residues could disallow the binding of particular sugar epimers through steric hindrance. For instance in L-arabinose-binding protein, because of the very close association of Trp16 with the B-face of the bound L-arabinose (Fig. 11-4), D-galactose, or D-fucose, it would prevent the binding of the C4 epimers of the three monopyranosides (Quiocho & Vyas, 1984, Quiocho et al., 1989). Similar discrimination exists for the C4 epimers of the lactose bound to ricin and enterotoxin (Rutenber & Robertus, 1991; Sixma et al., 1992).

There are two possible exceptions to the use of stacking interactions. Aromatic residues are not involved in the binding of monosaccharides to the catalytic site of glycogen phosphorylase and neuraminidase (Burmeister et al., 1992; Johnson et al., 1988). However, it is possible that an aromatic residue could participate in binding of longer oligosaccharide substrates at the catalytic site of phosphorylase (Johnson et al.,

1988). Although no aromatic residue makes an interface with a sugar unit of the oligosaccharide bound to C-type animal lectin, the imidazole ring side chain of a His residue, which could be considered a gratuitous substitute for an aromatic side chain, apposes the A-face of the mannose 8 unit (Weis et al., 1992).

Ordered Water Molecules in Protein–Carbohydrate Interactions

Carbohydrate and binding site residues undergo desolvation upon complex formation. However, with the exception of the complex of D-ribose-binding protein with sugar (Mowbray & Cole, 1992), desolvation is often not complete, and ordered water molecules are present in sugar-binding sites. Ordered water molecules contribute to binding affinity by mediating hydrogen bonds between carbohydrate and protein (see Figs. 11-1, 11-2, and 11-3) (Bourne et al., 1990b, 1992; Cygler et al., 1991; Johnson et al., 1988; Quiocho & Vyas, 1984; Vyas et al., 1988).

The well-refined structures of the complexes of the L-arabinose binding protein with three monosaccharide substrates have further revealed a novel role of ordered water molecules in modulating the affinity and specificity for the sugars (Quiocho et al., 1989). ABP binds L-arabinose, D-galactose, and D-fucose with K_d values of 0.098, 0.23, and 3.8 μM, respectively (Table 11-3). Thus, the addition at the C5 of L-arabinose of the $C(6)H_2OH$ group in D-galactose reduces the binding affinity by only twofold, but substitution with the apolar $C(6)H_3$ group in D-fucose decreases affinity by approximately 40-fold. These affinities are also reflected in the efficiency of active transport of these three sugars. Structures of complexes of ABP with arabinose (Fig. 11-2), galactose, and fucose refined at 1.7Å, 1.8Å, and 1.9Å resolution (Table 11-3), respectively, show in atomic detail how ordered bound-water molecules, coupled with localized changes, can govern substrate specificity and affinity (Quiocho, 1993; Quiocho et al., 1989). In all three complexes, the sugar ring is positioned identically in the binding site and each hydroxyl common to these sugars participates in identical hydrogen-bonding interactions. Two ordered water molecules in the site contribute further to tight binding of arabinose but create a much less favorable interaction with the

Table 11–3 Binding Energies and Structures of Complexes of ABP with Sugars

Sugar	K_d (μM)	ΔG (kcal mol^{-1})	$\Delta\Delta G$	Structure (Å)	Refinement (R-factor)	References
D-Galactose	0.23	−8.9	—	1.8	0.132	a, b, c
1-Deoxy-D-galactose	4.8	−7.1	−1.8	1.81	0.147	b
2-Deoxy-D-galactose	110	−5.3	−3.6	1.71	0.158	b
6-Fluoro-6-deoxy-D-galactose	18	−6.3	−2.6	1.8	0.183	b
6-Deoxy-D-galactose (D-fucose)	3.8	−7.2	−1.7	1.9	0.134	a, b, c
L-Arabinose	0.098	−9.4	—	1.7	0.137	a, b, c, d

[a]Quiocho et al. (1989).
[b]Vermersch et al. (1992).
[c]Miller et al. (1983).
[d]Quiocho & Vyas (1984).

methyl group of fucose. In addition, tight binding of galactose is attained by the replacement of one of the water molecules by its $-C(6)H_2OH$, coordinated with localized structural changes which include a shift and redirection of the hydrogen bonding interactions of the other water molecules.

In the structures of the lectin (LOL I) complexed with trisaccharide and octasaccharide, water molecules serve not only to mediate protein–carbohydrate interactions but also to stabilize the conformations of the oligosaccharides (Bourne et al., 1990b, 1992). Indeed, these structures are notable by the presence of a large number of water molecules. Since many of the water molecules are associated solely with the polysaccharide, they help stabilize the conformation of the bound oligosaccharides.

In the sugar-bound structures, the binding sites of Group I proteins contain many fewer ordered water molecules than those of Group II proteins.

Metal Ions in Carbohydrate Binding

A few proteins, especially lectins, have a metal ion in proximity to the bound carbohydrate (Bourne et al., 1990a,b, 1992; Delbaere et al., 1990; Derewenda et al., 1989). In addition to hydrogen bonds and van der Waals contacts, the metal participates in binding saccharides. In lectins (LOL I, GS4, Con A, and Pea) complexed with sugars, the $O\delta_1$ and $N\delta_2$ of an Asn residue coordinates to Ca^{2+} and donates a hydrogen bond to the OH4 hydroxyl of the bound monosaccharide (Bourne et al., 1990a,b, 1992; Delbaere et al., 1990; Derewenda et al., 1989; Rini et al., 1993). Despite being coordinated to the bound metal, saccharides bound to lectins have dissociation constants in the millimolar range.

In the structure of the mannose-binding protein (a C-type Ca^{2+}-dependent animal lectin) with oligosaccharide, the Ca^{2+} is coordinated to two carbohydrate hydroxyl groups as well as to protein residues (Weis et al., 1992).

Sugar Derivatives as Probes of Hydrogen-Bonding Interactions

Protein–carbohydrate interaction is an excellent system for evaluating the energetics of hydrogen bonds by ligand-binding studies coupled with X-ray crystallographic analysis. Two approaches can be employed in this evaluation—the use of engineered proteins accessible by site-directed mutagenesis and engineered synthetic ligands. Using site-directed mutagenesis, a laborious undertaking since it also includes protein purification, hydrogen bond acceptor/donor groups from the protein may be modified and their contributions inferred from the difference in the ligand-binding energy between native and mutant proteins. The use of engineered ligands is well suited to the study of sugars since a deoxy derivative can provide an assessment of the total hydrogen-bonding contribution of a sugar hydroxyl to the interaction with the native protein. Because fluorine or other halogens can accept (but not donate) hydrogen bonds, sugar derivatives in which the hydroxyl group has been replaced by halides are believed to probe the energetic contribution only of the donor component.

Interpretations of ligand-binding effects resulting from both approaches require at least a knowledge of the refined X-ray structure of the native protein. These interpretations tacitly assume that mutations of specific residues, in particular, are free from complications due to changes in the local environment or protein conformation. However, since this assumption is fraught with uncertainty, it is essential also to determine the X-ray structures of the complexes of mutant protein with natural ligands as well as those of the native protein with bound engineered ligands.

Because of the ease and simplicity of the analysis, we have employed engineered ligands (Quiocho, 1993; Vermersch et al., 1992) to dissect the contribution of hydrogen bonds to the stability of sugar binding by the L-arabinose-binding protein. As discussed above, ABP also binds D-galactose and D-fucose with dissociation constants shown in Table 11-3 (adapted from Quiocho, 1993). The structures of complexes with the sugars have been refined at very high resolution (Table 11-3), providing a solid basis for studies with engineered proteins and sugars. As the affinities are some of the tightest known for protein–sugar complexes, more sensitive and accurate determinations of the differential equilibrium binding affinities of sugar analogues may be obtained. The hydrogen bonding interaction between ABP and D-galactose (Fig. 11-2) forms the foundation of our studies.

To test the validity of using deoxy and fluoro-deoxy analogues as probes for hydrogen interactions, as well as to assess the contribution of OH1, OH2, and OH6, we have determined the K_d values of the complexes of ABP with 1-deoxy, 2-deoxy, and 6-fluoro-6-deoxy analogues of D-galactose (Table 11-3) (Quiocho, 1993; Vermersch et al., 1992). The structures of the complexes with the engineered sugars have also been refined to very high resolution (Table 11-3).

The major findings from this study are as follows (Quiocho, 1993; Vermersch et al., 1992). First, the binding of 1- and 2-deoxy derivatives provides a consistent measure of the overall and specific-binding energy associated with each hydroxyl group. As shown in Fig. 11-2, OH1 and OH2 of Gal form one and two hydrogen bonds, respectively, with ABP. The loss of the one hydrogen bond between the neutral OH1 and the charged Asp90 carboxylate group (Fig. 11-2), as reflected in the binding of the 1-deoxy-Gal, amounts to about 1.8 kcal mol^{-1} (Table 11-3). A similar value (per hydrogen bond) is observed with the 2-deoxy-Gal (Table 11-3); however, one hydrogen bond associated with the OH2 is neutral–neutral and the other is neutral–charged (e.g., Fig. 11-2).

Second, the refined structures of the complexes of ABP with the 1- and 2-deoxy sugar derivatives are virtually identical to that of the complex with Gal, strongly validating the use of these deoxy derivatives as probes of the energetics of sugar binding. It is important to note that no new water molecule has moved into the void created by "engineering out" OH1 and OH2. This finding dispels the belief by others that a water molecule might occupy the void (Batacharyya & Brewer, 1988; Street et al., 1986) and thus seriously complicate the assessment of the energetics of protein–carbohydrate interactions. In fact we find by modeling experiments that, with all atoms surrounding each void fixed in their refined positions, it is not possible to place a water molecule in the void without steric clashes.

Fersht and co-workers have popularized the use of the technique of site-directed mutagenesis to assess the strength of hydogen bonds in protein–ligand interactions (Fersht et al., 1985). This technique has determined the energies of hydrogen bonds between neutral and charge groups to be ≥ 3 kcal mol^{-1}, which is significantly greater than those obtained in our studies (Table 11-3).

Deoxy sugars allow a more realistic assessment of the total contribution of sugar hydroxyl binding energy. Two minor complications are associated with the use of modifed substrates as well as protein mutants to probe hydrogen-bonding interactions. Engineering out a hydrogen-bonding group also leads to the elimination of van der

Waals contacts originally associated with the group. The effect of a void in place of the position of a missing donor/acceptor group, such as a hydroxyl group of a sugar or of a Tyr residue, is unknown.

The third finding is that the 6-fluoro-6-deoxy-D-galactose is not suitable for directly assessing the participation of a hydroxyl solely as a hydrogen bond donor group. In Gal-ABP complex (Fig. 11-2), OH6 donates a hydrogen bond to the carboxylate of Asp 89 (the interaction that 6F-gal binding is designed to probe specifically) and accepts a hydrogen bond from Wat310. As shown in Table 11-3, the conversion of OH6 to F caused a far greater loss of binding energy (2.6 kcal mol^{-1}) than that observed for each hydrogen bond lost in the binding of 1- and 2-deoxy derivatives. Moreover, two structural changes accompany the binding of 6F-Gal. (1) Although the replacement of OH6 with F is believed to be sterically conservative, the creation of enough space, together with localized rearrangements of Asp89 and Wat310, has allowed the binding of a new water moelcule at a position close to that observed for Wat311 in the ABP-Ara and ABP-Fuc complexes (Quiocho et al., 1989). (Note that Wat311 must be expelled for steric reasons when Gal is bound [Quiocho et al., 1989].) Due to the binding of a new water molecule (also identified as Wat311) in the ABP-6F-Gal complex, two additional hydrogen bonds are formed relative to the ABP-Gal complex and (2) whereas an excellent hydrogen bond is formed between OH6 donor group and Asp89 carboxylate acceptor group in ABP-Gal complex (H-bond distance of 2.73Å) (Fig. 11-1), there is a repulsion (as to be expected) between the electronegative F6 and the negatively charged carboxylate of Asp89. The F6 to the caroxylate oxygen (OD2) distance is 3.25 Å. These two destabilizing structural changes are directly related to the fact that F, when compared to the oxygen of the OH group, is more electronegative. Moreover, the affinity of F (due to inductive effect) for protons that can be donated is satisfied additionally by the binding of a new water molecule. These properties of F, combined with our findings above, nullify the suggested utility by others (Batacharyya & Brewer, 1988; Glaudemans et al., 1989; Street et al., 1986) of fluoro derivatives in probing hydrogen-bonding interactions.

Summary

Well-refined structures of a variety of protein–carbohydrate complexes have contributed immensely to our understanding of the atomic interactions between proteins and carbohydrates. They clearly show the importance of polar and nonpolar surfaces for carbohydrate recognition and affinity. Proteins that bind carbohydrates have been grouped into two classes and distinct structural features associated with the carbohydrate-binding site of each class described. Moreoever, when compared with sites belonging to Group I proteins, the features of carbohydrate recognition in sites belonging to Group II proteins show (1) fewer hydrogen bonds and van der Waals contacts per sugar unit; (2) some unpaired polar groups of the sugar; (3) many fewer cooperative hydrogen bonds and bidentate hydrogen bonds; (4) more involvement of water molecules, that either hydrogen bond to sugars and between sugar units in polysaccharides or mediate the hydrogen bonding between protein and carbohydrates; and (5) much larger B-values (atomic thermal factors) than for carbohydrates bound to Group I proteins. The structures of the complexes indicate that the frequency of amino acid side chains involved in hydrogen bonding are in the following order: (Asp,

Asn) > Glu > (Arg, His, Trp, Lys) > (Tyr, Gln) > (Ser, Thr). Though very few in number, main chain polar groups (NH and CO) are also involved in hydrogen bonding of sugars. With only two or possibly three exception, all structures of protein–carbohydrate complexes determined thus far show stacking of aromatic residues against faces of sugars, mostly against B-faces of sugars.

Acknowledgments

Work in our laboratory was supported by the Howard Hughes Medical Institute, and grants from the NIH, the Welch Foundation, and the W. M. Keck Foundation. We also thank Tim Reynolds for assisting in the preparation of schematic drawings.

Cyclodextrins

Dominique Armspach, Giuseppe Gattuso,
Rainer Königer, and J. Fraser Stoddart

The cyclodextrins (CDs) are a family of cyclic oligosaccharides produced during the degradation of starch by cyclodextrin glucosyl transferases, a variety of amylases present in *Bacillus macerans* and related microorganisms. Cyclodextrins are composed of six, seven, eight, or more α-(1-4)-linked D-(+)-glucopyranose residues. The prefix used in naming cyclodextrins identifies the number of glucose units—α-cyclodextrin, β-cyclodextrin, and γ-cyclodextrin comprise six, seven, and eight glucose units, respectively (Fig. 12-1). Only the first three members of the series (i.e., α-cyclodextrin, β-cyclodextrin, and γ-cyclodextrin) possess rigid and well-defined cavities; accordingly, they show significant complexing abilities toward a wide range of substrates in aqueous solutions. For this reason, the higher homologues will not be considered in this review.

Historical Background

Just over 100 years ago, the isolation of cyclodextrins was reported for the first time by Villiers (1891), who recovered a small amount of a crystalline product from the bacteriological digestion of starch. This product was termed "cellulosine" by Villiers because of its perceived similarities with cellulose. The empirical formula was found to be a multiple of $(C_6H_{10}O_5)_2 \cdot 3\ H_2O$, which corresponds to crystalline β-cyclodextrin. Villiers also noted two distinct forms of crystalline "cellulosine," which are believed to have been the α- and β-cyclodextrins.

Twelve years later in 1903, while investigating the decay of foods, Schardinger noticed a bacterium which produced acetone and ethanol as by-products of its action on sugar- or starch-containing plant materials. This microorganism was named *Bacillus macerans* (from the Latin macerare: "to rot"). Upon closer investigation, this *Bacillus* strain was found to convert 25%–30% of an inoculated starch paste into cyclodextrins. Two types of cyclodextrins were isolated by Schardinger. They gave differently shaped crystals and differently colored products upon reaction with iodine. These two cyclodextrins, termed "crystallized dextrin α" and "crystallized dextrin β," were characterized by Schardinger as cyclic oligosacharides (Schardinger, 1904, 1911). In recognition of Schardinger's work in this field, cyclodextrins were referred to as "Schardinger

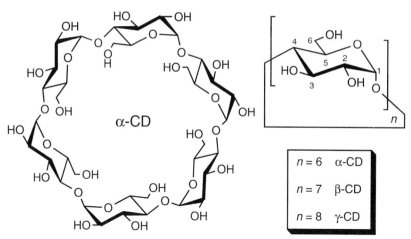

Figure 12–1. The structures of the cyclodextrins (CDs).

dextrins" well into the middle of this century. Indeed, this term is still in use today, albeit infrequently. Unfortunately, the cyclodextrin literature in the quarter of a century following Schardinger's last paper has contained little of lasting value (French, 1957). It was not until the late 1930s that the α-1,4 glycosidic linkage was accepted to be the only type of glycosidic linkage present in cyclodextrins. French & Rundle (1942) established the molecular weights of α- and β-cyclodextrin and, therefore, determined the number of α-1,4 linked D-glucopyranose units they contain—namely six and seven, respectively. The molecular structure determination of γ-cyclodextrin had to wait until the late 1940s, when it was finally confirmed as having eight α-1,4 linked D-glucopyranose units by Freudenberg & Cramer (1948) and French et al. (1950).

Sources of Cyclodextrins

Today's industrial production of cyclodextrins still relies on microorganisms (Sicard & Saniez, 1987). They provide the crude enzyme preparation that allows the conversion of starch into cyclodextrins. Since the late 1930s, when the first active, cell-free enzyme mixture was isolated from *Bacillus macerans*, several more microorganisms have been identified that also produce cyclodextrin glycosyl transferases. For the industrial preparation of cyclodextrins, the enzyme is isolated from the appropriate bacterium and allowed to act on a starch suspension.

Once a crude enzyme preparation had been isolated, it was quickly realized that its action is reversible (French et al., 1954), allowing the interconversion of α- and β-cyclodextrins in the presence of suitable cosubstrates.

This reversibility is used very effectively in the production process. To obtain high yields of β-cyclodextrin at the expense of α-cyclodextrin, a wide variety of selective precipitants can be added to the reaction mixture. For example, the addition of toluene will precipitate β-cyclodextrin, thereby shifting the equilibrium and resulting in the exclusive formation of β-cyclodextrin. This development was a great improvement on the earlier methods of separation developed by Schardinger and Freudenberg, who relied on the selective crystallization of the cyclodextrins and their acetylated derivatives.

A drawback of this selective precipitation method is the need to remove the precipitant after the separation has been achieved. Impurities introduced through this precipitation technique are believed to have led to inaccurate conclusions in past toxicity studies. Pure cyclodextrins can generally be isolated after steam distillation, which efficiently removes the precipitant from the cyclodextrin cavity.

As a result of this type of production research and ever increasing commercial demand, cyclodextrins are now industrially available in multiton quantities from several manufacturers around the world.

Physical Characteristics of Cyclodextrins

Cyclodextrins are colorless crystalline solids that are soluble in polar solvents. It is well established that the D-glucopyranose units are in the normal $^4C_1$ chair conformation. The overall shapes of the cyclodextrin molecules are those of truncated cones. Space-filling representations of α-, β-, and γ-cyclodextrins, which are based on their X-ray crystal structures, are shown in Figure 12-2. The more open top of the bucket-shaped α-cyclodextrin molecule (Fig. 12-3) bears the secondary 2- and 3-hydroxyl groups, whereas the primary 6-hydroxyl groups are located around the narrow rim of the bucket. The cavity interior is lined with the glucose ring oxygen atoms, as well as with the H3, H5, and, to a lesser extent, H6 protons. On the other hand, the H1, H2, and H4 protons are located on the outside surface of the molecule. A strong hydrogen

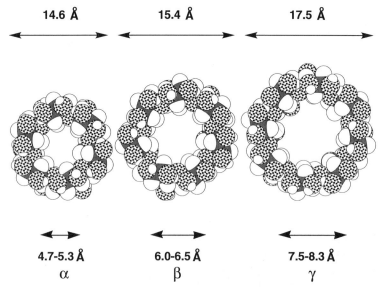

| 14.6 Å | 15.4 Å | 17.5 Å |

| 4.7-5.3 Å | 6.0-6.5 Å | 7.5-8.3 Å |
| α | β | γ |

Figure 12-2. Space-filling representation of the cyclodextrins viewed through the cavity from the secondary face.

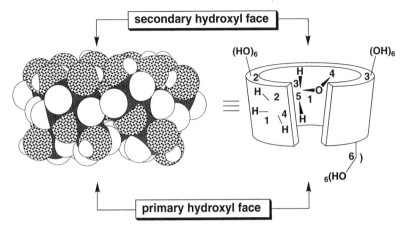

Figure 12–3. The bucket representation of α-cyclodextrin.

bonding network between the 2'-hydroxyl and 3-hydroxyl group of adjacent glucose units exists both in the solid state (Betzel et al., 1984; Le Bas & Rysanek, 1987; Saenger et al., 1983; Zabel et al., 1986) and in solution (Casu et al., 1968) (Fig. 12-4), with the 3-hydroxyl groups acting predominantly as hydrogen-bonding donors and the 2-hydroxyl groups acting mainly as hydrogen-bonding acceptors (Casu et al., 1968;

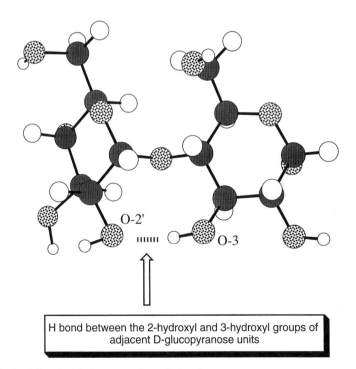

Figure 12–4. Disaccharide fragment of a cyclodextrin torus.

Christofides & Davies, 1982, 1983; Gillet et al., 1982; St-Jacques et al., 1976). The intramolecular hydrogen bonding further restricts the already limited torsion about the glycosidic bonds that connect the glucose residues. This situation appears to be the case for all naturally occurring cyclodextrins and is responsible for the rigidity of the molecules, which is an important prerequisite to efficient binding. According to ^{1}H NMR (Casu et al., 1968; St-Jacques et al., 1976) and ^{13}C NMR (Vincendon, 1981) spectroscopic studies in CD_3SOCD_3 solution, as well as X-ray crystallographic analysis in the solid state, the strength of these intramolecular hydrogen bonds increases in the order α-cyclodextrin $<$ β-cyclodextrin $<$ γ-cyclodextrin. The field of cyclodextrin research is already extensive and it is expanding all the time. A number of monographs and special journal issues (Bender & Komiyama, 1978; Duchêne, 1987; Stoddart, 1989; Szejtli, 1982a, 1988; Szejtli & Osa, 1996) are available, as are several reviews (Bergeron, 1977; Saenger, 1980, 1984) on general cyclodextrin chemistry. In addition, the proceedings of eight international symposia (Atwood et al., 1984; Atwood & Davies, 1987; Duchêne, 1990; Hedges, 1992; Huber & Szejtli, 1988; Osa, 1994; Szejtli, 1982b, 1996) on cyclodextrins give an excellent overview of the latest advances in research in the field.

Complex Formation

The most interesting feature of the cyclodextrins is that they act as promiscuous hosts toward a large variety of apolar aliphatic and aromatic substrates. In particular, the slightly more lipophilic interior of the cavity—compared to the more hydrophilic outside surface—has been held responsible for this inclusion phenomenon observed in aqueous solution. However, other species such as acids, amines, halides, and other small ions, as well as noble gases have been found to insert into the cyclodextrin cavities. Recently, it has even been shown that fullerenes can be complexed by γ-cyclodextrin (Andersson et al., 1992, 1994a,b; Boulas et al., 1994; Priyadarsini et al., 1994; Yoshida et al., 1994; Zhang et al., 1994).

A number of techniques have been employed to characterize cyclodextrin inclusion complexes. Proof that adduct formation, in the solid state, involves the presence of the guest inside the cavity has come from X-ray diffraction studies on cyclodextrin complexes crystallized from water. According to Saenger (McMullen et al., 1973), two modes of crystal packing have been observed for these complexes , that is, channel or cage structures. In channel-type structures, cyclodextrin molecules stack in such a way as to align their cavities linearly to form channels through the crystal (Fig. 12-5). These cyclodextrin stacks are stabilized by intermolecular hydrogen bonds, either between secondary hydroxyl groups on one face and primary hydroxyl groups on the other, or between the secondary and primary hydroxyl groups, leading to head-to-head or head-to-tail sequences, respectively. This hydrogen bonding between the stacked cyclodextrin molecules may either occur indirectly *via* a water molecule or directly through the hydroxyl groups of the cyclodextrins. In addition, cage-type structures form isolated cavities, which are blocked off on both rims by neighboring cyclodextrin molecules. This situation arises for two types of crystal packing (Fig. 12-5). First, the more commonly observed "herringbone" pattern, results from molecules packing in a criss–cross fashion. Second, a "brick-wall" pattern has been noted in a few cases, which contain offset layers in a head-to-tail array.

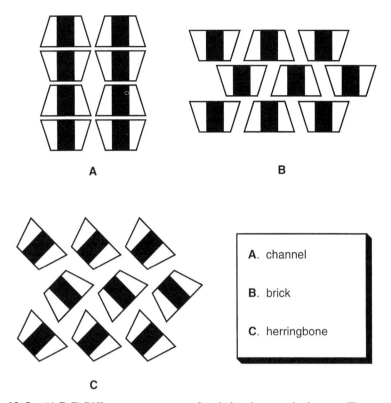

A

B

A. channel

B. brick

C. herringbone

C

Figure 12–5. (A,B,C) Different arrangements of cyclodextrin arrays in the crystalline state.

The elucidation of the solid state structures of cyclodextrin adducts not only provides structural information about the position of guest(s) in the cyclodextrin cavities, but also reflects the conformational changes occurring in the binding partners upon complexation. In solution, ^{1}H NMR spectroscopy has been recognized as the most versatile technique for studying the structures of inclusion complexes. Demarco & Thakkar (1970) first noticed that the values of the chemical shifts corresponding to the resonances of H3 and H5, protons, which line the interior of the cavity change as a result of the presence of an included guest. Similarly, the protons of the guest molecule which, upon complexation are in close contact with the interior of the cavity, are also strongly affected in relation to their chemical shifts. Thus, a comparison between the chemical shift changes of the different host or guest protons can be used (Schneider et al., 1988) to determine the disposition of the substrate(s) in the cavity of a cyclodextrin. As a result of the fast complexation–dissociation process on the ^{1}H NMR time scale, the observed signals correspond to the averaged chemical shifts of the bound and unbound species. If the changes in the chemical shifts of a particular resonance are sufficiently large, then the proton(s) corresponding to this resonance can be used as probe(s) for the determination of association constants (Bergeron et al., 1977). Structural information about the adducts can also be gained from intermolecular nuclear Overhauser effect (NOE) studies, including NOE differences measurements (Meister & Wenz, 1990) as well as two-dimensional (2D) ROESY experiments (Schneider et al., 1991).

Alternatively, ^{13}C NMR spectroscopy has been applied to the structural study of inclusion complexes both in solution (Bergeron & Channing, 1976) and in the solid state (Gidley & Bocieck, 1988; Hall & Lim, 1986; Ripmeester, 1988; Veregin et al., 1987). In addition, ^{1}H and ^{13}C NMR relaxation studies (Behr & Lehn, 1976; Lipkowitz et al., 1992) have shown that, for complexes formed with aromatic compounds, relatively weak couplings between host and guest species are observed, confirming the high mobility of the guest within the cyclodextrin host. There are many more applications of NMR spectroscopy to the investigation of solution state cyclodextrin complexes, including theoretical approaches (Inoue et al., 1987; Yamamoto et al., 1988). Comprehensive reviews are available in the literature (Bergeron, 1984; Yamamoto & Inoue, 1989). Various other physical methods have also been applied, often in conjunction with NMR spectroscopy and X-ray crystallography.

Sensse & Cramer (1969) first recognized that the complexation of an achiral, light-absorbing guest to a chiral, nonabsorbing cyclodextrin would give rise to induced Cotton effects in circular dichroism spectra. Since then, induced circular dichroism has been applied successfully to the structural elucidation of cyclodextrin—light-absorbing molecule host–guest complexes (Kobayashi & Opallo, 1990). In particular, the disposition of guest molecules in the cyclodextrin cavity has been inferred, based on the knowledge that electronic transitions parallel to the molecular axis of cyclodextrins produce positive Cotton effects while those normal to the axis show negative Cotton effects (Harata, 1979; Harata & Uedaira, 1975; Kobayashi, 1988; Shimizu et al., 1979, 1981, 1982). However, an induced circular dichroism spectrum can only be assigned to the correct geometry if the direction of the electronic transitions for a chromophoric guest is known. A general theory on induced circular dichroism involving cyclodextrins (Kodaka, 1993) has been proposed recently. It addresses not only included chromophoric residues, but also those that lie at the periphery of the cavity. In the particular systems this theory has been applied to so far, the predictions have been verified experimentally.

Other, less common techniques have also been used to study cyclodextrin complexes. They include ESR spectroscopy (Kotake & Janzen, 1989), which has a faster time scale than NMR spectroscopy so that signals for the bound and unbound species can be revealed, fluorescence spectroscopy (Minato et al., 1991), voltammetry (Isnin et al., 1991; Kaifer et al., 1989), conductometry (Palepu & Reinsborough, 1989), transport experiments (Poh & Chow, 1992), pH titrations (Eftink et al., 1989), mass spectrometry (Ashton et al., 1988), and so on. Microcalorimetric techniques (Danil De Namor, 1990; Eftink et al., 1989; Inoue et al., 1993) are of special interest as they provide values for the macroscopic thermodynamical parameters ($\Delta H°$ and $T\Delta S°$) for the binding of substrates.

The Mechanism of Inclusion

Although the driving forces involved in the formation of a cyclodextrin complex have been the subject of much debate, in recent years a general and realistic model (Tabushi et al., 1978), supported by experimental evidence has been proposed. A résumé of the proposals is given in this section.

The phenomenon of inclusion is a composite process which is best understood by considering the thermodynamics of the binding. Van der Waals interactions, which include dipole–dipole, dipole-induced dipole, and induced dipole–induced dipole terms, are the predominant attractive forces between a cyclodextrin host and a lipophilic substrate. The induced dipole-induced dipole interactions, also known as London dispersion interactions, comprise an attractive and a repulsive term. For an interatomic interaction, the sum of the two terms is described by the two parameter Lennard-Jones potential ($Ar^{-12} - Br^{-6}$) with r equal to the interatomic distance, with the parameters A and B describing the repulsive and attractive terms. The magnitude of the attractive term B is approximately proportional to the polarizability of the interacting atoms. For a series of phenyl acetates (Van Etten et al., 1967) as substrates, the relationship between the logarithm of the dissociation constant for the cyclodextrin complex and the molar refractivity of the guest was observed to be almost linear. The molar refractivity of a compound is linked to its polarizability: hence, the importance of van der Waals interactions in cyclodextrin inclusion complexes. This picture of a hydrophobic complex is fairly accurate if we consider the complex in the gas phase. The situation is much more complicated in solution. Figure 12-6 clearly shows the great complexity associated with apolar host–guest complexation in solution, as compared to the gas phase. Complexation in the gas phase proceeds essentially in a single step event (H + G → HG). It corresponds to step E in Figure 12-6, which describes the complexation between the guest in the gas phase and the externally solvated host. All other steps are solvent-related. In the thermodynamical process shown in Figure 12-6, step A describes the desolvation of the host cavity with transfer of the included water molecules (S) into the gas phase. In this process, the van der Waals and hydrogen bonding interactions between the included solvent molecules (S) and the host interior (H) ($\Delta H^{SH}_{VdW \& intermol} > 0$), as well as the interactions between the solvent molecules inside the cavity, are lost ($\Delta H^{SS}_{intermol} > 0$). On the other hand, the expulsion of water molecules is accompanied by a gain in entropy ($T\Delta S^{S}_{trans \& rot} > 0$). In step B, the solvating molecules are then transferred into the bulk, generating a loss of entropy ($T\Delta S'^{S}_{trans \& rot} < 0$) but a gain in cohesive interactions, mainly through hydrogen bonding with the bulk water ($\Delta H'^{SS}_{intermol} < 0$). Water molecules solvate apolar substrates by forming a cluster around the surface of the guest, reminiscent of the structure of ice. Step C describes the transfer of the guest from the cluster of water molecules into the gas sphere. As for the host (H), a loss of solvent-guest interactions ($\Delta H^{SG}_{VdW} > 0$) and a gain in entropy takes place upon desolvation ($T\Delta S^{G}_{trans \& rot} > 0$). In step D, the water cluster, held together by hydrogen bonding, breaks down and its components are transferred into the bulk with a gain in cohesive forces ($\Delta H''^{SS}_{intermol} < 0$) and a gain in entropy ($T\Delta S''^{S}_{trans \& rot} > 0$). Finally, step E involves attractive van der Waals interactions ($\Delta H^{HG}_{intermol} < 0$) and loss of translational and rotational entropy of one of the two through complex formation ($T\Delta S'^{G}_{trans \& rot} < 0$). Note that, for specific substrates, the inclusion complex can be stabilized further by hydrogen bonding (Matsui et al., 1970). Release of strain within the macrocyclic framework of the cyclodextrin has also been thought to participate in the binding. This hypothesis (Saenger et al., 1976) was based on the comparison between the X-ray crystal structures of hydrated α-cyclodextrin in which one glucose unit is distorted by rotation and the α-cyclodextrin macrocycle in complexes which possess the normal C_6 symmetry. However, more recent thermodynamical and theoretical investigations in the solution state have demonstrated that such a phenomenon counts for very little. On the contrary, energetically unfavorable conformational changes (Danil De Namor, 1990; Inoue et al., 1993; Tabushi et al., 1978), involving the reorganization of the original hydrogen bond network, usually take place upon complexation. This thermodynamical treatment of the binding event is also valid for less well-defined systems like micellar structures and

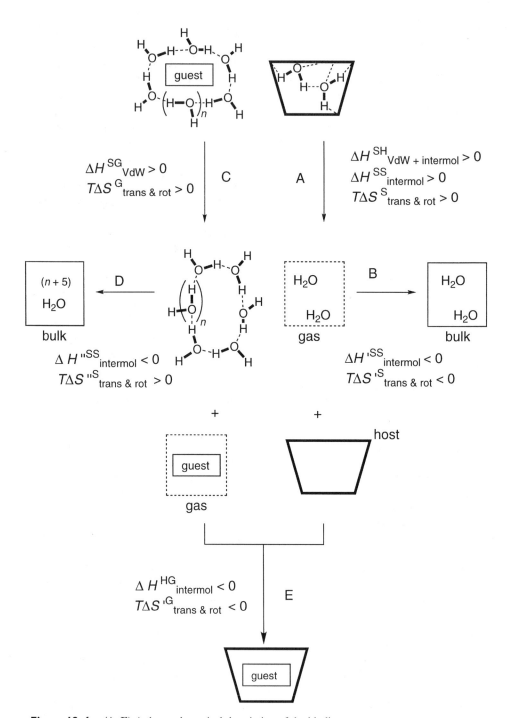

Figure 12–6. (A–E) A thermodynamical description of the binding event.

membranes (Ben-Naim, 1980; Ringsdorf et al., 1988; Tanford, 1980). In those systems, the associating monomers lose only a small amount of rotational and translational entropy, but virtually no changes in enthalpy ($\Delta H \sim 0$) take place. Here, the favorable desolvation entropies are not completely compensated by a large unfavorable term for monomer association, and thus a positive entropy is measured for the overall aggregation. On the other hand, host–guest complexation, as observed not only in cyclodextrins (Inoue et al., 1993; Danil De Namor, 1990) but also in water-soluble cyclophanes (Diederich, 1991), and in the hydrophobic pockets of enzymes and antibodies (Biltonen & Langerman, 1979; Ross & Subramanian, 1981), is essentially an enthalpy-driven process. The large enthalpic term and favorable entropies resulting from the release of "high energy water" into the bulk associated with the enthalpically favorable van der Waals interactions between matching binding partners (i.e., showing complementarity in shape and size) are not entirely compensated for by the losses in translational and rotational entropies. Interestingly, recent studies on cyclodextrin complexes with guest molecules containing both phenyl and naphthyl units (Schneider et al., 1991) have shown that this loss in entropy is minimal (i.e., the binding is optimal) when there is still enough room for high mobility of the guest within the cavity. The balance between the enthalpic and entropic terms determines the strength of the complex. It follows that the cavity size of the cyclodextrin has a profound effect on the mechanism of binding for a given substrate. For instance, Eftink and co-workers (Cromwell et al., 1985) have studied the interaction of 1-adamantanecarboxylate with the series of cyclodextrins by microcalorimetric titration at pH 8.5. They found that a 1:1 adduct is formed in the case of all cyclodextrins and that the strongest binding occurs with β-cyclodextrin. This observation is consistent with a near perfect match between the cavity of β-cyclodextrin and the diameter of 1-adamantanecarboxylate. The binding of 1-adamantanecarboxylate to β-cyclodextrin is almost entirely driven by an exothermic enthalpy change, whereas its binding to γ-cyclodextrin is characterized by an unusual and small endothermic ΔH° and a large positive ΔS°—the pattern observed for the classical hydrophobic effect related to loose binding.

Furthermore, numerous studies on cyclodextrin complexes have shown that α-cyclodextrin (5.7 Å diameter cavity) and β-cyclodextrin (7.8 Å diameter cavity) usually include one guest molecule or one hydrophobic moiety of a guest molecule (Bender & Komiyama, 1978; Duchêne, 1987; Stoddart, 1989; Szejtli, 1982a,b, 1988). It is not unusual, however, for γ-cyclodextrin to accommodate two guest molecules within its wider cavity—9.5 Å diameter cavity (Cromwell et al., 1985).

It transpires from these results that it is above all the complementarity in size and shape between a cyclodextrin and a guest molecule that controls

- the relative binding strength
- the shape
- the stoichiometry of a complex.

Applications of Cyclodextrins

Over the last few years, the price of cyclodextrins and some of their chemically modified derivatives has dropped dramatically. This drop in price, combined with their greatly increased availabilities on a multiton scale, has expanded tremendously the use of cyclodextrins in the commercial sphere.

Toxicity studies have established that orally administered cyclodextrins are harmless toxicologically. Studies on ^{14}C labeled cyclodextrins have demonstrated (Anderson et al., 1963) the incorporation of ^{14}C into tissues, as well as the presence of $^{14}CO_2$ in exhaled air. However, cyclodextrins are metabolized more slowly than starch. Some ill effects have been observed when cyclodextrins were administered in very high doses either subcutaneously or intravenously. This very low toxicity is crucial for many cyclodextrin applications, especially in the food, cosmetic, and pharmaceutical industries.

Cyclodextrins are used to achieve a wide range of different effects in commercial preparations (Saenger, 1980). Oils, volatile liquids, and even gases can form powders with cyclodextrins, thereby improving their handling, storage, and dispension. Furthermore, cyclodextrins can suppress unpleasant odors and tastes of certain food additives and pharmaceuticals, making them more acceptable to the consumer.

The stabilization of light- and oxygen-sensitive compounds can be achieved by complexation with cyclodextrin—a very striking example of this phenomenon is the stabilization of nitroglycerin, which can be manufactured into nonexplosive tablets upon complexation within β-cyclodextrin.

The water solubility of hydrophobic substances can also be increased using cyclodextrin complexes, thus improving, for example, the ease of application of certain agrochemicals or the bioavailability of drug molecules.

Cyclodextrins have become increasingly more important in analytical chemistry and separation technology (Li & Purdy, 1992). They have found applications in analytical luminescence spectrometry, where they can enhance signals observed in aqueous solutions. Furthermore, as a result of their chirality (Easton & Lincoln, 1996), they have been employed as chiral shift reagents in NMR spectroscopy, allowing the determination of optical purity. However, the greatest impact of cyclodextrins has been on chromatographic separations. They have found applications in affinity, thin layer, gas–liquid, and high performance liquid chromatography, as well as in electrophoresis. Addition of cyclodextrins to the stationary or mobile phases can significantly modify the separation obtained in certain systems.

In gas–liquid chromatography, many different chemically modified cyclodextrins have been employed with great success. The stationary phases containing cyclodextrins are easily prepared by coating the column with a cyclodextrin solution and allowing the solvent to evaporate. Greater difficulties were encountered in creating suitably immobilized cyclodextrin columns for high performance liquid chromatography. However, in 1985 Armstrong (Alak & Armstrong, 1986; Armstrong et al., 1986; Hinze et al., 1985; Seeman et al., 1988) developed a suitable resin bound cyclodextrin, which has become commercially available. These columns have been very successful for the separation of diastereoisomers and enantiomers, as well as constitutional isomers.

Many more applications of cyclodextrins exist and are being investigated. It is more than certain that the continuing research into cyclodextrin technology will lead to ever more practical applications for cyclodextrins and their chemically modified derivatives.

Chemical Modification of Cyclodextrins

Synthesizing derivatives of cyclodextrins has received much attention as a means of modifying their solubility and binding characteristics. The synthetic versatility of the hydroxyl group has led to the synthesis of a large number of cyclodextrin derivatives

(Croft & Bartsch, 1983). The type of substituent, as well as the constitutional iso-merism, may be varied to produce different chemically modified cyclodextrins. The most simply modified cyclodextrin is one which bears only a single substituent. However, three constitutional isomers are possible for these compounds, depending on which of the three hydroxyl groups accommodates the substituent. When two identical substituents are introduced, the number of isomers jumps to 27 in the case of α-cyclodextrin. For three identical substituents, this number increases dramatically to several hundred possible isomers. Higher or differing substitution leads to even larger numbers. These large numbers of possible isomers, and the difficulties associated with separating and characterizing them, has contributed to a proliferation of reports dealing with ill-defined mixtures of cyclodextrins.

Characterization is easiest for cyclodextrins in which each glucopyranose unit of the cyclodextrin is substituted in an identical fashion, that is, retaining the C_n symmetry of the cylcodextrin, where n is equal to the number of glucopyranose residues in the cyclodextrin. As a result of the degeneracy of the glucopyranose residues in these types of cyclodextrins, the ^1H NMR spectrum simplifies to show the protons as if they were associated with a single glucopyranose. However, even for monosubstituted cyclodextrins, this degeneracy is lost completely and the ^1H NMR signals have to be multiplied by 6, 7, or 8 corresponding to α-, β-, and γ-cyclodextrin, respectively. Very high field NMR spectroscopic investigations are necessary to characterize these unsymmetrical cyclodextrins satisfactorily (Ashton et al., 1995b; Spencer et al., 1987).

The synthesis of cyclodextrins containing different substituents on the three hydroxyl positions is theoretically feasible. The most reactive hydroxyl groups on a cyclodextrin are the primary hydroxyl functionalities at the 6-position. The secondary hydroxyl groups at the 2-position are only slightly less reactive, leading to poor selectivity. However, when sterically demanding substituents are being introduced, good selectivity for the primary hydroxyl groups can often be achieved. The 3-hydroxyl functionality is considerably less reactive as a consequence of the previously mentioned intramolecular hydrogen bond network. This difference becomes especially apparent in the relative resistance of the 3-hydroxyl group toward alkylation. This effect is found to be most pronounced in the β-cyclodextrin, followed closely by the α-cyclodextrin, with only poor selectivity occurring in the γ-cyclodextrin, where the differential reactivity of the two secondary hydroxyl groups is not very marked. In practice, it is still very difficult to obtain exclusive reaction at one position with no side reaction at other hydroxyl positions. Except for a few cases, enriched mixtures are generally obtained from which the desired product has to be isolated, usually by chromatography.

The methylation of cyclodextrins was one of the first chemical transformations performed (Irvine et al., 1924) to help elucidate their structures. More recently, selective alkylation has been used to affect the solubility of cyclodextrins in a variety of different solvents. Apart from exhaustive methylation, no selective one-step methylation procedure has been developed, so that the production of pure compounds involves extensive protection, chromatography, and deprotection (Ashton et al., 1995a; Spencer et al., 1987; Takeo, 1990). Partially methylated cyclodextrin mixtures are being produced commercially, as they show higher aqueous solubilities than their natural precursors. Other alkylation reactions, such as hydroxyalkylations and polymerizations, have produced cyclodextrins that have found applications in industry. Hydroxyalkylation affords very crude mixtures of modified cyclodextrins, which are highly water soluble, while the random cross-linking of cyclodextrins using epichlorohydrin yields an insoluble polymer, which is of interest in the field of separation technology.

Cyclodextrins also readily undergo acylations. Such reactions were used to good effect in some early methods (Freudenberg & Jacobi, 1935) for separating mixtures of natural cyclodextrins. Furthermore, the reversibility of acylations has made them useful tools in the purification of reaction mixtures by chromatography or as protecting groups for further synthetic modification of cyclodextrins.

The development of a large variety of different silylating agents in recent years has allowed for very selective protection of hydroxyl functionalities. Silyl groups, although only poorly investigated for protecting the hydroxyl groups on cyclodextrins, have afforded useful results for the selective protection at the 6 and 2,6-positions, particularly employing the *tert*-butyldimethylsilyl group (Ashton et al., 1995a; Fügedi, 1989; Icheln et al., 1996; Takeo et al., 1989).

Conventionally, the introduction of different functional groups has been achieved by converting one or more of the cyclodextrin hydroxyl functionalities into good leaving groups. Tosylation or halogenation procedures have been developed toward this end. A range of different functional groups have thus been introduced, affording a variety of amines or sulfides. Functionalizations of this type at the 6-position are reasonably straightforward. However, reactions at the 2- or 3-positions, which are more hindered and therefore react more slowly, lead to inversions of configuration at these chiral centers, yielding cyclodextrins containing sugar residues different from D-glucose. For these reasons, substitution at the 2- and 3-positions have not been investigated extensively.

The chemical derivatization of cyclodextrins have greatly modified their solubilities and complexation behaviors. Without these modifications, many of the industrial applications of cyclodextrins would not have been possible (Eastburn & Tao, 1994).

Novel Cyclodextrin Derivatives

Most of the chemical modifications described here have resulted from changes in the functional groups attached to the cyclodextrin torus. However, the core constitution of these cyclodextrin derivatives is essentially identical to those present in the natural products. Recent investigations have led, however, to the synthesis of novel cyclodextrin derivatives, which show significant changes in the conformation of the D-glucopyranose residues. In naturally occurring cyclodextrins, the D-glucopyranose residues adopt a $^4C_1$ conformation; however, it is well established that 3,6-anhydration locks a D-glucopyranose residue in a $^1C_4$ conformation. Base treatment (Fig. 12-7) of a cyclodextrin, whose primary 6-hydroxyl groups have been replaced by good leaving groups, should result in an intraresidue ring closure as a result of nucleophilic substitutions of the 3-hydroxyl groups expelling the leaving groups on the 6-positions. The fully derivatized per-3,6-anhydro-α-, β-, and γ-cyclodextrins have been synthesized recently by this approach (Ashton et al., 1991a,b, 1996b; Defaye & Gadelle, 1991; Yamamura et al., 1993, 1996; Yamamura & Fujita, 1991). As a consequence of this transformation, all the previously equatorial substituents adopt an axial position, resulting in the 2-hydroxyl group becoming oriented into the cavity of the cyclodextrin, thus reducing its diameter and making it relatively hydrophilic. Mass spectrometric binding studies (Ashton et al., 1996b; Yamamura et al., 1993) of group I metal ions with the per-3,6-anhydro-cyclodextrins have shown a size selectivity similar to that ex-

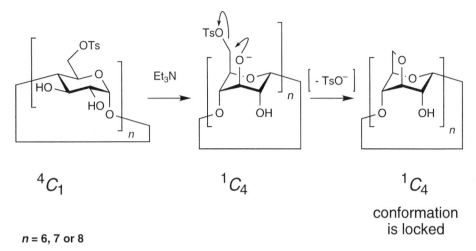

$^4C_1$

$^1C_4$

$^1C_4$

conformation
is locked

n = 6, 7 or 8

Figure 12–7. The 3,6-anhydration of cyclodextrins.

hibited by crown ethers, that is, the larger cations bind preferentially to the γ derivative while the smaller cations bind preferentially to the α derivative. Therefore, by changing the conformation of the D-glucopyranose residues, a reversal in the binding characteristics has been achieved to produce a cavity that binds hydrophilic guests in contrast to the original cyclodextrin cavity that complexes hydrophobic guests.

Other anhydro-cyclodextrin derivatives which have been reported recently include the per-2,3-epoxycyclodextrins (Coleman et al., 1992; Khan et al., 1992, 1996). These compounds are the result of introducing a leaving group on to the 2-position, followed by its intraresidue expulsion by the neighboring 3-hydroxyl group to form the manno-epoxide (Fig. 12-8). The ^{1}H NMR spectra of these compounds indicate a significant change in the conformation of the pyranose ring, as displayed by the change in the observed coupling constants for the individual protons. Opening of the epoxide ring in these per-2,3-manno-epoxy-cyclodextrins is expected to lead to cyclodextrin deriva-

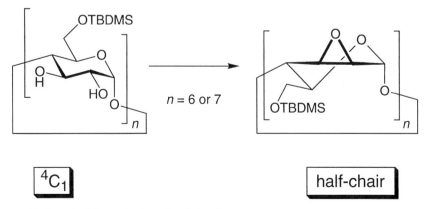

$^4C_1$

half-chair

Figure 12–8. The 2,3-epoxidation of cyclodextrins.

Table 12–1 Cyclodextrins and Their Analogues Obtained by the Total Synthesis Approach

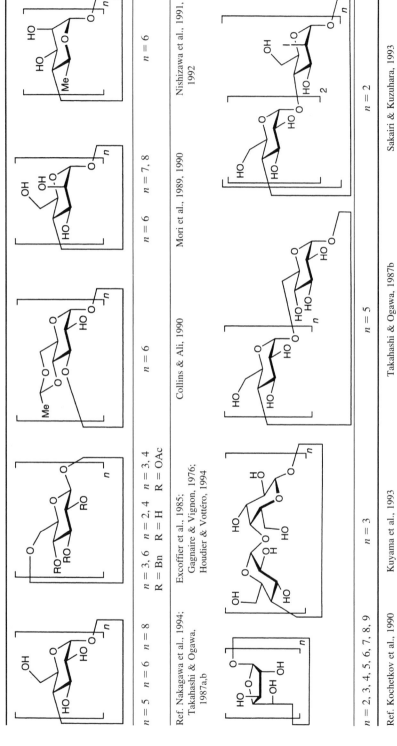

tives containing solely altrose residues in the $^1C_4$ conformation (Ikeda et al., 1990). It is hoped that these cyclodextrin derivatives will prove to be useful intermediates for further chemical modification.

The Synthesis of Cyclodextrins and Their Analogues

The first total synthesis of α- and γ-cyclodextrins from maltose was reported by Ogawa in 1987 in overall yields of 0.3 and 0.02%, respectively (Takahashi & Ogawa, 1987a,b). The synthetic strategy employed consisted of the synthesis of suitably protected linear oligosaccharides, following standard carbohydrate methodology. These linear oligosaccharides were then encouraged to undergo intramolecular cyclizations (Fig. 12-9) to yield α-cyclodextrin in 21% and γ-cyclodextrin in 8.4% yields. These yields are quite respectable, considering the untemplated cyclizations to form 30- and 40-membered rings, respectively. Interestingly, the synthesis of the unnatural cyclodextrin containing only five glucopyranose units has also been reported recently (Nakagawa et al., 1994).

The synthetic methodology lends itself to the incorporation of carbohydrate residues other than α-1,4 linked D-glucopyranose into the cyclodextrin ring. This approach has led to the synthesis of entirely unnatural cyclodextrins, for example, the α-1,4-linked cyclomannohexaose (Mori et al., 1989). Most surprisingly, the final cyclization step in this total synthesis was achieved in much higher yield (64%) than in the analogous reaction leading to the natural product (21%). Table 12-1 displays a variety of different cyclic oligosaccharides which have been obtained by total synthesis. Semisynthetic ap-

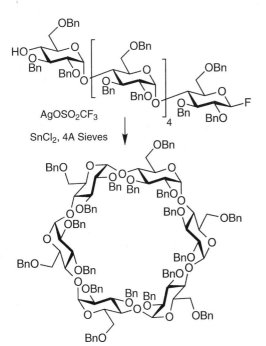

Figure 12–9. Cyclization of an oligosaccharide to form a protected α-cyclodextrin.

proaches have also led to a variety of new cyclodextrin analogues (Sakairi & Kuzuhara, 1992; Sakairi et al., 1991), in which novel sugar residues have been incorporated into the cyclodextrin ring. These syntheses rely on the selective opening of a cyclodextrin ring at a glycosidic bond. The resulting linear oligosaccharide is then extended by a sugar residue using standard carbohydrate chemistry methodology. Subsequent cyclization of the linear oligosaccharide yields a cyclodextrin derivative containing one (different) carbohydrate residue more than the cyclodextrin used as the starting material. Recently, a new methodology has been developed for the one-pot synthesis of families of cyclic oligosaccharides (Ashton et al., 1996a). This approach relies on the polycondensation/cycloglycosylation of a "disaccharide monomer" which bears glycosyl donor (cyanoethylidene) and acceptor (trityloxy) functions at the two ends of the same molecule. A cyclic hexasaccharide and octasaccharide, composed of alternating L-rhamnose and D-mannose residues, have been prepared by following this strategy, in 34% and 31% yields, respectively (Fig. 12-10). Furthermore, X-ray crystallography revealed that, in the solid state, the cyclic oligosaccharide assembles to form infinite stacks, which create nanotubes approximately 1 nm in diameter (Fig. 12-11).

Cyclodextrins as Catalysts

Cyclodextrins and some of their chemically modified derivatives can have profound effects on the outcome and kinetics of certain reactions. As a consequence, cyclodextrins have been investigated as catalysts for asymmetric reactions (Takahashi & Hattori, 1994) and as enzyme mimics (Bender & Komiyama, 1978; Breslow, 1994, 1995; Kirby, 1996). These effects are almost always the result of the formation of an inclusion complex between the cyclodextrin and at least one of the reagents.

In an early experiment, Cramer & Dietsche (1959) reported a rate enhancement in the hydrolysis of ethyl *para*-chloromandelate and other aromatic esters, upon addition of β-cyclodextrin. These observed rate enhancements are the result of the substrate binding in the cyclodextrin cavity, thus bringing the ester functionality into close proximity with the cyclodextrin hydroxyl groups. The complexed guest molecule is then oriented favorably to undergo a transesterification to the cyclodextrin (Fig. 12-12). This mechanism is consistent with the observations by Bender and co-workers (D'Souza & Bender, 1987; Van Etten et al., 1967) that *meta*-substituted phenyl esters experience a larger rate enhancement than their *para*-substituted analogues. In the case of the *meta*-substituted phenyl derivatives, the ester linkage is forced into close contact with the secondary hydroxyl groups on the cyclodextrin torus, leading to a facile transesterification reaction. The acylated cyclodextrin must then undergo hydrolysis to complete the catalytic cycle. To improve the turnover rate of the cyclodextrin, Bender derivatized β-cyclodextrin with an imidazolylbenzoic acid group, which is capable of catalyzing the hydrolysis of the acyl-cyclodextrin intermediate. Although this work has led to the synthesis of enzyme mimics, which are comparable in their catalytic activities to naturally occurring chymotrypsin, recently some doubt has been shed on the validity of the high catalytic accelerations associated with them (Menger & Ladika, 1987; Palmer et al., 1994).

The attachment singularly or multiply of other catalytic groups onto cyclodextrins, as demonstrated by Breslow (1986) for a variety of different reactions, has greatly extended and improved the scope of cyclodextrins as catalysts.

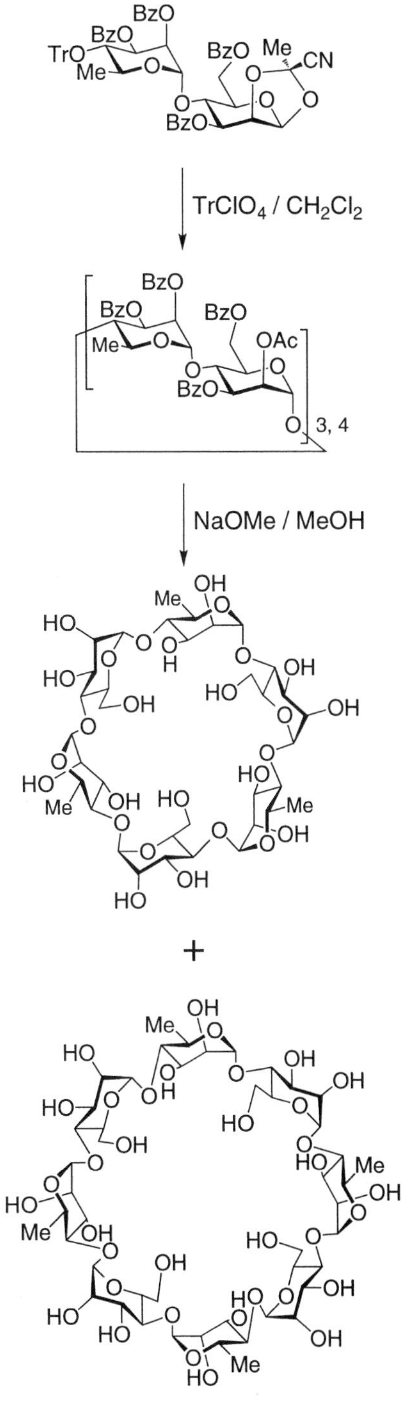

Figure 12–10. Polycondensation/cycloglycosylation of a "disaccharide monomer" for the preparation of L-rhamnose-D-mannose cyclic oligosaccharides.

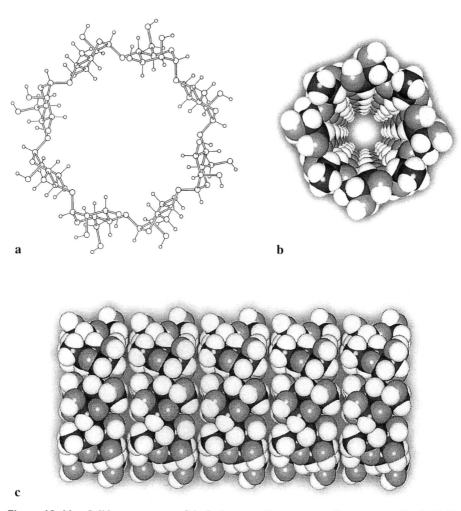

Figure 12–11. Solid state structure of the L-rhamnose-D-mannose cyclic octasaccharide. (**a**) Ball-and-stick representation. (**b**) Top-down and (**c**) side views of the nanotubular stacks.

Among others (Coates et al., 1990; Fujita et al., 1984; Jiang et al., 1994; Venema et al., 1994; Wang et al., 1994), Breslow and co-workers (Breslow, 1990, 1992; Breslow & Chung, 1990; Breslow & Zhang, 1992, 1994; Breslow et al. 1989) have recently reported the synthesis of cyclodextrin dimers which achieve very strong binding to specific substrates as well as high catalytic activities for phosphate ester hydrolysis. These ditopic receptors are capable of binding substrates with a much higher degree of control; furthermore, they were designed in such a way as to create strain in the ground state of the complexed molecule, yet to stabilize the geometry of the transition state. Consequently, this type of receptor forces the complexed molecule along the desired reaction pathway.

The investigations into cyclodextrins, as components for the construction of enzyme mimics, have been driven by the need to understand enzyme mechanisms and the de-

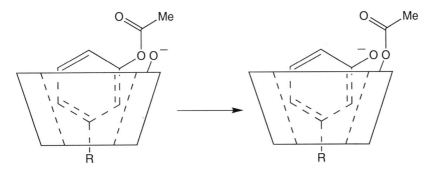

Figure 12–12. Catalytic hydrolysis of *meta*-substituted phenyl esters.

sire to design unnatural enzymes for any reaction of interest to an organic chemist. The recent reports on the syntheses of sophisticated enzyme mimics based on cyclodextrins holds much promise for future developments in this field.

The ability of cyclodextrins to direct aromatic substitution was first demonstrated by Breslow and co-workers (Breslow, 1986). In the presence of unmodified α-cyclodextrin, the chlorination of anisole with hypochlorous acid proceeded with a *para/ortho* selectivity of 96:4. This observed high selectivity is a result of the complexed anisole being shielded at the *ortho* position from electrophilic attack; only the *para* position of anisole is accessible in the complexed form (Fig. 12-13). Experiments with β-cyclodextrin resulted in similar effects being observed, if somewhat less pronounced. The lower regioselectivity observed in the case of β-cyclodextrin is believed to be a result of its larger cavity, which allows for more movement of the aromatic guest; this in turn, leads to less control in the regioselectivity of chlorination.

Upon closer examination of the reaction kinetics, it was found that the cyclodextrin takes part actively in the reaction sequence by rapidly and reversibly reacting, through its hydroxyl groups, with hypochlorous acid. This event leads to the formation of hypochlorite functionalities on the cyclodextrin rim, which can then deliver a chlorine atom to the bound substrate (Fig. 12-13), resulting in an increased rate of chlorination.

Other regioselective aromatic substitutions have been investigated by Komiyama and Hirai (Komiyama & Hirai 1983, 1984; Komiyama 1993), who have employed cyclodextrins successfully in the *para* formylation and carboxylation of phenols. Under

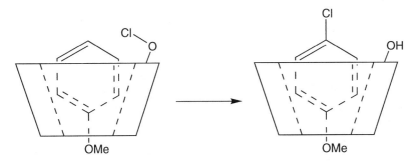

Figure 12–13. Regioselective catalytic chlorination of anisole.

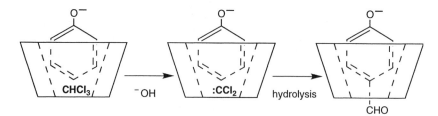

Figure 12–14. Regioselective formylation of phenols.

certain conditions, the *para* selectivity was found to be virtually absolute. This highly observed *para* selectivity has been rationalized by the formation of a ternary complex of β-cyclodextrin, chloroform, and the aromatic substrate (Fig. 12-14). The dichlorocarbene, formed from the complexed chloroform under basic conditions, is then trapped in a position in which it can only react with the *para* carbon atom of the bound phenol. As β-cyclodextrin can more easily accommodate two guests, it was found to be a more effective catalyst than α-cyclodextrin in the Komiyama systems.

More recently Komiyama has achieved good regioselective control in the hydroxymethylation of phenols with formaldehyde in the presence of hydroxypropyl β-cyclodextrin (Komiyama, 1989). It is believed that the hydroxypropyl groups aid in orienting the formaldehyde molecule in the cyclodextrin cavity through hydrogen bond formation. Interestingly, unmodified cyclodextrins show only slight increases in regioselectivity for this reaction.

Experiments with insoluble cyclodextrin polymers or immobilized cyclodextrins on solid supports have yielded promising results for extending this type of catalysis to the commercial production of *para*-substituted aromatics.

Cyclodextrins in Second Sphere Coordination

So far, only purely organic host–guest complexes have been discussed. Among the less well investigated species, which can undergo adduct formation with cyclodextrins, are certain transition metal complexes for which the cyclodextrin can act as a first (Fuchs et al., 1993; Klüfers & Schuhmacher, 1994) and second coordination sphere ligand (Colquhoun et al., 1986; Stoddart & Zarzycki, 1988).

One of the first transition metal complexes to be investigated for its binding in the cyclodextrin cavities was ferrocene and its derivatives. The binding of ferrocene with α-, β-, and γ-cyclodextrin has been investigated and different binding models have been proposed for each cyclodextrin. With α-cyclodextrin, a 2:1 sandwich complex is formed, where each of the two cyclopentadiene units of the ferrocene is capped with an α-cyclodextrin. This observation has since been confirmed by X-ray crystallography (Klingert & Rihs, 1991; Odagaki et al., 1990). These investigations revealed a ferrocene molecule trapped inside the cavity formed by two α-cyclodextrin molecules that are associated *via* their secondary faces in a head-to-head fashion. On the other hand, both β- and γ-cyclodextrins form 1:1 adducts with ferrocene. In the β-cyclodextrin complex, the C_5 symmetry axis of ferrocene is oriented (Kobayashi & Opallo, 1990) parallel to the symmetry axis of the cyclodextrin, whereas in the case of the γ-

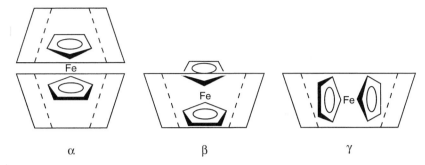

Figure 12–15. The proposed complexation of cyclodextrins with ferrocene.

cyclodextrin the ferrocene symmetry axis lies normal to the C_8 symmetry axis of the cyclodextrin (Fig. 12-15).

Apart from metallocenes and their functionalized derivatives, other organometallic complexes have also been investigated. Crystal structures of rhodium cyclooctadiene/α-cyclodextrin (Alston et al., 1985a), carboplatin/ α-cyclodextrin (Alston et al., 1985b,c, 1989), and platinum phosphine/β-cyclodextrin (Alston et al., 1988) complexes are recorded in the literature. In all these cases, a 1:1 binding stoichiometry between the cyclodextrin and the metal complex was observed. The cyclodextrin moiety binds the hydrophobic organic ligand, with additional stabilization coming from hydrogen bonding between the hydroxyl groups of the cyclodextrin and other, polar ligands on the metal complex. Further, binding between cyclodextrins and metal/heterocycle complexes has been observed and their physical characteristics have been investigated (Johnson & Reinsborough, 1994; Wylie & Macartney, 1993). The formation of inclusion complexes between cluster compounds and cyclodextrins has also been reported in the literature. Various cobalt carbonyl clusters have been shown to form 1:1 second sphere coordination complexes with γ-cyclodextrin (Harada et al., 1989; Shimida et al., 1992); no complexation has been observed with the α- or β-cyclodextrin. Furthermore, a crystal structure of a rheniumcarborane complex with α-cyclodextrin has been reported (Chetcuti et al., 1991) showing that the carborane anion only just penetrates the cyclodextrin cavity. More recently, it has even been demonstrated that cyclodextrins have the ability to solubilize hydrated lanthanide ions in aqueous solution (Yoshiro et al., 1994). It was found that the α-cyclodextrin solubilizes one mole equivalent of lanthanide ion, while the γ-cyclodextrin was able to solubilize two equivalents of lanthanide ion. This phenomenon is, however, rationalized by the association of the hydrated lanthanide with the hydroxyl functionalities of the cyclodextrin, rather than by inclusion complex formation.

The possibility that cyclodextrins may act as third sphere ligands has been demonstrated by the crystal structure of a potassium thiocyanate salt, which had been crystallized in the presence of a crown ether and γ-cyclodextrin to give a complex with 1:2:2 stoichiometry, respectively (Kamitori et al., 1987). The potassium ion was found to be sandwiched between two 12-crown-4 molecules, which, in turn, were sandwiched by two γ-cyclodextrin molecules. Cyclodextrins have even been employed as reaction vessels for ligand exchange reactions of metal complexes (Shimada et al., 1991). In this reaction, an alkyldicarbonylcyclopentadienyliron complex was included in a cyclodextrin cavity, where it underwent ligand exchange in a gas–solid system. For this type of reaction γ-cyclodextrin, by allowing better access to the included metal complex, proved to be more efficient than either α- or β-cyclodextrin.

Despite the very recent development of this field, the potential of these masked transition state metal complexes to act as sensors, delivery systems, and industrial catalysts is already being targeted.

Cyclodextrin Rotaxanes

Over the last few years, cyclodextrins have been used increasingly to construct molecular architectures (Wenz, 1994). The rigid cyclodextrin tori, combined with their abilities to incorporate guest molecules, have made cyclodextrins very useful materials for the construction of novel supramolecular structures. Some of the most fascinating systems of this type are rotaxanes and catenanes. These compounds contain mechanically interlocked components, either as threaded rings on a dumbbell or as two or more rings threaded through one another, reminiscent of links in a chain.

The first successful synthesis of a [2]rotaxane was reported in 1981 by Ogino and co-workers (Ogino, 1981; Ogino & Ohata, 1984). Treatment of the α,ω-diaminoalkanes in Figure 12-14 with α-cyclodextrin or β-cyclodextrin yields complexes which, after reaction with a dichloro cobalt complex, afford [2]rotaxanes as depicted in Figure 12-16. The terminal metal complexes are apparently large enough to prevent the cyclodextrin ring from slipping off the diaminoalkane moiety. This result is surprising, however, in that the synthesis was carried out in DMSO, a solvent which usually precludes the formation of cyclodextrin inclusion complexes. This concept of utilizing cyclodextrins and metal complexes to self-assemble rotaxanes has since been exploited in other systems (Yamanari & Shimura, 1983, 1984). Recently, Macartney and co-workers (Wylie & Macartney, 1990) have found that [2]rotaxanes are formed spontaneously in aqueous solution irrespective of the order of addition of α-cyclodextrin, the 1,1'-(α,ω-alkanediyl)-bis(4,4'-pyridylpyridinium) dicationic thread, and the pentacyanoferrate (II) stopper precursor depicted in Figure 12-17. Indeed, these [2]rotax-

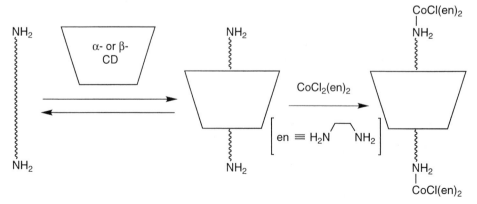

Figure 12–16. The synthesis of [2]rotaxanes containing α,ω-diaminoalkane threads and transition metal complexes as stoppers.

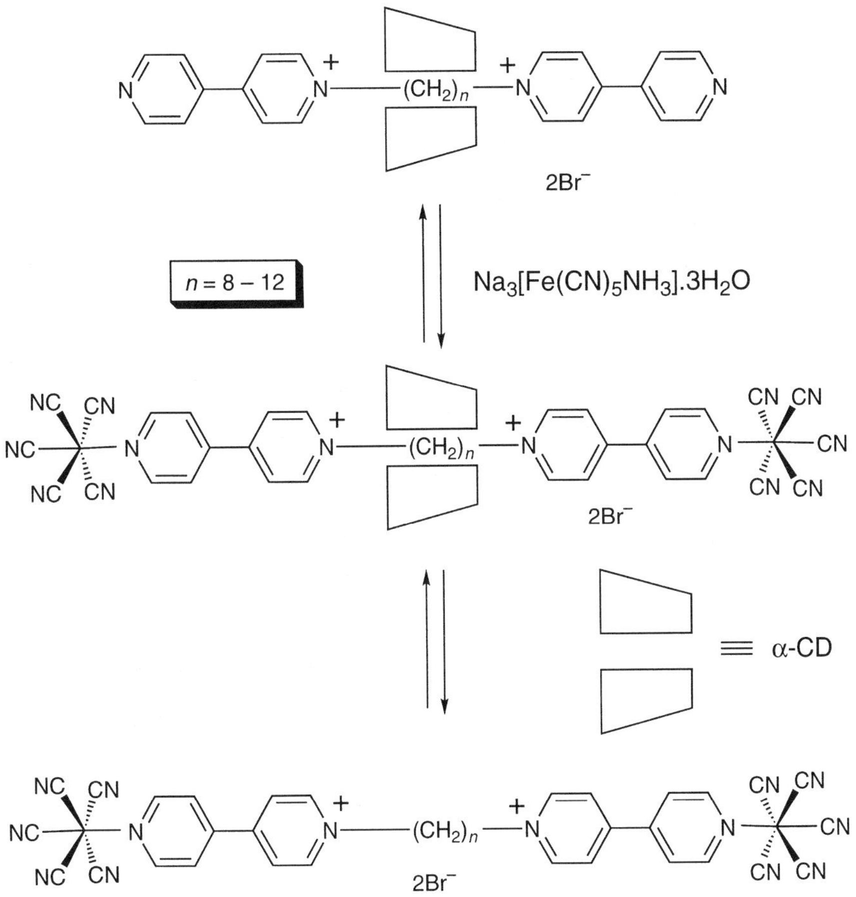

Figure 12–17. Macartney's [2]rotaxanes containing α-cyclodextrin.

anes are not only formed after adding solid $Na_3[Fe(CN)_5NH_3]\cdot3\ H_2O$ to an aqueous solution of α-cyclodextrin, but also when α-cyclodextrin is added directly to the pre-formed dumbbell-shaped component, implying slow dissociation of at least one of the $[Fe(CN)_5]^{3-}$ stoppers which allows formation of an intermediate semirotaxane and then its subsequent rapid recomplexation by $[Fe(CN)_5OH_2]^{3-}$ to give the [2]rotaxane. This system exemplifies the problems associated with the use of metal complexes as block-ing groups in rotaxanes. The kinetically labile complexes do not provide a full me-chanical blocking of dissociation.

Similarly, Rao & Lawrence (1990) have constructed a [2]rotaxane-like complex (Fig. 12-18) in 71% yield. The product precipitates upon addition of sodium tetraphenylborate to an aqueous solution of the complex of the bisammonium salt and per-2,6-O-methyl-β-cyclodextrin. Surprisingly, this "threaded molecular loop" can be eluted unchanged on cellulose with a mixture of benzene and acetone and its compo-sition determined by ^{1}H NMR and ^{13}C NMR spectroscopy. As expected, its FABMS shows only fragmentation. It seems likely, therefore, that the stabilization of this rotaxane arises from direct complexation between the biphenyl guest molecule and the per-2,6-O-methyl-β-cyclodextrin.

The gap between cyclodextrin pseudorotaxanes and true rotaxanes was bridged recently when Isnin & Kaifer (1991, 1993) reported the self-assembly of asymmetric zwitterionic [2]rotaxanes incorporating α-cyclodextrin. The alkyldimethyl(ferrocenylmethyl)ammonium cations in Figure 12-17 exhibit two different hydrophobic regions, and hence two types of binding interactions with cyclodextrins. α-Cyclodextrin binds preferentially to the alkyl chain while β-cyclodextrin engulfs the ferrocene subunit. The α-cyclodextrin complexes can, therefore, be considered as semirotaxane structures, since it is sterically impossible for the α-cyclodextrin bead to unthread or dissociate from the chain through the end occupied by the bulky dimethyl(ferrocenylmethyl)-ammonium group. Treatment of the complexes with the bulky water soluble amine, potassium 5-amino-2-naphthalensulfonate, and 1-(3-[dimethylamino]-propyl)-3-ethylcarbodiimide hydrochloride (EDC) as a water soluble coupling agent, affords the two possible isomeric zwitterionic [2]rotaxanes, in 15% yield (Fig. 12-19). The isomers have been separated by reverse phase TLC and the relative orientation of the α-cyclodextrin ring on the asymmetric dumbbell-shaped component has been established. Surprisingly, the isomers in which the secondary hydroxyl rim of the α-cyclodextrin faces the ferrocene unit are completely stable, while the other isomers undergo slow unthreading.

The presence of water is not necessarily a prerequisite for cyclodextrin rotaxane formation. Wenz et al. (1992) have taken advantage of the binding affinity of lipophilic

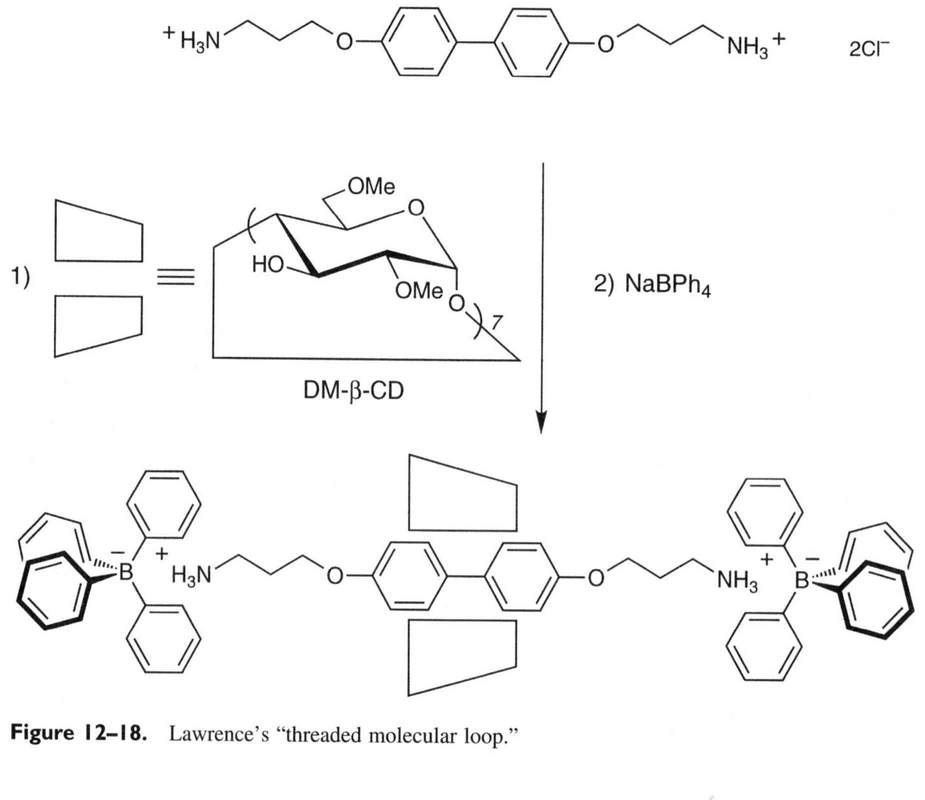

Figure 12–18. Lawrence's "threaded molecular loop."

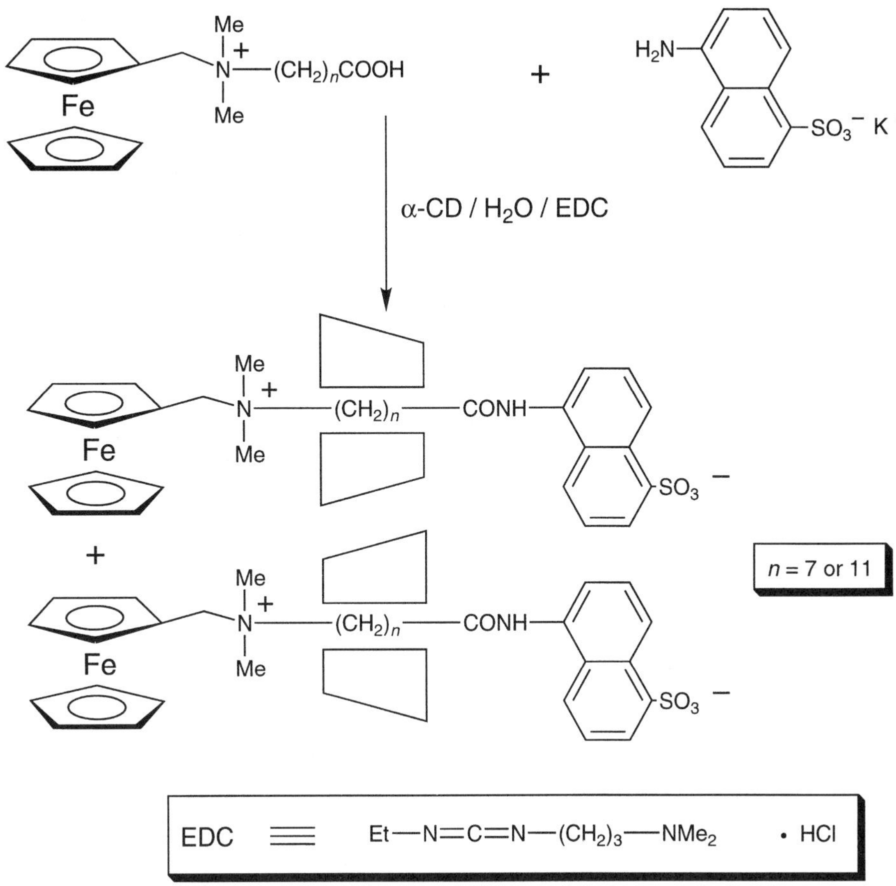

Figure 12–19. Kaifer's synthesis of isomeric zwitterionic [2]rotaxanes.

per-2,6-*O*-butyl-3-*O*-acyl-β-cyclodextrin for pyridinium cations to design and synthesize a [2]rotaxane (Fig. 12-20). ^{1}H NMR spectroscopic studies in acetonitrile indicate that, in the 1:1 complex formed between the guest and the lipophilic cyclodextrin, the cyclodextrin bead is indeed located around the pyridinium moiety and away from the trityl blocking group. It appears that stabilization of the complex arises not only from the coordinating power of the ester oxygen atoms towards the pyridinium cation, but also from hydrogen bonding between α-pyridinium protons and the same carbonyl oxygen atoms. The alkylation of the second pyridine moiety by bulky 7,7,7-triphenylheptyl 4-bromomethylenebenzoate affords the [2]rotaxane in 20% yield (Fig. 12-20). The [252]Cf–Pd mass spectrum of this [2]rotaxane clearly shows a pronounced signal at *m/z* 3293, which corresponds to the molecular ion of the intact [2]rotaxane. The ^{1}H NMR spectrum provides further evidence for the presence of a [2]rotaxane. The C_7 symmetry of the cyclodextrin bead is retained but the mirror symmetry of the dumbbell-shaped component is broken by the permanent threading of the host cyclodextrin, which is asymmetric along the axis of the dumbbell. Additionally, Steinbrunn & Wenz (1996) recently described the formation of water-

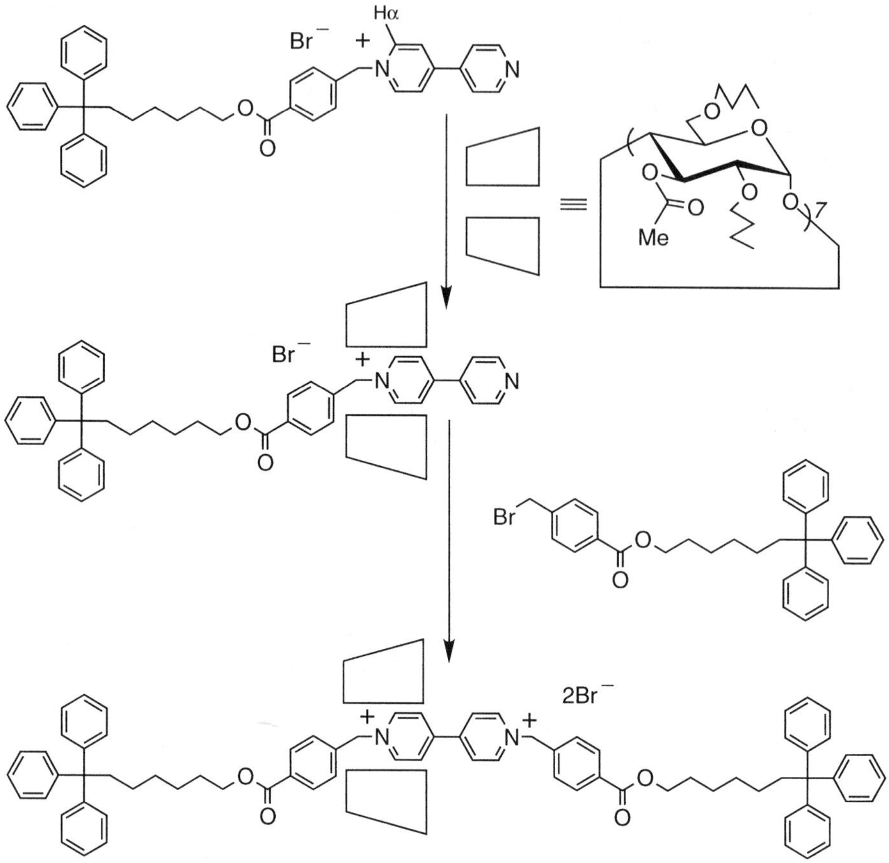

Figure 12–20. Incorporation of a lipophilic cyclodextrin into a [2]rotaxane.

soluble polyrotaxanes *via* polycondensation of α,ω-amino acids in the presence of α-cyclodextrin.

If cyclodextrins are able to form inclusion complexes with small linear molecules, there is no reason why they should not complex with polymeric ones. Indeed, Harada and co-workers (1990, 1993, 1994) have recently demonstrated that chains of poly(ethyleneglycol) (PEG) thread α-cyclodextrin beads in a necklace-like manner (Fig. 12-21). Interestingly, the view is now growing up in the scientific community that cyclodextrins might thread under conditions of equilibrium control with alternating orientations, where adjacent rings are matched head-to-head, tail-to-tail in order to optimize hydrogen bonding between the neighbouring cyclodextrin units (Stoddart, 1992). The same group in Japan (Harada et al., 1992, 1993b, 1994a) has now succeeded in capping the chain-ends of a poly(ethyleneglycol) bisamine (PEG-BA) of M_r 3450 with dinitrophenyl stoppers (Fig. 12-21). After extensive purification, a product whose molecular weight is in the region of 23,500 was obtained, commensurate with the threading of 20 α-cyclodextrin rings per polyrotaxane molecule. A further dimension was attained (Harada et al., 1993c) with the synthesis of polymeric molecular tubules based

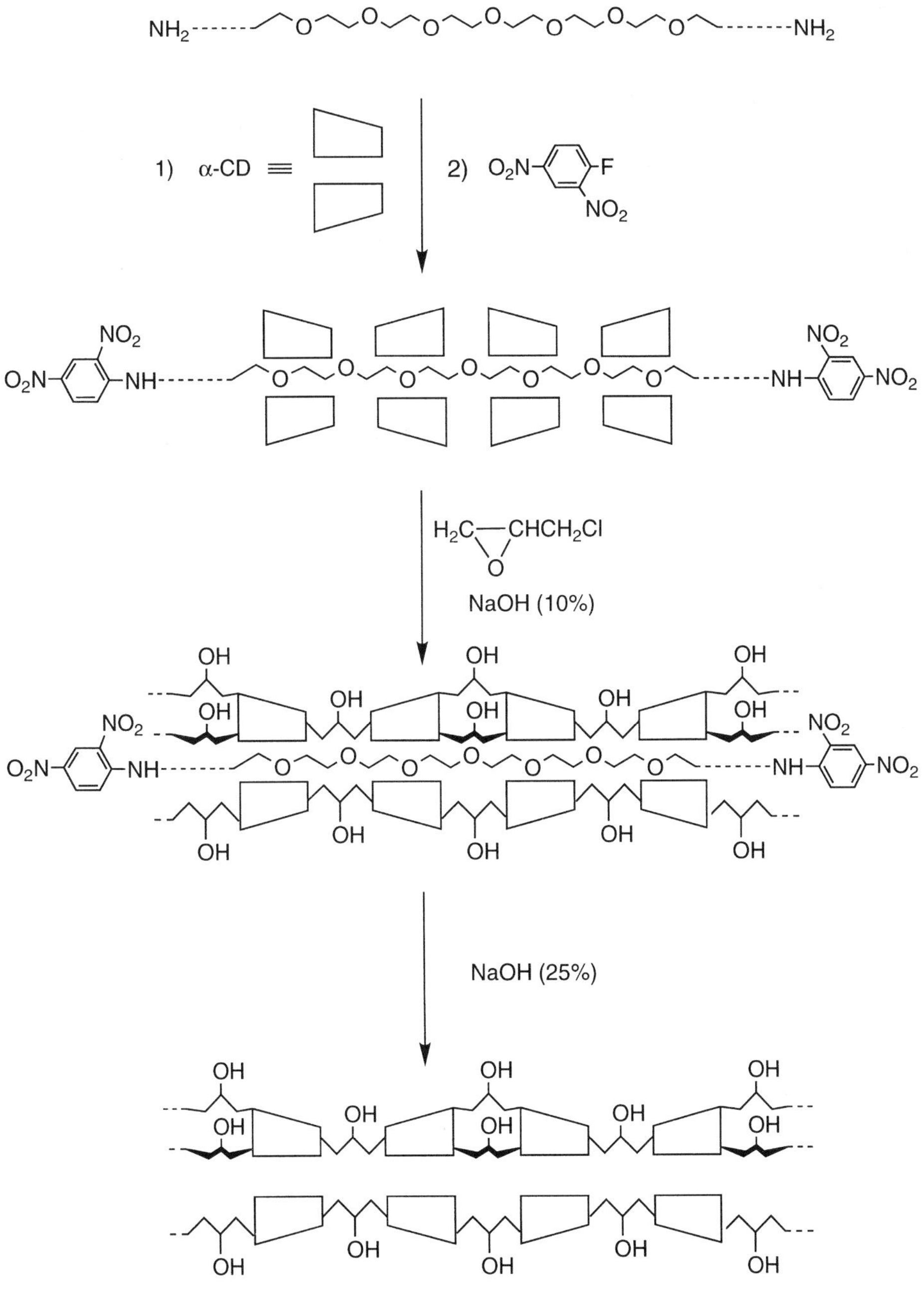

Figure 12–21. From polyrotaxanes to molecular tubes.

on α-cyclodextrin. In Harada's polyrotaxanes, the threaded cyclodextrins are well packed, essentially covering the entire length of the polyethyleneoxy chain. At this point, the tubular structure is already in place. One only needs to "glue" each cyclodextrin to its two neighbours in the polyrotaxanes. The formation of the desired tubules was apparently achieved by reaction with epichlorohydrin, removal of the capping groups, and subsequent release of the tubular cyclodextrin bead (Fig. 12-21). The average molecular weight of around 17,000 is consistent with the proposed structure. Further proof was provided by elemental analysis, and ^{1}H and ^{13}C NMR spectroscopic studies. The reaction is templated by the threaded chains, since freely dissolved α-cyclodextrin does not produce such tubular structures. However, no evidence for the proposed intrachain cross-linking pattern in both the [2]rotaxane and the molecular cylinder (Fig. 12-21) has been provided. It is, therefore, not unlikely that the product obtained is, in fact, a mixture of numerous products. Furthermore, the proposed head-to-head and tail-to-tail arrangement also remains uncertain, since full characterization of such polymeric structures is a particularly difficult task.

Quite independently, Wenz & Keller (1992) have demonstrated the threading of cyclodextrin rings onto the poly(iminoundecamethylene) chain and the poly-(iminotrimethylene-iminodecamethylene) chain (Fig. 12-22) in aqueous solution. Both ^{1}H NMR spectroscopy and viscosity measurements revealed that, for some of these compounds, the threading–dethreading equilibrium of the α-cyclodextrin with the polymer chain is reached only after 170 hr. Similarly, equilibrium dialysis of this polyrotaxane was far from complete after 2 weeks. The slow rates of association and dissociation are consistent with a mechanism of inclusion where each individual ring must find a chain terminus to thread and, once strung, must move toward the center of the polymer chain in order to make room for new rings. Reaction of the amino groups with nicotinoyl chloride then affords polyrotaxanes in which the nicotinoyl blocking groups are randomly located along the polymeric chain.

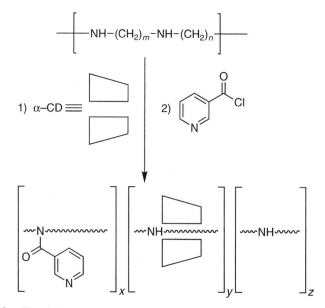

Figure 12–22. Wenz's heterogenous polyrotaxanes.

Apart from these main chain polyrotaxanes, in which the cyclodextrins are threaded onto the polymer backbone, the synthesis of side chain polyrotaxanes has also been reported (Born & Ritter, 1991). In these molecular architectures, cyclodextrin rotaxanes are attached as side chains to a main chain polymer backbone.

Catenated Cyclodextrins

Despite the first attempts by Lüttringhaus et al. at synthesizing catenated cyclodextrins as long ago as 1958, they are still very rare and only one successful synthesis (Armspach et al., 1993, 1994, 1995) of these species has been reported in the literature. Under

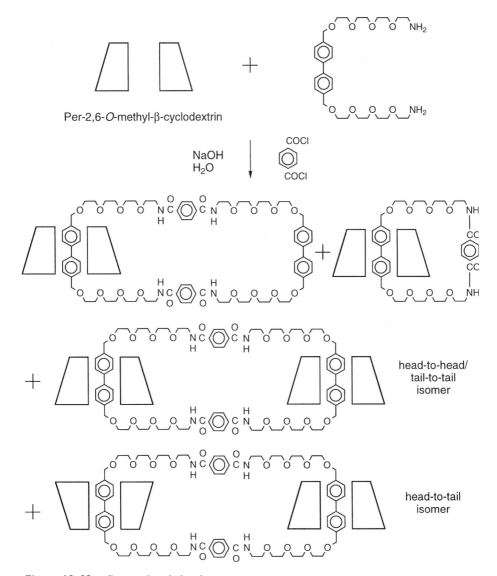

Figure 12–23. Catenated cyclodextrins.

aqueous conditions, the reaction of an aromatic diamine with terephthaloyl chloride in the presence of per-2,6-O-methyl-β-cyclodextrin afforded four different catenated cyclodextrins (Fig. 12-23). All four products were isolated and identified. X-ray crystallographic analysis of one of the products yielded the first crystal structure of a mechanically interlocked cyclodextrin. An interesting feature of these systems is their dynamic behavior. NMR spectroscopic investigations showed that the cyclodextrin rings are rapidly pirouetting about their symmetry axis. Similarly, rapid circumrotation of the dimeric cyclophane through the CD cavities is also indicated by dynamic NMR spectroscopy. Of particular stereochemical interest are the two isomeric [3]catenanes containing two cyclodextrin rings. These two species were only separable by high-pressure liquid chromatography (HPLC), but could be clearly identified as a result of their averaged symmetries, given the nature of the dynamic processes described above. Although structurally very similar, the head-to-tail [3]catenane and its head-to-head/tail-to-tail isomer possess distinctive local symmetries. Both ^{1}H NMR and ^{13}C NMR spectra, recorded in C_6D_6 and $CDCl_3$, respectively, indicate that, in the head-to-tail [3]catenane the synthetic macrocycle has averaged C_2 symmetry, whereas in the head-to-head/tail-to-tail isomer the synthetic macrocycle has D_2 symmetry. For example, the bitolyl methylene groups in the head-to-head/tail-to-tail isomer can be identified in the ^{13}C NMR spectrum ($CDCl_3$) as one degenerate signal, and yet as two distinct signals for these carbons in the head-to-tail isomer. Thus, a simple inspection of the ^{1}H and ^{13}C NMR spectra of the isomeric [3]catenanes allows the unequivocal assignment of their structures to be made and a comment to be recorded about the associated HPLC behavior of the compounds. Under the reverse phase elution conditions employed, it was noted that the head-to-tail isomers are always eluted ahead of the head-to-head/tail-to-tail ones.

Summary

Constructing molecular devices with cyclodextrins is a challenging contemporary area of cyclodextrin research. It combines many of the aspects of cyclodextrin chemistry such as selective binding, solubility, and their chemical modification, which are often addressed only in isolation. Judging from the recent interest in this area, many more exciting developments can be expected in the near future.

13

Biosynthesis of Bacterial and Mammalian Glycogen and Plant Starch Synthesis and Their Regulation

Jack Preiss

This chapter discusses comparative aspects of glycogen metabolism in bacteria and mammals, and starch metabolism in algae and higher plants. There are many similarities in the metabolism of plant starch and bacterial glycogen, especially in the mode of regulation of α1,4-glucan synthesis. Conversely, regulation of glycogen synthesis in mammals is different in that covalent modification as well as allosteric control is involved. The enzymology and biochemistry of the various enzymes in the animal, plant, and bacterial systems is also described.

Regulation of mammalian glycogen synthesis has been studied not only for attempting to understand glycogen synthesis itself but also for understanding the role of hormonal action and cellular regulation. A concise description of this important research area of cellular control is provided and includes *in vitro* experiments relevant to the physiological control processes.

There are some recent reviews on bacterial glycogen synthesis (Nelson & Pan, 1995; Preiss, 1996a; Preiss & Romeo, 1989, 1994) and on starch biosynthesis (Martin & Smith, 1995; Preiss, 1988, 1991, 1992, 1996b,c, 1997; Preiss & Sivak, 1996; Sivak & Preiss, 1995; Smith et al., 1995) which is discussed in more detail some of the areas presented in this chapter. References to the regulation of mamalian glycogen synthesis include Cohen (1986) and Roach (1986).

The chemical and physical aspects of the starch granule and its components amylose and amylopectin have been discussed in some recent excellent reviews (Hizukuri, 1995; Morrison & Karkalas, 1990) and on animal glycogen (Manners, 1991). Unfortunately, not as much information is available on the structures of bacterial and fungal glycogen (Manners, 1991). Yet, the available data does indicate that overall, the bacterial and fungal glycogen structures are quite similar to that of mammalian glycogen. A review on yeast glycogen structure is available (Manners, 1971).

Function of Glycogen and Starch in Nature

Glycogen in Animals

Glycogen is a primary energy source in animal cells. In a man weighing 70 kg, about 1% wet weight in skeletal muscle and 5% in liver can account for 90 g and 350 g of glycogen reserves, respectively. The storage of glucose as glycogen is important for mammalian homeostasis. There are also different functions for skeletal muscle and hepatic glycogen reserves. Liver rarely utilizes hepatic glycogen as its own energy source, but stores glycogen when excess glucose is available in the diet and utilizes it for maintaining blood glucose levels. A man weighing 70 kg uses 180 g of glucose per day for those tissues that can only utilize carbohydrate as a source of energy. About one-half of this is derived from hepatic glycogen. Therefore, hepatic glycogen synthesis and mobilization is dictated by blood glucose levels and is controlled by glucoregulatory hormones, primarily glucagon, insulin, and glucocorticoids. In contrast, in all other mammalian extrahepatic tissues, glycogen stores are utilized for specific functions of the tissue.

Although the glycogen content of skeletal muscle is large compared to the liver it is not directly available as a source of blood glucose. The lactate, derived from skeletal muscle glycogenolysis and glycolysis during exercise, at rest can be converted to glucose by the liver and kidney through gluconeogenesis, and then serve a source of about 10%–20% of the total blood glucose. Thus, muscle glycogen serves primarily as an energy source, broken down in response to contractile stress. Replenishment occurs when the diet is such that there is a concomitant increase in high blood glucose (hyperglycemia) and insulin.

Glycogen in Bacteria

Glycogen, found in many bacteria, usually accumulates in limited-growth conditions and in the presence of an excess carbon source (Dawes & Senior, 1973; Preiss, 1984, 1989; Preiss & Romeo, 1989). Glycogen accumulation has been shown to occur in the stationary phase of the growth cycle as a response to limitations of nitrogen, sulfur, or phosphate. Glycogen is not required for bacterial growth and glycogen-deficient mutants grow as well as the wild-type strains.

Biological functions of bacterial glycogen have been reviewed (Preiss, 1989). Under nongrowing conditions and when an endogenous carbon source is not available, glycogen is probably utilized for energy for the preservation of cell integrity. Bacteria require energy for maintenance under nongrowing conditions (references in Preiss, 1989). This "energy of maintenance" is most probably the energy required for processes such as turnover of proteins and RNA, for maintenance of motility, intracellular pH, chemotactic response and for osmotic regulation. The bacteria *Escherichia* (*E.*) *coli* and *Enterobacter* (*E.*) *aerogenes*, having glycogen, do not degrade their RNA and protein components in media devoid of a carbon source, while the glycogen-deficient bacteria immediately release NH_3 from their nitrogen-containing components (Preiss, 1989). It is suggested that glycogen aids in the preservation of cellular constituents undergoing turnover. Glycogen-containing *E. aerogenes*, *E. coli*, and *Streptococcus mitis* also survive better than organisms having no glycogen. Another function for glycogen has been suggested in various *Clostridia* species. These organisms accumulate glycogen up to 60% of their dry weight before or during initiation of sporulation and during spore formation, the glycogen is rapidly degraded (Mackey & Morris, 1971). Glycogen-deficient strains are poor spore formers, thus glycogen may serve as a source of carbon and energy for spore formation and maturation.

Although the above studies suggest a role for glycogen in bacterial survival, the issue is not clear cut. For example, glycogen-rich *Sarcina lutea* cells die faster when starved in phosphate buffer than cells with no polysaccharide (Burleigh & Dawes, 1967). Thus, further clarification of the role of glycogen in bacteria is required.

Starch in Plants

Starch is present in almost all green plants and in various types of plant tissues and organs, for example, leaves, roots, shoots, fruits, grains, and stems. The demonstration of the disappearance of starch in leaves either by exposure to low light or by extended exposure of the leaf in the dark (24–48 hrs) has been well documented as early as the nineteenth century (Sachs, 1887). Illumination of the leaf in bright light causes the reformation of starch granules in the chloroplast organelle. This is easily observed by iodine staining of the tissue (Edwards & Walker, 1983) or by light or electron microscopy (Badenhuizen, 1969). Starch accumulation is due to carbon fixation during photosynthesis; the starch formed in the light is degraded in the dark to products to be transported to other parts of the plant. These products are used as an energy source and the carbons are utilized for sucrose synthesis. Biosynthesis and degradation of starch in the leaf is, therefore, a dynamic process having diurnal fluctuations.

In fruit, storage organs, or seed, the synthesis of starch occurs during the development and maturation of the tissue (Jenner, 1979). Starch degradation in these tissues then occurs at the time of sprouting or germination of the seed or tuber, or ripening of the fruit, where it is used as a source both of carbon and energy. The degradative and biosynthetic processes in the storage tissues may, therefore, be temporally separated. However, there is a good possibility that during each phase of starch metabolism some turnover of the starch molecule does occur.

Synthesis of Glycogen and Starch

Glycogen Synthesis in Animals

The mammalian glycogen synthetic pathway from glucose-1-P is shown below in reactions 13-1–13-4.

$$UTP + \alpha\text{-glucose-1-P} \Leftrightarrow UDP\text{-glucose} + PPi \qquad \text{(Reaction 13–1)}$$

$$UDP\text{-glucose} + apo\text{-glycogenin} \rightarrow (glucosyl)_{10-11}\text{-glycogenin} \qquad \text{(Reaction 13–2)}$$

$$UDP\text{-glucose} + glucan \rightarrow \alpha\text{-1,4-glucosyl-glucan} + UDP \qquad \text{(Reaction 13–3)}$$

elongated α1,4-oligosaccharide chain-glucan $\rightarrow$
  α-1,4–α-1,6 branched-glucan   (Reaction 13–4)

First the synthesis of the glucosyl-nucleotide donor, uridine 5′-diphosphate (UDP)-glucose, (Reaction 13-1) is catalyzed by the enzyme, UDP-glucose (synthetase) pyrophosphorylase (EC 2.7.7.9; uridine 5′ triphosphate (UTP): α-D-glucose-1-phosphate uridylyltransferase). The equilibrium of the reaction towards UDP-glucose synthesis is slightly less than 1 but the other product of the reaction, pyrophosphate, is cleaved by the ubiquitous inorganic pyrophosphatase. Thus, formation of UDP-glucose is essentially irreversible in mammalian cells. Reaction 13-2 catalyzes a transfer of glucose from the sugar nucleotide to a protein, glycogenin, which is currently considered to be the protein that participates in the initiation of glycogen synthesis (Krisman & Barengo,

1975; Lomako et al., 1988; Pitcher et al., 1987). Rabbit muscle glycogenin is a 37 kDa self-glycosylating protein containing 332 amino acids (Campbell & Cohen, 1989) that glycosylates the hydroxyl group of its tyrosine residue 194 (Rodriguez & Whelan, 1985; Smythe et al., 1988). The reaction requires Mn^{2+} as a cofactor. Further auto-glycosylation occurs and the initial glucose residue is converted to a chain containing 7–11 glucosyl units linked by α-1,4 glucosidic bonds. This glucosylated protein can then serve as a primer (glycogen and lower molecular weight maltodextrins can also serve as primers) for Reaction 13-3 which is catalyzed by glycogen synthase (EC 2.4.1.11; UDP-glucose-glycogen 4-α-glucosyl transferase). Chain elongation of the glucosylated glycogenin must continue and most probably shortly thereafter, branching enzyme (EC 2.4.1.18; 1,4-α-D-glucan 6-α-[1,4-α-glucano]-transferase) catalyzes formation of α-1,6 branched chains (Reaction 13-4). The actual details of events between formation of the oligosaccharide–glycogenin primer and formation of the glycogen having molecular weight 10^7 kDa are presently not known. However, proglycogen, a lower molecular weight (~400 kDa) form of glycogen which is trichloroacetic acid-insoluble (in contrast to glycogen) has been isolated from astrocytes of newborn rat brain and is believed to be an intermediate fraction between glycogenin and glycogen (Alonzo et al., 1995b; Lomako et al., 1993). It is also postulated that there is a form of glycogen synthase, denoted proglycogen synthase, which glucosylates the glucosylated glycogenin to synthesize the proglycogen, and another form of glycogen synthase which is the enzyme forming the 10^7 kDa macroglycogen from proglycogen (Alonso et. al, 1995b; Lomako et al., 1991). The proglycogen synthase, as well as proglycogen, remains to be fully characterized and distinguished from the "classical glycogen synthase." Thus, in addition to the four reactions known for the initiation and synthesis of mammalian glycogen, other reactions converting glycogenin to glycogen are postulated. The reactions leading to glycogen synthesis are summarized in Figure 13-1.

Properties of the Mamalian Glycogen Biosynthetic Enzymes

The properties of the various enzymes involved in mamalian glycogen synthesis will be discussed below. However, the properties of the regulatory enzyme, glycogen synthase, will be discussed in the section dealing with regulation of mammalian glycogen synthesis.

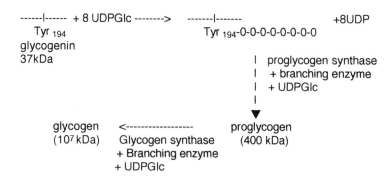

Figure 13–1. The biosynthesis of mammalian glycogen as described by Lomako et al. (1991) and Alonzo et al. (1995). Tyr 194-0-0-0-0-0-0-0-0 represents the glycogenin glucosylated at tyrosine residue 194 with 8 glucosyl residues.

UDP-glucose pyrophosphorylase is ubiquitous in nature and catalyzes the synthesis of UDP-glucose (Eq. 13-1). It was first detected in yeast (Munch-Peterson et al., 1953); the enzyme has now been isolated and characterized from microorganisms, plants, and mammalian tissue (Turnquist & Hansen, 1973). The enzyme has been obtained in highly purified form from calf liver, human liver, lamb, goat, rabbit liver, and human erythrocytes (see review by Turnquist & Hansen, 1973). A molecular weight of 480,000 was reported for the calf liver enzyme, and its subunit weight was 60,000. Thus, the reactive enzyme is an octamer of eight identical subunits. Other mammalian enzymes appear to have similar molecular weights. The UDP-glucose pyrophosphorylases have an absolute requirement for divalent cation, and magnesium is the best cation for supporting activity. Mn^{2+}, Ca^{2+}, and Ni^{2+} also support enzyme activity to some extent. The optimum pH is in the range of 7–9 and the equilibrium constant in the direction of UDP-glucose formation ranges from 0.15 to 0.34 for the animal, plant, and microorganismal enzymes studied (Turnquist & Hansen, 1973). The enzyme is highly specific for UDP-glucose. The calf and human liver enzymes can catalyze the pyrophosphorolysis of thymidine 5′-diphosphate (TDP)-glucose, cytidine 5′-diphosphate (CDP)-glucose, guanosine diphosphate (GDP)-glucose, UDP-galactose, UDP-xylose, and UDP-mannose to a small extent, 0.1%–2.2% (Knop & Hansen, 1970).

The reaction mechanism of UDP-glucose pyrophosphorylase has been studied with the enzyme from liver, erythrocytes, and *Acanthamoeba castel lani* (Turnquist & Hansen, 1973) and it appears to be an ordered BiBi mechanism. The nucleoside phosphate is both the first substrate to be added and the last product to be released in the mechanism.

$$
\begin{array}{c}
\text{UTP} + \text{E} \Leftrightarrow \text{E-UTP} + \text{Glc-l-P} \Leftrightarrow \text{E-UTP-Glc-l-P} \\
\uparrow \downarrow \qquad\qquad\qquad\qquad \text{(Equation 13–1)}\\
\text{E} + \text{UDPGlc} \Leftrightarrow \text{E-UDPGlc} + \text{PPi} \Leftrightarrow \text{E-UDPGlc-PPi}
\end{array}
$$

UDP-glucose is the most potent inhibitor of the animal UDP-glucose pyrophosphorylase. Thus, its concentration possibly exerts some regulation of the enzyme. The inhibition appears to be competitive with UTP. Roach et al. (1975) suggested that the concentration ratio of UDP-glucose to UTP may be the most important determinant of UDP-glucose pyrophosphorylase activity. No other regulatory phenomenon has been associated with the mammalian enzyme and because UDP-glucose functions not only in the synthesis of glycogen but also in the synthesis of other sugar nucleotides (e.g., UDP-galactose and UDP-glucuronic acid), it would be expected that the dominant regulation of glycogen synthesis would occur at the level of glycogen synthase.

Mammalian branching enzyme from rat liver (Krisman, 1962; Larner, 1953; Verhue & Hers, 1966) and rabbit skeletal muscle (Brown & Brown, 1966a,b; Gibson et al., 1971; Illingworth et al., 1961) has been studied in some detail. Larner (1953) showed that the enzyme catalyzed the formation of new (α-1-6)-α-D-glucosidic bonds with glycogen preparations containing average chain lengths of 11–21 glucosyl residues. Evidence was obtained for the rat liver (Verhue & Hers, 1966) and rabbit skeletal muscle (Gibson et al., 1971) enzymes, for the cleavage of linear chains of six or more glucosyl residues from the terminal portion of the outer chains of the α-1-4-α-D-glucan substrate, with the subsequent transfer and reattachment of the cleaved oligosaccharide portion in a 1-6-α-D-glucosidic linkage to the outer portions of the α-D glucan. The branching enzyme from rabbit skeletal muscle preferentially catalyzed the transfer of segments seven glucosyl units long when it acted on polysaccharides enlarged from UDP-glucose by rabbit muscle glycogen synthase action *in vitro* (Brown & Brown, 1966b). The liver and muscle enzymes are also active on amylose and amylopectin molecules (Brown & Brown, 1966a; Krisman, 1962). The rabbit muscle enzyme also catalyzed the formation of new 1-6-α-D-linkages in products formed from glucose-1-P with phosphorylase and also greatly stimulated the "unprimed" phosphorylase reaction (Illingworth et al., 1961). Branching enzyme in a combined system with mammalian glycogen synthase stimulated the rate of transfer of glucose units in the presence of a minimal quantity of glycogen primer (Brown & Brown, 1966a). This stimulation is thus very similar to that observed for bacterial branching enzyme on bacterial glycogen synthase activity and for plant branching enzyme on starch synthase activity. There is some information about the minimal size of the oligosaccharide chain transferred by the action of branching enzyme. More information on this is available for the plant and the bacterial enzymes and is discussed in those sections on the bacterial and plant branching enzymes. The amino acid residues involved in determining the specificity of size of oligosaccharide cleaved and transferred, and the interbranch distance between formation of the new branch points, are unknown for all the branching enzymes and remains to be determined in the future.

Glycogen storage disease ("type IV") is due to the absence of branching enzyme (Brown & Brown, 1966c, 1968). The glucan product isolated from the liver of these individuals is an amylopectin-type polysaccharide. Thus, some branches have been formed, suggesting that some branching activity is present at low levels. Purified rabbit muscle branching enzyme can cause further branching of this α-D-glucan (Gibson et al., 1971). The rabbit skeletal muscle enzyme has been purified to near homogeneity (Gibson et al., 1971).

The apparent molecular weight determined by sucrose density gradient centrifugation is about 92,000. Thus, it is similar to the molecular weights of plant and bacterial enzymes. The enzyme has a broad pH optimum in citrate buffer, 6.8–7.8, and was stimulated about twofold in 0.5 M sodium citrate at pH 7.0. The partially purified liver enzyme was also activated twofold by sodium citrate and ~1.7-fold by sodium borate (Krisman, 1962).

The eukaryotic branching enzyme genes that have been isolated are from *Saccharomyces cerevisiae* (Thon et al., 1992) and from a human hepatoma cell line (Thon et al., 1993). The yeast *GLC* 3 gene and deduced amino acid sequence was compared to the bacterial branching enzymes from *E. coli* (Baecker et al., 1986), *Synechococcus*, and *Bacillus stearothermophilus* (Kiel et al., 1990); there was only 8% identity and 42% similarity of the yeast branching enzyme with the prokaryotic sequences. The cDNA encoding the human branching enzyme could complement the yeast branching enzyme mutant, *glc* 3, and had a 67% identical amino acid sequence with the yeast branching enzyme (Thon et al., 1993). The human gene is located on chromosome 3.

Two reviews describing the discovery and characterization of *glycogenin* are available (Smythe and Cohen, 1991; Whelan, 1986). This protein, about 37 kDa, was found to be covalently bound to glycogen (Rodriguez & Whelan, 1985) and also associated with glycogen synthase even after extensive purification of the rabbit muscle glycogen synthase (Nimmo et al., 1976). Most of the glycogenin in rabbit muscle is considered to be covalently linked to glycogen. In liver, however, most of the glycogen and glycogenin are in free form and not associated with each other (Gannon & Nuttal, 1996). The first step in converting the apo-glycogenin, a primer for glycogen synthesis, involves an autoglucosylation of glycogenin Tyr 194 by UDP-glucose. The reaction requires either Mn^{2+} or Mg^{2+} for the autoglycosylation reactions. Up to 7–11 glucosyl units can be attached to the glucosyl-tyrosine residue. The complete sequence of the rabbit muscle glycogenin has been elucidated (Campbell & Cohen, 1989).

Glycogenin, when isolated and purified, contains a glycosylated tyrosine residue, leading to the question of whether the first glucosyl unit resulted from autoglucosylation by glycogenin or whether there was another enzyme responsible for the first glycosylation. The rabbit muscle glycogenin was expressed to a high level in *E. coli* (Alonzo et al., 1995c; Viskupic et al., 1992) and purified. Tyr194 was found to be glucosylated and contained from one to eight residues of glucose. Glycogenin could incorporate another 5 mol of glucose per mol of glycogenin if supplied with UDP-glucose. The Km value was 4.5 μM.

Isoamylase can remove the oligosaccharide chain from the tyrosine residue (Lomako et al., 1992) and pretreatment of the glucosylated glycogenin enhanced incorporation of labeled glucose from UDP-glucose. This suggested that glycogenin can self glucosylate its Tyr194 residue. More direct evidence was obtained by expressing the glycogenin in an *E. coli* mutant deficient in UDP-glucose pyrophosphorylase activity (Alonzo et al.,1994b), resulting in production of a carbohydrate-free glycogenin, apoglycogenin. When UDP-xylose + Mn^{+2} was incubated with the glycogenin, one mole of xylose was incorporated per mole of glycogenin. With UDP-glucose, an average of eight glucose chains are added per glycogenin. However, upon release of the glucose chains by isoamylase, the size of chains varied, with the predominant chains being in the 7–11-glucose units range. The production of a carbohydrate-free apo-glycogenin and its ability to self-glucosylate obviates the need to invoke a separate enzyme for the addition of the first glucose residue to Tyr194.

Thus, glycogenin can catalyze two different glycosylation reactions—first, glycosylation of a tyrosine hydroxyl group, and then further glucosylations to form α-1,4 glucosidic linkages. UDP and UTP were found to be effective inhibitors (Cao et al., 1993). Other pyrimidine sugar nucleotides could substitute for UDP-glucose as glucosyl donors (Alonzo et al., 1995a). The rate of glycosylation using CDP-glucose and TDP-glucose were 71% and 33% respectively, of that obtained with UDPGlc. Both ADP-glucose and GDP-glucose were inactive. Meezan et al. (1994) reported that UDP-xylose could also serve as a glycosyl donor with only one xylose molecule being transferred to glycogenin itself (Meezan, et al., 1994). No further chain growth could occur with either UDP-xylose or UDP-glucose. Other reactions were also catalyzed by glycogenin. Glycogenin could transfer glucose from UDP-glucose to exogenous substrates such as *p*-nitrophenyl-linked malto-oligosaccharides (Lomako et al., 1990), and also to tetradecyl β-D-maltoside, octyl β-D-maltoside, and dodecyl β-D-maltoside (Manzella et al., 1995).

If the recombinant glycogenin is mutated from Tyr to Phe or Thr at residue 194 the enzyme loses its ability to self glucosylate (Alonzo et al., 1994a, 1995c; Cao et al., 1993). However, the Phe194 and Thr194 mutants are still able to glycosylate with UDP-glucose, dodecyl β-D-maltoside (Alonzo et al., 1994a), and *p*-nitrophenyl-linked malto-oligosaccharides (Alonzo et al., 1995c). Also noted was the ability of mutant and normal glycogenins to hydrolyze UDP-glucose at rates similar to self-glucosylation rates of the normal enzyme. This hydrolysis is competitive with glucose transfer to *p*-nitrophenyl-linked maltoside (Alonzo et al., 1995c). Self-glucosylation, glucosylation of other acceptors, and hydrolysis all appear to be catalyzed by the same active center.

Glycogenin and the mutant proteins, Phe194 and Thr194, could also transfer glucose from UDP-glucose to maltose to form maltotriose (Cao et al., 1995b). However, no further conversion to a higher oligosaccharide occurred. Analysis of the crystal stucture by X-ray diffraction indicated that glycogenin existed as dimers (Cao et al., 1995b).

In *Saccharomyces cerevisiae* the genes Glg1p and Glg2p code for self-glucosylating proteins having amino acid sequences that are 33% and 34% identical, respectively, to that of rabbit muscle glycogenin in the N-terminal region of 258 amino acids (Cheng et al., 1995). Glg1p and Glg2p encode proteins of 618 and 380 amino acids, respectively, and have 55% sequence identity over their N-terminal 258 amino acids. Thus, they are larger than the muscle glycogenin which has 332 amino acids. The COOH termini of Glg1p and Glg2p are largely nonidentical in sequence except for two small segments of sequence similarity. Each contains a Tyr residue (Tyr232) in correspondence with Tyr194 of the rabbit muscle enzyme.

When the Glg1p and Glg2p genes are disrupted by homologous recombination separately, there was little effect on glycogen accumulation (Cheng et al., 1995). However, loss of both genes caused the almost complete loss of glycogen. The glycogen synthase activity was normal in this double mutant so lack of glycogen synthesis was not due to absence of glycogen synthase activity. If the rabbit muscle glycogenin was expressed in the double Glg1p, Glg2p mutant then glycogen synthesis was almost completely restored. Thus the mammalian glycogenin could complement the double mutant deficiency. These data indicated that the Glg1p and Glg2p genes were functioning in initiating glycogen synthesis and this report (Cheng et al., 1995) is the first *in vivo* evidence of the requirement of a glycogenin in the biosynthesis of glycogen in eukaryotes.

Starch Synthesis in Plants and Algae

The reactions of starch synthesis are similar to those indicated for mammalian glycogen synthesis except that the sugar nucleotide gucosyl donor is different and, as is shown later, the enzymatic site for regulation is also different. The reactions are indicated below.

$$ATP + \alpha\text{-glucose-1-P} \Leftrightarrow ADP\text{-glucose} + PPi \qquad \text{(Reaction 13–5)}$$

$$ADP\text{-glucose} + \alpha\text{-1,4 glucan} \rightarrow \alpha\text{-1,4-glucosyl-}\alpha\text{-1,4} \\ \text{glucan} + ADP \qquad \text{(Reaction 13–6)}$$

elongated α-1,4-oligosaccharide chain $\rightarrow$
    α-1,4-α-1,6 branched-glucan   (Reaction 13–7)
    (proamylopectin; phytoglycogen)

$$\text{proamylopectin} \rightarrow \text{amylopectin} + \text{proamylose} \qquad \text{(Reaction 13–8)}$$

$$\text{proamylose} \rightarrow \text{amylose} \qquad \text{(Reaction 13–9)}$$

The sugar nucleotide utilized for synthesis of the α-1,4 glucosidic linkages in amylose and amylopectin is ADP-glucose and not UDP-glucose. ADP-glucose synthesis is catalyzed by ADP-glucose (synthetase) pyrophosphorylase (2.7.7.27; ATP:α-D-glucose-1-phosphate adenylyltransferase) (Reaction 13-5). Reaction 13-6 is catalyzed by starch synthase (EC 2.4.1.21; ADP-glucose;1,4-α-D-glucan 4-α-glucosyltransferase). The reaction is the same as the glycogen synthase reaction except for the difference in glucosyl donor. Reaction 13-7 is catalyzed by branching enzyme (EC 2.4.1.18; 1,4-α-D-glucan 6-α-[1,4-α-glucano]-transferase) and the reaction is very similar to the one leading to the branching of glycogen (Reaction 13-4). However, the branch chains in amylopectin are longer (about 20 to 24 glucose units long) and there is less branching in amylopectin (~5% of the glucosidic linkages are α-1,6) as seen in glycogen (10–13 glucose units long and 10% of linkages are α-1,6). Thus, the starch branching enzymes may have different specificities with respect to chain transfer than those that branch glycogen or the interaction of the starch branching enzymes with the starch synthases may be different than the interaction of the mammalian branching enzymes with the respective glycogen synthase. Most probably the chain-elongating properties of the starch synthases are also different than those observed for the mammalian glycogen synthases and may account for some of the differences seen in amylopectin structure.

At least one other enzyme, a debranching enzyme, is involved in synthesis of the starch granule and its polysaccharide components amylose and amylopectin (James et al., 1995; Nakamura et al., 1996; Pan & Nelson, 1984). It has also been shown that there are isozymic forms of plant starch synthases (cited in reviews, Preiss, 1988; 1991; Preiss & Levi, 1980; and in recent literature, Denyer & Smith, 1992; Denyer et al., 1993, 1995; Hylton et al, 1996) and branching enzymes (cited in reviews, Preiss, 1988; 1991; Preiss & Levi, 1980; and in recent literature, Bhattacharyya et al., 1993; Burton et al., 1995; Guan & Preiss, 1993; Guan et al., 1994a,b; Mizuno et al., 1992). They are different gene products and seem to have different roles in the synthesis of the two polymers of starch, amylose, and amylopectin. For example, in many different plants as well in *Chlamydomonas reinhardtii* (Delrue et al., 1992; Dry et al., 1992; Nelson et al., 1978; Sano, 1984; Shure et al., 1983; Van der Leij et al., 1991; Visser et al., 1991) a granule-bound starch synthase is involved in the synthesis of amylose. Mutants defective in this enzyme, known as *waxy* mutants, give rise to starch granules having only amylopectin.

Reaction 13-6 was first described by Leloir et al. (1961) with UDP-glucose as the glycosyl donor, but it was later shown that ADP-glucose was more efficient in terms of maximal velocity and Km value (Recondo & Leloir, 1961). Leaf-starch synthases and the soluble starch synthases of reserve tissues are specific for ADP-glucose. In contrast, the starch synthases bound to the starch granule in reserve tissues also have some activity with UDP-glucose.

As indicated in a later section, there is also some evidence suggesting that initiation of starch synthesis in plants may also involve a protein acceptor similar to glycogenin.

Glycogen Synthesis in Bacteria

Sigal et al. (1964) showed that several UDP-glucose pyrophosphorylase-negative mutants of *E. coli* accumulated normal amounts of glycogen during growth in limiting nitrogen media. Thus, UDP-glucose could not be a glucosyl donor for glycogen synthesis. It was also reported in 1964 that extracts of several bacteria contained potent ADP-glucose: α-1,4-D-glucan-4-α-glucosyl transferase activities, which is known as the bacterial glycogen synthase reaction (Reaction 13-11, Greenberg & Preiss, 1964; Shen et al., 1964). These same bacterial extracts also contained ADP-glucose pyrophosphorylase (Reaction 13-10). Thus, the first two reactions in the ADP-glucose pathway of bacterial glycogen synthesis are similar to those present in plants.

About 10% of the linkages in bacterial glycogen are α-1,6. The formation of these linkages is catalyzed by an α- 1,4-glucan branching enzyme (Reaction 13-12) similar to that described for glycogen synthesis (Reaction 13-4) in animals. Branching enzyme activity has been detected in *E. coli* (Boyer & Preiss, 1977; Sigal et al., 1965), *Arthrobacter globiformis* (Zevenhuizen, 1964), *Salmonella typhimurium* (Steiner & Preiss, 1977), and *Streptococcus mitis* (Walker & Builder, 1971). The branching enzyme genes from *E. coli* (Baeker et al., 1986), *Streptomyces aureofaciens* (Homerova & Kormanec, 1994), *Bacillus stearothermophilus* (Kiel et al., 1991;Takata et al., 1994), *Bacillus caldolyticus* (Kiel et al., 1992) and cyanobacteria (Kiel et al.,1989, 1990) have been cloned.

$$\text{ATP} + \alpha\text{-glucose-1-P} \Leftrightarrow \text{ADP-glucose} + \text{PPi} \qquad \text{(Reaction 13–10)}$$

$$\text{ADP-glucose} + \text{glucan} \rightarrow \alpha\text{-1,4-glucosyl-glucan} + \text{ADP} \qquad \text{(Reaction 13–11)}$$

$$\text{elongated } \alpha 1,4\text{-oligosaccharide chain-glucan} \rightarrow$$
$$\alpha\text{-1,4-}\alpha\text{-1,6 branched-glucan} \qquad \text{(Reaction 13–12)}$$

Thus, the reactions seen for glycogen synthesis in bacteria are very similar to those that are involved in starch biosynthesis with respect to the use of ADP-glucose as the glucosyl donor. However, the plant systems have isozymes for their α-glucan synthases and branching enzymes. As of now isozymic forms of the glycogen synthase or branching enzymes have not been observed in bacteria. There have been some reports that there is a glycogenin in *E. coli* that is similar to the mammalian glycogenin involved in the initiation of glycogen synthesis. Barengo et al. (1975) reported the formation of a labeled TCA-insoluble fraction upon incubation of [^{14}C]UDP-glucose with *E. coli* extracts. The radioactivity was solubilized by α- amylase, suggesting that the label was an α-1,4 glucosyl oligosaccharide attached to a protein. Evidence was also presented to suggest that this labeled fraction was an intermediate in glycogen synthesis. Goldraij et al. (1993) have also isolated a 31 kDa protein presumably bound to *E. coli* glycogen. Whether this 31 kDa protein is indeed the bacterial glycogenin remains to be established. At present the glycogenin of bacteria is not well characterized. Moreover, the sugar nucleotide donor for glycogen synthesis is ADP-glucose and not UDP-glucose. As will be shown later mutants of *E. coli* defective in ADP-glucose PPase activity are deficient in glycogen. The involvement of UDP-Glc, if any, in glycogen synthesis is restricted to synthesis of a glucosylated glycogenin. However, as indicated before, *E. coli* mutants deficient in UDP-Glc PPase have normal glycogen levels. It is possible however, that the bacterial glycogenin may have a different sugar nucleotide glycosyl donor specificity.

Alternative Routes to Glycogen Synthesis in Bacteria

α-Glucans similar to glycogen have been shown to be formed either directly from sucrose (Reaction 13-12) or maltose (Reaction 13-13), or else from glucose-1-phosphate *via* the phosphorylase reaction in certain bacteria (Hestrin, 1960).

$$\text{sucrose} + (1,4\text{-}\alpha\text{-D-glucosyl})_n \Leftrightarrow$$
$$\text{D-fructose} + (1,4\text{-}\alpha\text{-D-glucosyl})_{n+1} \qquad \text{(Reaction 13–13)}$$

$$\text{maltose} + (1,4\text{-}\alpha\text{-D-glucosyl})_n \Leftrightarrow$$
$$\text{D-glucose} + (1,4\text{-}\alpha\text{-D-glucosyl})_{n+1}$$

(Reaction 13–14)

Amylosucrase, the enzyme catalyzing Reaction 13-13, is found in *Neisseria* strains that accumulate large amounts of glycogen-type polysaccharide when grown on sucrose (Hehre, 1951; Okada & Hehre, 1974). Amylosucrase, however, is limited to a few bacterial species and is active only when sucrose is in the grown media. *Neisseria* can metabolize exogenous sucrose but does not synthesize sucrose. Therefore, the observed accumulation of glycogen in *Neisseria* and in other microorganisms grown on carbon sources besides sucrose is not due to amylosucrase. Likewise, amylomaltase, the enzyme catalyzing Reaction 13-14, is induced along with a number of other enzymes when various strains of *E. coli*, *Streptococcus mutans*, *Aerobacter aerogenes* (Monod & Torriani, 1948; Palmer at al., 1973), *Streptococcus mitis* (Walker, 1966), *Dip/ococcus pneumoniae* (Lacks, 1968), and *Pseudomonas stutzeri* (Wöber, 1973) are grown on maltose or maltodextrins. The synthesis of amylomaltase, however, is repressed by glucose (Chao & Weathersbee, 1974) and its activity, therefore, cannot account for the synthesis of glycogen in organisms grown on glucose as a carbon source.

Maltodextrin phosphorylase and glycogen phosphorylase, as shown in Reaction 13-15, occur in many bacteria and can catalyze the synthesis or phosphorolysis of α-4-glucosidic linkages in α-1,4-glucans. But maltodextrin phosphorylase is induced only in the presence of maltodextrins, and for the microorganisms studied glycogen phosphorylase activity is insufficient to account for their rate of glycogen accumulation (Chen & Segel, 1968a,b; Khandelwal et al., 1973). Also, *E. coli* mutants deficient in maltodextrin phosphorylase accumulate maltodextrins, suggesting that the phosphorylase is involved in the degradation (phosphorolysis), and not the synthesis, of α-1,4-glucans (Schwartz, 1966).

$$\alpha\text{-}1,4\text{-(glucosyl)}_n + \text{Pi} \Leftrightarrow \alpha\text{-glucose-l-P} + \alpha\text{-}1,4\text{-(glucosyl)}_{n+1} \quad \text{(Reaction 13–15)}$$

Thus, practically all bacteria that accumulate glycogen synthesize their α-1,4-glucosidic linkages by the ADP-glucose pathway. Glycogen-deficient or glycogen-excess mutants of *E. coli* and of *S. typhimurium* (reviewed in Preiss, 1984, 1996a; Preiss & Romeo, 1989, 1994) have been isolated and are altered either in glycogen synthase activity or in ADP-glucose pyrophosphorylase activity, or both. A list of bacteria containing glycogen or the glycogen biosynthetic enzymes has been compiled in past reviews (Preiss, 1989; Romeo & Preiss, 1989) and indicates that glycogen accumulation is not restricted to any class of bacteria; it is present in gram-negative and -positive types, and even in archaebacteria. A recent report characterizes the ADP-Glc PPase of *Thermus caldophilus* Gk-24 (Ko et al., 1996).

Properties of the Bacterial and Plant α-1,4 Glucan Synthesizing Enzymes. Bacterial ADP-glucose Pyrophosphorylases: Structural and Enzymological Properties

The unique feature of the ADP-Glc PPases is that they are allosteric enzymes; their regulation *via* activation by glycolytic intermediates and inhibition by AMP, ADP, or inorganic phosphate is important for controlling the synthesis of both bacterial glycogen and plant starch. The regulatory properties of the ADP-Glc PPases will be discussed in a subsequent section dealing with regulation of bacterial glycogen synthesis. Other known properties of the enzyme are discussed here.

The molecular weights of the homogeneous enzymes from *E. coli* B (Haugen et al., 1974,1976), *R. rubrum* (Furlong & Preiss, 1969), and spinach leaf (Ribereau-Gayon & Preiss, 1971) have been estimated by sedimentation equilibrium ultracentrifugation analysis, and the molecular weights of partially purified enzymes from *Aeromnas hydrophila*, *Rhodospirillum molischianum*, *Rhodospirillum tenue*, *Rhodobacter spheroides*, *Rhodobacter gelatinosa*, *Rhodobacter viridis*, *Rhodospirillum fulvum*, *Rhodobacter acidophila*, *Rhodobacter glohiformis*, *Salmonella typhimurium*, and *Serratia marcescens* have been estimated from sucrose gradient ultracentrifugation data (Preiss & Walsh, 1981). A study of the subunit molecular weights of four bacterial enzymes indicate that they are in the range of 45,000–51,000, suggesting that the native enzymes from bacterial sources are tetramers of similar subunits. Of interest are the findings for the S. *marcescens* enzyme; two molecular weight species are seen: one has $Mr96,000$, the other 186,000 (Preiss et al., 1976). If the subunit molecular weight of the S. *marcescens* enzyme is similar to the *E. coli* enzyme, it probably exists as homo-tetrameric and homo-dimeric forms in equilibrium.

The substrate saturation curves of the *Rhodospirillum rubrum* ADP-glucose pyrophosphorylase are hyperbolic at low temperatures and in the presence of activator. Therefore, the reaction mechanism was investigated *via* initial kinetic studies (Paule & Preiss, 1971). Intersecting reciprocal plots were obtained, indicating a sequential kinetic mechanism. Product inhibition patterns eliminated all known sequential mechanisms except the ordered BiBi or Theorell–Chance mechanisms (Cleland, 1972). Small intercept effects suggested the existence of significant concentrations of central transistory complexes. Kinetic constants obtained in the study also favored the ordered BiBi mechanism. Moreover, equilibrium ATP-[^{32}P]pyrophosphate isotope exchange studies supported a sequential-ordered mechanism and also indicated that ATP is the first substrate to add and that ADP-glucose is the last product to dissociate from the enzyme (Paule & Preiss, 1971). A study of the binding of substrates and effectors of the *E. coli* B ADP-glucose pyrophosphorylase enzyme has been initiated (Haugen & Preiss, 1979).

Equilibrium dialysis studies showed that in the presence of 5 mM MgCl$_2$ and 1.5 mM fructose-1,6-bisphosphate, glucose-1-phosphate does not bind to the enzyme. However, ATP does bind, suggesting that the reaction mechanism of the *E. coli* B enzyme is similar to the *R. rubrum* enzyme in that there is ordered binding, with magnesium adenosine triphosphate (MgATP) binding first and glucose-1-phosphate binding second. Chromium adenosine triphosphate (CrATP), a potent inhibitor of many enzymes utilizing MgATP as a substrate (DePamphilis & Cleland, 1973; Janson & Cleland, 1974), is a potent competitive inhibitor of the *E. coli* ADP-glucose pyrophosphorylase (Haugen & Preiss, 1979). If this inactive analogue is present in the glucose-1-phosphate-binding equilibrium dialysis experiments, it is found that one mole of glucose-1-phosphate binds per mole of ADP-glucose pyrophosphorylase subunit.

Only two moles of MgATP or CrATP bind to the tetrameric protein in the absence of glucose-1-P. Thus MgATP sites appear to exhibit half-site reactivity (Lazdunski, 1972). This is in contrast to equilibrium dialysis experiments with ADP-glucose, in which 4 moles of this substrate were found to bind to 1 mole of the tetrameric protein (Haugen & Preiss,1979). In the presence of glucose-1-phosphate, however, 4 moles of CrATP bind to the tetrameric protein. Thus it appears that in the pyrophosphorylase reaction mechanism in the synthesis direction, 2 moles of MgATP initially bind to the tetrameric protein. This permits the binding of glucose-1-phosphate to its four binding sites on the tetrameric protein. Further binding of the next 2 moles of MgATP may then ensue, with concomitant catalysis occurring. A mechanism for enzyme catalysis is proposed that explains some of the kinetic and binding properties in terms of an asymmetry in the distribution of the conformational states of the four identical subunits (Haugen & Preiss, 1979).

The *E. coli* ADP-Glc PPase gene has been isolated and expressed (Okita et al., 1981) and the nucleotide sequence has been elucidated (Baecker et al., 1983). The calculated molecular weight of the deduced protein (Fig. 13-2) is 48,762, which is in excellent agreement with the experimentally determined approximate molecular weight of 50,000 (Haugen et al., 1974, 1976).

```
                                               FBP
 1     5     10    15    20    25    30    35    40    45    50
MVSLEKNDHLMLARQLPLKSVALILAGGRGTRLKDLTNKRAKPAVHFGGK

      55    60    65    70    75    80    85    90    95    100
FRIIDFALSNCINSGIRRMGVITQYQSHTLVQHIQRGWSFFNEEMNEFVD

            ATP,ADP-GLC
      105   110   115   120   125   130   135   140   145   150
LLPAQQRMKGENWYRGTADAVTQNLKIIRRYKAEYVVILAGDHIYKQDYS

                                            GLC-1-P
      155   160   165   170   175   180   185   190   195   200
RMLIDHVEKGARCTVACMPVPIEEASAFGVMAVDENDKIIEFVEKPANPP

      205   210   215   220   225   230   235   240   245   250
SMPNDPSKSLASMGIYVFDADYLYELLEEDDRDENSSHDFGKDLIPKITE

      255   260   265   270   275   280   285   290   295   300
AGLAYAHPFPLSCVQSDPDAEPYWRDVGTLEAYWKANLDLASVVPELDMY

      305   310   315   320   325   330   335   340   345   350
DRNWPIRTYNESLPPAKFVQDRSGSHGMTLNSLVSGGCVISGSVVVQSVL

      355   360   365   370   375   380   385   390   395   400
FSRVRVNSFCNLDSAVLLPEVWVGRSCRLRRCVIDRACVIPDGMVIGENA

      405   410   415   420   425   430
EEDARRFYRSEEGIVLVTREMLRKLGHKQER
```

Figure 13–2. The amino acid sequence of *E. coli* ADP-glucose pyrophosphorylase. Amino acids are represented by the one-letter amino acid code. The figure shows that Lys39 is involved in binding the activator, fructose-1,6-bis-P (FBP), Tyr114 is involved in binding ATP- and ADP-Glc, and Lys195 is involved in binding glucose-1-P.

Chemical modification and site-directed mutagenesis studies of *E. coli* ADP-Glc PPase have provided evidence for the location of the activator binding site (Parsons & Preiss, 1978a,b), the inhibitor binding site (Larsen & Preiss, 1986; Larsen at al., 1986), and the substrate binding site (Hill et al., 1991; Parsons & Preiss, 1978b). These experiments have used pyridoxal-P as the analogue for either the activator (fructose1,6-bisP) or the substrate (glucose-1-P). For an ATP analogue, the photoaffinity reagent 8-azido-ATP ($8N_3$ATP) proved to be a substrate for the *E. coli* enzyme (Lee & Preiss, 1986) whereas 8-azido AMP ($8N_3$AMP) was an effective inhibitor analogue (Larsen & Preiss, 1986; Larsen et al., 1986). Since the sequence of the *E. coli* ADP-Glc PPase gene *glgC* is known, the identification of the amino acid sequence about the modified residue permitted determination of the location of the modified residue in the primary structure of the enzyme. The amino acid residue involved in binding the activator was Lys39 and the amino acid involved in binding the adenine portion of the substrates (ADP-Glc and ATP) was Tyr114. Tyr114 was also the major binding site for the adenine ring of the inhibitor, AMP. Lys195 is protected from reductive phosphopyridoxylation by the substrate ADP-Glc: thus, it was proposed that it is also a part of the substrate binding site.

Figure 13-2 shows the amino acid sequence of the *E. coli* ADP-Glc PPase and the substrate and allosteric sites identified *via* chemical modification. The deduced amino acid sequence of the cloned *S. typhimurium* enzyme is known (Leung & Preiss, 1987b). There is about 80% identity of the nucleotide sequences between the *E. coli* and *S. typhimurium glgC* genes and a 90% identity in the amino acid sequence. Most of the changes are conservative. Amino acids that have been shown to be involved in substrate and allosteric effector binding, as well as those involved in maintaining allosteric function for the *E. coli* enzyme, are all conserved.

The *E. coli* enzyme has been crystallized (Mulichak et al., 1988) but the crystals are of poor diffraction quality and are sensitive to X-ray exposure damage. Thus, no three-dimensional structure of the enzyme is available at present.

Bacterial Glycogen Synthase

The *E. coli* B (Fox et al., 1976) and *E. coli* K12 (Cattaneo et al., 1979) glycogen synthases have been purified to homogeneity. The *E. coli* B enzyme is strongly absorbed to the hydrophobic resin 4-aminobutyl-Sepharose and is eluted only by solutions containing both 1 M KCl and 1 M maltose. The homogeneous enzyme has a specific activity of about 115 μmol of glucose transferred to glycogen per mg of protein per min (Cattaneo et al., 1978; Kawaguchi et al., 1978). Subunit molecular weight as determined by SDS-gel electrophoresis is 49,000 ± 2000, while sucrose density gradient centrifugation showed multiple aggregated forms of 98,000, 135,000, and 185,000 (Fox et al., 1976). Thus *E. coli* B glycogen synthase can exist as dimers, trimers, and tetramers. It should be noted that the bacterial glycogen synthase subunit is about one-half the size of the mammalian glycogen synthase subunit.

The bacterial glycogen synthase is unlike the mammalian glycogen synthase in two major respects. First, the enzyme exhibits no regulatory properties. It does not exist in both inactive and active forms, as does the mammalian glycogen synthase. Presently, there is no evidence for either phosphorylation or dephosphorylation of the enzyme, or for any other enzyme-catalyzed modification. The bacterial enzyme is not activated by glucose 6-phosphate or other glycolytic intermediates (Dietzler & Strominger, 1973;

Elbein & Mitchell, 1973; Greenberg & Preiss, 1965; Preiss & Greenberg, 1965). Second, the bacterial enzyme uses ADP-glucose as its physiological sugar nucleotide glucosyl donor (Greenberg & Preiss, 1964) rather than UDP-glucose, the physiological glucosyl donor of the mammalian glycogen–synthase reaction. Glucose-l-phosphate is totally inactive. In one report (Holmes & Preiss, 1979) the ratio of activities of ADP-glucose: UDP-glucose: CDP-glucose: GDP-glucose: TDP-glucose of the *E. coli* glycogen synthase reaction was shown to be 100:0.29:0.11:0.66:0.23. Thus, all other sugar nucleotides tested had less than 1% of the activity determined for ADP-glucose. Deoxy-ADP glucose, a nonphysiological analogue of ADP-glucose, has been found to be active to the same extent as ADP-glucose with the *Arthobacter viscosus* glycogen synthase (Greenberg & Preiss, 1965) as well as with the *E. coli* B enzyme (Preiss & Greenberg, 1965).

The Km values reported for ADP-glucose range from 20 μM for the *M. smegmatis* glycogen synthase (Elbein & Mitchell, 1973) and 25–35 μM for the *E. coli* B glycogen synthase (Fox et al., 1976) to 0.4 mM for the *Pasteurella pseudotuberculosis* glycogen synthase (Dietzler & Strominger, 1973). The Km for deoxy-ADP-glucose is 38 μM for *E. coli* B glycogen synthase and 27 μM for the *A. viscosus* enzyme. The glycogen synthase has been shown to be inhibited by ADP and very sensitive to inhibition by *p*-hydroxymercuribenzoate. ADP was competitive with ADP-glucose and its Ki for the *E. coli* system was 15 μM (Preiss & Greenberg, 1965).

Various α-glucans can serve as effective primers, namely, glycogen from animal and bacterial sources, amylose, and amylopectin. Lower molecular weight maltodextrins can also serve as primers. Maltotriose is quite effective as a primer, and maltotetraose has been identified as the immediate product. Maltose is effective as a primer only at high concentrations, and maltotriose is the immediate product. Glucose does not function as an acceptor of glucose in the bacterial glycogen synthase reaction.

The glycogen synthase reaction has been shown to be reversible (Fox et al., 1976). The formation of labeled ADP-glucose occurs from either [^{14}C]ADP or [^{14}C]glycogen. The ratio of ADP to ADP-glucose at equilibrium at 37°C has been found to vary threefold in the pH range 5.27–6.82. This suggests that in the formation of a new α-l,4-glucosidic bond, a proton is liberated. Thus, the glycogen synthase reaction in this pH range may be written as:

$$\text{ADP-glucose}^{2-} + (\text{glucosyl})_n \Leftrightarrow (\text{glucosyl})_{n+1} + \text{ADP}^{2-} \quad \text{(Equation 13–2)}$$

Since the pK_{al} of ADP^{2-} ionizing to $\text{ADP}^{3-} + H^+$ is about 6.4, then a proton would be liberated in varying amounts at pH 5.4 to 7.4 and in stoichiometric amounts above pH 7.4. If the pKa of ADP^{2-} is assumed to be 6.4 (Bock, 1960), then the ratio of ADP^{2-} to ADP-glucose could be determined and was calculated to be constant in the range of pH 5.27–6.82 and was 45.8 $\pm$ 4.5. The constancy of this equilibrium ratio suggests that ADP^{2-} is the reactive species in the reaction.

The equilibrium of the glycogen synthase reaction may be depicted as K_{eq} = [ADP] [G_n]/[ADP-glucose] [G_{n+1}] where K_{eq} is the equilibrium constant, G_n is the polyglucose primer molecule, and G_{n+1} is the polyglucose primer molecule where one glucosyl residue has been added. Since G_n, G_{n+1}, or even G_{n+x} can act as either acceptor or donors of glucosyl residues, K_{eq} would not be affected by the concentration of primer, and the ratio of ADP to ADP-glucose would be a measure of K_{eq}. Using 45.8 $\pm$ 4.5 as the K_{eq} value, the $\Delta F°$ of the reaction at 37°C may be calculated to be $-2.3 \pm$

0.06 kcal from the equation, $\Delta F° = -RT \ln K_{eq}$. If the $\Delta F°$ of hydrolysis of the α-1,4-glucosidic linkage formed is assumed to be -4 kcal, then the $\Delta F°$ of hydrolysis of the glucosidic linkage of ADP-glucose may be calculated to be -6.3 kcal.

The structural gene for glycogen synthase, *glgA*, has been cloned from both *E. coli* and *S. typhimurium* (Leung & Preiss, 1987a; Okita et al., 1981) and the nucleotide sequence of the *E. coli glgA* gene has been determined (Kumar et al., 1986). It consists of 1431 bp specifying a protein of 477 amino acids with a MW of 52,412.

The *E. coli* and *S. typhimurium* glycogen synthases are specific for the sugar nucleotide, ADP-glucose. Some chemical modification studies (Holmes & Preiss, 1982) have shown two distinct sulfydryl groups important for enzyme activity and protected by the primer, glycogen, and the substrate, ADP-Glc, respectively. The reactive sulfhydryl residues are probably located at or near the binding sites for the substrates, glycogen, and ADP-Glc.

An affinity analogue of ADP-Glc, adenosine diphosphopyridoxal (ADP-pyridoxal), was used to identify the ADP-Glc binding site (Furukawa et al., 1990). Incubation of the enzyme with the analogue plus sodium borohydride led to an inactivated enzyme. The degree of inactivation correlated with the incorporation of about 1 mol of analogue per mol of enzyme subunit for 100% inactivation. After tryptic hydrolysis one labeled peptide was isolated and the modified Lys residue was identified as Lys15. The sequence, Lys-X-Gly-Gly, where lysine is the amino acid modified by ADP-pyridoxal, has been found to be conserved in the mammalian glycogen synthase (Mahrenholz et al., 1988; Tagaya et al., 1985). As will be shown later, this amino acid sequence is also present in the plant starch synthases. The substrate analogue used for determination of the mammalian glycogen synthase was UDP-pyridoxal, the analogue for the substrate UDP-Glc.

Furukawa et al. (1993,1994) have performed site-directed mutagenesis experiments to determine structure–function relationships for a number of amino acids in the *E. coli* glycogen synthase. Substitution of other amino acids for Lys at residue 15 suggested that the Lys residue is involved mainly in binding the phosphate residue adjacent to the glycosidic linkage of the ADP-Glc and not in catalysis. The major effect on the kinetics of the mutants at residue 15 was the elevation of the Km of ADP-Glc, about 30- to 50-fold, when either Gln or Glu were the substituted amino acids. Substitution of Ala for Gly at residue 17 decreased the catalytic rate constant, k_{cat}, about three orders of magnitude compared to the wild-type enzyme. Substitution of Ala for Gly18 decreased the rate constant only 3.2-fold. The Km effect on the substrates, glycogen, and ADP-Glc, were minimal. It was postulated by the researchers (Furukawa et al., 1993) that the two glycyl residues in the conserved Lys-X-Gly-Gly sequence participated in the catalysis by assisting in maintaining the correct conformational change of the active site or by stabilizing the transition state.

Since there is still binding of the ADP-Glc and appreciable catalytic activity of the Lys15Gln mutant, the ADP-pyridoxal modification was repeated and in this instance about 30-fold higher concentration was needed for inactivation of the enzyme (Furukawa et al., 1994). The enzyme was maximally inhibited about 80% and tryptic analysis of the modified enzyme yielded one peptide containing the affinity analogue and having the sequence, Ala-Glu-Asn-modified Lys-Arg. The modified lysine was identified as Lys277. Site-directed mutagenesis of Lys277 to form a Gln mutant was done and the Km for ADP-Glc was essentially unchanged but k_{cat} was decreased 140-fold. It was concluded that Lys277 was more involved in the catalytic reaction than in substrate binding.

Bacterial Branching Enzyme

The structural gene of the *E. coli* branching enzyme (BE), *glgB*, has been cloned (Baecker et al., 1986) and its complete nucleotide and deduced amino acid sequences were determined and found to be consistent with the amino acid analysis of the pure protein, as well as with the molecular weight determined by SDS gel electrophoresis and with the amino acid sequence analyses obtained of the amino terminal and of the various peptides obtained *via* CNBr degradation. The gene consisted of 2181 bp specifying a protein of 727 amino acids having a MW of 84,231.

The relationship in amino acid sequences between that of branching enzyme (BE) and amylolytic enzymes such as α-amylase, pullulanase, glucosyltransferase, and cyclodextrin glucanotransferase was reported by Romeo & Preiss (1989), particularly at those sequences believed to be contacts between the substrate and the enzyme. Baba et al. (1991), reported that there was a marked conservation in the amino acid sequence of the four catalytic regions of amylolytic enzymes in maize endosperm BE I. As shown in Table 13-1, four regions that putatively constitute the catalytic regions of the amylolytic enzymes are conserved in the starch branching isoenzymes of maize endosperm, rice seed, potato tuber, and the glycogen branching enzymes of *E. coli*. Analysis of this high conservation in the α- amylase family has been pointed out and greatly expanded by Svensson (1994) and by Jesperson et al. (1993) with respect to sequence homology, but also in the prediction the $(\beta/\alpha)_8$-barrel structural domains with a highly symmetrical fold of eight inner, parallel β- strands, surrounded by eight helices, in the various groups of enzymes in the family. The $(\beta/\alpha)_8$-barrel structural domain was determined from the crystal stucture of some α- amylases and cyclodextrin glucanotransferases.

Table 13–1 Comparison of Primary Structures of Various Branching Enzymes with the Four Most Conserved Regions of the α-Amylase Family[a]

Branching Enzyme	Region 1	Region 2	Region 3	Region 4
Maise endosperm BE I	277 **DVVHSH**	347 **GFRFDGVTS**	402 **TVVAEDVS**	470 **CIAYAESHD**
Maize endosperm BE II	315 **DVVHSH**	382 **GFRFDGVTS**	437 **VTIGEDVS**	501 **CVTYAESHD**
Potato tuber BE	355 **DVVHSH**	424 **GFRFDGITS**	453 **VTMAEEST**	545 **CVTYAESHD**
Rice seed BE 1	271 **DVVHSH**	341 **GFRFDGVTS**	396 **TIVAEDVS**	461 **CVTYAESHD**
Rice seed BE 3	337 **DVVHSH**	404 **GFRFDGVTS**	459 **ITIGEDVS**	524 **CVTYAESHD**
E. coli glycogen BE	335 **DWVPGH**	400 **ALRVDAVAS**	453 **VTMAEEST**	517 **NVFLPLNHD**
B. subtilis α-amylase	100 **DAVINH**	171 **GFRFDAAKH**	204 **FQYGEILQ**	261 **LVTWVESHD**
B. sphaericus cyclodextrinase	238 **DAVFNH**	323 **GWRLDVANE**	350 **IIVGEVWH**	414 **SFNLLGSHD**

[a]The sequences have been derived from references referred to in the text. Only two examples of enzymes from the amylase family are compared. Svensson (1994) compares over 40 enzymes ranging from amylases, glucosidases, various α-1,6-debranching enzymes as well as four cases of branching enzymes. The invariant amino acid residues believed to be involved in catalysis are in bold letters.

The conservation of the putative catalytic sites of the α- amylase family in the glycogen and starch branching enzymes would be expected as the BE catalyzes two consecutive reactions in synthesizing α-1,6-glucosidic linkages by cleavage of an α-1,4-glucosidic linkage in an 1,4-α-D-glucan to form a nonreducing end oligosaccharide chain that is transferred to a C-6 hydroxyl group of the same or another 1,4-α-D-glucan. It would be of interest to know whether the eight highly conserved amino acid residues of the α- amylase family are also functional in branching enzyme catalysis. Further experiments such as chemical modification and analysis of the three-dimensional structure of the BE would be needed to determine the precise functions and nature of its catalytic residues and mechanism, these studies are discussed when the plant starch branching enzymes are reviewed.

Plant ADP-Glucose Pyrophosphorylases

The ADP-glucose pyrophosphorylases of higher plants, green algae, as well as the cyanobacteria are regulatory proteins under allosteric control and constitute the major site for regulation of starch synthesis. The enzymes are highly activated by 3-phosphoglycerate and inhibited by inorganic phosphate. The regulation of starch and cyanobacterial glycogen synthesis *via* regulation of the photosynthetic ADP-Glc PPase will be discussed in a later section. The structural properties of the plant ADP-Glc PPase is reviewed in this section.

The enzyme that has been studied in most detail with respect to kinetic properties and structure is that isolated from spinach leaf (Copeland & Preiss, 1981; Ghosh & Preiss, 1966; Preiss et al., 1967). However, the kinetic properties of the ADP-glucose pyrophosphorylases from other leaf extracts, barley, butter lettuce, kidney bean, maize, peanut, rice, sorghum, sugar beet, tobacco, and tomato are similar to those found for the spinach leaf enzyme (Sanwal et al., 1968).

The spinach leaf ADP-glucose pyrophosphorylase has been purified to homogeneity by preparative disc gel electrophoresis (Ribereau-Gayon & Preiss, 1971) and by hydrophobic chromatography (Copeland & Preiss, 1981). The spinach leaf enzyme has a molecular mass of 206,000 and is composed of two different subunits, having molecular masses of 51 and 54 kDa (Morell et al., 1987a,b; Preiss et al.,1987a,b). These subunits can be distinguished not only by differences in their molecular mass but also with respect to amino acid composition, amino terminal sequences, peptide patterns of the tryptic digests on high performance liquid chromatography (HPLC), as well as antigenic properties. The two subunits are, therefore, quite distinct and most probably the products of two genes. In contrast, bacterial ADP-Glc PPases including the cyanobacterial enzymes are homotetrameric, that is, composed of only one subunit, 50–55 kilodaltons in mass depending on the species (Preiss, 1984).

Other plant ADP-Glc PPases have been studied in detail and they also have been shown to be composed of two dissimilar subunits. The maize endosperm ADP-Glc PPase, which has a molecular mass of 230 kDa, reacts with the antibody prepared against the native spinach leaf enzyme in immunoblot experiments (Preiss et al., 1990). The enzyme is composed of subunits of 55 and 60 kDa, which seem to correspond to the spinach leaf 51 and 54 kDa subunits (Preiss et al., 1990).

The studies of the maize endosperm mutants *shrunken* 2 (*sh* 2) and *brittle* 2 (*bt* 2), which are ADP-Glc PPase activity deficient (reviewed by Preiss, 1988, 1991), are also relevant. In immunoblotting experiments and using antibodies raised against the native enzyme and each subunit of the spinach leaf enzyme, it was found that the mutant *bt* 2 endosperm lacks the 55 kDa subunit and that the mutant *sh* 2 endosperm lacks the 60 kDa subunit. These results (Preiss et al.,1990) strongly suggest that the maize endosperm ADP-Glc PPase is composed of two immunologically distinct subunits and that the *sh* 2 and *bt* 2 mutations cause reduction in ADP-Glc PPase activity through the lack of one of the subunits; the *sh* 2 gene would be the structural gene for the 60 kilodalton protein, while the *bt* 2 gene would be the structural gene for the 55 kDa protein. Consistent with this hypothesis was the isolation of an ADP-Glc PPase cDNA clone from a maize endosperm library (Barton et al., 1986) which hybridized with the small subunit cDNA clone from rice (Anderson et al., 1989). This maize ADP-Glc PPase cDNA clone was found to hybridize to a transcript present in maize endosperm but absent in *bt* 2 endosperm. Thus, the *bt* 2 mutant appears to be the structural gene of the 55 kDa subunit of the ADP-Glc PPase.

Hylton and Smith (1992) proposed the existence of not two, but four polypeptides of MW ~50 kDa for the ADP-glucose PPase of pea embryo, and a molecular mass for the holoenzyme of ~110 kDa. The authors, however, do not seem to have taken appropriate precautions to prevent proteolysis, and the data concerning the molecular mass of the holoenzyme are inadequate, as no data on enzyme activity are presented but only the presence of protein.

In short, data available so far indicate that both the seed and leaf ADP-glucose pyrophosphorylase are heterotetramers composed of two different subunits and that, on the basis of immunoreactivity and sequence data (Smith-White & Preiss, 1992), there is corresponding homology between the subunits in the leaf enzyme and with the subunits of reserve tissue enzyme.

Because the plant native ADP-Glc PPases are tetrameric and composed of two different subunits it is of interest to know why the two subunits are required for optimal catalytic activity. The enzyme must contain ligand-binding sites for the activator, 3PGA, and inhibitor, Pi, as well as catalytic sites for the two substrates, ATP and glucose-1-P; it is possible that that these sites could be located on different subunits. Chemical modification can be used to obtain information on the catalytic mechanism and on the catalytic site of the enzyme of interest. Chemical modification studies on the ADP-Glc PPase have involved the use of the following affinity labels, that can be used in radioactive form, depending on the nature of the experiment: (1) pyridoxal-5-phosphate (PLP), an analogue of 3-PGA, (2) the photoaffinity substrate analogues, 8-azido-ATP and 8-azido-ADP-Glc. When UV light (257 nm) irradiates 8-azido compounds, a nitrene is formed, which can react with electron-rich residues and inactivate the enzyme (Preiss et al., 1989), and (3) phenylglyoxal, for the identification of arginine residues.

These kind of studies have provided information on the catalytic and regulatory sites of the spinach ADP-Glc PPase, and on the role of the large and small subunits. In ADP-Glc PPase from *E. coli*, Lys195 has been identified as the binding site for the phosphate of glucose-1-P (Hill et al., 1991) and tyrosine residue 114 has been identified as involved in the binding of the adenosine portion of the other substrate, ATP (Lee & Preiss, 1986). The overall amino acid sequence identity of the *E. coli* enzyme when aligned with the plant and cyanobacterial ADP-Glc PPases ranges from 30% to 33% (Smith-White & Preiss, 1992). In contrast, there is greater sequence identity when the *E. coli* ATP and glucose-1-P binding sites (Table 13-2) are compared with the corresponding sequences of the plant and cyanobacterial enzymes. This suggests that those

Table 13–2 Conservation in Plant ADP-Glc PPases of the Glucose-1-Phosphate (Hill et al., 1991) and of the ATP Binding (Preiss & Romeo,1989) sites present in *E. coli* ADPGlc PPase[a]

Source	Glucose-1-P Site	ATP Site
Prokaryotes	195　　114	
E. coli	IIEFVEKP-AN	WYRGTADAV
S. typhimurium	**D*****-**	*********
Anabaena	V*D*S***KGE	*FQ*******
Synechocystis	*TD*S***QGE	*FQ*******
Plant Small Subunit		
A. thaliana (small subunit)	****A***KGE	*FQ*******
Maize endosperm 54 kDa	****A***KGE	*FQ*******
Potato tuber 50 kDa	****A***QGE	*FQ*******
Rice seed (small subunit)	*V**A***KGE	*FQ*******
Spinach leaf 51 kDa	****A***KGE	*FQ*******
Wheat endosperm (small subunit)	****A***KGE	*FQ*******
Plant Large Subunit		
Maize endosperm 60 kDa	VLQ*F***KGA	*FQ****SI
Potato tuber 51 kDa	VVQ*A***KGF	*FQ*******
Spinach leaf 54 kDa	VLS*S***KGD	*FQ*******
Wheat endosperm (large subunit)	VVQ*S*Q*KGD	*FR*****W

[a]References to these sequences for the plant ADP-Glc PPases are in Smith-White and Preiss, (1992). The sequences for the *Anabaena* enzyme is in Charng et al.(1992), for the *Synechocystis* enzyme in Kakefuda et al. (1992) and for the wheat endosperm small subunit in Ainsworth et al. (1993). The numbers 195 and 114 corresponds to Lys195 and Tyr114 of the *E. coli* enzyme and * Signifies the same amino acid as in the *E. coli* enzyme.

sequences are still important in the plant enzyme, and probably having the same function. Indeed, in a recent preliminary experiment with the potato tuber ADP-Glc PPase expressed in *E. coli* (Ballicora et al., 1995), site-directed mutagenesis of lysine residue K198 of the 50 kDa subunit (equivalent to the *E. coli* ADP-Glc PPase K195) to a glutamate residue increased the Km for glucose-1-P from 57 μM to over 31 mM without any perceptible change in the Km or Ka for the other substrates, Mg^{2+}, ATP or activator, 3PGA (Fu & Preiss, unpublished results). These results indicate an involvement of Lys198 of the plant ADP-Glc PPase in the binding of glucose-1-P. In the case of the putative ATP binding site, there is a phenylalanine residue instead of tyrosine in the corresponding sequences of the plant and cyanobacterial enzymes. It would be of interest to determine in future site-directed mutagenesis and chemical modification studies whether the WFQGTADAV region of the plant enzyme is indeed a portion of the ATP binding region or whether the conservative change of two amino acids in the sequence has affected the function of that portion of the protein.

The binding site for pyridoxal phosphate in the small subunit (Table 13-3) was isolated, revealing a lysine residue close to the C-terminus which may be important for 3-phosphoglycerate (3PGA) activation (Morell et al., 1988). When PLP is covalently bound, the plant ADP-Glc PPase no longer requires 3PGA for activation. Also, the covalent binding of PLP is prevented by the allosteric effectors 3PGA and Pi. These observations show that the modified enzyme no longer requires an activator for maximal activity and that the covalent modification is prevented by the presence of the allosteric effectors; they strongly indicate that the activator analogue, PLP, is binding at the activator site. Ball & Preiss (1994) also showed that three lysine residues of the spinach leaf large subunit are involved in or close to the binding site of pyridoxal-P and, presumably, of the activator, 3PGA (Table 13-3). The chemical modification of these Lys residues by pyridoxal-P was prevented by the presence of 3PGA during the reductive phosphopyridoxylation process and, in the case of the Lys residue of site 1 of the small subunit and site 2 of the large subunit, Pi also prevented them from being modified by reductive pyridoxylation.

Very similar results were otained with the *Anabaena* ADP-Glc PPase (Charng et al., 1994). Chemical modification of the enzyme with PLP resulted in the cyanobacterial enzyme no longer requiring activator for maximal activity, and the modification was prevented by 3PGA and Pi. The modified Lys residue was identified as Lys419 and the sequence adjacent to that residue is very similar to that observed for site 1 sequences of the higher plants. Site-directed mutagenesis of Lys 419 to either Arg, Ala, Gln, or Glu produced mutant enzymes having 25 to 150-fold lower apparent affinities than that of wild-type enzyme. No other kinetic constants such as Km for substrates and the inhibitor, Pi, were affected. Neither was the heat stability or the catalytic efficiency of the enzyme affected. These mutant enzymes, however, were still activated to a great extent at higher concentrations of 3PGA, suggesting that an additional site was involved in the binding of the activator. The Lys419Arg mutant was chemically modified with the activator analogue, PLP. Modification of Lys382 in the Arg mutant was observed and caused a dramatic alteration in the allosteric properties of the en-

Table 13–3 Plant and Cyanobacterial ADPglucose Pyrophosphorylase Activator Binding Sites[a]

	Activator Site 1	Activator Site 2
	419	382
Anabaena	SGIVVVLKNAVITDGTII	QRRAIIDKNAR
Synechocystis	NGIVVVIKNVTIADGTVI	IRRAIIDKNAR
Higher plants	Activator site 1, Small subunit	Activator site 2, Large subunit
Barley endosperm	SGIVTVIKDALLPSGTVI	ISNCIIDMNAR
Maize endosperm	GGIVTVIKDALLPSGTVI	IRNCIIDMNAR
Potato tuber	SGIVTVIKDALIPSGIII	IRKCIIDKNAK
Rice seed	SGIVTVIKDALLLAEQLY	
Spinach leaf	SGIVTVIKDALIPSGTVI	IKDAIIDKNAR
Wheat leaf		IKRAIIDKNAR
Wheat seed	SGIVTVIKDALLPSGTVI	IQNCIIDKNAR

[a]The sequences are listed in one letter code and were taken from Smith-White and Preiss (1992), and from references indicated in the text. The Lys residues that are in outline are those covalently modified by pyridoxal-P (the chemical modification prevented by 3PGA and Pi). In the case of the potato tuber enzyme the Lys residue was identified via site-directed mutagenesis experiments. The numbers 419 and 382 correspond to the Lys residues in the *Anabaena* ADPGlc PPase subunit. Site 1 is present in the small subunit of the plant ADP Glc PPase while site 2 is present in the large subunit.

zyme; this could be prevented by the presence of 3PGA or Pi during the chemical modification process. Therefore, Lys382 was identified as the additional site involved in the binding of the activator; as seen in Table 13-3 the adjacent sequence about Lys382 in the *Anabaena* enzyme is very similar to that seen for site 2 in higher plants, which is situated on the large subunit.

Site 1 corresponds to Lys440, the lysyl residue near the C-terminus that is phosphopyridoxylated in the spinach leaf small subunit (Morell et al., 1988); this corresponds to Lys468 in the rice seed small subunit and to Lys441 in the potato tuber ADP-Glc PPase small subunit. Site 2 is also situated close to the C-terminus, equivalent to Lys382 in the *Anabaena* ADP-Glc PPase and Lys404 of the potato tuber large subunit. Table 13-3 also shows that the amino acid sequence of the spinach leaf small subunit peptide containing the modified lysyl residue of site 1 is highly conserved in the rice seed, potato tuber, maize (Bai et al., 1990), barley (Thorbj⁻rnsen et al., 1996b), and wheat endosperm small subunits (Ainsworth et al., 1993), and the *Anabaena* (Charng et al., 1992) and *Synechocystis* (Kakefuda et al., 1992) ADP-Glc PPase subunits. Similarly, the amino acid sequence of site 2 of the spinach leaf large subunit is highly conserved in the large subunits of the potato tuber, maize and barley endosperm, as well as wheat seed (Olive et al., 1989) and wheat leaf ADP-Glc PPases (Olive et al., 1989).

Phenylglyoxal inactivation of the enzyme can be prevented by 3PGA or by Pi, evidence that one or more arginine residues are present in the allosteric sites of the spinach leaf enzyme; both subunits were labeled when [^{14}C]phenylglyoxal was used (Ball & Preiss, 1992). The location of the Arg residue(s) in the sequence is presently unknown but there is a possibility it may be close to the Lys residue at activator site 2.

cDNA clones encoding the putative mature forms of the large and small subunits of the potato tuber ADP-Glc PPase have been expressed together, using compatible vectors, in an *E. coli* mutant devoid of ADP-Glc PPase activity (Ballicora et al., 1995; Iglesias et al., 1993). The ADP-Glc PPase activity expressed was high and the enzyme displayed catalytic and allosteric kinetic properties very similar to the ADP-Glc PPase purified from potato tuber (Iglesias et al., 1993). Moreover, the enzyme activity was neutralized by antibody prepared against potato tuber but not by antibody prepared against the *E. coli* ADP-Glc PPase (Iglesias et al., 1993). This expression system is a very useful tool for performing site-directed mutagenesis to further characterize the allosteric function of the lysyl residues identified *via* chemical modification with pyridoxal-P of the spinach enzyme. Indeed in preliminary results, site-directed mutagenesis of Lys 441 of the potato ADP-Glc PPase small subunit to Glu and Ala resulted in mutant enzymes 30–83-fold lower in their affinity, respectively, for 3PGA (Preiss et al., 1996).

The ability to express cDNA clones representing the potato tuber small and large subunits together in *E. coli* (Ballicora et al., 1995) to obtain a highly active enzyme has also enabled us to express the subunits separately to determine their specific functions. It was found that the potato tuber small subunit when expressed alone had high catalytic activity, provided that the 3PGA concentration was increased to 20 mM. The 3PGA saturating concentration for the expressed transgenic or (normal) potato tuber heterotetrameric enzyme is 3 mM. It was found that the Ka of the transgenic enzyme in ADP-Glc synthesis was 0.16 mM while for the small subunit alone it was 2.4 mM. Thus, the small subunit by itself has about 15-fold lower affinity for the activator. Also, the small subunit is more sensitive to Pi inhibition than the transgenic heterotetrameric enzyme with an eight-fold lower Ki. The kinetics of 3PGA activation and the Pi inhibition were the main kinetic differences between the homotetrameric small subunit and the recombinant heterotetrameric ADP-Glc PPase. These results are consistent with those obtained for the *Arabidopsis thaliana* mutant ADP-Glc PPase lacking the large subunit (Lin et al., 1988b). In this case the enzyme had lower affinity for the activator and higher sensitivity towards Pi inhibition than the heterotetrameric normal enzyme (Li & Preiss, 1992).

The potato tuber large subunit expressed by itself had negligible activity. Thus, it seems that the dominant function of the small subunit is catalysis while the dominant function of the large subunit is modulation of the sensitivity of the small subunit to allosteric activation and inhibition.

A high degree of amino acid sequence identity is observed when comparing the sequences of corresponding ADP-Glc PPase subunits from different species, a result that could be predicted from the fact that the spinach leaf lower molecular weight subunit antibody reacts very well with the equivalent subunits of the enzymes from maize endosperm (Plaxton & Preiss, 1987; Preiss et al., 1990), rice seed (Anderson et al., 1989; Krishnan et al., 1986), *A. thaliana* (Lin et.al., 1988a), and potato tuber (Okita et al.,1990). The antibody for the lower molecular weight spinach leaf subunit does not react well with the higher molecular mass subunit of the ADP-Glc PPase of other species. Therefore, not much homology was expected between sequences of the small and large subunits, and the degree of identity between the large and small subunits (obtained by Edman degradation or deduced from nucleotide sequences of cDNAs or genomic DNA) is around 40%–60% (Smith-White & Preiss, 1992). Sequence analyses indicate a greater identity between the 54 kDa subunit of the spinach leaf enzyme, the subunit coded by the *sh*-2 gene from maize, and the subunit encoded by the cDNA insert, we7, from wheat endosperm (Olive et al.,1989) suggesting that the latter correspond to the large molecular weight subunit of the ADP-Glc PPase.

Because of the relatively low but certain homology between the two subunits of the ADP-Glc PPase, it is speculated that they originally arose from the same gene. The bacterial ADP-Glc PPase has been shown to be a homotetramer composed of only one subunit (Preiss, 1984). The cyanobacterial ADP-Glc PPase has 3PGA as an allosteric activator and Pi as an inhibitor, similar to the enzyme from higher plants (Levi & Preiss, 1976), but unlike the bacterial enzymes. Both bacterial (Preiss, 1984; Preiss & Romeo, 1989) and cyanobacterial (Iglesias et al., 1991) ADPglucose pyrophosphorylases are homotetrameric, unlike the higher plant enzymes, indicating that regulation by 3PGA and Pi is not related to the heterotetrameric nature of the higher plant enzyme. It is quite possible that during evolution there was duplication of the ADP-Glc PPase gene, and then divergence of the genes produced two different genes coding for the two peptides, both required for optimal activity of the native higher plant enzyme.

As previously indicated, one can tentatively assign catalytic function to the small subunit of the ADP-Glc PPase; this is consistent with the identity and similarity in sequence between the small subunits isolated from different plants and tissues. In the case of the large subunit in which amino acid sequences have lower similarity to what is observed for the small subunits, it is quite possible that the different large subunits lend different regulatory properties to the heterotetrameric ADP-Glc PPases of different species or tissues. This would make some sense given that the needs and amounts of starch required for each type of tissue are different. Thus, because sequences of the large subunits reflect their occurrences in different plant tissues, for example, leaf, stem, guard cells, tuber, endosperm, and root, (Smith-White & Preiss, 1992), it is possible that these sequence differences give the enzyme from each tissue different allosteric properties.

Starch Synthase

Starch synthase activity can be measured as the transfer of $[^{14}C]$glucose from ADP-glucose into a primer such as rabbit glycogen or amylopectin (Macdonald & Preiss, 1983, 1985).

As first shown in our laboratory, citrate stimulates a reaction in the absence of added maltodextrin primer which is due to small amounts of endogenous primer strongly bound to the enzyme (Boyer & Preiss, 1979; Ozbun et al., 1971; 1972; Pollock & Preiss, 1980). Citrate has been shown to increase the apparent affinity for the glucan primer (Boyer & Preiss, 1979; Pollock & Preiss, 1980). Thus, only a minute amount of endogenous glucan (e.g., 6 nmol of anhydroglucose units) is needed per reaction mixture for the citrate-stimulated "unprimed reaction."

Many questions regarding starch biosynthesis remain to be answered. The gaps in our understanding of starch biosynthesis arise partly from the difficulties inherent to the starch granule itself, which is insoluble in water, has a very complicated structure and is still the object of much research and speculation (Morrison & Karkalas, 1990).

In vivo, starch synthesis occurs by deposition on the granule surface by the concerted action of starch synthases and branching enzyme. Starch synthase activity is associated with the starch granules or in the supernatant of crude extracts. The elucidation of the roles of the multiple forms of starch synthase and branching enzyme in the biosynthesis of starch, and the determination of its fine structure requires purification and characterization of each isoform. In maize endosperm there are at least four starch synthases, two soluble (Ozbun et al., 1971) and at least two granule-bound (Macdonald & Preiss, 1985). The number of isoforms may vary with the plant species and the developmental stage, but those that have been studied more carefully seem to have a similar number of isoforms. Indeed, as in the case of pea embryo an isozyme of starch synthase, starch synthase II, can exist as a soluble protein and starch-granule bound (Edwards et al., 1996). The question remains whether the activities soluble and granule bound are both functional. Indeed, Mu-Forster et al. (1996) has reported that in maize endosperm, that more than 85% of the starch synthase I protein may be associated with the starch granule. This was determined by using antibody prepared against the starch synthase. However, no evidence was presented to indicate that the starch synthase I was active in the particulate stage. The cDNA clones that encode the two isozymes of granule-bound starch synthase of pea embryo are optimally expressed at different times during development (Dry et al., 1992); while isozyme II is expressed in every organ, isozyme I is not expressed in roots, stipules, or flowers (Dry et al., 1992).

Purification of the starch synthase and branching enzymes in large amounts and to high specific activity has proven to be difficult. Partly for this reason it has not been possible so far to find out how the enzymes interact to produce the two carbohydrates, amylose and amylopectin, that form the starch granule.

Genetic studies implicate one granule-bound starch synthase (GBSS) isoform in the synthesis of amylose. In *waxy* (*wx*) mutants there is virtually no amylose, GBSS activity is deficient (Nelson & Rines, 1962, Nelson et al. 1978; Shure et al., 1983) and the w*x* protein is missing. The final product of the *wx* locus is a protein of molecular weight 58 kDa associated with the starch granule.

Shure et al. (1983) prepared cDNA clones homologous to *wx* mRNA and, in subsequent experiments (Federoff et al.,1983) restriction endonuclease fragments containing part of the *wx* locus were cloned from strains carrying the different *wx* alleles to further characterize the controlling insertion elements activator (ac) and dissociation (ds). Excision of the ds element from the certain *wx* alleles produces two new alleles encoding for wx proteins with altered starch synthase activities (Wessler et al.,1986).

The DNA sequence of the *wx* locus of *Zea mays* was determined by analysis of both a genomic and an almost full length cDNA clone (Klösgen et al., 1986) and the *wx* locus from barley has been cloned and its DNA sequenced (Rhode et al.,1988). Amino acid sequences are also available for rice (Wang et al., 1990), potato (Van der Leij et al., 1991), cassava (Salehuzzaman et al., 1993), wheat (Ainsworth et al.,1993), and pea isozymes (Dry et al., 1992).

Table 13-4 compares three regions of the deduced amino acid sequences from the potato, cassava, maize, barley, wheat, rice, and isozyme I of pea embryo *wx* clones with the amino acid sequence for the *E. coli* glycogen synthase (Kumar et al., 1986), and the rice soluble starch synthase (Baba et al., 1993). Region 1 starts with the first 27 amino acids of the N-terminal of the *E. coli* glycogen synthase. Thirteen are identical to the amino acid sequences deduced for the plant *wx* proteins. Of particular significance is the sequence starting at residue Lys15 of the bacterial enzyme, . . . KTGGL . . . . The lysine in the bacterial glycogen synthase has been implicated in the binding of the substrate, ADP-glucose (Furukawa et al., 1990), on the basis of the chemical modification of that site by the substrate analogue ADP-pyridoxal. The similarity of sequences between the bacterial glycogen synthase, the soluble starch synthase and the wx protein provides further evidence that the *wx* gene is indeed the structural gene for the granule bound starch synthase.

There are two other regions of high conservation of the various GBSSs with the *E.coli* glycogen synthase. In region II, only one or two amino acids of the thirteen amino acids are different from the *E. coli* sequence and in region III, all the GBSSs are completely identical with respect to the amino acid sequence while the bacterial enzyme differs in only two of nine amino acids, an Arg for a Ser and an Ala for a Val.

The genetic evidence points to the *wx* locus as the structural gene for a starch synthase bound to the starch granule. However, direct biochemical evidence was lacking, mainly because of the difficulties involved in studying the proteins associated with starch. Starch was solubilized using amylases, and the starch proteins liberated into the supernatant were fractionated by chromatography on (diethylamino)ethyl (DEAE)-cellulose (Macdonald & Preiss, 1985). The GBSS I was clearly associated with the wx protein (recognized by its mobility on sodium dodecyl sulfate (SDS) polyacrylamide gels and its reaction with antibodies raised against the pure wx protein) throughout purification. The molecular mass of the GBSS I, determined by gel filtration or by sucrose density gradients, was about 59 kDa (Sivak et al., 1993).

Because of the failure to demonstrate that the wx protein from pea endosperm had starch synthase activity, Smith (1990) suggested that "the *waxy* protein of pea is not the major granule-bound starch synthase" and that a "re-examination, species by species, of the identity of the starch-granule-bound starch synthase . . ." may be required. Sivak et al. (1993), however, showed that starch extracted from developing embryos of pea contained starch synthase activity that was associated with the waxy pro-

Table 13–4 Conserved Regions of Amino Acid Sequences of the *E. coli* Glycogen Synthase and the Various Granule-Bound Starch Synthases[a]

	Region I
E.coli glycogen synthase	1MQVLHVCSEMFPLLKTGGLADVIGALP
Rice soluble starch synthase	20RSVVFVTGEASPYAKSGGLGDVCGSLP
Potato tuber wx protein	4MNLIFVGTEVGPWSKTGGLGDVLRGLP
Cassava wx protein	4MNLIFVGAEVGPWSKTGGLGDVLGGLP
Maize wx protein	5MNVVFVGAEMAPWSKTGGLGDVLGGLP
Barley wx protein	6MNLVFVGAEMAPWSKTGGLGDVLGGLP
Wheat wx protein	7MNLVFVGAEMAPWSKTGGLGDVLGGLP
Rice wx protein	6MNVVFVGAEMAPWSKTGGLGDVLGGLP
Pea wx protein I	1MSLVFVGAEVGPWSKTGGLGDVLGGLP

	Region II	Region III
E. coli glycogen synthase	372VPSRFEPCGLTQL	397RTGGLADTV
Rice soluble starch synthase	372MPSRFEPCGLNQL	397GTGGLRDTV
Potato tuber wx protein	397VPSRFEPCGLIQL	422STGGLVDTV
Cassava wx protein	398VPSRFEPCGLIQL	423STGGLVDTV
Maize wx protein	398VTSRFEPCGLIQL	423STGGLVDTV
Barley wx protein	396VTSRFEPCGLIQL	421STGGLVDTV
Wheat wx protein	410VTSRFEPCGLIQL	435STGGLVDTV
Rice wx protein	397VPSRFEPCGLIQL	422STGGLVDTV
Pea wx protein I	403IPSRFEPCGLIQL	428STGGLVDTV

[a]The numbers preceding the sequence indicates the residue number from the N-terminus in the sequence. The sequence in outline form, KTGGL, has been shown for the *E. coli* glycogen synthase to be involved in binding of the sugar nucleotide substrate (Furukawa et al., 1992). References to the other sequences may be obtained from Preiss and Sivak (1996).

tein. The molecular weight of the pea starch synthase is about 59 kDa, as determined by ultracentrifugation in sucrose density gradients. A pea granule-bound starch synthase preparation displayed a relatively high specific activity (over 10 μmol glucose incorporated per min per mg protein). This enzymatic fraction when subjected to SDS polyacrylamide gel electrophoresis (PAGE) migrated the same as the waxy protein; it gave a strong immunoblot with antibody prepared against the waxy protein either from pea embryo or maize and only the wx protein stain was visible (Sivak et al., 1993). Thus the immunological data indicated that the activity assayed by Sivak et al. (1993) was due to the granule bound starch synthase (Waxy protein) and not due to the truncated soluble starch synthase of 60 kDa detected by Edwards et al. (1996).

When the gene coding for the mature Waxy protein from maize kernel was expressed in *E. coli*, the recombinant protein had a molecular weight similar to the maize protein as determined by SDS PAGE, reacted with antibody raised against the plant protein and had starch synthase activity (Sivak et al., unpublished data).

Thus, the biochemical reexamination of starch synthase present in starch granules from two species, maize and pea, strengthens the genetic evidence supporting the role of the wx protein as a granule-bound starch synthase with a major role in the determination of amylose content of starch.

How the loss of GBSS causes the disappearance of amylose is still not clear. It has been shown by many experiments involving antisense RNA in potato (Kuipers et al., 1994; Visser et al., 1991) and in rice (Shimada et al., 1993), that disappearance of amylose correlated very well with the loss of wx gene expression. It is possible that the interior of the granule is devoid of branching enzyme or, if branching enzyme is in the granule itself, it is not appreciably active. The presence of an active chain elongating enzyme, that is, starch synthase, without an active branching enzyme present (in the presence or absence of some debranching activity), could lead to amylose formation. However, this situation may be more complicated since more than one isozyme of the GBSS has been found for a number of plants. Also, it is quite possible that the GBSS may also be involved in the initial formation of amylopectin near the exterior portion of the granule along with the soluble starch synthases. In *Chlamydomonas reinhardtii* a *wx* mutant deficient in GBSS was isolated (Delrue et al., 1992). In this mutant not only was the isolated starch deficient in amylose but also one of the amylopectin fractions, amylopectin II, was significantly lower. Amylopectin II has longer chains than the amylopectin I fraction as judged from the increase in λ_{max} of the glucan-I_2 complex. When GBSS is active, it would not be rate limiting and thus amylose and amylopectin are seen as normal components of the starch granule. When there is a loss of the major GBSS activity, then the rate of formation of the amylose and initial amylopectin structures would be limiting in the *wx* mutant and only the higher branched amylopectin I fraction would be present.

A normal amount of starch is made even if the GBSS activity is deficient, and this may suggest that there is sufficient activity of the minor granule-bound starch synthases present or that the soluble starch synthases activity may substitute to form sufficient unbranched chains allowing normal amylopectin synthesis to proceed at the same rate as in the wild type. However, these limiting amounts of GBSS isozyme activity are not sufficient to produce substantial amounts of amylose. The involvement of GBSS in amylopectin and amylose synthesis is discussed later.

A variety of plant systems have shown the presence of multiple forms of soluble starch synthases (SSS). Studies with barley, maize, and wheat endosperm, pea, rice, sorghum, and teosinte seeds, spinach leaf, and potato tuber extracts have indicated the presence of at least two major forms of SSS (reviewed in Preiss, 1988, 1991; Preiss & Levi, 1980; Preiss & Sivak, 1996), designated as types I and II. In most cases, SSS I elutes from an anion exchange column at lower salt concentration than SSS II.

Although starch synthase I (SSS I) has been partly purified from maize kernels (Boyer & Preiss, 1979; Pollock & Preiss, 1980), starch synthase II (SSS II), a less stable isoform, has been more difficult to purify.

The properties observed for the isoforms of maize endosperm tissue reflect the properties of the corresponding enzyme forms in other plant materials, and the properties of the starch synthase isozymes have been reviewed (Preiss & Sivak, 1996). The apparent affinity for ADP-glucose, measured by the Km, is similar for the two forms. The maximal velocity of the type I enzyme is greater with rabbit liver glycogen than with amylopectin, while the type II enzyme is less active with glycogen than with amylopectin. Citrate stimulation of the primed reaction is greater for type I than for type II. Both forms can use the oligosaccharides maltose and maltotriose as primers when present at high concentrations. Starch synthase I seemed to have more activity than SSS II with these acceptors.

The lower activity for SSS I with amylopectin as a primer as compared to glycogen, suggests that SSS I may prefer elongating the short exterior chains (A chains) that are more prevalent in glycogen than in amylopectin. The reverse may be true for SSS II, which could prefer the longer chains (B chains) seen in amylopectin. Differences were also noted in the apparent affinities with respect to primer. For example, the Km of the type I enzyme for amylopectin is nine times lower than that of the type II enzyme. It is worth noting that the type I enzyme is active without added primer in the presence of 0.5 M citrate while the type II enzyme is inactive under these conditions. Citrate decreases the Km of amylopectin for both types of enzymes; 160-fold for the type I enzyme and about 16-fold for the type II starch synthase with 0.5 M citrate.

The starch synthase isozymes in maize endosperm have different molecular masses. The GBSS isozyme I has a molecular mass of 60 kDa, that of GBSS II 95 kDa, the SSS I a molecular mass of 72 kDa, and SSS II, 95 kDa (reviewed in Preiss & Sivak, 1996). Mu et al. (1994) have reported the molecular mass of maize endosperm SSSI as 76 kDa, which is similar to the value reported previously for SSS I . These molecular mass values for the starch synthases are all higher than that of the E. coli glycogen synthase, which has a molecular weight of 52 kDa (Kumar et al., 1986). It appears that the maize endosperm SSS I and II are immunologically distinct (Macdonald & Preiss, 1985). Antibody prepared against maize endosperm SSS I showed very little reaction with SSS II in neutralization tests.

In summary, the maize SSS I and II seem to be distinct forms, as distinguished on the basis their physical, kinetic, and immunological properties. They are probably products of two different genes. Because of their different kinetic properties and different specificities with respect to primer activities they may have different functions in the formation of the starch granule. Purification of the isoforms to high specific activity and lack of interfering activities will facilitate the characterization of the isoforms with respect to primer specificity and interaction with isoforms of branching enzyme, supplying information about their role in vivo.

In rice, three isoforms of soluble starch synthase were separated by anion exchange chromatography which, in immunoblots, reacted with antibodies raised to the rice waxy protein (Baba et al., 1993). After affinity chromatography of the active fractions, amino terminal sequences were obtained for the protein bands of 55–57 kDa (separated by SDS PAGE) that cross-reacted weakly with serum raised against the rice waxy protein. It is worth noting that this experimental approach does not exclude the possibility that other soluble starch isoforms were present which did not cross-react with the antiserum, and the authors indicate that other results suggest that another soluble starch synthase isoform, with a molecular weight of 66 kDa, is also present in seed extracts.

Other forms of starch synthase may be present in plants. Recently, Marshall et al.

(1996) have reported the presence of a starch synthase, 140 kDa, in potato tubers which may account for 80% of the total soluble starch synthase activity. A cDNA representing the protein gene was isolated. Expression of an antisense mRNA caused a reduction of about 80% of the soluble starch synthase activity in the tuber extracts. Of interest was the finding that the severe reduction in activity had no effect on starch content or on the amylose/amylopectin ratio of the starch. However, there was a change in the morphology of the starch granules, suggesting an alteration in the starch structure. The specific change in structure causing the morphology change remains to be determined.

Baba et al. (1993) isolated cDNA clones coding for the putative soluble starch synthase from maize from an inmature rice seed library in λgt 11 using as probes synthetic oligonucleotides designed on the basis of the N-terminal amino acid sequences available. The insert of about 2.5 kb was sequenced and shown to code for a 1878-nucleotide open reading frame. Comparison with the corresponding amino terminal sequences led the authors to conclude that the protein is initially synthesized as a precursor, carrying a long transit peptide at the amino acid terminus and that the same gene would be expressed both in seeds and in leaves.

In order to understand the various functions of the different starch synthases Ball and his associates in Lille, France set out to isolate various mutants of *Chlamydomonas* deficient in starch synthase activities. They have been successful in isolating a soluble starch synthase II deficient mutant (Fontaine et al.,1993) and double mutants deficient both in GBSS and in SSS II (Maddelein et al., 1994). These studies have provided significant information on the function of both these enzymes and their involvement in amylopectin biosynthesis. The SSS II mutant contained only 20%–40% of the starch seen in the wild-type organism and the percent amylose of the total starch increased from 25% to 55%. This mutant also contained a modified amylopectin which had an increased amount of very short chains, 2–7 glucose units or degrees of polymerization (DP) and a concomitant decrease of intermediate-sized chains (8–60 DP). This suggested that the SSS II was involved in the synthesis or maintenance of the intermediate-sized chains (B chains?) in amylopectin. The higher amylose content could be explained because if an unbranched amylose-like intermediate were a precursor for amylopectin synthesis, it could not be utilized effectively in the SSS II mutant. It is quite possible that this amylose fraction may be more highly branched than the usual amylose. The absorption spectra of its I_2 complex has lower maximal wavelength than the wild-type amylose fraction, suggesting that more branching has occurred. The mutant amylose fraction may, therefore, have a greater amount of branched amylose intermediates on the route to amylopectin biosynthesis.

The double mutants defective in SSS II and a GBSS (Maddelein et al., 1994) had an even lower starch content, 2%–16% of the wild-type, and the amount of starch present was inversely correlated with the severity of the GBSS defect of the double mutant. The authors suggest that the GBSS is required to form the basic structure of the amylopectin and the effects of the GBSS absence are exacerbated due to the diminished SSS II activity. Of interest is that the SSSI may, in addition to a small amount of starch, synthesize a small water soluble polysaccharide. Analysis of both fractions suggest that they may be intermediate in structure between amylopectin and glycogen with respect to extent of branching.

These studies of the *Chlamydomonas* mutants by Ball and his colleagues have been quite informative in that they provide good evidence for involvement of the GBSS in amylopectin, as well as in amylose synthesis, and suggest that an important function for SSS II would be in its involvement in synthesis of the intermediate-sized (B) branches in amylopectin.

Branching Enzyme

Branching enzyme has been assayed in several different ways. An iodine assay is based on the decrease in absorbance of the glucan–iodine complex resulting from the branching of amylose or amylopectin by the enzyme. During incubation of the assay mixture containing amylose or amylopectin, aliquots are taken at intervals and iodine reagent is added (Boyer & Preiss, 1978a). For amylose, the decrease of absorbance is measured at 660 nm and, for amylopectin, at 530 nm. A unit of activity is defined as decrease in absorbance of 1.0 per min at 30°C at the defined wavelength.

The phosphorylase-stimulation assay (Boyer & Preiss, 1978a,b; Hawker et al., 1974) is based on the stimulation of the 'unprimed' (without added glucan) phosphorylase activity of the phosphorylase a from rabbit muscle as the branching enzyme present in the assay mixture increases the number of nonreducing ends available to the phosphorylase for elongation. One unit is defined as 1 μmol transferred from glucose-1-P per min at 30°C.

The branch-linkage assay (Takeda, et al., 1993) is the only assay that measures the number of branch chains formed by branching enzyme catalysis rather than an indirect effect of its action as in the two assays described above. The enzyme fraction is incubated with the substrate, NaBH$_4$-reduced amylose, the reaction is then stopped by boiling and the product is incubated with crystallized *Pseudomonas* isoamylase for debranching. Finally, the reducing power of the oligosaccharide chains transferred by the enzyme is measured by a modification of the Park-Johnson method. Reduction of amylose with borohydride gives about 2% of the reducing power of the nonreduced amylose, resulting in lower blanks.

Thus, the branching-linkage assay is the most quantitative assay for branching enzyme but amylolytic activity interferes most with this assay. The phosphorylase-stimulation assay is the most sensitive and the I$_2$ assay is not very sensitive but allows the testing of branching enzyme specificity with various α-1,4-dextrins and providing information on the possible role of the different branching enzyme isoforms. It may be best to employ all three assays when studying the properties of the branching enzymes. But, above all, if reliable information is being sought, the branching enzymes must be purified to the extent that all degradative enzymes are eliminated before studying its properties.

In maize endosperm there are three branching enzyme isoforms (Boyer & Preiss, 1978a; Guan & Preiss, 1993; Singh & Preiss, 1985). Reports on other tissues are consistent with the presence of more than one isoform, as in castor bean (Goldner & Beevers, 1989). BE I, IIa, and IIb purified from maize kernels (Boyer & Preiss, 1978a; Guan & Preiss, 1993; Singh & Preiss, 1985) no longer contained amylolytic activity (Guan & Preiss, 1993; Takeda et al., 1993). Molecular weights were 82,000 for isoform I and 80,000 for isoforms IIa and IIb (Boyer & Preiss, 1978a,b). Some progress has been made (Guan & Preiss, 1993; Takeda, et al., 1993) and some of the results using the different assays are summarized in Table 13-5.

Takeda et al. (1993) have analyzed the branched products made from amylose by each BE isoform. This was done by debranching the products of each reaction using isoamylase, followed by gel filtration. BE IIa and BE IIb are very similar in their affini-

Table 13–5 Specific Activities (units/mg protein) of Maize Endosperm Branching Enzyme Isoforms[a]

Branching Enzymes	BE I	BE IIa	BE IIb
Phosphorylase stimulation (a)	1196	795	994
Branching linkage assay (b)	2.6	0.32	0.14
Iodine stain assay (c)			
Amylose (c_1)	800	29.5	39
Amylopectin (c_2)	24	59	63
Ratio of activity			
a/b	460	2484	7100
a/c_1	1.5	27	25
a/c_2	49.8	13.5	15.8
c_2/c_1	0.03	2	1.6

[a]The units are for phosphorylase stimulation and branching linkage assays, mmol/min, and for the iodine stain assay, a decrease of one absorbance unit per min.

ties for amylose and the size of the chain transferred. When presented with amyloses of different average chain lengths, the three BEs have higher activity with the longer chain amylose, but while BE I could still catalyze the branching of an amylose of average chain length 197 with 89% of the activity shown with a chain length of 405, the activity of BE II dropped sharply with decreasing chain length. The study of the reaction products showed that the action of BE IIa and BE IIb results in the transfer of shorter chains than those transferred by BE I. The action of the isoforms on amylopectin has been studied by Guan & Preiss (1993); of the three isoforms, BE I had the highest activity in branching amylose (Table 13-5) and its rate of branching amylopectin was less than 5% of that with amylose. In contrast, the BE IIa and IIb isoforms effected the branch amylopectin at twice the rate they effected the branching of amylose, and catalyzed branching of amylopectin at six times the rate observed for BE I. These results are consistent with the results of Takeda et al. (1993) in suggesting that BE I catalyzes the transfer of longer branched chains and that BEIIa and IIb catalyze the transfer of shorter chains. Thus, it is quite possible that BEI may produce slightly branched polysaccharides which serve as substrates for enzyme complexes of BE II isoforms and starch synthases to synthesize amylopectin; BE II isoforms may play a major role in forming the short chains present in amylopectin. Also, BE I may be more involved in producing the more interior (B chains) chains of the amylopectin while BE IIa and BEIIb would be involved in forming the exterior (A) chains.

In potato tubers, Vos-Scheperkeuter et al. (1989) purified a single form of branching activity of 79 kDa molecular mass. Antibodies were prepared to the native potato enzyme and they were found to react strongly with maize BE I and very weakly with maize BE IIb. In neutralization tests the antiserum inhibited the activities of both the potato tuber BE and maize BE I. It was concluded that the potato branching enzyme shows a high degree of similarity to the maize BE I and to a lesser extent with the other maize BE.

However, whether potato tubers have two isoforms of branching enzyme such as BE I and BE II has not been determined. Borovsky et al. (1975) isolated from potato tubers a BE of molecular mass 85 kDa. This is close to the mass of 79 kDa found by Vos-Scheperkeuter et al. (1989). Recently, it has been claimed that branching enzymes of molecular mass of 97 and 103 kDa can be isolated (Blennow & Johansson, 1991; Khoßnoodi et al., 1993) and these results suggested that the previous lower molecular mass values of 79 and 85 kDa are the results of proteolysis during purification of the 103 kDa BE. Khoshnoodi et al. (1996) showed that limited proteolysis of the 103 kDa enzyme either with trypsin or chymotrypsin produced an enzyme, still fully active, having a molecular weight of 80 kDa. Indeed, up to four cDNA clones have been isolated for BE, one having molecular weight 91–99 kDa (Khoshnoodi et al., 1996; Koßmann et al., 1991; Poulsen & Kreiberg, 1993). All of these allelic clones have sequences similar to the BE I type. It is still not resolved whether or not the 97- and 80-kDa proteins could be the products of different allelic forms of the BE gene or different BE genes. Also, the *sbeIc* allele codes for a mature enzyme of 830 amino acids having a molecular weight of 95,180. The *sbeIc* BE protein product, expressed in *E. coli*, migrates as a 103 kDa protein during electrophetic analysis (Khoshnoodi et al., 1996).

It is of interest to note that BE isolated from other plants, bacteria, or mammals have molecular masses ranging from 75 to approximately 85 kDa. These molecular masses have been consistent with the molecular weights obtained from deduced amino acid sequences obtained from isolated genes or cDNA clones. Using Indica rice, Smyth (1988) resolved two or three peaks of BE by anion exchange chromatography. Although the molecular weight of the BE seemed to be only about 40,000, activity was determined using the iodine assay and a crude extract, and amylase contamination or possible proteolysis apparently were not taken into account.

Mizuno et al. (1992), has reported four forms of branching enzyme from immature rice seeds that were separated by chromatography on DEAE-cellulose chromatography. It seems that two of the forms, BE1 and BE 2 (the latter composed of BE 2a and BE 2b) were the major forms, while BE 3 and BE 4 were minor forms comprising less than 10% of the total branching enzyme activity. The molecular weight of the branching enzymes were BE 1, 82 kDa; BE 2a, 85 kDa; BE 2b, 82 kDa; BE 3, 87 kDa; BE 4a, 93 kDa; and BE 4b, 83 kDa. However, BE 1, 2a and b seem to be immunologically similar in their reaction to maize endosperm BE I antibody. Moreover, the rice seed BE 1, BE 2a, and BE 2b had very similar N-terminal amino acid sequences. All three BEs had two N-terminal sequences, TMVXVVEEVDHLPIT and VXVVEEVDHLPITDL. The latter sequence is very similar to the first but lacking just the first two N-terminal amino acids. Thus, although these activities eluted in separate fractions from the DEAE-cellulose column they seem to be the same protein on the basis of immunology and N-terminal sequences, although BE 2a is 3 kDa larger. Antibody raised against BE 3 reacted strongly against BE 3 but not toward BE 1 or BE 2a, 2b. Thus, as noted for maize endosperm, rice endosperm had essentially two different isoforms of BE.

Because of the many isoforms existing for the rice seed branching enzymes, Yamanouchi & Nakamura (1992) studied and compared the BEs from rice endosperm, leaf blade, leaf sheath, culm, and root. The BE activity could be resolved into two fractions, BE 1 and BE 2; both fractions were found in all tissues studied in different ratios of activity. The specific activity of the endosperm activity either on the basis of fresh weight or protein was 100–1000-fold greater than other tissues studied. On native gel electrophoresis, rice endosperm BE 2 could be resolved into two fractions, BE 2a and BE 2b. Upon electrophoresis of the other tissue BE 2 forms, only BE 2b was found. BE 2a was detected only in the endosperm tissue. It appears that there could be tissue-specific isoforms of BE in rice.

Three forms of branching enzyme from developing hexaploid wheat (*Triticum aestivum*) endosperm have been partially purified and characterized (Morell et al., 1997). Two forms are immunologically related to maize branching enzyme I and one form to maize BE II. The N-terminal sequences are consistent with these relationships. The wheat BE I_B gene is located on chromosome 7B while the wheat BE I_{AD} peptides genes are located on chromosomes 7A and D. The BE classes in wheat are expressed differentially during endosperm development in that BE II is expressed constitutively throughout the whole cycle while BE I_B and BE I_{AD} are expressed in late endosperm development.

There are some maize endosperm mutants that appear to increase the percent amylose of the starch granule. The normal starch granule contains about 25% of the polysaccharide as amylose with the rest as amylopectin. In contrast, *amylose extender* mutants may have as much as 55%–70% of the polysaccharide as amylose; they may have an amylopectin fraction with fewer branch points and with the branch chains longer in length compared to those of normal amylopectin. Results with the recessive maize endosperm mutant, amylose extender (*ae)*, suggest that *ae* is the structural gene for either branching enzyme IIa or IIb (Boyer & Preiss, 1978b, 1981) as the activity of BE I was not affected by the mutation. In gene dosage experiments, Hedman and Boyer (1982) reported a near linear relationship between increased dosage of the dominant *ae* allele and BE IIb activity. Since the separation of form IIa from IIb was not very clear, it is possible that the *ae* locus was also affecting the level of IIa.

Singh and Preiss (1985) concluded that although some homology exists between the three-starch branching enzymes, there are major differences in the structure of branching enzyme I when compared to IIa and IIb, as shown by its different reactivity with some monoclonal antibodies, and differences in amino acid composition and in proteolytic digest maps. It was also concluded (Singh & Preiss, 1985) that branching enzymes IIa and IIb are very similar and perhaps the product of the same gene. However, recent studies by Fisher et al. (1993, 1996) in analyzing 16 isogenic lines having independent alleles of the maize *ae* locus, suggest that BE IIa and BE IIb are encoded by separate genes and that the BE IIb enzyme is encoded by the *ae* gene. They isolated a cDNA clone labeled *Sbe* 2b, which had a cDNA predicted amino acid sequence at residues 58 to 65 identical with the N-terminal sequence of the maize BE IIb that they had purified (Fisher et al, 1993). Moreover, they did not detect in *ae* endosperm extracts any mRNA with the *Sbe* 2b cDNA clone. Some BE activity was observed in the *ae* extracts that had chromatographic behavior similar to BE IIa.

The finding that the enzyme defect in the *ae* mutant is BE IIb is consistent with the finding that BE II is involved in transfer of small chains *in vitro*. Besides having an increase in amylose, the altered amylopectin structure in the *ae* mutant consists of fewer and longer chains and a lesser number of total chains. In other words, there are very few short chains.

The wrinkled pea has a reduced starch level, about 66%–75% of that seen in the round seed. Whereas the amylose content is about 33% in the round form it is 60%–70% in the wrinkled pea seed. Edwards et al. (1988) measured the activities of several enzymes involved in starch metabolism in wrinkled pea at four different developmental stages. In this variety it was found that branching enzyme activity was, at its highest, only 14% of that seen for the round seed. The other starch biosynthetic enzymes and phosphorylase had similar activities in the wrinkled and round seeds. These results were confirmed by Smith (1988) who also showed that the r (*rugosus*) lesion (as found in the wrinkled pea of genotype rr) was associated with the absence of one isoform of branching enzyme. Edwards et al. (1988) proposed that the reduction in starch content observed in the mutant seeds is caused indirectly by the reduction in BE activity through an effect on the starch synthase. The authors suggested that, in the absence of branching enzyme activity, the starch synthase forms an α-1 $\rightarrow$ 4-glucosyl elongated chain which is a poor glucosyl acceptor (primer) for the starch synthase substrate, ADP-Glc, therefore decreasing the rate of α-1 $\rightarrow$ 4-glucan synthesis. Indeed, in a study of rabbit muscle glycogen synthase (Carter & Smith, 1978) it was found that continual elongation of the outer chains of glycogen caused it to become an ineffective primer, thus decreasing the apparent activity of the glycogen synthase. The observation that ADP-glucose in the wrinkled pea accumulates to higher concentrations than in round or normal pea, was considered evidence that activity of the starch synthase was restricted *in vivo*. Under optimal *in vitro* conditions, in which a suitable primer like amylopectin or glycogen is added, starch synthase activity in the wrinkled pea was equivalent to that found in the wild type.

Amylose extender mutants have been found in rice and studied (Mizuno et al., 1993); the alteration of the starch structure is very similar to that reported for the maize endosperm *ae* mutants. The defect is BE 3 isozyme and BE 3 of rice is more similar in amino acid sequence to maize BE II than to BE I (Guan et al., 1994b; Mizuno et al., 1993). Thus rice BE 3 may catalyze the transfer of small chains rather than long chains.

The r locus of pea seed has been cloned by using an antibody towards one of the pea branching enzyme isoforms and screening a cDNA library (Bhattacharyya et al.,1990). It appears that the branching enzyme gene in the wrinkled pea contains an 800 bp insertion, causing it to express an inactive branching enzyme. The authors indicated that the sequence of the 2.7 kb clone showed over 50% homology to the glycogen branching enzyme of E. coli (Baecker et al., 1986) and proposed that the cDNA that they had cloned corresponded to the starch branching enzyme gene of pea seed. The glg B gene sequence has been determined for a cyanobacterium (Kiel et al., 1990) and its deduced amino acid sequence has extensive similarity to the amino acid sequence (62% identical amino acids) in the middle of the E. coli protein. It appears, therefore, that branching enzymes in nature have extensive homology regardless of the degree of branching of their products, which is higher (about 10% α-1,6 linkages) in glycogen, the storage polysaccharide in enteric bacteria and in cyanobacteria, than in the amylopectin (about 5% α-1,6 linkages) present in higher plants.

cDNA clones of genes representing different isoforms of branching enzyme from various plants have been isolated from potato tuber (Khoshnoodi et al., 1996; Koßmann et al., 1991; Poulsen & Kreiberg, 1993), maize kernel (BE I; Guan et al, 1994a; BE II, Fisher et al., 1993, 1996; Guan et al., 1994b), cassava (Salehuzzaman et al., 1992), and rice seeds (branching enzyme I, Mizuno et al., 1992; Nakamura & Yamanouchi, 1992; branching enzyme 3, Mizuno et al., 1993). The cDNA clones of the maize BE I and BE II have been overexpressed in E. coli and purified (Guan et al, 1994a,b; Preiss et al., 1994). The transgenic enzymes had the same properties as seen with the natural maize endosperm BEs with respect to specific activity and specificity toward amylose and amylopectin (Guan & Preiss, 1993).

The data available on the localization of branching enzyme within the plastid have been obtained with potato (Kram et al., 1993) using antibodies raised against potato

BE and immunogold electron microscopy. The enzyme (which would be the equivalent of the BE I isoform of maize, as discussed above) was found in the amyloplast, concentrated at the interface between stroma and starch granule, rather than throughout the stroma, as it is the case with the ADP-glucose pyrophosphorylase (Kim et al., 1989). This would explain how amylose synthesis is possible when the enzyme responsible for its formation, that is, the Wx protein, one of the granule-bound starch synthases, is capable of elongating both linear and branched glucans. The spatial separation of branching enzyme and granule-bound starch synthase, even if only partial, would allow the formation of amylose without subsequent branching by the branching enzyme. However, even if spatial separation did not exist, starch crystallization would have the same effect, that is, would prevent further branching. Morell and Preiss (unpublished) found about 5% of the total branching enzyme activity associated with the starch granule after amylase digestion. Whether this branching enzyme was similar to the soluble branching enzymes was not determined, but Sivak (Preiss & Sivak; 1996), using SDS polyacrylamide electrophoresis, found among the proteins present in maize and pea starch of maize starch, a polypeptide of about 80 KDa that reacted with antibodies raised against maize BE II. It is worth noting, however, that small amounts of BE are expected to sediment with the starch granule because of its affinity for the polysaccharide. These results have been confirmed by Mu-Forster et al. (1996).

As indicated before (Table 13-1), four regions that constitute the catalytic regions of the amylolytic enzymes are conserved in the starch branching isoenzymes of maize endosperm, rice seed, potato tuber, and the glycogen branching enzymes of *E. coli* (Jesperson et al., 1993; Svensson, 1994).

Of interest is that the eight highly conserved amino acid residues of the α-amylase family are indeed also functional in branching enzyme catalysis. Preliminary experiments (Kuriki et al., 1996), in which amino acid replacements have been done by site-directed mutagenesis, do suggest that that the conserved Asp residues of regions 2 and 4 and the Glu residue of region 3 (Table 13-1, in bold letters) are important for BE II catalysis. Their exact functions, however, are unknown and further experiments such as chemical modification and analysis of the three-dimensional structure of the BE would be needed to determine the precise functions and nature of its catalytic residues and mechanism. Arginine residues are also important as suggested by chemical modification with phenylglyoxal (Cao & Preiss, 1996), as well as histidine residues as suggested by chemical modification studies with diethyl pyrocarbonate (Funane & Preiss, unpublished experiments). It would also be interesting to determine the regions of the C-terminus and N-terminus which are dissimilar in sequence and in size in the various branching isoenzymes. It may be these areas that are important with respect to BE preference with respect to substrate (amylose-like or amylopectin-like) as well as in size of chain transferred or the extent of branching.

Initiation of Starch Synthesis

Initiation of starch synthesis *via* a glucosyl protein, as seen for glycogen synthesis, is a viable hypothesis. Tandecarz and Cardini (1978) described a system which comprises at least two enzymatic reactions in which proplastid membranes from potato tuber glucosylate a membrane protein at a serine or threonine residue, using UDP-glucose to form a glucoprotein. This product, a 38 kDa protein, is used in turn as an acceptor for a long chain of glucoses added sequentially in a α-1,4 bond using either ADP-glucose or UDP-glucose as donors. This system has been further characterized and one of the enzymes has been purified (Moreno et al.,1986, 1987); the potato enzyme catalyzes its own glycosylation (Ardila & Tandecarz, 1992). The reaction requires Mn^{2+} and thus is similar to the self-glycosylation carried out by glycogenin. However, although the enzymatic formation of the glucosyl-protein has been demonstrated in maize endosperm (Rothschild & Tandecarz, 1994), not much information is available yet on the fate of the putative glucan protein.

The Site of Starch Synthesis in Plants: The Plastid

The site of starch synthesis in leaves and other photosynthetic tissues is the chloroplast (see Okita, 1992; Preiss, 1991). All of the starch biosynthetic enzymes are present solely in the chloroplast. The starch formed during the day is degraded at night and the carbon is utilized to synthesize sucrose. The seed imports carbon and energy from the source tissues in the form of sucrose. The site of starch synthesis in the seed is the amyloplast, a nonphotosynthetic organelle. Amyloplasts resemble chloroplasts in that they are enclosed by an envelope comprising two membranes (Badenhuizen, 1969) and in that they develop from proplastids.

Sucrose has to be converted to another product before it can be taken up by the amyloplast, because the inner envelope is practically impermeable to sucrose. Although the events that lead to the flow of carbon into starch have been fairly well established for photosynthetic tissues, the situation is far less clear for storage tissues. This is because the amyloplast, the organelle in which starch is stored in sink tissues, is even more fragile than the chloroplast.

Although it is possible that many of the reactions occurring in chloroplasts also occur in the amyloplast, direct extrapolation is not possible. Indeed, the metabolism of the amyloplast, which is dependent on the cytosol for carbon and energy, is bound to differ in many ways from that of the chloroplast, which generates ATP and fixes CO_2. Information regarding the amyloplasts can be obtained in a number of ways, for example, localization of the starch biosynthetic enzymes using immunocytochemical studies, measurement of enzyme activity in isolated amyloplasts, and measurement of uptake of labeled metabolites by isolated plastids.

To study the metabolism of a plastid it is essential to isolate active plastids in good yield that are intact, and free of cytosolic contamination and other organelles. If the isolated plastids are good enough, they will provide reliable information on the enzymes present therein, what metabolites they can take up, at what rate, whether transport of a particular metabolite is passive or active, for example.

Keeling et al. (1988) circumvented the problem by using a different approach: they supplied developing wheat endosperm with glucose or fructose ^{13}C-labeled in C-1 or C-6, and then examined the extent of redistribution of ^{13}C between carbons 1 and 6 in the starch glucosyl moieties. The redistribution was lower (12%–20%) that would have been expected if carbon flow into starch had been by the C_3 pathway *via* triosephosphate isomerase. The authors suggested that hexose monophosphates (rather than triose phosphates) were more likely to be the main source of energy and carbon for the amyloplast.

It appears that the major carbon transport system for the wheat grain amyloplast does not involve triose-P and that, most likely, it involves hexose-P. Entwistle et al.

(1988) found that wheat endosperm lacks significant amyloplastic fructose-1,6-bisphosphatase, an enzyme that would be required if a triose-P/Pi transport system (like the one in the chloroplast envelope) were required for starch synthesis. In search of a transport system capable of supplying carbon for starch synthesis in the wheat endosperm, Tyson and ap Rees (1988) incubated intact amyloplasts with different ^{14}C-labeled compounds, that is, glucose, glucose-1-P, glucose-6-P, fructose-6-P, fructose-1,6-bisP, dihydroxyacetone-P, and glycerol-P. Only glucose-1-P was incorporated into starch and this incorporation was dependent on the integrity of the amyloplast. These results are consistent with the results of Keeling et al. (1988). Direct import of six carbon compounds has been reported for amyloplasts of potato, fava beans (Viola et al., 1991), maize endosperm, and other tissues (Heldt et al., 1991). Hill and Smith (1991) reported that glucose 6-phosphate was the preferred metabolite for starch synthesis by pea embryo amyloplasts and that ATP was also required. In pea roots (Borchert et al., 1989, 1993) the Pi translocator is active with dihydroxyacetone-P, 3PGA, glucose-6-P and P-enol pyruvate.

Thus, there appears to be a great diversity between the various translocators that exchange Pi with phosphorylated compounds (for review see Flügge & Heldt, 1991; Heldt et al., 1991) and it seems that the major transport system for most reserve non-photosynthetic plant systems may be at the hexose-P level and not at the triose-P level. Studies during the past 5 years are consistent with this view. Glucose-6-P translocator was found in intact cauliflower-bud plastids (Neuhaus et al., 1993a,b; Batz et al., 1993) and in maize endosperm (Neuhaus et al., 1993a). A glucose-1-P translocator for storage tissue amyloplasts of potato suspension cultured cells has also been reported (Kosegarten & Mengel, 1994). In all these cases the uptake of the hexose-P into the amyloplast was much higher than that observed for dihydroxyacetone-P. Also if 6-week-old cut spinach leaves were incubated in a 50 mM glucose solution for over 4 days, a glucose-6-P transporter was induced in the chloroplasts (Quick et al., 1995). Fruit chloroplasts assimilate very little of the CO_2 that is fixed in the leaves but appear to import carbohydrates. The intact chloroplasts of green pepper fruits (Batz et al., 1995) and tomato fruit chloroplasts and chromoplasts (Schünemann & Borchert, 1994) also have systems that translocate hexose-phosphates. In the tomato systems the solubilized envelope proteins were reconstituted into liposomes and it was found that the leaf chloroplasts translocated with Pi, dihydroxyacetone-P, and 3PGA and had low activity with glucose-6-P or glucose-1-P. However, the fruit chloroplast and chloroplast envelope proteins in addition to good translocation with the triose-phosphates had good translocation activity with P-enol pyruvate, glucose-6-P, and glucose-1-P.

The properties of the glucose-6-P translocator of the cauliflower bud and maize endosperm plastids (Neuhaus et al., 1993a,b) and of the green pepper fruit chloroplast (Batz et al., 1995) are interesting. The translocator identified in the cauliflower-bud plastid is 31.6 kDa (Neuhaus et al., 1993a,b). Translocation of glucose-6-P is measured by its incorporation of its label into the plastid starch. This incorporation was stimulated 6–40-fold by the presence of ATP and 3PGA. The authors (Neuhaus et al., 1993b) interpret the effect due to the need for ATP and 3PGA for starch synthesis. They postulate that these compounds in the cytosol act as feed-forward signals for starch synthesis and are also translocated into the plastid and utilized for synthesis of ADP-glucose. 3PGA is the allosteric activator of ADP-Glc PPase, the enzyme synthesizing ADP-glucose.

The glycolytic scheme in the amyloplast may then take on a more important function than the one it has in the chloroplast in that it would contribute to the production of amyloplastic ATP. Thus, the concentration of 3PGA could be an indicator of the ATP supply and of the availability of carbon in the amyloplast. If this were true, the regulatory effect of 3PGA on the ADP-Glc PPase from nonphotosynthetic tissues (i.e., stimulation or reversal of its inhibition by P_i) would have a physiological role similar to the one it has for the leaf enzyme.

However, two recent reports have indicated that a significant portion of the ADP-Glc PPase activity may be present in the cytosol and not solely in the amyloplast. Before these reports it was generally thought that the ADP-Glc PPase was exclusively in the amyloplast as indicated in studies with soybean cell culture (MacDonald & ap Rees, 1983), wheat endosperm (Entwhistle & ap Rees, 1988), pea embryo (Smith, 1988), and oilseed rape embryos (Kang & Rawsthorne, 1994). In barley endosperm, isolation of intact amyloplasts was achieved with efficiencies ranging from 41% to 89% (Thornbjørnsen et al., 1996a,b). The amount of total endosperm activity of plastidial enzymes, recovered in the amyloplast, starch synthase, and alkaline pyrophosphatase, ranged from 13% to 17%, while the percent of total ADP-Glc PPase activity residing in these amyloplasts was 2.5%. On this basis, the amount of ADP-Glc PPase activity residing in the amyloplast was estimated to be 15% with the remainder present in the cytosol. On the basis of antibody studies two different isoforms of the ADP-Glc PPase were detected, one mainly in the cytosol and the other mainly plastid. The authors indicate however, that there is an excess of ADP-Glc PPase activity in the amyloplast to accommodate the starch synthetic rate and the function of the cytosolic ADP-Glc PPase to them at present is obscure. Moreover, the kinetic properties of the plastid ADP-Glc PPase have not been characterized.

In another report (Denyer et al., 1996), preparations enriched in maize endosperm plastids contained 24%–47% of the total activity of the plastid marker enzymes, starch synthase and alkaline pyrophosphatase, while containing only 3% of the total ADP-Glc PPase activity. On this basis it was estimated that more than 95% of the ADP-Glc PPase activity was nonplastid. Using antibodies prepared against the bt2 subunit of the maize endosperm ADP-Glc PPase it was demonstrated that most of the bt2 protein was confined to the supernatant, but was also present in the plastid. In bt 2 mutant kernels, the cytosolic bt2 protein was not detected but there was a plastidial form of ADP-Glc PPase observed. These data are somewhat different than what has been obtained by Miller and Chourey (1995) who detected the bt2 protein by immunogold labeling in the amyloplast. However, they could not eliminate the possibility that there was also a cytosolic ADP-Glc PPase.

In maize endosperm, if the above data are not artifactual, there may be many different possible routes for ADP-Glc synthesis. The carbon may be translocated into the plastid *via* a glucose-6-P translocator and then converted to ADP-Glc *via* plastidial phosphoglucomutase and ADP-Glc PPase catalysis (Neuhaus et al., 1993a). The results of Denyer et al. (1996) indicate that ADP-Glc synthesis in the amyloplast can be catalyzed *via* a plastidial ADP-Glc PPase, either containing the bt2 protein as one of its subunits, or by another ADP-Glc PPase isozyme. Since the major activity of ADP-Glc synthesis resides in the cytosol (Denyer et al., 1996), ADP-Glc synthesized in the cytosol must then be translocated into the plastid. An ADP-Glc transporter would logically be required. Thus far, no protein having those properties has been identified. ADP-Glc uptake by the *Acer pseudoplatanus* amyloplasts for starch synthesis has been reported (Pozueta-Romero et al., 1991). However, as shown by Borchert et al. (1993) and by Batz et al. (1994), the ADP-Glc transport may not be physiologically relevant. *In vitro*, ADP-Glc may be translocated *via* the ATP/ADP translocator, since both ADP and ATP, at concentrations lower than their physiological concentrations, effectively inhibit ADP-Glc uptake in the pea-root and cauliflower-bud amyloplasts. The hypothetical ADP-Glc transporter, if present, remains to be characterized.

There is a possibility that the *bt* 1 gene product maybe the ADP-Glc transporter. The *bt* 1 gene encodes a plastidial membrane-associated protein (Cao et al., 1995a,b; Sullivan & Kaneko, 1995) whose deduced amino acid sequence shows similarity to known adenine nucleotide transporters (Sullivan et al., 1991). The *bt* 1 mutant is starch deficient and shows a high level of ADP-Glc concentration in the endosperm compared to the normal endosperm (Shannon et al., 1996). Thus, this interesting Bt 1 protein remains to be studied in order to determine its exact function.

Much remains to be done to characterize the plastidial barley and maize endosperm ADP-Glc PPases with respect to their functions in starch synthesis and identification of the more important route for hexose incorporation into starch.

A Hypothesis Assigning Specific Roles for the Starch Synthase, Branching Isozymes, and Debranching Enzyme in *In Vivo* Synthesis of Amylopectin

The "basics" of starch biosynthesis, both in the leaves and in storage tissues, are described above. The regulation of the first enzyme in the pathway, the ADP-glucose pyrophosphorylase, is described below. As indicated, it has been possible to increase the starch content of potato tuber and tomato fruit; this is the first time that an increase in the accumulation of a useful natural product has been achieved by genetic transformation (Stark et al., 1992; Preiss et al., 1994). Many details, however, are still missing from the general picture so it is not possible to give a precise description of how the synthesis of the starch granule starts, how amylopectin and amylose are made, or why starch granules from different species differ in their size, number per cell, and composition.

The cluster model of amylopectin structure as postulated by Hizukuri (1986, 1995) is the currently accepted structure. A feature of the model is the clustering of the α-1,6 linkage branch points in certain regions of the amylopectin and the occurrence of B chains of varying sizes: B1, approximately 19 glucose units long, B2 with 41 units, B3 with 69 units, and B4, approximately 104–115 glucose units long. The number of B3 and B4 chains are few as compared to the number of B2 and B3 chains. The B1 chains extend only into one area of clusters while the B2, B3, and B4 chains extend into two, three, and four cluster areas of α-1,6 branch linkage areas, respectively. These areas are separated by 39–44 glucose units (Hizukuri, 1986).

What roles would the starch synthases (SS) and branching enzyme isozymes play in the formation of the crystalline starch granule and amylopectin structures? How is amylose formed? Why are starch granules from different species different in size and in the number per cell? These differences most probably are related to the SS specificities in chain elongation and in the size transfer of glucose chain units by BE and where the α-1,6 bond is formed after transfer. As indicated above, in the discussion of BEs (Guan & Preiss, 1993; Guan et al., 1995; Takeda et al., 1993;), BE I transfers long chains (DP 40–>100) while BE II only transfers shorter chains (DP 6–14). Amylose is the preferred substrate for BE I, while amylopectin is the preferred substrate for BE II. Thus, BE I may be more involved in synthesis of the interior B chains while BE II is involved in the synthesis of exterior A and B1 chains. These modes of action appear to occur *in vivo*, as BE I did transfer longer chains than BE II, and BE II transferred shorter chains when the isozymes were expressed in *E. coli* and maize (Guan et al., 1995).

Ball and associates have isolated various mutants of *Chlamydomonas* deficient in starch synthase activities a granule bound starch synthase (GBSS) deficient mutant, a SSS II-deficient mutant (Fontaine et al., 1993), and a double mutant deficient both in GBSS and in SSS II (Maddelein et al., 1994). As indicated in the discussion of the starch synthases, the SSS II mutant had only 20%–40% of the wild-type starch content and the amylose fraction of the starch increased from 25%–55%. This mutant also had a modified amylopectin with an increased amount of short chains of 2–7 DP, and a decrease of intermediate chains of 8–60 DP. This suggests that the SSS II is involved in the synthesis or maintenance of the intermediate chains (mainly B chains) in amylopectin. The higher amylose content could be explained because of the failure of the SSS II mutant to make extended chains.

Studies prior to those on the GBSS of *Chlamydomonas* showed that a deficiency of GBSS activity in plants such as maize, rice, barley, sorghum endosperms, and potato tuber resulted in a loss of the amylose fraction in the starch granule with little effect on the amylopectin fraction. Thus, GBSS was considered to play a major role in amylose synthesis. Moreover, the role of GBSS in amylose synthesis in potato plants was also demonstrated by transforming them to produce antisense RNA from a gene construct containing GBSS cDNA in reverse orientation. Total suppression of GBSS activity gives rise to tubers containing amylose-free starch (Visser et al., 1991). Once again, the amylopectin fraction did not appear to be affected.

The double mutants defective in SSS II and a GBSS in *C. reinhardtii* (Maddelein et al., 1994), however, had a starch content of only 2%–16% that of the wild type. The severity of the GBSS defect of the double mutant dictated the amount of starch present in the double mutant with an almost null mutant having very little starch. The authors suggested that GBSS is very important for synthesis of not only amylose, but also the internal structure of the amylopectin and the effect of GBSS deficiency is worsened by the diminished SSS II activity. These studies, using the *Chlamydomonas* mutant, provide evidence for the involvement of the GBSS not only in amylose but also in amylopectin synthesis, and suggest that a function for SSS II is in synthesis of the intermediate B branch chains in amylopectin.

From the above data on the BE and SS isozymes a possible route for amylopectin and amylose can be proposed and is shown in Figure 13-3. A reaction with the potential of being the initiating reaction for synthesis has been observed in potato tuber (Moreno et al., 1987) and maize endosperm (Rothschild & Tandecarz, 1994) and a transfer of glucose from UDP-Glc to serine or threonine residues has been observed. The resulting glucosylated 38 kDa protein can serve as a primer for the synthesis of starch *via* the starch synthase reactions. Whether there is an acceptor protein that could be glucosylated by ADP-Glc has not been demonstrated; the acceptor protein and reaction has not been characterized as well as the other starch biosynthetic enzymes.

After formation of the unbranched maltodextrin–protein primer of undetermined size, high rates of polysaccharide formation may occur at the surface of the develop-

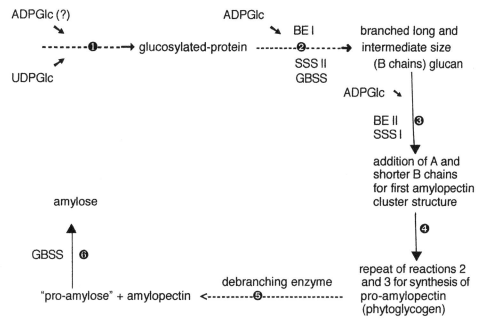

Figure 13–3. Proposed synthesis of amylose and amylopectin. Phase 1: initiation of α-glucan synthesis *via* synthesis of a glucoprotein primer for starch synthases. Phase 2: formation of the internal cluster structure of the ultimate amylopectin product by GBSS, SSS II, and BE I. Phase 3: formation of the external cluster stucture (exterior A and B chains) by SSS I and BE II. Phase 4: continuous repeat of phase 2 and 3 reactions to complete formation of a highly branched pro-amylopectin (phytoglycogen). Phase 5: debranching of pro-amylopectin to form amylopectin and the amylose primer, "pro-amylose." Phase 6: formation of amylose by GBSS-mediated elongation.

ing starch granule, where GBSS, SSS II, and branching enzyme I interact with the glucosylated protein primer to form a branched α-glucan containing both long and intermediate chains.

The postulation of phase 2 in Figure 13-3 is based on the studies of the polysaccharide structures observed in the *Chlamydomonas* SSS II and GBSS mutants as well as the *ae* mutants of rice (Mizuno et al., 1993) and maize (Boyer & Preiss, 1981), which are defective in BE II. BE II-deficient mutants have altered oligosaccharides with fewer branches and longer branched chains. In phase 3, SSS I and BE II are responsible for the synthesis of the A and exterior B chains to complete the first cluster region in the glucan. Continued synthesis in phase 4 is essentially a repeat of phases 2 and 3 to synthesize a highly branched α-glucan termed proamylopectin or phytoglycogen. This highly α-branched glucan is water soluble and noncrystalline. In phase 5, a debranching enzyme debranches the pro-amylopectin to form amylopectin which can crystallize. In phase 6, the chains liberated by debranching action of the pro-amylopectin (phytoglycogen) are used as primers by GBSS to form amylose. Amylose synthesis may occur only inside the starch granule and only GBSS would be involved because it may be the only starch synthase present at the site of amylose synthesis. Inside the granule, branching enzyme activity is quite restricted and, therefore, the amylose would be only slightly branched. Possibly the slight branching observed in the amylose fraction had occurred previously before the debranching phase, 5. Debranching of the proamylopectin may have liberated primer for GBSS that had some branch chains.

The reason for postulating a water-soluble pro-amylopectin is based on the existence of the *sugary* 1 mutation of maize endosperm, which contains reduced amounts of amylopectin and starch granules. The mutant accumulates about 35% of its dry weight as phytoglycogen, a highly branched water-soluble polysaccharide (Pan & Nelson, 1984). The *sugary* 1 mutation was shown to be deficient in debranching enzyme activity. The evidence that the *sugary* 1 mutation affects the structural gene for a debranching enzyme is supported by the isolation of a cDNA of the *su* 1 gene. Its deduced amino acid sequence is most similar to a bacterial isoamylase (James et al., 1995). It remains to be shown whether the *su* 1 gene product debranching enzyme activity is actually an isoamylase, a pullulanase, or an R enzyme (Hizukuri, 1995). Moreover, the specificity of the reaction needs to be studied with respect to the factors that determine which α-1,6 linkages are cleaved and which remain resistant to debranching action. It is quite possible that the crowding of the α-1,6 linkages in the cluster region causes some steric difficulties for the debranching of the linkages in the cluster region, but at present this is only a hypothesis.

These reactions do not have to occur in perfect sequence and the phases may have some overlap, for example, phases 2, 3, and 4 may overlap, and possibly even 5 and 6. However, the present evidence, such as intermediate products formed by starch mutants of *Chlamydomonas* and of higher plants, support the sequence of reactions shown in Figure 13-3 for amylopectin and amylose biosynthesis. Further experiments are certainly required to test the proposed scheme in Figure 13-3.

Regulation of Bacterial Glycogen and Starch Synthesis

The main regulatory site of bacterial glycogen synthesis and plant and algal starch synthesis is different from that found for mammalian glycogen synthesis. As indicated, the regulatory enzyme in mammalian systems is glycogen synthase. The differences for regulatory enzymes sites in the various systems are due to the specificity for ADP-glucose, the glucosyl donor in the bacterial and plant α-glucan systems and for UDP-glucose, the glucosyl donor for mammalian glycogen synthesis. The latter is also utilized for the synthesis of other sugar nucleotides, mainly UDP-galactose and UDP-glucuronate, precursors for the synthesis of several cellular constituents. Thus, the first unique reaction for mammalian glycogen synthesis is the glycogen synthase step, where both allosteric control and hormonal-mediated control is exerted. In bacteria and in plants, however, the only physiological function for ADP-glucose is as a donor of glucose for α-1,4-glucosyl linkages. It would be advantageous to conserve the ATP utilized for synthesis of the sugar nucleotide for the prokaryote and plant cell to regulate glucan synthesis at the level of ADP-glucose formation.

Over 50 ADP-Glc PPases (mainly bacterial but also plant) have been studied with respect to regulatory properties. In almost all cases, glycolytic intermediates activate ADP-Glc synthesis while AMP, ADP, or P_i are inhibitors. Glycolytic intermediates in the cell can be considered indicators of carbon excess; therefore, under conditions of limited growth with excess carbon in the media, accumulation of glycolytic intermediates would be signals for the activation of ADP-Glc synthesis. Thus, the enzyme seems to be modulated by the availability of ATP in the cell and the presence of glycolytic intermediates.

Activators and Inhibitors of ADP-Glc PPase

Based on the observed differences in specificity of metabolic intermediates activating the ADP-glucose pyrophosphorylases studied thus far, the enzyme may be classified into seven groups. These groups are listed in Table 13-6. The ADP-Glc PPases in the first group are isolated from bacteria of the genus *Rhodospirillum (R. rubrum, R. molis-chianum, R. tenue, R. fulvum,* and *R. photometricum)* and *Rhodocyclus purpureus* and are activated solely by pyruvate (Furlong & Preiss, 1969; Preiss & Greenberg, 1981; Preiss & Levi, 1978; Yung & Preiss, 1981).

A second group of prokaryotes that are capable of oxygenic photosynthesis are the cyanobacteria having ADP-Glc PPases activated by 3-phosphoglycerate (3PGA), the primary CO_2 fixation product of photosynthesis (Charng et al., 1992; Iglesias et al., 1991; Levi & Preiss, 1976). 3PGA is also the primary activator of the green algae (Iglesias et al.,1994; Sanwal & Preiss, 1967) and higher plant ADP-Glc PPases (Preiss, 1982, 1988, 1991; Preiss & Sivak, 1996). Thus the initial product of photosynthesis serves as an allosteric activator for the synthesis of a reserve product, glycogen, or starch. The specificity of the activation is the same whether the enzyme is from a plant incorporating CO_2 *via* the C3 pathway or the C4 pathway. The ADP-glucose pyrophosphorylases of *Chlorella pyrenoidosa, Chlorella vulgaris, Scenedesmus obliquus,* and *Chlamydomonas reinhardii* and of the cyanobacteria are also activated by 3-phosphoglycerate and to a lesser extent than by other glycolytic intermediates and inhibited by Pi. 3-Phosphoglycerate also increases the apparent affinity of all the

Table 13–6 Groups of Bacterial, Algal, and Plant ADP-Glucose Pyrophosphorylases Having Different Activator Specificities

Bacteria or Plant or Algae	Activator(s)
Rhodospirillum spp. (*R. fulvum, R. molischianum, R. photometricum, R. rubrum, R. tenue*), *Rhodocyclus purpureus*	Pyruvate
Aphanocapsa 6308, Synechococcus 6301, Synechocystis PCC6803, *Anabaena* PCC 7120, *Chlorella pyrenoidosa, Chlorella vulgaris, Chlamydomonas reinhardtii, Scenedesmus obliquus,* Plant tissues (leaf or reserve tissue)	3-P-glycerate
Agrobacterium tumefaciens, Arthrobacter viscosus, Chlorobium limicola, Chromatium vinosum, Rhodobacter spp. (*R. gelatinosa, R. blastica, R. capsulata, R. palustris), Rhodomicrobium vannielii*	Pyruvate Fructose-6-P
Rhodobacter spp. (*R. gelatinosa, R. globiformis, R. spheroides*)	Pyruvate Fructose-6-P Fructose 1,6-bis-P
Aeromonas hydrophilia, Micrococcus luteus, Mycobacterium smegmatis, Rhodopseudomonas viridis, Thermus caldophilus Gk-24	Fructose-6-P Fructose-1,6 bis-P
Citrobacter freundii, Edwardsiella tarda, Enterobacter aerogenes, Enterobacter cloacea, Escherichia aurescens, Escherichia coli, Klebsiella pneumoniae, Salmonella enteriditis, Salmonella typhimurium, Shigella dysenteriae	Fructose-1,6-bis-P Pyridoxal-5-P NADPH
Clostridium pasteurianum, Enterobacter hafniae, Serratia spp. (*S. liquifaceans, S. marcescens*)	None

substrates for the spinach leaf enzyme from 2–13-fold. ADP-glucose synthesis catalyzed by the spinach leaf enzyme is inhibited 50% by 22 μM Pi in the absence of activator at pH 7.5 (Ghosh & Preiss, 1966). In the presence of I mM 3-phosphoglycerate, 50% inhibition required 1.3 mM phosphate. Thus, the activator decreased sensitivity to Pi inhibition about 450-fold. However, Pi at 0.5 mM increased the concentration of 3-phosphoglycerate needed for activation.

Another group of ADP-Glc PPases having a different allosteric activator specificity is found in phototrophic bacteria belonging to the genus previously called *Rhodopseudomonas*, now called *Rhodobacter* and to *Rhodomicrobium* (*R. acidophila, R. capsulata, R. palustris*, and *R. vanniellii*) or to the families Chromatiaceae and Chlorobiaceae. They have as activators both fructose-6-phosphate and pyruvate (Eidels et al., 1970; Preiss & Levi, 1978). Included in this group are *Agrobacterium tumefaciens* (Eidels et al., 1970) and *Arthrobacter viscosus* (Shen & Preiss, 1964, 1965, 1966).

ADP-glucose PPases isolated from *R. sphaeroides, R. gelatinosa*, and *R. globiformis* (Greenberg et al., 1983; Preiss et al., 1980; Yung & Preiss, 1982) constitute a fourth group and are activated by fructose-1,6-diphosphate in addition to fructose-6-phosphate and pyruvate.

The ADP-glucose PPase from the anaerobic photosynthetic *Rhodobacter viridis* is unusual in that it is not activated by pyruvate but is activated by fructose 6-phosphate and fructose-1,6-diphosphate. ADP-glucose pyrophosphorylases with this activator specificity are the fifth group and are also found in nonphotosynthetic organisms such as *Aeromonas*, gram-negative facultative anaerobes, or from the gram-positive aerobic organisms *Micrococcus luteus* and *Mycobacterium smegmatis* (Lapp & Elbein, 1972; Yung et al., 1984).

The sixth group of ADP-Glc PPases are from the enteric bacteria (*Citrobacter freundii, Edwardsiella tarda, Escherichia coli, Escherichia aurescens, Enterobacter aerogenes, Enterobacter cloacae, Klebsiella pneumoniae, Salmonella enteritidis, Salmonella typhimurium*, and *Shigella dysenteriae*) and are activated by fructose-1,6-bisphosphate, NADPH, and pyridoxal phosphate (Creuzat-Sigal, 1972; Gentner et al., 1969; Govons et al., 1973; Preiss et al., 1966; Ribereau-Gayon et al., 1971; Steiner & Preiss, 1977).

However, there are some enteric organisms from the genus *Serratia* (*S. liquefaciens* and *S. marcescens*) and from *Enterobacter hafniae* containing an enzyme not activated by any of the metabolites tested (Preiss et al., 1976). Another ADP-glucose pyrophosphorylase with no apparent activator is that isolated from *Clostridium pasteurianum* (Robson et al., 1974). These enzymes may be classified as in the seventh group.

Overlapping Specificity of the ADP-Glc Pyrophosphorylase Activator Binding Site

In the seven classes the dominant activators are pyruvate, fructose-6-phosphate, 3-PGA, and fructose-1,6-bisphosphate. All the ADP-Glc PPases isolated from the photosynthetic anaerobic bacteria (except for *R. viridis*) are activated by pyruvate. A number are also activated by fructose-6-phosphate, and a few are activated by a third metabolite, fructose-1,6-bisphosphate. Fructose-1,6-bisphosphate is an activator of the enzyme from enteric organisms, as well as those from the Aeromonads, *M. luteus*, and *M. smegmatis*. However, fructose-6-phosphate, an effective activator of the ADP-glucose pyrophosphorylases in the three last-mentioned organisms, is not an activator for the enteric enzymes. 3-Phosphoglycerate, a highly effective activator for the enzyme from the cyanobacteria, green algae, and plant tissues, is a poor activator for the enteric enzymes. Conversely, fructose-1,6-bisphosphate is able to activate the plant leaf enzymes, but less effectively then 3-phosphoglycerate. Whereas 28 μM fructose-1,6-bisphosphate is required for 50% of maximal stimulation of the spinach leaf enzyme at pH 8.5, only 10 μM 3PGA is required (Ghosh & Preiss, 1966). Moreover, the maximum stimulation of V_{max} effected by fructose-1,6-bisphosphate (16-fold) is considerably less than that observed for 3-phosphoglycerate (58-fold) at pH 8.5. In contrast, fructose-1,6-bisphosphate stimulates 30-fold the rate of ADP-glucose synthesis catalyzed by *E. coli* B ADP-Glc PPase, while the same concentration of 3-phosphoglycerate gives only a 1.5-fold stimulation.

This overlapping of specificity for the activators in the various ADP-Glc PPase classes suggests that the activator sites for the different groups are similar or related. Indeed, if one looks at the deduced amino acid sequences of many of the ADP-Glc PPases (Preiss, 1996a; Smith-White & Preiss, 1992) one can see considerable similarity of amino acid sequences, particularly at the allosteric binding sites as well as the substrate binding sites. One may, therefore, speculate that mutation of the gene at the activator binding site of the ADP-Glc PPase occurred during evolution to modify the activator specificity. The likely pressure for change may have been the coordination or compatibility of the metabolite activator with the major carbon assimilation and dissimilation pathways of the organism.

Metabolites associated with energy metabolism, AMP, ADP, or inorganic phosphate, are inhibitors of the ADP-Glc PPases. The enteric ADP-glucose pyrophorphorylases (Gentner & Preiss, 1968; Preiss et al., 1966; Ribereau-Gayon et al., 1971), including those found in the genus *Serratia* (Preiss et al., 1976) and *Enterobacter hafniae* (Yung et al., 1984), are very sensitive to AMP inhibition. The plant, algal, and cyanobacterial enzymes are highly sensitive to inorganic phosphate (Charng et al., 1994; Ghosh & Preiss, 1966; Iglesias et al., 1991, 1994; Preiss & Levi, 1978; Preiss & Sivak, 1996; Sanwal & Preiss, 1967; Sanwal et al., 1968). Other ADP-glucose pyrophosphorylases are sensitive to either P_i, ADP, or AMP (Preiss, 1978). However, very little or negligible inhibition by the above three compounds at concentrations of 5 mM or less are observed for the enzyme isolated from the *Aeromonas* (Yung et al., 1984), M. *smegmatis* (Lapp & Elbein, 1972), *R. ruhrum* (Furlong & Preiss, 1969), and *R. molischianum* (Preiss & Greenberg, 1981). Thus, in most cases, ADP-Glc and glycogen synthesis proceeds only with the availability of ATP or if there is a high energy state in the cell.

The peptide domains involved in the inhibitor-binding sites remain to be identified. However, some data obtained with the *E. coli* enzyme (Haugen & Preiss, 1979; Kumar et al., 1988; Larsen & Preiss, 1986; Larsen et al., 1986; Parsons & Preiss, 1978a) strongly suggest that the inhibitor-binding site overlaps with both the activator and substrate ATP-binding sites.

Effect of Activators on Inhibition and Substrate Kinetics

Kinetic studies with various ADP-glucose pyrophosphorylases show that the presence of activator in reaction mixtures usually lowers the concentration of the substrates (ATP, glucose-1-phosphate, pyrophosphate, ADP-glucose, and the cationic activator Mg^{+2} [or Mn^{+2}]) required to achieve one-half of maximal velocity (Km or $S_{0.5}$) about

2–15-fold. The apparent affinity of the enzyme for the substrate is thus increased in the presence of activator. The activator also increases maximal velocity or k_{cat}, 2–60-fold, depending on the pH and the particular ADP-glucose pyrophosphorylase studied. The prime function of the allosteric activator, however, may be to reverse the sensitivity of the enzyme to inhibition by the inhibitors Pi, AMP, or ADP. The inhibitors are usually noncompetitive with the substrates. For many ADP-glucose pyrophosphorylases, the activator at relatively high concentrations can completely reverse the inhibition caused by AMP, Pi, or ADP. A well-studied system has been the *E. coli* B enzyme, where fructose-1,6-bisphosphate modulates the sensitivity to AMP inhibition (Gentner & Preiss, 1967, 1968). The $S_{0.5}$ (concentration giving 50% inhibition) for 5'-AMP is about 70 μM at 1.7 mM fructose-1,6-bisphosphate. With lower concentrations of activator, however, lower concentrations of AMP give 50% inhibition. At 60 μM fructose-1,6-bisphosphate only 3.4 μM AMP is needed to achieve 50% inhibition. Table 13-7 shows the interaction of the activator fructose-6-phosphate and fructose-1,6-bisP for the *R. viridis* ADP-Glc PPase, with the inhibitor inorganic phosphate. Fructose-6-phosphate stimulates the reaction rate 6-fold and fructose1,6-bisP, 3.5-fold (Greenberg et al., 1983); half-maximal stimulation ($A_{0.5}$) occurs at 180 μM for both activators. At low concentrations of activator (0.10 mM) there is about 90% inhibition by inorganic phosphate at 1 mM. However, this inhibition is almost completely overcome by increasing concentrations of the activator, fructose-6-phosphate. At 1 mM fructose-6-P, the inhibition is only 20%. Inorganic phosphate has two effects on the enzyme: it increases the $A_{0.5}$ value of fructose-1,6-bisP (from 180 μM to 550 μM in the presence of 1 mM Pi) and increases the sigmoidicity of the activation curve. The Hill $\tilde{n}$ increases from 1.0 in the absence of Pi to 2.0 in the presence of 1 mM Pi.

Similar interactions can be seen in Figure 13-4 with the potato tuber ADP-Glc PPase, which also show that large effects can be seen on the rate of ADP-glucose synthesis with relatively small changes in the 3PGA, the activator, and Pi concentrations, particularly at low concentrations of 3PGA where the activation is minimal in the presence of Pi. At 1.2 mM Pi and 0.2 mM 3PGA, ADP-Glc synthesis is inhibited over 95%. However, if the Pi concentration decreases 33% to 0.8 mM and the 3PGA con-

Table 13–7 Inorganic Phosphate Inhibition of the *Rhodobacter viridis* ADP-Glc PPase Achieved with Varying Concentrations of the Activator, Fructose-6-P or Fructose-1,6-bis-P

Activator	Inhibitor (conc.)	Relative V_0	$I_{0.5}$ (mM)	$A_{0.5}$ (mM)	$\tilde{n}$
None		0.19	0.36		1.7
Fructose-6-P, 0.1 mM		0.52	0.51		2.1
Fructose-6-P, 0.25 mM		0.8	0.88		2.7
Fructose-6-P, 1.0 mM		1.0	1.4		3.2
Fructose-1,6-P_2, 0.1 mM		0.45	0.54		2.4
Fructose-1,6-P_2, 0.25 mM		0.70	0.79		2.8
Fructose-1,6-P_2,	None			0.18	1.0
Fructose-1,6-P_2,	Pi, 0.5 mM			0.30	1.4
Fructose-1,6-P_2,	Pi, 1 mM			0.55	1.9

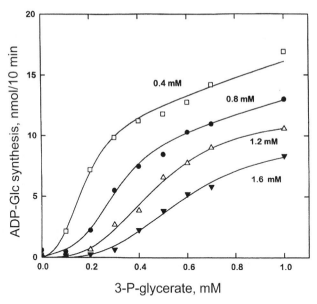

Figure 13–4. The effect of Pi and 3PGA on the rate of ADP-Glc synthesis catalyzed by potato tuber ADP-Glc PPase. 3PGA curve done in the presence of 0.4 mM Pi (■); 0.8 mM Pi (●); 1.2 mM Pi (△); and 1.6 mM Pi (▼).

centration increases 50% to 0.3mM there is an 8.5-fold increase in the rate of ADP-Glc synthesis. Conversely, at 0.4 mM 3PGA and 0.8 mM Pi. The rate of ADP-Glc synthesis is 7.5 nmol per 10 min. This is reduced to 2.2 nmol per 10 min if only the 3PGA concentration decreases 50% to 0.2 mM. If the Pi concentration increases as well to 1.2 mM (a 50% increase) then the synthetic rate is reduced to 0.65 nmol, which is a 91% reduction of ADP-Glc synthesis. The small changes in the effector concentrations give large effects in the synthetic rate due to the sigmoidal nature of the curves, particularly at the low concentrations of 3PGA. Evidence is presented below that indicates strongly that the ratio of activator/inhibitor modulates the activity of ADP-glucose pyrophosphorylase *in vivo* in bacteria and in plants, thereby regulating the synthesis of α-1,4 glucans in these systems.

In Vivo Evidence in Support of the ADP-Glucose Pyrophosphorylase Activator–Inhibitor Interaction Regulating α-1,4 Glucan Synthesis

Plant Systems. The leaf ADP-glucose pyrophosphorylase is highly sensitive to 3-phosphoglycerate, the primary CO_2 fixation product of photosynthesis, and to inorganic phosphate. Thus, it has been suggested that these compounds play a significant role *in vivo* in regulating starch biosynthesis in algae, in higher plants, and in the regulation of glycogen synthesis in cyanobacteria (Preiss, 1978, 1982, 1988, 1989, 1991, 1992, 1996a,b,c, 1997). For example, inorganic phosphate levels in leaves may decrease during photosynthesis because of photophosphorylation, and glycolytic intermediates increase in the chloroplast in the light. This situation would thus contribute to conditions necessary for optimal starch or glycogen synthesis *via* the increased rate of formation of ADP-glucose. In the light, the levels of ATP and reduced pyridine nucleotides are also increased, leading to the formation of sugar phosphates from 3-phosphoglycerate. In the dark, there is an increase in phosphate concentration with concomitant decreases in the levels of 3-phosphoglycerate, ATP, and reduced pyridine nucleotides. This would lead to inhibition of ADP-glucose synthesis and, therefore, of starch or glycogen synthesis.

Data correlating altered accumulation of starch or altered rates of starch synthesis due to changes of either cellular Pi or 3PGA levels have been discussed in previous reviews (Preiss, 1988, 1991; Preiss & Levi, 1980) and suggest that *in vivo*, 3PGA and Pi levels do affect starch synthetic rates *via* modulation of ADP-Glc PPase activity.

Recent evidence has also been obtained which suggests strongly that the regulatory effects seen for the plant and algal ADP-Glc PPase are important *in vivo* for starch synthesis. Kacser-Burns control analysis methods (Kacser & Burns, 1973; Kacser, 1987) vary enzyme activity, either by using mutants deficient in that enzyme or by varying the physiological conditions and correlating the effect of these changes on the rate of a metabolic process (e.g., starch synthesis). If the enzyme activity is rate limiting or important in controlling the metabolic process, then a large effect on that process should be seen. Conversely, if there is no or little effect, then the enzyme level of activity is considered not to be rate limiting for the metabolic process being measured. The effect is measured as a flux coefficient ratio. If the ratio correlates well for variation of enzyme activity with change in rate of the process measured, a correlation ratio close to one should be observed.

Analysis on photosynthate partitioning was done on starch synthesis in *Arabidopsis thaliana* (Neuhaus & Stitt, 1990; Neuhaus et al., 1989). The results showed that leaf ADP-Glc PPase is a major site of regulation for starch synthesis (Neuhaus & Stitt, 1990) and that regulation of the enzyme by 3PGA is an important determinant of the rate of starch synthesis *in vivo* (Neuhaus et al., 1989). *A. thaliana* mutant strains containing only 7% of the normal activity of ADPGlc PPase and a hybrid strain between the mutant and normal strain having 50%, exhibited 90% and 39% reduction in the starch synthetic rate respectively, at a high level of light, as compared to the wild type (Neuhaus & Stitt, 1990). Thus, there is fairly good correlation between the activity of the ADP-Glc PPase and the rate of starch synthesis. The flux-control coefficient was determined to be 0.64.

Despite the fairly high value seen for the ADP-Glc PPase flux control coefficient it is quite possible that it may be underestimated due to the allosteric properties of the enzyme. In flux-control analysis the maximal enzyme activity is measured. In the case of an allosteric enzyme the potential maximal enzyme activity may not be as critical as the allosteric effector concentrations that establish the enzyme activity. Therefore, for an allosteric enzyme a valid flux control coefficient based solely on potential maximum activity may not be observed. With ADP-Glc PPases, activation by 3PGA can be anywhere from 10- to 100-fold. Moreover, inhibition by the allosteric inhibitor Pi and variations in the [3PGA]/[Pi] ratio could cause greater fluctuations in the potential maximal activity. Thus, a flux coefficient control value analysis based on only the maximal activites of the *Arabidopsis thaliana* mutant enzymes and normal ADP-Glc PPases may underestimate the regulatory potential of the ADP-Glc PPase reaction.

In the experiments utilizing a *Clarkia xantiana* mutant deficient in leaf cytosolic phosphoglucoseisomerase (i.e., having only 18% of the activity seen in the wild type), lower sucrose synthetic rates and increased starch synthesis rates were observed (Neuhaus et al., 1989). The chloroplast 3PGA concentration increased about twofold, suggesting that the increase in starch synthetic rate measured in the mutant deficient in cytosolic phosphoglucoseisomerase was due to activation of the ADPGlc PPase by the increased 3PGA concentration and the 3PGA/Pi ratio.

Important evidence indicating that the *in vitro* activation of the ADP-Glc PPase is truly functional *in vivo* comes from isolation of a class of mutants in which the mutation affects the allosteric properties of the ADPGlc PPase directly. Such mutants were found easily for the bacteria *E. coli* and *Salmonella typhimurium* (Preiss, 1984;1996b; Preiss & Romeo, 1994). Similar mutants have been found for *Chlamydomonas (C.) rheinhardtii* and for maize endosperm. A significant finding was made by Ball et al. (1991) who isolated a starch-deficient mutant of *C. reinhardtii* in which the defect was shown to be in the ADP-Glc PPase; this mutant could not be activated effectively by 3PGA. The inhibition by Pi was similar to the wild type (Iglesias & Preiss, unpublished results). The starch deficiency was observed in the mutant whether the organism was grown photoautotrophically with CO_2 or in the dark with acetate as the carbon source. Thus, the allosteric mechanism seems to be operative for photosynthetic or nonphotosynthetic starch biosynthesis.

Another putative ADP-Glc PPase allosteric mutant from maize endosperm, which has 15% more dry weight (in addition to starch) than the normal endosperm, has been isolated and described by Giroux et al. (1996). The mutant allosteric ADP-Glc PPase was less sensitive to Pi inhibition than the normal enzyme.

The *Chlamydomonas* starch-deficient mutant and higher dry-weight maize endosperm mutant studies strongly suggest that the *in vitro* regulatory effects observed with the photosynthetic and nonphotosynthetic plant ADP-Glc PPases are highly functional *in vivo* and that ADP-Glc synthesis is rate-limiting for starch synthesis.

Thus, data continue to accumulate showing the importance of the plant ADP-Glc PPase in regulating starch synthesis, and that the allosteric effectors 3PGA and Pi are important *in vivo*, in photosynthetic as well as in nonphotosynthetic starch synthesis.

Bacterial Systems. Mutants of *E. coli* B (Govons et al., 1973; Preiss, 1973, 1978; Preiss & Walsh, 1981; Preiss et al., 1976), *E. coli* K12 (Cattaneo et al., 1969; Creuzat-Sigal, 1972), and *Salmonella typhimurium* LT-2 (Steiner & Preiss, 1977) containing normal activity levels of the glycogen biosynthetic enzymes but still affected in their ability to accumulate glycogen, have been isolated. They have been shown to contain ADP-glucose pyrophosphorylases altered in their regulatory properties.

Table 13-8 compares the maximum amounts of glycogen accumulated by four glycogen-excess mutants of *E.coli* B with the parent strain and the maximum amounts of glycogen accumulated by two *S. typhimurium* LT-2 glycogen-excess mutants with their parent strain. In minimal media having excess glucose, the rate of glycogen accumulated in SG5, CL1136, and 618 is 2-, 3.5-, and 3.7-fold greater, respectively, than that found in *E. coli* B (Preiss & Walsh, 1981). The amounts of glycogen synthesis are about 2-, 3.5-, and 3.5-fold greater for the mutants SG5, CL1136, and 618. The *S. typhimurium* LT-2 mutants JP51 and JP23 accumulate 67% and 25% more glycogen than the parent strain.

The levels of activity of the glycogen biosynthetic enzymes in the mutants and in the parent strains are equivalent and thus cannot account for the increased rate of accumulation of glycogen present in the mutants. Furthermore, the *E. coli* B mutant ADP-

Table 13-8 Comparison of Aallosteric Kinetic Constants of E. coli and S. typhimurium LT2 Allosteric Mutant ADP-Glc PPases and Their Glycogen Accumulation Amounts

Strain	Maximal Glycogen Accumulation[a] mg/gram cell	Fructose 1,6-bisP[b] ($A_{0.5}$, μM)	AMP[c] ($I_{0.5}$, μM)	Mutation	Reference
E. coli B	20	68	75		Govons et al., 1973
Mutant SG14	8.4	820	500	Ala44 → Thr	Preiss et al., 1975
					Meyer et al., 1993
Mutant SG5	35	22	170	Pro295 → Cys	Govons et al., 1973
					Meyer et al., 1992
Mutant 618	70	15	860	Gly336 → Asp	Cattaneo et al., 1969
					Kumar et al., 1989
Mutant CL1136	74	5	68	Arg67 → Thr	Preiss et al., 1976
					Ghosh et al., 1992
S. typhimurium					
LT2	12	95	110		Lehmann & Preiss, 1980
					Steiner & Preiss, 1977
Mutant JP23	15	Not activated	250		Steiner & Preiss, 1977
Mutant JP51	20	84	490		Steiner & Preiss, 1977

[a]The bacterial strains were grown in minimal media with 0.75% glucose; the data is expressed as maximal mg of anhydroglucose units per gram (wet wt) of cells in stationary phase.
[b]$A_{0.5}$ is the fructose-1,6-bisP giving 50% of maximal activation.
[c]$I_{0.5}$ is the AMP concentration required for 50% inhibition.

Glc PPases have approximately the same apparent affinities as the parent enzyme for the substrates ATP and glucose-1-P, and for the divalent cation activator Mg^{2+}. The affinities of the substrates for the S. *typhimurium* mutants have not been studied. Nevertheless, the modified properties of the E. coli and S. typhimurium mutant enzymes are their apparent affinities for the allosteric effectors.

The concentration of fructose-1,6-bisphosphate required for 50% of maximal activation ($A_{0.5}$) is about threefold less for the SG5 ADP-Glc PPase, 4.5-fold less for the 618 and 12-fold less for the CL1136 enzymes (Table 13-8). The apparent affinities for the activators NADPH and pyridoxal-5′ phosphate are also greater for the mutant ADP-glucose pyrophosphorylases than for the E. coli B enzyme.

The mutant ADP-glucose pyrophosphorylases are also less sensitive to the allosteric inhibitor AMP (Table 13-8). Higher concentrations of AMP are necessary to effect 50% inhibition of the mutant ADP-Glc PPases than are required for inhibition of the E. coli B enzyme. At 1.5 mM fructose-1,6-bisphosphate the mutant enzymes are less sensitive than the parent enzyme to inhibition. As would be expected, the CL1136 and SG5 ADP-Glc PPases have more activity than the E. coli B wild-type enzyme when assayed under equivalent conditions and energy charge values (Govons et al., 1973; Preiss et al., 1976). The energy charge is defined as [ATP] + 1/2 [ADP]/ [ATP] + [ADP] + [AMP] (Atkinson, 1970). At an energy charge of 0.7, the SG5 enzyme has about five times more activity than the E. coli B with 1.5 mM fructose-1,6-bisphosphate as the activator (Govons et al., 1973). Over a range of energy charge of 0.85–0.9 and with 0.75 mM fructose 1,6-bisP, that is, under approximately physiological conditions, the SG5 enzyme shows about twice as much activity as the *E. coli* B enzyme. The CL1136 ADP-Glc PPase is almost completely active in the energy charge range of 0.75–1.0 with fructose-1,6-bisP concentrations of 0.75 mM and higher (Preiss et al., 1976). Significant activity (10- to 25-fold of maximal activity) is seen even at an energy charge of 0.4 with fructose-1,6 bisP concentrations of 0.75 mM or higher. In the absence of the activator and at an energy charge value of 0.75, the activity of the CL1136 enzyme is almost 30% of the maximal activity. In contrast, the *E. coli* B ADP-glucose pyrophosphorylase activity is less than 4% of the maximal activity at an energy charge of 0.65, even in the presence of 3 mM fructose-1,6-bisP. At an energy charge level of 0.75 and with 3.0 mM fructose-1,6-bisphosphate, the *E. coli* B enzyme exhibits only 11% of its maximal activity. The CL1136 enzyme shows 98% of its maximal activity under these conditions. The physiological energy charge range in various *E. coli* strains is found to be 0.85–0.91 (Dietzler et al., 1974a; Swedes et al., 1975). The physiological concentrations of fructose-bisP range between 0.71 and 3 mM (Dietzler et al., 1973, 1974a). Under these conditions, the CL1136 enzyme shows maximal activity and is not sensitive either to change in energy charge or to fluctuations in fructose-1,6-bisP. In contrast, the *E. coli* B enzyme is very sensitive to these ranges of fructose-1,6-bisP concentration and energy charge (Govons et al., 1973; Preiss et al., 1976). These studies suggest strongly that the increased accumulation of glycogen in the mutants SG5 and CL1136 is due to alterations of the ADP glucose pyrophosphorylases that cause a greater affinity for the activators and a lower affinity for the inhibitor. The correlation of the relative sensitivities of *E. coli* B and SG5 and CL1136 ADP-glucose pyrophosphorylases to AMP inhibition and fructose-1,6-bisP activation with the increased rates of glycogen accumulation is in agreement with the expressed view that the cellular levels of the allosteric activators and inhibitors of ADP-glucose pyrophosphorylase modulate the rate of synthesis and accumulation of glycogen in the cell.

The *S. typhimurium* ADP-Glc PPase mutants are of some interest because their altered kinetic properties are different than seen for the *E. coli* allosteric mutants. As indicated in Table 13-8, both activator and inhibitor constants ($A_{0.5}$ and $I_{0.5}$) are affected by the mutation in *E. coli* mutants SG5, 618, and CL1136. In contrast the *S. typhimurium* LT-2 mutant JP51 enzyme has an $A_{0.5}$ value for fructose-1,6 bisphosphate similar to that of the parent strain ADP-Glc PPase. However, the JP51 enzyme has about a five-fold higher $I_{0.5}$ value for AMP than the parent enzyme, thus suggesting that the lesser sensitivity to AMP inhibition is the reason for higher accumulation of glycogen in mutant JP51. Mutant JP23 enzyme, in contrast to the parent strain enzyme, is not activated by the fructose bis-P; that is, it is not dependent on fructose-1,6 bisphosphate for full activity (Steiner & Preiss, 1977). The mutant enzyme also is less sensitive than the parent strain enzyme to AMP inhibition (Table 13-8) and in addition cannot be inhibited to an extent more than 60% of its original activity at higher concentrations of AMP, either in the presence or absence of fructose-1,6-bisP. If the levels of AMP in cells are assumed to be 0.15 ± 0.05 mM, then the JP23 ADP-glucose pyrophosphorylase activity could range from 50% to 60% of its total activity. Under the same conditions in the presence of 1 mM fructose-1,6 bisphosphate and range of AMP concentration indicated above, the *S. typhimurium* LT-2 enzyme would exhibit from 5% to 50% of its total activity (Steiner & Preiss, 1977). Since JP23 crude extracts have 50% of the activity of the extracts of LT-2, the JP23 enzyme activity would be 0.7–4.9 times as active as the LT-2 enzyme over this concentration range of AMP. Mutant JP23 could therefore accumulate more glycogen than LT-2, based solely on the alteration of its ADP-glucose pyrophosphorylase to sensitivity to AMP inhibition.

Dietzler et al. (1973, 1974a) have shown that when the growth of E. coli W4597(K) ceased because of NH_4^+ limitation, glycogen accumulation rates increased about 3.3- to 4.2-fold. There was also a decrease of fructose-1,6 bisphosphate of about 76% (from 3.1 to 0.71 mM) and an increase in the energy charge from 0.74 in exponential phase to 0.87 in stationary phase. The total concentration of the adenylate pool (ATP + AMP + ADP) in exponential and stationary phase was 3 mM. Although the fructose-1,6-bisphosphate concentration decreased 76% from exponential to stationary phase, Dietzler et al. (1973, 1974a) concluded that this decrease in the concentration of allosteric activator was more than offset by the increase in energy charge. The conclusion of Dietzler et al. (1974a) was based on the data of Govons et al. (1973) that showed a threefold increase in E. coli B ADP-gluclose pyrophosphorylase activity when energy charge increased from 0.74 to 0.87, with a concomitant decrease of fructose-1,6-bisphosphate concentration from 1.5 to 0.5 mM. The total adenosine nucleotide pool was 2 mM in these in vitro experiments.

In various E. coli strains grown in nitrogen-limiting minimal media, with glucose as the carbon source, the total adenine nucleotide concentration is about 3 mM (Dietzler et al., 1974a,b; Lowry et al., 1971), the fructose-1,6-bisP concentration ranges from 0.71 to 3.2 mM (Lowry et al., 1971; Dietzler et al., 1973, 1974a,b), the mass action ratio of the adenylate kinase reaction ranges from 0.37 to 0.69, and energy charge values lie between 0.74 and 0.9. While glucose-1-phosphate concentrations in E.coli cells have not been determined, they have been calculated from the published glucose-6-phosphate values reported for E. coli HFr 139 (Lowry et al., 1971) and E.coli strains W4597(K) or G34 (416) using the assumption that the phosphoglucomutase reaction is at equilibrium in vivo (K_{equil} = 17 at 37°C). Determinations of the glucose-1-phosphate concentration range from 39 to 45 μM.

Using the above values of fructose-1,6-bisP, adenine pool concentrations, and an adenylate kinase mass action of 0.45, the response of the E. coli B ADP-glucose pyrophosphorylase to energy charge was determined (Preiss, 1978; Preiss et al., 1976). The E. coli B enzyme shows only 4% of its maximal velocity at an energy charge value of 0.75 with 0.75 mM fructose-1,6-bisphosphate. This increases to 37% of maximal velocity at energy charge 0.9. With 3 mM fructose-1,6-bisphosphate, the E. coli B enzyme activity is 11% of its maximal velocity at energy charge of 0.75, and this increases to 78% maximal velocity at energy charge of 0.9. The rate of ADP-Glc synthesis in exponential phase E. coli B extracts is sevenfold greater than the glycogen accumulation rate and about 10-fold greater in stationary phase cell extracts (Preiss et al., 1975, 1976). Thus only 10%–14% of the maximal ADP-glucose synthetic activity is needed to account for the observed glycogen accumulation rates. Glycogen degradation rates in E. coli appear to be two orders of magnitude less than synthetic rates (Holme & Palmstierna, 1956). Thus, glycogen accumulation rates may be considered to be determined solely by glycogen synthetic activity. Taking into account that the calculated glucose-1-phosphate concentration is at less than saturating concentrations for the ADP-Glc PPase in E. coli B, the percentage of maximal activity can be calculated at various charge values and fructose-1,6-bisphosphate concentrations in the physiological range. Sufficient E. coli B enzyme activity to account for the glycogen accumulation rate in exponential phase would be seen at energy charges of 0.87

with 0.75 mM fructose-1,6-bisphosphate or at 0.85 with 1.5–3.0 mM fructose-1,6-bisphosphate. In stationary phase, sufficient ADP-glucose synthesis rates are also observed at energy charge values of 0.8 with 2.5–3.0 mM fructose-1,6-bisP, 0.85 with 1.5 mM, or 0.86 with 0.75 mM (Preiss, 1978; Preiss et al., 1976). Thus, the calculated ADP-glucose-synthesizing activity in *E. coli* B cells within the physiological range of energy charge and fructose-1,6-bisphosphate levels is sufficient to account for the observed glycogen accumulation rates. The fructose-1,6-bisP concentrations and energy charge values during growth of *E. coli* B are presently unknown. In contrast to *E. coli* W4597(K), Lowry et al. (1971) have shown that the fructose-1,6-bisP concentrations increase slightly from 2.6 to 3.2 mM in another strain of *E. coli* K-12, Hfr 139, when it reaches stationary phase in the presence of excess glucose and limiting nitrogen. The energy charge remains essentially constant. The foregoing calculations also have assumed no compartmentation of the metabolites in relation to the possible compartmentation of glycogen biosynthetic enzymes. Third, the effects of other cations, anions, and other metabolites on ADP-glucose pyrophosphorylase activity have not been considered. Nevertheless, the correlation of ADP-glucose pyrophosphorylase activity with the known fructose-1,6-bisphosphate concentrations and energy charge values with the observed rates of glycogen accumulation appears to be quite good.

An interesting and important finding has been made by Dietzler et al. (1974b, 1975). Two mutants of E. *coli*, W4597(K) and G34, were grown under five different nutrient conditions that gave a 10-fold range in the rate of glycogen accumulation in stationary phase. This was obtained by varying the carbon or the nitrogen source of the media. The different rates of glycogen accumulation found in the various nutrient conditions is related linearly to the square of the fructose-1,6-bisP concentration in the bacteria. The ATP concentrations in the bacteria were the same and were independent of the composition of the media. Thus, a relationship exists between the concentration of the cellular fructose-1,6-bisP and the rate of glycogen accumulation. Dietzler et al. (1975) fitted these data to the Hill equation and obtained an $A_{0.5}$ of 0.82 mM and a Hill slope value ñ of 2.08. These values are in agreement with the *in vitro* kinetic values obtained with the *E. coli* B ADP-Glc PPase at energy charge 0.85 (Govons et al., 1973: $A_{0.5}$, 0.68 mM; ñ, 2.0).

The above data suggest that the physiological activator for the *E. coli* ADP-glucose pyrophosphorylase is fructose-1,6-bisP. This view is also supported by studies with mutant SG14 (Preiss et al., 1971, 1975). Mutant SG14 accumulates glycogen at 28% the rate of *E. coli* B (Table 13-8) and contains about 23%–25% of the ADP-glucose-synthesizing activity of *E. coli* B (Preiss et al., 1975). The activity is still sixfold greater than that required for the observed rate of glycogen accumulation in SG14. The concentrations of ATP and Mg^{2+} required for 50% of maximal activity ($S_{0.5}$) are four- to fivefold higher for the SG14 enzyme than the *E. coli* B enzyme. Whereas the $S_{0.5}$ values for ATP and Mg^{2+} are 0.38 and 2.3 mM, respectively, for the *E. coli* B enzyme in the presence of 1.5 mM fructose-1,6-bisphosphate, the $S_{0.5}$ values for ATP and Mg^{2+} are 1.6 and 10.8 mM, respectively, for the SG14 ADP-Glc PPase in the presence of saturating fructose-1,6-bisP concentration (4.0 mM). Reports in the literature indicate that the ATP level in growing *E. coli* is approximately 2.4 mM (Dietzler et al., 1974a,b; Lowry et al., 1971) and the Mg^{2+} level is about 25–40 mM (Silver, 1969). Therefore, the SG14 ADP-glucose pyrophosphorylase would be essentially saturated with respect to these substrates. The apparent affinities ($S_{0.5}$) for glucose-l-phosphate for the *E. coli* B and SG14 enzymes are about the same (Preiss et al., 1975). The major difference between the SG14 and *E. coli* B ADP-Glc PPases appears to lie in their sensitivities toward activation and inhibition. About 12-fold as much fructose-1,6-bisphosphate is needed for 50% maximal stimulation of the SG14 ADP-glucose pyrophosphorylase ($A_{0.5} = 0.82$ mM) as for half-maximal stimulation of the *E. coli* B enzyme. The $A_{0.5}$ for pyridoxal phosphate for the SG14 enzyme (0.44 mM) is about 25-fold higher than the $A_{0.5}$ observed for the *E. coli* B enzyme. Pyridoxal phosphate and fructose-1,6-bisphosphate stimulate ADP-glucose synthesis catalyzed by the *E. coli* B enzyme to about the same extent. However, the stimulation of the SG14 ADP-glucose pyrophosphorylase seen with pyridoxal phosphate is only half that elicited by fructose-1,6-bisphosphate (Preiss et al., 1975). Another notable difference is that NADPH does not stimulate the SG14 enzyme. Compounds structurally similar to NADPH, such as l-pyrophosphorylribose-5-phosphate and 2'-PADPR, which are capable of activating the *E. coli* B ADP-glucose pyrophosphorylase, do not activate the SG14 enzyme. Since the apparent affinity of the SG14 enzyme for its activators is considerably lower than that observed for the *E. coli* ADP-glucose pyrophosphorylase, it was an unexpected finding that SG14 is capable of accumulating glycogen even at 28% the rate observed for the parent strain. This rate is accounted for by the relative insensitivity of the SG14 enzyme to inhibition by AMP (Preiss et al., 1971, 1975). The SG14 ADP-glucose pyrophosphorylase is much less sensitive than the parent strain enzyme to AMP inhibition in the concentration range of 0–0.2 mM. At a saturating concentration of fructose-1,6-bisP for the SG14 enzyme (4.0 mM), only 7% inhibition of the SG14 enzyme is observed at 0.2 mM AMP. The same concentration of AMP gives 40% inhibition of the *E. coli* B enzyme. At a concentration of fructose-1,6-bisP (1.5 mM) that gives 80% of maximal velocity for the SG14 enzyme, 0.2 mM 5'-AMP causes 80% and 33% inhibition of the *E. coli* B and SG14 enzymes, respectively. A decrease in fructose-1,6-bisphosphate to 1.0 or 0.5 mM further increases the sensitivity of the *E. coli* B ADP-glucose pyrophosphorylase activity to inhibition, while at these fructose-bisP concentrations the sensitivity of the SG14 ADP-glucose pyrophosphorylase to AMP remains the same or becomes less than that observed at 1.5 mM fructose-bisP. At concentrations of 0.5–1.0 mM of fructose-bis-P, the *E. coli* B enzyme is inhibited 90% or more by 0.2 mM AMP and the inhibition of the SG14 enzyme ranges from 12% to 30%. Although the SG14 enzyme has a lower apparent affinity for its activators, it is also less sensitive to AMP inhibition. The two effects on activation and inhibition appear to compensate for each other and to allow SG14 to accumulate glycogen at about 28% the rate of the parent strain (Preiss et al., 1975). The data obtained from kinetic studies of the SG14 ADP-glucose pyrophosphorylase suggest that fructose-1,6-bisP is the most important physiological activator of the *E. coli* ADP-Glc PPase. This is based on the observation that NADPH is not an activator of the SG14 enzyme and that the concentration of pyridoxal phosphate needed for activation of the enzyme ($A_{0.5} = 0.44$ mM) is considerably higher than the concentration reported to be present in *E. coli* B. The concentration of pyridoxal phosphate is 24–48 μM (Dempsey, 1972), and most of this metabolite is probably protein bound in the cell and unavailable for activation of the ADP-Glc PPase (Dempsey & Arcement, 1971). The concentration of fructose-1,6-bisphosphate in *E. coli* is approximately 0.71–3.2 mM (Dietzler et al., 1973, 1974a,b; Lowry et al., 1971), and the $A_{0.5}$ of SG14 ADP-Glc PPase is 0.82 mM. The concentration of fructose-bisP in the *E. coli* cell is, therefore, sufficient for activation of ADP-glucose synthesis at the rates required for the observed glycogen accumulation rate in SG14.

Genetic Regulation of Bacterial Glycogen Synthesis

Since 1987, information pertaining to the genetic regulation of glycogen synthesis in *E. coli* has accumulated. The location of the various structural genes required for glycogen synthesis on the *E. coli* chromosome and their nucleotide sequences has been determined. Moreover, it has also been shown that cyclic AMP (cAMP), guanosine 5′-diphosphate 3′-diphosphate, and the cAMP receptor protein are strong positive regulators of the expression of the ADP-Glc PPase (*Glg* C) and glycogen synthase (*Glg*A) genes, and that a protein gene (*Csr* A) product, is a negative regulator of expression. Genetic regulation of the level of the glycogen biosynthetic enzymes does play an important role in determining the ultimate level of glycogen synthesis and accumulation under different physiological conditions. The various phenomena involved in genetic regulation of bacterial glycogen synthesis have been summarized in a number of reviews (Preiss & Romeo, 1989, 1994; Preiss, 1996a).

Regulation of Mammalian Glycogen Synthesis

As seen above, the site of regulation of glycogen synthesis in bacteria and starch synthesis in plants is at the ADP-Glc PPase step and is different from that found for mammalian glycogen synthesis. The regulatory enzyme in mammalian systems is glycogen synthase (Reaction 13-3). The difference in regulatory sites in the various systems may be based on the difference in specificity for the glucosyl donor, ADP-glucose, for the bacterial and plant α-D-glucan systems. As shown in Figure 13-5, UDP-glucose is utilized for the synthesis of other sugar nucleotides and glycosyl substituents required for the synthesis of many cellular constituents. Therefore, the first unique reaction for mammalian glycogen synthesis, after synthesis of the glucosylated acceptor protein glycogenin, is the glycogen synthase step where both allosteric control and covalent modification control are exerted. In contrast, in bacteria and plants the only known function for ADP-glucose is for synthesis of α-1,4-D-glucosyl bonds in bacterial glycogen and starch. Thus, the prokaryote and plant cells regulate α-glucan synthesis at the level of ADP-glucose formation so as to conserve ATP utilized for synthesis of the sugar nucleotide. The findings of regulation occurring at the glycogen synthase step in the mam-

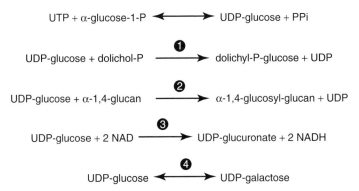

Figure 13–5. Some reactions of UDP-glucose in mammalian cells.

malian systems and at the ADP-Glc PPase step in bacteria and plants are consistent with the concept that major regulation of a biosynthetic pathway does occur at the first unique step of the pathway. One can also recognize the need in mammalian systems for a type of regulation that involves an efficient, rapid off–on type of control of glycogen synthesis that permits synthesis of glycogen when carbohydrate or carbon is plentiful in the diet but prevents synthesis and permits degradation of glycogen to occur during muscular contraction or during starvation. Such a mechanism involves covalent modification of the enzyme catalyzing the limiting reaction of the process in order to produce either inactive or active forms of the enzyme.

Regulation of Glycogen Synthase by Phosphorylation–Dephosphorylation

Mammalian glycogen synthase has been purified from several sources, for example, skeletal and cardiac muscle, liver, adipose, and kidney (Brown & Larner, 1971; Issa & Mendicino, 1973; Killilea & Whelan, 1976; Lin & Segal, 1973; Miller et al., 1975; Nakai & Thomas, 1975; Soderling et al., 1970). Moreover, cDNA representing the structural gene of the enzyme has been isolated (human muscle, Browner et al., 1989; rabbit muscle, Zhang et al., 1989; rat liver, Bai et al., 1990; human liver, Nuttal et al., 1994; yeast, Farkas et al., 1990, 1991) and the rabbit skeletal muscle cDNA has been expressed both in bacterial (Zhang et al., 1993) and COS cells (Skurat et al., 1993, 1994, 1996). The deduced amino acid sequences of the rabbit and human muscle glycogen synthases have 97% identity while the rat and human liver enzymes have 92% similarity. The two yeast glycogen synthases have an 80% identity in sequence but only a 50% overall identity to the muscle glycogen synthases. The amino acid similarity between human muscle and human liver was only 60%, being least similar in the N- and C-terminal regions of the protein. The human and rat liver enzymes are truncated by 32–34 amino acids compared to the rabbit and human muscle glycogen synthases.

Glycogen synthase from the above sources appear to be composed of identical or similar subunits of MW 80,000–85,000. The native forms from liver or adipose are aggregates of two identical subunits, whereas that from muscle contains four. Glycogen synthase exists in at least two forms; a phosphorylated form, arising from covalent modification by phosphorylation of serine residues by ATP, and a dephosphorylated form, which can be obtained from the phosphorylated form by phosphatase action. This is shown in Figure 13-6 and the forms were distinguished by their sensitivity to al-

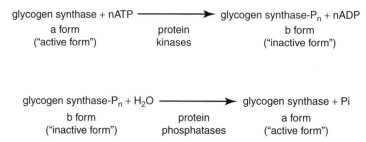

Figure 13–6. Covalent modification of glycogen synthase *via* phosphorylation by protein kinases and dephosphorylation by protein phosphatases.

losteric activation by glucose-6-P. The forms were originally noted as the a (or I) (un-phosphorylated) form and b (or D) (phosphorylated) form. The b form was dependent on the presence of glucose-6-P for activity, whereas the a form did not require glucose-6-P for activity. The a and b forms can also be distinguished on the basis of Km for the substrate, UDP-glucose. The b form usually has a higher Km (lower apparent affinity) than the a form. Thus it is apparent that the a form is the active form of the enzyme while the b, phosphorylated form is an inactive form of the glycogen synthase. The a/b ratio, that is, activity in absence of glucose-6-P/activity in presence of glucose-6-P, has been used as a measure of the state of phosphorylation of the enzyme. The a form was recognized as the primary active species within the cell, especially as modulated by hormonal control. The cAMP-dependent protein kinase was shown to catalyze phosphorylation and consequently inactivation of this enzyme. Subsequent studies also indicated that on each subunit of the glycogen synthase there were multiple and unique phosphorylation sites. Originally, Smith et al. (1971) reported that, if partially purified enzyme was incubated extensively with ATP and subsequently purified, the inactive glycogen synthase contained 6 mol of phosphate per subunit. Although this information did not initially gain acceptance, Embi et al. (1982) and Picton et al. (1982a), showed that there are seven serine phosphorylation sites on the rabbit skeletal muscle glycogen synthase. Currently, the observed number of potential phosphorylation sites found *in vivo* are 9; and more than 10 sites can be phosphorylated *in vitro* (Roach, 1990). Table 13-9 shows the various *in vivo* phosphorylation sites and the major protein kinases that are involved in the phosphorylation of those sites.

Phosphorylation of these sites is catalyzed by more than six kinases (Cohen et al., 1982; Roach, 1990; Roach & Larner, 1976, 1977). Phosphorylation at the distinct sites synergistically inactivates the enzyme; however, the effects observed may vary depending on the conditions used to assay the glycogen synthase. The effects of multisite phosphorylation would depend on the concentration of effectors used, such as glucose-6-P or the substrate, UDP-glucose. For example, with preparations of enzymes containing 0.27–3.49 mol of alkali-labile phosphate per glycogen synthase subunit, Roach and Larner (1976) reported that the $A_{0.5}$ (concentration of activator, glucose-6-P, needed for 50% of maximal activation) and $S_{0.5}$ (concentration of substrate, UDP-glucose, having 50% of maximal velocity) varied with phosphate content, from 3.3 μM to 2.7 mM, and from 0.75 mM to at least 60 mM, respectively. Both parameters increased with phosphate content. The greatest absolute change occurred at values greater than 2 mol of phosphate bound per enzyme subunit. Plots of activity versus glucose-6-P became more sigmoidal with increasing enzyme phosphate content. Activation by glucose-6-P was related primarily to modulation of UDP-glucose affinity. Several inhibitors, such as ATP, ADP, AMP, UDP, and Pi, had increasing effects with enzyme of increasing alkali-labile phosphate content. These investigators presented a scheme in which glycogen synthase activity is sigmoidally and inversely dependent on phosphorylation state. The hormonal effects of insulin were counteractive to those of epinephrine, glucagon, and so on, respectively decreasing or increasing the extent of phosphorylation. Inhibitors such as UDP, ATP, and AMP accentuate the inhibition by phosphorylation, whereas the activator glucose-6-P and substrate UDP-glucose diminished the extent of such inhibition.

As seen in Table 13-9, there are two phosphorylation sites at the N-terminal region of the rabbit skeletal muscle glycogen synthase. The remaining seven are situated in the C-terminal region. The cAMP-dependent protein kinase preferentially phosphory-

Table 13–9 *In vivo* Phosphorylation Sites in Rabbit Muscle Glycogen Synthase

Phosphorylation Site

Residue	Name	Protein Kinases
7	2	cAMP-dependent protein kinase Calmodulin-dependent protein kinase II Phosphorylase kinase Protein kinase C
10		Casein kinase I
640	3a	Glycogen synthase kinase 3 cAMP-dependent protein kinase
644	3b	Glycogen synthase kinase 3
648	3c	Glycogen synthase kinase 3
652	4	Glycogen synthase kinase 3 cAMP-dependent protein kinase
656	5	Casein kinase II
697	1a	cAMP-dependent protein kinase Protein kinase C
710	1b	cAMP-dependent protein kinase Calmodulin-dependent protein kinase II

This table is derived from Roach (1990). The contributions of the identification of the glycogen synthase phosphorylation sites and the glycogen synthase kinases by many investigators is acknowledged by Roach (1990). The sites listed in this table are only those sites that are known to be labeled *in vivo*. The protein kinases listed for the sites are those that phosphorylate the enzyme *in vitro*.

lates three sites, 1a, 1b, and 2 (Embi et al., 1982). The cAMP-dependent kinase can also phosphorylate sites 3a and 4 but at a much slower rate (Sheorain et al., 1985) and thus is not considered to be as important as glycogen synthase kinase for those sites. There are overlapping specificities among the different protein kinases for site 2 as phosphorylase kinase (Roach, 1986), calmodulin-dependent protein kinase, and protein kinase C also can phosphorylate this site (see Cohen, 1986).

Of interest is the phosphorylation of site 5 catalyzed by casein kinase II. Picton et al. (1982b) showed that dephosphorylation at site 5 did not alter the regulatory kinetics of rabbit muscle glycogen synthase. Nor did rephosphorylation of site 5 by casein kinase II (also called glycogen synthase kinase 5) affect the activity of the glycogen synthase. In other words, the glycogen synthase phosphorylated at site 5 did not depend to any extent on glucose-6-P for maximal activity. However, the presence of phosphate at site 5 was necessary for the phosphorylation of sites 3a, 3b, and 3c (Rylatt et al., 1980) by glycogen synthase kinase 3, which did increase the dependency of the glycogen synthase activity for glucose-6-P. The site 5 phosphate appears to be highly stable, as it is resistant to dephosphorylation by the rabbit muscle protein phosphatases, but can be removed by potato acid phosphatase. Phosphorylation of sites 1a, 1b, or 2 did not require the presence of phosphate at site 5.

This observation was confirmed and extended by Fiol et al. (1987, 1990). They synthesized a peptide corresponding to the rabbit muscle glycogen synthase amino acid sequence containing sites 3a, 3b, 3c, 4, and 5. Synergism between casein kinase II phosphorylation of site 5 and phosphorylation by glycogen synthase kinase 3 (GSK-3) of sites 3a, b, c, and 4 of the synthesized peptide were seen. Indeed, phosphorylation of site 5 was obligatory for the phosphorylation of the four sites at 3 and 4 by glycogen synthase kinase 3. As seen in Table 13-10, the GSK-3 sites were regularly spaced every fourth residue in the motif, SXXXS(P). What was also found is that the phosphorylations by GSK-3 were ordered. First site 4 was phosphorylated, then site 3c, 3b, and finally 3a (Fiol et al., 1990). This was shown clearly by synthesizing a series of peptides in which sites 3a, 3b, 3c, and 4 were replaced, one at a time, with alanine. The peptide-containing alanine at site 4 could not be phosphorylated at the site 3 residues even though it was phosphorylated at site 5 by GSK-3. Also it was observed that GSK-3 would not phosphorylate serine residues at sites closer to the amino terminus than the site containing the alanine residue. When alanine replaced serine at site 3b, only sites 4 and 3c were phosphorylated. With alanine substituted at site 3c, only site 4 was phosphorylated. Thus, the multiple phosphorylation by GSK-3 was of an obligate order, first 4 then 3c, 3b, and finally 3a. Most probably GSK-3 recognizes the motif-SXXXS(P) and this may explain the need for GSK-5 to phosphorylate site 5. The sequential formation of new recognition sequences, SXXXS(P) at the sites 5, 4, 3c, 3a, and 3b would explain the ordered phosphorylation.

This interdependency of GSK-3 with GSK-5 has been defined as hierarchal phosphorylation (Roach, 1990). Another example of the hierarchal phosphorylation is seen with cAMP-dependent protein kinase and casein kinase I. The cAMP-dependent protein kinase enhances phosphorylation of the glycogen synthase by casein kinase I (Flotow & Roach, 1989). The phosphorylation by casein kinase was at serine residue 10 (Zhang et al., 1989). Synthetic peptides based on the four phosphorylated regions in the muscle glycogen sequence (residues 694–707, 706–733, 1–14, and 636–662) were synthesized and phosphorylated. Casein kinase could not phosphorylate the unphosphorylated peptides but if cAMP-dependent kinase phosphorylated peptides 694–707, 706–733, and 1–14, all three peptides were easily phosphorylated by the casein kinase I (Flotow et al., 1990). The greatest stimulation was seen with peptide 1–14. In the case of peptide 1–14 the phosphorylation site was at Ser10 and in the case of

Table 13–10 The Amino Acid Sequence of the Regions Corresponding to the Phosphorylated Sites of Rabbit Muscle Glycogen Synthase

N-Terminal phosphorylation sites

       7   10

  PLSRTLS VSS LPGL----

     (site 2)

C-Terminal phosphorylation sites

     640   644   648   652   656 RYPRPAS VPPS PSLS RHSS PHCS

EDEEEPRDGLPEEDGERYDEDEEAAKD

   (sites 3a 3b   3c   4   5)

           697   710

RRNIRAPQWPRRAS CTSSSGGSKRSNS VDTSSLSTPSEP-------

     (sites 1a   1b)

the of the peptides 694–707 and 706–733, the phosphorylated residue was Thr713. However, the rate of phosphorylation was 20–40 times greater at Ser10 than Thr713. Moreover, whereas Ser10 has been demonstrated to be phosphorylated *in vivo*, Thr713 has not (Roach,1990). Thus, the physiologically important site for phosphorylation by casein kinase I is considered to be Ser10 and this phosphorylation is considered to be dependent on an initial phosphorylation of Ser7 (site 2) by cAMP- dependent protein kinase.

Of interest is the phosphorylation of peptide 636–662, the peptide encompassing glycogen synthase phosphorylation sites 3a, b, c, 4, and 5 (Flotow et al., 1990). Using the peptides in which alanine was substituted for serine at the different phosphorylation sites, it was shown that serine residues 646 and 651 (Table 13-10) were phosphorylated by casein kinase I and this phosphorylation was enhanced significantly by prior phosphorylation of the sites 3a, b, c, 4, and 5 (Flotow et al., 1990). However, it is still not clear whether these sites are phosphorylated *in vivo* and whether their phosphorylation by casein kinase I substantially affects the glycogen synthase activity. These studies, carried out mainly by Roach's group, also indicate that the recognition of the serine phosphorylation site by casein kinase I is -S(P)-XXS- and for GSK-3 is SXXX-S(P).

The Effect of Phosphorylation on Glycogen Synthase Activity and Relative Effects of Phosphorylation on Different Sites

In Vitro Studies. Phosphorylation of the glycogen synthase activity leads to decreased activity but phosphorylations at different sites have different effects. Little or no inactivation is seen with phosphorylation at sites 5, 1a, and 1b, while site 2 phosphorylation gives moderate inactivation. A most potent inactivation is seen with phosphorylation by GSK-3 at sites 3a, b, and c (Cohen, 1986; Roach, 1986). Although site 5 phosphorylation doesn't cause any change in activity, it is functional in regulation as sites 3 cannot be phosphorylated unless there is an initial phosphorylation of site 5. Phosphorylation of sites 1a, 1b, and 2 by cAMP-dependent protein kinase leads only to a partial inactivation of the rabbit muscle glycogen synthase, and phosphorylation of site 2 did not decrease activity in rabbit (Wang et al.,1986a) or rat liver glycogen synthase (Huang et al., 1983). However, casein kinase I phosphorylation of Ser10, proceeding after phosphorylation of sites 1a, 1b, and 2 in rabbit muscle glycogen synthase and 2 in liver glycogen synthase, causes a total inactivation of the muscle (Wang et al., 1986a) and liver enzymes (Huang et al., 1983). It should be pointed out that sites 1a and 1b are absent in rat liver (Bai et al., 1990; Wang et al., 1986a,b) and human liver glycogen synthase (Nuttall et al., 1994).

Thus, the secondary phosphorylations by casein kinase I of Ser10, and by GSK-3 of sites 3a, b, c, and 4 have greater effects on the activity of glycogen synthase than the primary phosphorylations by cAMP-dependent kinase (sites 2, 1a, and 1b) and glycogen synthase kinase-5; there may be two different routes to inactivating glycogen synthase, the GSK-3 sites, and the casein kinase site.

It should also be mentioned that the above studies were done with a recombinant glycogen synthase expressed in *E. coli* (Zhang et al., 1993), and the same results were seen with phosphorylation site peptide analogues. This included the dependency of phosphorylation by GSK-3 on the prior phosphorylation casein kinase II, the potent inactivation by GSK-3, the partial inactivation by phosphorylation by cAMP-dependent kinase, the stimulation of phosphorylation by casein kinase I by prior phosphorylation by phosphorylase kinase, and the greater inactivation of glycogen synthase after the combined phosphorylation by casein kinase I and cAMP-dependent protein kinase.

Further studies of the recombinant glycogen synthase expressed in *E. coli* allowed Wang and Roach (1993) to generate mutant forms of the rabbit muscle glycogen synthase at the GSK-3 phosphorylation sites S640A, S644A, and S648A (sites 3a, 3b, and 3c, respectively). All three mutants had high ∓ glucose-6-P ratios of activity (0.8–0.9). Phosphorylation of the mutants was carried out using GSK-3 and casein kinase II. The mutants phosphorylated at sites 5 and 4 (mutant S648A) and at sites 3c, 4, and 5 (mutant S644A) had full activity. When sites 3b, 3c, 4, and 5 (mutant S640A) were phosphorylated the activity ratio decreased modestly to approximately 0.6 to 0.7. When all sites were phosphorylated in the recombinant enzyme, the activity ratio decreased to 0.1. The results of this study demonstrated that phosphorylation site 3a and to a lesser extent 3b, correlated with the inactivation of the glycogen synthase. The apparent affinity constant for the activator, glucose-6-P, was 2.4 μM and only increased appreciably, to 24 μM, when site 3a was phosphorylated with the other 4 sites.

In Vivo Studies. At least for the rabbit muscle glycogen synthase, the above *in vitro* studies clearly showed that the important phosphorylation sites for inactivation were sites 2, 2a, 3a, and 3b. However, it is of interest to know the important phosphorylation sites under hormonal control in the *in vivo* systems. For the rabbit muscle enzyme intravenous insulin administration doubles the ∓ glucose-6-P activity ratio with a decrease in phosphorylation of all the sites (Cohen, 1986; Lawrence & Zhang, 1994; Parker at al., 1983). Epinephrine, which increases the phosphate content of the glycogen synthase, increases the phosphorylation at practically all the phosphorylation sites (Parker et al., 1982, 1983; Poulter et al., 1988; Sheroain et al., 1984).

With respect to the liver glycogen synthase, the hormones glucagon, vasopressin, and epinephrine all increased the phosphorylation of the peptide regions containing the site 2 and 3 phosphorylation sites (Akatsuka et al., 1985; Roach, 1986, 1990). Glucagon-promoted phosphorylation of the casein kinase I site now referred to as site 2a or equivalent liver Ser residue to muscle glycogen synthase residue, Ser10.

To study in detail the role of individual phosphorylation sites in the regulation of the rabbit muscle glycogen synthase, the enzyme was overexpressed in COS M9 cells (Skurat et al.,1993, 1994). The activity ratio of ∓ glucose-6-P was found to be very low, ~0.01, indicative of a high level of phosphorylation. Ser to Ala mutations were introduced singly, or in combinations, at the nine known phosphorylation sites; it was found that no single Ser to Ala mutation caused a substantial increase in the activity ratio (Skurat et al., 1994). It was shown that simultaneous mutations were needed at both regions of sites 2, the N-terminal region, and sites 3, the C-terminal region. The most effective combinations were mutations at site 3a (Ser 640) or site 3b (Ser644) together with site 2 (Ser7). Double mutants, Ser640Ala-Ser7Ala, Ser644Ala-Ser7Ala, and Ser10Ala (site 2a)-Ser640Ala, gave activity ratios of 0.59, 0.25, and 0.21, respectively. These results were consistent with site 2 phosphorylation being a prerequisite for phosphorylation of site 2a.

In contrast to the *in vitro* results was the observation that mutation of site 5 (Ser656), although affecting phosphorylation at the various sites 3 and 4, did not give any increase in the activity ratio (Skurat et al., 1994). This was interpreted to mean that sites 3a and 3b in COS cells may be phosphorylated by an alternative pathway not dependent on phosphorylation of site 5. Nevertheless, the COS cell data did show that the most important sites in regulation of glycogen synthase were sites 2, 2a, 3a, and 3b, and this conclusion was consistent with the *in vitro* data.

This system was studied further and it was shown that phosphorylation of sites 3a and 3b occurred even though mutations were made at sites 5, 4, and 3c (Skurat & Roach, 1995). Thus, phosphorylation of sites 3a and 3b may occur *via* protein kinases other than GSK-3. Evidence supporting this view has recently been obtained. Skurat and Roach (1996) mutagenized amino acid residues close to the phosphorylation sites, 3a and 3b, that might be important for a protein kinase to recognize and phosphory-late sites 3a and 3b. These were arginine 637 and proline 645. The mutants made were Arg637Gln and Pro645Ala, a double mutant, R637Q, S644A (site 3b), and two triple mutants, S7A (site 2), R637Q, S644A, and S7A, S644A, P645A. In addition, the ser-ine residue of sites 3c, 4, and 5 of these mutants were mutagenized to alanine to avoid possible phosphorylation of sites 3a and 3b by GSK-3. Mutation of Arg637 to Gln eliminated phosphorylation of site 3a, suggesting that Arg637 may be important for another protein kinase to recognize site 3a. The mutant Pro645Ala also eliminated phos-phorylation of site 3b, suggesting a possible involvement of a "proline-directed pro-tein kinase." Either mutation alone did not substantially increase the activation ratio meaning that phosphorylation at either site plus phosphorylation at sites 2 and 2a pro-duced totally inactive enzyme. However, the triple mutant, S7A, R637Q, S644A was active with an activity ratio of 0.62, while S7A, S644A, P645A had an activity ratio of 0.21 (Skurat & Roach, 1996). The results also indicate that in the COS cells sites 2, 2a, 3a, and 3b are all important for regulation of glycogen synthase and, most sig-nificantly, suggest that phosphorylation of sites 3a and 3b can be phosphorylated in-dependently by distinct protein kinases. Thus, three protein kinases are proposed for the phosphorylation of sites 3a and 3b (Skurat & Roach, 1996) and are shown in Figure 13-7.

Thus, multiple mechanisms involving at least three protein kinases may occur in the cell in the regulation of sites 3 of glycogen synthase, and two protein kinases for reg-ulation of sites 2. The reason for redundant mechanisms of inactivation of glycogen synthase is unknown at present, but could be the mechanism for integrating messages of a number of hormonal and signal transduction pathways. It should be pointed out, however, that the *in vivo* studies have been done in COS cells and whether the same phenomena, particularly the unidentified protein kinases Pk_x and Pk_y, are relevant to control of glycogen synthase in skeletal muscle remains to be established.

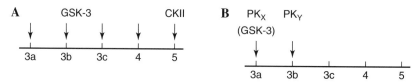

Figure 13–7. (A) Phosphorylation of sites 3a, 3b, 3c, 4 by glycoggen synthase kinase 3 and phos-phorylation of site 5 by casein kinase II. These phosphorylations occur in sequence with site 5 first, and then phosphorylation of 4, 3c, 3b, and finally, 3a. (B) Phosphorylation of site 3a by the putative protein kinase, PK_x and site 3b by the putative protein kinase, PK_y. These phosphorylations are in-dependent of each other and do not require prior phosphorylation of sites 3c, 4, and 5.

Other interesting experiments done in the COS cell system indicated that overexpression of the glycogen biosynthetic enzymes, glycogen synthase, glycogenin, branching enzyme, and UDP-glucose pyrophosphorylase alone did not lead to increased or over-accumulation of glycogen (Skurat et al., 1996). However, if the glycogen synthase mutant S7A, S640A was overexpressed, then there was approximately a 2- to 2.5-fold increase in glycogen levels. This would suggest that the glycogen synthase activity was rate limiting. However the co-overexpression of the mutant glycogen synthase with the UDP-glucose pyrophosphorylase or with glycogenin led to even greater glycogen accumulation: another 34%–70% increase. Thus, with over expression of the hyperactive glycogen synthase, the synthesis of UDP-glucose or the synthesis of the glycosylated acceptor protein may become somewhat rate limiting.

Mechanism of Insulin Stimulation of Mammalian Skeletal Muscle Glycogen Synthesis by Activation of Protein Phosphatase 1 and Inactivation of Glycogen Synthase Kinase-3

Protein Phosphatase 1

Because the glycogen synthase is phosphorylated at multiple sites per subunit, it would be of interest to know whether one or several protein phosphatases catalyze dephosphorylation of these distinct sites. Four protein phosphatases are known and have been reviewed (Ballou & Fischer 1986; Cohen, 1989; Cohen & Cohen, 1989) and the reader is referred to these references for background on the structures and biochemical properties of these phosphatases. Of these, protein phosphatase 1 which mainly hydrolyzes the phosphate of the β-subunit of phosphorylase kinase has potent activity on the phosphate sites 1a, 2, 2a, 3a, 3b, 3c, and 4 of glycogen synthase, and is the principal enzyme in dephosphorylating glycogen synthase since both the phosphatase and glycogen synthase are usually bound to the glycogen particle. Protein phosphatase 2A, which has greater activity on the α-subunit phosphates of phosphorylase kinase than protein phosphatase 1, also has activity on the above glycogen synthase phosphate sites.

It is of interest that protein phosphatase 1, MW 37 kDa, is associated with the glycogen particles *in vivo* when it is complexed to a 160 kilodalton protein referred to as the G subunit (Hubbard & Cohen, 1989a; Strålfors et al., 1985). When protein phosphatase 1 is associated with the G subunit, it binds to the glycogen particle and is far more active enzymatically under physiological conditions in dephosphorylating glycogen synthase, glycogen phosphorylase, and phosphorylase kinase (enzymes that also bind to the glycogen particle) than when it is free (Hubbard & Cohen, 1989c). The G subunit is phosphorylated by cAMP-dependent protein kinase *in vitro* (Hubbard & Cohen, 1989b; Strålfors et al., 1985) and in response to epinephrine *in vivo* (Dent et al., 1990a,b; MacKintosh et al., 1988). Phosphorylation occurs at two serine sites, 1 and 2, separated by 18 residues; the phosphorylation of both sites causes a dissociation of the phosphatase from the G subunit (Hubbard & Cohen, 1989b). The phosphatase is approximately 5- to 8-fold less active than the complex in dephosphorylating glycogen synthase and phosphorylase present in the glycogen particle; this lowering of the protein phosphatase activity is one way epinephrine stimulates glycogen breakdown and inhibits glycogen synthesis.

The dissociation of subunit G from the phosphatase correlates with phosphorylation of site 2 and not site 1 (Hubbard & Cohen, 1989b). Reassociation of G subunit with the phosphatase occurs with dephosphorylation of site 2 by protein phosphatase 2A under conditions where site 1 still retains the phosphate residue, as it is more resistant to dephosphorylation by protein phosphatase 2A. Thus, inactivation of the protein phosphatase is due to phosphorylation of site 2 and not site 1.

In response to insulin administration, Parker et al. (1983) showed that the phosphate released from glycogen synthase is mainly from sites 3a, 3b, and 3c, suggesting that insulin caused the inhibition of GSK-3 or activated protein phosphatase 1. As will be shown, both phenomena, activation of protein phosphatase 1 and inhibition of GSK-3, do occur due to insulin.

Dent et al. (1990a) showed that the G protein–phosphatase complex with phosphate mainly at site 1, had protein phosphatase activity associated with the glycogen particle, and approximately 2.5- to 3-fold higher activity than the dephosphorylated protein phosphatase. Moreover, a protein kinase was isolated from rabbit skeletal muscle that phosphorylated the G subunit at site 1 but not at site 2. This protein kinase also phosphorylated ribosomal protein S6 in vitro. This protein kinase activity was increased approximately twofold within 15 min after insulin administration, which is the same time at which glycogen synthase activity is increased (Parker et al., 1983). The phosphorylation of the G subunit increased the phosphatase activity about 2.8-fold on glycogen synthase and phosphorylase kinase. Likewise, the in vivo stimulation of phosphorylation of site 1 and not site 2 by insulin administration was also demonstrated. As shown in Figure 13-8, Dent et al. (1990b) proposed that the interaction of insulin with its

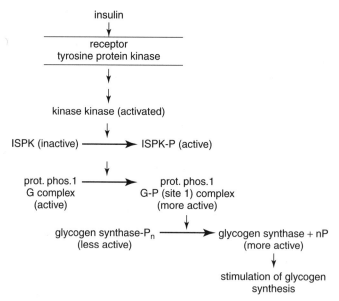

Figure 13–8. Activation of glycogen synthase by insulin as proposed by Dent et al. (1990b). Attachment of insulin to its receptor tyrosine protein kinase activates a series of protein kinases which finally activate a Ser/Thr protein kinase phosphorylating the protein phosphatase 1–G protein complex, making the protein phosphatase activity more active. In turn, this phosphatase dephosphorylates the glycogen synthase at sites 3a, 2, and 2a making the glycogen synthase more active.

membrane-bound receptor activated its tyrosine protein kinase, leading to an activation of a protein serine/threonine kinase-kinase which in turn phosphorylates and makes active a kinase labeled as the insulin-stimulated kinase (ISPK). This ISPK then phosphorylates site 1 of the G subunit, making the protein phosphatase 1-G complex more active. The active protein phosphatase then dephosphorylates the phosphate residues from sites 3a, b, and c, thus activating glycogen synthesis.

Inactivation of Glycogen Synthase Kinase-3

The accumulated evidence also indicates that insulin can induce inactivation of glycogen synthase kinase-3 in many different cells (Cross et al.,1994; Welsh & Proud, 1993; Welsh et al., 1994). The inactivation appears to be a phosphorylation catalyzed by protein kinase B, also known as Akt/RAC (Bellacosa et al., 1991; Coffer & Woodgett, 1991; Cross et al., 1995), which is regulated by an activated phosphatidylinositol (PI3) kinase (Franke et al., 1995). Inactivation of the GSK-3 by protein kinase B is accompanied by phosphorylation of serine 9 of the GSK-3β and serine 21 of the GSK-3α isozymes *in vitro* or *in vivo* (Cross et al., 1995). Although other protein kinases are known to phosphorylate and inactivate GSK-3 (Cross et al., 1995), it is believed that the activation by insulin of glycogen synthase *via* inactivation of GSK-3 is due to activation of protein kinase B. Insulin stimulation of L6 myotube cells caused a 10-fold increase in protein kinase B activity and decreased GSK-3 activity by 40%–50%. The half-time for activation of protein kinase B was one min; this was slightly faster than the half-time inhibition of GSK-3, which was 2 min. The inhibition of GSK-3 was reversed by incubation with protein phosphatase 2A.

Other studies done in whole cells also indicate the importance of GSK-3 in regulating glycogen synthase activity (Eldar-Finkelman et al., 1996). Mutants of GSK-3 at Ser9 were made (Ser9Ala and Ser9Glu). These mutants could not be phosphorylated and their GSK-3 activity made inactive by phosphorylation. The wild-type, normal, and mutant enzymes were expressed in 293 cells and their activity was determined. Cells expressing the S9A mutant, WT, and S9E mutant GSK-3 had 2.6-, 1.8-, and 2.0-fold higher GSK-3 activity, respectively, as compared with control cells. The higher activity of the S9A mutant suggested Ser9 as a key regulatory site for GSK-3 inactivation. However, substitution of glutamic acid for serine could not mimic the inactivation caused by the negative ion, phosphate, on Ser9. The effects of expressing the WT and GSK-3 mutants in the 293 cells on glycogen synthase–glucose-6-P/+ glucose-6-P activity ratio was measured. A 50% reduction in the activity ratio was seen for the cells having the S9A mutant while a 20%–30% decrease was observed in cells having the WT and S9E mutant. Thus, evidence was obtained that activation of GSK-3 is sufficient to inhibit glycogen synthase in intact cells and supports a physiological role for GSK-3 in regulating glycogen synthase.

In summary, insulin stimulates glycogen synthesis by activating protein phosphatase 1 and by inactivating glycogen synthase kinase-3. Thus, the inactivated glycogen synthase becomes active and at the same time phosphorylase a and phosphorylase kinase become inactive due to their dephosphorylation by activated protein phosphatase 1. This results in an overall increase in glycogen synthesis and decreased glycogen degradation.

Recently, three lines of transgenic mice were generated in which the rabbit skeletal muscle glycogen synthase was overexpressed in mouse skeletal muscle (Manchester et al., 1996). The glycogen synthase expressed was the glycogen synthase sites 2 and 3a mutant (Ser7Ala, Ser640Ala) so that the overexpression of the synthase could not be inactivated by phosphorylation. The glycogen synthase activity was increased by as much as 10-fold with concomitant increases of up to fivefold in glycogen content. The levels of UDP-glucose decreased markedly, consistent with the increase in glycogen synthase activity. Levels of the glycogen degradative enzyme, phosphorylase, increased up to threefold but the insulin-sensitive glucose transporter either remained the same or decreased. Thus, increasing active glycogen synthase caused an increase in glycogen, supporting the conclusion that activation of glycogen synthase contributes to the increased accumulation of glycogen in response to insulin.

References

Chapter 1

Angyal, S. J. (1968) *Aust. J. Chem. 21*, 2737–2746.

Angyal, S. J. (1969) *Angew. Chem. Int. Ed. Engl. 8*, 157–166.

Angyal, S. J. (1984) *Adv. Carbohydr. Chem. Biochem. 42*, 15–68.

Angyal, S. J. (1991) *Adv. Carbohydr. Chem. Biochem. 49*, 19–35.

Aspinall, G. O. (Ed.) (1982) in *The Polysaccharides*, Academic Press, New York.

Berdy, J. (1980) *CRC Handbook of Antibiotic Compounds*, Vol. 2, pp 35, CRC Press, Boca Raton, FL.

Booth, G. E., & Ouellete J. (1966) *J. Org. Chem. 31*, 544–546.

Cahn, R. S. (1964) *J. Chem. Ed. 41*, 116–125.

Cahn, R. S., Ingold, C. K., & Prelog, V. (1956) *Experientia 12*, 81–94.

Cahn, R. S., Ingold, C. K., & Prelog, V. (1966) *Angew. Chem. Int. Ed. Engl. 5*, 385–415.

Capon, B. (1969) *Chem. Rev. 69*, 407–498.

Clarke, R. J., Coates, J. M. & Lincoln, S. F. (1988) *Adv. Carbohydr. Chem. Biochem. 46*, 205–249.

Deslongchamps, P. (1983) in *Stereoelectronic Effects in Organic Chemistry*, Pergamon Press, Oxford, England.

Drew, H. D. K., & Haworth, W. N. (1926) *J. Chem. Soc.*, 2303–2310.

Edward, J. T. (1955) *Chem. Ind.* (London), 1102–1104.

El Khadem, H. S. (Ed.) (1982) in *Anthracycline Antibiotics*, Academic Press, New York.

El Khadem, H. S., Horton, D., & Page, T. F., Jr., (1968) *J. Org. Chem. 33*, 734–740.

Eliel, E. L., Allinger, N. L., Angyal, S. J., & Morrison, G. A. (1965) in *Conformational Analysis*, Wiley-Interscience, New York, pp 381.

Ferrier, R. J. & Collins P. M. (1972) in *Monosaccharide Chemistry*, Penguin Books, Harmondsworth, Middlesex, England.

Fischer, E. (1891) *Chem. Ber. 24*, 2683.

Fischer, E. (1907) *Chem. Ber. 40*, 102.

Garg, H. G., & Lyon, N. B. (1991) *Adv. Carbohydr. Chem. Biochem. 49*, 239–261.

Haines, A. H. (1976) *Adv. Carbohydr. Chem Biochem. 33*, 11–109.

Haines, A. H. (1981) *Adv. Carbohydr. Chem Biochem. 39*, 13–70.

Haworth, W. N. (1929) in *The Constitution of Sugars*, Edward Arnold and Co., London.

Horton, D., & Pigman, W. (Ed.) (1970) in *The Carbohydrates, Chemistry and Biochemistry*, pp 217–1099, Academic Press, New York.

Horton, D., & Wander, J. D. (1969) *Carbohydr. Res. 10*, 279–288.

Horton, D., & Wander, J. D. (1970) *Carbohydr. Res. 15*, 271–284.

Hudson, C. S. (1948) *Adv. Carbohyd. Chem. 3*, 1–22.

Jeffrey, G. A., & Fasiska, E. J. (1972) *Carbohydr. Res. 21*, 187–199.

Jeffrey, G. A., & Kim, H. S. (1970) *Carbohydr. Res. 14*, 207–216.

Jochims, J. C., Taigel, G., Seegler, A., Lutz, P., & Driesen, H. E. (1967) *Tetrahedron Lett.*, 4363–4369.

Karplus, M. (1963) *J. Am. Chem. Soc. 85*, 2870–2871.

Kennedy, J. F., & White, C. A. (1983) in *Bioactive Carbohydrates: In Chemistry, Biochemistry and Biology*, pp 132–142, John Wiley & Sons.

Kochetkov, N. K., & Smirnova, G. P. (1986) *Adv. Carbohydr. Chem. Biochem. 44*, 387–438.

Kuettner, K. E., & Kimura, J. H. (1985) *J. Cell Biochem. 27*, 327–336.

Lemieux, R. U. (1971) *Pure Appl. Chem. 25*, 527–548.

Lemieux, R. U., & Chu, N. J. (1958) *Abstr. Papers Amer. Chem. Soc. Meeting 133*, 31N.

Lemieux, R. U., & Morgan A. R. (1965) *Can. J. Chem. 43*, 2205–2213.

Lemieux, R. U., & Stevens, J. D. (1966) *Can. J. Chem. 44*, 249–262.

Mills, J. A. (1956) *Adv. Carbohydr. Chem.. 10*, 1–53.

Montreuil, J. (1982) in *Comprehensive Biochemistry*, Vol. 19B, Part II, Neuberger, A. (Ed.), pp 1–188, Elsevier, Amsterdam.

Reeves, R. E. (1949) *J. Am. Chem. Soc. 71*, 215–217.

Reeves, R. E. (1951) *Adv. Carbohydr. Chem. 6*, 107–134.

Rinehart, Jr., K. L. & Suami, T. (1980), *Aminocyclitol Antibiotics*, ACS Symposium Series 125. American Chemical Society, Washington, DC.

Romers, C., Altona, C., Buys, H. R., & Havinga, E. (1969) in *Topics of Stereochemistry*, Vol. 4, Eliel, E. L., & Allinger, N. L. (Eds.), pp 39, Wiley-Interscience, New York.

Rosanoff, M. A. (1906) *J. Am. Chem. Soc. 28*, 114–121.

Rudrum, M., & Shaw, D. F. (1965) *J. Chem. Soc.*, 52–57.

Ruoslathi, E. (1989) *J. Biol. Chem. 264*, 13,369–13,372.

Schaffer, R. (1970) in *The Carbohydrates, Chemistry and Biochemistry*, Horton, D. & Pigman, W. (Eds.), pp 69–111, Academic Press, New York.

Schardinger, F. (1911) *Zentrabl. Bakteriol. Abt. II 29*, 188–197.

Schwarz, J. C. P. cited in L. Hough & A. C. Richardson (1967) in *Rodd's Chemistry of Carbon Compounds*, Vol I, Part F, Coffey S. (Ed.), p 90 ff, Elsevier, Amsterdam.

Sharon, N. (1975) in *Complex Carbohydrates. Their Chemistry, Biosynthesis, and Functions*, Addison-Wesley Publishing Co., Reading, MA.

Sharon, N., & Lis, H. (1982) in *The Proteins*, Third Ed., Vol. 5, Neurath, H. & Hills, H. L. (Eds.), pp 1–144, Academic Press, New York.

Stoddart, J. F. (1971) in *Stereochemistry of Carbohydrates*, Wiley-Interscience New York.

Sugihara, J. M. (1953) *Adv. Carbohydr. Chem. 8*, 1–44.

Sweet, F., & Brown, R. K. (1968) *Can. J. Chem. 46*, 1543–1548.

Umezawa, H. (Ed.) (1978) in "Index of Antibiotics from Actinomycetes," Vol. 2, University Park Press, Baltimore, MD.

Chapter 2

Agosta, W. C., & Foster, A. M. (1971) *J. Chem. Soc., Chem. Commun.* 433–434.

Aitken, R. A. (1991) in *Organic Reaction Mechanisms* (Knipe, A. C., & Watts, W. E., Eds.) pp 207–227, J. Wiley, Chichester.

Ayral-Kaloustian, S., & Agosta, W. C. (1982) *J. Org. Chem. 47*, 284–287.

Bethell, D. (1969) *Adv. Phys. Org. Chem. 7*, 153–209.

Bozó, E., & Vasella, A. (1992) *Helv. Chim. Acta 75*, 2613–2633.

Bozó, E., & Vasella, A. (1994) *Helv. Chim. Acta 77*, 745–753.

Briner, K. (1995) D. Phil. Thesis, University of Zürich.

Briner, K., Bernet, B., Maloisel, J.-L., & Vasella, A. (1994) *Helv. Chim. Acta 77*, 1969–1984.

Briner, K., & Vasella, A. (1989) *Helv. Chim. Acta 72*, 1371–1382.

Briner, K., & Vasella, A. (1990) *Helv. Chim. Acta 73*, 1764–1778.

Briner, K., & Vasella, A. (1992) *Helv. Chim. Acta 75*, 621–637.

Brook, A. G., & Dillon, P. J. (1969) *Can. J. Chem. 47*, 4347–4351.

Collins, P. M., Oparache, N. N., & Whitton, B. R. (1974) *J. Chem. Soc., Chem. Commun.*, 292–292.

Crawford, R. J., & Raap, R. (1965a) *Can. J. Chem. 43*, 126–132.

Crawford, R. J., & Raap, R. (1965b) *Can. J. Chem. 43*, 356–368.

Crooks, J. E. (1975) in *Proton-Transfer Reactions* (Caldin, E., & Gold, V., Eds.) pp 155–177, Chapman and Hall, London.

Czechtizky, W. (1996) D. M. Sc., ETH Zürich.

Du, X.-M., Fan, H., Goodman, J. L., Kesselmayer, M. A., Krogh-Jespersen, K., LaVilla, J. A., Moss, R. A., Shen, S., & Sheridan, R. S. (1990) *J. Am. Chem. Soc. 112*, 1920–1926.

Eigen, M. (1963) *Angew. Chem. 75*, 489–508.

El-Saidi, M., Kassam, K., Pole, D. L., Tadey, T., & Warkentin, J. (1992) *J. Am. Chem. Soc. 114*, 8751–8752.

Fraser, R. R., Kaufmann, M., Morand, P., & Govil, G. (1969) *Can. J. Chem. 47*, 403–409.

Frey, H. M. (1966) *Adv. Photochem. 4*, 225–256.

Frey, O., Hoffmann, M., Wittmann, V., Kessler, H., Uhlmann, P., & Vasella, A. (1994) *Helv. Chim. Acta 77*, 2060–2069.

Friedrich, K., Jansen, U., & Kirmse, W. (1985) *Tetrahedron Lett. 26*, 193–196.

Glaser, R., Chen, G. S., & Barnes, C. L. (1993) *J. Org. Chem. 58*, 7446–7455.

Graßberger, M. (1971) in *Organische Borverbindungen*, Chemische Taschenbücher Vol. 15, pp 77–85, Verlag Chemie, Weinheim.

Heydt, H., & Regitz, M. (1990) in *Methoden der Organischen Chemie (Houben-Weyl)* (Regitz, M., Ed.) Vol. E19b, pp 1628–1682, Georg Thieme Verlag, Stuttgart.

Hoffmann, R. W. (1971) *Angew. Chem. 83*, 595–630.

Hoffmann, R. W., Lilienblum, W., & Dittrich, B. (1974) *Chem. Ber. 107*, 3395–3407.

Hönig, H., & Weidmann, H. (1979) *Carbohydr. Res. 73*, 260–266.

Hordvik, A. (1974) *Acta Chem. Scand., Ser. B 28*, 261–262.

James, V. J., Stevens, J. D., & Moore, F. H. (1978) *Acta Crystallogr., Sect. B 34*, 188–193.

Jeffrey, G. A., & Saenger, W. (1991) *Hydrogen Bonding in Biological Structures*, Springer-Verlag, Berlin.

Jennings, B. M., & Liu, M. T. H. (1976) *J. Am. Chem. Soc. 98*, 6416–6417.

Jochims, J. C., & Kobayashi, Y. (1976) *Tetrahedron Lett. 24*, 2065–2068.

Jones, S. W., Scheinmann, F., Wakefield, B. J., Middlemiss, D., & Newton, R. F. (1986) *J. Chem. Soc., Chem. Commun.* 1260–1261.

Kirmse, W. (1963) *Liebigs Ann. Chem. 666*, 9–16.

Kirmse, W. (1971) *Carbene Chemistry*, Academic Press, New York.

Kirmse, W., & Kund, K. (1989) *J. Am. Chem. Soc. 111*, 1465–1473.

Kirmse, W., & Kund, K. (1990) *J. Org. Chem. 55*, 2325–2332.

Kirmse, W., Lelgemann, R., & Friedrich, K. (1991) *Chem. Ber. 124*, 1853–1863.

Kirmse, W., Loosen, K., & Sluma, H.-D. (1981) *J. Am. Chem. Soc. 103*, 5935–1696.

Kirmse, W., Van Chiem, P., & Schurig, V. (1985) *Tetrahedron Lett. 26*, 197–200.

Kolthoff, I. M., & Chantooni, M. K. (1969) *J. Am. Chem. Soc. 91*, 4621–4625.

Kuhn, C.-S., Glaudemans, C. P. J., & Lehmann, J. (1989) *Liebigs Ann. Chem.* 357–360.

Kuhn, C.-S., Lehmann, J., Jung, G., & Stefanovic, S. (1992) *Carbohydr. Res. 232*, 227–233.

Kuhn, C.-S., Lehmann, J., & Steck, J. (1990) *Tetrahedron 46*, 3129–3134.

Kurz, G., Lehmann, J., & Thieme, R. (1985) *Carbohydr. Res. 136*, 125–133.

Lehmann, J., Steck, J., Weiser, W., Boos, W., & Wrissenberg, S. (1988) *Carbohydr. Res. 184*, 113–120.

Lehmann, J., & Thieme, R. (1986) *Liebigs Ann. Chem.*, 525–532.

Lemieux, R. U., Kendriks, K. B., Stick, R. V., & James, K. (1975) *J. Am. Chem. Soc. 97*, 4056–4062.

Li, C., & Vasella, A. (1993a) *Helv. Chim. Acta 76*, 197–210.

Li, C., & Vasella, A. (1993b) *Helv. Chim. Acta 76*, 211–221.

Linden, A., Vasella, A., & Witzig, C. (1992) *Helv. Chim. Acta 75*, 1572–1577.

Liu, M. T. H. (1982) *Chem. Soc. Rev. 11*, 127–140.

Liu, M. T. H., & Ramakrishnan, K. (1977) *Tetrahedron Lett. 36*, 3139–3142.

Liu, M. T. H., & Stevens, I. D. R. (1987) in *Chemistry of Diazirines* (Liu, M. T. H., Ed.) Vol. 1, pp 111–160, CRC Press, Inc., Boca Raton, Florida.

Liu, M. T. H., Tencer, M., & Stevens, I. D. R. (1986) *J. Chem. Soc., Perkin Trans. 2*, 211–214.

Lopusinski, A., Bernet, B., Linden, A., & Vasella, A. (1993) *Helv. Chim. Acta 76*, 94–112.

Maloisel, J.-L., & Vasella, A. (1992) *Helv. Chim. Acta 75*, 1491–1514.

Mangholz, S. E. (1997) D. Phil. Thesis, University of Zürich, *in preparation.*

Mangholz, S. E., & Vasella, A. (1991) *Helv. Chim. Acta 74*, 2100–2111.

Mangholz, S. E., & Vasella, A. (1995) *Helv. Chim. Acta 78*, 1020–1035.

Moss, R. A. (1980) *Acc. Chem. Res. 13*, 58–64.

Moss, R. A. (1989) *Acc. Chem. Res. 22*, 15–21.

Moss, R. A., & Jones, M., Eds. (1973) *Carbenes*, Vol. 1, Wiley-Interscience, New York.

Moss, R. A., & Jones, M., Eds. (1975) *Carbenes*, Vol. II., Wiley-Interscience, New York.

Moss, R. A., & Jones, M. (1978) in *Reactive Intermediates* (Jones, M., & Moss, R. A., Eds.) Vol. 1, pp 69–116, J. Wiley, New York.

Moss, R. A., & Jones, M. (1981) in *Reactive Intermediates* (Jones, M., & Moss, R. A., Eds.) Vol. 2, pp 59–134, J. Wiley, New York.

Moss, R. A., & Jones, M. (1985) in *Reactive Intermediates* (Jones, M., & Moss, R. A., Eds.) Vol. 3, pp 69–116, J. Wiley, New York.

Moss, R. A., Shen, S., Hadel, L. M., Kmiecik-Lawrynowicz, G., Wlostowska, J., & Krogh-Jespersen, K. (1987) *J. Am. Chem. Soc. 109*, 4341–4349.

Moss, R. A., Shen, S., & Wlostowski, M. (1988) *Tetrahedron Lett. 29*, 6417–6420.

Moss, R. A., & Zdrojewski, T. (1990) *J. Phys. Org. Chem. 3*, 694–696.

Moss, R. A., Zdrojewski, T., Krogh-Jespersen, K., Wlostowski, M., & Matro, A. (1991) *Tetrahedron Lett. 32*, 1925–1928.

Muddasani, P. R., Bernet, B., & Vasella, A. (1994a) *Helv. Chim. Acta 77*, 334–350.

Muddasani, P. R., Bozó, E., Bernet, B., & Vasella, A. (1994b) *Helv. Chim. Acta 77*, 257–290.

Padwa, A., & Hornbuckle, S. F. (1991) *Chem. Rev. 91*, 263–309.

Park, Y. J., Kim, H. S., & Jeffrey, G. A. (1971) *Acta Crystallogr., Sect. B 27*, 220–227.

Pirrung, M. C., Chang, V. K., & DeAmicis, C. V. (1989) *J. Am. Chem. Soc. 111*, 5824–5831.

Praly, J.-P., Di Stefano, C., Descotes, G., & Faure, R. (1994) *Tetrahedron Lett. 35*, 89–92.

Praly, J.-P., El Kharraf, Z., & Descotes, G. (1990) *Tetrahedron Lett. 31*, 4441–4442.

Rajamannar, T. (1990) *Internal Report, University of Zürich.*

Ratcliffe, A. J., & Fraser-Reid, B. (1990) *J. Chem. Soc., Perkin Trans. 1*, 747–750.

Schmidt, R. R. (1986) *Angew. Chem. 98*, 213–236.

Schmidt, R. R., Behrendt, M., & Toepfer, A. (1990) *Synlett* 694–696.

Schmitz, E., & Ohme, R. (1961) *Chem. Ber. 94*, 2166–2173.

Sheridan, R. S., Moss, R. A., Wilk, B. K., Shen, S., Wlostowski, M., Kesselmayer, M. A., Subramanian, R., Kmiecik-Lawrynowicz, G., & Krogh-Jespersen, K. (1988) *J. Am. Chem. Soc. 110*, 7563–7564.

Somsak, L., Praly, J.-P., & Descotes, G. (1992) *Synlett* 119–120.

Spedding, H. (1961) *J. Chem. Soc.* 3617–3621.

Takahashi, Y., & Vasella, A. (1992) *Helv. Chim. Acta 75*, 1563–1571.

Tichy, M. (1965) *Adv. Org. Chem. 5*, 115–298.

Uhlmann, P., Harth, E., Naughton, A. B., & Vasella, A. (1994a) *Helv. Chim. Acta 77*, 2335–2340.

Uhlmann, P., Nanz, D., Bozó, E., & Vasella, A. (1994b) *Helv. Chim. Acta 77*, 1430–1440.

Uhlmann, P., & Vasella, A. (1992) *Helv. Chim. Acta 75*, 1979–1994.

Uhlmann, P., & Vasella, A. (1994) *Helv. Chim. Acta 77*, 1175–1192.

Vasella, A. (1991) *Pure Appl. Chem. 63*, 507–518.

Vasella, A. (1993) *Pure Appl. Chem. 65*, 731–752.

Vasella, A., Baudin, G., & Panza, L. (1991a) *Heteroatom Chem. 2*, 151–161.

Vasella, A., Dhar, P., & Witzig, C. (1993a) *Helv. Chim. Acta 76*, 1767–1778.

Vasella, A., Uhlmann, P., Waldraff, C. A. A., Diederich, F., & Thilgen, C. (1992) *Angew. Chem. 104*, 1383–1385.

Vasella, A., & Waldraff, C. A. A. (1991) *Helv. Chim. Acta 74*, 585–593.

Vasella, A., Witzig, C., & Husi, R. (1991b) *Helv. Chim. Acta 74*, 1362–1372.

Vasella, A., & Witzig, C. (1995) *Helv. Chim. Acta 78*, 1971–1982.

Vasella, A., Witzig, C., Waldraff, C., Uhlmann, P., Briner, K., Bernet, B., Panza, L., & Husi, R. (1993b) *Helv. Chim. Acta 76*, 2847–2875.

Waldraff, C. (1995) D. Phil. Thesis, University of Zürich.

Wittmann, V. & Kessler, H. (1993) *Angew. Chem. Int. Ed. Engl. 32*, 1091–1093.

Yates, P., & Crawford, R. J. (1966) *J. Am. Chem. Soc. 88*, 1463–1464.

Yates, P., & Loutfy, R. O. (1975) *Acc. Chem. Res. 8*, 209–216.

Zapata, A., Bernet, B., & Vasella, A. (1996) *Helv. Chim. Acta 79*, 1169–1189.

Chapter 3

Alais, J., & David, S. (1990) *Carbohydr. Res. 201*, 69–77.

Alais, J., & Veyrieres, A. (1990) *Carbohydr. Res. 207*, 11–31.

Andersson, F., Fügedi, P., Garegg, P. J., & Nashed, M. (1986) *Tetrahedron Lett. 27*, 3919–3922.

Apparu, M., Blanc-Muesser, M., Defaye, J., & Driguez, H. (1981) *Can. J. Chem. 59*, 314–320.

Araki, Y., Watanabe, K., Kuan, F.-H., Itoh, K., Kobayachi, N., & Ishido, Y. (1984) *Carbohydr. Res. 127*, C5–C9.

Arcamone, F., Penco, S., Redaelli, S., & Hanessian, S. (1976) *J. Med. Chem. 19*, 1424–1425.

Auge, C., Warren, C. D., & Jeanloz, R. W. (1980) *Carbohydr. Res. 82*, 85–95.

Barresi, F., & Hindsgaul, O. (1991) *J. Am. Chem. Soc. 113*, 9376–9377.

Baumann, H., Lönn, H., & Lönngren, J. (1983) *Carbohydr. Res. 114*, 317–321.

Birberg, W., Fügedi, P., Garegg, P. J., & Pilotti, Å. (1989) *J. Carbohydr. Chem. 8*, 47–57.

Bochkov, A. F., & Zaikov, G. E. (1979) *Chemistry of the O-Glycosidic Bond*, Chapter 2, Pergamon Press, Oxford.

Bock, K., & Pedersen, C. (1974) *J. Chem. Soc., Perkin Trans. 2*, 293–297.

Bock, K., & Pedersen, C. (1976) *Acta Chem. Scand., Ser. B 30*, 727–732.

Bock, K., & Pedersen, C. (1983) *Adv. Carbohydr. Chem. Biochem. 41*, 27–66.

Bock, K., & Pedersen, H. (1988) *Acta Chem. Scand., Ser. B 42*, 75–85.

Bock, K., & Thogersen, H. (1982) *Annu. Rep. NMR Spectrosc. 13*, 1–57.

Bock, K., Fernandez-Bolanos Guzmán, J., & Refn, S. (1992) *Carbohydr. Res. 232*, 353–357.

Bock, K., Pedersen, C., & Pedersen, H. (1984) *Adv. Carbohydr. Chem. Biochem. 42*, 193–225.

Bommer, R., Kinzy, W., & Schmidt, R. R. (1991) *Liebigs Ann. Chem.* 425–433.

Bommer, R., & Schmidt, R. R. (1989) *Liebigs Ann. Chem.* 1107–1111.

Boons, G.-J. (1996a) *Tetrahedron 52*, 1095–1121.

Boons, G.-J. (1996b) *Contemp. Org. Syn. 3*, 173–200.

Brauns, D. H. (1923a) *J. Am. Chem. Soc. 45*, 833–835.

Brauns, D. H. (1923b) *J. Am. Chem. Soc. 45*, 2381–2390.

Brigl, P., Gronemeier, K., & Schulz, A. (1939) *Ber. 72*, 1052–1059.

Cerny, M., & Pacak, J. (1961) *Collect. Czech. Chem. Commun. 26*, 2084–2086.

Cerny, M., Vrkoc, J., & Stanek, J. (1959) *Collect. Czech. Chem. Commun. 24*, 64–69.

Cerny, M., Zachystalova, D., & Pacak, J. (1961) *Collect. Czech. Chem. Commun. 26*, 2206–2211.

Chittenden, G. J. F. (1992) *Carbohydr. Res. 242*, 297–301.

Chiu, S.-H. L., & Anderson, L. (1976) *Carbohydr. Res. 50*, 227–228.

Collins, P. M., & Ali, M. H. (1990) *Tetrahedron Lett. 31*, 4517–4520.

Collins, P. M., Manro, A., Opara-Mottah, E. C., & Ali, M. H. (1988) *J. Chem. Soc., Chem. Commun.* 272–274.

Danishefsky, J. J., Gervay, J., Peterson, J. M. McDonald, F. E., Koseki, K., Oriyama, T., Griffith, D. A, Wong, C.-H., & Dumas, D. P. (1992a) *J. Am. Chem. Soc. 114*, 8329–8331.

Danishefsky, S. J., Koseki, K., Griffith, D. A., Gervay, J., Peterson, J. M., McDonald, F. E., & Oriyama, T. (1992b) *J. Am. Chem. Soc. 114*, 8331–8333.

Danishefsky, S. J., McClure, K. F., Randolph, J. T., Ruggeri, R. B. (1993) *Science 260*, 1307–1309.

Ding, Y., & Liu, Y. (1991) *Carbohydr. Res. 209*, 306–310.

Dmitriev, B. A., Chernyak, A. Y., & Bairamova, N. E. (1975) *Bull. Acad. Sci. USSR Div. Chem. Sci.. 24*, 132–137.

Dolle, R. E., & Nicolaou, K. C. (1985) *J. Am. Chem. Soc. 107*, 1691–1694.

Douglas, S. P., Whitfield, D. M., & Krepinsky, J. J. (1991) *J. Am. Chem. Soc. 113*, 5095–5097.

Douglas, S. P., Whitfield, D. M., & Krepinsky, J. J. (1993) *J. Carbohydr. Chem. 12*, 131–136.

Douglas, S. P., Whitfield, D. M., & Krepinsky, J. J. (1995) *J. Am. Chem. Soc. 117*, 2116–2117.

Duchaussoy, P., Lei, P. S., Petitou, M., Sinaÿ, P., Lormeau, J. C., & Choay, J. (1991) *Bioorg. Med. Chem. Lett. 1*, 99–1104

Edward, J. T. (1955) *Chem. Ind.* (London), 1102–1104.

Evans, D. A., Truesdale, L. K., Grimm, K. G., & Nesbitt, S. L. (1977) *J. Am. Chem. Soc. 99*, 5009–5017.

Excoffier, G., Gagnaire, D., Utille, J. P., & Vignon, M. (1972) *Tetrahedron Lett.*, 5065–5068.

Ferrier, R. J., Hay, R. W., & Vethaviyasar, N. (1973) *Carbohydr. Res. 27*, 55–61.

Fischer, E. (1893) *Ber. 26*, 2400.

Fischer, E. (1895) *Ber. 28*, 1145–1167.

Fischer, E., & Armstrong, E. F. (1901) *Ber. 34*, 2885–2908.

Fischer, E., & Delbrük, K. (1909) *Ber. 42*, 1476–1482.

Flowers, H. M. (1972) *Methods Carbohydr. Chem. 6*, 474–480.

Fodor, S. P. A., Read, J. L., Pirrung, M. C., Stryer, L., Lu, A. T., & Solas, D. (1991) *Science 251*, 767–773.

Fraser-Reid, B., Konradsson, P., Mootoo, D. R., & Udodong, U. E. (1988) *J. Chem. Soc., Chem. Commun.* 823–825.

Fraser-Reid, B., Udodong, U. E., Wu, Z., Ottosson, H., Merritt, J. R., Rao, C. S., Roberts, C., & Madsen, R. (1992) *Synlett*, 927–942.

Fraser-Reid, B., Wu, Z., Udodong, U. E., & Ottosson, H. (1990) *J. Org. Chem. 55*, 6068–6070.

Frechet, J. M. & Schuerch, C. (1971) *J. Am. Chem. Soc. 93*, 492–496.

Friesen, R. W., & Danishefsky, S. (1989) *J. Am. Chem. Soc. 111*, 6656–6660.

Frush, H. L., & Isbell, H. S. (1941) *J. Res. Natl. Bur. Stds. 27*, 413–428.

Fügedi, P., & Garegg, P. J. (1986) *Carbohydr. Res. 149*, C9–C12.

Fügedi, P., Garegg, P. J., Lönn, H., & Norberg, T. (1987) *Glyconjugate J. 4*, 97–108.

Garegg, P. J. (1992) *Acc. Chem. Res. 25*, 575–580.

Garegg, P. J., & Helland, A.-C. (1993) *J. Carbohydr. Chem. 12*, 105–117.

Garegg, P. J., & Ossowski, P. (1983) *Acta Chem. Scand., Ser. B 37*, 249–250.

Geysen, H. M., Meloen, R. H., & Barteling, S. J. (1984) *Proc. Natl. Acad. Sci. U.S.A. 82*, 3998–4002.

Greene, T. W., & Wuts, P. G. M. (1991) *Protective Groups in Organic Synthesis*, Second Ed., John Wiley & Sons, Inc. New York.

Greilich, U., Zimmermann, P., Jung, K.-H., & Schmidt, R. R. (1993) *Liebigs Ann. Chem.* 859–864.

Griffith, D. A., & Danishefsky, S. J. (1992) *J. Am. Chem. Soc. 113*, 5863–5864.

Gross, H., Farkas, I., & Bognár, R. (1978) *Z. Chem. 18*, 201–210.

Grundler, G., & Schmidt, R. R. (1984) *Liebigs Ann. Chem.* 1826–1847.

Grundler, G., & Schmidt, R. R. (1985) *Carbohydr. Res. 135*, 203–218.

Gunther, W., & Kunz, H. (1992) *Carbohydr. Res. 228*, 217–241.

Guthrie, R.D., Jenkins, A., & Roberts, G.A.F. (1973) *J. Chem. Soc., Perkin Trans. 1*, 2414–2417.

Halcomb, R. L., & Danishefsky, S. J. (1989) *J. Am. Chem. Soc. 111*, 6661–6666.

Hall, L. D., Manville, J. F., & Bhacca, N. S. (1969) *Can. J. Chem. 47*, 1–17.

Hällgren, C., & Widmalm, G. (1993) *J. Carbohydr. Chem. 12*, 309–333.

Handlon, A. L., & Fraser-Reid, B. (1993) *J. Am. Chem. Soc. 115*, 3796–3797.

Hanessian, S. (1968) in *Advances in Chemistry Series*, No. 74, p 159 ff, ACS, Washington, DC.

Hanessian, S. Ed. (1997) *Preparative Carbohydrate Chemistry*, Marcel Dekker, New York.

Hanessian, S., Bacquet, C., & Lehong, N. (1980) *Carbohydr. Res. 80*, C17–C22.

Hanessian, S., & Banoub, J. (1977) *Carbohydr. Res. 53*, C13–C16.

Hanessian, S., & Plessas, N. R. (1969) *J. Org. Chem. 34*, 1053–1058.

Hanessian, S., Ponpipom, M. M., & Lavallee, P. (1972) *Carbohydr. Res. 24*, 45–56.

Hasegawa, A., Ishida, H., Nagahama, T., & Kiso, M. (1993) *J. Carbohydr. Chem. 12*, 703–718.

Hasegawa, A., Nakamura, J., & Kiso, M. (1986) *J. Carbohydr. Chem. 5*, 11–19.

Hasegawa, A., Nagahama, T., Ohki, H., & Kiso, M. (1992) *J. Carbohydr. Chem. 11*, 699–714.

Hasegawa, A., Ohki, H., Nagahama, T., Ishida, H., & Kiso, M. (1991) *Carbohydr. Res. 212*, 277–281.

Hashimoto, S., Hayashi, M., & Noyori, R. (1984) *Tetrahedron Lett. 25*, 1379–1382.

Hashimoto, S.-I., Yanagiya, Y., Honda, T., Harada, H., & Ikegami, S. (1992) *Tetrahedron Lett. 33*, 3523–3526.

Hayashi, M., Hashimoto, S., & Noyori, R. (1984) *Chem. Lett.* 1747–1750.

Helferich, B., Bohn, E., & Winkler, S. (1930) *Ber. 63*, 989–998.

Helferich, B., & Gootz, R. (1929) *Ber. 62*, 2505–2507.

Helferich, B., & Gootz, R. (1931) *Ber. 64*, 109–114.

Hermkens, P. H. H., Ottenheijm, H. C. J., & Rees, D. (1996) *Tetrahedron 52*, 4527–4554.

Hermkens, P. H. H., Ottenheijm, H. C. J., & Rees, D. C. (1997) *Tetrahedron 53*, 5643–5678.

Horton, D. (1963) *Methods Carbohydr. Chem. 2*, 433–437.

Hronowski, L. J. J., Szarek, W. A., Hay, G. W., Krebs, A., & Depew, W. T. (1989) *Carbohydr. Res. 190*, 203–218.

Hudson, C. S. (1924) *J. Am. Chem. Soc. 46*, 462–477.

Igarashi, K. (1977) *Adv. Carbohydr. Chem. Biochem., 34*, 243–283.

Ishida, H., Ohta, Y., Tsukada, Y., Kiso, M., & Hasegawa, A. (1993) *Carbohydr. Res. 246*, 75–88.

Ishikawa, T., & Fletcher, H. G. (1969) *J. Org. Chem. 34*, 563–571.

Ito, Y., Numata, M., Sugimoto, M., & Ogawa, T. (1989) *J. Am. Chem. Soc. 111*, 8508–8510.

Ito, Y., & Ogawa, T. (1987) *Tetrahedron Lett. 28*, 4701–4704.

Iversen, T., & Bundle, D. R. (1981) *J. Org. Chem. 46*, 5389–5393.

Iversen, T., & Bundle, D. R. (1982) *Can. J. Chem. 60*, 299–303.

Jain, R. K., & Matta, K. L. (1992) *Carbohydr. Res. 226*, 91–100.

Jain, R. K., Piskorz, C. F., & Matta, K. L. (1993) *Carbohydr. Res. 243*, 385–391.

Joshi, S., Rodebaugh, R., Merritt, J. R., Geysen, H. M., & Fraser-Reid, B. (1997) Abstracts *Am. Chem. Soc. Natl. Mtg.*, San Francisco, April 13–18.

Jünnemann, J., Thiem, J., & Pedersen, C. (1993) *Carbohydr. Res. 249*, 91–94.

Kahne, D., Walker, S., Cheng, Y., & Engen, D. V. (1989) *J. Am. Chem. Soc. 111*, 6881–6882.

Kameyama, A., Ishida, H., Kiso, M., & Hasegawa, A. (1990) *Carbohydr. Res. 200*, 269–285.

Kartha, K. P. R., Kameyama, A., Kiso, M., & Hasegawa, A. (1989) *J. Carbohydr. Chem. 8*, 145–158.

Khan, S. H., & Matta, K. L. (1993) *J. Carbohydr. Chem. 12*, 335–348.

Khan, S. H., & O'Neill, R. A., Eds. (1996) *Modern Methods in Carbohydrate Synthesis*, Harwood Academic Publishers, Amsterdam.

Kihlberg, J., Eichler, E., & Bundle, D. R. (1991) *Carbohydr. Res. 211*, 59–75.

Kihlberg, J., Leigh, D. A., & Bundle, D. R. (1990) *J. Org. Chem. 55*, 2860–2863.

Kinzy, W., & Löw, A. (1993) *Carbohydr. Res. 245*, 193–218.

Kinzy, W., & Schmidt, R. R. (1987) *Carbohydr. Res. 166*, 265–276.

Kirby, A. J. (1983) *The Anomeric Effect and Related Stereoelectronic Effects at Oxygen*, pp. 709, Springer-Verlag, New York.

Kobayashi, M., Yamazaki, F., Ito, Y., & Ogawa, T. (1990) *Carbohydr. Res. 201*, 51–67.

Kochetkov, K., Klimov, E. M., Malysheva, N. N., & Demchenko, A. V. (1991) *Carbohydr. Res. 212*, 77–91.

Kochetkov, K., Zhuin, V. M., Klimov, E. M., Matlysheva, N. N., Makarova, Z. G., & Ott, A. Y. (1987) *Carbohydr. Res. 164*, 241–254.

Koenigs, W., & Knorr, E. (1901) *Ber. 34*, 957–981.

Konradsson, P., Mootoo, D. R., McDevitt, R. E., & Fraser-Reid, B. (1990) *J. Chem. Soc., Chem. Commun.* 270–272.

Konradsson, P., Roberts, C., & Fraser-Reid, B. (1991) *Rec. Trav. Chim. Pays-Bas 110*, 23–24.

Korytnyk, W., & Mills, J. A. (1959) *J. Chem. Soc.* 636–649.

Kovac, P., & Edgar, K. J. (1992) *J. Org. Chem. 57*, 2455–2467.

Kreuzer, M., & Thiem, J. (1986) *Carbohydr. Res. 149*, 347–361.

Kronzer, F. J., & Schuerch, C. (1973) *Carbohydr. Res. 27*, 379–390.

Kuntz, H., & Sager, W. (1985) *Helv. Chim. Acta 68*, 283–287.

Kuyama, H., Nakahara, Y., Nukada, T., Ito, Y., Nakahara, Y., & Ogawa, T. (1993) *Carbohydr. Res. 243*, C1–C7.

Lemieux, R. U. (1963a) *Methods Carbohydr. Chem 2*, 221–222.

Lemieux, R. U. (1963b) *Methods Carbohydr. Chem. 2*, 223–224.

Lemieux, R. U. (1964) in *Molecular Rearrangements* (De Mayo, P., Ed.), Vol. 2, p 709 ff, Interscience, New York.

Lemieux, R. U., Hendricks, K. B., Stick, R. V., & James, K. (1975) *J. Am. Chem. Soc. 97*, 4056–4062.

Lemieux, R. U., & Huber, G. (1956) *J. Am. Chem. Soc. 78*, 4117–4119.

Lemieux, R. U., James, K., & Nagabhushan, T. L. (1973a) *Can. J. Chem. 51*, 42–47.

Lemieux, R. U., James, K., & Nagabhushan, T. L. (1973b) *Can. J. Chem. 51*, 48–52.

Leontein, K., Nilsson, M., & Norberg, T. (1985) *Carbohydr. Res. 144*, 231–240.

Liang, R., Yan, L., Loebach, J., Ge, M., Uozumi, Y., Sekanina, K., Horan, N., Glidersleeve, J., Thompson, C., Smith, A., Biswas, K., Still, W.C., & Kahne, D. (1996) *Science 274*, 1520–1522.

Liptak, A., Jodal, I., Harangi, J., & Nanasi, P. (1983) *Acta Chim. Hung. 113*, 415–422.

Lönn, H. (1985) *Carbohydr. Res. 139*, 105–113.

Lucas, H., Basten, J. E. M., Konradsson, P., & van Boeckel, C. A. A. (1993) *Angew. Chem. Int. Ed. Engl. 32*, 434–436.

Lucas, H., Basten, J. E. M., van Dinther, T. G., Meuleman, D. G., van Aelst, S. F., & van Boeckel, C. A. A. (1990) *Tetrahedron Lett. 46*, 8207–8228.

Marino-Albernas, J.-R., Harris, S. L., Varma, V., & Pinto, B. M. (1993) *Carbohydr. Res. 245*, 245–257.

Marousek, V., Lucas, T. J., Wheat, P. E., & Schuerch, C. (1978) *Carbohydr. Res. 60*, 85–96.

Martin, T. J., Brescello, R., Toepfer, A., & Schmidt, R. R. (1993) *Glyconjugate J. 10*, 16–25.

Matsumoto, T., Katsuki, M., Jona, H., & Suzuki, K. (1989) *Tetrahedron Lett. 30*, 6185–6188.

Matsumoto, T., Katsuki, M., & Suzuki, M. (1988b) *Tetrahedron Lett. 29*, 6935–6938.

Matsumoto, T., Maeta, H., Suzuki, K., & Tsuchihashi, G. (1988a) *Tetrahedron Lett. 29*, 3567–3570.

Matsuzaki, Y., Ito, Y., & Ogawa, T. (1992) *Tetrahedron Lett. 33*, 4025–4028.

Matsuzaki, Y., Nunomura, S., Ito, Y., Sugimoto, M., Nakahara, Y., & Ogawa, T. (1993) *Carbohydr. Res. 242*, C1–C6.

Merrifield, R. B. (1963) *J. Am. Chem. Soc. 85*, 2149–2154.

Merritt, J. R., & Fraser-Reid, B. (1992) *J. Am. Chem. Soc. 114*, 8334–8336.

Michael, A. (1879a) *Am. Chem. J. 1*, 305–312.

Michael, A. (1879b) *Compt. Rend. 89*, 355.

Michael, A. (1885) *Am. Chem. J. 6*, 336–340.

Micheel, F., & Klemer, A. (1952) *Chem. Ber. 85*, 187–188.

Micheel, F., Klemer, A., Baum, G., Ristic, P., & Zumbülte, F. (1955) *Chem. Ber. 88*, 475–479.

Micheel, F., Klemer, A., Nolte, M., Nordiek, H., Tork, L., & Westermann, H. (1957) *Chem. Ber. 90*, 1612–1617.

Mootoo, D. R., Konradsson, P., Udodong, U. E., & Fraser-Reid, B. (1988) *J. Am. Chem. Soc. 110*, 5583–5584.

Mori, M., Ito, Y.,& Ogawa, T. (1990) *Carbohydr. Res. 195*, 199–224.

Mukaiyama, T., Hashimoto, Y., & Shoda, S. (1983) *Chem. Lett.*, 935–938.

Mukaiyama, T., Murai, Y., & Shoda, S. (1981) *Chem. Lett.*, 431–432.

Murase, T., Ishida, H., Kiso, M., & Hasegawa, A. (1988) *Carbohydr. Res. 184*, C1–C4.

Nakahara, Y., & Ogawa, T. (1987) *Tetrahedron Lett. 28*, 2731–2734.

Newth, F. H., & Phillips, G. O. (1953) *J. Chem. Soc.* 2896–2900.

Nicolaou, K. C., Caulfield, T., Kataoka, H., & Kumagawa, T. (1988) *J. Am. Chem. Soc. 110*, 7910–7911.

Nicolaou, K. C., Caulfield, T. J., Kataoka, H., & Stylianides, N. A. (1990) *J. Am. Chem. Soc. 112*, 3693–3695.

Nicolaou, K. C., Chucholowski, A., Dolle, R. E., & Randall, J. L. (1984a) *J. Chem. Soc., Chem. Commun.* 1155–1156.

Nicolaou, K. C., Dolle, R. E., Papahatjis, D. P., & Randall, J. L. (1984b) *J. Am. Chem. Soc. 106*, 4189–4192.

Nicolaou, K. C., Hummel, C. W., & Iwabuchi, Y. (1992) *J. Am. Chem. Soc. 114*, 3126–3128.

Nicolaou, K. C., Seitz, S. P., & Papahatjis, D. P. (1982) *J. Am. Chem. Soc. 105*, 2430–2434.

Nicolaou, K. C., Winssinger, N., Pastor, J., & DeRoose, F. (1997) *J. Am. Chem. Soc. 119*, 449–450.

Nilsson, M., & Norberg, T. (1990) *J. Carbohydr. Chem. 9*, 1–14.

Nilsson, S., Lönn, H., & Norberg, T. (1991) *J. Carbohydr. Chem. 10*, 1023–1048.

Noyori, R., & Hayashi, M. (1987) *Chem. Lett.*, 57–60.

Nukada, T., Kitajima, T., Nakahara, Y., & Ogawa, T. (1992) *Carbohydr. Res. 228*, 157–170.

Numata, M., Sugimoto, M., Shibayama, S., & Ogawa, T. (1988) *Carbohydr. Res. 174*, 73–85.

Ogawa, T., Kitajima, T., & Nukada, T. (1983a) *Carbohydr. Res. 123*, C8–C11.

Ogawa, T., & Matsui, M. (1977) *Carbohydr. Res. 54*, C17–C21.

Ogawa, T., & Nukada, T. (1985) *Carbohydr. Res. 136*, 135–152.

Ogawa, T., & Takahashi, Y. (1985) *Carbohydr. Res. 138*, C5–C9.

Ogawa, T., Nakabayashi, S., & Kitajima, T. (1983b) *Carbohydr. Res. 114*, 225–236.

Ogawa, T., Nakabayashi, S., & Sasajima, K. (1981) *Carbohydr. Res. 95*, 308–312.

Ogawa, T., Sugimoto, M., Kitajima, T., Sadozai, K. K., & Nukada, T. (1986) *Tetrahedron Lett. 27*, 5739–5742.

Olah, G. A., Welch, J. T., Vankar, Y. D., Nojima, M., Kerekes, I., & Olah, J. A. (1979) *J. Org. Chem. 44*, 3872–3881.

Pacsu, E. (1928) *Ber. 61*, 1508–1513.

Pacsu, E., & Wilson, E. J., Jr. (1939) *J. Am. Chem. Soc. 61*, 1930–1931.

Paulsen, H. (1982) *Angew. Chem. Int. Ed. Engl. 21*, 155–173.

Paulsen, H., Heume, M., & Nürnberger, H. (1990) *Carbohydr. Res. 200*, 127–166.

Paulsen, H., & Höffgen, E. C. (1993) *Liebigs Ann. Chem.* 531–541.

Paulsen, H., & Lebuhn, R. (1984a) *Carbohydr. Res. 125*, 21–45.

Paulsen, H., & Lebuhn, R. (1984b) *Carbohydr. Res. 130*, 85–101.

Paulsen, H., & Lebuhn, R. (1983) *Liebigs Ann. Chem.*, 1047–1072.

Paulsen, H., & Lockhoff, H. (1981) *Chem. Ber. 114*, 3102–3114.

Paulsen, H., Meinjohanns, F., Reck, F., & Brockhausen, I. (1993) *Liebigs Ann. Chem.*, 721–735.

Paulsen, H., & Peters, T. (1987) *Carbohydr. Res. 165*, 229–249.

Paulsen, H., Richter, A., Sinnwell, V., & Stenzel, W. (1978) *Carbohydr. Res 64*, 339–364.

Paulsen, H., van Dorst, J. A. L. M., Reck, F., & Meinjohanns, E. (1992) *Liebigs Ann. Chem.* 513–521.

Pedersen, C., & Refn, S. (1978) *Acta Chem. Scand., Ser. B 32*, 687–689.

Peters, T., & Bundle, D. R. (1989) *Can. J. Chem. 67*, 491–496.

Petitou, M., Jaurand, G., Derrien, M., Duchaussoy, P., & Choay, J. (1991) *Bioorg. Med. Chem. Lett. 1*, 95–98.

Pinto, B. M., Morissette, D. G., & Bundle, D. R. (1987) *J. Chem. Soc., Perkin Trans. 1*, 9–14.

Posner, G. H., & Haines, S. R. (1985) *Tetrahedron Lett. 26*, 5–8.

Pozsgay, V., & Jennings, H. J. (1987) *Tetrahedron Lett. 28*, 1375–1376.

Prabhanjan, H., Aoyama, K., Kiso, M., & Hasegawa, A. (1992) *Carbohydr. Res. 233*, 87–99.

Prabhanjan, H., Kiso, M., & Hasegawa, A. (1991) *Carbohydr. Res. 211*, C1–C5.

Rademann, I., & Schmidt, R. R. (1996) *Tetrahedron Lett. 37*, 3989–3990.

Raghavan, S., & Kahne, D. (1993) *J. Am. Chem. Soc. 115*, 1580–1581.

Randolph, J. T., McClure, K. F., & Danishfesky, S. J. (1995) *J. Am. Chem. Soc. 117*, 5712–5719.

Rao, C. S., Ratcliffe, A. J., & Fraser-Reid, B. (1993) *J. Chem. Soc., Perkin Trans. 1*, 1207–1211.

Ratcliffe, A. J., & Fraser-Reid, B. (1990) *J. Chem. Soc., Perkins Trans. 1* 747–750.

Ratcliffe, A. J., Konradsson, P., & Fraser-Reid, B. (1990) *J. Am. Chem. Soc. 112*, 5665–5667.

Ratcliffe, A. J., Konradsson, P., & Fraser-Reid, B. (1991) *Carbohyd. Res. 216*, 323–335.

Rio, S., Beau, J.-M., & Jacquinet, J.-C. (1993) *Carbohydr. Res. 244*, 295–313.

Roberge, J. Y., Beebe, X., & Danishefsky, S. J. (1995) *Science 269*, 202–204.

Rosenbrook, W., Jr., Riley, D. A., & Lartey, P. A. (1985) *Tetrahedron Lett. 26*, 3–4.

Roy, R., Andersson, F. O., & Letellier, M. (1992) *Tetrahedron Lett. 33*, 6053–6056.

Sadozai, K. K., Kitajima, T., Nakahara, Y., Ogawa, T., & Kobata, A. (1986a) *Carbohydr. Res. 152*, 173–182.

Sadozai, K. K., Nukada, T., Ito, Y., Nakahara, Y., Ogawa, T., & Kobata, A. (1986b) *Carbohydr. Res. 157*, 101–123.

Sato, S., Ito, Y., Nukada, T., Nakahara, Y., & Ogawa, T. (1987) *Carbohydr. Res. 167*, 197–210.

Sato, S., Mori, M., Ito, Y., & Ogawa, T. (1986) *Carbohydr. Res. 155*, C6–C10.

Schlubach, H. H. (1926) *Ber., 59*, 840–844.

Schlubach, H. H., & Gilbert, R. (1930) *Ber. 63*, 2292–2297.

Schmidt, R. R. (1986) *Angew. Chem. Int. Ed. Engl. 25*, 212–235.

Schmidt, R. R. (1989) *Pure Appl. Chem. 61*, 1257–1270.

Schmidt, R. R., & Michel, J. (1981) *Angew. Chem. Int. Ed. Engl. 21*, 72–73.

Schmidt, R. R., & Michel, J. (1985) *J. Carbohydr. Chem. 4*, 141–169.

Schmidt, R. R., Michel, J., & Roos, M. (1984) *Liebigs Ann. Chem.*, 1343–1357.

Schmidt, R.R., & Stumpp, M. (1983) *Liebigs Ann. Chem.* 1249–1256.

Schwarz, J. C. P. (1973) *J. Chem. Soc., Chem. Commun.*, 505–508.

Sharp, V. E., & Stacey, M. (1951) *J. Chem. Soc.* 285–288.

Sim, M. M., Kondo, H., & Wong, C.-H. (1993) *J. Am. Chem. Soc. 115*, 2260–2267.

Sinaÿ, P. (1978) *Pure Appl. Chem. 50*, 1437–1452.

Sinaÿ, P., & Pougny, J. R. (1976) *Tetrahedron Lett.* 4073–4076.

Sliedregt, L. A. J. M., Zegelaar-Jaarsvald, K., van der Marel, G. A., & van Boom, J. H. (1993) *Synlett*, 335–337.

Smid, P., de Ruiter, G. A., van der Marel, G. A., Rombouts, F. M., & van Boom, J. H. (1991) *J. Carbohydr. Chem. 10*, 833–849.

Spijker, N. M., Westerduin, P., & van Boeckel, C. A. A. (1992) *Tetrahedron Lett. 48*, 6297–6316.

Srivistava, G., & Hindsgaul, O. (1991) *J. Carbohydr. Chem. 10*, 927–933.

Stanek, J., Cerny, M., Kocourek, J., & Pacák, J. (1963) *The Monosaccharides*, Chapter 5, Academic Press, New York.

Stork, G., & Kim, G. (1992) *J. Am. Chem. Soc. 114*, 1087–1088.

Szarek, W.A., Grynkiewicz, G., Doboszewski, B., & Hay, G. W. (1984) *Chem. Lett.*, 1751–1754.

Szurmai, Z., Kerékgyártó, J., Harang, J., & Liptak, A. (1987) *Carbohydr. Res. 164*, 313–325.

Takahashi, Y., & Ogawa, T. (1987) *Carbohydr. Res. 164*, 277–296.
Taubken, N., Sauerbrei, B., & Thiem, J. (1993) *J. Carbohydr. Chem. 12*, 651–667.
Thatcher, G. R. J., Ed. (1993) *The Anomeric Effect and Associated Stereoelectronic Effects*, ACS Symposium Series # 539.
Thiem, J., Karl, H., & Schwenter, J. (1978) *Synthesis* 696–698.
Toshima, K., & Tatsuta, K. (1993) *Chem. Rev. 93*, 1503–1531.
Tsuchiya, T. (1990) *Adv. Carbohydr. Chem. Biochem. 48*, 91–277.
Udodong, U. E., Madsen, R., Roberts, C., & Fraser-Reid, B. (1990) *J. Am. Chem. Soc. 115*, 7886–7887.
Udodong, U. E., Rao, C. S., & Fraser-Reid, B. (1992) *Tetrahedron 48*, 4713–4724.
Urban, F. J., Moore, B. S., & Breitenbach, R. (1990) *Tetrahedron Lett. 31*, 4421–4424.
Veeneman, G. H., Brugghe, H. F., van den Elst, H., & van Boom, J. H. (1990a) *Carbohydr. Res. 195*, C1–C4.
Veeneman, G. H., & van Boom, J. H. (1990a) *Tetrahedron Lett. 31*, 27–30.
Veeneman, G. H., & van Boom, J. H. (1990b) *Tetrahedron Lett. 31*, 275–278.
Veeneman, G. H., van Leeuwen, S. H., & van Boom, J. H. (1990b) *Tetrahedron Lett. 31*, 1331–1334.
Veeneman, G. H., van Leeuwen, S. H., Zuurmond, H., & van Boom, J. H. (1990c) *J. Carbohydr. Chem. 9*, 783–796.
Vliegenthart, J. F. G., Dorland, L., & van Halbeck, H. (1983) *Adv. Carbohydr. Chem. Biochem. 41*, 209–374.
Voznij, Y. V., Kalicheva, I. S., & Galoyan, A. A. (1984) *Bioorg. Khim. 10*, 1256.
Wessel, H. P., Mayer, B., & Englert, G. (1993) *Carbohydr. Res. 242*, 141–151.
Weygand, F., & Ziemann, H. (1962) *Liebigs Ann. Chem.*, 179–198.
Winstein, S., & Buckles, R. E.(1942) *J. Am. Chem. Soc. 64*, 2780–2786.
Wolfrom, M. L., & Thomson, A. (1963) *Methods Carbohydr. Chem. 2*, 427–430.
Wulff, G., & Röhle, G. (1974) *Angew. Chem. Int. Ed. Engl. 13*, 157–170.
Yamazaki, F., Kitajima, T., Nukada, T., Ito, Y., & Ogawa, T. (1990a) *Carbohydr. Res. 201*, 15–30.
Yamazaki, F., Sato, S., Nukada, T., Ito, Y., & Ogawa, T. (1990b) *Carbohydr. Res. 201*, 31–50.
Yamanoi, T., Nakamura, K., Takeyama, H., Yanagihara, K., & Inazu, T. (1993) *Chem. Lett.* 343–346.
Yoshida, M., Uchimura, A., Kiso, M., & Hasegawa, A. (1993) *Glycoconjugate J. 10*, 3–15.
Zehavi, T., & Patchornik, A. (1973) *J. Am. Chem. Soc. 95*, 5673–5677.
Ziegler, T., Eckhardt, E., & Birault, V. (1993) *J. Org. Chem. 58*, 1090–1099.
Zimmermann, P., Bommer, R., Bär, T., & Schmidt, R. R. (1988) *J. Carbohydr. Chem. 7*, 435–452.
Zuurmond, H. M., van der Marel, G. A., & van Boom, J. H. (1991) *Rec. Trav. Chim. Pays-Bas 110*, 301–302.
Zuurmond, H. M., van der Meer, P. H., van der Klein, P. A. M., van der Marel, G. A., van Boom, J. H. (1993) *J. Carbohydr. Chem. 12*, 1091–1103.

Chapter 4

Amvam-Zollo, R. H. (1983) Dissertation, Université d'Orléans.
Barresi, F., & Hindsgaul, O. (1995) *Mod. Synth. Methods*, Vol. 7, 281–330.
Boons, G.-J. (1995) *Contemp. Org. Synth. 3*, 173–200.
Danishefsky, S. J., McClure, K. F., Randolph, J. T., & Ruggeri, R. B. (1993) *Science 260*, 1307–1309.
David, S., Auge, C., & Gautheron, C. (1991) *Adv. Carbohydr. Chem. Biochem. 49*, 175–237.
Douglas, S. P., Whitfield, D. M., & Krepinsky, J. J. (1991) *J. Am. Chem. Soc. 113*, 5095–5097.

Ferrier, R. J., Hay, R. W., & Vethaviyasar, N. (1973) *Carbohydr. Res. 27*, 55–61.

Ferrier, R. J., & Haines, S. R. (1984) *Carbohydr. Res. 127*, 157–161.

Fraser-Reid, B., Konradsson, P., Mootoo, D. R., & Udodong, U. (1988) *J. Chem. Soc., Chem. Commun.* 823–825.

Frechet, J. M. J., & Schuerch, C. (1972) *Carbohydr. Res. 22*, 399–412.

Fügedi, P., Lipták, A., & Nanasi, P. (1982) *Carbohydr. Res. 107*, C5–C8.

Fügedi, P., & Garegg, P. J. (1986) *Carbohydr. Res. 149*, C9–C12.

Fügedi, P., Garegg, P. J., Lönn, H., & Norberg, T. (1987) *Glycoconjugate J. 4*, 97–108.

Gait, M. J., (Ed.) (1984) *Oligonucleotide Synthesis*, IRL Press, Oxford.

Garegg, P. J., & Hällgren, C. (1992) *J. Carbohydr. Chem. 11*, 425–443.

Groneberg, R. D., Miyazaki, T., Stylianides, N. A., Schulze, T. J., Stahl, W., Schreiner, E. P., Suzuki, T., Iwabuchi, Y., Smith, A. L., & Nicolaou, K. C. (1993) *J. Am. Chem. Soc. 115*, 7593–7611.

Halcomb, R. L., & Danishefsky, S. J. (1989) *J. Am. Chem. Soc. 111*, 6661–6666.

Hanessian, S., & Guindon, Y. (1980) *Carbohydr. Res. 86*, C3–C6.

Hanessian, S., Bacquet, C., & Lehong, N. (1980) *Carbohydr. Res. 80*, C17–C22.

Hanessian, S., Ed. (1997) Preparative Carbohydrate Chemistry, Marcel Dekker Inc., New York.

Hashimoto, S., Hayashi, M., & Noyori, R. (1984) *Tetrahedron Lett. 25*, 1379–1380.

Ichikawa, Y., Liu, J. L.-C., Shen, G.-J., & Wong, C.-H. (1991) *J. Am. Chem. Soc. 113*, 6300–6302.

Ichikawa, Y., Lin, Y.-C., Dumas, D. P., Shen, G.-J., Garcia-Junceda, E., Williams, M. A., Bayer, R., Ketchanm, C., Walker, L. E., Paulson, J. C., & Wong, C.-H. (1992a) *J. Am. Chem. Soc. 114*, 9283–9298.

Ichikawa, Y., Look, G. C., & Wong, C.-H. (1992b) *Anal. Biochem. 202*, 215–238.

Igarashi, K. (1977) *Adv. Carbohydr. Chem. Biochem. 34*, 243–283.

Ito, Y., & Ogawa, T. (1987) *Tetrahedron Lett. 28*, 2723–2726.

Janrand, G., Beau, J.-M., & Sinaÿ, P. (1981) *J. Chem. Soc., Chem. Commun.*, 572–573.

Jiang, Z.-H., & Schmidt, R. R. (1992) *Liebigs Ann. Chem.*, 975–982.

Kahne, D., Walker, S., Cheng, Y., & Van Engen, D. (1989) *J. Am. Chem. Soc. 111*, 6881–6882.

Kameyama, A., Ishida, H., Kiso, M., & Hasegawa, A. (1991) *J. Carbohydr. Chem. 10*, 549–560.

Kent, S. B. H. (1988) *Annu. Rev. Biochem. 547*, 957–990.

Koenigs, W., & Knorr, E. (1901) *Chem. Ber. 34*, 957–981.

Kondo, H., Ichikawa, Y., & Wong, C.-H. (1992) *J. Am. Chem. Soc. 114*, 8748–8750.

Lee, M. D., Ellestad, G. A., & Borders, D. B. (1991) *Acc. Chem. Res. 24*, 235–243.

Lemieux, R. U., & Haymi, J. I. (1965) *Can. J. Chem. 43*, 2162–2174.

Lemieux, R. U., & Morgan, A. R. (1965) *Can. J. Chem. 43*, 2190–2198.

Lemieux, R. U., & Driguez, H. (1975a) *J. Am. Chem. Soc. 97*, 4063–4069.

Lemieux, R. U., & Driguez, H. (1975b) *J. Am. Chem. Soc. 97*, 4069–4076.

Lönn, H. (1985) *Carbohydr. Res. 139*, 105–113.

Martin, T. J., & Schmidt, R. R. (1992) *Tetrahedron Lett. 33*, 6123–6126.

Matsuzaki, Y., Ito, Y., & Ogawa, T. (1992) *Tetrahedron Lett. 33*, 6343–6346.

Mootoo, D. R., Konradsson, P., Udodong, U., & Fraser-Reid, B. (1988) *J. Am. Chem. Soc. 110*, 5583–5584.

Mukaiyama, T., Murai, Y., & Shoda, S. (1981) *Chem. Lett.*, 431–432.

Nicolaou, K. C., Bockovich, N. J., & Carcanague, D. R. (1992a) *J. Am. Chem. Soc. 114*, 8701–8702.

Nicolaou, K. C., Caulfield, T. J., Kataoka, H., & Stylianides, N. A. (1990a) *J. Am. Chem. Soc. 112*, 3693–3695.

Nicolaou, K. C., Dolle, R. E., Papahatjis, D. P., & Randall, J. L. (1984) *J. Am. Chem. Soc. 106*, 4189–4192.

Nicolaou, K. C., Dolle, R. E., Papahatjis, P. P., & Randall, J. L. (1985) *J. Am. Chem. Soc. 107*, 5556–5558.

Nicolaou, K. C., Groneberg, R. D., Miyazaki, T., Stylianides, N. A., Schulze, T. J., & Stahl, W. (1990b) *J. Am. Chem. Soc. 112*, 8193–8195.

Nicolaou, K. C., Hummel, C. W., & Iwabuchi, Y. (1992b) *J. Am. Chem. Soc. 114*, 3126–3128.

Nicolaou, K. C., Hummel, C. W., Pitsinos, E. N., Nakada, M., Smith, A. L, Shibayama, K., & Saimot, H. (1992c) *J. Am. Chem. Soc. 114*, 10082–10084.

Nicolaou, K. C., Seitz, S. P., & Papahatjis, D. P. (1983) *J. Am. Chem. Soc. 105*, 2430–2434.

Paulsen, H. (1982) *Angew. Chem. Int. Ed. Engl. 21*, 155–173, and references cited therein.

Paulsen, H., Heume, M., & Nürnberger, H. (1990) *Carbohydr. Res. 200*, 127–166.

Posner, G. H., & Haines, S.R. (1985) *Tetrahedron Lett. 26*, 5–8.

Pougny, J.-R, Jacquinet, J.-C, Nassr, M., Cushet, D., Milat, M.-L., & Sinaÿ, P. (1977) *J. Am. Chem. Soc. 99*, 6762–6763.

Schmidt, R. R., & Michel, J. (1980) *Angew. Chem. Int. Ed. Engl. 19*, 731–732.

Schmidt, R. R., & Michel, J. (1984) *Tetrahedron Lett. 25*, 821–824.

Schmidt, R. R., & Michel, J. (1985) *J. Carbohydr. Chem. 4*, 141–169.

Schmidt, R. R. (1986) *Angew. Chem. Int. Ed. Engl. 25*, 212–235, and references cited therein.

Sharon, N., & Lis, H. (1993) *Scientific Am. 268*, 82–89.

Sim, M. M., Kondo, H., & Wong, C.-H. (1993) *J. Am. Chem. Soc. 115*, 2260–2267.

Sinaÿ, P. (1991) *Pure & Appl. Chem. 63*, 519–528.

Suzuki, K., Maeta, H., & Matsumoto, T. (1989) *Tetrahedron Lett. 30*, 4853–4856, and references cited therein.

Thiem, J., Karl, H., & Schwentner, J. (1978) *Synthesis*, 696–698.

Toone, E. J., Simmon, E. S., Bednarski, M. D., & Whitesides, G. M. (1989) *Tetrahedron 45*, 5365–5422.

Toshima, K., & Tatsuta, K. (1993) *Chem. Rev. 93*, 1503–1531.

Veeneman, G. H., & van Boom, J. M. (1990a) *Tetrahedron Lett. 31*, 275–278.

Veeneman, G. H., van Leeuwen, S. H., & van Boom, J. H. (1990b) *Tetrahedron Lett. 31*, 1331–1334.

Veeneman, G. H., Notermans, S., Liskamp, R. M. J., van der Marel, G. A., & van Boom, J. H. (1987) *Tetrahedron Lett. 28*, 6695–6698.

Weygand, F., Ziemann, H., & Bestmann, H. J. (1958) *Chem. Ber. 91*, 2534–2537.

Wolfrom, M. L., & Groebke, W. (1963) *J. Org. Chem. 28*, 2986–2988.

Chapter 5

Aiyar, J., Danishefsky, S. J., & Crothers, D. M. (1992) *J. Am. Chem. Soc. 114*, 7552–7554.

Bailly, C., & Waring, M. J. (1995) *J. Am. Chem. Soc. 117*, 7311–7316.

Banville, D. L., Keniry, M. A., Kam, M., & Shafer, R. H. (1990a) *Biochemistry 29*, 6521–6534.

Banville, D. L., Keniry, M. A., Kam, M., & Shafer, R. H. (1990b) *Biochemistry 29*, 9294–9304.

Behr, D., Honikel, K., & Harmann, G. (1969) *Eur. J. Biochem. 9*, 82–92.

Bergman, R. G. (1973) *Acc. Chem. Res. 6*, 25–31.

Bifulco, G., Galeone, A., Gomez-Paloma, L., Nicolau, K. C., & Chazin, W.J. (1996) *J. Am. Chem. Soc. 118*, 8817–8824.

Chatterjee, M., Cramer, K. D., & Townsend, C. A. (1994) *J. Am. Chem. Soc. 116*, 8819–8820.

Chatterjee, M., Mah, S. C., Tullius, T. D., & Townsend, C. A. (1995) *J. Am. Chem. Soc. 117*, 8074–8082.

Chiang, S.-Y., Welch, J., Rauscher, F. J., & Beerman, T. A. (1994) *Biochemistry 33*, 7033–7040.

Darby, N., Kim, C.U., Salaün, J. A., Shelton, K. W., Takada, S., & Masamune, S. (1971) *J. Chem. Soc., Chem. Commun.*, 1516–1517.

Dedon, P. C. & Goldberg, I. H. (1992) *Chem. Res. Toxicol. 5*, 311–332.

De Voss, J. J., Hangeland, J. J., & Townsend, C. A. (1990a) *J. Am. Chem. Soc. 112*, 4554–4556.

De Voss, J. J., Townsend, C. A., Ding, W.-D., Morton, G. O., Ellestad, G. A., Zein, N., Tabor, A. B., & Schreiber, S. L. (1990b) *J. Am. Chem. Soc. 112*, 9669–9670.

Ding, W.-D., & Ellestad, G. A. (1991) *J. Am. Chem. Soc. 113*, 6617–6620.

Doyle, T. W. & Borders, D. B. (Eds.) (1995) *Ene-diyne Antibiotic Antitumor Agents*, Dekker: New York.

Fox, K. R. & Howarth, M. R. (1985) *Nucleic Acid Res. 13*, 8695–8714.

Gao, X. & Patel, D. J. (1989) *Biochemistry 28*, 751–762.

Gao, X., Mirau, P., & Patel, D. J. (1992) *J. Mol. Biol. 223*, 259–279.

Goldberg, I. H. (1986) in *Basic Life Sciences: Mechanisms of DNA Damage and Repair* (Simic, M. G., Grossman, L., & Upton, A. C. Eds.), p. 231, Vol. 38, Plenum, NY.

Goldberg, I. H. (1991) *Acc. Chem. Res. 24*, 191–198.

Graddon, D. P. (1969) *Coord. Chem. Rev. 4*, 1–28.

Hangeland, J. J., De Voss, J. J., Heath, J. A., Townsend, C. A., Ding, W., Ashcroft, J. S., & Ellstad, G. A. (1992) *J. Am. Chem. Soc. 114*, 9200–9202.

Hawley, R. C., Kiessling, L. L., & Schreiber, S. (1989) *Proc. Natl. Acad. Sci. U. S. A. 86*, 1105–1109.

Hayasaka, T. & Inoue, Y. (1969) *Biochemistry 8*, 2342–2347.

Hecht, S. M. (1986) *Acc. Chem. Res. 19*, 383–391.

Heinemann, U., Alings, C., & Lauble, H. (1990) in *Structure & Methods*, Vol. 3: DNA & RNA (Sarma, R. H., & Sarma, M. H., Eds) pp. 39–53, Adenine Press, NY.

Hendrickson, W. A. (1990) 10th International Biophysics Congress, Vancouver, British Columbia.

Hurley, L. H., Reynolds, V. L., Swenson, D. H., Petzold, G. L., & Scahill, T. A. (1984) *Science 226*, 843–844.

Ikemoto, N., Kumar, R. A., Ling, T.-T., Ellestad, G. A., Danishefsky, S. J., & Patel, D. J. (1995) *Proc. Natl. Acad. Sci. U.S.A. 92*, 10506–10510.

Itzhaki, L., Weinberger, S., Livnah, N., & Berman, E. (1990) *Biopolymers 29*, 481–489.

Jones, R. R., & Bergmann, R. G. (1972) *J. Am. Chem. Soc. 94*, 660–661.

Kamiyama, M. (1968) *J. Biochem. 63*, 566–672.

Kaziro, Y., & Kamiyama, M. (1967) *J. Biochem. (Tokyo) 62*, 424–429.

Kersten, W., Kersten, H., & Szybalski, W. (1966) *Biochemistry 5*, 236–244.

Koschel, K., Hartmann, G., Kersten, W., & Kersten, H. (1966) *Biochem. Z. 344*, 76–86.

Kumar, R. A., Ikemoto, N., & Patel, D. J. (1997) *J. Mol. Biol. 265*, 182–201.

Langley, D. R., Doyle, T. W., & Beveridge, D. L. (1994) *Tetrahedron 50* (5), 1379–1396.

Lee, M. D., Ellstad, G. A., & Borders, D. B. (1991) *Acc. Chem. Res. 24*, 235–243.

Li, T., Zeng, Z., Estevez, V. A., Baldenius, K. U., Nicolaou, K. C., & Joyce, G. F. (1994) *J. Am. Chem. Soc. 116*, 3709–3715.

Lown, J. W. (1988) Anthracycline and Anthracenedione-Based Anticancer Agents, Section 2. *Bioact. Mol., 6*.

Mah, S. C., Townsend, C. A., & Tulius, T. D. (1994) *Biochemistry 33*, 614–621.

Mayer, J., & Sondheimer, F. (1966) *J. Am. Chem. Soc. 88*, 602–603.

Nicolaou, K. C., Keiichi, A., Komatsu, H., Smith, B. M., Li, T., Egan, M. G., & Gomez-Paloma, L. (1995) *Angew. Chem. Int. Ed. Engl. 34*, 576–578.

Nicolaou, K. C., Smith, B. M., Ajito, K., Komatsu, H., Gomez-Paloma, L., & Tor, Y. (1996) *J. Am. Chem. Soc. 118*, 2303–2304.

Nicolaou, K. C., Tsay, S.-C., Suzuki, T., & Joyce, G. F. (1992) *J. Am. Chem. Soc. 114*, 7555–7559.

Paloma, L. G., Smith, J. A., Chazin, W. J., & Nicolaou, K. C. (1994) *J. Am. Chem. Soc. 116*, 3697–3708.

Shibata, M., Tanabe, K., Hamada, Y., Nakazawa, K., Miyaba, A., Hitomi, H., Miyamoto, M., & Mizuno, K. (1960) *J. Antibiot., Ser. B 13*, 1–4.

Silva, D. J., & Kahne, D. (1993) *J. Am. Chem. Soc. 115*, 7962–7970.

Silva, D. J., Goodnow, R., & Kahne, D. (1993) *Biochemistry 32*, 463–471.

Silva, D. J., Kahne, D., & Kraml, C. M. (1994) *J. Am. Chem. Soc. 116*, 2641–2642.

Skarbek, J. D., & Speedie, M. K. (1981) in *Antitumor Compounds of Natural Origin: Chemistry and Biochemistry* (Aszalos, A., Ed.), pp. 191–235, Vol. 1, CRC Press, FL.

Snyder, J. P. (1990) *J. Am. Chem. Soc. 112*, 5367–5369.

Stankus, A., Goodisman, J., & Dabrowiak, J. (1992) *Biochemistry 31*, 9310–9318.

Stubbe, J., & Kozarich, J. W. (1987) *Chem. Rev. 87*, 1107–1136.

Sugiura, Y., Uesawa, Y., Takahashi, Y., Kuwahara, J., Golik, J., & Doyle, T. W. (1989) *Proc. Natl. Acad. Sci. U.S.A. 86*, 7672–7676.

Sugiura, Y., & Matsumoto, T. (1993) *Biochemistry 32*, 5548–5553.

Uesugi, M., & Sugiura, Y. (1993) *Biochemistry 32*, 4622–4627.

van Dyke, M. W., & Dervan, P. B. (1983) *Biochemistry 22*, 2373–2377.

Walker, S., Gange, D., Gupta, V., & Kahne, D. (1994a) *J. Am. Chem. Soc. 116*, 3197–3206.

Walker, S., Landovitz, R., Ding, W.-D., Ellestad, G. A., & Kahne, D. (1992) *Proc. Natl. Acad. Sci. U.S.A. 89*, 4608–4612.

Walker, S., Murnick, J., & Kahne, D. (1993) *J. Am. Chem. Soc. 115*, 7954–7961.

Walker, S., Valentine, K. G., & Kahne, D. (1990) *J. Am. Chem. Soc. 112*, 6428–6429.

Walker, S., Yang, D., Gange, D., & Kahne, D. (1991) *J. Am. Chem. Soc. 113*, 4716–4717.

Walker, S. L., Andreotti, A. H., & Kahne, D. E. (1994b) *Tetrahedron 50*, 1351–1360.

Wakisaka, G., Uchino, H., Nakamura, T., Sotobayashi, H., Shirakawa, S., Adachi, A., & Sakurai, M. (1963) *Nature 198*, 385–386.

Ward, D. C., Reich, E., & Goldberg, I. H. (1965) *Science 149*, 1259–1263.

Waring, M. J. (1981a) *Annu. Rev. Biochem. 50*, 159–192.

Waring, M. J. (1981b) in *The Molecular Basis of Antibiotic Action* (Gale, E. F., Cundliffe, E., Reynolds, P., Richmond, E., & Waring, M. J., Eds.), pp. 258–401, 2nd Ed., Wiley, London.

Yanagi, K., Privé, G., & Dickerson, R. E. (1991) *J. Mol Biol. 217*, 201–214

Yu, L, Salzberg, A. A., Dedon, P. C. (1995) *Bioorg. Med. Chem. 3*, 729–741.

Zein, N., Sinha, A.M., McGahren, W. J., & Ellestad, G. A. (1988) *Science 240*, 1198–1201.

Zimmer, C., & Wähnert, U. (1986) *Prog. Biophys. Mol. Biol. 47*, 31–112.

Chapter 6

Aisaka, K., Tamura, S., Arai, Y., & Uwajima, T. (1987) *Biotechnol. Lett. 9*, 633–637.

Alefounder, P. R., Baldwin, S. A., Perham, R. N., & Short, N. (1989) *J. Biochem. 257*, 529–534.

Andersson, F. O., Claesson, B., & Samuelsson, B. (1990) *J. Org. Chem. 55*, 4699–4704.

Auge, C., David, S., Gautheron, C., Malleron, A., & Cavaye, B. (1988) *New J. Chem. 12*, 733–744.

Auge, C., David, S., Mathieu, C., & Gautheron, C. (1984) *Tetrahedron Lett. 25*, 1467–1470.

Auge, C., & Gautheron, C. (1987) *J. Chem. Soc., Chem. Commun.*, 859–860.

Auge, C., & Gautheron, C. (1988) *Tetrahedron Lett. 29*, 789–790.

Auge, C., Gautheron, C., David, S., Malleron, A., Cavaye, B., & Bouxom, B. (1990) *Tetrahedron 46*, 201–214.

Auge, C., Gautheron, C., & Pora, H. (1989) *Carbohydr. Res. 193*, 288–293.

Barbas, C. F., III, Wang, Y.-F., & Wong, C.-H. (1990) *J. Am. Chem. Soc. 112*, 2013–2014.

Baumann, H., Brisson, J.-R., Michon, F., Pon, R., & Jennings, H. J. (1993) *Biochemistry 32*, 4007–4013.

Bednardski, M. D., Chenault, H. K., Simon, E. S., & Whitesides, G. M. (1987) *J. Am. Chem. Soc. 109*, 1283–1285.

Bednarski, M. D., Crans, D. C., DiCosimo, R., Simon, E. S., Stein, P. D., & Whitesides, G. M. (1988) *Tetrahedron Lett. 29*, 427–430.

Bednarski, M. D., Simon, E. S., Bischofberger, N., Fessner, W.-D., Kim, M.-J., Lees, W., Saito, T., Waldmann, H., & Whitesides, G. M. (1989) *J. Am. Chem. Soc. 111*, 627–635.

Bednarski, M. D., Waldmann, H. J., & Whitesides, G. M. (1986) *Tetrahedron Lett. 27*, 5807–5810.

Belasco, J. G., & Knowles, J. R. (1983) *Biochemistry 22*, 122–129.

Bender, H. (1980) *Carbohydr. Res. 78*, 133–145.

Berliner, L. J., Davis, M. E., Ebner, K. E., Beyer, T. A., & Bell, J. E. (1984) *Mol. Cell. Biochem. 62*, 37–42.

Berliner, L. J., & Robinson, R. D. (1982) *Biochemistry 21*, 6340–6343.

Beyer, T. A., Sadler, J. E., Rearick, J. I., Paulson, J. C., & Hill, R. L. (1981) *Adv. Enzymol. 52*, 23–175.

Bielfeldt, T., Peters, S., Meldal, M., Bock, K., & Paulsen, H. (1992) *Angew, Chem. Int. Edn. Engl. 31*, 857–859.

Bischofberger, N., Waldmann, H., Saito, T., Simon, E. S., Lees, W., Bednarski, M. D., & Whitesides, G. M. (1988) *J. Org. Chem. 53*, 3457–3465.

Blacklow, R. S., & Warren, L. (1962) *J. Biol. Chem. 237*, 3520–3526.

Bolte, J., Demuynck, C., & Samaki, H. (1987) *Tetrahedron Lett. 28*, 5525–5528.

Boren, T., Falk, P., Roter, K. A., Larson, G., & Normark, S. (1993) *Science 262*, 1892–1895.

Borysenko, C. W., Spaltenstein, A., Straub, J. A., & Whitesides, G. M. (1989) *J. Am. Chem. Soc. 111*, 9275–9276.

Brockhausen, I., Carver, J., & Schacter, H. (1988) *Biochem. Cell. Biol. 66*, 1134–1151.

Brockhausen, I., Hull, E., Hindsgaul, O., Schachter, H., Shah, R. N., Michnick, S. W., & Carver, J. P. (1989) *J. Biol. Chem. 264*, 11211–11221.

Card, P. J., & Hitz, W. D. (1984) *J. Am. Chem. Soc., 106*, 5384–5350.

Card, P. J., Hitz, W. D., & Ripp, K. G. (1986) *J. Am. Chem. Soc. 108*, 158–161.

Carlson, D. M. (1987) *Pure Appl. Chem. 59*, 1489–1492.

Carver, J. P. (1993) *Pure Appl. Chem. 65*, 763–770.

Chen, L., Dumas, D. P., & Wong, C.-H. (1992) *J. Am. Chem. Soc. 114*, 741–748.

Choay, J., Lormeau, J.-C., Petitou, M., Sinaÿ, P., & Fareed, J. (1981) *Ann. N. Y. Acad. Sci. 370*, 664–669.

Chou, W.-C., Fotsch, C., & Wong, C.-H. (1995) *J. Org. Chem. 60*, 2916–2917.

Clark, R. S., Banerjee, S., & Coward, J. K. (1990) *J. Org. Chem. 55*, 6275–6285.

Comb, D. G., & Roseman, S. (1960) *J. Biol. Chem. 235*, 2529–2537.

Conradt, H. S., Buensch, A., & Brossmer, R. (1984) *FEBS Lett. 170*, 295–300.

Conradt, H. S., Egge, H., Peter-Katalinic, J., Reiser, W., Siklosi, T., & Schaper, K. (1987) *J. Biol. Chem. 262*, 14600–14605.

Crout, D. H. G., MacManus, D. A., Ricca, J.-M., Singh, S., Crithley, P., & Gibson, W. T. (1992) *Pure Appl. Chem. 64*, 1079–1084.

Crowell, D. N., Anderson, M. S., & Raetz, C. R. H. (1986) *J. Bacteriol. 168*, 152–159.

Cygler, M., Rose, D. R., & Bundle, D. R. (1991) *Science 253*, 442–445.

David, S., Auge, C., & Gautheron, C. (1991) *Adv. Carbohydr. Chem. Biochem. 49*, 175–237.

Dedonder, R. (1966) *Methods Enzymol. 8*, 500–505.

Deijl, C. M., & Vliegenthart, J. F. G. (1983) *Biochem. Biophys. Res. Commun. 111*, 668–674.

Demuynck, C., Bolte, J., Hecquet, L., & Dalmas, V. (1991) *Tetrahedron Lett. 32*, 5085–5088.

Demuynck, C., Fisson, F., Bennani-Baiti, I., Samaki, H., & Mani, J.-C. (1990) *Agr. Biol. Chem. 54*, 3073–3078.

Denarie, J., & Cullimore, J. (1993) *Cell 74*, 951–954.

Dreyer, M. K., & Schulz, G. E. (1993) *J. Mol. Biol. 231*, 549–553.

Drueckhammer, D. G., Durrwachter, J. R., Pederson, R. L., Crans, D. C., Daniels, L., & Wong, C.-H. (1989) *J. Org. Chem. 54*, 70–77.

Durrwachter, J. R., Drueckhammer, D. G., Nozaki, K., Sweers, H. M., & Wong, C.-H. (1986) *J. Am. Chem. Soc. 108*, 7812–7818.

Durrwachter, J. R., & Wong, C.-H. (1988) *J. Org. Chem. 53*, 4175–4181.

Liu, K. K.-C., & Wong, C.-H. (1992) *J. Org. Chem. 57*, 4789–4791.

Livingston, B. D., Robertis, E. M. D., & Paulson, J. C. (1990) *Glycobiology 1*, 39–44.

Look, G. C., Fotsch, C. H., & Wong, C.-H. (1993a) *Acc. Chem. Res. 26*, 182–190.

Look, G. C., Ichikawa, Y., Shen, G.-J., Cheng, G.-J., & Wong, C.-H. (1993b) *J. Org. Chem. 58*, 4326–4330.

Look, G. C., & Wong, C.-H. (1992) *Tetrahedron Lett. 33*, 4523–4256.

Low, M. G., & Saltiel, A. R. (1988) *Science 239*, 268–275.

Lowary, T. L., & Hindsgaul, O. (1993) *Carbohydr. Res. 249*, 163–195.

Lowe, J. B. (1991) *Seminars Cell Biol. 2*, 289–307.

Lowe, J. B., Stoolman, L. M., Nair, R. P., Larson, R. D., Berhend, T. L., & Marks, R. M. (1990) *Cell 63*, 475–484.

McCoy, R. D., Vimr, E. R., & Troy, F. A. (1985) *J. Biol. Chem. 260*, 12695–12699.

McDowell, W., Grier, T. J., Rasmussen, J. R., & Schwarz, R. T. (1987) *Biochem. J. 248*, 523–531.

Moradian, A., & Benner, S. A. (1992) *J. Am. Chem. Soc. 114*, 6980–6987.

Moris-Varas, F., Qian, X.-H., & Wong, C.-H. (1996) *J. Am. Chem. Soc. 118*, 7647–7652.

Morris, A. J., & Tolan, D. R. (1993) *J. Biol. Chem. 268*, 1095–1100.

Myles, D. C., Andrulis, P. J., III, & Whitesides, G. M. (1991) *Tetrahedron Lett. 32*, 4835–4838.

Nicolaou, K. C., & Dai, W.-M. (1991) *Angew. Chem. Int. Ed. Engl. 30*, 1387–1416.

Nilsson, K. G. I. (1987) *Carbohydr. Res. 167*, 95–103.

Nilsson, K. G. I. (1988a) *Carbohydr. Res. 180*, 53–59.

Nilsson, K. G. I. (1988b) *TIBTECH 6*, 256–264.

Nishida, Y., Wiemann, T., Sinwell, V., & Thiem, J. (1993a) *J. Am. Chem. Soc. 115*, 2536–2537.

Nishida, Y., Wiemann, T., & Thiem, J. (1992) *Tetrahedron Lett. 33*, 8043–8046.

Nishida, Y., Wiemann, T., & Thiem, J. (1993b) *Tetrahedron Lett. 34*, 2905–2906.

Nunez, H. A., & Barker, R. (1980) *Biochemistry 19*, 489–495.

Oehrlein, R., Hindsgaul, O., & Palcic, M. M. (1993) *Carbohydr. Res. 244*, 149–159.

Oths, P. J., Hayer, R. M., & Floss, H. G. (1990) *Carbohydr. Res. 198*, 91–100.

Ozaki, A., Toone, E. J., von der Osten, C. H., Sinskey, A., & Whitesides, G. M. (1990) *J. Am. Chem. Soc. 112*, 4970–4971.

Palcic, M. M., & Hindsgaul, O. (1991) *Glycobiology 1*, 205–209.

Palcic, M. M., Srivastave, O. P., & Hindsgaul, O. (1987) *Carbohydr. Res. 159*, 315–324.

Palcic, M. M., Venot, A. P., Ratcliffe, R. M., & Hindsgaul, O. (1989) *Carbohydr. Res. 190*, 1–11.

Paulson, J. C. (1989) *TIBS 14*, 272–276.

Paulson, J. C. (1992) in *Adhesion: Its Role in Inflammatory Disease*, (Harlan, J., & Liu, D., Eds.), p 19, W. H. Freeman, New York.

Paulson, J. C., & Colley, K. J. (1989) *J. Biol. Chem. 264*, 17615–17618.

Pederson, R. L., Kim, M.-J., & Wong, C.-H. (1988) *Tetrahedron Lett. 29*, 4645–4648.

Pederson, R. L., Liu, K. K.-C., Rutan, J. F., Chen, L., & Wong, C.-H. (1990) *J. Org. Chem. 55*, 4897–4901.

Petrie, C. R., Sharma, M., Simmons, O. D., & Korytnyk, W. (1989) *Carbohydr. Res. 186*, 326–334.

Phillips, M. L., Nudelman, E., Gaeta, F. C. A., Perez, M., Singhal, A. K., Hakomori, S., & Paulson, J. C. (1990) *Science 250*, 1130–1132.

Racker, E. (1952) *J. Biol. Chem. 196*, 347–365.

Racker, E. (1961) in *The Enzymes*, Vol. 5, (Boyer, P. D., Lardy, H., & Myrback, K., Eds.), p 397, Academic Press, New York.

Rathbone, E. B., Hacking, A. J., & Cheetham, P. S. J. (1986) U. S. Patent 4,617,269.

Ray, P. H. (1980) *Bacteriol. 141*, 635–644.

Ray, P. H. (1982) *Methods Enzymol. 83*, 525–530.

Rietschel, E. T., & Brade, H. (1992) *Sci. Amer. 267*, 54–61.

Rose, I. A., & Warms, J. V. B. (1985) *Biochemistry 24*, 3952–3957.

Rosen, O. M., Hoffee, P., & Horecker, B. L. (1965) *J. Biol. Chem. 240*, 1517–1524.

Russell, R. N., & Liu, H.-W. (1991) *J. Am. Chem. Soc. 113*, 7777–7778.

Sabesan, S., & Paulson, J. C. (1986) *J. Am. Chem. Soc. 108*, 2068–2080.

Saenger, W. (1980) *Angew, Chem. Int. Ed. Engl. 19*, 344–362.

Schachter, H. (1986) *Biochem. Cell. Biol. 64*, 163–181.

Schanbacher, F. L., & Ebner, K. E. (1970) *J. Biol. Chem. 245*, 5057–5061.

Schenkman, S., Man-Shiow, J., Hart, G. W., & Nussenzweig, V. (1991) *Cell 65*, 1117–1125.

Schmid, W., & Whitesides, G. M. (1990) *J. Am. Chem. Soc. 112*, 9670–9671.

Scholz, D., Bednarik, K., Ehn, G., Neruda, W., Janzek, E., Loibner, H., Briner, K., & Vasella, A. (1992) *J. Med. Chem. 35*, 2070–2074.

Schultz, M., Waldmann, H., Vogt, W., & Kunz, H. (1990) *Tetrahedron Lett. 31*, 867–868.

Schumann, R. R., Leong, S. R., Flaggs, G. W., Gray, P. W., Wright, S. D., Mathison, J. C., Tobias, P. S., & Ulevitch, R. J. (1990) *Science 249*, 1429–1431.

Schuster, M., Wang, P., Paulson, J. C., & Wong, C.-H. (1994) *J. Am. Chem. Soc. 116*, 1135–1136.

Schwarz, R.-T., & Datema, R. (1982) *Adv. Carbohydr. Chem. Biochem. 40*, 287–379.

Schwarzmann, G., & Sandhoff, K. (1990) *Biochemistry 29*, 865–871.

Schwelberger, H. G., Kohlwein, S. D., & Paltauf, F. (1989) *Eur. J. Biochem. 180*, 301–308.

Serianni, A. S., Cadman, E., Pierce, J., Hayes, M. L., & Barker, R. (1982) *Methods Enzymol. 89*, 83–92.

Simon, E. S., Bednarski, M. D., & Whitesides, G. M. (1988) *J. Am. Chem. Soc. 110*, 7159–7163.

Simon, E. S., Grabowski, S., & Whitesides, G. M. (1989) *J. Am. Chem. Soc. 111*, 8920–8921.

Simon, E. S., Grabowski, S., & Whitesides, G. M. (1990) *J. Org. Chem. 55*, 1834–1841.

Srivastava, G., Alton, G., & Hindsgaul, O. (1990) *Carbohydr. Res. 207*, 259–276.

Srivastava, G., Hindsgaul, O., & Palcic, M. M. (1993) *Carbohydr. Res. 245*, 137–144.

Srivastava, G., Kaur, K. J., Hindsgaul, O., & Palcic, M. M. (1992) *J. Biol. Chem. 267*, 22356–22361.

Sugai, T., Shen, G.-J., Ichikawa, Y., & Wong, C.-H. (1993) *J. Am. Chem. Soc. 115*, 413–421.

Sygusch, J., Beaudy, D., & Allaire, M. (1987) *Proc. Natl. Acad. Sci. U.S.A. 84*, 7846–7850.

Takaoka, Y., Kajimoto, T., & Wong, C.-H. (1993) *J. Org. Chem. 58*, 4809–4812.

Thiem, J., & Treder, W. (1986) *Angew. Chem. Int. Ed. Engl. 25*, 1096–1097.

Thiem, J., & Wiemann, T. (1990) *Angew. Chem. Int. Ed. Engl. 29*, 80–82.

Tomlinson, S., de Carvalho, L. P., Vanderkerckhove, F., & Nussenzweig, V. (1991) *Glycobiology 2*, 549–551.

Treder, W., Thiem, J., & Schlingmann, M. (1986) *Tetrahedron Lett. 27*, 5605–5608.

Turner, N. J., & Whitesides, G. M. (1989) *J. Am. Chem. Soc. 111*, 624–627.

Uchida, Y., Tsukada, Y., & Sugimori, T. (1985) *Agr. Biol. Chem. 49*, 181–187.

Unverzagt, C., Kunz, H., & Paulson, J. C. (1990) *J. Am. Chem. Soc. 112*, 9308–9309.

van Brunt, J. (1986) *Bio/Technology 4*, 780–782.

Vanderkerckhove, F., Schenkman, S., de Carvalho, L. P., Tomlinson, S., Kiso, M., Yoshida, M., Hasegawa, A., & Nussenzweig, V. (1992) *Glycobiology 2*, 541–548.

van Rooijen, R. J., van Schalkwijk, S., & de Vos, W. M. (1991) *J. Biol. Chem. 266*, 7176–7181.

Vara, J. A., & Hutchinson, C. R. (1988) *J. Biol. Chem. 263*, 14992–14995.

Varki, A. (1993) *Glycobiology 3*, 97–130.

von der Osten, C. H., Barbas, C. F., III, Wong, C.-H., & Sinskey, A. J. (1989a) *Mol. Microbiol. 3*, 1625–1637.

von der Osten, C. H., Sinskey, A. J., Barbas, C. F., Jr., Pederson, R. L., Wang, Y.-F., & Wong, C.-H. (1989b) *J. Am. Chem. Soc. 111*, 3924–3927.

Vyplel, H., Scholz, D., Macher, I., Schindlmaier, K., & Schutze, E. (1991) *J. Med. Chem. 34*, 2759–2767.

Waldmann, H., Gygax, D., Bednarski, M. D., Shangraw, W. R., & Whitesides, G. M. (1986) *Carbohydr. Res. 157*, C4–C7.

Wallenfels, K., Foeldi, B., Niermann, H., Bender, H., & Linder, D. (1978) *Carbohydr. Res. 61*, 359–368.

Walsh, C. T. (1989) *J. Biol. Chem. 264*, 2393–2396.
Waltz, G., Arufto, A., Kalanus, W., Bevilacqua, M., & Seed, B. (1990) *Science 250*, 1132–1135.
Wang, P., Shen, G.-J., Wang, Y.-F., Ichikawa, Y., & Wong, C.-H. (1993a) *J. Org. Chem. 58*, 3985–3990.
Wang, Y.-F., Dumas, D. P., Wong, C.-H. (1993b) *Tetrahedron Lett. 34*, 403–406.
Weiss, W. I., Drickamer, K., & Hendrickson, W. A. (1992) *Nature 360*, 127–134.
Winchester, B., & Fleet, G. W. J. (1992) *Glycobiology 2*, 199–210.
Williams, M. A., Kitagawa, H., Datta, A. K., Paulson, J. C., & Jamieson, J. C. (1995) *Glycoconjugate J. 12*, 755–761.
Wong, C.-H., Druekhammer, D. G., & Sweers, H. M. (1988) in *Fluorinated Carbohydrates: Chemical and Biochemical Aspects*, ACS Symposium Series 374, (Taylor, N. F., Ed.), pp 29–42, American Chemical Society, Washington, DC.
Wong, C.-H., Haynie, S. L., & Whitesides, G. M. (1982) *J. Org. Chem. 47*, 5416–5418.
Wong, C.-H., Haynie, S. L., & Whitesides, G. M. (1983a) *J. Am. Chem. Soc. 105*, 115–117.
Wong, C.-H., Ichikawa, Y., Krach, T., Gautheron-Le Narvor, C., Dumas, D. P., & Look, G. C. (1991a) *J. Am. Chem. Soc. 113*, 8137–8145.
Wong, C.-H., Mazenod, F. P., & Whitesides, G. M. (1983b) *J. Org. Chem. 48*, 3493–3497.
Wong, C. H., Schuster, M., Wang, P., & Sears, P. (1993) *J. Am. Chem. Soc. 115*, 5893–5901.
Wong, C.-H., Shen, G.-J., Pederson, R. L., Wang, Y.-F., & Hennen, W. J. (1991b) *Methods Enzymol. 202*, 591–620.
Wong, C.-H., Wang, R., & Ichikawa, Y. (1992) *J. Org. Chem. 57*, 4343–4344.
Wong, C.-H., & Whitesides, G. M. (1983) *J. Org. Chem. 48*, 3493–3497.
Yuasa, H., Hindsgaul, O., & Palcic, M. M. (1992) *J. Am. Chem. Soc. 114*, 5891–5892.
Zehavi, U., & Herchman, M. (1984) *Carbohydr. Res. 133*, 339–342.
Zehavi, U., Sadeh, S., & Herchman, M. (1983) *Carbohydr. Res. 124*, 23–34.
Ziegast, G., & Pfannemuller, B. (1987) *Carbohydr. Res. 160*, 185–204.
Ziegler, T., Straub, A., & Effenberger, F. (1988) *Angew. Chem. Int. Ed. Engl. 27*, 716–717.

Chapter 7

Akasaka, K., Imoto, T., Shibata, S., & Hatano, H. (1975) *J. Magn. Reson. 18*, 328–343.
Altona, C. (1996) *Vicinal Coupling Constants and Conformation of Biomolecules*, in Encyclopedia of NMR, Wiley, pp 4909–4923.
Altona, C., & Sundaralingam, M. (1972) *J. Am. Chem. Soc. 94*, 8205–8212.
Ando, I., & Webb, G. A. (1983), Theory of NMR Parameters, Academic Press, New York, 83–113.
Angelotti, T., Krisko, M., O'Connor, T., & Serianni, A. S. (1987) *J. Am. Chem. Soc. 109*, 4464–4472.
Angyal, S. J. (1969) *Angew. Chem. Intl. Ed. Engl. 8*, 157–166.
Angyal, S. J. (1991) *Adv. Carbohydr. Chem. Biochem. 498*, 19–35.
Anteunis, M., & Danneels, D. (1975) *Org. Magn. Reson. 7*, 345–348.
Bandyopadhyay, T., Wu, J., & Serianni, A. S. (1993) *J. Org. Chem. 58*, 5513–5517.
Bandyopadhyay, T., Wu, J., Stripe, W.A. Carmichael, I., & Serianni, A. S. (1997) *J. Am. Chem. Soc. 119*, 1737–1744.
Barker, R., & Serianni, A. S. (1986) *Acc. Chem. Res. 19*, 307–313.
Bax, A., & Davis, D. G. (1985) *J. Magn. Reson. 65*, 355–360.
Bax, A., Freeman, R., & Kempsell, S. P. (1980) *J. Am. Chem. Soc. 102*, 4849–4851.
Bax, A., Freeman, R., & Morris, G. (1981) *J. Magn. Reson. 42*, 164–168.
Bax, A., & Subramanian, S. (1986) *J. Magn. Reson. 67*, 565–569.
Bax, A., & Summers, M. F. (1986) *J. Am. Chem. Soc. 108*, 2092–2094.
Berry, J. M., Hall, L. D., Welder, D. G., & Wong, K. F. (1979) *Anomeric Effect: Origin and Consequences*, (Szarek, W. A., & Horton, D., Eds.) p 7, ACS Symposium Series 87, American Chemical Society, Washington, DC.

Biamonti, C., Rios, C.B., Lyons, B. A., & Montelione, G. T. (1994) *Adv. Biophys. Chem. 4*, 51–20.

Bock, K., & Duus, J. Ø. (1994) *J. Carbohydr. Chem. 13*, 513–543.

Bock, K., Lundt, I., & Pedersen, C. (1973) *Tetrahedron Lett.* 1037–1040.

Bock, K., & Pedersen, C.(1974) *J. Chem. Soc. Perkin II* 293–297.

Bock, K., & Pedersen, C.(1975) *Acta Chem. Scand. B29*, 258–264.

Bock, K., & Pedersen, C.(1977) *Acta Chem. Scand. B31*, 354–358.

Bock, K., & Pedersen, C. (1983) *Adv. Carbohydr. Chem. Biochem. 41*, 27–66.

Bodenhausen, G., & Ernst, R. R. (1982) *J. Am. Chem. Soc. 104*, 1304–1309.

Bodenhausen, G., & Freeman, R. (1977) *J. Magn. Reson. 28*, 471–476.

Bossennec, V., Firmin, P., Perly, B., & Berthault, P. (1990) *Magn. Res. Chem. 28*, 149–155.

Bothner-by, A., & Castellano, S. (1964) *J. Chem. Phys. 41*, 3863.

Braunschweiler, L., & Ernst, R. R. (1983) *J. Magn. Reson. 53*, 521–528.

Church, T. J., Carmichael, I., & Serianni, A. S. (1996) *Carbohydr. Res. 280*, 177–186.

Church, T. J., Carmichael, I., & Serianni, A. S. (1997) *J. Am. Chem. Soc. 119*, 8946–8964.

Cumming, D. A., & Carver, J. P. (1987) *Biochemistry 26*, 6664–6676.

Cyr, N., Ritchie, R. G. S., Spotswood, T. M., & Perlin, A. S. (1974) *Can. J. Spectrosc. 19*, 190–193.

Dais, P., & Perlin, A. S. (1987) *Adv. Carbohydr. Chem. Biochem. 45*, 125–168.

Davis, D. G., & Bax, A. (1985) *J. Am. Chem. Soc. 107*, 2820–2821.

Duker, J. M., & Serianni, A. S. (1993) *Carbohydr. Res. 249*, 281–303.

du Penhoat, C. H., Imberty, A., Roques, N., Michon, V., Mentech, J., Descotes, G., & Perez, S. (1991) *J. Am. Chem. Soc. 113*, 3720–3727.

Emsley, L., Dwyer, T. J., Spielmann, H. P., & Wemmer, D. E. (1993) *J. Am. Chem. Soc. 115*, 7765–7771.

Engelsen, S. B., Perez, S., Braccini, I., & du Penhoat, C. H. (1994) 3rd International Satellite Symposium, Carbohydrate Research Center, University of Toronto, Val Morin, Quebec, Canada, Poster #9.

Ernst, R. R., Bodenhausen G., & Wakaun, A. (1987) *Principles of Nuclear Magnetic Resonance in One and Two Dimensions*, Clarendon Press, Oxford.

Fesik, S. W. (1991) *J. Med. Chem. 34*, 2937–2945.

Fesik, S. W., Eaton, H. L., Olejniczak, E. T., Zuiderweg, E., McIntosh, L. P., & Dahlquist, F. W. (1990) *J. Am. Chem. Soc., 112*, 886–888.

Fesik, S. W., & Zuiderweg, E. R. P. (1990) *Quart. Rev. Biophys. 23*, 97–131.

Garrett, E. C., & Serianni, A. S. (1990) *Carbohydr. Res. 208*, 23–35.

Haasnoot, C. A. G., de Leeuw, F. A. A. M., & Altona, C. (1980) *Tetrahedron 36*, 2783–2792.

Hamer, G. K., Balza, F., Cyr, N. and Perlin, A. S. (1978) *Can. J. Chem. 56*, 3109–3116.

Harris, R. K. (1986) *Nuclear Magnetic Resonance Spectroscopy, Longnan Scientific and Technical*, Essex, England, pp. 217–218.

Hayes, M. L., Pennings, N. J., Serianni, A. S., & Barker, R. (1982a) *J. Am. Chem. Soc. 104*, 6764–6769.

Hayes, M. L., Serianni, A. S., & Barker, R. (1982b) *Carbohydr. Res. 100*, 87–101.

Hines, J. V., Landry, S. M., Vanani, G., & Tinoco, Jr., I. (1994) *J. Am. Chem. Soc. 116*, 5823–5831.

Hines, J. V., Vanani, G., Landry, S. M., & Tinoco, Jr., I. (1993) *J. Am. Chem. Soc. 115*, 11002–11003.

Homans, S. W. (1990) *Prog. NMR Spectros. 22*, 55–81.

Isbell, H. S., & Pigman, W. (1969) *Adv. Carbohydr. Chem. Biochem. 24*, 13–65.

Jardetzky, O. (1980) *Biochim. Biophys. Acta 621*, 227–232.

Juaristi, E., & Cuevas, G. (1995) *The Anomeric Effect*, CRC Press, Boca Raton, FL.

Karplus, M. (1959) *J. Chem. Phys. 30*, 11–15.

Kellenbach, E. R., Remerowski, M. L., Eib, D., Boelens, R., van der Marel, G. A., van den Elst, H., van Boom, J. H., & Kaptein, R. (1992) *Nucleic Acids Res. 20*, 653–657.

Kessler, H., Gehnke, M., & Griesinger, C. (1988) *Angew. Chem. Int. Ed. Engl. 27*, 490–536.

King-Morris, M. J., & Serianni, A. S. (1986) *Carbohydr. Res. 154*, 29–36.

King-Morris, M. J., & Serianni, A. S. (1987) *J. Am. Chem. Soc. 109*, 3501–3508.

Kirby, A. J. (1983) The Anomeric Effect and Related Stereoelectronic Effects at Oxyen, Springer, New York.

Kline, P. C., Huang, S.-G., Hayes, M. L., Barker, R., & Serianni, A. S. (1990) *Can. J. Chem. 68*, 2171–2182.

Kline, P. C., & Serianni, A. S. (1988) *Magn. Reson. Chem. 26*, 120–123.

Kline, P. C., & Serianni, A. S. (1990a) *J. Am. Chem. Soc. 112*, 7373–7381.

Kline, P. C., & Serianni, A. S. (1990b) *Magn. Reson. Chem. 28*, 324–330.

Kline, P. C., & Serianni, A. S. (1992) *J. Org. Chem. 57*, 1772–1777.

Lancelot, G., Chanteloup, L., Beau, J.-M., & Thuong, N. T. (1993) *J. Am. Chem. Soc. 115*, 1599–1600.

Lemieux, R. U. (1963) in *Molecular Rearrangements*, (de Mayo, P., Ed.), p. 713, Wiley-Interscience, New York.

Lemieux, R. U. (1971) *Pure Appl. Chem. 25*, 527–548.

Lemieux, R. U., Koto, S., & Voisin, D. (1979) in *Anomeric Effect: Origin & Consequences*, (Szarek, W.A., & Horton, D., Eds.), pp. 17–29, ACS Symposium Series 87, American Chemical Society, Washington, DC.

Lemieux, R. U., Kullnig, R. K., Bernstein, H. J., & Schneider, W. G. (1957) *J. Am. Chem. Soc. 79*, 1005–1006.

Lemieux, R. U., Nagabhushan, T. L., & Paul, B. (1972) *Can. J. Chem. 50*, 773–776.

Liu, Q., & Brady, J. W. (1996) *J. Am. Chem. Soc. 118*, 12276–12286.

London, R. E. (1980) *Magnetic Resonance in Biology* (Cohen, J. S., Ed.) John Wiley & Sons, Vol. 1, pp 1–69.

London, R. E., Matwiyoff, N. A., & Mueller, D. D. (1975) *J. Chem. Phys. 63*, 4442–4449.

Lyerla, Jr., J. R., & Levy, G. C. (1974) *Topics in Carbon-13 NMR Spectroscopy*, (Levy, G. C. Ed.), pp. 79–148, John Wiley & Sons, New York.

Macura, S., & Ernst, R. R. (1980) *Mol. Phys. 41*, 95.

Majumdar, A., & Hosur, R. V. (1992) *Prog. NMR Spectrosc. 24*, 109–158.

Maple, S. R., & Allerhand, A. (1987) *J. Am. Chem. Soc. 109*, 3168–3169.

Mareci, T. H., & Freeman, R. (1982) *J. Magn. Reson. 48*, 158–163.

Marino, J. P., Schwalbe, H., Glaser, S. J., & Griesinger, C. (1996) *J. Am. Chem. Soc. 118*, 4388–4395.

Marshall, J. L. (1983) *Carbon-Carbon and Carbon-Proton NMR Couplings: Applications to Organic Stereochemistry and Conformational Analysis*, Verlag Chemie International, Dearfield Beach, FL.

Martin, M. L., Delpuech, J.-J., & Martin, G. J. (1980 *Practical NMR Spectroscopy*) Heyden and Son, Ltd., London, pp. 122–124.

Mega, T. L., & Van Etten, R. L. (1988) *J. Am. Chem. Soc. 110*, 6372–6376.

Müller, L., Kumar, A., & Ernst, R. R. (1975) *J. Chem. Phys. 63*, 5490–5491.

Mulloy, B., Frenkiel, T. A., & Davies, D. B. (1988) *Carbohydr. Res. 184*, 39–46.

Nicholson, L. K., Kay, L. E., Baldisseri, D. M., Arango, J., Young, P. E., Bax, A., & Torchia, D. A. (1992) *Biochemistry 31*, 5253–5263.

Nikonowicz, E. P., & Pardi, A. (1992) *J. Am. Chem. Soc. 114*, 1082–1083.

Nuzillard, J.-M., & Freeman, R. (1994) *J. Magn. Reson. Series A, 110*, 262–265.

Pardi, A., & Nikonowicz, E. P. (1992) *J. Am. Chem. Soc. 114*, 9202–9203.

Patt, S. L., & Shoolery, J. N. (1982) *J. Magn. Reson. 46*, 535–539.

Perrin, C. L., & Armstrong, K. B. (1993) *J. Am. Chem. Soc. 115*, 6825–6834.

Perrin, C. L., Armstrong, K. B., & Fabian, M. A. (1994) *J. Am. Chem. Soc. 116*, 715–722.

Pfeffer, P. E., Valentine, K. M., & Parrish, F. W. (1979) *J. Am. Chem. Soc. 101*, 1265–1274.

Piantini, U., Sorensen, O. W., & Ernst, R. R. (1982) *J. Am. Chem. Soc. 104*, 6800–6801.

Pierce, J., Serianni, A. S., & Barker, R. (1985) *J. Am. Chem. Soc. 107*, 2448–2456.

Pigman, W., & Isbell, H. S. (1968) *Adv. Carbohydr. Chem. Biochem. 23*, 11–57.

Podlasek, C. A., & Serianni, A. S. (1994) *J. Biol. Chem. 269*, 2521–2528.

Podlasek, C. A., Stripe, W. A., Carmichael, I., Shang, M., Basu, B., & Serianni, A. S. (1996) *J. Am. Chem. Soc. 118*, 1413–1425.

Podlasek, C. A., Wu, J., Stripe, W. A., Bondo, P. B., & Serianni, A. S. (1995) *J. Am. Chem. Soc. 117*, 8635–8644.

Poppe, L. (1993) *J. Am. Chem. Soc. 115*, 8421–8426.

Poppe, L., & van Halbeek, H. (1992) *J. Am. Chem. Soc. 114*, 1092–1094.

Rao, V. S. R., Balaji, P. V., & Qasba, P. K. (1995) *Glycobiology 5*, 273–279.

Risley, J. M., & Van Etten, R. L. (1979) *J. Am. Chem. Soc. 101*, 252–253.

Ritchie, R. G. S., Cyr, N., Korsch, B., Koch, H. J., & Perlin, A. S. (1975) *Can. J. Chem. 53*, 1424–1433.

Rosevear, P. R., Nunez, H. A., & Barker, R. (1982) *Biochemistry 21*, 1421–1431.

Saenger, W. (1984) *Principles of Nucleic Acid Structure*, pp. 51–104, Springer-Verlag, New York.

Salzner, U., & Schleyer, P. von Rague' (1993) *J. Am. Chem. Soc., 115*, 10231–10236.

Schmidt, R. K., Karplus, M., & Brady, J. W. (1996) *J. Am. Chem. Soc. 118*, 541–546.

Schmitz, U., Zon, G., & James, T. L. (1990) *Biochemistry 29*, 2357–2368.

Schwarcz, J. A., & Perlin, A. S. (1972) *Can. J. Chem. 50*, 3667–3676.

Schwarcz, J. A., Cyr, N., & Perlin, A. S. (1975) *Can. J. Chem. 53*, 1872–1875.

Serianni, A. S. (1992) in *Glycoconjugates: Composition, Structure and Function*, (Allen, H. J., & Kisailus, E. C., Eds.), pp. 71–102, Marcel Dekker, Inc., New York.

Serianni, A. S. (1994) in *NMR of Biological Macromolecules*, (Stassinopoulou, C. I., Ed.), pp. 293–306, Springer-Verlag, Berlin.

Serianni, A. S., & Barker, R. (1979) *Can. J. Chem. 57*, 3160–3167.

Serianni, A. S., & Barker, R. (1982) *J. Magn. Reson. 49*, 335–340.

Serianni, A. S., & Barker, R. (1984) *J. Org. Chem. 49*, 3292–3300.

Serianni, A. S., Bondo, P. B., & Zajicek, J. (1996) *J. Magn. Reson., Ser. B 112*, 69–74.

Serianni, A. S., Clark, E. L., & Barker, R. (1979a) *Carbohydr. Res. 72*, 79–91.

Serianni, A. S., Hunez, H. A., & Barker, R. (1980) *J. Org. Chem. 45*, 3329–3341.

Serianni, A. S., Pierce, J., Huang, S.-G., & Barker, R. (1982) *J. Am. Chem. Soc. 104*, 4037–4044.

Serianni, A. S., Pierce, J., & Barker, R. (1979b) *Biochemistry 18*, 1192–1199.

Serianni, A. S., & Podlasek, C. A. (1994) *Carbohydr. Res. 259*, 277–282.

Serianni, A. S., Wu, J., & Carmichael, I. (1995) *J. Am. Chem. Soc. 117*, 8645–8650.

Shaka, A. J., & Keeler, J. (1987) *Prog. NMR Spectrosc. 19*, 47–129.

Shaka, A. J., Keeler, J., & Freeman, R. (1983) *J. Magn. Reson. 53*, 313–340.

Shimizu, H. (1962) *J. Chem. Phys. 37*, 765–778.

Snyder, J. R., Johnston, E. R., & Serianni, A. S. (1989) *J. Am. Chem. Soc. 111*, 2681–2687.

Snyder, J. R., & Serianni, A. S. (1986) *J. Org. Chem. 51*, 2694–2702.

Snyder, J. R., & Serianni, A. S. (1987a) *Carbohydr. Res. 163*, 169–188.

Snyder, J. R., & Serianni, A. S. (1987b) *Carbohydr. Res. 166*, 85–99.

Snyder, J. R., & Serianni, A. S. (1988) *Carbohydr. Res. 184*, 13–25.

Snyder, J. R., & Serianni, A. S. (1991) *Carbohydr. Res. 210*, 21–38.

Spoormaker, T., & de Bie, M. J. A. (1978) *Rec. Trav. Chim. Pays-Bas 97*, 85–87.

Summers, M. F., Marzilli, L. G., & Bax, A. (1986) *J. Am. Chem. Soc. 108*, 4285–4294.

Takagi, S., & Jeffrey, G. A. (1978) *Acta Crystallogr., Sect. B 34*, 2006–2010.

Tran, V., & Brady, J. W. (1990) *Biopolymers, 29*, 961–976.

Tvaroska, I., & Bleha, T. (1989) *Adv. Carbohydr. Chem. Biochem. 47*, 45–123.

Tvaroska, I., Hricovini, M., & Petrakova, E. (1989) *Carbohydr. Res. 189*, 359–362.

van Beuzekom, A. A., de Leeuw, F. A. A. M., & Altona, C. (1990) *Magn. Reson. Chem. 28*, 68–74.

Varela, O., Zunszain, P. A., Cicero, D. O., Baggio, R. F., Vega, D. R., & Garland, M. T. (1996) *Carbohydr. Res. 280*, 187–196.

Vliegenthart, J. F. G., Dorland, L., & van Halbeek, H. (1983) *Adv. Carbohydr. Chem. Biochem. 41*, 209–374.

Vold, R. L., Waugh, J. S., Klein, M. P., & Phelps, D. E. (1968) *J. Chem. Phys. 48*, 3831–3832.

Vuorinen, T., & Serianni, A. S. (1990a) *Carbohydr. Res. 207*, 185–210.

Vuorinen, T., & Serianni, A. S. (1990b) *Carbohydr. Res. 209*, 13–31.
Walker, T. E., London, R. E., Whaley, T. W., Barker, R., & Matwiyoff, N. A. (1976) *J. Am. Chem. Soc. 98*, 5807–5813.
Wijmenga, S. S., Hallenga, K., & Hilbers, C. W. (1989) *J. Magn. Reson. 84*, 634–642.
Woessner, D. E. (1962) *J. Chem. Phys. 37*, 647–654.
Wolfe, S. (1972) *Acc. Chem. Res. 5*, 102–111.
Wu, G. D., Serianni, A. S., & Barker, R. (1983) *J. Org. Chem. 48*, 1750–1757.
Wu, J., Bondo, P. B., Vuorinen, T., & Serianni, A. S. (1992) *J. Am. Chem. Soc. 114*, 3499–3505.
Wu, J., & Serianni, A. S. (1992) *Carbohydr. Res. 226*, 209–219.
Wu, J., & Serianni, A. S. (1994) *Biopolymers 34*, 1175–1186.
Wu, J., Serianni, A. S., & Vuorinen, T. (1990) *Carbohydr. Res. 206*, 1–12.
Wüthrich, K. (1986) *NMR of Proteins and Nucleic Acids*, John Wiley and Sons, New York.
Zefirov, N. S., Samoshin, V. V., Subbotin, O. A., Baranenkov, V. I., & Wolfe, S. (1978) *Tetrahedron 34*, 2953–2959.
Zhu, G., Renwick, A., & Bax, A. (1994) *J. Magn. Reson. Series A 110*, 257–261.

Chapter 8

Aruffo, A., Stamenkovic, I., Melnick, M., Underhill, C. B., & Seed, B. (1990) *Cell 61*, 1303–1313.
Aruffo, A., Kolanus, W., Walz, G., Fredman, P., & Seed, B. (1991) *Cell 67*, 35–44.
Atherton, A. & Born, G. V. (1972) *J. Physiol. 222*, 447–474.
Bajorath, J., Stenkamp, R., & Aruffo, A. (1993) *Protein Sci. 2*, 1798–1810.
Baumheter, S., Singer, M. S., Henzel, W., Hemmerich, S., Renz, M., Rosen, S. D., & Lasky, L.A. (1993) *Science 262*, 436–438.
Berg, E. L., McEvoy, L. M., Berlin, C., Bargatze, R. F., & Butcher, E. C. (1993) *Nature 366*, 695–698.
Bernstein, F. C., Koetzle, T. F., Williams, G. J., Meyer, E. E., Jr., Brice, M. D., Rodgers, J. R., Kennard, O., Shimanouchi, T., & Tasumi, M. (1977) *J. Mol. Biol. 112*, 535–542.
Bevilacqua, M. P., Pober, J. S., Mendrick, D. L., Cotran, R. S., & Gimbrone, M. A., Jr. (1987) *Proc. Natl. Acad. Sci. U.S.A. 84*, 9238–9242.
Bevilacqua, M. P., Stengelin, S., Gimbrone Jr., M. A., & Seed, B. (1989) *Science 243*, 1160–1165.
Bevilacqua, M. P. (1993) *Annu. Rev. Immunol. 11*, 767–804.
Bevilacqua, M. P., & Nelson, R. M. (1993) *J. Clin. Invest. 91*, 379–387.
Blundell, T. L., & Johnson, M. S. (1993) *Protein Sci. 2*, 877–883.
Bourne, Y., & Cambillau, C. (1993) in *Water and Biological Macromolecules* (Westhof, E., Ed.) pp 321–337, CRC, Boca Raton, FL.
Bowen, B. R., Nguyen, T., & Lasky, L. A. (1989) *J. Cell Biol. 109*, 421–427.
Bowie, J. U., & Eisenberg, D. (1993) *Current Opinion Struct. Biol. 3*, 437–444.
Braaten, B. A., Spangrude, G. J., & Daynes, R. A. (1984) *J. Immunol. 133*, 117–122.
Briskin, J. M., McEvoy, L. M., & Butcher, E. C. (1993) *Nature 363*, 461–464.
Brown, T. A., Bouchard, T., St. John, T., Wayner, E., & Carter, W. G. (1991) *J. Cell Biol. 113*, 207–221.
Bruccoleri, R. E., Haber, E., & Novotny, J. (1988) *Nature 335*, 564–568.
Bundle, D. R. & Young, N. M. (1992) *Current Opinion Struct. Biol. 2*, 666–673.
Camerini, D., James, S. P., Stamenkovic, I., & Seed, B. (1989) *Nature 342*, 78–82.
Carlos, T., Kovach, N., Schwartz, B., Rosa, M., Newman, B., Wayner, E., Benjamin, C., Osborn, L., Lobb, R., & Harlan, J. (1991) *Blood 77*, 2266–2271.
Carter, W. G., & Wayner, E. A. (1988) *J. Biol. Chem. 263*, 4193–4201.
Chothia, C., & Lesk, A. M. (1986) *EMBO J. 5*, 823–826.

Collins, T., Williams, A., Johnston, G. I., Kim, J., Eddy, R., Shows, T., Gimbrone, M. A., Jr., & Bevilacqua, M. P. (1991) *J. Biol. Chem. 266*, 2466–2473.

Corral, L., Singer, M. S., Macher, B. A., & Rosen, S. D. (1990) *Biochem. Biophys. Res. Commun. 172*, 1349–1356.

Dobrina, A., Schwartz, B. R., Carlos, T. M., Ochs, H. D., Beatty, P. G., & Harlan, J. M. (1989) *Immunology 67*, 502–508.

Doré, M., Korthuis, R. J., Granger, D. N., Entman, M. L., & Smith, C. W. (1993) *Blood 82*, 1308–1316.

Dougherty, G. J., Lansdorp, P. M., Cooper, D. L., & Humphries, R. K. (1991) *J. Exp. Med. 174*, 1–5.

Drickamer, K. (1992) *Nature 360*, 183–186.

Erbe, D. V., Wolitzky, B. A., Presta, L. G., Norton, C. R., Ramos, R. J., Burns, D. K., Rumberger, J. M., Narasinga Rao, B. N., Foxall, C., Brandley, B. K., & Lasky, L. A. (1992) *J. Cell Biol. 119*, 215–227.

Erbe, D. V., Watson, S. R., Presta, L. G., Wolitzky, B. A., Foxall, C., Brandley, B. K., & Lasky, L. A. (1993) *J. Cell Biol. 120*, 1227–1235.

Etzioni, A., Frydman, M., Pollack, S., Avidor, I., Phillips, M. L., Paulson, J. C., & Gershoni-Baruch, R. (1992) *New England J. Med. 327*, 1789–1792.

Foxall, C., Watson, S. R., Dowbenko, D., Fennie, C., Lasky, L. A., Kiso, M., Hasegawa, A., Asa, D., & Brandley, B. K. (1992) *J. Cell Biol. 117*, 895–902.

Gallatin, W. M., Weissman, I. L., & Butcher, E. C. (1983) *Nature 304*, 30–34.

Geng, J. G., Bevilacqua, M. P., Moore, K. L., McIntyre, T. M., Prescott, S. M., Kim, J. M., Bliss, G. A., Zimmerman, G. A., & McEver, R. P. (1990) *Nature 343*, 757–760.

Goelz, S. E., Hession, C., Goff, D., Griffiths, B., Tizard, R., Newman, B., Chi-Rosso, G., & Lobb, R. (1990) *Cell 63*, 1349–1356.

Goetinck, P. F., Stirpe, N. S., Tsonis, P. A., & Carlone, D. (1987) *J. Cell Biol. 105*, 2403–2408.

Goldstein, L. A., Zhou, D. F., Picker, L. S., Minty, C. N., Bargatze, R. F., Ding, J. F., & Butcher, E. C. (1989) *Cell 56*, 1063–1072.

Graves, B. J., Crowther, R. L., Chandran, C., Rumberger, J. M., Li, S., Huang, K.-S., Presky, D. H., Familletti, P. C., Wolitzky, B. A., & Burns, D. K. (1994) *Nature 367*, 532–538.

Greer, J. (1991) *Methods Enzymol. 202*, 239–252.

Grober, J. S., Bowen, B. L., Ebling, H., Athey, B., Thompson, C. B., Fox, D. A., & Stoolman, L. M. (1993) *J. Clin. Invest. 91*, 2609–2619.

Gundel, R. H., Wegner, C. D., Torcellini, C. A., Clarke, C. C., Haynes, N., Rothlein, R., Smith, C. W., & Letts, L. G. (1991) *J. Clin. Invest. 88*, 1407–1411.

Günthert, U., Hofmann, M., Rudy, W., Reber, S., Zöller, M., Haußmann, I., Matzku, S., Wenzel, A., Ponta, H., & Herrlich, P. (1991) *Cell 65*, 13–24.

Harlan, J. M., Winn, R. K., Vedder, N. B., Doerschuk, C. M., & Rice, C. L. (1992) in *Adhesion: Its Role in Inflammatory Disease* (Harlan, J. M., & Liu, D. Y., Eds.) pp. 117–150, W. H. Freeman & Company, New York.

Hattori, R., Hamilton, K. K., Fugate, R. D., McEver, R. P., & Sims, P. J. (1989) *J. Biol. Chem. 264*, 7768–7771.

Haynes, B. F., Liao, H.-X., & Patton, K. L. (1991) *Cancer Cells 3*, 347–350.

Haynes, B. F., Telen, M. J., Hale, L. P., & Denning, S. M. (1989) *Immunol. Today 10*, 423–428.

Herrlich, P., Zöller, M., Pals, S. T., & Ponta, H. (1993) *Immunol. Today 14*, 395–399.

Hollenbaugh, D., Bajorath, J., Stenkamp, R., & Aruffo, A. (1993) *Biochemistry 32*, 2960–2966.

Huet, S., Groux, H., Caillou, B., Valentin, H., Prieur, M., & Bernard, A. (1989) *J. Immunol. 143*, 798–801.

Imai, Y., True, D. D., Singer, M. S., & Rosen, S. D. (1990) *J. Cell Biol. 111*, 1225–1232.

Imai, Y., Lasky, L. A., & Rosen, S. D. (1993) *Nature 361*, 555–557.

Jackson, D. G., Buckley, J., & Bell, J. I. (1992) *J. Biol. Chem. 267*, 4732–4739.

Jalkanen, S., Bargatze, R. F., de los Toyos, J., & Butcher, E. C. (1987) *J. Cell Biol. 105*, 983–990.

Jalkanen, S., & Jalkanen, M. (1992) *J. Cell Biol. 116*, 817–825.

Johnston, G. I., Cook, R. G., & McEver, R. P. (1989) *Cell 56*, 1033–1044.

Jones, T. A., & Thirup, S. (1986) *EMBO J. 5*, 819–822.

Kuijpers, T. W., Hoogerwerf, M., van der Laan, L. J. W., Nagel, G., van der Schoot, C. E., Grunert, F., & Roos, D. (1992) *J. Cell Biol. 118*, 457–466.

Larsen, E., Celi, A., Gilbert, G. E., Furie, B. C., Erban, J. K., Bonfanti, R., Wagner, D. D., & Furie, B. (1989) *Cell 59*, 305–312.

Larsen, E., Palabrica, T., Sajer, S., Gilbert, G. E., Wagner, D. D., Furie, B. C., & Furie, B. (1990) *Cell 63*, 467–474.

Larsen, G. R., Sako, D., Ahern, T. J., Shaffer, M., Erban, J., Sajer, S. A., Gibson, R. M., Wagner, D. D., Furie, B. C., & Furie, B. (1992) *J. Biol. Chem. 16*, 11104–11110.

Lasky, L. A. (1992) *Science 258*, 964–969.

Lasky, L. A., Singer, M. S., Dowbenko, D., Imai, Y., Henzel, W. J., Grimley, C., Fennie, C., Gillett, N., Watson, S. R., & Rosen, S. D. (1992) *Cell 69*, 927–938.

Lasky, L. A., Singer, M. S., Yednock, T. A., Dowbenko, D., Fennie, C., Rodriguez, H., Nguyen, T., Stachel, S., & Rosen, S. D. (1989) *Cell 56*, 1045–1055.

Lawrence, M. B., & Springer, T. A. (1991) *Cell 65*, 859–873.

Ledbetter, J. A., Rabinovitch, P. S., Hellström, I., Hellström, K. E., Grosmaire, L. S., & June, C. H. (1988) *Eur. J. Immunol. 18*, 1601–1608.

Leeuwenberg, J. F. M., Von Asmuth, E. J. U., Jeunhomme, T. M. A. A., & Buurman, W. A. (1990) *J. Immunol. 145*, 2110–2114.

Ley, K., Cerrito, M., & Arfors, K.-E. (1991a) *Am. J. Physiol. 260*, 1667–1673.

Ley, K., Gaehtgens, P., Fennie, C., Singer, M. S., Lasky, L. A., & Rosen, S. D. (1991b) *Blood 77*, 2553–2555.

Lowe, J. B., Stoolman, L. M., Nair, R. P., Larsen, R. D., Berhend, T. L., & Marks, R. M. (1990) *Cell 63*, 475–484.

Lowe, J. B., Kukowska-Latallo, J. F., Nair, R. P., Larsen, R. D., Marks, R. M., Macher, B. A., Kelly, R. J., & Ernst, L. K. (1991) *J. Biol. Chem. 266*, 7467–7477.

Ma, X.-L., Weyrich, A. S., Lefer, D. J., Buerke, M., Albertine, K. H., Kishimoto, T. K., & Lefer, A. M. (1993) *Circulation 88*, 649–658.

Mayadas, T. N., Johnson, R. C., Rayburn, H., Hynes, R. O., & Wagner, D. D. (1993) *Cell 74*, 541–554.

McEver, R. P., Beckstead, J. H., Moore, K. L., Marshall-Carlson, L., & Bainton, D. F. (1989) *J. Clin. Invest. 84*, 92–99.

Miyake, K., Underhill, C. B., Lesley, J., & Kincade, P. W. (1990) *J. Exp. Med. 172*, 69–75.

Moore, K. L., Varki, A., & McEver, R. P. (1991) *J. Cell Biol. 112*, 491–499.

Mulligan, M. S., Varani, J., Dame, M. K., Lane, C. L., Smith, C. W., Anderson, D. C., & Ward, P. A. (1991) *J. Clin. Invest. 88*, 1396–1406.

Mulligan, M. S., Polley, M. J., Bayer, R. J., Nunn, M. F., Paulson, J. C., & Ward, P. A. (1992) *J. Clin. Invest. 90*, 1600–1607.

Mulligan, M. S., Paulson, J. C., De Frees, S., Zheng, Z.-L., Lowe, J. B., & Ward, P. A. (1993) *Nature 364*, 149–151.

Needham, L. K., & Schnaar, R. L. (1993) *Proc. Natl. Acad. Sci. U.S.A. 90*, 1359–1363.

Ord, D. C., Ernst, T. J., Zhou, L.-J., Rambaldi, A., Spertini, O., Griffin, J., & Tedder, T. F. (1990) *J. Biol. Chem. 265*, 7760–7767.

Patel, K. D., Zimmerman, G. A., Prescott, S. M., McEver, R. P., & McIntyre, T. M. (1991) *J. Cell Biol. 112*, 749–759.

Peach, R. J., Hollenbaugh, D., Stamenkovic, I., & Aruffo, A. (1993) *J. Cell Biol. 122*, 257–264.

Picker, L. J., Kishimoto, T. K., Smith, C. W., Warnock, R. A., & Butcher, E. C. (1991a) *Nature 349*, 796–799.

Picker, L. J., Warnock, R. A., Burns, A. R., Doerschuk, C. M., Berg, E. L., & Butcher, E. C. (1991b) *Cell 66*, 921–933.

Pober, J. S., Bevilacqua, M. P., Mendrick, D. L., Lapierre, L. A., Fiers, W., & Gimbrone, M. A., Jr. (1986) *J. Immunol. 136*, 1680–1687.

Postigo, A. A., Garcia-Vicuña, R., Diaz-Gonzalez, F., Arroyo, A. G., Landázuri, M. O., Chi-Rosso, G., Lobb, R. R., Laffon, A., & Sánchez-Madrid, F. (1992) *J. Clin. Invest.* *89*, 1445–1452.

Quiocho, F. A. (1989) *Pure Appl. Chem. 61*, 1293–1306.

Rosen, S. D., Singer, M. S., Yednock, T. A., & Stoolman, L. M. (1985) *Science 228*, 1005–1007.

Rosen, S. D. (1993) *Immunology 5*, 237–247.

Sako, D., Chang, X.-J., Barone, K. M., Vachino, G., White, H. M., Shaw, G., Veldman, G. M., Bean, K. M., Ahern, T. J., Furie, B., Cumming, D. A., & Larsen, G. R. (1993) *Cell 75*, 1179–1186.

Schleiffenbaum, B., Spertini, O., & Tedder, T. F. (1992) *J. Cell Biol. 119*, 229–238.

Screaton, G. R., Bell, M. V., Jackson, D. G., Cornelis, R. B., Gerth, U., & Bell, J. I. (1992) *Proc. Natl. Acad. Sci. U.S.A. 89*, 12160–12164.

Shimizu, Y., Van Seventer, G. A., Siraganian, R., Wahl, L., & Shaw, S. (1989) *J. Immunol. 143*, 2457–2463.

Shimizu, Y., Shaw, S., Graber, N., Gopal, T. V., Horgan, K. J., Van Seventer, G. A., & Newman, W. (1991) *Nature 349*, 799–802.

Siegelman, M. H., van de Rijn, M., & Weissman, I. L. (1989) *Science 243*, 1165–1172.

Siegelman, M. H., Cheng, I. C., Weissman, I. L., & Wakeland, E. K. (1990) *Cell 61*, 611–622.

Sippl, M. J. (1990) *J. Mol. Biol. 213*, 859–883.

Sippl, M. J., & Weitckus, S. (1992) *Proteins 13*, 258–271.

Skinner, M. P., Lucas, C. M., Burns, G. F., Chesterman, C. N., & Berndt, M. C. (1991) *J. Biol. Chem. 266*, 5371–5374.

Spangrude, G. J., Braaten, B. A., & Daynes, R. A. (1984) *J. Immunol. 132*, 354–362.

St. John, T., Meyer, J., Idzerda, R., & Gallatin, W. M. (1990) *Cell 60*, 45–52.

Stamenkovic, I., Amiot, M., Pesando, J. M., & Seed, B. (1989) *Cell 56*, 1057–1062.

Stamenkovic, I., Aruffo, A., Amiot, M., & Seed, B. (1991) *EMBO J. 10*, 343–348.

Stamper, H. B., Jr., & Woodruff, J. J. (1976) *J. Exp. Med. 144*, 828–833.

Stenberg, P. E., McEver, R. P., Shuman, M. A., Jacques, Y. V., & Bainton, D. F. (1985) *J. Cell Biol. 101*, 880–886.

Stoolman, L. M., & Rosen, S. D. (1983) *J. Cell Biol. 96*, 722–729.

Stoolman, L. M., Tenforde, T. S., & Rosen, S. D. (1984) *J. Cell Biol. 99*, 1535–1540.

Stoolman, L. M., Yednock, T. A., & Rosen, S. D. (1987) *Blood 70*, 1842–1850.

Sugama, Y., Tiruppathi, C., Janakidevi, K., Andersen, T. T., Fenton, J. W., II & Malik, A. B. (1992) *J. Cell Biol. 119*, 935–944.

Sy, M. S., Guo, Y.-J., & Stamenkovic, I. (1991) *J. Exp. Med. 174*, 859–866.

Sy, M.-S., Guo, Y.-J., & Stamenkovic, I. (1992) *J. Exp. Med. 176*, 623–627.

Tarone, G., Ferracini, R., Galetto, G., & Comoglio, P. (1984) *J. Cell Biol. 99*, 512–519.

Tedder, T. F., Isaacs, C. M., Ernst, T. J., Demetri, G. D., Adler, D. A., & Disteche, C. M. (1989) *J. Exp. Med. 170*, 123–133.

Thomas, L., Byers, H. R., Vink, J., & Stamenkovic, I. (1992) *J. Cell Biol. 118*, 971–977.

Tiemeyer, M., Swiedler, S. J., Ishihara, M., Moreland, M., Schweingruber, H., Hirtzer, P., & Brandley, B. K. (1991) *Proc. Natl. Acad. Sci. U.S.A. 88*, 1138–1142.

Tyrrell, D., James, P., Rao, N., Foxall, C., Abbas, S., Dasgupta, F., Nashed, M., Hasegawa, A., Kiso, M., Asa, D., Kidd, J., & Brandley, B. K. (1991) *Proc. Natl. Acad. Sci. U.S.A. 88*, 10372–10376.

Tölg, C., Hofmann, M., Herrlich, P., & Ponta, H. (1993) *Nucleic Acids Res. 21*, 1225–1229.

von Andrian, U. H., Chambers, J. D., McEvoy, L. M., Bargatze, R. F., Arfors, K.-E., & Butcher, E. C. (1991) *Proc. Natl. Acad. Sci. U.S.A. 88*, 7538–7542.

von Andrian, U. H., Berger, E. M., Ramezani, L., Chambers, J. D., Ochs, H. D., Harlan, J. M., Paulson, J. C., Etzioni, A., & Arfors, K. E. (1993) *J. Clin. Invest. 91*, 2893–2897.

Vyas, N. K. (1991) *Current Opinion Struct. Biol. 1*, 732–740.

Walz, G., Aruffo, A., Kolanus, W., Bevilacqua, M., & Seed, B. (1990) *Science 250*, 1132–1135.

Watson, M. L., Kingsmore, S. F., Johnston, G. I., Siegelman, M. H., Le Beau, M. M., Lemons, R. S., Bora, M. S., Howard, T. A., Weissman, I. L., McEver, R. P., & Seldin, M. F. (1990a) *J. Exp. Med. 172*, 263–272.

Watson, S. R., Imai, Y., Fennie, C., Geoffroy, J. S., & Rosen, S. D. (1990b) *J. Cell Biol. 110*, 2221–2229.

Watson, S. R., Fennie, C., & Lasky, L. A. (1991) *Nature 349*, 164–166.

Webb, D. S. A., Shimizu, Y., Van Seventer, G. A., Shaw, S., & Gerrard, T. L. (1990) *Science 249*, 1295–1297.

Weis, W. I., Kahn, R., Fourme, R., Drickamer, K., & Hendrickson, W. A. (1991) *Science 254*, 1608–1615.

Weis, W. I., Drickamer, K., & Hendrickson, W. A. (1992) *Nature 360*, 127–134.

Weyrich, A. S., Ma, X.-L., Lefer, D. J., Albertine, K. H., & Lefer, A. M. (1993) *J. Clin. Invest. 91*, 2620–2629.

Winn, R. K., Liggitt, D., Vedder, N. B., Paulson, J. C., & Harlan, J. M. (1993) *J. Clin. Invest. 92*, 2042–2047.

Wodak, S. J., & Rooman, M. J. (1993) *Current Opinion Struct. Biol. 3*, 247–259.

Yang, B., Zhang, L., & Turley, E. A. (1993) *J. Biol. Chem. 268*, 8617–8623.

Yang, B., Yang, B. L., Savani, R. C., & Turley, E. A. (1994) *EMBO J. 13*, 286–296.

Yednock, T. A., Butcher, E. C., Stoolman, L. M., & Rosen, S. D. (1987a) *J. Cell Biol. 104*, 725–731.

Yednock, T. A., Stoolman, L. M., & Rosen, S. D. (1987b) *J. Cell Biol. 104*, 713–723.

Yuen, C.-T., Bezouska, K., O'Brien, J., Stoll, M., Lemoine, R., Lubineau, A., Kiso, M., Hasegawa, A., Bockovich, N. J., Nicolaou, K. C., & Feiz, T. (1994) *J. Biol. Chem. 269*, 1595–1598.

Zhou, Q., Moore, K. L., Smith, D. F., Varki, A., McEver, R. P., & Cummings, R. D. (1991) *J. Cell Biol. 115*, 557–564.

Chapter 9

Ahlers, J., Müller, W., Reichert, A., Ringsdorf, H., & Venamer, J. (1990) *Angew. Chem. Int. Ed. Engl. 29*, 1269–1285.

Albrecht, T. R., & Quate, C. F. (1987) *J. Appl. Phys. 62*, 2599–2602.

Alving, C. R., & Richards, R. L. (1983) (Ostro, M. J., Ed.), pp. 209–287, Dekker, New York.

Andersson, M., & Oscarson, S. (1992) *Bioconjugate. Chem. 4*, 246–249.

Andresz, H., Richter, G. C., & Pfannemueller, B. (1978) *Makromol. Chem. 179*, 301.

Azzam, R. M. A., & Bashara, N. M. (1977) *Ellipsometry and Polarized Light*, North Holland Publishing Company, Amsterdam.

Bader, H., van Wagenen, R., Andrade, J. D., & Ringsdorf, H. (1984) *J. Coll. Interf. Sci. 101*, 246–249.

Bain, C. D., & Whitesides, G. M. (1989) *Adv. Mater. 4*, 110.

Bangham, A. D., Standish, M. M., & Watkins, J. C. (1965a) *J. Mol. Biol. 13*, 238–252.

Bangham, A. D., Standish, M. M., & Weissmann, G. J. (1965b) *J. Mol. Biol. 13*, 253–259.

Barth, E., Myrvik, Q. M., Wagner, W., & Gristina, A. G. (1989) *Biomaterials 10*, 325–328.

Bater, E. A., & Wilchek, M. (1989) *TIBS 14*, 408–412.

Behm, R. J., & Hoesler, W. (1986) in *Chem. Phys. Solid Surf.*, (Vanselow, R., & Tong, S. Y., Eds.) Vol. 5, pp. 361–411, CRC Press, Cleveland, OH.

Bertozzi, C. R., & Bednarski, M. D. (1992) *J. Am. Chem. Soc. 114*, 5543–5546.

Bigelow, W. C., Pickett, D. L., & Zisman, J. (1946) *J. Colloid Sci. 1*, 513–538.

Boren, T., Falk, P., Roth, K. A., Larson, G., & Normark, S. (1993) *Science 262*, 1892–1895.

Bretscher, M. S. (1985) *Sci. Am. 252*, 100–109.

Buckland, R. M. (1986) *Nature 320*, 557–558.

Byramova, N. E., Mochalova, L. V., Belyanchiov, J. M., Matrosovich, M. N., & Bovin, N. V. (1991) *J. Carbohydr. Chem. 10*, 691–700.

Cemel, A., Fort, T., & Lando, J. B. (1972) *J. Polym. Sci. 10*, 2061–2083.

Charych, D. H., Nagy, J. O., Spevak, W., & Bednarski, M. D. (1993) *Science 261*, 585–588.

Chernyak, A. Y., Kononov, L. O., & Kochetkov, N. K. (1991) *Carbohydr. Res. 216*, 381–398.

Curatolo, W. A., Yau, A. O., & Small, D. M. (1978) *Biochemistry 17*, 5740–5744.

Day, D., & Ringsdorf, H. (1985) *J. Polym. Sci.: Polym. Lett. Ed. 16*, 205.

Deamer, D. W. (1984a) *Liposome Technology*, Vol. I, pp. 29–35, CRC, Boca Raton, FL.

Deamer, D. W. (1984b) *Liposome Technology*, Vol. I, 109 ff, CRC, Boca Raton, FL.

Dean, B., Ogushi, H., Cai, S., Otsuji, E., Tashiro, K., Hakomori, S., & Toyokuni, T. (1993) *Carbohydr. Res. 245*, 175–192.

DeGennes, P. G. (1985) *Rev. Mod. Phys. 57*, 827–863.

Ebara, Y., & Okahata, Y. (1994) *J. Am. Chem. Soc. 116*, 11209–11212.

Fowkes, F. M. Z. (1964) *Contact Angles, Wettability and Adhesion*, Vol. 43, American Chemical Society, Washington, DC.

Gaines, G. L. (1966) *Insoluble Monolayers at the Liquid–Gas Interface*, p 386 ff, John Wiley, New York.

Gamian, A., Chomik, M., Laferriere, C., & Roy, R. (1991) *Can. J. Microbiol. 37*, 233–237.

Golocchenko, J. A. (1986) *Science 232*, 48–53.

Ghosh, P., & Bachhawat, B. K. (1980) *Biochim. Biophys. Acta*, Vol. 632, 562–572.

Green, N. M. (1975) *Adv. Protein Chem. 29*, 85–133.

Gregoriadis, G., Ed. (1984) *Liposome Technology*, Vol. II, CRC Press, Boca Raton, FL.

Gristina, A. G. (1987) *Science 237*, 1588–1595.

Halcomb, R. L., Huang, H., & Wong, C.-H. (1994) *J. Am. Chem. Soc. 116*, 11315–11322.

Hampton, R. Y., Holz, R. W., & Goldstein, I. J. (1980) *J. Biol. Chem. 255*, 6766–6771.

Hansma, P. K., & Tersoff, J. (1987) *J. Appl. Phys. 61*, R1–R23.

Haussling, L., Ringsdorf, H., Schmitt, F. J., & Knoll, W. (1991) *Langmuir 7*, 1837–1840.

Helferich, B., & Hoffman, H. J. (1952) *Chem. Ber. 85*, 175–180.

Helm, C. A., Schmitt, F. J., Israelachvili, J. N., & Knoll, W. (1991) *Makromol. Chem. Symp. 46*, 103–111.

Hope, M. J., Bally, M. B., Mayer, L. D., Janoff, A. S., & Cullis, P. R. (1986) *Chem. Phys. Lipids 40*, 89–107.

Horejsi, V., & Kocourek, J. (1974) *Methods Enzymol. 34*, 361–367.

Horejsi, V., Smoleck, P., & Kocourek, J. (1978) *Biochim. Biophys. Acta 538*, 293–298.

Horejsi, V., Ticha, M., & Kocourek, J. (1977a) *Biochim. Biophys. Acta 449*, 290–300.

Horejsi, V., Ticha, M., & Kocourek, J. (1977b) *Biochim. Biophys. Acta 449*, 301–308.

Huang, L. (1993) *Liposomes*, (Ostro, M. J., Ed.), pp. 87–124, Dekker, New York.

Hub, H., Hupfer, B., & Ringsdorf, H. (1980) *Springer Ser. Chem. Phys. 11*, 253.

Humphries, G. M., & McConnell, H. M. (1977) *Proc. Natl. Acad. Sci. U.S.A. 74*, 3537–3541.

Israelachvili, J. N. (1985) *Intermolecular and Surface Forces*, Academic Press, London.

Juliano, R. L. (1983) in *Liposomes*, (Ostro, M. J., Ed.), pp. 53–86, Dekker, New York.

Juliano, R. L., & Stamp, D. (1976) *Nature 261*, 235–237.

Kallin, E. (1994) U*se of Aminoalditols and Glycosylamines in Neoglycoconjugate Synthesis*, in Neoglycoconjugates: Preparation and Applications, Lee, Y. C. and Lee, R. T., Eds., pp. 199–223, Academic Press, New York.

Kennedy, J. F. (1988) *Carbohydrate Chemistry*, Oxford University Press, Oxford, England.

Ketis, N. V., Girdleston, J., & Grant, C. W. M. (1980) *Proc. Natl. Acad. Sci. U.S.A. 77*, 3788–3790.

Kirby, C., & Gregoriadis, G. (1984) *BioTechnology*, 979–984.

Knight, C. G., (1981) *Liposomes: From Physical Structure to Therapeutic Applications*, Elsevier, Amsterdam.

Kobayashi, K., Kobayashi, A., Tobe, S., & Akaike, T. (1994) *Carbohydrate Containing Polystyrenes*, in Neoglycoconjugates: Preparation and Applications, Lee, Y. C. and Lee, R. T., Eds., pp. 261–284, Academic Press, New York.

Kojima, N., Handa, K., Newman, W., & Hakomori, S.-I. (1992) *Biochem. Biophys. Res. Commun. 189*, 1686–1694.

Kraska, B., & Mester, L.(1978) *Tetrahedron Lett. 46*, 4583–4586.

Langmuir, I. (1939) *Proc. Roy. Soc. London, Ser. A 170*, 1–39.

Laschewsky, A., Ringsdorf, H., & Schmidt, G. (1988) *Polymer 29*, 448–456.

Lee, R. T., & Lee, Y. C. (1982) *Methods Enzymol. 83*, 299–305.

Lee, R. T., & Lee, Y. C. (1994a) *Enhanced Biochemical Affinities of Multivalent Neoglycoconjugates*, in Neoglycoconjugates: Preparation and Applications, Lee, Y. C. and Lee, R. T., Eds., pp 23–50, Academic Press, New York.

Lee, Y. C., & Lee, R. T., Eds. (1994b) *Neoglycoconjugates*: *Preparation and Applications*, Academic Press, New York.

Lelkes, P. I. (1984) in *Liposome Technology*, (Gregoriadis, E., Ed.), Vol. 1, 51–65, CRC Press, Boca Raton, FL.

Lemieux, R. U., Bundle, D. R., & Baker, D. A. (1980), *U.S. Patent 4.238.473.*

Luna, E. J., & Hitt, A. L. (1992) *Science 258*, 955–963.

MacDonald, R. C., MacDonald, R. I., Menco, B. P. M., Takeshita, K., Subbarao, N. K., & Hu, L. R. (1991) *Biochim. Biophys. Acta 1061*, 297–303.

Marti, O., Ribi, H. O., Drake, B., Albrecht, T. R., Quate, C. F., & Hansma, P. K. (1988) *Science 239*, 50–52.

Mastendrea, M., & Bednarski, M. (1990) *Sugar Coated Semiconductors: Model Surfaces to Study Biological Adhesion*, Vol. 174, Proceedings of the Materials Research Society, Pittsburgh, PA.

Mate, C. M., Lorenz, M. R., & Novotny, V. J. (1989) *J. Chem. Phys. 90*, 7550–7555.

Matrosovich, M. N., Mochalova, L. V., Marinina, V. P., Biramova, N. E., & Bovin, N. V. (1990) *FEBS Lett. 272*, 209–212.

Mayer, L. D., Hope, M. J., & Cullins, P. R. (1986) *Biochim. Biophys. Acta 858*, 161–168.

Mayer, L. D., Hope, M. J., Cullis, P. R., & Janoff, A. S. (1985) *Biochim. Biophys. Acta 817*, 193–196.

Merritt, K. (1984) *Biomaterials 5*, 47–53.

Mo, H., Van Damme, E. J. M., Peumans, W. J., & Goldstein, I. J. (1994) *J. Biol. Chem. 269*, 7666–7673.

Nagy, J. O., Wang, P., Gilbert, J. H., Schaefer, M. E., Hill, T.G., & Callstrom, M. R. (1992) *J. Med. Chem. 35*, 4501–4502.

New, R. R. C., Ed. (1990) *Liposomes*: *A Practical Approach*, IRL, Oxford.

Nishimura, S.-I., Matsuoka, K., & Kurita, K. (1990) *Macromol. 23*, 4182–4194.

Olson, F., Hunt, C. A., Szoka, F. A., Vail, W. J., & Papahadjopoulos, D. (1979) *Biochim. Biophys. Acta 557*, 9–23.

Ostro, M. J., Ed. (1983) *Liposomes*, Dekker, New York.

Ostro, M. J., Ed. (1987) *Liposomes: From Biophysics to Therapeutics*, Dekker, New York.

Parce, W. J., Henry, N., & McConnell, H. M. (1978) *Proc. Natl. Acad. Sci. U.S.A. 75*, 1515–1518.

Paulsen, H., & Hoffgen, E. C. (1991) *Tetrahedron Lett. 32*, 2747–2750.

Phillips, M. L., Nudelman, E., Gaeta, F. C. A., Perez, M., Singhal, A. K., Hakomori, S.-I., & Paulson, J. C. (1990) *Science 250*, 1130–1132.

Pockels, A. (1893) *Nature 48*, 152.

Quate, C. F. (1973) *Phys. Today 26*, 1.

Rando, R. R., & Bangerter, F. W. (1979) *J. Supramol. Struct. 11*, 295–309.

Raz, A., Bucana, C., Fogler, W. E., Poste, G., & Fidler, I. J. (1981) *J. Cancer Res. 41*, 487–494.

Redwood, W. R., Jansons, V., & Patel, B. C. (1975) *Biochem. Biophys. Acta* Vol. 406, 347–361.

Reimer, P., Weissleder, R., Wittenberg, J., & Brady, T. J. (1992) *Radiology 182*, 565–569.

Ribi, H. O., Ludwig, D. S., Mercer, K. L., Schoolnik, G. K., & Kornberg, R. D. (1988) *Science 239*, 1272–1276.

Roche, A. C., Maget-Dana, R., Obrenovitch, A., Hildenbrand, K., Nicolau, C., & Monsigny, M. (1977) *FEBS Lett. 93*, 91–96.

Romanowska, A., Meunier, S., Tropper, F. D., Laferriere, C. A., & Roy, R. (1994) *Methods Enzymol. 242*, 90–101.

Roy, R., Andersson, F. O., Harms, G., Kelm, S., & Schauer, R. (1992a) *Angew. Chem. Int. Ed. Engl. 31*, 1478–1481.

Roy, R., Pon, R. A., Tropper, F. D., & Andersson, F. O. (1993a) *J. Chem. Soc., Chem. Commun.*, 264–266.

Roy, R., & Tropper, F. D. (1988) *J. Chem. Soc., Chem. Commun.*, 1058–1060.

Roy, R., Tropper, F. D., Morrison, T., & Boratynski, J. (1991) *J. Chem. Soc., Chem. Commun.*, 536–538.

Roy, R., Tropper, F. D., & Romanowska, A. (1992b) *Bioconjugate Chem. 3*, 256–261.

Roy, R., Tropper, F. D., & Romanowska, A. (1992c) *J. Chem. Soc., Chem. Commun.*, 1611–1613.

Roy, R., Tropper, F. D., Williams, A. J., & Brisson, J.-R. (1993b) *Can. J. Chem. 71*, 1995–2006.

Roy, R., Zanini, D., Meunier, S. J., & Romanowska, A. (1993c) *J. Chem. Soc., Chem. Commun.*, 1869–1872.

Samuelson, L., Miller, P., Galotti, D., Marx, K. A., Kumar, S., Tripathy, S., & Kaplan, D. (1992) *Thin Solid Films 210*, 796–798.

Schmitt, F. J., Weisenhorn, A. L., Hansma, P. K., & Knoll, W. (1991) *Makromol. Chem. 46*, 133–143.

Schnaar, R. L., & Lee, Y. C. (1975) *Biochemistry 14*, 1525–1541.

Schnaar, R. L., Weigel, P. H., Kuhlenschmidt, M. S., Schmell, E., Lee, Y. C., & Roseman, S. (1978) *J. Biol. Chem. 253*, 7940–7951.

Schnaar, R. L., Weigel, P. H., Roseman, S., & Lee, Y. C. (1982) *Methods Enzymol. 83*, 306–310.

Sharon, N. & Lis, H. (1993) *Sci. Am. 268*, 82–89.

Sharom, F. J., Barratt, D. G., & Grant, C. W. (1977) *Proc. Natl. Acad. Sci. U.S.A.*, 2751–2755.

Sigal, G. B., Mammen, M., & Whitesides, G. M. (1996) *J. Am. Chem. Soc. 118*, 3789–3800.

Slama, J. S., & Rando, R. R. (1980) *Biochemistry 19*, 4595–4600.

Sparks, M. A., Williams, K. W., & Whitesides, G. M. (1993) *J. Med. Chem. 36*, 778–783.

Spevak, W., Nagy, J. O., Charych, D. H., Schaefer, M. E., Gilbert, J. H., & Bednarski, M. D. (1993) *J. Am. Chem. Soc. 115*, 1146–1147.

Sukenik, C. N., Balachander, N., Culp, L. A., Lewandowski, K., & Merritt, K. (1990) *J. Biomed. Mat. Res. 24*, 1307–1323.

Surolia, A., Bachhawat, B. K., & Podder, S. K. (1975) *Nature 257*, 802–804.

Swalen, J. D., Allara, D. L., Andrade, J. D., Chandross, E. A., Garoff, S., Israelachvili, J., McCarthy, T. J., Murray, R., Pease, R. F., Rabolt, J. F., Waynne, K. J., & Yu, H. (1987) *Langmuir 3*, 932–950.

Szoka, F., & Papahadjopoulos, D. (1980) *Annu. Rev. Biophys. Bioeng. 9*, 467–508.

Szoka, F., Jacobson, K., Dezko, Z., & Papahadjopoulos, D. (1980) *Biochim. Biophys. Acta 600*, 1–18.

Tanford, C. (1980) T*he Hydrophobic Effect: Formation of Micelles & Biological Membranes*, 2nd edition, Wiley, New York.

Tieke, B. (1985) *Adv. Polym. Sci. 71*, 79–151.

Ticha, M., & Kocourek, J. (1991) *Carbohydr. Res. 213*, 339–342.

Tropper, F. D., Romanowska, A., & Roy, R. (1994) *Methods Enzymol. 242*, 257–271.

Varki, A. (1993) *Glycobiology 3*, 97–130.

Wattenbarger, M. R., Graves, D. J., & Lauffenburger, D. A. (1990) *Biophys. J. 57*, 765–777.

Weigel, P. H., Schmell, E., Lee, Y. C., & Roseman, S. (1978a) *J. Biol. Chem. 253*, 330–333.

Weigel, P. H., Schnaar, R. L., Kuhlenschmidt, M. S., Schmell, E., Lee, R. T., Lee, Y. C., & Roseman, S. (1978b) *J. Biol. Chem. 254*, 10830–10838.

Weinstein, J. N., Yoshikami, S., Henkart, P., Blumenthal, R., & Hagins, W. A. (1977) *Science 195*, 489–492.

Weisz, O. A., & Schnaar, R. L. (1991) *J. Cell. Biol. 115*, 485–493.

Whitesides, G. M., & Laibinis, P. E. (1990) *Langmuir 6*, 87–96.

Whitesides, G. M., & Ferguson, G. S. (1988) *Chemtracts: Organic Chemistry 1*, 171–187.

Wiley, D. C., & Skehel, J. J. (1987) *Annu. Rev. Biochem. 56*, 365–394.

Wu, P., Tin, G. W., & Baldeshwieler, J. D. (1981) *Proc. Natl. Acad. Sci. U.S.A. 78*, 2033–2037.

Young, T. (1805) *Philosophical Trans. Roy. Soc. London 95*, 65.

Zheng, Z., Tsurruoka, T., Tsuji, T., & Hakomori, S.-I. (1992) *Biochem. Biophys. Res. Commun. 186*, 1397–1402.

Chapter 10

Ahmed, R., & Gray, D. (1996) *Science, 272*, 54–60.

Altman, E., & Bundle, D. R. (1994) *Methods Enzymol. 247*, 243–253.

Amzel, L. M., Poljak, R. J., Saul, F., Varga, J. M., & Richards, F. F. (1974) *Proc. Natl. Acad. Sci. U.S.A. 71*, 1427–1430.

Arakatsu, Y., Ashwell, G., & Kabat, E. A. (1966) *J. Immunol. 97*, 858–866.

Arepali, S. R., Glaudemans, C. P. J., Daves, G. D., Kovac, P., & Bax, A. (1995) *J. Magn. Reson. 106*, 195–198.

Arevalo, J., Stura, E. A., Taussig, M. J., & Wilson, I. A. (1993) *J. Mol. Biol. 231*, 103–118.

Ashwell, G. (1972) *Methods Enzymol., 28*, 219–222.

Auzanneau, F-I., & Bundle, D. R. (1993a) *Can. J. Chem. 71*, 534–548.

Auzanneau, F-I., & Bundle, D. R. (1993b) *Carbohydr. Res. 247*, 195–209.

Auzanneau, F-I., Hanna, H. R., & Bundle, D. R. (1993) *Carbohydr. Res. 240*, 161–181.

Avery, O. T., & Goebel, W. F. (1931) *J. Exp. Med. 54*, 437–447.

Bannett, A. D., Bensinger, W. I., Raja, R., Baquero, A., & McAlack, R. F. (1987) *Transplantation 43*, 909–910.

Baumann, H., Altman, E., & Bundle, D. R. (1993) *Carbohydr. Res. 247*, 347–354.

Bax, A., Summers, M. F., Egan, W., Guirgis, N., Schneerson, R., Robbins, J. R., Ørskov, F., Ørskov, I., & Vann, W. F. (1988) *Carbohydr. Res. 173*, 53–64.

Beierbeck, H., Delbaere, L. T. J., Vandonselaar, M., & Lemieux, R. U. (1994) *Can. J. Chem. 72*, 463–470.

Beierbeck, H., & Lemieux, R. U. (1990) *Can. J. Chem. 68*, 820–827.

Bensinger, W. I., Baker, D. A., Buckner, C. D., Clift, R. A., & Thomas, E. D. (1981) *New Eng. J. Med. 304*, 160–162.

Bensinger, W. I., Buckner, C. D., Clift, R. A. Williams, B. M., Banaji, M., & Thomas E. D.(1987) *Transplant. Proc. 19*, 4605–4608.

Berek, C., Griffiths, G. M., & Milstein, C. (1985) *Nature 316*, 412–418.

Bhat, T. N., Bentley, G. A., Fischmann, T. O., Boulot, G., & Poljak, R. J. (1990) *Nature, 347*, 483–485.

Bjorkman, P. J., Saper, M. A., Samraoui, B., Bennett, W. S., Strominger, J. L., & Wiley, D. C. (1987a) *Nature 329*, 506–512.

Bjorkman, P. J., Saper, M. A., Samraoui, B., Bennett, W. S., Strominger, J. L., & Wiley, D. C. (1987b) *Nature 329*, 512–518.

Björndal, H., Hellerqvist, C. G., Lindberg, B., & Svensson, S. (1970) *Angew. Chem. Intl. Ed. Engl., 9*, 610–619.

Bock, K. (1983) *Pure Appl. Chem. 55*, 605–622.

Bock, K. (1987) *Pure Appl. Chem. 59*, 1447–1456.

Bock, K., Frejd, T., Kihlberg, J., & Magnusson, G. (1988) *Carbohydr. Res. 176*, 253–270.

Bock, K., Josephson, S., & Bundle, D. R. (1982) *J. Chem. Soc., Perkin Trans. 2*, 59–70.

Bock, K., & Lemieux, R. U. (1982) *Carbohydr. Res. 100*, 63–74.

Bock, K., & Meldal, M. (1984a) *Acta Chem. Scand. B 38*, 71–77.

Bock, K., & Meldal, M. (1984b) *Acta Chem. Scand. B 38*, 255–266.

Bock, K., Meldal, M., Bundle, D. R., Iversen, T., Garegg, P., Norberg, T., Lindberg, A. A., & Svenson, S. B. (1984a) *Carbohydr. Res. 130*, 23–34.

Bock, K., Meldal, M., Bundle, D. R., Iversen, T., Pinto, B. M., Garegg, P. J., Kvanström, I., Norberg, T., Lindberg, A. A., & Svenson, S. B. (1984b) *Carbohydr. Res. 130*, 35–53.

Borén, H. B., Garegg, P. J., & Wallin N.-H. (1972) *Acta Chem. Scand. 26*, 1082–1086.

Bott, R. R., Navia, M. A., & Smith, J. L. (1982) *J. Biol Chem. 257*, 9883–9886.

Bourne, Y, Rougés, P, & Cambillau, C. (1992) *J. Biol. Chem. 267*, 197–203.

Bourne, Y., van Tilbeurgh, H., & Cambillau, C. (1993) *Current Opinion Struct. Biol., 3*, 681–686.

Branden, C., & Tooze, J. (1991) *Introduction to Protein Structure*, Garland Publishing Inc., New York & London.

Brandts, J. F., Lin, L.-N, Wiseman, T., Williston, S., & Yang, C. P. (1990) *Am. Lab., (Shelton Conn.)*, 30–41.

Braun, D. G., Eichmann, K., & Krause, R. M. (1969) *J. Exp. Med. 129*, 809–830.

Braun, D. G., Kjems, E., & Cramer, M. (1973) *J. Exp. Med. 138*, 645–658.

Brisson, J.-R., Baumann, H., Imberty, A., Perez, S., & Jennings, H. J. (1992) *Biochemistry 31*, 4996–5004.

Brummell, D. A., Sharma, V. P., Anand, N. N., Bilous, D., Dubuc, G., Michniewicz, J., MacKenzie, C. R., Sadowska, J., Sigurskjold, B. W., Sinnott, B., Young, N. M., Bundle, D. R., & Narang, S. A.(1993) *Biochemistry 32*, 1180–1187.

Bundle, D. R. (1986) in *Protein-Carbohydrate Interactions in Biological Systems*, (Lark, D. L., Ed.) Academic Press, London, pp. 165–171.

Bundle, D. R. (1989) *Pure Appl. Chem. 61*, 1171–1180.

Bundle, D. R. (1990) *Topics Current Chem. 154*, 1–37.

Bundle, D. R. (1997) in *Glycosciences: Status and Perspectives* (Gabius, H.-J., & Gabius, S., Eds.) Chapman & Hall, Weinheim, Germany, pp. 311–331.

Bundle, D. R., Altman, E., Auzanneau, F.-I., Baumann, H., Eichler, E., & Sigurskjold, B. W. (1994c) *The Alfred Benzon Symposium No. 36, Complex Carbohydrates in Drug Research*, (Bock, K., & Clausen, H., Eds.) Copenhagen, Munksgaard, Copenhagen, pp. 168–181.

Bundle, D. R., Baumann, H., Brisson, J.-R., Gagne, S. M., Zdanov, A., & Cygler, M. (1994b) *Biochemistry, 33*, 5183–5192.

Bundle, D. R., Cherwonogrodzky, J. W., Gidney, M. A. J., Meikle, P. J., Perry, M. B., & Peters, T. (1989) *Infect. Immun., 57*, 2829–2836.

Bundle, D. R., Eichler, E., Gidney, M. A. J., Meldal, M., Ragauskas, A., Sigurskjold, B. W., Sinnott, B., Watson, D. C., Yaguchi, M., & Young, N. M. (1994a) *Biochemistry 33*, 5172–5182.

Bundle, D. R., Gidney, M. A. J., Josephson, S., & Wessel, H.-P. (1983) *Am. Chem. Soc. Symp. Ser.*, 49–63.

Bundle, D. R., Gidney, M. A. J., Kassam, N., & Rahman, A. F. R. (1982) *J. Immunol., 129*, 678–682.

Bundle, D. R., Gidney, M. A. J., Perry, M. B., Duncan, J. R., & Cherwonogrodzky, J. W. (1984) *Infect. Immun., 46*, 389–393.

Bundle, D. R., & Josephson, S (1979a) *Can. J. Chem., 57*, 662–668.

Bundle, D. R., & Josephson, S. (1979b) *J. Chem. Soc., Perkin Trans. 1*, 2736–2739.

Bundle, D. R., & Josephson, S. (1980) *Carbohydr. Res., 80*, 75–85.

Bundle, D. R., & Sigurskjold, B. W. (1994) *Methods Enzymol., 242*, 288–305.

Bundle, D. R., & Young, N. M. (1992) *Current Opinion Struct. Biol. 2*, 666–673.

Carlin, N. I. A., Bundle, D. R., & Lindberg, A. A. (1987a) *J. Immunol., 138*, 4419–4427.

Carlin, N. I. A., Gidney, M. A. J., Lindberg, A. A., & Bundle, D. R. (1986) *J. Immunol. 137*, 2361–2366.

Carlin, N. I. A., Lindberg, A. A., Bock, K., & Bundle, D. R. (1984) *Eur. J. Biochem., 139*, 189–194.

Carlin, N. I. A., Svenson, S. B., & Lindberg, A. A. (1987b) *Microbiol. Pathogenesis 2*, 171–183.

Caroff, M., Bundle, D. R., & Perry, M. B. (1984) *Eur. J. Biochem. 139*, 195–200.

Carver, J. P., Michnick, S. W., Imberty, A., & Cumming, D. A. (1991) *Ciba Found. Symp. 158*, 6–26.

Chervenak, M. C., & Toone, E. J. (1994) *J. Am. Chem. Soc. 116*, 10533–10539.

Cho, J.-W., & Troy, F. A. (1994) *The Alfred Benzon Symposium No. 36, Complex*

Carbohydrates in Drug Research, (Bock, K., & Clausen, H., Eds.) Copenhagen, Munksgaard, Copenhagen, pp. 260–279.

Christofides, J. C., Davies, D. B., Martin, J. A., & Rathbone, E. B. (1986) *J. Am. Chem. Soc. 108*, 5738–5743.

Clore, G. M., & Gronenborn, A. M. (1982) *J. Magn. Reson. 48*, 402–417.

Clore, G. M., & Gronenborn, A. M. (1983) *J. Magn. Reson. 53*, 423–442.

Colman, P. M. (1988) *Adv. Immunol. 43*, 99–132.

Conelly, M. L. (1983) *J. Appl. Crystallogr. 16*, 548–558.

Cooper, D. K. C., Ye, Y., Niekrasz, M., Kehoe, M., Martin, M., Neethling, F. A., Kosanke, S., DeBault, L. E., Worsley, G., Zuhdi, N., Oriol, R., & Romano, E. (1993) *Transplantation 56*, 769–777.

Cotton, R. G. H., Secher, D. S., & Milstein, C. (1973) *Eur. J. Immunol. 3*, 135–140.

Cromer, R., Spohr, U., Khare, D. P., LePendu, J., & Lemieux, R. U. (1992) *Can. J. Chem. 70*, 1511–1530.

Cumming, D. A., & Carver, J. P. (1987) *Biochemistry 26*, 6664–6676.

Cygler, M., Rose, D. R., & Bundle, D. R. (1991) *Science 253*, 442–445.

Davies, D. R., & Chacko, S. (1993) *Acc. Chem. Res. 26*, 421–427.

Davies, D. R., & Metzger, H. (1983) *Annu. Rev. Immunol. 1*, 87–117.

Davies, D. R., Padlan, E.A., & Sheriff, S. (1990) *Annu. Rev. Biochem. 59*, 439–473.

Davies, D. R., Padlan, E. A., & Sheriff, S. (1990) *Annu. Rev. Biochem. 59*, 439–473.

Delbaere, L. T. J., Vandonselaar, M., Prasad, L., Quail, J. W., Nikrad, P. V., Pearlstone, J. R., Carpenter, M. R., Smillie, L. B., Spohr, U., & Lemieux, R. U. (1990) *Can. J. Chem. 68*, 1116–1121.

Delbaere, L. T. J., Vandonselaar, M., Prasad, L., Quail, J. W., Wilson, K. S., & Dauter, Z. (1993) *J. Mol. Biol. 230*, 950–965.

Deng, S.-J., MacKenzie, C. R., Sadowska, J., Michniewicz, J., Young, N. M., Bundle, D. R., & Narang, S. (1994) *J. Biol. Chem. 269*, 9533–9538.

Du, M.-H., Spohr, U., & Lemieux, R. U. (1994) *Glycoconjugate J. 11*, 443–461.

Dunitz, J. D. (1994) *Science 264*, 670.

Edmundson, A. B., Ely, K. R., Girling, R. L., Abola, E. E., Schiffer, M., Westholm, F. A., Fausch, M. D., & Deutsch, H. F. (1974) *Biochemistry, 13*, 3816–3827.

Ekborg, G., Eklind, K.,Garegg, P. J., Gotthammar, B., Carlsson, H. E., Lindberg, A. A., & Svenungsson, B. (1977) *Immunochemistry 14*, 153–157.

Ekborg, G., Garegg, P. J., & Gotthammar, B. (1975) *Acta Chem. Scand. B 29*, 765–771.

Evans, S. V., Sigurskjold, B. W., Jennings, H. J., Brisson, J. R., To, R., Altman, E., Frosch, M., Weisgerber, C., Kratzin, H., Klebert, S., Vaesen, M., Bitter-Suermann, D., Rose, D. R., Young, N. M., & Bundle, D. R. (1995) *Biochemistry 34*, 6737–6744.

Fersht, A. R. (1987) *Trends Biochem. Sci., 12*, 301–304.

Finkelstein, A. V., & Janin, J. (1989) *Protein Eng. 3*, 1–3.

Frosch, M., Görgen, I., Boulnois, G. J., Timmis, K. N., & Bitter-Suermann, D. (1985) *Proc. Natl. Acad. Sci. U.S.A. 82*, 1194–1198.

Glaudemans, C. P. J. (1991) *Chem. Rev. 91*, 25–33.

Glaudemans, C. P. J., & Kovac, P. (1985) *Mol. Immunol. 22*, 651–653.

Glaudemans, C. P. J., & Kovac, P. (1988) *Am. Chem. Soc. Symp. Ser. 374*, 78–108.

Glaudemans, C. P. J., Kovac, P., & Nashed, E. M. (1994) *Methods Enzymol. 247*, 305–322.

Glaudemans, C. P. J., Lerner, L., Daves, G. D., Jr., Kovac, P., Venable, R., & Bax, A. (1990) *Biochemistry 29*, 10906–10911.

Goding, J. W. (1983) *Monoclonal Antibodies: Principles and Practice*, Academic Press, Inc., Harcourt Brace Jovanovich, Sydney, Australia.

Goebel, W. F. (1939) *J. Exp. Med. 69*, 353–364.

Goebel, W. F. (1940) *J. Exp. Med. 72*, 33–48.

Goebel, W. F., & Avery, O. T. (1931) *J. Exp. Med. 54*, 431–436.

Grootenhuis, P. D. J., & van Boeckel, C. A. A. (1991) *J. Am. Chem. Soc. 113*, 2743–2747.

Ha, S. N., Giammona, A., Field, M., & Brady, J. W. (1988) *Carbohydr. Res. 180*, 207–221.

Haber, E. (1970) *Fed. Proc. 29*, 66–71.

Hakomori, S.-I. (1981) *Sem. Hematol. 18*, 39–62.

Hakomori, S.-I. (1989) *Adv. Cancer Res. 52*, 257–331.

Hanna, H. R., & Bundle, D. R. (1993) *Can. J. Chem. 71*, 125–134.

Heidelberger, M., Kabat, E. A., & Mayer, M. (1942) *J. Exp. Med. 75*, 35–47.

Hellerqvist, C. G., Larm, O., Lindberg, B., Holme, T., & Lindberg, A.A. (1969a) *Acta Chem. Scand. 23*, 2217–2222.

Hellerqvist, C. G., Lindberg, B., Pilotti, A., & Lindberg, A. A. (1970a) *Acta Chem. Scand. 24*, 1168–1174.

Hellerqvist, C. G., Lindberg, B., Samuelsson, K., & Lindberg, A. A. (1971) *Acta Chem. Scand. 25*, 955–961.

Hellerqvist, C. G., Lindberg, B., Svensson, S., Holme, T., & Lindberg, A. A. (1968) *Carbohydr. Res. 8*, 43–55.

Hellerqvist, C. G., Lindberg, B., Svensson, S., Holme, T., & Lindberg, A. A. (1969b) *Carbohydr. Res., 9*, 237–241.

Hellerqvist, C. G., Lindberg, B., Svensson, S., Holme, T., & Lindberg, A. A. (1970b) *Carbohydr. Res. 14*, 17–26.

Hellström, I., Garrigues, H. J., Garrigues, U., Hellström, K. E. (1990) *Cancer Res.* 50, 2183–2190.

Hilschmann, N., & Craig, L. C. (1965) *Proc. Natl. Acad. Sci. U.S.A. 53*, 1403–1409.

Himmelspach, K., Westphal, O., & Teichmann, B. (1971) *Eur. J. Immunol. 1*, 106–112.

Hindsgaul, O., Khare, D. P., Bach, M., & Lemieux, R. U. (1985) *Can. J. Chem. 63*, 2653–2658.

Homans, S. W. (1990) *Biochemistry 29*, 9110–9118.

Honsik, C. J., Jung, G., & Reisfeld, R. A. (1986) *Proc. Natl. Acad. Sci. U.S.A. 83*, 7893–7897.

Hricovini, M., Shah, R. N., & Carver, J. P. (1992) *Biochemistry 31*, 10018–10023.

Imberty, A., & Perez, S. (1994) *Glycobiology 4*, 351–366.

Imberty, A., Hardman, K. D., Carver, J. P., & Perez, S. (1990) *Glycobiology 1*, 631–642.

Iversen, T., & Bundle, D. R. (1982) *Carbohydr. Res. 103*, 29–40.

Jaton, J.-C., & Braun, D. G. (1972) *Biochem. J. 130*, 539–546.

Jaton, J.-C., Huser, H., Braun, D. G., Givol, D., Pecht, I., & Schlessinger, J. (1975) *Biochemistry 14*, 5312–5315.

Jeffrey, G. A., & Sänger, W. (1991) in *Hydrogen Bonding in Biological Structures*, Springer-Verlag, Berlin.

Jeffrey, P. D., Bajorath, J., Chang, C. Y., Yelton, D., Hellstrom, I., Hellstrom, K. E., & Sheriff, S. (1995) *Nature Struct. Biol. 2*, 466–471.

Jencks, W. P. (1981) *Proc. Natl. Acad. Sci. U.S.A. 78*, 4046–4050.

Jennings, H. J. (1983) *Adv. Carbohydr. Chem. Biochem. 41*, 155–208.

Johnson, L. N., Cheetham, J., McLaughlin, P. J., Acharya. K. R., Barford, D., & Phillips, D. C. (1988) *Current Topics Microbiol. Immunol. 139*, 81–134.

Jolley, M. E., Glaudemans, C. P. J., Rudikoff, S., & Potter, M. (1974) *Biochemistry 13*, 3179–3184.

Jolley, M. E., Rudikoff, S., Potter, M., & Glaudemans, C. P. J. (1973) *Biochemistry 12*, 3039–3044.

Jones, J. K. N., & Perry, M. B. (1957) *J. Am. Chem. Soc., 79*, 2787–2793.

Jörbäck, H. J. A., Svenson, S. B., & Lindberg, A. A. (1984) *J. Immunol. 133*, 950–957.

Josephson, S., & Bundle, D. R. (1979) *Can. J. Chem. 57*, 3073–3079.

Josephson, S., & Bundle, D. R. (1980) *J. Chem. Soc., Perkin Trans. 1*, 297–301.

Kabat, E. A. (1962) *Fed. Proc. 21*, 694–701.

Kabat, E. A. (1966) *J. Immunol. 97*, 1–11.

Kabat, E. A. (1973) *Adv. Chem. Ser. 101*, 334–361.

Kabat, E. A. (1976) *Structural Concepts in Immunology & Immunochemistry*, Second Ed., Holt, Rinehart, & Winston, New York.

Kabat, E. A. (1980) *Membranes, Receptors, and the Immune Response*, Alan R. Liss, Inc. New York, pp. 1–46.

Kabat, E. A. (1983) *Annu. Rev. Immunol. 1*, 1–32.

Kabat, E. A. (1988) *Annu. Rev. Immunol. 6*, 1–24.

Kabat, E. A., Liao, J., Burzynska, M. H., Wong, T. C., Thˉgerson, H., & Lemieux, R. U. (1981) *Mol. Immunol. 18*, 873–881.

Kabat, E. A., & Mayer, M. M. (1961) *Experimental Immunochemistry*; Second Ed., C. C. Thomas, Springfield, Illinois.

Kabat, E. A., Wu, T. T., Perry, H. M., Gottesman, K. S., & Foeller, C. (1991) *Sequences of Proteins of Immmunological Interest*, Fifth Ed., National Institutes of Health, Bethesda.

Karlsson, K.-A. (1991) *Trends Pharmacol. Sci. 12*, 265–272.

Karlsson, R., Michaelsson, A., & Mattson, L., (1991) *J. Immunol. Methods 145*, 229–240.

Kay, L. E. (1995) *Current Opinion Struct. Biol. 5*, 674–681.

Kihlberg, J., & Bundle, D. R. (1991) *Carbohydr. Res., 216*, 67–78.

Kihlberg, J., Eichler, E., & Bundle, D. R. (1991) *Carbohydr. Res. 211*, 59–75.

Kihlberg, J., Hultgren, S. J., Normark, S., & Magnusson, G. (1989) *J. Am. Chem. Soc. 111*, 6364–6368.

Kimball, J. W., Pappenheimer, A.M., Jr., & Jaton, J.-C. (1971) *J. Immunol. 106*, 1177–1184.

Klein, J. (1982) *Immunology*, Wiley, New York.

Köhler, G., & Milstein, C. (1975a) *Nature 256*, 495–497.

Köhler, G., & Milstein, C. (1975b) *Eur. J. Immunol. 6*, 511–519.

Krause, R. M. (1970) *Adv. Immunol. 12*, 1–56.

Kronis, K. A., & Carver, J. P. (1985) *Biochemistry 24*, 834–840.

Landsteiner, K. (1945) *The Specificity of Serological Reactions*, Harvard University Press, Cambridge.

Lemieux, R. U. (1978) *Chem. Soc. Rev. 7*, 423–452.

Lemieux, R.U. (1989) *Chem. Soc. Rev. 18*, 347–374.

Lemieux, R. U. (1993) *Carbohydrate Antigens, ACS Symposium Series 519*, (Garegg, J., & Lindberg, A. A., Eds.) American Chemical Society, Washington, D.C.

Lemieux, R. U. (1994) *The Alfred Benzon Symposium No. 36, Complex Carbohydrates in Drug Research*, (Bock, K., & Clausen, H., Eds.), Copenhagen, Munksgaard, Copenhagen, pp. 188–197.

Lemieux, R. U. (1996) *Acc. Chem. Res. 29*, 373–380.

Lemieux, R. U., Baker, D. A., & Bundle, D. R. (1977) *Can. J. Biochem. 55*, 507–512.

Lemieux, R. U., Bock, K., Delbaere, L. T. J., Koto, S., & Rao, V. S. (1980) *Can. J. Chem. 58*, 631–653.

Lemieux, R. U., Bundle, D. R., & Baker, D. A. (1975a) *J. Am. Chem. Soc. 97*, 4076–4083.

Lemieux, R. U., Cromer, R., & Spohr, U. (1988a) *Can. J. Chem. 66*, 3083–3098.

Lemieux, R. U., Delbaere, L. T. J., Beierbeck, H., & Spohr, U. (1991) *Ciba Found. Symp. 158*, pp 231–248, Wiley, Chichester.

Lemieux, R. U., & Driguez, H. (1975a) *J. Am. Chem. Soc. 97*, 4063–4069.

Lemieux, R. U., & Driguez, H. (1975b) *J. Am. Chem. Soc. 97*, 4069–4075.

Lemieux, R. U., Du, M.-H., & Spohr, U. (1994) *J. Am. Chem. Soc. 116*, 9803–9804.

Lemieux, R. U., Hendriks, K. B., Stick, R. V., & James, K. (1975b) *J. Am. Chem. Soc. 97*, 4056–4062.

Lemieux, R. U., Hindsgaul, O., Bird, P., Narasimhan, S., & Young, W.W., Jr. (1988b) *Carbohydr. Res. 178*, 293–305.

Lemieux, R. U., Koto, S., & Voisin, D. (1979a) *ACS Symposium Series 87*, 17–29.

Lemieux, R. U., LePendu, J., & Hindsgaul, O. (1979b) *J. Antibiot. 32*, S21–31.

Lemieux, R. U., Szweda, R., Paszkiewicz-Hnatiw, E., & Spohr, U. (1990) *Carbohydr. Res. 205*, c12–c17.

Lemieux, R. U., Venot, A. P., Spohr, U., Bird, P., Mandel, G., Morishima, N., Hindsgaul, O., & Bundle, D. R. (1985) *Can. J. Chem. 63*, 2664–2668.

Lemieux, R. U., Wong, T. C., Liao, J., & Kabat, E. A. (1984) *Mol. Immunol. 21*, 751–759.

Levy, R., Assulin, O., Scherf, T., Levitt, M., & Anglister, J. (1989) *Biochemistry 28*, 7168–7175.

Lindberg, A. A., Wollin, R., Bruse, G., Ekwall, E., & Svenson, S. B. (1983) *Am. Chem. Soc. Symp. Ser. 231*, 83–118.

Lloyd, K. O., Kabat, E. A., & Licerio, E. (1968) *Biochemistry 7*, 2976–2990.

Lönngren, J., Goldstein, I. J., & Niederhuber, J. E. (1976) *Arch. Biochem. Biophys. 175*, 661–669.

Lüderitz, O. (1970) Angew. *Chem. Intl. Ed. Engl. 9*, 649–750.

Lüderitz, O., Westphal, O., & Staub, A. M. (1966) *Bacteriol. Rev. 30*, 192–255.

Lüderitz, O., Westphal, O., Staub, A. M., & Le Minor, L. (1960) *Nature, 188*, 556–558.

Lüderitz, O., Westphal, O., Staub, A. M., & Nikaido, H. (1971) in *Microbial Toxins Weinbaum* (Kadis, G. S., & Ajl, S. J., Eds.) Vol. 4, pp. 145–233.

Luk, J. M. C., & Lindberg, A. A. (1991) *J. Clin. Microbiol. 29*, 2424–2433.

Mäda, H., Schmidt-Kessen, E. J., & Jaton, J.-C. (1977) *Biochemistry 16*, 4086–4089.

Mäkelä, O., Peterfy, F., Outschoorn, I. G., Richter, A. W., & Seppälä, I. (1984) *Scand. J. Immunol. 19*, 541–550.

Mäkelä, O., Seppälä, I., & Pelkonen, J. (1985) *Eur. J. Biochem. 15*, 827–833.

McBroom, C. R., Samanen, C. H., & Golstein, I. J. (1972) *Methods Enzymol. 28*, 212–219.

Meikle, P. J., Young, N. M., & Bundle, D. R. (1990) J. *Immunol. Methods 132*, 255–261.

Metzger, H. (1970) *Adv. Immunol. 12*, 57–116.

Michon, F., Brisson, J. R., & Jennings, H. J. (1987) *Biochemistry 26*, 8399–8405.

Milstein, C. (1986) *Science 231*, 1261–1268.

Mollison, P. L. (1983) in *Blood Transfusion in Clinical Medicine*, 7th Ed., Blackwell, London.

Moreno, C., Hale, C., Hewett, R., & Cussell, D. (1979) *Eur. J. Immunol. 9*, 916–919.

Mulligan, M.S., Paulson, J. C., De Frees, S., Zheng, Z.-L., Lowe, J. B., & Ward, P. A. (1993) *Nature 364*, 149–151.

Murray-Rust, P., Stallinger, W. C., Monti, C. T., Preston, R., & Glusker, J. P. (1983) *J. Am. Chem. Soc. 105*, 3206–3214.

Nashed, E. M., Perdomo, G. R., Padlan, E. A., Kovac, P., Matsuda, T., Kabat, E. A., & Glaudemans, C. P. J. (1990) *J. Biol. Chem. 265*, 20699–20707.

Nelson, P.W., Helling, T. S., Shield, C. F., Beck, M., & Bryan, C. F. (1992) *Am. J. Surgery 164*, 541–545.

Neurohr, K. J., Bundle, D. R., Young, N. M., & Mantsch, H. H. (1982b) *Eur. J. Biochem. 123*, 305–310.

Neurohr, K. J., Mantsch, H. H., Young, N. M., & Bundle, D. R. (1982a) *Biochemistry 21*, 498–503.

Nikrad, P. V., Beierbeck, H., & Lemieux, R. U. (1992) *Can. J. Chem. 70*, 241–253.

Nisonoff, A. (1982) *Introduction to Molecular Immunology*, Sunderland, Sinauer Associates, Massachusetts.

Norberg, T., Svenson, S. B., Bock, K., & Meldal, M. (1985) *FEMS Microbiol. Lett. 28*, 171–176.

Nossal, G. J. V., & Ada, G. L. (1971) *Antigens, Lymphoid Cells and the Immune Response*, Academic Press, New York.

Novotny, J., Bruccoleri, R., Newall, J., Murphy, D., Haber, E., & Karplus, M. (1983) *J. Biol. Chem. 258*, 14433–14437.

Odaka, A., Kim, J. I. L., Takahashi, H., Shimada, I., & Arata, Y. (1992) *Biochemistry 31*, 10686–10691.

Oomen, R., Young, N. M. & Bundle, D. R. (1991) *Protein Eng. 4*, 427–433.

Oriol, R., Opelz, G., Chun, C., & Terasaki, P. I. (1980) *Transplantation, 29*, 397–400.

Oriol, R., Ye, Y., Koren, E., & Cooper, D. K. C. (1993) *Transplantation 56*, 1433–1442.

Padlan, E. A., & Kabat, E. A. (1988) *Proc. Natl. Acad. Sci. U.S.A. 85*, 6885–6889.

Padlan, E. A., Segal, D. M., Spande, T. F., Davies, D. R., Rudikoff, S., & Potter, M. (1973) *Nature New Biol. 245*, 165–167.

Paul, W. E. (1989) *Fundamental Immunology*, Raven Press, New York.

Paulsen, H. (1982) Angew. *Chem. Intl. Ed. Engl. 21*, 155–173.

Perlmutter, R. M., Hansburg, D., Briles, D. E., Nicolotti, R. A., & Davie, J. M. (1978) *J. Immunol. 121*, 566–572.

Perry, M. B., & Bundle, D. R. (1990) *Infect. Immun. 58*, 1391–1395.
Perry, M. B., Bundle, D. R., Gidney, M. A. J., & Lior, H. (1988) *J. Clin. Microbiol. 26*, 2391–2394.
Peters, T., Brisson, J.-R., & Bundle, D. R. (1990) *Can. J. Chem. 68*, 979–988.
Peters, T., & Pinto, B. M. (1996) *Current Opinion Struct. Biol. 6*, 710–720.
Pincus, J. H., Jaton, J.-C., Bloch, K. J., & Haber, E. (1970a) *J. Immunol. 104*, 1143–1148.
Pincus, J. H., Jaton, J.-C., Bloch, K. J., & Haber, E. (1970b) *J. Immunol. 104*, 1149–1154.
Pinto, B. M., & Bundle, D. R. (1983) *Carbohydr. Res. 124*, 313–318.
Pinto, B. M., & Bundle, D. R. (1984) *Carbohydr. Res. 133*, 333–338.
Pinto, B. M., Reimer, K. B., Morissette, D. G., & Bundle, D. R. (1989) *J. Org. Chem. 54*, 2650–2656.
Pinto, B. M., Reimer, K. B., Morissette, D. G., & Bundle, D. R. (1990a) *J. Chem. Soc., Perkin Trans. 1*, 293–299.
Pinto, B. M., Reimer, K. B., Morissette, D. G., & Bundle, D. R. (1990b) *Carbohydr. Res., 196*, 156–166.
Poljak, R. J. (1978) *CRC Critical Rev. Biochem., 5*, 45–84.
Poppe, L., & van Halbeek, H. (1991) *J. Am. Chem. Soc. 113*, 363–365.
Poppe, L., & van Halbeek, H. (1992) *J. Am. Chem. Soc. 114*, 1092–1094.
Poppe, L., Stuike-Prill, R., Meyer, B., & van Halbeek, H. (1992) *J. Biomol. NMR 2*, 109–136.
Potter, M. (1972) *Physiol. Res. 52*, 631–719.
Potter, M., Mushinsky, E. B., Rudikoff, S., Glaudemans, C. P. J., Padlan, E. A., & Davies, D. R. (1979) *Ann. Immunol. (Inst. Pasteur) 130 C*, 263–271.
Quiocho, F. A. (1986) *Annu. Rev. Biochem. 55*, 287–315.
Quiocho, F. A. (1988) *Current Topics Microbiol. Immunol. 139*, 135–148.
Quiocho, F. A. (1989) *Pure Appl. Chem. 61*, 1293–1306.
Raffa, R. B., & Porreca, F. (1989) *Life Sci. 44*, 245–258.
Ragupathi, G., Park, T. K., Zhang, S., Kim, I. J., Garber, L., Adluri, S., Lloyd, K. O., Danishefsky, S. J., & Livingston, P.O. (1997) *Angew. Chem. Int. Ed. Engl. 36*, 125–127.
Riechmann, L., & Weill, M. (1993) *Biochemistry 32*, 8848–8855.
Rietschel, E. T., Brade, L., Lindner, B., & Zahringer, U. (1992) in *Bacterial Endotoxic Lipopolysaccharides* (Morrison, D.C., & Ryan, J. L., Eds.) Vol. 1, CRC Press, Boca Raton, Florida, pp. 3–41.
Rini, J. M. (1995) *Annu. Rev. Biophys. Biomol. Struct. 24*, 551–577.
Rini, J. M., Schulze-Gahmen, U., & Wilson, I. A. (1992) *Science 255*, 959–965.
Rivera-Sagredo, A., Solis, D., Diaz-Maurino, T., Jimenez-Barbero, J., & Martin-Lomas, M. (1991) *Eur. J. Biochem. 197*, 217–228.
Roitt, I. M., Brostoff, J., & Male, D. K. (1985) *Immunology*, C.V. Mosby, St. Louis, MO.
Romano, E. L., Soyano, A., Montano, R. F., Ratcliffe, M., Olson, M., Suarez, G., Martínez, N., & Worsley, G. (1994) *Vox Sang. 66*, 194–199.
Rose, D. R., Cygler, M., To, R. J., Przybylska, M., Sinnott, B., & Bundle, D. R. (1990) *J. Mol. Biol. 215*, 489–492.
Rose, D. R., Przybylska, M., To, R. J., Kayden, C. S., Oomen, R. P., Vorberg, E., Young, N. M., & Bundle, D. R. (1993) *Protein Sci. 2*, 1106–1113.
Rutherford, T. J., Partridge, J., Weller, C. T., & Homans, S. W. (1993) *Biochemistry 32*, 12715–12724.
Scherf, T., Hiller, R., Naider, F., Levitt, M., & Anglister, J. (1992) *Biochemistry 31*, 6884–6897.
Schwartz, B. A., & Gray, G. R. (1977) *Arch. Biochem. Biophys. 181*, 542–549.
Seppälä, I., & Mäkelä, O. (1989) *J. Immunol. 143*, 1259–1264.
Sher, A., & Tarikas, H. (1971) *J. Immunol. 106*, 1227–1233.
Shibata, S., Goldstein, I. J., & Baker, D. (1982) *J. Biol. Chem. 25*, 9324–9329.
Sierks, M. R., Bock, K., Refn, S., & Svensson, B. (1992) *Biochemistry 31*, 8972–8977.
Sigurskjold, B. W., Altman, E., & Bundle, D. R. (1991) *Eur. J. Biochem. 197*, 239–246.
Sigurskjold, B. W., & Bundle, D. R. (1992) *J. Biol. Chem. 267*, 8371–8376.

Spohr, U., & Lemieux, R. U. (1988) *Carbohydr. Res. 174*, 211–237.

Spohr, U., Hindsgaul, O., & Lemieux, R. U. (1985a) *Can. J. Chem. 63*, 2644–2652.

Spohr, U., Morishima, N., Hindsgaul, O., & Lemieux, R. U. (1985b) *Can. J. Chem., 63*, 2659–2663.

Spohr, U., Paszkiewicz-Hnatiw, E., Morishima, N., & Lemieux, R. U. (1992a) *Can. J. Chem. 70*, 254–271.

Spohr, U., Petrakova, E., & Lemieux, R. U. (1992b) *Can. J. Chem. 70*, 233–240.

Stanfield, R. L., Fieser, T. M., Lerner, R. A., & Wilson, I. A. (1990) *Science 248*, 712–719.

Stirm, S., Staub, A. M., Leluc, B., Mayer, H., Lüderitz, O., & Westphal (1966) *Biochem. Zeit. 344*, 401–412.

Street, I. P., Armstrong, C. R., & Withers, S. G. (1986) *Biochemistry 25*, 6021–6027.

Stuike-Prill, R., & Meyer, B. (1990) *Eur. J. Biochem., 194*, 903–919.

Suh, S. W., Bhat, T. N., Navia, M. A., Cohen, G. H., Rao, D. N. Rudikoff, S., & Davies, D. R. (1986) *Proteins 1*, 74–80.

Takahashi, H., Suzuki, E., Shimada, I, & Arata, Y. (1992) *Biochemistry 31*, 2464–2468.

Takahashi, K., Yagisawa, T., Sonda, K., Kawaguchi, H., Yamaguchi, Y., Toma, H., Agishi, T., & Ota, K. (1993) *Trans. Proc. 25*, 271–273.

Thøgersen, H., Lemieux, R. U., Bock, K., & Meyer, B. (1982) *Can. J. Chem. 60*, 44–57.

Thurin, J. (1988) *Current Topics Microbiol. Immunol. 139*, 59–79.

Tonegawa, S. (1983) *Nature 302*, 575– 581.

Trail, P. A., Willner, D., Lasch, S. J., Henderson, A. J., Hofstead, S., Casazza, A. M., Firestone, R. A., Hellström, I., & Hellström, K. E. (1993) *Science 261*, 212–215.

Vadhan, R., Cordon, C. C., Carswell, E., Mintzer, D., Dantis, L., Duteau, C., Templeton, M. A., Oettgen, H. F., Old, L. J., & Houghton, A. (1988) *J. Clin. Oncol. 6*, 1636–1648.

van Boeckel C. A. A., & Petitou, M. (1993) *Angew. Chem. Intl. Ed. Engl. 32*, 1671–1690.

Vandonselaar, M., Delbaere, L. T. J., Spohr, U., & Lemieux, R. U. (1987) *J. Biol. Chem. 262*, 10848–10849.

Varki, A. (1993) *Glycobiology 3*, 97–130.

Vermersch, P. S., Tesmer, J. J. G., & Quiocho, F. A. (1992) *J. Mol. Biol. 226*, 923–929.

von Itzstein, M., Wu, W.-Y., Kok, G. B., Pegg, M. S., Dyason, J. C., Jin, B., Phan, T. V., Smythe, M. L., White, H. F., Oliver, S. W., Colman P. M., Varghese, J. N. Ryan, D. M., Woods, J. M., Bethell, R. C., Hotham, V. J., Cameron, J. M., & Penn, C. R. (1993) *Nature 363*, 418–423.

Vorberg, E., & Bundle, D. R. (1990) *J. Immunol. Methods 132*, 81–89.

Vyas, M. N., Vyas, N. K., Meikle, P. J., Sinnott, B., Pinto, B. M., Bundle, D. R., & Quiocho, F. A. (1993) *J. Mol. Biol. 231*, 133–136.

Watkins, W. M. (1980) *Adv. Human Genet. 10*, 1–136.

Watson, K. A., Mitchell, E. P. Johnson, L. N., Son, J. C., Bichard, C. J. F., Orchard, M. G., Fleet, G. W. J., Oikonomakos, N. G., Leonidas, D. D., Kontou, M., & Papageorgioui, A. (1994) *Biochemistry 33*, 5745–5758.

Weimar, T., Harris, S. L., Pitner, J. B., Bock, K., & Pinto, B. M. (1995) *Biochemistry 34*, 13672–13680.

Weiner, S. J., Kollman, P. A., Nguyen, D. T., & Case, D. A. J. (1986) *Comput. Chem. 7*, 230–252.

Wessel, H. P., & Bundle, D. R. (1983) *Carbohydr. Res. 124*, 301–311.

Westphal, O. (1952) Angew. *Chem. 64*, 314.

Westphal, O., & Lüderitz, O. (1960) *Angew. Chem. 72*, 881–891.

Wilbrandt, R., Tung, K. S. K., Deodhar, S. D., Nakamoto, S., & Kolff, W. J. (1969) *Am. J. Clin. Pathol. 51*, 15–23.

Wilson, I. A., & Stanfield, R. L. (1994) *Current Opinion Struct. Biol. 4*, 857–867.

Wilson, I. A., & Stanfield, R. L. (1995) *Nature Struct. Biol. 2*, 433–436.

Wiseman, T., Williston, S., Brandts, J. F., & Liu, L.-N. (1989) *Anal. Biochem. 179*, 131–137.

Wu, T. T., & Kabat, E. A. (1970) *J. Exp. Med. 132*, 211–250.

Young, W.W., Jr., Johnson, H. S., Tamura, Y., Karlsson, H. S., Larson, G. Parker, J. M.

R., Khare, D. P., Spohr, U., Baker, D. A., Hindsgaul, O., & Lemieux, R. U. (1983) *J. Biol. Chem. 258*, 4890–4894.

Zdanov, A., Li, Y., Bundle, D. R., Deng, S.-J., MacKenzie, C. R., Narang S. A., Young, N. M., & Cygler, M. (1994) *Proc. Natl. Acad. Sci. U.S.A. 91*, 6423–6427.

Zeigler, T., Pavliak, V., Lin, T.-H., Kovac, P., & Glaudemans, C. P. J. (1990) *Carbohydr. Res. 204*, 167–186.

Zidovetzki, R., Blatt, Y., Schepers, G., & Pecht, I. (1988) *Mol. Immunol. 25*, 379–383.

Chapter 11

Aleshin, A., Golubev, A., Firsov, L. M., & Honzatko, R. B. (1992) *J. Biol. Chem. 267*, 19291–19298.

Batacharyya, L., & Brewer, C.F. (1988) *Eur. J. Biochem. 176*, 235–238.

Bauer, A. J., Rayment, I., Frey, P. A., & Holden, H. M. (1992) *Proteins 12*, 372–381.

Becker, J. W., & Reeke, G. N., Jr., (1991) *Trans. Am. Crystallogr. Assoc. 25*, 37–50.

Bennet, W. A., & Steitz, T. A. (1980) *J. Mol. Biol. 140*, 183–209.

Bossart-Whitaker, P., Carson, M., Babu, Y. S,. Smith, C. D., Laver, W. G., & Air, G.M. (1993) *J. Mol. Biol. 232*, 1069–1083.

Bourne, Y., Rouge, P., & Cambillau, C. (1990b) *J. Biol. Chem. 265*, 18161–18165.

Bourne, Y., Rouge, P., & Cambillau, C. (1992) *J. Biol. Chem. 267*, 197–203.

Bourne, Y., Roussel, A., Frey, M., Rouge, P., Fontecilla-Camps, J.-C., & Cambillau, C. (1990a) *Proteins 8*, 365–376.

Buisson, G., Duee, E., Haser, R., & Payan, F. (1987) *EMBO J. 6*, 3903–3916.

Burmeister, W. P., Ruigrok, R. W. H., & Cusak, S. (1992) *EMBO J. 11*, 49–56.

Cygler, M., Rose, D. R., & Bundle, D. R. (1991) *Science 253*, 442–445.

Dai, J.-B., Liu, Y., Ray, W. J., Jr., & Konno, M. (1992) *J. Biol. Chem. 267*, 6322–6337.

Deisenhofer, J. (1981) *Biochemistry 20*, 2361–2370.

Delbaere, L. T. J., Vandonselaar, M., Prasad, L., Quial, J. W., Pearlstone, J. R., Carpenter, M. R., Smillie, L. B., Nikrad, P. V., Sphor, U., & Lemieux, R. U. (1990) *Can. J. Chem. 68*, 1116–1121.

Derewenda, Z., Yariv, J., Helliwell, J. R., Kalb, A. J. (Gilboa), Dodson, E. J., Papiz, M. Z., Want, T., & Campbell, J. (1989) *EMBO J. 8*, 2189–2193.

Fersht, A. R., Shi, J.-P., Knill-Jones, J., Lowe, D. M., Wilkinson, A. J., Blow, D. M., Brick, P., Carter, P., Waye, M. M. Y., & Winter, G. (1985) *Nature 314*, 235–238.

Glaudemans, C. P. J., Kovac, P., & Rao, A. S. (1989) *Carbohydr. Res. 190*, 252–270.

Goldsmith, E., Sprang, S., & Fletterick, R. (1982) *J. Mol. Biol. 156*, 411–427.

Goldsmith, E. J., Fletterick, R. J., & Withers, S. G. (1987) *J. Biol. Chem. 262*, 1449–1455.

Goldsmith, E. J., Sprang, S. R., & Fletterick, R. J. (1991) *Trans. Am. Crystallogr. Assoc. 25*, 51–63.

Hecht, H. J., Kalisz, H. M., Hendle, J., Schmid, R. D., & Schomburg, D. (1993) *J. Mol. Biol. 229*, 153–172.

Ito, N., Phillips, S. E. V., Stevens, C., Ogel, Z. B., McPherson, M. J., Keen, J. N., Yadav, K. D. S., & Knowles, P. F. (1991) *Nature 350*, 87–90.

Johnson, L. N., Acharya, K. R., Jorson, M. D., & McLaughlin, P. J. (1990) *J. Mol. Biol. 211*, 645–661.

Johnson, L. N., Cheetham, J., McLaughlin, P. J., Acharya, K. R., Barford, D., & Phillips, D. C. (1988) *Curr. Topics Microbiol. Immunol. 139*, 81–134.

Johnson, L. N., Snape, P., Martin, J. L., Acharya, K. R., Barford, D., & Oikonomakos, N. G. (1993) *J. Mol. Biol. 232*, 253–267.

Juy, M., Amit, A. G., Alzari, P. M., Poljak, R. J., Claeyssens, M., Beguin, P., & Aubert, J.-P. (1992) *Nature 357*, 89–91.

Katzin, B. J., Collins, E. J., & Robertus, J. D. (1991) *Proteins 10*, 251–259.

Klein, C., & Schulz, G.E. (1991) *J. Mol. Biol. 217*, 737–750.

Levitt, M., & Perutz, M. F. (1988) *J. Mol. Biol. 201*, 751–754.

Liao, D., Kapadia, G., Ahmed, H., Vasta, G. R., & Herzberg, O. (1994) *Proc. Natl. Acad. Sci. U.S.A. 91*, 1428–1432.

Liao, D. I., Kapadia, G., Reddy, P., Saier, M. H., Reizer, J., & Herzberg, O. (1991) *Biochemistry 30*, 9583–9594.

Loris, R., Steyaert, J., Maes, D., Lisgarten, J., Pickergill, R., & Wyns, L. (1993) *Biochemistry 32*, 8772–8781.

Martin, J. L., Johnson, L. N., & Withers, S. G. (1990) *Bicohemistry 29*, 10745–10757.

Matsuura, Y., Kusunoki, M., Harada, W., & Kakudo, M. (1984) *J. Biochem. (Tokyo) 95*, 697–702.

Mikami, B., Hehre, E. J., Sato, M., Katsube, Y., Hirose, M., Morita, Y., & Sacchettini, J. C. (1993) *Biochemistry 32*, 6836–6845.

Miller, D. M., III., Olson, J. S, Pflugrath, J. W., & Quiocho, F. A. (1983) *J. Biol. Chem. 258*, 13665–13672.

Mowbray, S. L., & Cole, L. B. (1992) *J. Mol. Biol. 225*, 155–175.

Mowbray, S. L., Smith, R. L., & Cole, L. B. (1990) *Receptor 1*, 41–45.

Phillips, D. C. (1966) *Sci. Am. 215*, 78–90.

Quiocho, F. A. (1986) *Annu. Rev. Biochem. 55*, 287–315.

Quiocho, F.A. (1989) *Pure. Appl. Chem. 61*, 1293–1306.

Quiocho, F. A. (1990) *Phil. Trans. Royal Soc. Lond. 326*, 341–351.

Quiocho, F. A. (1993) *Biochemical Society Trans.*, Vol. 21, 442–448.

Quiocho, F. A., & Vyas, N. K. (1984) *Nature 310*, 381–386.

Quiocho, F. A., Vyas, N. K., & Spurlino, J. C. (1991) *Trans. Am. Crystallogr. Assoc. 25*, 23–35.

Quiocho, F.A., Wilson, D. K., & Vyas, N. K. (1989) *Nature 340*, 404–407.

Rini, J. M., Hardman, K. D., Einspahr, H., Suddath, F. L., & Carver, J. P. (1993) *J. Biol. Chem. 268*, 10126–10132.

Rose, D. R. Przybylska, M., To, R. J., Kayden, C. S., Oomen, R. P., Vorberg, E., Young, N. M., & Bundle, D. R. (1993) *Prot. Sci. 2*, 1106–1113.

Rouvinen, J., Bergfors, T., Terri, T., Knowles, J. K. C., & Jones, T. A. (1990) *Science 249*, 380–386.

Rutenber, E., Katzin, B. J., Ernst, S., Collins, E. J., Mlsna, D., Ready, M. P., & Robertus, J. D. (1991) *Proteins 10*, 240–250.

Rutenber, E., & Robertus, J. D. (1991) *Proteins 10*, 260–269.

Shaanan, B., Lis, H., & Sharon, N. (1991) *Science 254*, 862–866.

Sharff, A.J. Rodseth, L. E., & Quiocho, F. A. (1993) *Biochemistry 32*, 10553–10559.

Sixma, T. K., Pronk, S. E., Kalk, K. H., van Zanten, B. A. M., Berghuis, A. M., & Hol, W. G. J. (1992) *Nature 355*, 561–564.

Sprang, S. R., Goldsmith, E. J., Fletterick, R. J., Withers, S. G., & Madsen, N. B. (1982) *Biochemistry 21*, 5364–5371.

Spurlino, J. C., Lu, G.-Y., & Quiocho, F. A. (1991) *J. Biol. Chem. 266*, 5202–5219.

Stein, P. E., Boodhoo, A., Tyrell, G. J., Brunton, J. L., & Read, R. J. (1992) *Nature 355*, 748–750.

Street, I. P., Armstrong, C. R., & Wither, S. G. (1986) *Biochemistry 25*, 6021–6027.

Strynadka, N. C. J., & James, M. N. J. (1991) *J. Mol. Biol. 220*, 401–424.

Vermersch, P. S., Tesmer, J. J. G., Lemon, D. D., & Quiocho, F. A. (1990) *J. Biol. Chem. 265*, 16592–16603.

Vermersch, P. S., Tesmer, J. J. G., Lemon, D. D., & Quiocho, F. A. (1992) *J. Mol. Biol. 226*, 923–929.

Vyas, N. K. (1991) *Current Opinion Struct. Biol. 1*, 732–740.

Vyas, N. K., Vyas, M. N., & Quiocho, F. A. (1988) *Science 242*, 1290–1295.

Vyas, N. K., Vyas, M. N., & Quiocho, F. A. (1991) *J. Biol. Chem. 266*, 5226–5237.

Vyas M. N., Vyas, N. K., & Quiocho, F. A. (1994) *Biochemistry 33*, 4762–4768.

Weis, W., Brown, J. H., Cusak, S., Paulson, J. C., Skehel, J. J., & Wiley, D. C. (1988) *Nature 333*, 426–431.

Weis, W. I., Drickamer, K., & Hendrickson, W. A. (1992) *Nature 360*, 127–134.
Whitlow, M., Howard, A. J., Finzel, B. C., Poulous, T. L., Winborne, E., & Gilliland, G. L. (1991) *Proteins 9*, 153–173.
Wright, C. S. (1990) *J. Mol. Biol 215*, 635–651.
Wright, C. S. (1992) *J. Biol. Chem. 267*, 14345–14352.
Zou, J., Flocco, M. M., & Mowbray, S. L. (1993) *J. Mol. Biol. 233*, 739–752.

Chapter 12

Alak, A., & Armstrong, D. W. (1986) *Anal. Chem. 58*, 582–584.
Alston, D. R., Ashton, P. R., Lilley, T. H., Stoddart, J. F., & Zarzycki, R. (1989) *Carbohydr. Res. 192*, 259–281.
Alston, D. R., Lilley, T. H., & Stoddart, J. F. (1985a) *J. Chem. Soc., Chem. Commun.*, 1600–1602.
Alston, D. R., Slawin, A. M. Z., Stoddart, J. F., & Williams, D. J. (1985b) *Angew. Chem. Int. Ed. Engl. 24*, 786–787.
Alston, D. R., Slawin, A. M. Z., Stoddart, J. F., & Williams, D. J. (1985c) *J. Chem. Soc., Chem. Commun.*, 1602–1604.
Alston, D. R., Slawin, A. M. Z., Stoddart, J. F., Williams, D. J., & Zarzycki, R. (1988) *Angew. Chem. Int. Ed. Engl. 27*, 1184–1185.
Anderson, G. H., Robbins, F. J., Domingues, F. J., Moores, R. G., & Long, C. L. (1963) *Toxicol. Appl. Pharmacol. 5*, 257–266.
Andersson, T., Nilsson, K., Sundahl, M., Westman, G., & Wennerström, O. (1992) *J. Chem. Soc., Chem. Commun.*, 604–606.
Andersson, T., Sundahl, M., Westman, G., & Wennerström, O. (1994a) *Tetrahedron Lett. 35*, 7103–7106.
Andersson, T., Westman, G., Wennerström, O., & Sundahl, M. (1994b) *J. Chem. Soc., Perkin Trans. 2*, 1097–1101.
Armspach, D., Ashton, P. R., Ballardini, R., Balzani, V., Godi, A., Moore, C. P., Prodi, L., Spencer, N., Stoddart, J. F., Tolley, M. S., Wear, T. J., & Williams, D. J. (1995) *Chemistry 1*, 33–35.
Armspach, D., Ashton, P. R., Moore, C. P., Spencer, N., Stoddart, J. F., Wear, T. J., & Williams, D. J. (1993) *Angew. Chem. Int. Ed. Engl. 32*, 854–858.
Armspach, D., Ashton, P. R., Spencer, N., Stoddart, J. F., & Williams, D. J. (1994) *Pestic. Sci. 41*, 232–235.
Armstrong, D. W., Ward, T. J., Armstrong, R. D., & Bees, T. E. (1986) *Science 232*, 1132–1135.
Ashton, P. R., Boyd, S. E., Gattuso, G., Hartwell, E. Y., Königer, R., Spencer, N., & Stoddart, J. F. (1995a) *J. Org. Chem. 60*, 3898–3903.
Ashton, P. R., Brown, C. L., Menzer, S., Nepogodiev, S. A., Stoddart, J. F., & Williams, D. J. (1996a) *Chem. Eur. J. 2*, 580–591.
Ashton, P. R., Ellwood, P., Staton, I., & Stoddart, J. F. (1991a) *Angew. Chem. Int. Ed. Engl. 30*, 80–81.
Ashton, P. R., Ellwood, P., Staton, I., & Stoddart, J. F. (1991b) *J. Org. Chem. 56*, 7274–7280.
Ashton, P. R., Gattuso, G., Königer, R., Stoddart, J. F., & Williams, D. J. (1996b) *J. Org. Chem. 61*, 9553–9555.
Ashton, P. R., Hartwell, E. Y., Philp, D., Spencer, N., & Stoddart, J. F. (1995b) *J. Chem. Soc., Perkin Trans. 2*, 1263–1277.
Ashton, P. R., Stoddart, J. F., & Zarzycki, R. (1988) *Tetrahedron Lett. 29*, 2103–2106.
Atwood, J. L., & Davies, J. E. D., Eds. (1987) in *Proceedings of the Third International Symposium on Cyclodextrins, J. Incl. Phenom. 1*, 29–98.
Atwood, J. L., Davies, J. E. D., & Osa, T. Eds. (1984) in *Proceedings of the Second International Symposium on Cyclodextrins, J. Incl. Phenom. 3–4*.

Behr, J. P., & Lehn, J.-M. (1976) *J. Am. Chem. Soc. 98*, 1743–1747.

Ben-Naim, A. (1980) in *Hydrophobic Interactions*, Plenum Press, New York.

Bender, M. L., & Komiyama, M. (1978) in *Cyclodextrin Chemistry*, Springer-Verlag, Berlin.

Bergeron, R., & Channing, M. A. (1976) *Bioorg. Chem. 5*, 437.

Bergeron, R., Channing, M. A., Gibeily, G. J., & Pillor, P. M. (1977) *J. Am. Chem. Soc. 99*, 5146–5151.

Bergeron, R. J. (1977) *J. Chem. Educ. 54*, 204–207.

Bergeron, R. J. (1984) in *Inclusion Compounds* (Atwood, J. L., Davies, J. E. D., & MacNicol, D. D. Eds.) Vol. 3, pp. 391–443, Academic Press, London.

Betzel, C., Saenger, W., Hingerty, B. E., & Brown, G. M. (1984) *J. Am. Chem. Soc. 106*, 7545–7557.

Biltonen, R. L., & Langerman, N. (1979) *Methods Enzymol. 61*, 287–318.

Born, M., & Ritter, H. (1991) *Makromol. Chem., Rapid Commun. 12*, 471–476.

Boulas, P., Kutner, W., Jones, M. T., & Kadish, K. M. (1994) *J. Phys. Chem. 98*, 1282–1287.

Breslow, R. (1986) *Adv. Enzym. Relat. Areas Mol. Biol. 58*, 1–60.

Breslow, R. (1990) *Pure Appl. Chem. 62*, 1859–1866.

Breslow, R. (1992) *Israel J. Chem. 32*, 23–30.

Breslow, R. (1994) *Pure Appl. Chem. 66*, 1573–1582.

Breslow, R. (1995) *Acc. Chem. Res. 28*, 146–153.

Breslow, R., & Chung, S. (1990) *J. Am. Chem. Soc. 112*, 9659–9660.

Breslow, R., Greenspoon, N., Guo, T., & Zarzycki, R. (1989) *J. Am. Chem. Soc. 111*, 8296–8297.

Breslow, R., & Zhang, B. (1992) *J. Am. Chem. Soc. 114*, 5882–5883.

Breslow, R., & Zhang, B. (1994) *J. Am. Chem. Soc. 116*, 7893–7894.

Casu, B., Reggiani, M., Gallo, G. G., & Vigevani, A. (1968) *Tetrahedron 24*, 803–821.

Chetcuti, P. A., Moser, P., & Rihs, G. (1991) *Organometallics 10*, 2895–2897.

Christofides, J. C., & Davies, D. B. (1982) *J. Chem. Soc., Chem. Commun.*, 560–562.

Christofides, J. C., & Davies, D. B. (1983) *J. Am. Chem. Soc. 105*, 5099–5105.

Coates, J. H., Easton, C. J., van Eyk, S. J., Lincoln, S. F., May, B. L., Whalland, C. B., & Williams, M. L. (1990) *J. Chem. Soc., Perkin Trans. 1*, 2619–2620.

Coleman, A. W., Zhang, P., Ling, C.-C., Mahuteau, J., Parrot-Lopez, H., & Miocque, M. (1992) *Supramol. Chem. 1*, 11–14.

Collins, P. M., & Ali, M. H. (1990) *Tetrahedron Lett. 31*, 4517–4520.

Colquhoun, H. M., Stoddart, J. F., & Williams, D. J. (1986) *Angew. Chem. Int. Ed. Engl. 25*, 487–507.

Cramer, F., & Dietsche, W. (1959) *Chem. Ber. 92*, 1739–1747.

Croft, A. P., & Bartsch, R. A. (1983) *Tetrahedron 39*, 1417–1474.

Cromwell, W. C., Bystrom, K., & Eftink, M. R. (1985) *J. Phys. Chem. 89*, 326–332.

Danil De Namor, A. F. (1990) *Pure Appl. Chem. 62*, 2121–2127.

Defaye, J., & Gadelle, A. (1991) *Angew. Chem. Int. Ed. Engl. 30*, 78–80.

Demarco, P. V., & Thakkar, A. L. (1970) *J. Chem. Soc., Chem. Commun.*, 2–4.

Diederich, F. (1991) in *Cyclophanes* (Stoddart, J. F., Ed.) pp. 246–263, The Royal Society of Chemistry, Cambridge.

D'Souza, V. T., & Bender, M. L. (1987) *Acc. Chem. Res. 20*, 146–152.

Duchêne, D., Ed. (1987) in *Cyclodextrins and their Industrial Uses*, Editions de la Santé, Paris.

Duchêne, D., Ed. (1990) in *Minutes of the Fifth International Symposium on Cyclodextrins*, Editions de la Santé, Paris.

Eastburn, S. D., & Tao, B. Y. (1994) *Biotech. Adv. 12*, 325–339.

Easton, C. J., & Lincoln, S. F. (1996) *Chem. Soc. Rev.*, 163–170.

Eftink, M. R., Andy, M. L., Bystrom, K., Perlmutter, H. D., & Kristol, D. S. (1989) *J. Am. Chem. Soc. 111*, 6765–6772.

Excoffier, G., Paillet, M., & Vignon, M. (1985) *Carbohydr. Res. 135*, C10–C11.

French, D. (1957) *Adv. Carbohydr. Chem. 12*, 189–260.

French, D., Knapp, D. W., & Pazur, J. H. (1950) *J. Am. Chem. Soc. 72*, 5150–5152.

French, D., Levine, M. L., Norberg, E., Nordin, P., Pazur, J. H., & Wild, G. M. (1954) *J. Am. Chem. Soc. 76*, 2387–2390.

French, D., & Rundle, R. E. (1942) *J. Am. Chem. Soc. 64*, 1651–1653.

Freudenberg, K., & Cramer, F. (1948) *Z. Naturforsch. 3b*, 464.

Freudenberg, K., & Jacobi, R. (1935) *Liebigs Ann. Chem. 518*, 102–108.

Fuchs, R., Habermann, N., & Klüfers, P. (1993) *Angew. Chem., Int. Ed. Engl. 32*, 852–854.

Fügedi, P. (1989) *Carbohydr. Res. 192*, 366–369.

Fujita, K., Ejima, S., & Imoto, T. (1984) *J. Chem. Soc., Chem. Commun.*, 1277–1278.

Gagnaire, D., & Vignon, M. (1976) *Carbohydr. Res. 51*, 140–144.

Gidley, M. J., & Bocieck, S. M. (1988) *J. Am. Chem. Soc. 110*, 3820–3829.

Gillet, B., Nicole, D. J., & Delpuech, J.-J. (1982) *Tetrahedron Lett. 23*, 65–68.

Hall, L. D., & Lim, T. K. (1986) *J. Am. Chem. Soc. 108*, 2503–2510.

Harada, A., & Kamachi, M. (1990) *J. Chem. Soc., Chem. Commun.*, 1322–1323.

Harada, A., Li, J., & Kamachi, M. (1992) *Nature 356*, 325–327.

Harada, A., Li, J., & Kamachi, M. (1993a) *Macromolecules 26*, 5698–5703.

Harada, A., Li, J., & Kamachi, M. (1993b) *Nature 364*, 516–518.

Harada, A., Li, J., & Kamachi, M. (1994a) *J. Am. Chem. Soc. 116*, 3192–3196.

Harada, A., Li, J., & Kamachi, M. (1994b) *Macromolecules 27*, 4538–4543.

Harada, A., Li, J., Nakamitsu, T., & Kamachi, M. (1993c) *J. Org. Chem. 58*, 7524–7528.

Harada, A., Shimada, M., & Takahashi, S. (1989) *Chem. Lett.* 275–276.

Harata, K. (1979) *Bull. Chem. Soc. Jpn. 52*, 1807–1812.

Harata, K., & Uedaira, H. (1975) *Bull. Chem. Soc. Jpn. 48*, 375–378.

Hedges, A. R. (1992) *Minutes of the Sixth International Symposium on Cyclodextrins*, Editions de Santé, Paris.

Hinze, W. L., Riehl, T. E., Armstrong, D. W., DeMond, W., Alak, A., & Ward, T. (1985) *Anal. Chem. 57*, 237–242.

Houdier, S., & Vottéro, P. J. A. (1994) *Angew. Chem., Int. Ed. Engl. 33*, 354–356.

Huber O. & Szejtli J., Eds. (1988) *Proceedings of the Fourth International Symposium on Cyclodextrins*, Kluwer, Dordrecht.

Icheln, D., Gehrcke, B., Piprek, Y., Mischnick, P., König, W. A., Dessoy, M. A., & Morel, A. F. (1996) *Carbohydr. Res. 280*, 237–250.

Ikeda, H., Nagano, Y., Du, Y., Ikeda, T., & Toda, F. (1990) *Tetrahedron Lett. 31*, 5045–5048.

Inoue, Y., Hakushi, T., Liu, Y., Tong, L.-H., Shen, B.-J., & Jin, D.-S. (1993) *J. Am. Chem. Soc. 115*, 475–481.

Inoue, Y., Kitagawa, M., Hoshi, H., Sakurai, M., & Chûjô, R. (1987) *J. Incl. Phenom. 5*, 55–58.

Irvine, J. C., Pringsheim, H., & Macdonald, J. (1924) *J. Chem. Soc. 125*, 942–947.

Isnin, R., & Kaifer, A. E. (1991) *J. Am. Chem. Soc. 113*, 8188–8190.

Isnin, R., & Kaifer, A. E. (1993) *Pure Appl. Chem. 65*, 495–498.

Isnin, R., Salam, C., & Kaifer, A. E. (1991) *J. Org. Chem. 56*, 35–41.

Jiang, T., Sukumaran, D. K., Soni, S.-D., & Lawrence, D. S. (1994) *J. Org. Chem. 59*, 5149–5155.

Johnson, M. D., & Reinsborough, V. C. (1994) *J. Solution Chem. 23*, 185–193.

Kaifer, A. E., Quintela, P. A., & Schuette, J. M. (1989) *J. Incl. Phenom. 7*, 107–115.

Kamitori, S., Hirotsu, K., & Higuchi, T. (1987) *J. Am. Chem. Soc. 109*, 2409–2414.

Khan, A. R., Barton, L., & D'Souza, V. T. (1992) *J. Chem. Soc., Chem. Commun.* 1112–1114.

Khan, A. R., Barton, L., & D'Souza, V. T. (1996) *J. Org. Chem. 61*, 1112–1114.

Kirby, A. J. (1996) *Angew. Chem., Int. Ed. Engl. 35*, 707–724.

Klingert, B., & Rihs, G. (1991) *J. Inc. Phenom. 10*, 255–265.

Klüfers, P., & Schuhmacher, J. (1994) *Angew. Chem., Int. Ed. Engl. 33*, 1863–1865.

Kobayashi, N. (1988) *J. Chem. Soc., Chem. Commun.*, 918–919.

Kobayashi, N., & Opallo, M. (1990) *J. Chem. Soc., Chem. Commun.*, 477–479.

Kochetkov, N. K., Nepogod'ev, S. A., & Backinowsky, L. V. (1990) *Tetrahedron 46*, 139–150.

Kodaka, M. (1993) *J. Am. Chem. Soc. 115*, 3702–3705.

Komiyama, M. (1989) *J. Chem. Soc., Perkin Trans. 1*, 2031–2034.

Komiyama, M. (1993) *Prog. Polym. Sci. 18*, 871–898.

Komiyama, M., & Hirai, H. (1983) *J. Am. Chem. Soc. 105*, 2018–2021.

Komiyama, M., & Hirai, H. (1984) *J. Am. Chem. Soc. 106*, 174–178.

Kotake, Y., & Janzen, E. G. (1989) *J. Am. Chem. Soc. 111*, 5138–5140.

Kuyama, H., Nukada, T., Nakahara, Y., & Ogawa, T. (1993) *Tetrahedron Lett. 34*, 2171–2174.

Le Bas, G., & Rysanek, N. (1987) in *Cyclodextrins and their Industrial Uses* (Duchêne, D., Ed.) pp. 105–130, Editions de la Santé, Paris.

Li, S., & Purdy, W. C., (1992) *Chem. Rev. 92*, 1457–1470.

Lipkowitz, K. B., Raghothama, S., & Yang, J. (1992) *J. Am. Chem. Soc. 114*, 1554–1562.

Lüttringhaus, A., Cramer, F., Prinzbach, H., & Henglein, F. M. (1958) *Liebigs Ann. Chem. 613*, 185–198.

Matsui, Y., Naruse, H., Mochida, K., & Date, Y. (1970) *Bull. Chem. Soc. Jpn. 43*, 1909.

McMullen, R. K., Saenger, W., Fayos, J., & Mootz, D. (1973) *Carbohydr. Res. 31*, 37–46.

Meister, A., & Wenz, G. (1990) in *Minutes of the Fifth International Symposium on Cyclodextrins* (Duchêne, D., Ed.) pp. 188–191, Editions de la Santé, Paris.

Menger, F. M., & Ladika, M. (1987) *J. Am. Chem. Soc. 109*, 3145–3146.

Minato, S., Osa, T., & Ueno, A. (1991) *J. Chem. Soc., Chem. Commun.*, 107–108.

Mori, M., Ito, Y., & Ogawa, T. (1989) *Carbohydr. Res. 192*, 131–146.

Mori, M., Ito, Y., Uzawa, J., & Ogawa, T. (1990) *Tetrahedron Lett. 31*, 3191–3194.

Nakagawa, T., Ueno, K., Kashiwa, M., & Watanabe, J. (1994) *Tetrahedron Lett. 35*, 1921–1924.

Nishizawa, M., Imagawa, H., Kan, Y., & Yamada, H. (1991) *Tetrahedron Lett. 32*, 5551–5554.

Nishizawa, M., Imagawa, H., Kubo, K., Kan, Y., & Yamada, H. (1992) *Synlett.*, 447–448.

Odagaki, Y., Hirotsu, K., Higuchi, T., Harada, A., & Takahashi, S. (1990) *J. Chem. Soc., Perkin Trans. 1*, 1230–1231.

Ogino, H. (1981) *J. Am. Chem. Soc. 103*, 1303–1304.

Ogino, H., & Ohata, K. (1984) *Inorg. Chem. 23*, 3312–3316.

Osa, T. (1994) *Proceedings of the Seventh International Symposium on Cyclodextrins*, Publ. Office Acad. Soc., Tokyo, Japan.

Palepu, R., & Reinsborough, V. C. (1989) *Can. J. Chem. 67*, 1550–1553.

Palmer, D. R. J., Buncel, E., & Thatcher, G. R. J. (1994) *J. Org. Chem. 59*, 5286–5291.

Poh, B.-L., & Chow, Y. M. (1992) *J. Incl. Phenom. 14*, 85–90.

Priyadarsini, K. I., Mohan, H., Tyagi, A. K., & Mittal, J. P. (1994) *J. Phys. Chem. 98*, 4756–4759.

Rao, T. V. S., & Lawrence, D. S. (1990) *J. Am. Chem. Soc. 112*, 3614–3615.

Ringsdorf, H., Schlarb, B., & Wenzmer, J. (1988) *Angew Chem., Int. Ed. Engl. 27*, 113–158.

Ripmeester, J. A. (1988) *J. Incl. Phenom. 6*, 31–40.

Ross, P. D., & Subramanian, S. (1981) *Biochemistry 20*, 3096–3402.

Saenger, W. (1980) *Angew. Chem. Int. Ed. Engl. 19*, 344–362.

Saenger, W. (1984) in *Inclusion Compounds* Vol. 2, pp. 231–259, (Atwood, J. L., Davies, J. E. D., & MacNicol, D. D., Eds.), Academic Press, London.

Saenger, W., Betzel, C., Hingerty, B., & Brown, G. M. (1983) *Angew Chem., Int. Ed. Engl. 22*, 883–884.

Saenger, W., Noltemeyer, M., Manor, P. C., Hingerty, B., & Klav, B. (1976) *Bioorg. Chem. 5*, 187.

Sakairi, N., & Kuzuhara, H. (1992) *J. Chem. Soc., Chem. Commun.*, 510–512.

Sakairi, N., & Kuzuhara, H. (1993) *J. Chem. Soc., Chem. Commun.*, 1874–1875.

Sakairi, N., Wang, L.-X., & Kuzuhara, H. (1991) *J. Chem. Soc., Chem. Commun.* 289–290.

Schardinger, F. (1904) *Wien. Klin. Wochenschr. 17*, 207–209.

Schardinger, F. (1911) *Zentralbl. Bakteriol. Parasitenkd. Infektionskr. Hyg. II 29*, 188–197.

Schneider, H.-J., Blatter, T., & Simova, S. (1991) *J. Am. Chem. Soc. 113*, 1996–2000.

Schneider, H.-J., Kramer, R., Simova, S., & Schneider, U. (1988) *J. Am. Chem. Soc. 110*, 6442–6448.

Seeman, J. I., Secor, H. V., Armstrong, D. W., Timmons, K. D., & Ward, T. J. (1988) *Anal. Chem. 60*, 2120–2127.

Sensse, K., & Cramer, F. (1969) *Chem. Ber. 102*, 509–521.

Shimada, M., Harada, A., & Takahashi, S. (1991) *J. Chem. Soc., Chem. Commun.*, 263–264.

Shimida, M., Harada, A., & Takahashi, S. (1992) *J. Organomet. Chem. 428*, 199–205.

Shimizu, H., Kaito, A., & Hatano, M. (1979) *Bull. Chem. Soc. Jpn. 52*, 2678–2684.

Shimizu, H., Kaito, A., & Hatano, M. (1981) *Bull. Chem. Soc. Jpn. 54*, 513–519.

Shimizu, H., Kaito, A., & Hatano, M. (1982) *J. Am. Chem. Soc. 104*, 7059–7065.

Sicard, P. J., & Saniez, M.-H. (1987) in *Cyclodextrins and Their Industrial Uses* (Duchêne, D., Ed.) pp. 75–103, Editions de la Santé, Paris.

Spencer, C. M., Stoddart, J. F., & Zarzycki, R. (1987) *J. Chem. Soc., Perkin Trans. 2*, 1323–1336.

Steinbrunn, M. B., & Wenz, G. (1996) *Angew. Chem., Int. Ed. Engl. 35*, 2139–2141.

St-Jacques, M., Sundarajan, P. R., Taylor, K. J., & Marchessault, R. H. (1976) *J. Am. Chem. Soc. 98*, 4386–4391.

Stoddart, J. F. (1992) *Angew. Chem., Int. Ed. Engl. 31*, 846–848.

Stoddart, J. F., Ed. (1989) *Carbohydr. Res. 192*.

Stoddart, J. F., & Zarzycki, R. (1988) *Recl. Trav. Chim. Pays-Bas 107*, 515–528.

Szejtli, J. (1982a) in *Cyclodextrins and their Inclusion Complexes*, Akadémiai Kiado, Budapest.

Szejtli, J. (1988) in *Cyclodextrin Technology*, Kluwer Academic, Dordrecht.

Szejtli, J. (1996) *Proceedings of the Eight International Symposium on Cyclodextrins*, Kluwer Academic, Norwell, MA.

Szejtli, J., Ed. (1982b) in *Proceedings of the First International Symposium on Cyclodextrins* Reidel, Dordrecht.

Szejtli, J., & Osa, T., Eds. (1996) *Comprehensive Supramolecular Chemistry*, Vol. 3, Elsevier, Oxford.

Tabushi, I., Kiyosuke, Y., Sugimoto, T., & Yamamura, K. (1978) *J. Am. Chem. Soc. 100*, 916–919.

Takahashi, K., & Hattori, K. (1994) *J. Inc. Phenom. 17*, 1–24.

Takahashi, Y., & Ogawa, T. (1987a) *Carbohydr. Res. 169*, 127–149.

Takahashi, Y., & Ogawa, T. (1987b) *Carbohydr. Res. 164*, 277–296.

Takeo, K. (1990) *Carbohydr. Res. 200*, 481–485.

Takeo, K., Mitoh, H., & Uemura, K. (1989) *Carbohydr. Res. 187*, 203–211.

Tanford, C. (1980) in *The Hydrophobic Effect, Formation of Micelles and Biological Membranes*, Second Edition, Wiley, New York.

Van Etten, R. L., Sebastion, J. F., Clowes, G. A., & Bender, M. L. (1967) *J. Am. Chem. Soc. 89*, 3242–3253.

Venema, F., Baselier, C. M., van Dienst, E., Ruël, B. H. M., Feiters, M. C., Engbersen, J. F. J., Reinhoudt, D. N., & Nolte, R. J. M. (1994) *Tetrahedron Lett. 35*, 1773–1776.

Veregin, R. P., Fyfe, C. A., Marchessault, R. H., & Taylor, M. G. (1987) *Carbohydr. Res. 160*, 41–56.

Villiers, A. (1891) *C. R. Seances Acad. Sci. 112*, 536–538.

Vincendon, M. (1981) *Bull. Soc. Chim. Fr. II*, 129–134.

Wang, Y., Ueno, A., & Toda, F. (1994) *Chem. Lett.*, 167–170.

Wenz, G. (1994) *Angew. Chem. Int. Ed. Engl. 33*, 803–822.

Wenz, G., & Keller, B. (1992) *Angew. Chem. Int. Ed. Engl. 31*, 197–199.

Wenz, G., Von der Bey, E., & Schmidt, L. S. (1992) *Angew Chem., Int. Ed. Engl. 31*, 783–785.

Wylie, R. S., & Macartney, D. H. (1990) *J. Am. Chem. Soc. 114*, 3136–3138.

Wylie, R. S., & Macartney, D. H. (1993) *Inorg. Chem. 32*, 1830–1837.

Yamamoto, Y., & Inoue, Y. (1989) *J. Carbohydr. Chem. 8*, 29–46.

Yamamoto, Y., Onda, M., Takahashi, Y., Inoue, Y., & Chûjô, R. (1988) *Carbohydr. Res. 182*, 41–52.

Yamamura, H., Ezuka, T., Kawase, Y., Kawai, M., Butsugan, Y., & Fujita, K. (1993) *J. Chem. Soc., Chem. Commun.*, 636–637.

Yamamura, H., & Fujita, K. (1991) *Chem. Pharm. Bull. 39*, 2505–2508.

Yamamura, H., Nagaoka, H., Kawai, M., Butsugan, Y., & Einaga, H. (1996) *J. Chem. Soc., Chem. Commun.*, 1069–1070.

Yamanari, K., & Shimura, Y. (1983) *Bull. Chem. Soc. Jpn. 56*, 2283–2289.

Yamanari, K., & Shimura, Y. (1984) *Bull. Chem. Soc. Jpn. 57*, 1596–1603.

Yoshida, Z.-I., Takakuma, H., Takekuma, S.-I., & Matsubara, Y. (1994) *Angew Chem., Int. Ed. Engl. 33*, 1597–1599.

Yoshiro, M., Miyama, S., Takarada, T., & Komiyama, M. (1994) *J. Inc. Phenom. 17*, 393–397.

Zabel, V., Saenger, W., & Mason, S. A. (1986) *J. Am. Chem. Soc. 108*, 3664–3673.

Zhang, D.-D., Liang, Q., Chen, J.-W., Li, M.-K., & Wu, S.-H. (1994) *Supramol. Chem. 3*, 235–239.

Chapter 13

Akatsuka, A., Singh, T. J., Nakabayashi, H., Lin, M. C., & Huang, K. P. (1985) *J. Biol. Chem. 260*, 3239–3242.

Ainsworth, C., Clark, J., & Balsdon, J. (1993) *Plant Molec. Biol. 22*, 67–82.

Alonzo, M. D., Lagzdins, E. J., Lomako, J., Lomako, W. M., & Whelan, W. J. (1995a) *FEBS Lett. 359*, 110–112.

Alonzo, M. D., Lomako, J., Lomako, W. M., & Whelan, W. J. (1994a) *FEBS Lett. 342*, 38–42.

Alonzo, M. D., Lomako, J., Lomako, W. M., Whelan, W. J., & Preiss, J. (1994b) *FEBS Lett. 352*, 222–226.

Alonzo, M. D., Lomako, J., Lomako, W. M., & Whelan, W. J. (1995b) *FASEB J. 9*, 1126–1137.

Alonzo, M. D., Lomako, J., Lomako, W. M., & Whelan, W. J. (1995c) *J. Biol.Chem. 270*, 15315–15319.

Anderson, J. M., Hnilo, J., Larson, R., Okita, T. W., Morell, M., & Preiss, J. (1989) *J. Biol. Chem. 264*, 12238–12242.

Ardila, F. J., & Tandecarz, J. S. (1992) *Plant Physiol. 99*, 1342–1347.

Atkinson, D. E. (1970) *in The Enzymes*, Third Ed., Vol. 1, (Boyer, P.D., Ed.), pp. 461–489, Academic Press, NY.

Baba, T., Kimura, K., Mizuno, K., Etoh, H., Ishida, Y., Shida, O., & Arai, Y. (1991) *Biochem. Biophys. Res. Commun. 181*, 87–94.

Baba, T., Nishihara, M., Mizuno, K., Kawasaki, T., Shimada, H., Kobayashi, E., Ohnishi, S., Tanaka, K., & Arai, Y. (1993) *Plant Physiol. 103*, 565–573.

Badenhuizen, I. P. (1969) *The Biogenesis of Starch Granules in Higher Plants*, Appleton-Century Crofts, NY.

Baecker, P. A., Furlong, C. E., & Preiss, J. (1983) *J. Biol. Chem. 258*, 5084–5088.

Baecker, P. A., Greenberg, E., & Preiss, J. (1986) *J. Biol. Chem 261*, 8738–8743.

Bai, G., Zhang, Z., Werner, R., Nuttall, F. Q., Tan, A. W. H., & Lee, E. Y. C. (1990) *J. Biol. Chem. 265*, 7843–7848.

Ball, K. L., & Preiss, J. (1992) *J. Prot. Chem. 11*, 231–238.

Ball, K. L., & Preiss, J. (1994) *J. Biol. Chem. 269*, 24706–24711.

Ball, S., Marianne, T., Dirick, L., Fresnoy, M., Delrue, B., & Decq, A. (1991) *Planta 185*, 17–26.

Ballicora, M. A., Laughlin, M. J., Fu, Y., Okita, T. W., Barry, G. F., & Preiss, J. (1995) *Plant Physiol. 109*, 245–251.

Ballou, L. M., & Fischer, E. H. (1986) in *The Enzymes* (Boyer, P. D., & Krebs, E. G., Eds.) Vol. 17, Third Ed., pp. 312–361, Academic Press, San Diego, CA.

Barengo, R., Flawia, M., & Krisman, C. R. (1975) *FEBS Lett. 53*, 274–278.

Barton, C., Yang, L., Galvin, M., Sengupta-Gopalan, C., & Borelli, T. (1986) in *Regulation of Carbon and Nitrogen Reduction and Utilization in Maize* (Shannon, J. C., Knievel, D. P., & Boyer, C. D., Eds.), American Society of Plant Physiologists, pp 363–365, Rockville, MD.

Batz, O., Maaß, U., Heinrichs, G., Scheibe, & Neuhaus, H. E. (1994) *Biochim. Biophys. Acta 1200*, 148–154.

Batz, O., Scheibe, R., & Neuhaus, H. E. (1993) *Biochem. J. 294*, 15–17.

Batz, O., Scheibe, R., & Neuhaus, H. E. (1995) *Planta 196*, 50–57.

Bellacosa, A., Testa, J. R., Staal, S. P., & Tsichlis, P. N. (1991) *Science 254*, 274–277.

Bhattacharyya, M., Martin, C., & Smith, A. M. (1993) *Plant Mol. Biol. 22*, 525–531.

Bhattacharyya, M. K., Smith, A. M., Noel-Ellis, T. H., Hedley, C., & Martin, C.(1990) *Cell 60*, 115–122.

Blennow, A., & Johansson, G. (1991) *Phytochemistry 30*, 437–444.

Bock, R. M. (1960) in *The Enzymes* (Boyer, P. D., Lardy, H., & Myrback, K., Eds.), Second Ed., Vol. 2, pp. 3–38, Academic Press, NY.

Borchert, S., Grosse, H., & Heldt, H. W. (1989) *FEBS Lett. 253*, 183–186.

Borchert, S., Harborth, J., Schünemann, D., Hoferichter, P., & Heldt, H. W. (1993) *Plant Physiol. 101*, 303–312.

Borovsky, D., Smith, E. E., & Whelan, W. J. (1975) *Eur. J. Biochem. 59*, 615–625.

Boyer, C., & Preiss, J. (1977) *Biochemistry 16*, 3693–3699.

Boyer, C., & Preiss, J. (1979) *Plant Physiol. 64*, 1039–1042.

Boyer, C. D., & Preiss, J. (1978a) *Carbohydrate Res. 61*, 321–334.

Boyer, C. D., & Preiss, J. (1978b) *Biochem. Biophys. Res. Commun. 80*, 169–175.

Boyer, C. D., & Preiss, J. (1981) *Plant Physiol. 67*, 1141–1145.

Brown, B. I., & Brown, D. H. (1966a) in *Methods in Enzymology* (Neufeld, E. F., & Ginsburg, V., Eds.), Vol. 8, pp. 395–403, Academic Press, NY.

Brown, B. I., & Brown, D. H. (1966c) *Proc. Natl. Acad. Sci. U.S.A.* 56, 725–729.

Brown, B. I., & Brown, D. H. (1968) in *Carbohydrate Metabolism and Its Disorders* (Dickens, F., Randle, P. J., & Whelan, W. J., Eds.), Vol. 2, pp. 123–150, Academic Press, London.

Brown, D. H., & Brown, B. I. (1966b) *Biochim. Biophys. Acta 130*, 263–266.

Brown, N. E., & Larner, J. (1971) *Biochim. Biophys. Acta 242*, 69–80.

Browner, M. F., Nakano, K., Bang, A.G., & Fletterick, R. J. (1989) *J. Biol. Chem.* 86, 1443–1447.

Burleigh, I. G., & Dawes, E. A. (1967) *Biochem. J.* 102, 236–250.

Burton, R. A., Bewley, J. D., Smith, A. M., Bhattacharyya, Tatge, H., Ring, S., Bull, V., Hamilton, D. O., & Martin, C. (1995) *Plant J.* 7, 3–15.

Campbell, D. G., & Cohen P. (1989) *Eur. J. Biochem.* 185, 119–125.

Cao, H., & Preiss, J. (1996) *J. Prot. Chem.* 15, 291–304.

Cao, H., Sullivan, T. D., Boyer, C. D., & Shannon, J. C. (1995a) *Physiol. Plant. 95*, 176–186.

Cao, Y., Mahrenholz, A. M, Depaoli-Roach, A. A., & Roach, P. (1993) *J. Biol. Chem. 268*, 14687–14693.

Cao, Y., Steinrauf, L. K., & Roach, P. (1995b) *Arch. Biochem. Biophys. 319*, 293–298.

Carter, J., & Smith, E. E. (1978) *Carbohydr. Res. 61*, 395–406.

Cattaneo, J., Chambost, J. P., & Creuzat-Sigal, N. (1978) *Arch. Biochem. Biophys. 190*, 85–96.

Cattaneo, J., Damotte, M., Sigal, N., Sanchez-Medina, G., & Puig, J. (1969) *Biochem. Biophys. Res. Commun. 34*, 694–701.

Cattaneo, J., Magnan, M., & Bigliardi, J. (1979) *Arch. Biochem. Biophys. 196*, 449–458.

Chao, J., & Weathersbee, C. J. (1974) *J. Bacteriol. 117*, 181–188.

Charng, Y.-Y., Iglesias, A. A., & Preiss, J. (1994) *J. Biol. Chem. 269*, 24107–24113.

Charng, Y.-Y., Kakefuda, G., Iglesisas, A. A., Buikema, W. J., & Preiss, J. (1992) *Plant Mol. Biol. 20*, 37–47.

Chen, G. S., & Segel, I. H. (1968a) *Arch. Biochem. Biophys. 127*, 164–174.

Chen, G. S., & Segel, I. H. (1968b) *Arch. Biochem. Biophys. 127*, 175–186.

Cheng, C., Mu, J., Farkas, I., Huang, D., Goebl, M. G., & Roach, P. J. (1995) *Mol. Cell. Biol. 15*, 6632–6640.

Cleland, W. W. (1972) in *The Enzymes* (Boyer, P. D., Ed.) Vol 2, Third Ed., pp. 1–65, Academic Press, San Diego, CA.

Coffer, P. J., & Woodgett, J. R. (1991) *Eur. J. Biochem. 201*, 475–481.

Cohen, P. (1986) in *The Enzymes* (Boyer, P. D., & Krebs, E. G., Eds.) Vol. 17, Third Ed., pp. 461–497, Academic Press, San Diego, CA.

Cohen, P. (1989) *Annu. Rev. Biochem. 58*, 453–508.

Cohen, P., & Cohen, P. T. (1989) *J. Biol. Chem. 264*, 21435–21438.

Cohen, P., Yellowlees, D., Aitken, A., Donella-Deana, A., Hemmings, B. A., & Parker, P. J. (1982) *Eur. J. Biochem. 124*, 21–35.

Copeland L., & Preiss, J. (1981) *Plant Physiol. 68*, 996–1002.

Creuzat-Sigal, N., Latil-Damotte, M., Cattaneo, J., & Puig, J. (1972) in *Biochemistry of the Glycosidic Linkage* (Piras, R., & Pontis, H.G., Eds.), pp. 647–680, Academic Press, NY.

Cross, D. A., Alessi, D. R., Cohen, P., Andjelkovich, M., & Hemmings, B. A. (1995) *Nature 378*, 785–789.

Cross, D. A., Alessi, D. R.,Vandenheede, J. R., McDowell, H. E., Hundal, H. S., & Cohen, P. (1994) *Biochem J. 303*, 21–26.

Dawes, E. A., & Senior, P. J. (1973) *Adv. Microbiol. Physiol. 10*, 135–266.

Delrue, B., Fontaine, T., Routier, F., Decq, A., Wieruszeski, J.-M., van den Koornhuyse, N., Maddelein, M.-L., Fournet, B., & Ball, S. (1992) *J. Bacteriol. 174*, 3612–3620.

Dempsey, W. B. (1972) *Biochim. Biophys. Acta 264*, 344–353.

Dempsey, W. B., & Arcement, L. J. (1971) *J. Bacteriol. 107*, 580–582.

Dent, P., Campbell, D. G., Caudwell, F. B., & Cohen, P. (1990a) *FEBS Lett. 259*, 281–285.

Dent, P., Lavoinne, A., Nakielny, S., Caudwell, F. B., Watt, P., & Cohen, P. (1990b) *Nature 348*, 302–308.

Denyer, K., Dunlap, F., Thornbjørnsen, T., Keeling, P., & Smith, A. M. (1996) *Plant Physiol. 112*, 779–785.

Denyer, K., Hylton, C. M., Jenner, C.F., & Smith, A. M. (1995) *Planta 196*, 256–265.

Denyer, K., Sidebottom, Hylton, C. M., & Smith, A. M. (1993) *Plant J. 4*, 191–198.

Denyer, K., & Smith, A. M. (1992) *Planta 186*, 609–617.

DePamphilis, M. L., & Cleland, W. W. (1973) *Biochemistry 122*, 3714–3724.

Dietzler, D. N., Lais, C. J., & Leckie, M. P. (1974a) *Arch. Biochem. Biophys. 160*, 14–25.

Dietzler, D. N., Leckie, M. P., & Lais, C. J. (1973) *Arch. Biochem. Biophys. 156*, 684–693.

Dietzler, D. N., Leckie, M. P., Lais, C. J., & Magnani, J. L. (1974b) *Arch. Biochem. Biophys. 162*, 602–606.

Dietzler, D. N., Leckie, M. P., Lais, C. J., & Magnani, J. L. (1975) *J. Biol. Chem. 250*, 2383–2387.

Dietzler, D. N., & Strominger, J. L. (1973) *J. Bacteriol. 113*, 946–952.

Dry, I., Smith, A., Edwards, A., Bhattacharyya, M., Dunn, P., & Martin, C. (1992) *Plant J. 2*, 193–202.

Edwards, J., Green, J. H., & aP Rees, T. (1988) *Phytochemistry 27*, 1615–1620.

Edwards, A., Marshall, J., Denyer, K., Sidebottom, C., Visser, R.G., Martin, C., & Smith, A. M. (1996) *Plant Physiol. 112*, 89–97.

Edwards, G., & Walker, D. A. (1983) In *C₃,C₄: Mechanisms of Cellular and Environmental Regulation, Photosynthesis*, pp. 204–207, University of California Press, Berkeley, CA.

Eidels, L., Edelman, P. L., & Preiss, J. (1970) *Arch. Biochem. Biophys. l40*, 60–74.

Elbein, A. D., & Mitchell, M. (1973) *J. Bacteriol. 113*, 863–873.

Eldar-Finkelman, H., Argast, G. M., Foord, O., Fischer, E. H., & Krebs, E. G. (1996) *Proc. Natl. Acad. Sci. U.S.A. 93*, 10228–10233.

Embi, N., Parker, P. J., & Cohen, P. (1982) *Eur. J. Biochem. 115*, 405–413.

Entwistle, G., & ap Rees, T. (1988) *Biochem. J. 255*, 391–396.

Entwistle, G., Tyson, R. H., & aP Rees, T. (1988) *Phytochemistry 25*, 2033–2039.

Farkas, I., Hardy, T. A., DePaoli-Roach, A. A., & Roach, P. J. (1990) *J. Biol. Chem. 265*, 20879–20886.

Farkas, I., Hardy, T. A., Goebl, M. G., & Roach, P. J. (1991) *J. Biol. Chem. 266*, 15602–15607.

Fedoroff, N., Wessler, S., & Shure, M. (1983) *Cell 35*, 235–242.

Fiol, C., Mahrenholz, A. M., Wang, A., Roeske, R. W., & Roach, P. J. (1987) *J. Biol. Chem. 262*, 14042–14048.

Fiol, C., Wang, A., Roeske, R. W., & Roach, P.J. (1990) *J. Biol. Chem. 265*, 6061–6065.

Fisher, D. K., Boyer, C. D., & Hannah, L. C. (1993) *Plant Physiol. 102*, 1045–1046.

Fisher, D. K., Gao, M., Kim, K.-N., Boyer, C. D., & Guiltinan, M. J. (1996) *Plant Physiol. 110*, 611–619.

Flotow, H., Graves, P. R.,Wang, A., Fiol, C. J., Roeske, R.W., & Roach, P. J. (1990) *J. Biol. Chem. 265*, 14264–14269.

Flotow, H., & Roach, P. J. (1989) *J. Biol. Chem. 264*, 9126–9128.

Flügge, U. I., & Heldt, H. W. (1991) *Annu. Rev. Plant Physiol. Plant Mol. Biol. 42*, 129–144.

Fontaine, T., D'Hulst, C., Maddelein, M.-L., Routier, F., Pepin, T. M., Decq, A., Wieruszeski, J.-M. Delrue, B., Van Den Koornhuyse, N., Bossu, J.-P., Fournet, B., & Ball, S. (1993) *J. Biol. Chem. 268*, 16223–16230.

Fox, J, Kawaguchi, K., Greenberg, E., & Preiss, J. (1976) *Biochemistry 15*, 849–856.

Franke, T. F., Yang, S. I., Chan, T. O., Datta, K., Kazlauskas, A., Morrison, D. K., Kaplan, D. R., & Tsichlis, P. N. (1995) *Cell 81*, 727–736.

Furlong, C. E., & Preiss, J. (1969) *J. Biol. Chem. 244*, 2539–2548.

Furukawa, K., Tagaya, M., Inouye, M., Preiss J., & Fukui, T. (1990) *J. Biol. Chem. 265*, 2086–2090.

Furukawa, K., Tagaya, M., Tanazawa, K., & Fukui, T. (1993) *J. Biol. Chem. 268*, 23837–23842.

Furukawa, K., Tagaya, M., Tanazawa, K., & Fukui, T. (1994) *J. Biol. Chem. 269*, 868–871.

Gannon, M. C., & Nuttall, F. Q. (1996) *Trends Glycosci. Glycotechnol. 8*, 183–194.

Gentner, N., Greenberg, E., & Preiss, J. (1969) *Biochem. Biophys. Res. Commun. 36*, 373–380.

Gentner, N., & Preiss, J. (1967) *Biochem. Biophys. Res. Commun. 27*, 417–423.

Gentner, N., & Preiss, J. (1968) *J. Biol. Chem. 243*, 5882–5891.

Ghosh, H. P., & Preiss, J. (1966) *J. Biol. Chem. 241*, 4491–4505.

Ghosh, P., Meyer, C., Remy, E., Peterson, D., & Preiss, J. (1992) *Arch. Biochem. Biophys. 296*, 122–128.

Gibson, W. B., Brown, B. I., & Brown, D. H. (1971) *Biochemistry 10*, 4253–4262.

Giroux, M. J., Shaw, J., Barry, G., Cobb, B.G., Greene, T., Okita, T., & Hannah, L.C. (1996) *Proc. Nat. Acad. Sci., U.S.A. 93*, 5824–5829.

Goldner, W., & Beevers, H. (1989) *Phytochemistry 28*, 1809–1812.

Goldraij, A., Miozzo, M. C., & Curtino, J. A. (1993) *Biochem. Mol. Biol. Internat. 30*, 453–460.

Govons, S., Gentner, N., Greenberg, E., & Preiss, J. (1973) *J. Biol. Chem. 248*, 1731–1740.

Greenberg, E., & Preiss, J. (1964) *J. Biol. Chem. 239*, 4314–4315.

Greenberg, E., & Preiss, J. (1965) *J. Biol. Chem. 240*, 2341–2348.

Greenberg, E., Preiss, J. E., VanBoldrick, M., & Preiss, J. (1983) *Arch. Biochem. Biophys. 220*, 594–604.

Guan, H., Kuriki, T., Sivak, M., & Preiss, J. (1995) *Proc. Nat. Acad. Sci. U.S.A. 92*, 964–967.

Guan, H. P., Baba, T., & Preiss, J. (1994a) *Plant Physiol. 104*, 1449–1453.

Guan, H. P., Baba, T., & Preiss, J. (1994b) *Cell. Mol. Biol. 40*, 981–988.

Guan, H. P., & Preiss, J. (1993) *Plant Physiol. 102*, 1269–1273.

Haugen, T., & Preiss, J. (1979) *J. Biol. Chem. 254*, 127–136.

Haugen, T. H., Ishaque, A., Chatterjee, A. K., & Preiss, J. (1974) *FEBS Lett. 42*, 205–208.

Haugen, T. H., Ishaque, A., & Preiss, J. (1976) *J. Biol. Chem. 251*, 7880–7885.

Hawker, J. S., Ozbun, J. L., Ozaki, H., Greenberg, E., & Preiss, J. (1974) *Arch. Biochem. Biophys. 160*, 530–551.

Hedman, K. D., & Boyer, C. D. (1982) *Biochem. Genet. 20*, 483–492.

Hehre, E. J., (1951) *Adv. Enzymol. 11*, 297–337.

Heldt, H. W., Flügge, U.-I., & Borchert, S. (1991) *Plant Physiol. 95*, 341–343.

Hestrin, S. (1960) in *The Bacteria* (Gunsalas, I., & Stanier, R. Y., Eds.), Vol. 3, pp. 373–388, Academic Press, NY.

Hill, L. M., & Smith, A. M. (1991) *Planta 185*, 91–96.

Hill, M. A., Kaufmann, K., Otero, J., & Preiss, J. (1991) *J. Biol. Chem. 266*, 12455–12460.

Hizukuri, S. (1986) *Carbohydr. Res. 147*, 342–347.

Hizukuri, S. (1995) in *Carbohydrates in Food* (Eliasson, A-C., Ed.), pp. 347–429, Marcel Dekker, Inc. NY.

Holme, T., & Palmstierna, H. (1956) *Acta Chem. Scand. 10*, 578–586.

Holmes, E., & Preiss, J. (1979) *Arch. Biochem. Biophys. 196*, 436–448.

Holmes, E., & Preiss, J. (1982) *Arch. Biochem. Biophys. 216*, 736–740.

Homerova, D., & Kormanec, J. (1994) *Biochim. Biophys. Acta 1200*, 334–336.

Huang, K. P., Akatsuka, A., Singh, T. J.,& Blake, K. R. (1983) *J. Biol. Chem. 258*, 7094–7101.

Hubbard, M. J., & Cohen, P. (1989a) *Eur. J. Biochem. 180*, 457–465.

Hubbard, M. J., & Cohen, P. (1989b) *Eur. J. Biochem. 186*, 701–709.

Hubbard, M. J., & Cohen, P. (1989c) *Eur. J. Biochem. 186*, 711–716.

Hylton, C., & Smith, A. M. (1992) *Plant Physiol. 99*, 1626–1634.

Hylton, C. M., Denyer, K., Keeling, P. L., Chang, M.-T., & Smith, A. (1996) *Planta, 198*, 230–237.

Iglesias, A. A., Barry, G. F., Meyer, C., Bloksberg, L., Nakata, P. A., Greene, T., Laughlin, M. J., Okita, T. W., Kishore, G. M., & Preiss, J. (1993) *J. Biol. Chem. 268*, 1081–1086.

Iglesias, A. A., Charng, Y.-Y., Ball, S., & Preiss, J. (1994) *Plant Physiol. 104*, 1287–1294.

Iglesias, A. A., Kakefuda, G., & Preiss, J. (1991) *Plant Physiol. 97*, 1187–1195.

Illingworth, B., Brown, D. H., & Cori, C. F. (1961) *Proc. Natl. Acad Sci. U.S.A. 46*, 469–478.

Issa, H. A., & Mendicino, J. (1973) *J. Biol. Chem. 248*, 685–696.

James, M. G., Robertson, D. S., & Meyers, A. M. (1995) *Plant Cell 7*, 417–429.

Janson, C. A., & Cleland, W. W. (1974) *J. Biol. Chem. 249*, 2572–2574.

Jenner, C. F. (1979) in *Encyclopedia of Plant Physiology New Series* (Tanner, W., & Loewus, F.A., Eds.), Vol. 13A, pp. 700–746, Springer-Verlag, Berlin.

Jesperson, H. M., Macgregor, E. A., Henrissat, B., Sierks, M. R., & Svensson, B. (1993) *J. Prot. Chem. 12*, 791–805.

Kacser, H. (1987) In *The Biochemistry of Plants* (Davies, D. D., Ed.) Vol. 11, pp. 39–67, Academic Press, NY.

Kacser, H., & Burns, J. A. (1973) *Symp. Soc. Exp. Biol. 27*, 65–107.

Kakefuda, G., Charng, Y.-Y., Iglesias, A. A., McIntosh, L., & Preiss, J. (1992) *Plant Physiol. 99*, 359–361.

Kang, F., & Rawsthorne, S. (1994) *Plant J. 6*, 795–805.

Kawaguchi, K., Fox, J., Holmes, E., Boyer, C., & Preiss, J. (1978) *Arch. Biochem. Biophys. 190*, 385–397.

Keeling, P. L., Wood, J. R., Tyson, R. H., & Bridges, I. G. (1988) *Plant Physiol. 87*, 311–319.

Khandelwal, R. L., Spearman, T. N., & Hamilton, I. R. (1973) *Arch. Biochem. Biophys. 154*, 295–305.

Khoshnoodi, J., Blennow, A., Ek, B., Rask, L., & Larsson, H. (1996) *Eur. J. Biochem. 242*, 132–138.

Khoshnoodi, J., Ek, B., Rask, L., & Larsson, H. (1993) *FEBS 332*, 132–138.

Kiel, J. A. K. W., Boels, J. M., Beldman, G., & Venema, G. (1990) *Gene 89*, 77–84.

Kiel, J. A. K. W., Boels, J. M., Beldman, G., & Venema, G. (1991) *Mol. Gen. Genet. 230*, 136–144.

Kiel, J. A. K. W., Boels, J. M., Beldman, G., & Venema, G. (1992) *DNA Seq. Map. 3*, 221–232.

Kiel, J. A. K. W., Elgersma, H. S. A., Beldman, G., Vossen, J. P. M. J., & Venema, G. (1989) *Gene 78*, 9–17.

Killilea, S. D., & Whelan, W. J. (1976) *Biochemistry 15*, 1349–1355.

Kim, W. T., Francheschi, V. R., Okita, T. W., Robinson, N. L., Morell, M., & Preiss, J. (1989) *Plant Physiol. 91*, 217–220.

Klösgen, R. B., Gierl, A., Schwartz-Sommer, Z., & Saedler, H. (1986) *Mol. Gen. Genet. 203*, 237–244.

Knop, J. K., & Hansen, R. G. (1970) *J. Biol. Chem. 245*, 2499–2504.

Ko, J. H., Kim, C. H., Lee, D.-S., & Kim, Y. S. (1996) *Biochem. J. 319*, 977–983.

Kosegarten, H., & Mengel, K. (1994) *Physiol. Plant. 91*, 111–120.

Koßmann, J., Visser, R. G. F., Müller-Röber, Willmitzer, L., & Sonnewald, U. (1991) *Mol. Gen. Genet. 230*, 39–44.

Kram, A. M., Oostergetel, G. T., & Van Bruggen, E. F. J. (1993) *Plant Physiol. 101*, 237–243.

Krishnan, H. B., Reeves, C. D., & Okita, T. W. (1986) *Plant Physiol. 81*, 642–645.

Krisman, C. R. (1962) *Biochim. Biophys. Acta 65*, 307–315.

Krisman, C. R., & Barengo, R. (1975) *Eur. J. Biochem. 52*, 117–123.

Kuipers, A. G. J., Jacobsen, E., & Visser, R. G. F. (1994) *Plant Cell 6*, 43–52.

Kumar, A., Larsen, C. E., & Preiss, J. (1986) *J. Biol. Chem. 26l*, 16256–16259.

Kumar, A., Tanaka, T., Lee, Y. M., & Preiss, J. (1988) *J. Biol. Chem. 263*, 14634–14639.

Kuriki, T., Guan, H., Sivak, M., & Preiss, J. (1996) *J. Prot. Chem. 15*, 305–313.

Lacks, S. (1968) *Genetics 60*, 685–706.

Lapp, D., & Elbein, A. D. (1972) *J. Bacteriol. 112*, 327–336.

Larner, J. (1953) *J. Biol. Chem. 202*, 491–503.

Larsen, C. E., Lee, Y. M., & Preiss, J. (1986) *J. Biol. Chem. 261*, 15402–15409.

Larsen, C. E., & Preiss, J. (l986) *Biochemistry 25*, 4371–4376.

Lawrence, J. C., & Zhang, J. N. (1994) *J. Biol. Chem. 269*, 11595–11600.

Lazdunski, N. (1972) *Curr. Top. Cell. Regul. 6*, 267–310.

Lee, Y. M., & Preiss, J. (1986) *J. Biol. Chem. 261*, l058–l064.

Lehmann, M., & Preiss, J. (1980) *J. Bacteriol. 143*, 120–127.

Leloir, L. F., deFekete, M. A. R., & Cardini, C.E. (1961) *J. Biol. Chem. 236*, 636–641.

Leung, P., & Preiss, J. (l987a) *J. Bacteriol. l69*, 4349–4354.

Leung, P., & Preiss, J. (l987b) *J. Bacteriol. l69*, 4355–4360.

Levi, C., & Preiss, J. (1976) *Plant Physiol. 58*, 753–756.

Levitski, A., & Koshland, D.E., Jr. (1976) *Curr. Top. Cell. Regul. 10*, 1–40.

Li, L., & Preiss, J. (1992) *Carbohydr. Res. 227*, 227–239.

Lin, T. P., Caspar, T., Somerville, C., & Preiss, J. (1988a) *Plant Physiol. 86*, 1131–1135.

Lin, T. P., Caspar, T., Somerville, C., & Preiss, J. (1988b) *Plant Physiol. 88*, 1175–1181.

Lin D. C., & Segal, H. L. (1973) *J. Bio/. Chem. 248*, 7007–7011.

Lomako, J., Lomako, W. M., & Whelan, W.J. (1988) *FASEB J. 2*, 3097–3103.

Lomako, J., Lomako, W. M., & Whelan, W.J. (1990) *FEBS Lett. 264*, 13–16.

Lomako, J., Lomako, W. M., & Whelan, W.J. (1991) *FEBS Lett. 279*, 223–228.

Lomako, J., Lomako, W. M., & Whelan, W.J. (1992) *Carbohydr. Res. 227*, 331–338.

Lomako, J., Lomako, W. M., Whelan, W. J., Dombro, R. S., Neary, J. T., & Norenberg, M. D. (1993) *FASEB J. 7*, 1386–1393.

Lowry, O. H., Carter, J., Ward, J. B., & Glaser, L. (1971) *J. Biol. Chem. 246*, 6511–6521.

MacDonald, F. D., & ap Rees, T. (1983) *Biochim. Biophys. Acta 755*, 81–89

MacDonald, F. D., & Preiss, J. (1983) *Plant Physiol. 73*, 175–178.

MacDonald, F. D., & Preiss, J. (1985) *Plant Physiol. 78*, 849–852.

Mackey, B. M., & Morris, J. G. (1971) *J. Gen Microbiol. 66*, 1–13.

MacKintosh, C., Campbell, D. G., Hiraga, A., & Cohen, P. (1988) *FEBS Lett. 234*, 189–194.

Maddelein, M.-L., Bellanger, F., Delrue, B., Libessart, N., D'Hulst, C., Van Den Koornhuyse, N., Fontaine, T., Wieruszeski, J.-M. Decq, A., & Ball, S. (1994) *J. Biol Chem. 269*, 25150–25157.

Mahrenholz, A. M., Wang, Y., & Roach, P. J. (1988) *J. Biol. Chem. 263*, 10561–10567.

Manchester, J., Skurat, A. V., Roach, P., Hauschka, S. D., & Lawrence, J. C. (1996) *Proc. Natl. Acad. Sci. U.S.A. 93*, 10707–10711.

Manners, D. J. (1971) in *The Yeasts*, (Rose, A. H., & Harrison, J. A., Eds.), Vol. 2, pp. 419–439, Academic Press, NY.

Manners, D. J. (1991) *Carbohydr. Polymers 16*, 37–82.

Manzella, S. M., Rodén, L., & Meezan, E. (1995) *Glycobiology 5*, 263–271.

Marshall, J., Sidebottom, C., Debet, M., Martin, C., Smith, A. M., & Edwards, A. (1996) *Plant Cell 8*, 1121–1135.

Martin, C., & Smith A. M. (1995) *Plant Cell 7*, 971–985.

Meezan, E., Ananth, S., Manzella, S., Campbell, P., Siegal, S., Dillion, D. J., & Roden, L. (1994) *J. Biol. Chem. 269*, 11503–11508.

Meyer, C. R., Ghosh, P., Nadler, S., & Preiss, J. (1992) *J. Bacteriol. 174*, 4509–4512.

Meyer, C. R., Ghosh, P., Remy, E., & Preiss, J. (1993) *Arch. Biochem. Biophys. 302*, 64–71.

Miller, M. E., & Chourey, P. (1995) *Planta 197*, 522–527.

Miller, R. E., Miller, E. A., Fredholm, B., Yellin, J. B., Eichner, R. D., Mayer, S. E., & Steinberg, D. (1975) *Biochemistry 14*, 2481–2488.

Mizuno, K., Kawasaki, T., Shimada, H., Satoh, H., Kobayashi, E., Okamura, S., Arai, Y., & Baba, T. (1993) *J. Biol. Chem. 268*, 19084–19091.

Mizuno, K., Kimura, K., Arai, Y., Kawasaki, T., Shimada, H., & Baba, T. (1992) *J. Biochem. 112*, 643–651.

Monod, J., & Torriani, A. M. (1948) *C.R. Acad. Sci. Paris 227*, 240–242.

Morell, M., Bloom, M., Larsen, R., Okita, T. W., & Preiss, J. (1987b) in *Plant Gene Systems and their Biology*. (Key, J. L., & McIntosh, L., Eds.), pp. 227–242, Alan R. Liss, Inc., NY.

Morell, M., Bloom, M., & Preiss, J. (1988) *J. Biol. Chem. 263*, 633–637.

Morell, M. K., Blennow, A., Kosar-Hashemi, B., & Samuel, M. S. (1997) *Plant Physiol. 113*, 201–208.

Morell, M. K., Bloom, M., Knowles, V., & Preiss, J. (1987a) *Plant Physiol. 85*, 185–187.

Moreno, M., Cardini, C. E., & Tandecarz, J. S. (1986) *Eur. J. Biochem. 157*, 539–545.

Moreno, M., Cardini, C. E., & Tandecarz, J. S. (1987) *Eur. J. Biochem. 162*, 609–614.

Morrison, W. R., & Karkalas, J. (1990) in *Methods in Plant Biochemistry* (Dey, P.M., Ed.) pp. 323–352, Academic Press, Ltd., London.

Mu, C., Harn, C., Ko, Y-T., Singletary, G. W., Keeling, P. L., & Wasserman, B. P. (1994) *Plant J. 6*, 151–159.

Mu-Forster, C., Huang, R., Powers, J. R., Harriman, R. W., Knight, M., Singletary, G. W., Keeling, P., & Wasserman, B. P. (1996) *Plant Physiol. 111*, 821–829.

Mulichak, A. M., Skrzypczak-Jankum, E., Rydel, T. J., Tulinsky, A., & Preiss, J. (1988) *J. Biol. Chem. 263*, 17237–17238.

Munch-Petersen, A., Kalckar, H. M., Cutole, E., & Smith, E. E. B. (1953) *Nature 172*, 1036–1037.

Nakai, C., & Thomas, J. A. (1975) *J. Biol. Chem. 250*, 4081–4086.

Nakamura, Y., & Yamanouchi, H. (1992) *Plant Physiol. 99*, 1265–1266.

Nakamura, Y., Umemoto, T., Takahata, Y., Komae, K., Amano, E., & Satoh, H. (1996) *Physiol. Plant. 97*, 491–498.

Nelson, O., & Pan, D. (1995) *Annu. Rev. Plant Physiol. Plant Mol. Biol. 46*, 475–496.

Nelson O. E., Chourey, P. S., & Chang, M. T. (1978) *Plant Physiol. 62*, 383–386.

Nelson, O. E., & Rines, H. W. (1962) *Biochem. Biophys. Res. Commun. 9*, 297–300.

Neuhaus, H. E., Batz, O., Thom, E., & Scheibe, R. (1993a) *Biochem. J. 296*, 395–401.

Neuhaus, H. E., Henrichs, G., & Scheibe, R. (1993b) *Plant Physiol. 101*, 573–578.

Neuhaus, H. E., Kruckeberg, A. L., Feil, R., & Stitt, M. (1989) *Planta 178*, 110–122.

Neuhaus, H. E., & Stitt, M. (1990) *Planta 182*, 445–454.

Nimmo, H. G., Proud, C. G., & Cohen, P. (1976) *Eur. J. Biochem. 68*, 21–30.

Nuttall, F. Q., Gannon, M. C., Bai, G., & Lee, E. Y. C. (1994) *Arch. Biochem. Biophys. 311*, 443–449.

Salehuzzaman, S. N. I. M., Jacobsen, E., & Visser, R. G. F. (1992) *Plant Mol. Biol. 20,* 809–819.

Salehuzzaman, S. N. I. M., Jacobsen, E., & Visser, R. G. F. (1993) *Plant Mol. Biol. 23,* 947–962.

Sano, Y. (1984) *Theor. Appl. Genet. 68,* 467–473.

Sanwal, G., Greenberg, E., Hardie, J., Cameron, E., & Preiss, J. (1968) *Plant Physiol. 43,* 417–427.

Sanwal, G. G., & Preiss, J. (1967) *Arch. Biochem. Biophys. 119,* 454–469.

Schünemann, D., & Borchert, S. (1994) *Bot. Acta 107,* 461–467.

Schwartz, M. (1966) *J. Bacteriol. 92,* 1083–1089.

Shannon, J. C., Pien, F.-M., & Lui, K.-C. (1996) *Plant Physiol. 110,* 835–843.

Shen, L., Ghosh, H. P., Greenberg, E., & Preiss, J. (1964) *Biochim. Biophys. Acta 89,* 370–372.

Shen, L., & Preiss, J. (1964) *Biochem. Biophys. Res. Commun. 17,* 424–429.

Shen, L., & Preiss, J. (1965) *J. Biol. Chem. 240,* 2334–2340.

Shen, L., & Preiss, J. (1966) *Arch. Biochem. Biophys. 116,* 375–390.

Sheroain, V. S., Corbin, J. D., & Soderling, T. R. (1985) *J. Biol. Chem. 260,* 1567–1572.

Sheroain, V. S., Juhl, H., Bass, M., & Soderling, T. R. (1984) *J. Biol. Chem. 259,* 7024–7030.

Shimada, H., Tada, Y., Kawasaki, T., & Fujimura, T. (1993) *Theor. Appl. Genet. 86,* 665–672.

Shure, M., Wessler, S., & Federoff, N. (1983) *Cell 35,* 225–233.

Sigal, N., Cattaneo, J., & Segel, I. H. (1964) *Arch. Biochem. Biophys. 108,* 440–451.

Sigal, N., Cattaneo, J., Chambost, J. P., & Favard, A. (1965) *Biochim. Biophys. Res. Commun. 20,* 616–620.

Silver, S. (1969) *Proc. Nat. Acad. Sci. U.S.A. 62,* 764–771.

Singh, B. K., & Preiss, J. (1985) *Plant Physiol. 78,* 849–852.

Sivak, M. N., & Preiss, J. (1995) in *Seed Development and Germination* (Kigel, J., & Galili, G., Eds.), pp. 139–168, Marcel Dekker, NY.

Sivak, M. N., Wagner, M., & Preiss, J. (1993) *Plant Physiol. 103,* 1355–1359.

Skurat, A. V., Cao, Y., & Roach, P. J. (1993) *J. Biol. Chem. 268,* 14701–14707.

Skurat, A. V., Peng, H.-W., Chang, H.-Y., Cannon, J. F., & Roach, P. J. (1996) *Arch. Biochem. Biophys. 328,* 283–288.

Skurat, A. V., & Roach, P. J. (1995) *J. Biol. Chem. 270,* 12491–12497.

Skurat, A. V., & Roach, P. J. (1996) *Biochem. J. 313,* 45–50.

Skurat, A. V., Wang, Y., & Roach, P. J. (1994) *J. Biol. Chem. 269,* 25534–25542.

Smith, A. (1990) *Planta 182,* 599–604.

Smith, A. M. (1988) *Planta 175,* 270–279.

Smith, A. M., Denyer, K., & Martin, C. (1995) *Plant Physiol. 107,* 673–677.

Smith, C. H., Brown, N. E., & Larner, J. (1971) *Biochim. Biophys. Acta 242,* 81–88.

Smith-White, B., & Preiss, J. (1992) *J. Mol. Evol. 34,* 449–464.

Smyth, D. A. (1988) *Plant Sci. 57,* 1–8.

Smythe, C., Caudwell, F.B. Ferguson, M., & Cohen, P. (1988) *EMBO J. 7,* 2681–2686.

Smythe, C., & Cohen, P. (1991) *Eur. J. Biochem. 200,* 625–631.

Soderling, T. R., Hickenbottom, J. P., Reiman, E., Hunkeler, F. L., Walsh, D. A., & Krebs, E. G. (1970) *J. Biol. Chem. 245,* 6317–6328.

Stark, D. M., Timmerman, K. P., Barry, G. F., Preiss, J., & Kishore, G. M. (1992) *Science 258,* 287–292.

Steiner, K. E., & Preiss, J. (1977) *J. Bacteriol. 129,* 246–253.

Strålfors, P., Hiraga, A., & Cohen, P. (1985) *Eur. J. Biochem. 149,* 295–303

Sullivan, T., & Kaneko, Y. (1995) *Planta 196,* 477–484.

Sullivan, T., Strelow, L. I., Illingworth, C. A., Phillips, C. A., & Nelson, O. E. (1991) *Plant Cell 3,* 1337–1348.

Svensson, B. (1994) *Plant Molec. Biol. 25,* 141–157.

Swedes, J. S., Sedo, R. J., & Atkinson, D. E. (1975) *J. Biol. Chem. 250,* 6930–6938.

Tagaya, M., Nakano, K., & Fukui, T. (1985) *J. Biol. Chem. 260,* 6670–6676.

Takata, H., Takaha, T., Kuriki, T., Okada, S., Takagi, M., & Imanaka, T. (1994) *Appl. Environ. Microbiol. 60*, 3096–3104.

Takeda, Y., Guan, H. P., & Preiss, J. (1993) *Carbohydr. Res. 240*, 253–263.

Tandecarz, J. S., & Cardini, C. E. (1978) *Biochim. Biophys. Acta 543*, 423–429.

Thon, V. J., Khalil, M., & Cannon, J. F. (1993) *J. Biol. Chem. 268*, 7509–7513.

Thon, V. J., Vigneron-Lesens, C., Marianne-Pepin, T., Montreuil, J., Decq, A., Rachez, C., Ball, S. G., & Cannon, J. F. (1992) *J. Biol. Chem. 267*, 15224–15228.

Thornbjørnsen, T., Villand, P., Denyer, K., Olsen, O.-A., & Smith, A. (1996a) *Plant J. 10*, 243–250.

Thornbjørnsen, T., Villand, P., Kleczkowski, L. A., & Olsen, O.-A. (1996b) *Biochem. J. 313*, 149–154.

Turnquist, R. L., & Hansen, R. G. (1973) *in The Enzymes*, (Boyer, P. D., Ed.), Vol. 8, Third Ed., pp. 51–71, Academic Press, NY.

Tyson, R. H., & ap Rees, T. (1988) *Planta 175*, 33–38.

Van der Leij, F. R., Visser, R. F. G., Ponstein, A. S., Jacobsen, E., & Feenstra, W. J. (1991) *Mol. Gen. Genet. 228*, 240–248.

Verhue, W., & Hers, H. G. (1966) *Biochem. J. 99*, 222–227.

Viola, R., Davies, H. V., & Chudeck, A. R. (1991) *Planta 183*, 202–208.

Viskupic, E., Cao, Y., Zhang, W., Cheng, C., Depaoli-Roach, A. A., & Roach, P. (1992) *J. Biol. Chem. 267*, 25759–25763.

Visser, R. F. G., Somhorst, I., Kuipers, G. J., Ruys, N. J., Feenstra, W. J., & Jacobsen, E. (1991) *Mol. Gen. Genet. 225*, 289–296.

Vos-Scheperkeuter, G. H., de Wit, J. G., Ponstein, A. S., Feenstra, W. J., & Witholt, B. (1989) *Plant Physiol. 90*, 75–84.

Walker, G. J. (1966) *Biochem. J. 101*, 861–872.

Walker, G. J., & Builder, J. E. (1971) *Eur. J. Biochem. 20*, 14–21.

Wang, Y., Bell, A., Hermodson, M. A., & Roach, P. J. (1986a) *J. Biol. Chem. 261*, 1609–1615.

Wang, Y., Camici, M., Lee, F. T., Ahmad, Z., Depaoli-Roach, A. A., & Roach, P. J. (1986b) *Biochim. Biophys. Acta 888*, 225–236.

Wang, Y., & Roach, P. J. (1993) *J. Biol. Chem. 268*, 23876–23880.

Wang, Z.-Y, Wu, Z.-L., Xing, Y.-Y., Zheng, F.-G., Guo, X.-L., Zang, W.-G., & Hong, M.-M. (1990) *Nucleic Acids Res. 18*, 5898.

Welsh, G. I., Foulstone, E. J., Young, S. W., Tavar'e, J. M., & Proud, C. G. (1994) *Biochem. J. 303*, 15–20.

Welsh, G. I., & Proud, C. G. (1993) *Biochem. J. 294*, 625–629.

Wessler, S. R., Baran, G., Varagona, M., & Dellaporta, S. L. (1986) *EMBO J. 5*, 2427–2432.

Whelan, W. J. (1986) *BioEssays 5*, 136–140.

Wöber, G. (1973) *Hoppe-Seyler's Z. Physiol. Chem. 354*, 75–82.

Yamanouchi, H., & Nakamura, Y. (1992) *Plant Cell Physiol. 33*, 985–981.

Yung, S.-G., Paule, M. Beggs, R., Greenberg, E., & Preiss, J. (1984) *Arch. Microbiol. 138*, 1–8.

Yung, S-G., & Preiss, J. (1981) *J. Bacteriol. l47*, 101–109.

Yung, S-G., & Preiss, J. (1982) *J. Bacteriol. 151*, 742–749.

Zevenhuizen, L. P. T. M. (1964) *Biochim. Biophys. Acta 81*, 608–611.

Zhang, W., Browner, M. F., Fletterick, R. G., DePaoli-Roach, A. A., & Roach, P. J. (1989) *FASEB J. 3*, 2532–2536.

Zhang, W., DePaoli-Roach, A. A., & Roach, P. J. (1993) *Arch. Biochem. Biophys. 304*, 219–225.

Index